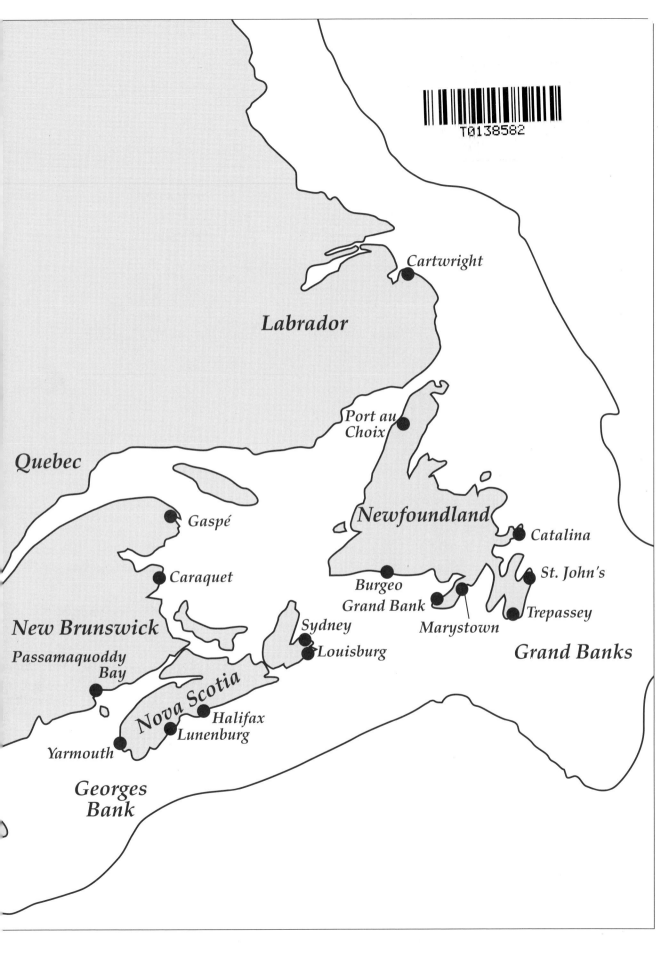

Managing
Canada's
Fisheries:

from early days to the year 2000

 Septentrion

Les éditions du Septentrion thanks the Canada Council for the Arts for
its support. We are also grateful for the financial support of the
Government of Canada through its Book Publishing Industry
Development Program (BPIDP).

Legal Deposit – 2006
Bibliothèque et Archives nationales du Québec
Library and Archives Canada
ISBN 2-89448-523-9

Les éditions du Septentrion
1300, avenue Maguire
Sillery, Quebec
G1T 1Z3

Distribution in Canada:
McGill-Queen's University Press
c/o Georgetown Terminal Warehouses
34 Armstrong Avenue
Georgetown, Ontario
L7G 4R9

Printed in Canada by Friesens

Published by Les éditions du Septentrion in co-operation with Fisheries
and Oceans Canada, and Public Works and Government Services
Canada.

 Printed on
recycled paper

CONTENTS

Maps

Tables, graphs, and statistical summaries

ACRONYMS AND INITIALISMS

A.C.O.A.	Atlantic Canada Opportunities Agency
A.D.B.	Atlantic Development Board
A.D.M.	assistant deputy minister
A.F.A.	Atlantic Fishermen's Association
A.F.A.P.	Atlantic Fisheries Adjustment Program
A.F.E.R.P.	Atlantic Fisheries Early Retirement Program
A.G.A.P.	Atlantic Groundfish Assistance Plan
A.G.L.R.P.	Atlantic Groundfish Licence Retirement Program
A.H.F.M.C.	Atlantic Herring Fishermen's Marketing Co-operative
A.R.D.A.	Agricultural and Rural Development Act
B.N.A.	British North America
C.A.F.S.A.C.	Canadian Atlantic Fisheries Scientific Advisory Committee
C.A.W.	Canadian Auto Workers
C.C.G.	Canadian Coast Guard
C.C.P.F.H.	Canadian Council of Professional Fish Harvesters
C.F.A.R.	Canadian Fisheries Adjustment and Recovery
C.I.D.A.	Canadian International Development Agency
C.P.R.	Canadian Pacific Railway
C.S.A.	Canadian Sablefish Association
C.S.C.	Canadian Saltfish Corporation
C.S.O.	community service officer
C.S.U.	Canadian Seamen's Union
D.F.O.	Department of Fisheries and Oceans
D.I.A.N.D.	Department of Indian Affairs and Northern Development
D.O.E.	Department of Environment
D.R.E.E.	Department of Regional Economic Expansion
D.R.I.E.	Department of Regional Industrial Expansion
E.A.	enterprise allocation
E.C.	European Community
E.F.F.	Eastern Fishermen's Federation
E.I.	Employment Insurance
E.U.	European Union
F.A.D.S.	Federal Aquaculture Development Strategy
F.A.O.	Food and Agriculture Organization of the United Nations
F.C.C.	Fisheries Council of Canada
F.D.A.	Fisheries Development Act
F.F.A.W.	Fish, Food and Allied Workers Union
F.F.M.C.	Freshwater Fish Marketing Corporation
F.F.T.	factory freezer trawler
F.M.S.	Fisheries and Marine Service
F.O.R.A.C.	Fisheries and Oceans Research Advisory Council
F.O.R.D.	Federal Office of Regional Development
F.P.A.	Fishermen's Protective Association
F.P.C.	fish protein concentrate
F.P.S.B.	Fisheries Prices Support Board
F.P.U.	Fishermen's Protective Union
F.R.B.	Fisheries Research Board
F.R.C.C.	Fisheries Resource Conservation Council
F.S.R.S.	Fishermen and Scientists Research Society
F.V.I.P.	Fishing Vessel Insurance Program

G.A.A.P.	General Adjustment Assistance Program
G.L.R.P.	Groundfish Licence Retirement Program
G.P.S.	global positioning system
H.B.C.	Hudson's Bay Company
H.R.D.C.	Human Resources Development Canada
I.C.C.A.T.	International Commission for the Conservation of Atlantic Tunas
I.C.N.A.F.	International Convention for the Northwest Atlantic Fisheries
I.C.O.D.	International Centre for Ocean Development
I.D.S.	Industrial Development Service
I.F.A.W.	International Fund for Animal Welfare
I.F.M.P.	Integrated Fishery Management Plan
I.J.C.	International Joint Commission
I.N.P.F.C.	International North Pacific Fisheries Commission
I.P.H.C.	International Pacific Halibut Commission
I.P.S.F.C.	International Pacific Salmon Fisheries Commission
I.Q.	individual quota
I.T.Q.	individual transferable quota
I.V.Q	individual vessel quota
J.P.A.	joint project agreement
L.T.A.	long-term agreement
M.F.U.	Maritime Fishermen's Union
M.P.	member of Parliament
M.S.Y.	maximum sustainable yield
N.A.C.F.I.	North American Council on Fishery Investigations
N.A.F.E.L.	Newfoundland Associated Fish Exporters Limited
N.A.F.O.	Northwest Atlantic Fisheries Organization
N.A.S.C.O.	North Atlantic Salmon Conservation Organization
N.C.A.R.P.	Northern Cod Adjustment and Recovery Plan
N.F.D.A.	Newfoundland Fisheries Development Authority
N.N.F.C.	Northern Native Fishing Corporation
O.S.Y.	optimum sustainable yield
P.A.I.T.	Program for the Advancement of Industrial Technology
P.F.R.C.C.	Pacific Fisheries Resource Conservation Council
P.R.F.C.A.	Prince Rupert Fishermen's Co-operative Association
Q.U.F.	Quebec United Fishermen
R.A.P.	Regional Advisory Process
R.C.M.P.	Royal Canadian Mounted Police
R.D.G.	regional director-general
S.E.P.	Salmonid Enhancement Program
S.I.A.P.	Shipyard Industry Assistance Program
St. F.X.	St. Francis Xavier University
S.U.F.	Society of United Fishermen
T.A.C.	total allowable catch
T.A.G.S.	The Atlantic Groundfish Strategy
T.A.P.	Temporary Assistance Program

U.F.A.W.U.	United Fishermen and Allied Workers' Union
U.I.	Unemployment Insurance Program
U.M.F.	United Maritime Fishermen
U.N.C.L.O.S.	United Nations Conference on the Law of the Sea
U.N.F.A.	United Nations Fish Agreement

PREFACE AND ACKNOWLEDGEMENTS

This narrative outlines commercial fishery management in Canada from pre-Contact days to the year 2000. Such a wide-ranging survey can make no attempt to treat individual episodes in full detail. Even if one tried, the complexities of the fishery often make it difficult to offer final judgement on the ins, outs, justifications, and consequences of management actions.

The Table of Contents makes clear the framework of the book. Its six parts include: The Fisheries Before Confederation; 1867-1914—Bringing Law and Order to the Fishery; 1914-1945—Short Booms, Long Depression; 1945-1968—The Age of Development; 1968-1984—Comprehensive Management Begins; and 1984-2000—Making the New System Work. The chapters within those divisions, from Confederation on, first treat national and international matters and then the various fishery regions, to the extent they can be disentangled. Since the book deals mainly with federal management, coverage of freshwater fisheries diminishes in the 20th century, when provincial governments carried out much of that work.

The narrative shows some recurring themes, such as the contest between development and conservation. But I find no all-inclusive and inexorable pattern, and I have learned to distrust fishery formulas. I do, however, offer some modest conclusions in the final chapter.

One clear aspect of fisheries history is the effort and dedication that have gone into fisheries management over the years since Confederation, by federal officials and others. When I mentioned this book to former fisheries minister Roméo LeBlanc, he remarked that "nothing has ever captured the excitement of that department." Neither will this recitation of main events, but I hope it will serve as a general guide to the difficult and unremitting struggle to protect fishery resources and use them responsibly.

A fair chunk of history has gone by since I started working on this book. Ken Johnstone, author of *The Aquatic Explorers*, a history of the Fisheries Research Board of Canada, had commenced a fisheries-management history before his untimely death in 1978. Around the same time, I left the fisheries department for the private sector. Ken Lucas, Senior Assistant Deputy Minister for federal fisheries at the time, and J. Cam Stevenson of the scientific publishing side of the department, backed me in 1979 to restart the history project, which was approved by minister Roméo LeBlanc. Because of unforeseen events, over the next few years I could devote only a fraction of my time to the history. I researched the broad picture of Canadian fisheries management, and wrote a preliminary narrative up to about 1960. After rejoining the Department of Fisheries and Oceans (D.F.O.) in 1987, I published a partial history in 1991, in D.F.O.'s manuscript report series, under the title *Fisheries Management in Canada, 1880-1910.* But circumstances continued to delay revision and completion of the larger project.

Years went by, as I worked in other capacities. Late in 2000, when Jack Stagg, then Associate Deputy Minister of D.F.O. and himself a historian, saw the 1880-1910 publication, he sponsored, and deputy minister Wayne Wouters approved, my completing the book. This was a wonderful opportunity to work full-time at filling out the picture of fisheries history. I researched, wrote, and revised over the next two and a half years, before retiring from D.F.O. The text remains as completed in 2003 and reflects my state of knowledge then, although proofreading, translation, and so on took further work after that time. Mr. Wouters's successors, especially Larry Murray, gave continued support, and Jean-Claude Bouchard, during his time as Associate Deputy Minister, was of invaluable help.

In fact, I have received assistance of one kind or another from all levels of the department and from others outside. The catalogue of those I must thank is long, with Linda MacMillan and Denise Charron at the head of the list, Linda for lining up authorizations, translation and other publishing-related contracts, and much more, and Denise for arranging design, photographs (those not otherwise credited in the book came through her good offices), and other necessities of production. Micheline Gilbert and Ted Gale also assisted in the early stages, and Micheline Steals typed in a multitude of revisions. Gord McWilliams of Carisse Graphic Design carried out the design work, Sharon Stewart proofread the manuscript, and Lexi-tech International provided translation.

The several dozen people with whom I did tape-recorded interviews, mostly in the early 1980's, are too many to list in full. But fishery figures such as Wilfred Templeman, Joey Smallwood, and Alfred Needler on the Atlantic, and Jimmy Sewid, Homer Stevens, and Cliff Levelton on the Pacific, exemplify the knowledge and engagement that I found among interviewees in general. I also inherited some interviews from Ken Johnstone, as well as his early partial narrative.

Many people read all or parts of my manuscript. I am especially grateful to Professor Shannon Ryan, Memorial University of Newfoundland, and to Peter Rider, Atlantic Provinces historian and Curator at the Canadian Museum of Civilization. W.C. MacKenzie, Ron MacLeod, Bob Applebaum, Scott Parsons, Ralph Halliday, and Pat Chamut, all now retired from federal fisheries, also commented on long sections of the manuscript. Brian Richman generously shared historical information. Other expert readers of sections, chapters, or passages included Henry Lear, Arthur May, Greg Peacock, Doug Pezzack, Jake Rice, Max Stanfield, and Robert Steinbock.

I thank the many others, inside and outside D.F.O., who gave me facts and opinions, served as sounding boards on one subject or another, or otherwise helped me. To list even the half of them would be unwieldy, but I will mention the following: Dennis Brock, Richard Cashin, Jerry Conway, Lynn Dolan, Dave Dunn, Charles Friend, Kate Glover, Jean-Eudes Haché, Jon Hansen, Tim Hsu, Jim Jones, Trevor Kenchington, Brian Lester, Charles Maginley, Alain Meuse, Don Pepper, Barry Rashotte, Victor Rabinovitch, Paul Steele, Ralph Surette, Steve Tilley, Bernard Vezina, and my mother, father, and other citizens of Campobello Island, N.B. For photographs, I got further help from Bernard Collin, Paul Richer, Michel Thérien, Norwood Whynot, and members of the Small Craft Harbours branch. The late Kevin McVeigh took the cover photograph and many others. Geof Thompson provided computerized data for the east and west coast illustrations inside the front cover. All those unmentioned, including many D.F.O. librarians, deserve equal thanks.

Printed sources for this narrative form a wide mélange: books, interviews, annual reports, speeches and press releases, reports and briefing notes, the fishery and general press, and a good deal of "gray literature" such as D.F.O. and industry leaflets. Among books, I relied often on the works, noted in the bibliography, by Harold Innis, H.Y. Hind, D.W. Prowse, Moses Perley, Shannon Ryan, Scott Parsons, a companion volume by Scott Parsons and Henry Lear, Margaret Beattie Bogue, A.B. McCullough, Cicely Lyon, and finally Geoff Meggs, who also shared supplementary information. The trade journal *Canadian Fishing Report*, which I published 1979-1984, yielded significant data. A large amount of information for recent decades came to me through informal conversations and other forms of osmosis. For the 1990's, D.F.O. management plans and other information posted on the Internet became an important source.

Statistical information, unless otherwise attributed, mostly came from D.F.O.'s statistical Web-site and printed material, including annual reports, the late lamented *Annual Statistical Review*, and miscellaneous documents. I thank Kieth Brickley, formerly of D.F.O., for special help. A smaller fraction of the unattributed data came from Statistics Canada, often from its historical Web-site. Tables in the latter part of the book for what I call "main-income fishermen," those for whom fishing is the single biggest source of income, came from Revenue Canada data.

I should note that methods of gathering or presenting statistical data sometimes changed over time. For example, D.F.O.'s landings statistics after 1972 include marine plants and miscellaneous, and after the 1960's generally use live or "nominal" weight of the fish. Another notable change took place in the "conversion factor" used to back-calculate live weight from dried saltfish. While old departmental figures derive from a lower conversion factor, in my graphs relating to salt cod I have used 4.88 to 1, a ratio employed by the Northwest Atlantic Fisheries Organization (N.A.F.O.). Finally, when N.A.F.O. and its predecessor I.C.N.A.F. (International Commission for the Northwest Atlantic Fisheries) present landings statistics, they generally mean finfish such as cod and herring, while omitting shellfish such as lobster.

Common types of Atlantic fishing gear appear on pages 222 and 236; most of these standard methods also find use on the Pacific. A Pacific troller appears on page 204. The book includes a list of acronyms and a glossary. Some notes on terminology: "fishery officer" can mean either the Fishery Officer of today or officials in the past with similar duties but different titles, such as Inspector. I use "fisherman" rather than "fisher" because that is the historical usage, and because female fishers, in my experience, prefer to be called fishermen. As for Aboriginal peoples, I most often use "Native," but sometimes "Indian" where that usage seems best to suit the context. Finally, up until the 1940's, I tend to distinguish "boats" from the larger decked "vessels"; afterwards, reflecting industry and department usage, boats and vessels are often synonymous.

The maps are for illustration only, and the same could be said of the text, which does not aim to present legally accurate information, but only a reasonable guide to what happened in fisheries management. The outline offered is mine, to the best of my ability. Many aspects could use more research (for example, the pre-Confederation origins of the Fisheries Act), and I know readers knowledgeable in fisheries will find some items missing or mistaken. That being said, I and those who helped me have taken every care with this narrative, and the publisher, the Department of Fisheries and Oceans, and the author can accept no responsibility for errors or omissions. We do however welcome suggested amendments for any future edition.

Finally, I thank Marie McCormack, my wife, not only for her help and support, but also for putting up with the growing number of historical file boxes, now more than 200 of them, that have co-occupied our dwellings over the past quarter-century.

PART 1:
THE FISHERIES BEFORE CONFEDERATION
CHAPTER 1.
The Aboriginal fishery

For millennia before the Vikings lighted on Canada's shores, hundreds of thousands of Aboriginal people depended on fishing for survival. Fish abounded. Spawning salmon could mass so closely in the water that they pushed many fish high and dry on the bank. But poverty could arise amid plenty. Fish could be absent from rivers or coves for much of the year, and when spawning, could be scarcely edible. Natural cycles could wipe out expected runs. Air-drying or smoking fish for the winter took long weeks of work; even then, wild animals might steal the caches of dried or frozen fish. As game and fish fluctuated, most Aboriginal groups moved from place to place. The quest for food never ended, and scarcities kept the population in check.[1]

Fish wove themselves into religion and culture. "The unity of the universe meant that all living beings were related—indeed, were 'people,' some of whom were human—and had minds.... So did some objects that the Western world considers to be inanimate; for example, certain stones, under certain conditions, could be alive or inhabited by minds."[2]

The First Peoples used almost all the basic fishing techniques of today. They made nets from kelp, roots, plants, and caribou thongs to entangle or entrap fish. They developed many varieties of hook and spear. They often attracted fish with torches. And they made great use of river and tidal weirs, particularly in the extensive fishery of the Pacific coast.[3]

Pacific fishery used "modern" methods

The Pacific Northwest supported perhaps as many as 200,000 people. Less migratory than most Native groups, they constructed wooden lodges which could last for generations. They took large quantities of salmon—by one estimate, as many as 17.5 million salmon a year.[4] Besides these and other fishes, they took sea mammals including sea otters, sea lions, seals, and whales. They dried and stored fish for the winter, and followed "grease trails" into the interior to trade the oil of the eulachon, or candlefish. Bands also exchanged such goods as blubber, oil, herring spawn, and dried fish and clams among themselves.[5]

Weirs, which partly barred rivers and streams, held great significance. (A "weir" generally means a barrier that lets water pass through but confines fish.) When Alexander Mackenzie, the first known man to cross the continent, reached the Pacific "from Canada, by land" in 1793, he asked if he could examine a weir in the upper reaches of the Bella Coola River. The Indians felt it was too important to show him. From a distance, Mackenzie marvelled at its ingenuity. The Native people embedded small trees in the river in an intricate fashion, reaching about four feet above water level and barring nearly two-thirds of the stream, to trap the salmon attempting to leap over.

Intricate wooden fish traps in the Stikine River district, British Columbia, photographed ca. 1900. (Library and Archives Canada, PA-186340)

The Indians also trolled and jigged for Pacific cod, halibut, and salmon. Some tribes speared or clubbed salmon migrating upriver by night. They sometimes used trawl-like dragnets slung between two canoes. Other methods included bagnets, gillnets, dipnets, and such devices as the Nootka's rake-like gaff, a long pole with a row of bone spikes, which they would draw through shoals of herring or eulachon.[6]

A federal fisheries report for 1886 described methods used in the black cod fishery, including a longline with many hooks. (Presumably ancient to the Haida, the longline technology was still new and controversial on the Atlantic coast.) The Haida used lines of giant kelp that they bleached, stretched, dried, smoked, and knotted together. They would attach 75–100 hooks to a line, spacing them about two feet apart. The hemlock hook's special shape prevented it from getting snagged on the rocky bottom around the Queen Charlotte Islands. Between the hook's two wooden arms, the fisherman would insert a baited stick. When the biting fish dislodged the stick, the two ends of the hook sprang together to hold it, without penetrating its flesh. The stick floated to the surface as a signal to the fisherman. Finally, the Haida fisherman would release clever slipknots holding the heavy stones he used for sinkers. This made hauling back an easier task.[7]

Every April and May, however uncertain the weather, the Nuu'chah'nulth (Nootka) of Vancouver Island put to sea in dugout canoes to hunt whales. To the harpoon they attached a line bearing sealskins or fish bladders filled with air. When first struck, the whale would dive, carrying the air-buoys with it. Each time it resurfaced, the Indians would throw more harpoons. Finally the air-buoys would keep the whale on the surface until it died.

The Pacific peoples also harpooned sea otters and clubbed sea lions. More elaborately, they lured seals within arrow-shot by wearing wooden masks, covering their bodies with branches, and imitating the actions of a basking seal.[8]

The Atlantic fishery

When Jacques Cartier in 1534 sailed into the Strait of Belle Isle, Port au Choix had already been a major fishing and sealing point for thousands of years. Cartier noted Indians in birchbark canoes catching seals in the strait, and others fishing mackerel with nets in Gaspé Bay.[9] Seven decades later, describing his 1606–1607 sojourn at Port Royal on the Bay of Fundy, Marc Lescarbot wrote:

> The savages do make a hurdle, or weir, that crosseth the brook, which they hold almost up straight, propped against wooden bars, arch-wise, and leave there a space for the fishes to pass, which space they stop when the tide doth retire, and all the fish is found stayed in such a multitude that they suffer it to be lost. And as for the dolphins, sturgeons, and salmons, they take them after that manner, or do strike them with harping-irons, so that these people are happy. For there is nothing in the world so good as these fresh meats.

Atlantic peoples used many other methods: spears, jigs, set-lines with baited hooks of bone or hardened wood, gillnets to entangle fish, and seine nets generally dragged from shore or under the ice to encircle fish. They would carry torches in canoes at night, to attract and spear salmon, sturgeon, or eels.

In many places, of course, they could gather shellfish by hand. Lescarbot described great quantities of mussels, lobsters, and crabs being taken without net or boat, along with cockles, sea urchins, and scallops "twice as big as oysters." The Mi'kmaq and Maliseet of the Maritimes took small whales as well as seals, walrus, sturgeon, and deep-sea swordfish.[10]

Inuit depend on sea mammals

Most of the Arctic Inuit lived most of the time on the coast, moving inland in spring to hunt caribou or fish in inland lakes. They depended heavily on fish and sea mammals, using hooks, spears, stone weirs, and other means to capture them. They would sometimes set lures, such as a bear's tooth, at holes in the ice, and spear the fish that approached. To capture seals, they sometimes used a kind of square seine made of baleine.[11]

Like the Nootka on the Pacific, the Inuit caught whales by using air-buoys to tire them and bring them to the surface. They also used a toggling technology. To a spear they would attach a sepa-

---

Lescarbot on the Native fishery

Marc Lescarbot in *Nova Francia* (1606) wrote about the return of fish after the winter:

[T]hose citizens of the sea, after the gusts and furious storms be past, they come to enlarge themselves through the salted fields: they skip, they trample, they make love, they approach to the shore, and come to seek the refreshing of fresh water. And then our said savages, that know the rendezvous of every one and the time of their return, go to wait for them in good devotion to bid them welcome. The smelt is the first fish of all that presents himself in the spring. ... There be certain brooks where such schools of these smelts do come that for the space of five or six weeks one might take of them sufficient to feed a whole city. There be other brooks where after the smelt cometh the herring, with like multitude.... The pilchards do come in their season, in such abundance that sometimes ... in less than the space of an hour we had taken enough of them to serve us for three days. ... The dolphins, sturgeons, and salmons do get to the head of the river in the said Port Royal, where such quantity of them are that they carried away the nets which we had laid for them.... In all places fish aboundeth there in like manner....

An 1813 depiction of torch-light fishing in North America. (Artist: John Heaviside-Clark. Library and Archives Canada, C-41915)

rable headpiece, to which they would tie a line. When the hunter speared the animal, the spear handle came loose, while the toggle embedded itself in the whale's flesh. The Inuit could then use the line to capture the whale.

According to Native historian Olive Dickason, the Inuit whaling technology was, from the 13th to the 17th century, the most advanced in the world. When "combined with European deep-sea ships, that technology led to the efflorescence of world-wide whaling."[12]

Inuit also exploited the walrus. And they depended particularly on seals, not only for food and clothing but also for seal oil to burn in stone lamps. They "dreaded most a sea-goddess reputed to control the weather and to regulate the supply of seals."[13]

Native conservation worked for many centuries

Although less dependent on fish than coastal peoples, inland groups made good use of them. Among agricultural tribes such as the Iroquois, women generally tilled the fields in summer so that the men could fish. The least dependent on fish were the peoples of the prairies, where lakes were scarce, rivers muddy, and fish less plentiful.[14]

The Amerindians lacked the paraphernalia of modern fisheries management, with multiple rules controlling size of nets and gear, type or volume of catch, seasons, fishing permits, and so on. Fish

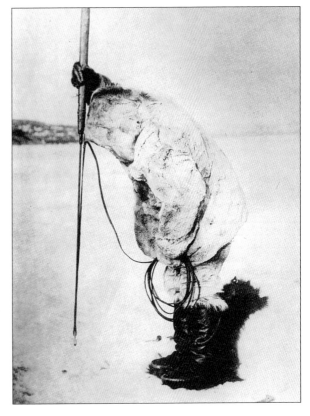

Seal hunting in the 20th century. (Library and Archives Canada, C-73178)

Inuit men spearing fish at a fish trap, photographed 1916. (Library and Archives Canada, PA-165664)

were plentiful, conservation problems few. Still, in the Pacific Northwest, some customs, such as taboos against throwing refuse into a salmon stream, had a management significance. Diane Newell writes that "for every group, traditional laws governed use and application of fishing technology...." For example, weirs would span only part of a river, and the downstream groups would open their barricades for the benefit of upstream users and escapement to the spawning beds.[15]

The Indians closely observed salmon life stages. The Nuu'chah'nulth stocked salmon streams by transplanting eggs. Experts could predict the timing of runs and the potential effects of weather and sea conditions.[16]

Rules of usage could be entwined with spirituality. Many tribes had rituals, myths, and taboos concerning the salmon. Solemn ceremonies took place at the arrival of the salmon run; these may have helped escapement to the spawning grounds by delaying the fishery. No one could throw refuse into a salmon river, lest the salmon spirit resent it. Among the Bella Coola people, according to Roderick Haig-Brown, the penalty for throwing

refuse during the salmon run was death.[17] Reflecting the importance of the sea, the Haida reserved nearly all their sacrifices for the ocean spirits, especially the killer whales.

Newell notes that "... all systems of resource management in aboriginal Canada relied on communal property arrangements ..."[18] Building a large weir, for example, might require effort from the whole community. The great anthropologist Diamond Jenness wrote, "Every man contributed his labour to the building and maintenance of the weir or pound, and every man was entitled to his share of the booty. ... At the weirs each man retained whatever fish he caught, but allowed no family to remain in want."[19]

Even so, individual or group rights sometimes manifested themselves. In a Pacific village, a higher position in the social hierarchy might carry not only a title but also other privileges, such as ownership of a certain fish-weir. Indian families or clans are said to have owned specific fishing places over hundreds of years, gaining them by inheritance or marriage, and sometimes leasing them out to others. As late as the Hell's Gate landslide on the

Fraser River in 1913, federal fisheries engineers seeking to repair the damage had to deal with Indians who owned certain fishing-rocks.

As Diane Newell notes, the Aboriginal salmon-management system in British Columbia "sustained yields for several thousand years." It not only produced large harvests, but also "assured everyone adequate stocks of fish over the long term."[20]

Europeans disrupt the fishing life

It is well known that in the centuries after Columbus, Cabot, and Cartier, the Europeans brought new diseases and enormous dislocation to the First Peoples. It is less often remarked that the newcomers sometimes interfered with food supplies from the fishery, the disruption getting worse as time went on. Whale and walrus hunting in the Arctic in the late 19th century caused widespread starvation among the Inuit.[21] Diamond Jenness noted in the 1930's that after decades of European influence "over large parts of the Arctic and sub-Arctic the Eskimo are now worse clad, and more ill-nourished, than in the days of their isolation."

Contact with the Europeans ... revolutionized the economic conditions in every part of the Dominion.... In British Columbia the canneries that sprang up at the mouths of the Columbia, Fraser, and Skeena rivers depleted the salmon on which the Indians had depended for their daily food. ... Throughout the whole country, indeed, there was a serious diminution in the food resources of the tribes that depended on fishing and hunting, all the tribes, that is, except the Iroquois and some of their Algonkian neighbours who cultivated maize. No longer was each tribe a self-contained and self-supporting unit, but ... all alike found themselves inextricably enmeshed in the economic system forced upon them from without.[22]

CHAPTER 2.
1500–1713: Fisheries spread settlement in North America

As explorers rounded Africa and crossed the Atlantic, the seaward-looking regions of western Europe moved into the forefront of world history. The ocean fostered both conquest and commerce, with much of the trade deriving from the vast stocks of cod off North America. England and France would become the two great fishing powers in the northwest Atlantic, using the fishery to feed their population, increase their commerce, and build up their naval strength.

But they fished differently, which had great consequences. The French had ample supplies of salt, which enabled salting fish at sea. The English, with less salt, depended more on shore-drying of fish. The shore fishery helped England, though smaller than France, to outpace her in settling North America.

Waters swarm with fish

A well-known account of Cabot's first voyage in 1497 said that the sea he crossed "is swarming with fish, which can be taken not only with the net but in baskets let down with a stone, so that it sinks in the water. ... [Cabot's companions] say that they could bring so many fish that this kingdom would have no further need of Iceland, from which place comes a very great quantity of the fish called stockfish."[1] (Stockfish were cod dried without salting, a product associated with Scandinavia.) Another description said that the sea off the new land "yeeldeth plenty of fish, and those very great, as seales, and those which commonly we call salmons: there are soles also above a yard in length: but especially there is great abundance of that kind of fish which the Savages call baccalaos [cod]."[2]

The Grand Banks alone covered more than 30,000 square miles. As vessels moved on to coastal waters, they still reported great abundance. A ship off the Magdalen Islands, late in the 1500's, caught 250 cod in an hour with only four hooks. Voyagers off Maine in the same era reported catching fish as fast as the hooks came down; "some they measured to be five foot long and three foot about." Others marvelled at the amount of fish on the Labrador coast.[3] (The term "Labrador," probably of Portuguese derivation, in those days took in the mainland side of the Strait of Belle Isle and much of the North Shore of the Gulf of St. Lawrence.)

In the 1600's, the governor of New York wrote that nearby waters had six-foot lobsters. Other reports told of "codfish as big as a man" (indeed, as late as 1895, a six-foot cod weighing 211 pounds was caught off Massachusetts).[4] Fishermen caught whales just by rowing offshore. A voyager to New England wrote of the "infinite multitudes" of mackerel.[5] The governor of Acadia wrote that the sea was "paved with salmon." Even allowing for tall tales, modern estimates have confirmed that salmon abundance was prodigious.[6] Other species equally abounded, some for generations to come. Even at the end of the 19th century, Atlantic coast herring would sometimes leave spawn on the beaches two to three feet deep, now unheard of.[7] In Labrador in the 1860's, a single haul of a beach-seine could take 4,000 or 5,000 cod. Beach-seines, which depend on fish being abundant enough that schools come almost to the shore, remained common into the 20th century.

Shore fishery requires great space and labour

Europe was hungry for fish from the new grounds. The European religious calendar, which then called for some 150 meatless days a year, accentuated demand.[8] In the first half of the 1500's, fishermen from many countries would carry on, as described by the Newfoundland historian D.W. Prowse:

> ... a great free trade; oils and wines, and fruits of France, Spain, and Portugal, were exchanged for English cutlery and West of England cordage, cloth-hats, caps, and hosiery. ... Of the French and Portuguese, some fished on the banks and brought their fish home green, but the majority met in St. John's every year, spring and autumn. From this harbour they spread themselves out, north and south, to carry on the shore fishery, each nationality going together in small companies of from four to six ships; returning to St. John's as a rendezvous each nation's ships sailed home together in a convoy.[9]

The first European fishery took place near the shores of Newfoundland, Cape Breton, and the Strait of Belle Isle. Men fished from the main vessel or from smaller boats brought along with them. Usually, three men fished to a boat, and two dried the fish on shore. They would head, gut, and split the cod, removing the backbone, then salt them, pile them in layers to squeeze out the moisture liberated by the salt, and let them drain for days. As well, they dropped the cod livers into a vat, where they would gradually render into oil. After the cod had taken the salt, workers would wash the fish, pile them in stocks, and finally spread them to dry in the open air.

Drying could take several days or even weeks, depending on outdoor conditions, with men turning over the cod at intervals and otherwise tending them. Sometimes dried on the beach itself, cod more commonly went onto flakes—raised wooden platforms. As the fishery grew, millions of fish needed laying out flat. Flakes took up great amounts of shore space; conflicts arose as vessels vied for "fishing rooms."[10]

Marc Lescarbot, in *Nova Francia* (1606), described operations thus:

> There is in Newfoundland and in Bacaillos [Cape Breton] great number of ports, where ships lie at anchor for three months. At the very break of day the mariners do go two or three leagues off in the sea to take their lading. They have every

one filled their shallop [open boat] by one or two o'clock in the afternoon, and do return into the port, where being, there is a great scaffold built on the seashore, whereon the fish is cast, as one cast sheaves of corn through a barn window. There is a great table whereon the fish cast is dressed, as above said. After six hours they are turned, and so sundry times. Then all is gathered and piled together; and again at the end of eight days put to the air. In the end, being dried, it is laid up. But there must be no fogs when it is a-drying, for then it will rot; not too much heat, for it would become red; but a temperate and windy weather.[11]

Many variations of the drying procedure evolved over time, producing different cures for different markets. Salted, dried fish not only fed populations on the land, but also provided an ideal, compact, non-spoiling

View of a fishing stage at Newfoundland in the early 1700's, showing catching, cleaning, extracting of oil, and drying. (Library and Archives Canada, C-3686)

food for naval and other seamen in the great fleets of western Europe.

Bank fishery requires more salt

By the mid-1500's, some vessels began making trips only as far as the offshore banks, returning without touching the coast. The shorter distance let some vessels make two voyages in a season. While the shore fishery took place mainly in June, July, and August, the bank fishery (also known as the green fishery) enjoyed a much longer season. But bank fish required more salt than shore-dried fish.[12]

Lescarbot wrote that in the French bank fishery,

Fifteen or twenty (more or less) mariners have every one a line ... of forty or fifty fathoms long, at the end whereof is a hook baited, and a lead of three pounds weight to bring it to the bottom; with this implement they fish their cods, which are so greedy that no sooner let down but as soon caught, where good fishing is. The fish being drawn a-ship-board, there are boards in form of narrow tables along the ship where the fish is dressed. There is one that cutteth off their heads and casteth them commonly in the sea; another cutteth their bellies and garbelleth them, and sendeth back to his fellow the biggest part of the backbone which he cutteth away. That done, they are put into the salting-tub for four and twenty hours: then they are laid up; and in this sort do they work continually (without respecting the Sunday, which is the Lord's day) for the space of almost three months, their sails down, until the lading be fully made. And because the poor mariners do endure there some cold among the fogs, specially them that be most hasty, which begin their voyage in February: from thence cometh the saying that 'it is cold in Canada'.[13]

In the French bank fishery, a 100-ton vessel would have a crew of 15–18 men and provisions for six months. The fishermen fished directly from outboard stagings along one side of the vessel, on which each fisherman placed a half hogshead reaching to his waist. Pieces of herring or cod entrails served as bait. "The catch might vary from nothing to between 25 and 200, or, exceptionally, 350 or 400 a day, about the limit of a fisherman's capacity."

Boys took the fish to those who dressed them. Salters then salted them in layers and left them for three or four days to drain. This treatment was then repeated, and the fish were ready to be barrelled and taken to France. Although wet-salted fish from the bank fishery would later get dried on shore, the cure was inferior to shore-drying at the outset.[14]

Canada's first oil industry

The Europeans sought oil as well as fish. Lescarbot noted that

of the livers of cods our Newfoundland men do make oils, casting those livers into barrels set in the sun, where they melt of themselves. There is great traffic made in Europe of the oil of the fish of Newfoundland. And for this only cause many go to the fishing of the whale, and of the hippopotamuses which they call the beast with the great tooth, or the *morses* [walrus].

Lescarbot described whaling methods near Tadoussac on the St. Lawrence, a well-known habitat of white whales (beluga). For rendering the oil, "They cut her in pieces, and in great kettles do seethe the fat, which melteth itself into oil, wherewith they may fill 400 hogsheads, sometimes more and sometimes less according to the greatness of the beast, and of the tongue commonly they draw five—yea, six—hogsheads full of train [oil]."[15] Fish and whale oil, and later seal oil, found many uses in Britain and Europe, for example in soap-making and leather tanning, and, of growing importance, as fuel for household and city-street lamps.[16]

Fishery creates many related jobs

Then as now, the fishing industry created a great deal of related work for boatbuilders, provisioners, and others. Fishing vessels themselves might carry sailmakers, blacksmiths, and carpenters. Vessels would leave home with large quantities of salt and other supplies such as bread, beef, pork, beer, tar, and candles, along with fishing supplies. The fishermen worked on a share system.

Leaving home to spend several months in the shore fishery, working hard in all weather, staying aboard the boat or in a shack on the beach, with a dull diet, few books (for those who could read), and no female companionship, must have been hard enough. The bank fishery was worse, with the drinking water limited and foul-smelling, the food monotonous, and the bedding unchanged. The vessel tossed and turned without end, often in the fog, with the captain none too sure of latitude, having hardly a concept of longitude, and half-lost half the time. But the thousands of fishermen left no accounts of their life for posterity.[17]

Europeans joust for supremacy

Portuguese, Spanish fisheries flourish and fade

Portugal had a seafaring nature, weak agriculture, cheap salt, and a Catholic population obliged to eat fish often. All this fostered a northwest Atlantic fishery from at least 1506. "From Aveiro alone sixty vessels went every year to Newfoundland, and later, in 1550, the number increased to 150." Portuguese fishermen explored harbours from Cape Breton to Labrador, but concentrated on the Avalon Peninsula, and left many place names in Newfoundland.[18] It appears, however, that the Portuguese fishery in the 1500's was smaller than that of the French, Spanish, and English.[19] By the 1600's, the Portuguese fishery was dying out, to regenerate only in the 1800's.

Early in the 1500's, Spain, enriched by shiploads of gold which the conquistadors brought home from Latin America, enjoyed great strength and sea power. Taking over the Spanish side of Navarre in 1516, she gained intrepid Basque fishermen. A 1553 account reported a Spanish fishing fleet numbering 200 ships and 6,000 men in the northwest Atlantic. In the 1570's, typically more than 100 vessels (probably 50–60 tons each) would fish cod at Newfoundland, and 20 or 30 would chase whales.[20]

By the 1580's, relations between Spain and England had deteriorated. In 1583, Sir Humphrey Gilbert extended England's overlordship in coastal Newfoundland. In following years, English ships often attacked the Spanish. In 1588, Spain sent a magnificent Armada against England, and failed miserably. In 1603, Spain and England signed a peace treaty, which would last throughout most of the 1600's.

Spain's fishery continued at a reduced rate. In 1625, the Spanish Basque port of San Sebastien was still sending 41 ships with 295 shallops and 1,475 men to the westward.[21] But Spain was declining and her fleet fading into insignificance. Spain's weakness opened up a rich salt-cod market for the French and English in Iberia, where salt cod became a prized dish with thousands of recipes.

Basques take cod and whales

Though one speaks of the Spanish and French fleets, often their fishermen were better known as Basques. Even after Navarre, on the Bay of Biscay, fell under the rule of France and Spain, the Basques retained their distinctiveness. French and especially Spanish Basques frequented the North American coast in large numbers in the 1500's. Their traces linger in such place names as Port aux Basques, Trepassey, Placentia, Santa Maria (St. Mary's Bay), Port au Choix (Portochova), and Renews.[22]

Besides fishing cod, Basques dominated the whale fishery. Archaeological work has documented operations at a Basque whaling station at Red Bay, Newfoundland, which employed an estimated 2,000 men fishing and processing. Using open boats of 20-odd feet in length, they harpooned right whales, attaching drogues to the end of the line to hold back the whale's flight.[23]

In the last half of the 1500's, Dutch, French, and English vessels joined the whale fisheries, and some also fished walruses. But by the end of the 1500's, the number of right whales—the easiest ones to catch from small boats—around Newfoundland dropped sharply. The Basque presence faded with them, in the first half of the 1600's. As for the walrus, by the end of the

The Basque whale fishery was part of an international whale fishery that grew in different areas of the North Atlantic. This depiction shows fishing of whales and killing of bears at Greenland, 1790. (Library and Archives Canada, C-111499)

1600's fishing vessels had badly depleted them in the Gulf of St. Lawrence.[24]

No care for conservation

Whales and walrus were the first major examples of stock depletion in the northwest Atlantic. Fishermen no doubt noticed increasing scarcities and shook their heads, but no one seems to have considered limiting the fishery. Scattered fishery regulations already existed for streams and rivers. But the ocean was swarming with fish and whales, and if they got scarce locally, one could always move on to find more.

Anyway, who would do the controlling? Strong nations espoused freedom of the seas. Elizabeth I of England told the Spanish ambassador in 1580 that "the use of the sea and air is common to all; neither can a title to the ocean belong to any people or private persons, forasmuch as neither nature nor public use and custom permit any possession thereof." In 1609, the Dutch scholar Hugo Grotius, in his book *Mare Liberum*, noted that "the forest is easily exhausted of wild animals and the river of fish, but such a contingency is impossible in the case of the sea." Grotius's articulation of freedom for navigation and fishing would have lasting influence.

The fleets become huge

More and more vessels fished the northwest Atlantic grounds. They ranged from less than 40 to more than 100 tons. Anthony Parckhurst's account in 1578 gave a total of 380 ocean-crossing vessels from England, Spain, Portugal, and France, without counting the small boats they carried with them.

As England and France took dominance, their fleets varied with the fortunes of war and commerce. Although statistics were few and happenstance, yet it is clear the numbers were large. Some reports in the 1600's estimated that at various times, France and England each sent as many as 20,000 men to the northwest Atlantic. Even if to be cautious one cuts those estimates and posits only a force of 20,000 men by both countries combined, that would still be a large number, fishing cod almost entirely.[26]

French fishery spreads widely

In commerce, culture, and population, France in the 1500's held a clear lead over England. This strength soon expressed itself in the fishery.

In his voyages of 1534 and 1535, Jacques Cartier found French fishermen already present at both mouths of the Gulf of St. Lawrence. Voyaging into the Gulf and the St. Lawrence River, Cartier discovered great abundance of fish and whales, including white whales, or beluga, at the mouth of the Saguenay. His reports opened the way for a fast-growing fishery.

France's northwest Atlantic fishery developed in a widespread way. By 1544, the French had a presence at Cape Breton, Gaspé, and Anticosti, and their vessels reached eastern and northeastern Newfoundland.

With her supplies of cheap salt, France pursued not only the dry fishery but also the wet-salt fishery of the offshore banks, whence vessels could return without landing in Newfoundland. Parckhurst's account in

---

The multinational fleet

The English captain Anthony Parckhurst described the Newfoundland fleet in 1578 as follows:

[S]ince my first travell being but 4.yeeres, [the English] are increased from 30.sayle to 50. ... I am informed that there are above 100.saile of Spaniards that come to take Cod (who make all wet, and do drie it when they come home) besides 20. or 30. more that come from Biskaie to kill Whale for Traine [oil]. These be better appoynted for shipping and furniture of munition, than any nation saving the Englishmen, who commonly are lords of the harbors where they fish ... As touching their tunnage, I thinke it may be neere five or six thousand tunne. But of Portugals there are not lightly above 50.saile, and they make all wet in like sorte, whose tunnage may amount to three thousand tuns, and not upwards. Of the French nation and Britons [Bretons], are about one hundred and fiftie sailes, the most of their shipping is very small, not past fortie tunnes, among which some are great and reasonably well appointed, better than the Portugals, and not so well as the Spaniards, and the burden of them may be some 7000. tunne. Their shipping is from all parts of France and Britaine, and the Spaniards from most parts of Spaine, the Portugals from Aviero and Vianna, and from 2. or 3. ports more. The trade that our nation hath to Island [Iceland] maketh, that the English are not there in such numbers as other nations.

As touching the kinds of Fish beside Cod, there are Herrings, Salmons, Thornebacks, Plase, or rather wee should call them Flounders, Dog Fish, and another most excellent of taste called of us a Cat, Oisters, and Muskles; in which I have found pearles above 40. in one Muskle, and generally all have some, great or small. ... there are also other kinds of Shel-fish, as limpets, cockles, wilkes, lobsters, and crabs: also a fish like a Smelt which commeth on shore, and another that hath the like propertie, called a Squid....[25]

1578 counted 150 French and Bretons at Newfoundland. In 1599, it was said that about 100 French vessels went to the banks, some making two voyages. There were reports of 600 French ships in 1611, and of 400 French on the banks in 1630.

England's takeover of the Avalon Peninsula in the later 1500's pushed many French vessels elsewhere, creating a stronger presence at Nova Scotia, notably Canso, the Bay of Fundy, and Maine, as well as in the Gulf of St. Lawrence. French vessels took cod, walrus, and seals at the Magdalen Islands, and also fished at Prince Edward Island, the Caraquet Islands, Miscou, the Gaspé, and the North Shore of the Gulf. Fishing vessels frequently traded in furs as well; indeed, the fur trade grew out of the fishery. France's growing economy provided capital for fleet expansion. Complex ownership arrangements existed: sometimes the master owned the vessels, sometimes he owned part and various shareholders divided the revenue.[27]

French fleet peaks in 1600's

The French distant-water fleet stayed strong in the 1600's. Nicolas Denys noted in 1669 that the French dry fishery typically had 100–150 vessels, manned by Basques and people of La Rochelle, Brittany, and Bordeaux. A 200-ton vessel in the dry fishery carried about 50 men with provisions for eight or nine months; the catch would average 200,000 fish. Channel ports sent 200–250 vessels to the green fishery on the banks. A single banking vessel of 200 tons, with a smaller crew of 25 men, would produce 45,000–50,000 fish.

In 1678–1689, with the English embroiled in internal conflicts ending in the "Glorious Revolution" and greater parliamentary power, the French fishery reached its peak. In 1678, it was estimated that the French had 300 vessels and 20,000 men in the fishery, with about 60 large vessels around Placentia. Some accounts give even larger estimates. Ships in the shore fishery, depending on their size, could carry from four to as many as 20 boats.

At the end of the 1600's, with France fighting various wars, her northwest Atlantic fleet dropped to about 100. In 1710, during the War of the Spanish Succession, the fleet declined further, to 50 or 60 vessels. But this was still a significant number, and France in the next century would boost her fleet by subsidies.[28]

French settlements remain weak and scattered

Despite France's widespread fishery, her shore settlements in North America proceeded only slowly. Permanent habitation began in 1604, when Samuel de Champlain, Pierre Du Gua de Monts, and their followers overwintered at St. Croix Island in southern New Brunswick. Champlain moved the next year to Port Royal, in Nova Scotia's Annapolis Basin, then left Acadia for Quebec in 1608. When he died there in 1635, his settlement counted only a few hundred people.

Meanwhile, Acadia (which then meant mainly the territory surrounding the Bay of Fundy, although the term would expand) passed back and forth between rival governors, falling at times under English control. French settlement took place chiefly in the Annapolis Basin, Minas Basin, and Isthmus of Chignecto areas, the population reaching about 2,000 by the early 1700's.[29] The Acadians mainly farmed their pleasant land, although fishing was significant, and a fishing company was chartered in 1682. Local authorities at times issued fishing licences to New England vessels, but this was largely a pretence; they had no enforcement fleet to keep the English from fishing nearby waters or drying fish on shore.[30]

Fishery concessions fail to take hold

In the 1600's, France gave fishery concessions to various parties, especially in the Gulf of St. Lawrence, but these suffered from the lack of ports on the ocean. When Nicolas Denys in the 1640's set up a fishing post at Miscou, others seized it. His later concession of lands from Cape Breton to Gaspé never became strong. New Englanders mounted raids on Acadia, where they dried fish without much interference, and they also attacked Quebec. The French sometimes tried to control English fishermen by licences, and on one occasion seized eight English vessels.[31]

Outside the concessions, in free-fishing zones in the southern Gulf, good ports were scarce. French fishing ships fought over beach space. Officials in France wrote elaborate regulations spelling out the number of boats per ship and the amount of beach space per boat, favouring fishing ships over the few residents. It may be that the concessions and the bureaucratic rigmarole discouraged some who would otherwise have settled. In any case, none of the French coastal concessions in the Maritimes built up a truly strong resident fishery. A fishing company based at Cape Breton ultimately failed.

In 1662, the French authorities set up a settlement at Placentia, their first real base on Newfoundland. This fishery organized itself with small boats supplying bait, especially herring from Cape St. Mary's, for larger vessels. In 1681, French ships as large as 200–400 tons frequented Placentia, with an estimated 100 ships fishing the area from St. Mary's, on the southern Avalon Peninsula, to St. Pierre. Placentia survived raids from the English but never grew populous. As late as 1713, Newfoundland had only some 180 French people.[32]

Instead, New France took strongest root far up the St. Lawrence River. In 1663, Louis XIV, the Sun King, put New France under direct control. France sent soldiers, encouraged settlement, and set up an administration at Quebec. Jean Talon became Intendant of New France in 1665; the *Jesuit Relations* described his fisheries activities in a hopeful account:

The first thoughts of Monsieur Talon, Intendant for the King in this country, were to exert himself with tireless activity to seek out the means for rendering this country prosperous.

He was so successful in this that fisheries of all kinds are in operation; the rivers, being very rich in fish, such as salmon, brill, perch, sturgeon, and—without leaving the stream, even—herring and cod, which are prepared both fresh and dried, and the sale of which in France is very profitable. This year, trial has been made of these fisheries by shallops that have been sent out, and [they] have yielded large returns.

Of similar nature is the seal-fishery, which furnishes the whole country with oil; and yields a great surplus that is sent to France and to the Antilles. ... The white-whale fishery, which they hope to make successful with little expense, will yield oils of higher grade for manufacturing purposes, and in even greater quantity.

The commerce which Monsieur Talon proposes to carry on with the Island of the Antilles will be one of this country's chief resources; and already, to ascertain its profitableness, he is this year shipping to those islands fresh and dried cod-fish, salted salmon, eels, peas, both green and white, fish-oil, staves, and boards—all produced in this country.

But as permanent fisheries are the soul, and form the chief maintenance of commerce, he intends to establish them as soon as possible; and, to attain this end, he purposes forming some sort of company to plant the first of these and bear their initial expense.

This passage ends with a sentence on fisheries reflecting the chronic optimism of fishing enterprises: "In a year or two they will yield marvellous profits."[33]

To the degree it existed, fisheries management—still an unknown term—concerned itself mainly with questions of trade, development, and sovereignty. In 1664, France placed import duties on cod, to provide a protected market for her fleet. A regulation in 1669 allowed inhabitants of Canada to export cod to France; another of 1685 exempted New France's trade with the West Indies from duties.

In an early conservation regulation, the French authorities in 1684 forbade the use of the jigger, apparently because in taking one cod it could damage others, and when wounded fish fled others would follow. This pronouncement failed to resolve the situation; controversy about the jigger continued, with protestations that it had damaged the fishery on the Labrador coast.[34]

Resident fishery remains weak

Despite some official encouragement, New France's resident fishery failed to become Talon's "soul and chief maintenance of commerce." Vessels from England and New England were now plying between Europe, North America, the West Indies, and Africa, exchanging cargoes of rum, saltfish, slaves, and other goods. But New France found it difficult to develop such a "triangle trade." Fishing grounds were scattered, and many areas faced longer hauls to the markets. The winter ice afflicting northeast Newfoundland, the Gulf, and the St. Lawrence River shortened their trading seasons. Plentiful French salt helped the bank fleet and made shore-drying less necessary, thus pulling people seaward from the coast, while the fur trade pulled others to the interior. In English areas, fishing and settlement tended to reinforce each other; in New France, they held each other back.

Even up the St. Lawrence River, settlement was sparse. As late as 1663, when Boston alone had 14,300 people, all of New France had only 2,500. Far more French had gone to the West Indies, which had 15,000 French people and 12,000 slaves.[35] As historians often note, the French despite their far-flung explorations were less emigration-minded than the English. French policy after 1666 gave little encouragement to emigration.[36] And the Canadian climate discouraged settlement.

England promotes the fishery

In *The Wealth of Nations* (1776), Adam Smith wrote that:

> To increase the shipping and naval power of Great Britain by the extension of the fisheries of our colonies is an object which the legislature seems to have had almost constantly in view. These fisheries upon this account have had all the encouragement which freedom can give them and they have flourished accordingly.[37]

England's concern to build up the fishery was apparent early on. An act in 1548 forbade the naval forces from exacting levies of money or fish from the Newfoundland fishing fleet. Another law under Henry VII set a penalty of ten pounds for buying fish at sea or from a foreign port. Under Elizabeth I (1558–1603), even though the Tudors had turned England away from Roman Catholicism, a law obliged citizens to eat fish rather than meat on Wednesdays and Saturdays, and also exempted vessels carrying fish out of England from customs dues. This legislation seems to have speeded the growth of the fishery. England also encouraged the merchant marine.

Foreign legislation also gave a backhanded boost to the English fishery in North America. At first, English vessels fished and traded at Iceland, exchanging manufactured goods for stockfish, which they sold mainly to continental Europe. Anthony Parckhurst reported in 1578 that the English fleet in Newfoundland numbered only 50 sail, because of the popular trade with Iceland. But in 1580 Denmark, which controlled Iceland, imposed a licence fee against English vessels fishing there. This helped turn English attention to the northwest Atlantic, with Bristol and West Country (southwestern England) interests taking the lead.[38]

England sticks to the shore

With limited access to salt, and with experience in unsalted stockfish, the English specialized in shore-drying. They developed a hard, dry, non-perishable cure that traded well in the all-important Mediterranean market. For the dry cure, the English worked in areas of less humidity, chiefly the Avalon Peninsula. They at first tended to pick sites near the Portuguese, from whom they could get salt. The dry fishery depended on wood and shore space, and the English fishermen spread out along the coast to build stages, flakes, and dwellings. Their strong presence in their chosen areas helped scatter their French rivals to other areas.[39]

In their areas of strength, the English began to take a degree of overlordship. Although fewer in numbers at first, they often had larger vessels, and in southern Newfoundland were usually the "admirals" who kept rough order among vessels of whatever nation. The custom was usually that the first captain to arrive in an area became its admiral for the season. (The French had the same practice.) In 1634, the "Western Charter," issued by the Privy Council, formalized the practice.

On the Avalon Peninsula, the English had good harbours that were close to their overseas markets, and where their operations could build up a critical mass in a fairly confined space. As in New England, fishing and settlement encouraged one another. The English population was growing on both sides of New France, which had no strong coastal settlements.

England operates a "free fishery"

England's Newfoundland fishery took off in the late 1500's. Vessels had at first supplied the home market. Then wars disrupted France's fishery for a period; she needed imports. So did Spain, especially after British attacks weakened her fishing fleet, and the Armada turned into a disaster. Iberia became a strong, long-lasting market for the growing English fleet. The fish trade with Europe and the Mediterranean helped lay the foundation for England's widespread trading system.[40]

England in this period often granted trading monopolies, for example to the East India company, chartered in 1600, or to the Hudson's Bay Company later in the century. But the growing fishery by West Country interests developed independently, as many people surged into it. Investors put money into large vessels in both fishing and trading ventures.

English fishery booms

In the early 1590's, the English began sending vessels into the Gulf of St. Lawrence, initially after walrus, then whales.[41] But mainly they stuck to the near-shore fishery at Newfoundland. By the early 1600's, it was booming.

One report says that some 500 vessels left England every April. Besides bringing their own cargo back in September, these fishing crews would load trading vessels, which would take fish to the Mediterranean and there load other goods for England or North America. The English freighters coming to Newfoundland picked up the name "sack ships," apparently from the "sec" (dry) wine they would carry.

A 1615 report counted about 250 English sail on the Newfoundland coast, averaging perhaps 60 tons with a crew of 20, for a total of 5,000 fishermen. A small barque of 30 tons would carry two boats. A larger 100-ton ship with 40 men could operate eight three-man boats. Besides those 24 fishermen, there would be "7 skilled headers and splitters, 2 boys to lay the fish on the table, 3 to salt fish, 3 to pitch salt on land and to wash and dry fish."

With each of the eight boats catching an average 25,000 fish, the take from spring to fall would be 200,000 dried fish, plus 12 tons of cod oil and 10,000 green (salted but undried) fish. The vessel might take the fish to France, Iberia, Italy, or the British Isles themselves. The master and ship's company, the owners, and the victuallers would each get about one-third of the proceeds.

Another report, in 1621, had England's total overseas fishery employing 200 ships and 10,000 men, mostly at Newfoundland, with a few in New England. By 1634, accounts put the number at more than 18,000 men. In 1637, there were said to be 500 English ships, and in 1640 about 250 ships and 20,000 men.[42]

Carrying trade cuts into fishing fleet

Market and other problems weakened the English fleet from the late 1620's on. The Civil Wars of the 1640's further reduced the West Country fleet at Newfoundland to 100 vessels. As well, the carrying trade was attracting vessel operators out of the fishery. Rather than catching their own fish, more vessels were buying cured fish from small Newfoundland-based boats.[43]

Although the first great cod rush had cooled, the fishery remained important. With the navy and the carrying trade, it provided a third pillar of England's maritime empire. Especially after mid-century, English traders increased their commerce with Brazil and the West Indies; the rapid growth of slavery there and in Virginia created a growing demand for the poorer "refuse" grades of fish. The English would deliver fish, other goods, and slaves from Africa to southern areas, and bring back sugar, tobacco, and rum.[44]

England curbs the competition

In New England, the fast-growing colonial fleet began cutting into the English trade. England also faced carrying-trade competition in Europe, particularly from the Dutch (whose own maritime strength had grown with the herring fishery).

To protect their carrying trade, the English in 1651–1673 passed several trade and "Navigation" laws. Only ships owned in England or her colonies could carry goods from Asia, Africa, or America into England or her dependencies. Other European countries could carry their own goods to England, but could not pick up goods from elsewhere to bring there. These measures helped the English take much of the carrying trade away from the previously dominant Dutch. The Navigation Acts prompted a war with the Dutch, which the English won, taking supremacy over the Dutch at sea.

Under these laws, only vessels from England itself had full trading freedom. English colonies faced certain duties even when trading with one another, and could ship certain articles such as tobacco and sugar only to England. (They sometimes got preferential treatment for their goods, just as England sometimes got a partial or total monopoly in the colonies.) Continental European goods destined for the colonies were supposed to go first to England, though New England gained a partial exemption from this rule.[45] The British restricted both trade and manufacturing in their overseas lands, their perpetual idea being that the colonies should produce raw materials, Britain the manufactured goods, with colonial markets closed to anything but British and colonial goods.

Adjusted from time to time, and becoming ever more complicated, the Navigation Acts in their general approach prevailed until the middle of the 19th century. They reflected the mercantile theory, which came to dominance in the 1600's. Gold and silver were the basic wealth; countries should seek a favourable balance of trade by promoting exports and curbing imports. Economic theories come and go, and no one today espouses mercantilism in its original form. Still, the Navigation Acts accompanied, whether or not they caused, England's rise to maritime supremacy.

England encouraged the fishery itself not only as a source of wealth, but also as a basis of naval support. Laws exempted the fishery from duties on materials used (1660) and from all taxes (1663).[46] Meanwhile, to protect her own fishing and trading fleet, England curbed settlement in Newfoundland, to avoid colonial competition. But New England was already too strong for such measures.

Thomas Wesley McLean, a 20th-century Canadian historical artist, depicted early fishing craft used in the Gulf of St. Lawrence and along the Atlantic coast. (Library and Archives Canada, C-69713)

The fishery builds New England

Reports by voyagers in the early 1600's drew fishing vessels to New England. Captain John Smith wrote that for the colonies, the fishery was "their mine and the sea the source of those silvered streames of all their vertue."

The Plymouth Colony planted in 1620 harboured high expectations. Sir Ferdinando Gorges, who became Treasurer of the Council for New England, in 1620 obtained a charter covering the territory between 40 and 48 degrees (roughly from present-day New Jersey to Newfoundland). No one was to visit the coast without obtaining a licence from the New England Council; fishermen were forbidden to land or procure wood to build stages.

West Country vessels fishing off New England resisted enforcement, their representatives complaining to Parliament. Gorges's charter faded away in any case. Free fishing prevailed for ships from England, as for the residents.

But fishermen from England found it hard to compete with the fast-growing colonial fishery. New England fishermen first used small boats, fishing nearby grounds in warmer months, and drying the catch ashore. A winter fishery developed especially after 1630, taking cod that came near the shore to spawn. Complementing the summer fishery, it gave New England a base for settlement, agriculture, year-round industry, and trade.[47]

Laws favour fishery, demand quality

The government of Massachusetts (whose territory then included Maine) favoured the fishery. An act in 1639 exempted fishermen from military duty and exempted fishing vessels from taxation for seven years.[48] The government kept an eye on quality. In 1641, Massachusetts brought in rules to standardize pickled-fish containers, and to appoint gaugers to ensure sufficient fish per barrel. In 1651, the gaugers also began inspecting for spoilage and quality.[49] This seems to have been the first fish inspection service in North America. Each town was to choose a proper person as inspector, who was required to "swear by thye living God" that he would well and truly carry out his duties. Another law in 1692 stipulated production by drying only. The odd regulation addressed conservation; for example, Plymouth colony in 1670 set a closed season for mackerel. Shad, salmon, mackerel, and other river and estuarine fisheries were important.[50]

New Englanders begin to roam

New England fishermen soon developed an offshore bank fishery. Small decked vessels would go to nearer banks including George's and Jeffrey's, salting the fish at sea and drying them ashore. Soon larger vessels such as two-masted ketches were travelling further afield. New England fishermen reached Nova Scotia and Newfoundland, fishing and trading. Besides cod, they were soon fishing walrus and seals at Sable Island and as far away as Newfoundland.[51]

Whaling started from Long Island in the 1640's, perhaps earlier. Slow-swimming right whales yielded high quantities of oil. People would keep watch, row off to harpoon the whales, and tow them ashore to render them down for fuel oil. There were also various uses for baleen, the flat, flexible, water-filtering, food-collecting plates of horn-like material in the whale's mouth. Further east, Nantucket islanders started whaling in 1672, and gradually took the lead. New Englanders would continue chasing right whales throughout the 17th and 18th centuries.[52]

Meanwhile, England's distant-water fishery at New England declined. Possibly 40 or 50 vessels a year went to New England in the 1620's. By 1637, the number was down to 15. Local competition probably contributed; as well, the carrying trade was attracting vessels out of the fishery. After about 1660, the British ships abandoned New England in favour of more northern waters.[53]

For New England's fisherman-traders, Newfoundland, with few material goods, provided an early market. After about 1645, New Englanders began supplying them with foodstuffs such as flour, sugar, corn, and beef, and other goods such as lumber, despite England's customs and navigation rules. Complaints of the time held that Newfoundland was "a magazine of contraband goods," and that with all the rum available from New England or English vessels, the "fishers grow debauched."[54]

By 1650, New England was also trading with Virginia, England, Holland, France, Portugal, and Spain. The growth of slavery in the West Indies hiked the demand for poor-quality fish, and this became New

Depiction by Thomas Wesley McLean. (Library and Archives Canada, C-69715)

England's primary codfish market. Vessels might fish in the summer off Nova Scotia, then in winter make a couple of trading voyages to the West Indies, bringing back sugar, molasses, and rum, goods which further aided their Newfoundland trade.[55]

New England becomes a powerhouse

By 1662, Boston had an estimated 300 vessels doing long-distance trading. New England's fishing and trading fleet kept growing. In 1670, there were as many as 30 New England shallops fishing at Port Rossignol alone in Acadia, and 12–15 large vessels at La Have. By 1700, New England had well over 200 larger ships, and Boston and Charleston were clearing a thousand vessels a year. In 1708, it was claimed that 300 New England vessels had been on the coast of Acadia. And in 1713, an estimated 300 vessels took part in the trade between New England, Nova Scotia, Newfoundland, and England.[56]

Fishing, with its complementary activities of lumbering, boatbuilding, shipping, and trading, had become the foundation of New England, as later generations recognized. The codfish appeared on currency, became the Massachusetts state symbol, and was depicted on automobile licence plates for many years in the 20th century. Remembering the colony's origins, the legislators of Massachusetts still keep a "sacred cod," a wooden model, in the State House.

Newfoundland colonizes against the law

England allowed free fishing, rather than monopolies, at both Newfoundland and New England. Although exerting some control through the Navigation

Acts and other laws, she let New England grow. England treated Newfoundland differently, constricting her freedom.

The English government at first supported colonizing attempts, none of which took strong hold. The island became a fishing station, which England's West Country interests wanted to use solely for the benefit of their own distant-water fishery. The West Country believed in free fishing—but only for their boats, not for Newfoundland residents. Parliament began to oppose settlement more than support it. English policies, although not always strictly applied, were enough to retard settlement and stunt the colony's economic and political growth.

Chartered companies fade away

As early as the 1500's, some fishermen may have stayed for the winter as caretakers. Sporadic attempts followed to colonize through chartered companies. John Guy of Bristol and his associates in 1610 received a charter for the territory between Cape Bonavista and Cape St. Mary's. The charter carefully specified that "there is no intent of depriving [fishermen thereabouts] of their former right of fishing."[57] Still, Guy in 1611 set some regulations affecting the Newfoundland fishery. Among other things, these forbade throwing anything harmful into the harbours (there being a common complaint about large stones used for pressing fish being thrown overboard), set out the amount of beach space that "admirals" could get, and stipulated that fishing stages be left alone.

Fishing interests from the West Country argued against "planters" (settlers) and sometimes attacked them, burning down their buildings. Confusion over rights encouraged piracy and theft. Guy's colony gradually lost strength and faded away. Other attempts at orderly settlement, including those of Calvert (Lord Baltimore) and Vaughan, at Ferryland and Renews, left no lasting impact.[58]

Far from London, fishing ships made sport of the law. Captains would race to reach the harbours first and become "admiral." Crews might steal the boats and salt others had left there through the winter, or burn the planters' stages and mills. They would fish on Sundays, and ruin anchorages by dumping stones in the harbours. By wasteful cutting of wood for stages and buildings, and "rinding" the bark of trees to cover and protect fish on the flakes, they destroyed many woodlands of the Avalon Peninsula. "By 1600, the east coast of Newfoundland—probably much of the shoreline between Ferryland and Bonavista—was becoming an ecological mess."[59]

The Western Charter

The Western Charter, issued in 1633–1634 by the Privy Council, tried to bring some order, while protecting West Country interests. It regularized the authority of the fishing admirals. The first captain to enter a harbour would be admiral, but would be able to claim only the shore space needed, with a small extra space to reward his first arrival. Free fishing would remain but under stronger regulations, governing such matters as protection of fishing stages, bait, and bait seining.[60]

In a final attempt at organized colonization, David Kirke in 1637 received a charter allowing him to take over Baltimore's colony at Ferryland, and giving him authority to tax all foreigners buying fish in Newfoundland, by taking part of their production (this produced French complaints to King Charles II of England). But Kirke's charter had strict limits. All houses between Cape Race and Bonavista had to be six miles inshore. The planters could fish, but with no monopoly. The charter protected free fishing by the English fleet. Even so, the West Country interests found Kirke too aggressive in pressing his privileges, and got him recalled in 1651.[62] Future settlement occurred in a disorganized way, as fishermen from the fishing ships took to the new land.

Crew members, boat-keepers settle on shore

Although fishing-ship crews had at first fished from the vessels, soon the men began fishing from small boats brought over with them or built on shore. They got used to the coves and bays. When the ships left for the winter, the captains often left crews behind to cut wood for flakes, stages, boats, and houses and to look after the fishing premises. A few chose to stay permanently and even to fish for themselves, often building their houses in the safety of creeks and coves where the fishing ships were unable to anchor. "These caretakers, plus the occasional deserters from the fishing ships and the remnants of several largely unsuccessful colonization attempts, became the first permanent European residents."[63] Early residents sold their fish to New England traders and English sack ships.

Beginning in the 1640's, fishing ships sometimes brought along other fishermen known as "bye-boat keepers," who kept their own small boats at Newfoundland. Bye-boat keepers would bring their own crews to work with them for the season. Over time, the bye-boat keepers, like the caretakers, contributed to settlement. Wives came over as well. By about 1650, the English-dominated areas of eastern Newfoundland, running from Cape Bonavista to Trepassey, had about 1,500 permanent residents, including 350 women and children, scattered among 30–40 settlements. These residents employed another 1,000 fishing "servants" who came over for the season.[64]

"[R]esidents often returned to the West of England to spend the winter," writes Shannon Ryan, historian of the saltfish trade, "and bye boat keepers often spent a winter on the island. Similarly, fishing ship captains sometimes remained in Newfoundland to look after

Excerpts from the Western Charter of 1634

Charles, by the Grace of God, King of England, Scotland, France, and Ireland, Defender of the Faith, and so forth....

... Whereas, the region or country, called Newfoundland ... which we hold, and our people have many years resorted to those parts, where, and on the coasts adjoining, they employed themselves in fishing, whereby a great number of our people have been set on work, and the navigation and mariners of our realm have been much increased ... until of late some of our subjects of the realm of England planting themselves in that country, and there residing and inhabiting, have imagined that for wrongs or injuries done there, either on the shore or in the sea adjoining, they cannot be here impeached ... [and] our subjects resorting thither injure one another and use all manner of excess, to the great hindrance of the voyage and common damage of this realm...

1st. If any man on the land there shall kill another, or if any shall secretly or forcibly steal the goods of any other in the value of forty shillings, he shall be forthwith apprehended and arrested, detained, and brought prisoner into England....

2d. That no ballast, prestones, or any thing else hurtful to the harbours, be thrown out to the prejudice of the said harbours....

3d. That no person whatever, either fisherman or inhabitant, do destroy, deface, or any way work any spoil or detriment to any stage, cook-room, flakes, spikes, nails, or any thing else that belongeth to the stages whatsoever....

4th. That, *according to the ancient custom*, every ship, or fisher that first entereth a harbour in behalf of the ship, be Admiral of the said harbour, wherein, for the time being, he shall receive only so much beech and flakes, or both, as is needful for the number of boats that he shall use, with an overplus only for one boat more than he needeth, as a privilege for his first coming....

5th. That no person cut out, deface, or in any way alter or change the marks of any boats or train-fats, whereby to defraud the right owners....

6th. That no person do diminish, take away, purloin, or steal any fish, or train, or salt....

7th. That no person set fire in any of the woods of the country, or work any detriment or destruction to the same, by *rinding of the trees*....

8th. That no man cast anchor or aught else hurtful, which may breed annoyance, or hinder the haling of seines for bait ....

9th. That no person rob the nets of others out of any drift, boat, or drover for bait ... nor rob or steal any of their nets....

10th. That no person do set up any tavern for selling of wine, beer, or strong waters, cyder, or tobacco, to entertain the fishermen; because it is found that by such means they are debauched, neglecting their labours, and poor ill-governed men not only spend most part of their shares before they come home, upon which the life and maintenance of their wives and children depend, but are likewise hurtful in divers other ways....[61]

that end of their company's trade, and other captains and mates often became bye boat keepers in times of depression."[65]

Why they settled

Despite the harshness of climate and life, the island had attractions: beauty, lots of fish, wood for heating and building, berries to pick and animals to trap, and freedom from authority. Aided by the growing trade with New England, residents might actually acquire more material goods than at home. Some Newfoundland fishermen undertook a second migration, finding passage on the trading vessels to New England itself.

Back in England, some argued that the country would do better to expand settlement at Newfoundland and improve its governance. England would still be able to use the fishery as a nursery for seamen. But this view generally lost out to West Country representations, for their own fishery and against a resident fishery.

Despite the rules of the Western Charter, little order existed. The fishing admirals were said to behave as kings and tyrants, abusing the fishermen. Contemporary accounts told of every house being a tavern. The English captains often forced the planters to buy wine and brandy along with their purchases of salt. There might be 100–200 men drunk in the small settlement of St. John's on the Sabbath.[66]

Planters gain minimal concessions

Gradually, the reality of growing settlement brought some concessions. The first governor appointed by the English government, John Treworgie, from 1653 to 1660 kept better order. New regulations appeared, aiming to protect trees and stages, and to prevent planters taking up too much beach room with their stages and flakes. Although Treworgie gave some protection to planters, West Country interests and government policy still opposed them. In 1661, the increase of bye-boat keepers brought a re-statement of rules in the Western Charter, together with an important addition to prevent settlement: "All owners of ships trading to Newfoundland [are] forbidden to carry any persons not of ships Company or such as are to plant or do intend to settle there."[68]

In 1671 and 1675, England passed more regulations against settlers, repeating such items as no settling within six miles of the shore, and even trying to force settlers to leave Newfoundland. Men from the fishing ships made new attacks against planters' property, which by now was often substantial. But in 1677, new laws made clear that planters could keep their houses and stages. And in 1680, more such regulations let planters live near the shore.[69]

English fleet fluctuates; Newfoundland fleet grows

The planters' fishery was gaining on the original distant-water fishery. A 1675 report put the English settlers in Newfoundland at 1,655 men using 277 boats and curing 69,000 quintals of merchantable fish, worth more than one-third the production value of the fishing ships. The England-based fishing fleet was still substantial—a 1676 report noted a total of 125 fishing vessels—but declined during disturbances at home. In 1684, for example, it included only 43 fishing ships at Newfoundland, with 1,489 men and 294 boats. The 304 resident boats outnumbered them, and were taking nearly as much cod as the fishing ships.[70]

The English fleet was fluctuating according to the trends of war and commerce. It rebounded by 1700, when 171 fishing ships operated 800 boats and the inhabitants 764 boats; declined during the War of the Spanish Succession (1702–1713); and rose again with the Treaty of Utrecht (1713), which confirmed English possession of Newfoundland.

While the English fishery rose and fell, that of Newfoundland trended upward. Residents hired fishing servants, often Irish lads, who came over for the fishing season. Sometimes these helpers settled in the new land. Residents numbered some 2,000 early in the 1700's.[71]

The West Country kept fighting the planters' encroachments on their accustomed fishing "rooms" (shore frontage with flakes and equipment).[72] In 1699, Parliament passed the Newfoundland Act, reaffirming the primary rights of the fishing ships to the shores and fisheries of Newfoundland. Still, the same act gave some comfort to settlers. Planters could keep buildings they had erected before 1685, as well as those erected after that year on seashore not used by fishing ships. The act also encouraged bye-boat keepers, allowing them to build houses and fishing rooms to be held as their own.[73]

At the beginning of the 1700's, then, Newfoundlanders had at least some recognition of their settlement. Even so, to try settling on the island was to enter a battle of nerves. For example, the founders of Twillingate had to sneak off in 1700 to start setting up in the new outport.[74] Residents had only limited rights of ownership, little law, no roads, few of the normal institutions of government, commerce, and agriculture—little but their fishing and their trade, via foreign carriers, with New England and the Old World. They lived on the shore of an island wilderness that they had colonized against the law.

Fish conquers fur

By the early 1700's, the English dominated at Newfoundland and from Maine to Georgia, on land and sea. Fishermen from England and New England were nipping at the edges of New France, sometimes drying fish in French territory, sometimes attacking French settlements. When in 1713 the Treaty of Utrecht ended the War of the Spanish Succession, England gained uncontested possession of Hudson Bay, Newfoundland, and mainland Nova Scotia. Thus the British took a major bite out of New France. Fifty years later, they would swallow the whole territory.

Why did France, the leading country in Europe, with a navy, merchant fleet, and fishery comparable to England's and more military strength overall, lose

North America? Partly because she had a smaller resident fishery, and therefore planted fewer people in the New World.

France by 1713 had made one of the greatest land grabs in world history. Her explorers and traders had pursued the Great Lakes system to its uttermost reaches, travelled down the Mississippi, and founded Louisiana. A handful of people had claimed the heart of the continent. But on the coast, France had only a skeleton crew. Her shore fishery remained small for reasons already noted—reluctance to emigrate; the pull of farming and the interior fur trade for those who did migrate; the abundant supplies of salt, which lessened the need for shore-drying; the ice that yearly afflicted the Gulf and the St. Lawrence River; the greater distance from markets; the excessive regulation, which impeded growth; and the monopolies and charters, which probably held back the fishery more than they helped it.

The English by contrast espoused free fishing, and specialized in shore-drying. They occupied shores at Newfoundland and New England that were closer to markets and, in New England's case, were ice-free and able to support a winter fishery, which completed a cycle of year-round employment. The fishery fostered the complementary trades of lumbering, shipping, and trading. Even in Newfoundland, where the British government opposed planters, the fishery drew settlers anyway. By 1700, the Newfoundland population had reached about 2,000. It had grown as much, against government opposition, as the Acadian population, also about 2,000 at that point, had grown with ostensible government encouragement. With English-speaking settlers far outnumbering French on the Atlantic coast, it was easy for negotiators of the Treaty of Utrecht to assign Nova Scotia to Britain.

The same population trends would continue, with English growth far exceeding French, despite high birth rates in New France. As late as the mid-1700's, the French numbered only some 65,000 in total; of them, the 10,000 by now residing on the Atlantic were mostly Acadian farmers tucked up in the Bay of Fundy. The British colonists concentrated along the coasts numbered about a million, more than ten times the strength of New France.[75]

The shore fishery was a prime factor in building British coastal dominance. That dominance aided the take-overs of Acadia, Newfoundland, Hudson Bay, and finally the fur-trading heartland of Quebec. In colonial North America, fish conquered fur.

Management mostly missing

One can view the elements of fisheries management as the following:

- Understanding the resource;

- Providing a basic system of laws, administration, and enforcement for conservation, protection, and sovereignty;

- Using the fish, which has such sub-elements as:

  - Setting a purpose, and relating the fishery to economic, social, or other goals such as sovereignty;

  - Controlling the degree of exploitation, to serve both conservation and the interests of business or pleasure;

  - Controlling access and allocation: that is, who gets the fish;

  - Ensuring decent handling of fish and quality of products;

  - Developing fisheries and markets; and,

  - Dealing with governance: that is, who makes the decisions, and how.

Which elements were evident by 1713? In the realm of understanding, people made limited observations of the habits and migrations of fish, and noticed at least some peculiarities of their biology.[76] But there was no systematic study. As for administration, there were glimmerings of laws and enforcement, as in the rough justice of the fishing admirals and the licences imposed by several authorities. But there was no thorough approach, and little thought of conservation in the sea fisheries.

France and England tried to use the fish for their best advantage, linking the fishery to goals of commerce and sea power, and passing many laws to encourage development. "Who gets the fish?" was already a vital question, reflected in charters, monopolies, and licences, and in various conflicts on the waters and shores. But in general, fisheries management existed not as an independent area, but as a by-product of political support for economic and military goals.

CHAPTER 3.
1713–1791: Rival powers defend, subsidize fisheries

The period 1713–1791 saw the map of North America change almost into its present form. New France disappeared, the United States took shape, and a string of British North American (B.N.A.) colonies foreshadowed Canadian Confederation.

In France and Great Britain (as it became after 1707), the distant-water fisheries remained important, partly because governments still believed in them as a nursery of seamen. In North America, the marine economy remained the strength of New England. Newfoundland lived entirely off the fishery. And the industry was growing in the Maritimes and Quebec.

But fishery economics were worsening. England and especially France resorted to subsidies to keep their overseas fleets strong. So did private-enterprise New England, and eventually the Maritimes. Rudimentary enforcement came into place for sea fisheries. And conservation surfaced as a worry in the river and estuary fisheries—especially for salmon, a species that wrote many laws.

At the outset of the period, France and New France still loomed large. Despite losing Newfoundland and mainland Nova Scotia in the 1713 Treaty of Utrecht, France retained Cape Breton, Île Saint-Jean (Prince Edward Island), and the mainland shores of the Gulf of St. Lawrence. As part of the treaty, Britain allowed the French to catch and dry fish in roughly the northern half of Newfoundland. As well, a large France-based fleet still fished the banks. In Newfoundland, there were few British settlers. In Nova Scotia, there were almost none; Acadians remained the majority.

On the coast, New England and the rest of the 13 colonies south of Nova Scotia dwarfed everything to the north, in terms of population. But the future Americans were still penned in between the Atlantic coast and the Appalachian Mountains, beyond which lay huge territories populated by Indians and claimed by France. On the coast and inland, the contest for North America was still open.

Newfoundland residents grow in strength

In Newfoundland, with the French gone from Placentia, the 2,000 or so English settlers had the run of the island. But in northern areas, they still had to contend with French fishermen. And they remained under the thumb of West Country interests.

British fishing ships and their small boats still dominated the shores of Newfoundland. Numbering less than a hundred during the War of the Spanish Succession, the fishing ships gained strength after the peace of 1713. But their shore fishery soon met a major setback. "A series of bad fishing years had begun with the severe and prolonged winter of 1713–1714. It had chilled the water along the coast, and it had been followed by 'the worst season for many years'."[1]

West Country interests responded by sending smaller ships to catch fish on the offshore banks and bring them ashore for curing by others. Vessels of 40–100 tons would carry seven to 12 men, fishing the banks from Trinity south to Trepassey. They found that "every fish brings its own bait with it to catch another with for by opening the maw you are always stock'd with fresh bait" from fish the cod had swallowed. The British vessels took some residents out as fishermen.[2] This was the first systematic offshore fishery with shore processing in Newfoundland. The fleet thereafter divided into coastal operations and bankers.

On the coastal side, a contemporary description of the fishing ships' boat fishery said that "men's food is beef, fish, pease, &c. Beer brewed with molasses and spruce. Go out of harbours in shallops, seven men and five men in a boat; catch fish with hook and line, first part of year their bait is muscles and lances; about middle of June bait is capeling, squid, and fresh herring, and end of year they fish with herring only—nets purposely for taking the sort of bait." On the fishing ships, wages displaced shares early in the century.[3]

The British fishing ships kept coming: in 1750, they numbered 93 fishing and sack ships, with about 1,600 men and 200 boats. Bye-boat men who operated their own small boats were also numerous; in 1751, about 550 of them employed more than 3,800 servants.[4] Among British vessels, the bank fishery was taking the lead; of the strong fleet in 1769, about two-thirds of the 354 vessels fished the banks.[5] Meanwhile, the fishing ships were losing ground on shore.

Residents begin to dominate fishery

Newfoundland was beautiful and had lots of fish, but life could be hard. In 1714, an observer noted that Newfoundland had about 500 families, but "their condition ... is more to be pitied than that of slaves and negroes."[6] That same year, the residents were reported to run some 360 boats, and the bye-

Vessel in the dry fishery, illustrated by the French academician Henri-Louis Duhamel du Monceau. (*Traité général des pesches* by Duhamel du Monceau, 1772, vol. 2, section 1, part 2, plate XIV, fig. 1)

boat keepers another 133 boats. These fleets produced about one-third as much fish as the 106 fishing ships with their 441 boats.[7]

As the fishing ships shifted more to the banks, emboldened residents took over some of their fishing rooms on the shore. The encroaching residents pushed fishing ships towards outlying ports,[8] causing new conflicts. In 1718, the *Report of the Lords Commissioners for Trade and Plantations to His Majesty* spoke of the need to remove inhabitants from the island.[9] By 1719, there were about 2,300 residents, with new settlements sprouting. The residents depended on hired helpers, who came from England or Ireland for the season. By 1750, about 850 fishing plantations located in hundreds of coves and harbours, and with a resident population of a few thousand, depended on some 5,400 migrant fishing servants.[10]

Settlers spread up the northeast coast, where the season was shorter and fish sold at a lower price. The settlers had to dismantle stages yearly to protect them from ice. Still, fish were plentiful; to handle them, eight men would fish to a boat, compared with six or seven in the south of the island.

While hook and line dominated the cod fishery, fishermen used nets for bait fish such as herring and capelin. Later in the century, they sometimes used beach-seines to catch cod schooling close to shore. Some people charged that cod-seines were destructive, because "a great quantity of small fish,

... after being inclosed in the sean (and not worth the attention of the person who hauls them) are left to rot."[11]

The truck system takes hold

As the resident population rose, some merchants, often English or English-backed, became dominant in the fishing communities. Complaints arose at mid-century that merchants were cornering the supply of provisions and selling them to the residents at exorbitant prices, "by which means they keep them poor and in debt, and dependent upon them."[12] The credit or "truck" system became strong. Merchants first extended credit to their own "servants," then to local boat-owning "planters" who made their own fish and sold it to the merchants.

A 1765 account stated the following:

> These merchants, store-keepers and boat keepers in order to secure the produce of the labour of the poor inhabitants to themselves, press their goods upon them in advance for that product, so that they contract debts without a possibility of paying them.... The inhabitants under these conditions of oppression and deprived of every view of bettering their condition, become abandoned to that dissolute way of life ... and remain under a slavish servitude.[13]

21

New Englanders bring goods, take people

As population grew, visiting traders enjoyed a sizeable traffic. Some came from Europe, but New England was taking the lead. Vessels trading for fish would bring rum, molasses, salt, pork, cider, clothes, and other supplies. Business kept increasing: in 1716, there were 31 trading ships from New England in Newfoundland; in 1774, there were 175. New Englanders often ignored British duties, their trade taking on the nature of a smuggle.

Many Newfoundland fishermen migrated aboard trading vessels to New England. Bye-boat keepers sometimes encouraged their servants to emigrate, since it saved their masters from paying their passage home. In the year 1717 alone, an estimated 1,300 fishermen emigrated from Newfoundland to New England.[14]

Government lags behind

By the late 1760's, the population numbered 11,000 or 12,000, if one included the 5,000–7,000 migratory servants. Residents now operated more than 1,200 boats, nearly 500 in Conception Bay alone. By the 1770's, residents were producing more than 300,000 quintals of fish, exceeding the production by the still-numerous British fishing ships.

Law and order took hold by fits and starts. From early in the century, naval governors exerted more authority. A resident governor took office from 1729. The fishing admirals retained authority over fishery matters until late in the century, to no one's satisfaction but their own: a 1751 report said they concerned themselves only with their own fishery, and Governor Palliser wrote in 1764 that "for the most part, they are ignorant, illiterate men...."[15]

Overall, Newfoundland remained backward, far behind New England. The home government sometimes took a disdainful attitude towards the fishing outpost. A privy council report in 1765 noted that the conflict between settling and fishing meant that neither was getting anywhere. There were complaints about the amount of rum drunk at Newfoundland, and the uselessness of the more and more numerous Irish settlers, who by 1750 outnumbered the English in St. John's.[16] Governor Palliser in 1765 wrote that the resident population never became good fishermen or seamen, and would in war likely join the enemy. Criticisms of the residents accompanied praise for the fishing ships. The renowned Edmund Burke wrote in 1766 that "the most valuable trade we have in the world is that with Newfoundland."[17]

New England fishery expands northward

While Newfoundland crept forward, New England and her southern neighbours were speeding ahead. By 1713, the colonies stretching from Maine to Georgia were vigorous, with their economies expanding landward. But the sea was still the centre of commerce, with New England the strongest force.

In the first quarter of the century (probably 1714 in Gloucester), New Englanders developed the schooner, a lean, fast, fore-and-aft rigged vessel that would dominate the northwest Atlantic for two centuries. New England schooners were most often two-masters, many of them around 65 feet long, with rounded bow and raised afterdeck.[18]

After 1720, with mainland Nova Scotia now British, New Englanders increased their offshore and northern fishery. Some fishermen settled in Nova Scotia, particularly at Canso. Many schooners made five trips annually to Sable Island Bank, Brown's Bank, and other banks near Cape Sable, and to Georges Bank, nearer home. They might trade along the way, for example, swapping fish at Canso for European goods brought by sack ships from England. New Englanders dried their fish on shore for export, mainly to the British and foreign West Indies.

With fishing, shipbuilding, and trading reinforcing one another, different areas built up their fleets. In 1741, Gloucester had about 70 schooners on the Grand Banks. Marblehead operated about 160 schooners. Massachusetts now had about 400 such vessels, and many ketches, shallops, and undecked boats.[19]

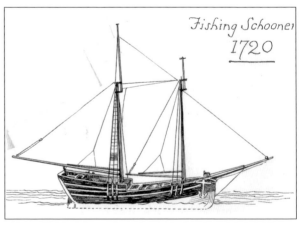

Schooner, 1720, depicted by Thomas Wesley McLean. (Library and Archives Canada, C-69714)

The growth continued. By 1765–1775, according to one estimate, the Massachusetts fishing fleet averaged 665 vessels of 40 tons, with 4,405 men. About 350 larger vessels of 70–180 tons, with an average crew of eight men, carried the fish to market.[20] Boston alone had 600 fishing and trading vessels in 1770. Fishery exports from Massachusetts to Europe and the West Indies were worth an estimated $1.25 million. The strength of the marine trade encouraged manufacturing as well.[21]

New Englanders sail around restrictions

New Englanders traded wherever they could, sometimes including New France. For example, a vessel might exchange provisions and tar in Newfoundland for refuse fish, take this to West Indies slave-owners, and exchange it for cotton, sugar, or molasses. After sailing home, the traders might make the molasses into rum, for sale to slave-traders, Indians, Newfoundlanders, or Africans. Or, traders might buy fish in Newfoundland (their own New England cure being less "merchantable" for Europe), take it to the Mediterranean, and load a cargo in England, Holland, or the Baltic for the voyage home. Sometimes they would sell the ship itself in England or Europe, take passage home, and build another.

Great Britain, following mercantile policies of the day, tried to control New England's trade for its own advantage. In 1733, Britain passed the Molasses Act, to restrain New England's trade with the non-British West Indies. New Englanders often sailed around such laws. But they were an irritant which would bring conflict.

Whalers comb the seas

New England's drive and ambition manifested themselves also in her whaling fleets, which combed the Atlantic and Pacific oceans. As right whales became scarcer, Nantucket Islanders about 1712 began taking sperm whales further offshore. Their spermaceti yielded better oil. At first whalers took the blubber ashore for rendering. In the 1750's they began rendering the blubber at sea. By the second half of the century, New Englanders were producing large numbers of brightly-burning spermaceti candles, and exporting them to Europe and the West Indies.

Attacking a whale. Another is being flensed beside the vessel. (Duhamel du Monceau, *Traité général des pesches*, 1772, part 2, vol. 3, section X, plate 1, fig. 3)

Whaling was already big business elsewhere. Davis Strait had since about 1720 become an international whaling ground. Dutch whalers dominated at first. The British government, wanting more whale oil for street lights and industrial uses, in 1733 began subsidizing whalers. The British fleet by the end of the century far surpassed the Dutch.

American whalers went to Davis Strait by 1732, and later to Baffin Bay and other Arctic areas. From the 1770's, New England whalers made trips around Cape Horn to the Pacific. As Edmund Burke told British parliamentarians at the time of the American Revolution: "No sea but what is vexed by their fisheries. No climate that is not a witness to their toils. … [And all this is done by] a people who are still, as it were, but in the gristle, and not yet hardened into the bone, of manhood."[22]

Meanwhile, American whalers also fished the northwest Atlantic. In the 1730's, fishermen mainly from Nantucket operated out of Canso to take whales on the Grand Banks.[23] Americans also fished the northern shores. One account around 1770 reported a hundred New England sloops and schooners, of 50–100 tons, chasing whales and codfish in the Gulf of St. Lawrence.

For their dominions north of New England, British authorities in 1764 passed an act to encourage the whale fishery. Duties on whale fins were reduced until 1770. The measures may have helped the growth of the whale fishery in Nova Scotia, where Halifax had several vessels by 1784.[24] Some whalers from Nantucket Island migrated in 1785 to the future town of Dartmouth, across the harbour from Halifax.

Wiping out the walrus

New Englanders took walruses at the Magdalen Islands, some 1,990 of them in 1765,[25] provoking complaints from Governor Palliser of Newfoundland. Their walrus fishery also caused concern at the Island of Saint John (Prince Edward Island). After that colony got a legislative assembly in 1769, its first law in 1770 was "An Act for the better regulating the carrying on [of] the Sea Cow Fishery on the Island of St. John." The act imposed a licence for hunting sea cows and set an October–November season, with fines for hunting without a licence or out of season, or for preventing sea cows from landing.

Problems continued. The Island's Governor Patterson noted in 1774 that "the Whale, Sea Cow, and Cod Fisheries … are now carried on in a very bad manner by Vessels from New England, and not to the twentieth part of the Extent, nor by much to that advantage they are capable of."[26] In addition, French vessels killed walrus by the hundreds at Miscou in the Gulf of St. Lawrence.

Walrus had been fairly plentiful at the Magdalen Islands, Sable Island, and south Newfoundland, numbering perhaps 250,000.[27] But over-exploitation soon led to commercial extinction.

Maritimes fishery slowly rises

Nova Scotia lagged far behind New England. For decades after the Treaty of Utrecht in 1713, the British authorities largely ignored the new colony. The population consisted of a few merchants and troops, and some fishermen mainly at Canso. Every summer, hundreds of fishermen from New England would congregate at Canso, making it an important centre for fishing and trading. Otherwise, Nova Scotia was almost a phantom colony. But from mid-century on, wars, government efforts, and the commercial drive of settlers built a substantial colony, a much smaller New England.

Lunenburg takes root

By mid-century, the British were making some effort at colonization. In 1753, to farm the land and strengthen their claims to Acadia, they brought to a Nova Scotia harbour some 1,453 Protestant Germans and Swiss. Over time, Lunenburg settlers spread out to the surrounding area, mixing to some degree with the New Englanders and other settlers.

Within a few years, the Lunenburg Germans (or "the Dutchmen," as other Nova Scotia fishermen sometimes called them) started a small "vessel" fishery, travelling to the banks or to nearby shore-fishing grounds.[28] They would in future demonstrate an innate drive that made them a high-line town. Of all Atlantic provinces, Nova Scotia has had the most far-ranging fisheries, and Lunenburgers have often been the spearhead.

Nova Scotia enacts quality standards

Other settlers were starting to arrive. After the Seven Years' War ended in 1763, many "pre-Loyalists" from New England moved in. Some went to vacated Acadian farmlands, others to such ocean-facing areas as Cape Sable and Yarmouth. Settlers from the British Isles joined them. In 1763, after the fall of New France, Nova Scotia gained what are now New Brunswick and Prince Edward Island. Nova Scotia's population doubled from about 8,000 in 1763 to more than 17,000 in 1775, more than half of them New Englanders,[29] who brought with them a marked attitude of independence and enterprise.

Nova Scotia had fish, lumber, and a geographical position comparable to New England's for the triangle trade with Europe and the West Indies. Enterprise naturally turned to the sea. From the 1750's on, there are reports of vessel construction in Nova Scotia.[30] By 1766, the province had 367 boats, 119 schooners, and three square-rigged ships.[31]

In 1751 and again in 1757, the Nova Scotia government provided bounties for dry and pickled fish.

("Pickled fish" refers to fish salted in brine; often, herring or mackerel.) In 1762, the government set standards regarding the quality of fish to be exported and the size of the containers.[32]

As the vessel-owning diarist Simeon Perkins recorded from the new settlement of Liverpool, not all prospered. The Nova Scotians suffered competition from Newfoundland and the Channel Islanders who took over many of the French fisheries in the Gulf. Perkins wrote of people being "in poor circumstances." The marine trade was still relatively small, while competing New England vessels filled the horizon.[33]

French fishery stays strong

France at the outset of the 1713–1791 period still operated a huge fishery on the banks and shores of the northwest Atlantic. The Treaty of Utrecht stated that "it shall be allowed to the subjects of France, to catch fish and to dry them on land" in certain areas of Newfoundland. This "French Shore" was to run from Cape Bonavista on Newfoundland's northeast coast up to the tip of the Great Northern Peninsula, and down the western side of the peninsula as far as Pointe Riche, near present-day Port au Choix. France also held the Gulf, Île Saint-Jean (Prince Edward Island), and Cape Breton. To offset the loss of mainland Nova Scotia, the French started building a great fortress at Louisbourg, and encouraged settlement on Cape Breton.

In 1719, it was said that 500 ships left France for the northwest Atlantic, with 200 of the largest engaged in the bank fishery. Around 1740, according to one perhaps exaggerated estimate, the French dry fishery employed 414 ships and 24,500 men in such places as Cape Breton, Gaspé, and Newfoundland. The bank fishery employed another 150 ships and 3,000 men.[34]

Cape Breton fishery grows

With no base in Newfoundland after 1713, many Placentia French moved to Cape Breton, where the new fortifications at Louisbourg promised protection. The island's population increased to some

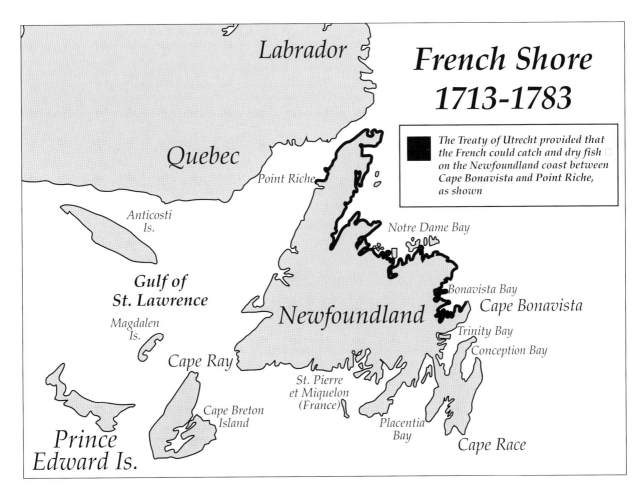

French Shore 1713-1783

The Treaty of Utrecht provided that the French could catch and dry fish on the Newfoundland coast between Cape Bonavista and Point Riche, as shown

Labrador

Quebec

Point Riche

Anticosti Is.

Gulf of St. Lawrence

Magdalen Is.

Cape Ray

Notre Dame Bay

Newfoundland

Bonavista Bay

Cape Bonavista

Trinity Bay

Conception Bay

St. Pierre et Miquelon (France)

Cape Breton Island

Placentia Bay

Cape Race

Prince Edward Is.

4,000 by the 1750's. Settlement also increased at Île Saint-Jean.

Cape Breton's fishery grew strong. A report in 1739 said that the French had 500 shallops in different parts of the island. These would employ three men fishing and two on shore curing. A shallop could produce about 300 quintals from April to September, with different cures for different markets in France, Spain, Portugal, and Italy, where French fish fetched a better price than English fish "by reason of the care that is taken in curing it, and of the method they have in sorting and sizing it for the proper markets." A winter fishery took place on the Atlantic side, primarily by small boats, with larger vessels coming from France to buy product or to fish for themselves. And the fishery was increasing:

> Some of those ships men fish in shallops but most in scooners from twenty to Forty tons who go to the Isle of Sable Bank, Bank Quero, St. Peters Banks and all the Banks on the coast of Nova Scotia, and catch their Fish there and make great Part of them in the English uninhabited ports on that coast; in [1720], the French fishery consisted at most of about Fifteen or twenty sail of ships, Few shallops and no scooners but have gradually increased ever since; besides those French ships that load for Europe, there are yearly above Thirty sail of large sloops, scooners and Briggs load for the French settlements in the West Indies.

Two decades later, a British assessment in 1762 found that "the dry Fishery upon the coast of Cape Breton and Acadia from Cape Breton to the River St. Lawrence was chiefly carried on by the Inhabitants, who employed in the several Parts about 800 Boats, each Boat having 4 men at an average, and catching 300 quintals of Fish; also about 60 sloops, schooners etc. each at an average carrying ten men, and catching 800 quintals of Fish."[35]

New France trails New England

Outside of Cape Breton, New France's resident fishery remained small. But some marine expansion took place. In 1731, the French government gave a bounty for ships built in New France; this created a bit of a building boom, with ten vessels ranging from 40 to 100 tons built in 1732. Between 1720 and 1740, interests around Quebec built around 200 ships all told, using them to trade with Cape Breton and the French West Indies.[36]

Commerce was growing, particularly at Quebec, Montreal, and Louisbourg. The latter port did some trading with Canso and New England, despite official restrictions. Merchants around Quebec invested in fishing and fur-trading operations at Gaspé and along the North Shore of the Gulf, which at some point became Labrador. A seal fishery developed on the North Shore, using elaborate systems of nets among the rocks and coves.

But New France lagged far behind New England. The Governor and Intendant of New France reported in 1737 that "extreme poverty is the rule in Canada."[37] Officialdom itself got in the way of progress, through over-regulation and sometimes corruption. For most people, the subsistence farm remained the basis of life.

British expel Acadians

On the Atlantic coast, the main group of French-speakers consisted of Acadian farmers in Nova Scotia. Back in 1650, there had been fewer than 500 of them. By about 1750, they numbered somewhere between 10,000 and 13,000, an amazing increase coming through birth rather than immigration. Acadian settlement had grown particularly at the Annapolis Basin, Minas Basin, and the Isthmus of Chignecto. Smaller French settlements took hold elsewhere, on both sides of the Bay of Fundy, at Cape Sable, and from Cape Breton all along the southern Gulf of St. Lawrence shore.[38]

In 1745, during the War of the Austrian Succession, New Englanders without much difficulty took the great fortress of Louisbourg. Peace negotiations in 1748 gave Louisbourg back to the French. But in 1749, the British founded Halifax as a naval base, provincial capital, and counterweight to Louisbourg. Some worried Acadians began moving to French-controlled areas at Île Saint-Jean and the Gulf of St. Lawrence shore.

Hostilities resumed in 1754, when Virginians clashed with French in the interior of North America. In 1755, angered by the "neutral Acadian" refusal to swear loyalty, the British expelled some 6,000 men, women, and children from the Nova Scotia peninsula. Some evaded the Deportation, escaping to French-controlled areas in New Brunswick and as far as Gaspé and the St. Lawrence Valley. They helped boost the population of Île Saint-Jean, which was increasing to several thousand.

In 1756, the Seven Years' War broke out. The British under Wolfe captured Louisbourg in 1758; now they could dominate the Gulf. More Acadian deportations followed from Cape Breton, the mainland, and Île Saint-Jean; that island's population dropped to a few hundred. By 1763, only about three or four thousand French people remained in all Acadia.[39] In future years many exiles would work their way back, only to find their best lands taken over.

Native peoples, Nova Scotia governors sign peace treaties

In the 1750's and 1760's, the British governors of Nova Scotia made several treaties of peace and

French fishing station in the 1700's. (From Duhamel du Monceau, *Traité général des pesches*, 1772, vol. 2, section 1, part 2, plate XVIII)

friendship with Mi'kmaq, Maliseet, and Passamaquoddy tribes. Although expressing Native "submission to His Majesty in the most perfect, ample and solemn manner," the treaty of 1760 also contained provisions that bound the Crown. Nearly two and a half centuries later, the Supreme Court of Canada's *Marshall* decision in 1999 would state that treaties of the day conveyed certain commercial fishing rights.

Still, Native peoples on the Atlantic in the decades and centuries after 1760 played no major part in the commercial fishery as developed by the newcomers. They took part here and there on a small scale. In a few cases they were more prominent, as in the Nova Scotia porpoise fishery, which died out in the 20th century, and the fishery in northern Labrador.

British take Canada; France retains fishing rights

The British captured Quebec in 1759, and New France soon fell. With the Peace of Paris in 1763, the Atlantic coast became British from the Arctic to the Gulf of Mexico. (The Peace of Paris also put an official end to any Spanish claim to fishing and drying rights in Newfoundland.)[40]

France kept only the islands of St. Pierre and Miquelon, just off the Burin Peninsula on the south coast of Newfoundland. Besides this important speck of land, France kept her fishing and drying position in northern Newfoundland, as laid out by the Treaty of Utrecht. France further kept "the liberty of fishing in the gulph of St. Lawrence, on condition that the subjects of France do not exercise the said fishery but at the distance of three leagues from all the coasts belonging to Great Britain... [and] at the distance of fifteen leagues from the coasts of the island of Cape Breton...."[41]

The latter rule was presumably to keep the French from interfering with the shore fishery by British ships and settlers. Although little was heard of this rule in following years, it seems to have been the first use of offshore zones in Canada. Previously, nothing officially stopped a French vessel from fishing close to a British shore, except the practical obstacle that the French might be unable to use shore space for drying.

France turns to subsidies

The French distant-water fishery remained considerable. Government buttressed it with subsidies. In 1767, ships fishing that part of the French Shore between Bonavista and Cape St. John (further north in Notre Dame Bay) got a bounty of 500 livres (pounds) each. The next year saw higher bounties for bigger vessels: 750 livres for those with 40–60 men, and 1,000 livres for those with more than 60 men. France also started paying a bounty of 25 sols per quintal on cod exported to the West Indies.[42] Table 3-1 shows the French Newfoundland fishery in 1765 to have more ships than the British, and nearly as much production.

Table 3-1. French and English cod fisheries, 1765 (as presented by H.Y. Hind, 1877).

		French	English
No. of	ships	339	293
"	men	14,952	17,876
"	tonnage	40,795	31,621
"	boats	1,765	1,823
"	seines	617	....
	Quintals of fish caught	488,790	522,512
	Hhds. of oil	6,840	....
	Stages	109	1,005
	Tuns of oil	1,760	2,384³/₄
	Tierces of salmon	......	1,172
	Sea Cows (Madelaine Island)	......	1,190
	Tons of oil	......	125
	Value of seal oil taken last winter	....	£5,109

The English figures include 9,976 inhabitants of Newfoundland.[43]

Mainland map changes

Inland, the Royal Proclamation of 1763 created the Province of Quebec, covering only a fragment of the old New France. Newfoundland took over the greater part of Labrador, the Gulf of St. Lawrence, Anticosti Island, and the Magdalen Islands. Nova Scotia got Cape Breton, Prince Edward Island, and the New Brunswick of today.

The huge lands west of rivers draining into the Atlantic (roughly, the Appalachians) became Indian territory. Thus, the Proclamation of 1763 irritated British subjects in Quebec, by reducing their territory, and in New England and the other seaboard colonies, by blocking their westward expansion.

The Native peoples, ostensibly protected by the huge territory set aside for them, actually became more vulnerable. Before, when France and England were still contesting North America, they had been able to a degree to play balance-of-power politics and negotiate advantages for themselves. With the fall of New France, they lost this position.

Quebec saw an influx of New England and British merchants. These new residents complained about the province's reduced size. In 1774, the Quebec Act re-enlarged the province. Inland, Quebec now took in part of the Indian territory south of the Great Lakes, between the Mississippi and Ohio rivers. This handover of western lands to the northern province angered New England and other seaboard colonies. On the coast, Quebec got back Anticosti Island, the Magdalen Islands, and all of Labrador except Hudson's Bay Company lands.

Quebec's population, aided by a francophone birth rate of legendary robustness and the influx of 20,000 anglophones, shot up from 65,000 in 1760 to more than 160,000 by 1791. This was far more than the combined population of the Maritimes and Newfoundland.

Channel Islanders set up in Gulf

After the Peace of Paris in 1763, many exiled Acadians made their way back. In the Bay of Fundy area, New Englanders had taken over their best farmlands; Acadians now set up elsewhere. In the Pubnico–Argyle–Wedgeport area of southwest Nova Scotia, where they earlier had a fishing tradition, Acadians took strongly to the fishery. In St. Mary's Bay, they first turned more to lumbering, but later developed a vigorous fishery. Others settled in the Isle Madame area, off Cape Breton. But the main population base of the Acadians now became the shores of the Gulf of St. Lawrence, especially in New Brunswick. A large proportion of men gravitated to the fishery. They had little money at first, and only small boats.

Who would dominate the Gulf fishery, opened up to British subjects by the Peace of Paris in 1763? There were several other groups at hand: Quebec interests, Nova Scotians from that still-tiny colony, Newfoundlanders, and New Englanders. The Yankees quickly increased their fishing and trading. In the Chaleur Bay region, rum sometimes helped persuade local fishermen to sell their dried cod to New Englanders.[44]

But a strong new force arrived on the coastal scene: merchants from Jersey and the other Channel Islands, English possessions lying near the French coast. The bilingual Channel Islanders moved quickly into the francophone Gulf, where Cape Breton and the Gaspé Peninsula had the fish-drying conditions suitable for a high-grade, light-salted product.[45]

In 1763, Jacques Robin petitioned for a grant at the mouth of the Miramichi River, and he later began hiring Acadians who had escaped the Deportation or returned after it. Some Indians also fished for the Robins. Soon the Robin interests were dealing in cod, cod and whale oil, salmon, and sometimes furs. They set up establishments at Carleton and Paspébiac on the Gaspé, Arichat and Cheticamp on Cape Breton, and elsewhere. In 1777, it was estimated that the Gaspé fishery employed an average 12 vessels a year, and exported 16,000 quintals of fish.

The Janvrin family set up on the Magdalens after 1782. Other Channel Island concerns (sometimes called Jersey houses) established themselves on the Gulf shores.[46] Although their individual fortunes varied, collectively the Jersey houses took firm root. They would hold French-Canadian fishermen in an iron grip for a century.

Newfoundland: Palliser takes over

After the Seven Years' War, the British government wanted to reaffirm the fishery as the "nursery of seamen." Sir Hugh Palliser, serving as governor of Newfoundland 1764–1768, and later as a member of Parliament in London, became the imperial strongman of the fishery. The Newfoundland historian D.W. Prowse summed him up thus:

Palliser has been highly praised in our histories; in some respects he is entitled to our gratitude; the bounty [incentive subsidy] for the fishery ... is undoubtedly due to his exertions. ...

The Governor had only one great fault—beyond his own circumscribed vision he could see no horizon; ... the one narrow insular idea of the age pervaded his official mind, that it should be a fishing colony, used for one great purpose only in his eyes, supplying men for the Navy. ... Every other consideration, every attempt to promote settlement, cultivation, and civilisation, must be ruthlessly swept aside. ... He could see clearly enough that settlement could not be prevented, so he abused the Colony and the colonist.[47]

From 1763 to 1774, Newfoundland governors also controlled Labrador (the term then including part of the North Shore of the Gulf of St. Lawrence), Anticosti, and the Magdalen Islands. Palliser moved quickly against colonial competition.

The French regime had granted charters to residents of New France for fishing and sealing rights along the Labrador. After the Seven Years' War, the new British masters of New France did the same for their own settlers, even extending the grants somewhat. In 1765, Palliser revoked these charters. This action caused consternation among the English merchants of Quebec, who had just put money into the fishing concessions. Newfoundland-based merchants moved in, protected by Palliser, who erected fortifications at Chateau Bay, on the northern end of the Strait of Belle Isle, and sent a detachment of troops to oversee matters on the Labrador.[48]

Palliser also had the French and New England fleets to deal with. On the Treaty Shore of northern Newfoundland, the French still made good use of their fishing and drying rights. In 1774, 273 French ships went to Newfoundland, with 1,455 boats and 12,367 men, producing 215,000 quintals of cured fish and 3,153 hogsheads of oil.[49] Palliser enforced the rules strictly, giving the French no extra leeway. In Palliser's view, according to Prowse, "all disputes [regarding the French fishery] were to be decided by English authorities alone. ... [The treaty] gave no right whatever to the French to catch salmon, to trade or traffic; they were only to fish for codfish, and dry them on land; they were not even permitted to cut spars or to build boats." When some Frenchmen captured a whale at Great Orange Harbour, Palliser had it taken away and sold.[50]

Regulations in 1766 admitted the other colonies, including New England, to the fishery, but distance and regulations hampered their operations.[51] Palliser kept tight watch against smuggling, and viewed their fishery warily—especially the whale fishery. He complained to the governor of Massachusetts that he had made peace with the "Esquemeaux" of the Labrador; but then, "some New England vessels contrary to the orders I have published went to the Northward, and robbed, plundered, and murdered some of their old men, women and children, who they left at home, so I expect some mischief will happen this year; revenge being their declared principle."[52]

Cod fishery yields high production

Favoured by government, the British ship fishery in 1771 reached a high point with 369 ships, although they dropped back to 254 in 1774. The number of bankers now varied from about one-third to two-thirds of the fleet. As well, bye-boat keepers were still fishing. In 1774, their 518 boats outnumbered the 451 operated by the fishing ships.

But the Newfoundland settlers were overtaking the British. The resident fishery had nearly 1,500 boats in 1774, more than the fishing ships and bye-boat keepers put together.[53] That same year the fishing ships brought over, besides their own crews, nearly 5,000 passengers to help in the Newfoundland fishery. Ireland supplied the most, England placed second, and the Isle of Jersey a poor third.[54] By the 1780's, residents were producing half or more of the total British catch, and their share was rising.[55]

After more than two centuries of exploitation, Newfoundland waters were still producing huge catches. Total production by British and

Newfoundlanders in the peak year of 1788 came to 949,000 quintals, or about 106 million pounds of product. Let us say that three-quarters was dried codfish. At a conversion factor of 4.88 (the number now used for such calculations), the live-weight equivalent would be about 390 million pounds, or 177,000 tonnes. The wet-salted fish, at a conversion factor of 2.7, would bring the live-fish equivalent to more than 200,000 tonnes, a major amount, without even counting the French or American fishery.

Palliser's Act benefits British, shuts out others

Leaving Newfoundland in 1768, Palliser became a member of Parliament in England. There he master-minded what may be seen either as the last great act of self-interest by British interests or as the first major attempt at fisheries regulation in British North American fisheries. "Palliser's Act" of 1775 set out to favour and subsidize British fishing ships at Newfoundland, keep the nursery of seamen productive, shut out the Americans, provide trade goods for Newfoundlanders from Great Britain, and stop the growth of settlement.

The fishery was to be exclusively for British vessels and Newfoundland residents. Colonists outside Newfoundland—including New Englanders, Quebecers, and Nova Scotians—were excluded. Coming in a period of British–American tension, this exclusion caused great indignation in New England. British fishing ships owned in England and over 50 tons received bounties to fish on the banks. Incentives also applied for the whale fishery. Fishing vessels could bring provisions from overseas free of duty (this would help the residents who had lost their New England suppliers). And sealskins and oil were to be free of duty, apparently a move to help this developing fishery.[56]

The act once again constrained property rights of residents. But it held some benefits for fishing-ship crews. Operators had to make agreements in writing with their crews, and fishing servants had the first rights to proceeds from fish and oil for their wages.[57]

Noting the benefits to West Country interests, a group of "Merchants, Boat-keepers, and Principal Inhabitants of St. John's Petty Harbour and Tor Bay" asked for bounties of their own. They also wanted something done about the wasting of small fish in cod-seines, the rinding (bark-removal) of trees to cover flakes and huts, and the destruction of birds on northern islands (the birds were used by the inhabitants for food and bait, but were being destroyed by fishing crews who sold the feathers).

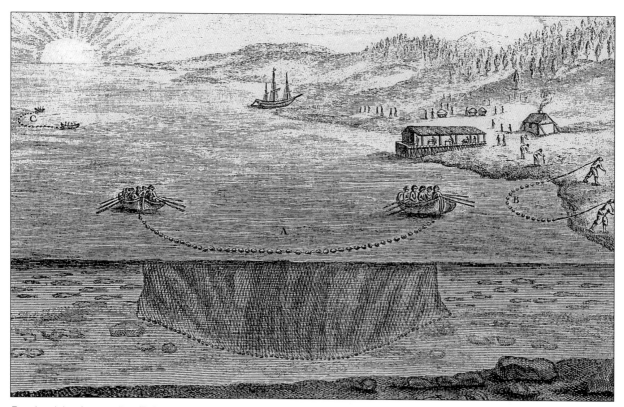

Beach seining for capelin. (Duhamel du Monceau, *Traité général des pesches*, 1772, vol. 2, section 1, part 2, plate XII, fig. 1)

The petitioning merchants asked that all shop-keepers be obliged to operate a fishing shallop. This recommendation arose because the now-established merchants, who both operated boats and sold goods to their fishermen, disliked competition from smaller storekeepers who operated no boats. The merchants in effect thought they deserved a captive market for their goods, in light of the "very great" and "enormous" wages they paid their servants.[58]

Some changes would come to pass. A modification of Palliser's Act in 1786 continued bounty payments for ten more years (in the end, they lasted until 1803). The 1786 amendment also required that fishing vessels increase their net mesh size from three and a half inches to four inches, to protect small fish. This provision, for cod-seines used near the beach, was one of the first sea-fishery conservation regulations. An accompanying regulation stipulated that birds valuable for food or bait were not to be destroyed for their feathers.[59]

British fishery declines

What effect did Palliser's Act have? Innis states that "attempts to restrict colonization and increase the fishing ships by legislation failed."[60] That is true in the longer term. But for the short run, Palliser's Act kept more vessels coming from Britain than would have made the voyage otherwise. Indeed, the migratory fleet reached an all-time peak of 389 vessels in 1788.

If Palliser's Act helped, who can say by how much? Fishery regulation has always been difficult to evaluate, given the welter of resource, market, and other variables. Regulators have generally responded to pressure and paid less attention to analyzing the results. Even when they try, the fluctuating factors often defy analysis.

In any case, the British migratory fishery for 1790–1792 remained sizeable, the fleet averaging 260 vessels. Then other factors intervened. Cold water temperatures and adverse trade conditions worked against the British in the early 1790's. The bank fishery out of St. John's, still mainly British, dropped from 140 vessels in 1788 to 70 in 1792. On top of that, from 1793 until 1814, the wars between Great Britain and France further curtailed both countries' fisheries, to the benefit of Newfoundlanders. By 1795, the number of British fishermen had dropped to 1,400, compared with 5,200 in 1784, and the bye-boat fishery had practically disappeared. Resident fishermen, meanwhile, had increased from 5,100 to 7,100.[61]

Seal fishery builds up

During the 1713–1791 period, the seal fishery, Newfoundland's first great home-grown fishery, was slowly gathering force. It developed as residents moved northward along the coast, closer to the pup-ping grounds. Twillingate and Fogo were selling seal oil by 1738.

Newfoundlanders at first took seals near the shore and in narrow places with nets and ice skiffs. Although erratic, this fishery produced important quantities of oil, used for lighting, soap, and other purposes. As time went on the pelts gained importance; they were used for leather by hat-, shoe-, saddle-, and trunk-makers in Britain and elsewhere. By 1791–1792, perhaps 7,000 seals were caught at Bonavista.[62]

The growing cod fishery and the seal fishery encouraged new settlement. From somewhere around 2,000 people in 1700, including fewer than 500 French, the population rose to possibly 19,000 people by 1789. By 1791, residents and visitors operated nearly 1,400 fish stages.[63] The resident schooner fleet built up, with both smaller (25–35 ton) and larger (50–75 ton) vessels. Small shipbuilding expanded in Trinity and Harbour Grace. The growing competition from the resident fleet helped to weaken the British fishing ships, already declining with the wars.

The residents were gaining ground despite obstacles. In 1786 the British privy council, still fighting the old battle against settlement, had advised against any increase of waterfront buildings or strengthening of property rights in Newfoundland. Yet shore property and resident influence increased, and the British government slowly responded. Acts of Parliament in 1791 and 1792 set up courts in Newfoundland, replacing the law of the fishing admirals.[64]

Wars, tariffs hinder American fishery

When Newfoundlanders in the late 1700's were just beginning to assert themselves as an entity, New Englanders had long since said goodbye to the British.

Back in the 1760's, after the Seven Years' War and the fall of New France, Britain had tried to impose more controls on the American colonies. The Proclamation of 1763 and the Quebec Act curtailed their western expansion. London tried to hike taxes to pay for the Seven Years' War; the colonies resisted. Imperial regulations set new import duties, which authorities tried to collect with a new strictness.

As bad feelings mounted, Britain closed the port of Boston to trade in 1774. Other laws, including Palliser's Act, restricted New England trade and prohibited fishing off the northern provinces, the fishery thus playing its part in provoking revolution. New Englanders retaliated by forbidding their residents to supply English vessels. Britain then passed the Prohibitory Act, forbidding all nations to trade with the American colonies. The Declaration of

Independence followed in 1776, and the Revolutionary War began, ending in American nationhood.[65]

Post-Revolution treaty creates "American shore"

In negotiating the Peace of Versailles in 1783, Britain was hardly dealing from strength. The United States got lands stretching west to the Mississippi and north to roughly the present border. And they regained northern fishing privileges. The Americans could fish anywhere off the British colonies, with sea fishing a right and coastal fishing a liberty. Shore-drying was a liberty applying only in mainland Nova Scotia and present-day New Brunswick, the Magdalens, and Labrador (defined as running east from Mont Joli, a point on the North Shore adjacent to the eastern end of Anticosti Island). Even in those places, it could take place only in unsettled areas.

Fishery excerpt from the Treaty of Versailles

It is agreed that the people of the United States shall continue to enjoy unmolested the right to take fish of every kind on the Grand Bank, and on all the other banks of Newfoundland; also in the Gulf of St. Lawrence, and at all other places in the sea where the inhabitants of both countries used at any time heretofore to fish. And also that the inhabitants of the United States shall have liberty to take fish of every kind on such part of the coast of Newfoundland as British fishermen shall use (but not to dry or cure the same on that island) and also on the coasts, bays and creeks of all other of His Britannic Majesty's dominions in America; and that the American fishermen shall have liberty to dry and cure fish in any of the unsettled bays, harbours and creeks of Nova Scotia, Magdalen Islands and Labrador, so long as the same shall remain unsettled; but so soon as the same or either of them shall be settled, it shall not be lawful for the said fishermen to dry or cure fish at such settlements, without a previous agreement for that purpose with the inhabitants, proprietors or possessors of the ground.

The French Shore shifts

France in 1778 had come to the aid of the rebelling American colonists, and the peace treaty involved French interests. The French Shore where the French had both fishing and drying rights shifted counterclockwise. Instead of starting at Cape Bonavista, it would now start at Cape St. John, a more northern point of Notre Dame Bay. This shift opened up space for the northward-pressing English settlers. The French Shore running up and around the Great Northern Peninsula now extended all the way down to Cape Ray, the island's southwestern tip. The French thus gained space on the southwest coast of the island which later became significant.

France still believed in the fishery. More bounties came into play in 1785, for dried fish carried to the West Indies and Europe. New duties appeared against foreign fish. But the French fishery was declining, from 431 ships in 1769 to only 86 in 1786. After the French Revolution in 1789, the fleet shrank further to only 46 ships by 1792.[66]

Britain blocks American trade with colonies

While agreeing to American fishing and drying in British North America, the British at first blocked New England's fisheries market in the British West Indies. They also banned American vessels from trading with the B.N.A. colonies. Only British ships could trade back and forth. This reining in of New England would give new opportunities to fishermen and traders in the Maritimes.[67]

The various American states were now independent of Britain, but were far from thoroughly united. Irritated by British post-war policies, they sought collective strength. Indeed, "the difficulty of securing united action in measures of retaliation against British policy contributed to the movement for the adoption of the American federal constitution," which was ratified in 1788.[68]

Thus, the fishery in its progress had built up New England's economy; fishery irritants had helped to prompt the Revolution; and now post-Revolution fishery reversals helped shape the Constitution of the United States of America.

New England takes to subsidies

Concerned about her fish trade, and suffering from British restrictions, New England promptly applied bounties. From 1789, subsidies supported dried fish and pickled fish, and the government placed duties on imports. In 1792, the bounties changed, to subsidize operations rather than production. Vessels of 5–20 tons received $1 per ton; those of 20–30 tons got $2.50 per ton. These bounties later increased. Vessels could also buy salt and fishing gear in foreign ports without paying duty.[69]

The subsidies accelerated New England's natural enterprise. From 1786 to 1790, an estimated 539 ships with about 3,400 men exported about a quarter-million quintals of dried fish yearly. Local fish-

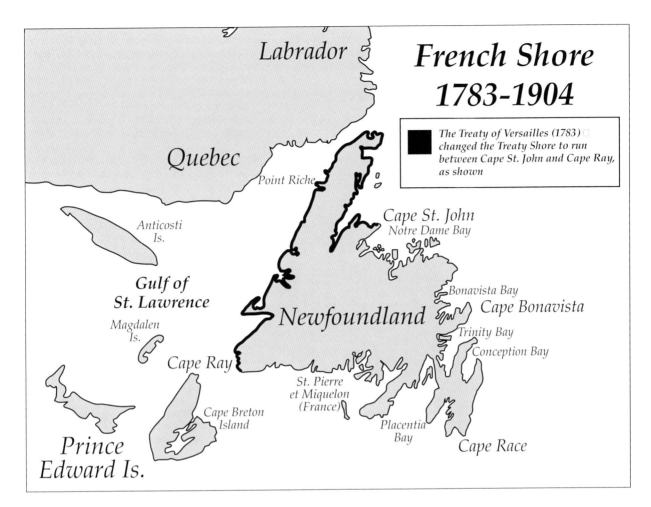

French Shore 1783-1904

The Treaty of Versailles (1783) changed the Treaty Shore to run between Cape St. John and Cape Ray, as shown

eries expanded—mackerel, herring, clams, and lobsters. Trade resumed in 1794 with the British West Indies, and grew with Europe. Off the northern colonies, New England built up a huge fishing effort.

Nova Scotians try to block New Englanders

Meanwhile, the growing colony of Nova Scotia was trying to stave off American fishermen. After the American Revolution, the arrival of some 20,000 Loyalists, disbanded soldiers, and refugees redoubled Nova Scotia's population. British authorities in 1784 carved off the separate provinces of Cape Breton (until 1820) and New Brunswick.

Although New England vessels were forbidden to trade with the northern colonies, or to dry fish in settled areas, they were skilful in skirting regulations. They needed watching. In 1786, George Leonard was appointed as Nova Scotia's superintendent of trade and fisheries, and he soon had four deputies. Leonard seems to have been the first high official in the B.N.A. colonies appointed for such a post. Later his authority was extended to all the Maritimes and Newfoundland. Leonard patrolled

vigorously in his armed brig, The *Earl of Moira*, though American vessels outnumbered him by many hundreds, fishing, trading, and smuggling.

In 1787, Leonard tried to counter the practice of Americans drying their own fish, ordering that they must send their fish ashore in vessels belonging to the King's subjects, and use help from the crews of those vessels in drying. In 1793, 40 or 50 American vessels were charged with throwing offal overboard, a practice forbidden by Nova Scotian authorities. And in what appears to be the first form of local licensing in the Maritimes, fishermen wanting permits to fish needed to have a boat built in the province, to swear allegiance to the King, and to pay $2.[70]

Merchants continued to complain about Americans fishing, and about their smuggling lower-priced goods to trade for fish and other local items. In the 1750's, Nova Scotia had subsidized production. Now in 1786, no doubt in response to commercial representations, the colony put bounties on the operation of vessels. These applied at first to vessels over 40 tons, later only to vessels over 75 tons.[71] While trying to fend off the Americans, Nova

Scotians were also endeavouring to take over their trade wherever possible, including Newfoundland and the West Indies.

New Brunswick and Prince Edward Island take shape

As New Brunswick became a separate province in 1784, some 14,000 Loyalists were joining the much smaller number of Acadians. With abundant forests, New Brunswick swept into the fishing–shipping–trading marine economy, emphasizing vessels and the timber trade. It was said that in the decade following 1783, New Brunswick built 93 square-rigged vessels and 71 sloops and schooners.[72]

The Island of Saint John fell under Nova Scotia's control after 1763, became a separate colony in 1769, and changed its name to Prince Edward Island in 1799. Under absentee proprietors, settlement was slow, the population reaching 4,000 by 1798. Most looked towards farming rather than fishing. Some people saw the fishery's potential, but it would be slow to rise.[73]

Management: Regulations start from the shore

In the 1700's and later, sea-fishery management in the sense of conservation was slow to emerge. Rather, governments tried to ensure access to fishing grounds for their vessels, and to encourage the trade through such means as tariffs and subsidies. Sometimes they set minimal quality standards. The sea fishery still seemed limitless.

Conservation began with river and shore fisheries. As settlement grew, more fishing took place for salmon, herring, shad, alewives, and other species. Gradually, these items entered into commerce. Local populations of fish got thinner, and people took notice. As conservation concerns arose, regulations slowly came into place, especially for salmon.

Salmon fishery becomes commercial

New England exported salmon in the 1600's, and Nova Scotia at least from 1773. By 1789, Nova Scotia was producing 10,000 barrels of mackerel, salmon, and herring, along with 20,000 quintals of cod, 1,500 barrels of whale and fish oil, and 10,000 pounds of whalebone (baleen).[74] The Nova Scotian fish trade was still small compared with New England's, but growing.

Great salmon rivers poured into the Gulf of St. Lawrence, and in the 1700's the French exported small amounts to the home country. After the Seven Years' War and the arrival of the Channel Islanders, the Robin interests and others began exporting salmon. A fishery developed at the Restigouche,

although Charles Robin complained in 1787 that after being speared by the Indians, the fish were poor quality and fit only for the West Indies.[75] Salmon fisheries also sprang up in the Bay of Fundy, where a Massachusetts trading firm—Simonds, Hazen and White—received a licence for a fishery at Saint John. Their success with salmon and other species encouraged new settlers to move in.[76]

In Newfoundland one George Skeffington, backed by New England capital, in 1723 got a 21-year monopoly for his salmon fisheries in certain harbours north of Cape Bonavista. He sold the salted salmon to Spain and Italy. By 1757, several operators were exporting pickled salmon. The fishery spread onwards into Notre Dame Bay, and reached into Labrador by the end of the century. Fishermen used weirs and nets of four-inch mesh. By 1786, about a dozen salmon fisheries existed, producing many hundred tierces (a size of cask between a barrel and a hogshead).

The Native peoples had often taken salmon with spears, or weirs of brush or stone, and colonists did the same. But nets became more and more common. Production increased sharply in the 1700's. R.W. Dunfield, a historian of Atlantic salmon, has calculated that in British North America, catches between 1762 and 1784 ranged between three and eight million pounds annually (roughly 1,300–3,700 tonnes), with as many as a million salmon taken in some years.[77]

Salmon spawn new regulations

Salmon were the great inspiration of fisheries management, because people could see their beauty, vigour, and vulnerability. As the colonists fished down the original abundance, a wave of regulations came into place. Over time, management techniques similar to those used for salmon—regulating seasons, gear, size, and so forth—would move into the sea fisheries.

Dunfield has listed many early regulations. In New England, mill-dams as well as fishing caused great damage to salmon. Various localities passed laws requiring fish passages, appointing overseers, and putting limits on the fishery. In Nova Scotia, a 1763 regulation provided that justices would "annually, at the first Sessions, ... regulate the river fishery; persons transgressing regulations to forfeit £10, one half to the poor, and the other to the informer, to be recovered in the Court of Record. Act to continue two years." This seems to have been the source of a long-lasting provision in the Fisheries Act, whereby those reporting the offence got part of the fine.

A 1770 regulation made it illegal to throw fish offal into the sea within three leagues of shore (possibly because fishermen believed that dead and rotted fish scare others away). And in 1775, justices received authority to appoint fishery overseers, thus making an occasional practice more general.[78]

Rules could vary according to county or local wishes. In Saint John, New Brunswick, the city council controlled the harbour fishery, and a tradition took root of issuing fish lots by lottery. In Prince Edward Island, the legislature in 1780 reacted to falling abundance with an "act to regulate the salmon, salmon trout, and eel fishery." This set a salmon season of January 15 to September 30, giving the salmon some time off from running the fishery gauntlet.

In the Maritimes, notably New Brunswick, fishing rights tended to go along with property rights. If you bought land on a riverbank, you owned the fishery at that place. In 1765, the governor of Nova Scotia, then including New Brunswick, granted land along the Miramichi to Messrs. Davidson and Cort, two Scots with the idea of setting up a salmon fishery. The authorities let them do so, so long as they also cleared land and encouraged colonization. By the mid-1770's they were exporting up to 850,000 pounds a year. But their private fishery caused problems for other people.[79]

On August 4, 1785, Benjamin Marston, then Sheriff of the Miramichi, wrote to the secretary of the new province of New Brunswick pleading for the government to impose salmon fishery regulations.

> The salmon fishery on this river is an object of great importance and worthy the attention of the government. ... Fisheries are uncertain in their annual produce, but the great falling off from what used to be caught in this river when Davidson and Cort first got their grant must be imputed to the destructive mode of catching the fish, which is by nets principally...
>
> I forgot to mention another principal evil consequent on the setting of the cross nets, which is, the depriving of all above of an equal chance in fishing. This injury falls chiefly upon the Indians whose fishing places are above the Grant of D. & C., in both branches....[80]

Perhaps such protestations had an effect. In 1786, after the influx of Loyalists, both New Brunswick and Nova Scotia passed laws similar to those of Massachusetts to keep rivers free of encumbrances. Though these laws lapsed, others followed, including Sunday closures on salmon fishing, penalties of up to 30 days in jail, a closed season (August 30 to April 1) on the Miramichi and Restigouche, net-length regulations on the Miramichi, and requirements for fishways at dams.[81] And in Newfoundland, the governor around 1774 issued, according to Prowse, "an admirable set of regulations" for the salmon fishery.[82]

These early rules in the B.N.A. colonies got the regulatory ball rolling, but the laws sometimes lapsed. Besides, enforcement was weak, conservation education minimal, and compliance seemingly poor. The ordinary person who lacked river property had little incentive to obey the rules that benefitted riparian owners. Complaints would rise about "proprietors of the salmon fisheries" who enjoyed them without expense, and expected the public to pay for their protection.[83]

Subsidies a chief form of sea-fishery "management"

In the sea fishery, rudimentary forms of enforcement were beginning, as exemplified by George Leonard. But trade, not conservation, was the uppermost idea. The various fish-quality regulations of the day were trying to protect trade. International disputes over fishing rights and customs duties stemmed from economic motives. And subsidies had by 1791 become a chief form of intervention in the fisheries.

In the 1500's and 1600's, strong fleets had grown up without subsidies. In the 1700's, fishery resources and market demand remained fundamentally strong. Why then did sea-fishery subsidies become common?

Jockeying for position on the ocean explained part of it. Britain encouraged the cod and whale fisheries for both economic and military purposes. The argument that the fishery was the "nursery of seamen" for the navy still carried force in other countries as well. France applied bounties vigorously from 1767, after losing New France, to protect her place on the oceans. New England used them from 1789, when her fishing and trading fleet was weak from war. Future president John Adams called the fishery "a nursery of seamen and a source of naval power."[84] And Nova Scotia applied subsidies from 1751, when active colonization started; it was natural to want to build up the fleet.

But below all these reasons, the fishery generally lacked the air of prosperity. Reports were emerging of the poor condition of fishermen, especially in Newfoundland. One imagines that in each country, vessel owners were lobbying for assistance.

Fishery alone rarely created prosperity

In general on the Atlantic coast, prosperity tended to stem not from the fishery alone, but from a mixture of fishing, shipping, shipbuilding, and trading, especially where manufacturing and agriculture existed as well. At one extreme was New England, with all those elements strong; at the other was Newfoundland, with all of them weak except the fishery itself. If fish alone created prosperity, Newfoundland would have been the richest of all; but it was the poorest.

The owners of sizeable vessels could use them for fishing, freighting, or trading. But small-boat fishermen, the great majority in British North America, could only fish and sell to the local buyer, while also growing food for subsistence.

Although rich fishermen were scarce, the occupation never had a shortage of people. It might take decades to become a true master, knowing all there was to know about a boat, local waters, navigation, catching and handling fish, and dealing with people. But starting the process was easy. Fishing required no formal education, only minimal investment for small-boat owners, and none for crewmen. Alternative employment was scarce; the fishery was in many instances the employer of last resort for the poor and uneducated.

With fish thronging on the coast, one needed only a small boat to get into the fishery. It appears likely that easy entry worked against prosperity. One suspects that, just as in the 20th century, whenever a fishery started to make money additional vessels would enter, perhaps with some effect on the resource but with more on the market, where buyers could pick and choose from competing sellers.

Yet the fishery despite low incomes and high competition had its attractions. There was the pride, skill, and independence of running a boat. Fishermen knew every part of the boat and rigging; they might even build the craft. They knew how to get around in the fog, the sound of the surf in different coves, the signs of fish, a thousand tricks of the trade. And there were the beauty of the coast, the feeling of being on the water, the chance of a high-line catch, the company of men and their stories, the sharing of a small community where people know one another over a lifetime.

CHAPTER 4.
1791–1848: Fisheries expand, regulation lags

The period began with the creation of Upper and Lower Canada in 1791 and ended with responsible government in 1848. A wide sweep of colonies—the two Canadas, New Brunswick, Prince Edward Island, Nova Scotia, Cape Breton, and Newfoundland—was growing stronger. As new immigrants followed the Loyalists, the population increased in every colony—doubling, tripling, quadrupling, or more. The British distant-water fishery faded away. The resident fishery in British North America came into its own, though beset by American and French rivals. The marine economy flourished, with the fishery expanding, shipyards booming, and a trading fleet criss-crossing the oceans of the world.

Much of the impetus came from the Anglo-French Wars (1793–1815) and the Anglo-American War (1812–1814), which curtailed fisheries by France, England, and New England. But after 1815, foreign competition rebounded. Roving American fleets developed the purse-seine, the mackerel jig, the longline, and bigger and better schooners. Other fisheries besides groundfish became sizeable. The Americans and French dominated the vessel fishery off British North America, and both had rights to cure fish on shore.

The B.N.A. colonies defended their fishery position vigorously. They made use of trade regulations and subsidies, and tried to maintain the new three-mile limit through British naval help and their own efforts. Like their American rivals, they did more fishing for herring and mackerel. And Newfoundland's seal fishery became a major industry.

The Atlantic coast showed many signs of vigour and prosperity. Few prospects so please the eye as sailing vessels, busy docks, and people working by the water in a thriving commerce. Even today, east coast people will recount stories from their forebears about the old days, with "a forest of masts" in the bay, and the wharves so numerous that one could walk around the harbour without touching land.

Even so, the mainland colonies were beginning to look more to the continent than to the coasts for growth. As the British system of imperial trade preferences came to an end in the 1840's, anxious politicians and business leaders hankered for co-operative arrangements with their American rivals. The B.N.A. colonies would soon strike a free-trade deal with the United States, the fishery providing the centrepiece.

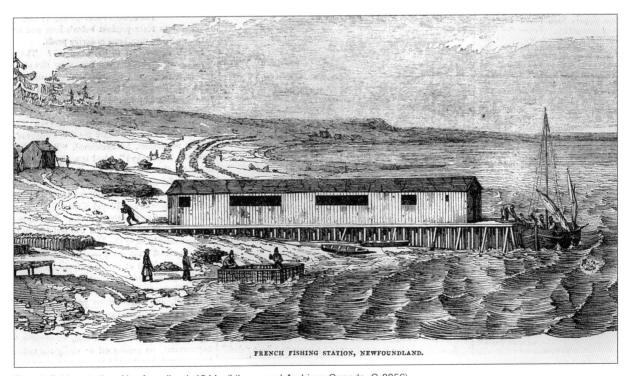

FRENCH FISHING STATION, NEWFOUNDLAND.

French fishing station, Newfoundland, 1844. (Library and Archives Canada, C-8856)

By mid-century, anxiety was rising about fishery conservation, mainly in the freshwater and shore fisheries. The individual colonies passed local laws, none very effective. All told, the fishery was growing, but so were its worries.

The Maritimes and Quebec: growing fishery, growing problems

In the period 1791-1848, the Maritimes grew faster than ever before or since. Nova Scotia's population rose ninefold, from an estimated 30,000 in 1790 to 276,800 in 1851. New Brunswick went from some 4,500 people in 1775 to 193,800 in 1851. In 1770, Prince Edward Island might have had a thousand settlers, mostly French; by 1848, it had 62,700.

But central Canada, with more and better land, was growing even faster. In Lower Canada (Quebec), the population rose from perhaps 60,000 in 1760 to 890,300 in 1851—not far from double the combined population of the Maritimes.[1] Upper Canada also began a prolonged upsurge, as Loyalists and other North American and British settlers poured in.

British wars strengthened the Atlantic economy. The French–English wars lasting from 1793 to 1815 not only held back the fishing fleets of Britain and France, but also blocked Britain's supplies of timber from the Baltic. Britain had to turn more to the colonies, with their virgin forests and huge fishery resources. Meanwhile, British-American conflicts cut down the American fishery off the Maritimes, giving the provinces a chance to supply New England's markets in the West Indies.

Although forestry led in New Brunswick and farming in Prince Edward Island, still the fishery was the most pervasive sector of the Maritime economy. The towns and settlements springing up along the coast might farm their own food, mill their own lumber, and set up a carpenter and blacksmith shop. But they needed rope, salt, and many manufactured goods from elsewhere. Their main trading commodity was fish, cured on flakes or pickled in barrels along the waterfront. Every area had its fishing fleet (though small in the case of P.E.I.) and merchant establishments.

Roads were few and poor. Many goods had to go by boat. Versatile vessels might both fish and carry. Nova Scotia developed a sizeable fleet, including larger vessels. By 1801, Liverpool alone had a ship of 200 tons, 14 brigs, 25 schooners, and a sloop. Maritime traders had by then made strong links with the West Indies and Europe. In Newfoundland, they largely replaced the traditional New England trade.[2]

Nova Scotia subsidizes the fishery

Frequent bounties helped the Nova Scotia fleet compete with the subsidized New England and French fleets. In 1800, fishermen of Ketch Harbour, near Halifax, petitioned for bounties because of their "Poverty and Distress." A legislative committee referred to the "annihilated Fisheries" of the county. In 1802, Nova Scotia provided a bounty of one shilling a quintal on cured cod. In 1806–1808, and again in 1815 and 1818, the colony gave bounties on salt imports.[3] In 1806–1807, the province tried bounties both by vessel tonnage and by production, for the Labrador fishery and West Indies trade. Also in 1806-1807, the imperial government paid bounties on exports of Newfoundland and British–American saltfish, herring, mackerel, and salmon.

Complaints arose about bounties going to exporters rather than to people actually in the fisheries. The system was extended and modified in 1810 and 1811. The Liverpool diarist, Simeon Perkins, in 1811 sent his crew on a short fishing trip "to make up four months to entitle us to the bounty."[4]

Shipbuilding grows

Fishing and trading, especially the timber trade, helped shipbuilding to grow. With Baltic supplies of timber blocked, London enacted preferential tariffs for colonial timber. Shipyards and boatyards sprang up in many places. While Nova Scotia tended to build smaller vessels for coasting, fishing, and the carrying trade, New Brunswick built many larger vessels for transporting timber. On both banks of the Miramichi, shipyards extended for 20 kilometres. All around the Atlantic coast, vessels went down the ways by the hundreds.

British fleet declines, American grows

Meanwhile, the British distant-water fishery in the northwest Atlantic was getting weaker. The wars beginning in 1793, the consequent loss of Spanish market, and the withdrawal of bounties in 1803 cut sharply into the British fleet. In 1806, few ships went to Newfoundland. The Avalon Peninsula from Bay Bulls to Trepassey had formerly supported more than 200 mainly English-owned bankers; in 1807, it had almost none. Sack ships dwindled to only 20 in 1804. Newfoundland's exports were increasing, but with the resident fishery now providing nearly all the catch.[5]

As the British fishery faded, the American grew. Back in 1783, the Peace of Paris had given the Americans fishing liberty all along the coast of

British North America, and shore-drying liberty in mainland Nova Scotia, New Brunswick, the Magdalens, and Labrador east. Offsetting that advantage, the British had restricted American trade with the B.N.A. colonies.

If the British thought the trade restrictions would hold back American fishing, they were wrong. Subsidies and natural drive boosted the New England fleet, which traded heavily with the foreign, and from 1794, the British West Indies. Great fleets fished off British North America. An estimate for the years 1790–1810 had more than 1,200 vessels a year going north, an average 584 to the banks off New England and Nova Scotia, and 648 to Chaleur Bay and Labrador. Complaints arose that New England schooners at Labrador were driving out British fishermen. The Americans were exporting saltfish to Spain and France, as well as to their main market in the Caribbean.

Ships going to Chaleur Bay and Labrador made one fare a year, and employed on average 5,800 men and boys. American vessels were also active at the Magdalens, and around this time began fishing the St. George Bay bank off western Newfoundland. This part of the fleet "brought home an average of 648,000 quintals of fish worth $5 a quintal and 20,000 barrels of oil."

These were large quantities. As noted earlier, a quintal meant 112 pounds of product; to get the live-weight equivalent, one multiplies by about three for wet-salted fish, and by nearly five for dry-salted fish. Depending on their proportion of wet and dry salt, the Americans must have been catching more than 100,000 tonnes (2,205 pounds to a tonne) from the Gulf and Labrador alone, plus their large catches from the banks.[6]

The French, British, and Canadian fleets made the total catch much higher. The fleet of the late 1700's had only sailing vessels, small boats, and handlines, the least effective method. Yet they took major catches, which reflects both hard work and an enormous abundance in the water.

Restrictions help Maritimes traders, smugglers

With Britain restricting American trade, Maritimers could fill the gap by increasing their own trade. But to meet the demands of the West Indies and Newfoundland, Maritime traders needed American goods. Accordingly, colonial governments relaxed some of the restrictions against American goods. More importantly, smugglers often made light of the remaining laws. "The Passamaquoddy Islands [Campobello, Deer Island, and Indian Island in New Brunswick] became the centre of so great an informal commerce that the statistics of trade were rendered utterly meaningless."[7] There and elsewhere, Maritimers traded in the shade, loading fish or, frequently, gypsum (plaster of Paris) dug in the upper Bay of Fundy, into American vessels.

When the Americans in 1794 regained their market in the British West Indies by Jay's Treaty in 1794, the Maritimes' trade with the West Indies took a drop. Merchant houses in Halifax complained about Americans underselling them. Nova Scotians demanded new restrictions on New England and new bounties at home.[8] Then war came once again to the Maritimes' aid.

Blockades, embargoes end in war

In 1806, Napoleon ordered all peoples under his sway, including those in the French West Indies, to have no commerce with the British. Any ships attempting to trade would be lawful prizes of war. Britain retaliated in kind. Now each power was blocking access to the other's possessions in the West Indies and elsewhere, which reduced American trade.

Then American legislation itself helped bottle up the U.S. fleet. Britain had controlled the seas since the Battle of Trafalgar in 1805. High-handed British naval forces searched American vessels for British deserters, and sometimes forced Americans into their naval service. President Thomas Jefferson struck back in 1807 with an Embargo Act forbidding foreign commerce. Jefferson wanted to starve the British into reason. But the British made do without American goods. The embargo boomeranged, hurting Jefferson's own people. American exports fell in a single year to one-fifth their previous volume. Although Jefferson softened his approach, difficulties continued.

The whole confused episode helped the Maritimes. With the new hostilities, American fishing on the Labrador and Newfoundland coasts dropped sharply. This in turn opened markets for Maritime and Newfoundland fish. Nova Scotian exporters sold more fish to the West Indies and overseas.

On the trade side, Passamaquoddy and other smugglers continued taking fish and other goods across the foggy B.N.A.–U.S.A. border.[9] In addition, the Maritime authorities got Britain to open free ports in Nova Scotia and New Brunswick (including Halifax, Shelburne, St. Andrews, and Saint John), where they could receive and tranship goods regardless of embargoes. Maritime traders gained further advantage in 1811, when Britain partly relaxed the Navigation Laws; now colonial ships could bring Mediterranean goods to British North America without stopping at England to pay customs duties. Out of all this, Maritimers built up a great trade, partly with Europe but especially with the West Indies.[10]

Meanwhile, London and Washington were drifting into armed conflict. When war commenced in 1812, Britain contended that Americans had thereby lost all fishing privileges off British North America. Now Maritimers had almost a monopoly on fishing. They again increased their fish trade with the West Indies, where Britain's naval power kept Americans out. And they still kept smuggling, especially at Passamaquoddy Bay, with New Englanders, who took little interest in the war.[11]

Post-war patrols capture Americans

The War of 1812–1814 cemented America's national identity, helped build Canada's, and began a long peace between the two countries. But for Maritime fishing businesses, peace meant returning to the reality of enormous American competition.

The Americans maintained that they should get their fishing privileges back. The British thought differently, and seized many U.S. vessels in the years 1815–1818. In June 1815, H.M.S. *Jaseur* captured eight American vessels and sent them into Halifax as prizes. The captain warned several other vessels not to come within 60 miles of the coast; the imperial authorities later disavowed this forerunner of the 200-mile limit. In 1816, the British stationed H.M.S. *Menai*, with 64 guns, in the Bay of Fundy. In June 1817, H.M.S. *Dee* seized 20 U.S. vessels lying with nets set at Cape Negro and the Ragged Islands, Nova Scotia; the captain sent them to Halifax for adjudication.[12]

1818 convention defines American fishing

Following American protests, the United Kingdom (the political title from 1800) and the United States set out to draft a general treaty of commerce, including fisheries. The resulting convention was signed in London on October 20, 1818.

The previous agreement of 1783 had let the Americans fish anywhere off British North America, right up to the beaches. Now, the 1818 Convention set up a three-mile limit—perhaps the first such legal limit.[13] There would be no American fishing within three marine miles of "any of the Coasts, Bays, Creeks or Harbours of His Britannic Majesty's Dominions in America"—except for certain defined areas. They could now fish at the Magdalen Islands, Labrador above Mont Joli, the west coast of Newfoundland, and that island's southwest coast between Ramea and Cape Ray. This meant a major loss to the Americans, compared with 1783, and a

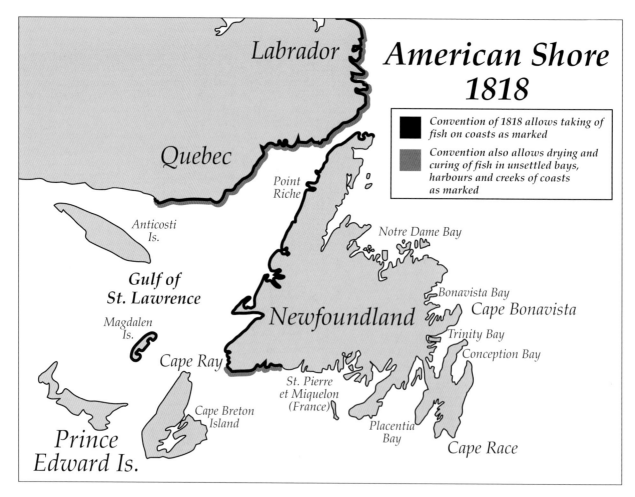

American Shore 1818

■ Convention of 1818 allows taking of fish on coasts as marked

▨ Convention also allows drying and curing of fish in unsettled bays, harbours and creeks of coasts as marked

Labrador

Quebec

Point Riche

Anticosti Is.

Notre Dame Bay

Gulf of St. Lawrence

Bonavista Bay

Cape Bonavista

Magdalen Is.

Newfoundland

Trinity Bay

Conception Bay

Cape Ray

St. Pierre et Miquelon (France)

Cape Breton Island

Placentia Bay

Cape Race

Prince Edward Is.

major gain for B.N.A. colonists in the Maritimes and Gaspé, and on the North Shore west of Mont Joli.

As for drying fish on shore, the 1783 agreement had let the Americans use unsettled bays in Nova Scotia and New Brunswick, the Magdalens, and Labrador. Now they lost all those areas except Labrador, but they picked up the southwest coast of Newfoundland between Ramea and Cape Ray. As before, when bays became settled, the Americans were to ask the residents' permission to cure fish on shore. There was one more point: the Americans could in all areas come to shore for shelter, to repair damages, to buy wood, or to get water, and for "no other purpose whatever."

All told, the 1818 Convention protected the Maritimes and Gaspé, channelling American fishing effort to the north, and opening southwest Newfoundland to shore-drying by the Americans. The Newfoundlanders were the only British North Americans to yield anything. But they also gained from the three-mile limit now protecting the east and most of the south coast. Newfoundland also continued to play host, under the 1763 Peace of Paris, to the French fleet, which could fish and cure on the west and part of the northeast coast. (The French and American shores overlapped on the west coast.)

The Convention left gaps that would bedevil fishing relations. What was a bay? Did the three-mile limit extend from a line drawn headland-to-headland, as the British later contended, or from the shore itself, as the Americans contended? The latter interpretation would open any bay more than six miles wide, such as Placentia Bay or the Bay of Fundy, to the Americans. And what about fishery management: did the American "liberties" of fishing allow them to ignore colonial regulations?

The treaty immediately drew attacks from colonists and the British press, for giving too much to the Americans. The New England fleet reappeared in force. The hundreds of American vessels at Orphan Bank in the Gulf of St. Lawrence drew blame for depleting inshore cod stocks at Chaleur Bay and Gaspé.[14]

Americans dominate bank and shore fisheries

The Americans saw Canada as the treasure-house of fish. In 1822, John Quincy Adams, then secretary of state and later president, wrote that "the portion of the fisheries to which we are entitled, even within the British territorial jurisdiction, is of great importance to this Union. To New England it is the most valuable of earthly possessions." And again:

> The shores, the creeks, the inlets of the Bay of Fundy, the Bay of Chaleurs, and the Gulf of St. Lawrence, the Straits of Bellisle, and the coast of Labrador, appear to have been designed by the God of Nature as the great ovarium of fish;—the inexhaustible repository of this species of food,

---

Partial text of the London Convention, 1818, between the United States and Great Britain

Whereas differences have arisen respecting the liberty claimed by the United States for the Inhabitants thereof to take, dry, and cure fish, on certain Coasts, Bays, Harbours, and Creeks, of His Britannic Majesty's Dominions in America, it is agreed between the High Contracting Parties, that the inhabitants of the said United States shall have, for ever, in common with the Subjects of His Britannic Majesty, the liberty to take fish of every kind on that part of the southern coast of Newfoundland, which extends from Cape Ray to the Ramea Islands, on the western and northern Coast of Newfoundland, from the said Cape Ray to the Quirpon Islands, on the Shores of the Magdalen Islands, and also on the Coasts, Bays, Harbours, and Creeks, from Mount Joly, on the southern Coast of Labrador [this mainland point was roughly opposite the eastern end of Anticosti Island], to and through the Straits of Belle Isle, and thence northwardly, indefinitely, along the Coast, without prejudice, however, to any of the exclusive rights of the Hudson Bay Company; and that the American fishermen shall also have liberty, for ever, to dry and cure Fish in any of the unsettled Bays, Harbours, and Creeks, of the southern part of the Coast of Newfoundland, hereabove described, and of the Coast of Labrador; but so soon as the same, or any portion thereof, shall be settled, it shall not be lawful for the said Fishermen to dry or cure Fish at such portion so settled, without previous agreement for such purpose with the inhabitants, proprietors, or possessors of the ground. And the United States hereby renounce, for ever, any liberty heretofore enjoyed or claimed by the inhabitants therefor, to take, dry, or cure Fish on, or within three marine miles of, any of the Coasts, Bays, Creeks, or Harbours of His Britannic Majesty's Dominions in America, not included within the above-mentioned limits; provided, however, that the American Fishermen shall be admitted to enter such Bays or Harbours for the purpose of shelter, and of repairing damages therein, of purchasing wood, and of obtaining water, and for no other purpose whatever. But they shall be under such restrictions as may be necessary to prevent their taking, drying, or curing Fish therein, or in any other manner whatever abusing the privileges hereby reserved to them.

not only for the supply of the American, but of the European continent. At the proper season, to catch them in endless abundance, little more of effort is needed than to bait the hook and pull the line, and occasionally even this is not necessary. In clear weather, near the shores, myriads are visible and the strand is at times almost literally paved with them.[15]

At Newfoundland, the British distant-water fishery had nearly vanished during the wars. Peace with the Americans in 1814 and the French in 1815 brought no major resurgence. The fishery was losing its supposed importance as the great nursery of sailing-vessel seamen; steam and steel were starting to require specially trained sailors. Britain was leading the Industrial Revolution; entrepreneurs were putting their money elsewhere than the fishery, where colonial competition was strong. In 1817, fewer than 50 fishing ships came over from Britain to Newfoundland. In 1823, there were only 15, making only 34,000 quintals, compared with 750,000 by the resident boat fishery.[16]

The British fleet still operated. The Labrador fishery was growing, and in 1829, it was said that England and the Jersey Islands sent 80 vessels and 4,000 men. Even at mid-century, Jersey houses were still bringing ships and boats across to work the Gulf of St. Lawrence.[17] Still, the traditional fishing-ship fleet was but a fading shadow of its former self. The British government would mainly let the fleet alone, to die off by itself. Meanwhile, France and New England would both make vigorous use of subsidies. Their fleets would compete with the B.N.A. colonies for a century.

New Englanders had a substantial fishery close to home, fishing Georges Bank for cod, haddock, halibut, and mackerel, the latter for bait and food. Further north, Americans were occupying B.N.A. banks and coasts, sometimes sneaking inside the three-mile limit. A vessel might in spring make trips to the nearer banks, then go to Labrador for cod and the Gulf of St. Lawrence for mackerel, then finish the year with a trip to the banks in November. U.S. vessels often picked up Nova Scotia fishermen for their trips north, a typical wage being $20 a month. They also bought herring and other bait from local fishermen.[18]

On the Labrador, a British naval report in 1820 noted 530 American sail, mostly schooners with a few brigs and sloops; this was more than ten times the 49 vessels of British, mainly Newfoundland, origin.[19] The U.S. fleet carried an estimated 5,830 men.[20] In 1829, of a reported 2,108 vessels and 24,110 men on the Labrador, the United States had 1,500 vessels and 15,000 men. In 1843, it was said that 700–800 American sail passed through the Strait of Canso annually, returning to the United States with nearly half a million quintals of fish from British waters, and buying many supplies in the strait.[21]

Vessels on the Labrador often moved along from harbour to harbour, following north the capelin, which attracted the cod and also provided bait. Vessels would anchor, then launch three or four boats to fish with hook and line or with seines. Although Americans had fish-drying privileges on the Labrador, often they salted their fish in bulk to take home.

The Grand Banks were also attracting more vessels. In the years 1830–1850, Marblehead alone typically sent 50–100 vessels, of 50–70 tons, twice yearly to the Grand Banks for handlining. Everywhere off the B.N.A. coast, the Americans were strong. In 1851, about 1,000 American vessels fished in Canada, with about 15,000 men. By comparison, Nova Scotia in the previous decade had only an estimated 10,000 fishermen, mostly in small open boats, rather than vessels.[22]

U.S. regains market in British West Indies, subsidizes fishery

The U.S. fishery during the wars had suffered losses in the European trade. After 1814, New England never re-won a major place in the Spanish market, where Newfoundland and Norway dominated. But New England could sell to a growing home market and to the non-British West Indies. In 1823, American pressure got the British West Indies reopened to trade by American vessels and goods. Fish was first excluded, but that restriction vanished in 1830. Americans could now send trading vessels to British colonies and the United Kingdom itself. Americans flourished in the West Indies trade (which, however, declined somewhat after 1833, when the British abolished slavery in their possessions).[23]

Bounties and duties helped the Americans. In 1819, the government set significant bounties at $3.50 or $4.00 per vessel ton, depending on vessel and crew size. With vessels probably averaging more than 60 tons, a typical subsidy could easily amount to $250. This equalled the value of 40–50 quintals, roughly five per cent of vessel production. Subsidization would continue until 1866.

Various duties applied against foreign products for much of the period: for example, $1.00 a quintal on dried or smoked fish—a substantial amount, since the full value of a quintal was only $5 or $6. Other typical duties were $2 a barrel on salmon, $1.50 a barrel on mackerel, and $1 a barrel on other pickled fish.[24] The United States, despite its deserved reputation for energy and private enterprise, was quick to use the power of the state for its fishery.

Mackerel and herring fisheries grow

New Englanders had long fished mackerel. Back in the 1600's, they had drag-seined mackerel from the shore, handlined them, and later trolled them. But the fishery remained modest. Then, in the 1800's, the huge American groundfish fleet branched into mackerel as a complement. Around 1815, Massachusetts fishermen developed an efficient mackerel jig. This helped create a great fishery ranging from Delaware to the coasts of Canada, said to employ nearly 2,000 vessels by the 1830's.

Fishing pressure or natural fluctuations soon made mackerel scarce off Gloucester. About 1834, Americans started fishing them in the Gulf of St. Lawrence, a fishery that lasted to the 1870's. As time went on, more Canadians worked in the American fleet, which at its peak employed some 10,000 men. American catches fluctuated widely; in some years U.S. interests imported large quantities from British North America.[25]

From the 1830's to the 1880's, New Englanders also fished a lot of herring off British North America, and bought herring from local fishermen. Their fishing methods included torching (attracting herring by flames), gillnets, and weirs.[26] Herring were useful both for bait and for food. In 1839, for example, nearly 150 American schooners of 60–80 tons "made" pickled herring at the Magdalens.[27]

Americans develop the purse-seine

By 1826, if not earlier, American fishermen invented the purse-seine, still the best means of catching large volumes of pelagic species (fish such as herring, capelin, mackerel, and tuna, which live near the surface and often school densely).[28] Fishermen had long used seines (sometimes called drag- or beach-seines) to enclose fish schooling close to shore. The fishermen would run a net behind them, often between two points of land, to cut off their escape. A cork-line held up the net's top; a lead-line sank its lower edge to the bottom. The men would pull ropes and net to bring in the seine and its fish.

The purse-seine transformed the process. The fishermen added metal rings at the bottom of the net, with a line rove through them. After encircling the fish, they pulled the line to cinch up the bottom of the net as a purse-string does a purse. The net hung like a floating bowl in the water; the fish had no escape. No longer did the fishermen need to stay in shoal water; there was no need of the bottom. Boats could purse-seine far at sea.

In deeper waters, groundfish generally swim too deep for a purse-seine. But for pelagic species, which live near the surface, the purse-seine would become a major technology, revolutionizing the mackerel fishery. The purse-line technology gradually replaced the mackerel jig, and spread from

mackerel and menhaden to the herring fishery. The Americans began using steam engines to power seines, both purse and conventional. By the 1860's and 1870's, some New Englanders were protesting the purse-seine, as they had earlier protested the mackerel jig, on conservation grounds.[29] In 1878, a Gloucester vessel sailed to Norway to purse-seine for mackerel. A Norwegian newspaper wrote:, "[I]t is obvious that the Norwegian fishermen will have to discard their old mode of fishing, and to have recourse to the American fishing method if they do not want to lose all the advantages enjoyed till now."[30]

Fresh-fish fishery becomes prominent

As cities grew and transportation improved, American fishermen built up their fresh-fish trade. In 1837, U.S. operators began carrying live inshore fish to Boston in smacks, and shipping them by rail. The fresh-fish fishery turned first to inshore stocks, then to Georges Bank. By mid-century, vessels were taking ice (cut from frozen ponds) to Georges and bringing back iced haddock.

They also iced halibut, the homely queen of groundfish, valued both for its taste and its slowness to spoil. By about 1850, the Gloucester fleet alone had about 60 halibut vessels. The winter fishery on Georges Bank in 1846 counted 29 vessels on Georges; by 1873, there were some 250. But even by mid-century, halibut were getting scarce. Many American vessels switched from halibut to cod, on Georges and the Grand Bank.[31]

New England whale fishery expands

In the whale fishery, small numbers of American vessels continued to fish on the Atlantic. From 1796 to 1807, a dozen or so craft fished whales off the south coast of Newfoundland, the Hermitage– Despair– Fortune area, until the American–British conflicts stopped them. But from about 1800, most New England whalers found they could do better in the Pacific. Vessels sailed around the Horn on multi-year voyages. By 1848, the American whaling fleet had more than 700 vessels, and had penetrated the Pacific Arctic.

Meanwhile, North American industry, which had demanded whale oil, was about to replace it. In the 1840's, Abraham Gesner of Nova Scotia developed cheap kerosene, or "coal oil," distilled from coal or petroleum. This became the standard lighting fuel. Soon Ontario and Pennsylvania entrepreneurs began drilling oil wells to produce kerosene. Although whale products were still in demand, petroleum soon replaced whale oil as the main lubricant and illuminant. (The inventor and geologist Gesner also took an interest in the fisheries. His 1847 book on New Brunswick noted the overbearing

The whale fishery, ca. 1850-1870. (Currier and Ives, Library and Archives Canada, C-32708)

energy of the Americans, the relative backwardness of the colonial fishery, and the increased worries about damage to fish stocks.)[32]

French subsidize strong fishery

The B.N.A. colonies faced fishery competition from overseas as well as New England. At the outset of the 1791–1848 period, the French Revolution and Napoleonic Wars practically destroyed the French distant-water fishery. Bounties were suspended from 1793, and their revival in 1802 had little effect.

But after the Napoleonic Wars ended in 1815, France strengthened tariffs and granted new bounties: 50 francs a man on St. Pierre and Miquelon and Newfoundland coast vessels, and 15 francs a man on vessels fishing the North Sea and Grand Banks. Exports also got bounties. From time to time, the bounties increased. Eventually they were estimated to equal the whole cost of catching and curing fish. "There is common agreement among contemporary observers that the system of bounties allowed France to become and remain an important saltfish exporter."[33]

The measures helped the fleet rebound. By 1830, the fleet was said to employ 300–400 vessels or about 12,000 men on the banks. Another 300 vessels worked in the shore fishery, each employing about 50 men and five boys. The French fishery was

big business again. The French government controlled operations closely, using a draw to assign vessels to particular harbours.[34]

French fleet pioneers the longline

Until the 19th century, the northwest Atlantic cod fisheries used mainly handlines to fish from the ship or boat. They also used beach-seines along the shore; it was said in the 1800's that the French cod-seines had ruined their northern fishery.[35] With handlines, fishermen could use only a few hooks.

Now the French changed that, by introducing the longline, or bultow, with many baited hooks attached to a "groundline" running close to the sea bottom. Dieppe fishermen first developed the longline in the second half of the 18th century. After the Napoleonic wars, it became widespread. With the longline, a single small boat could fish hundreds of hooks, and a vessel carrying a number of boats could fish thousands.

It was common from the early days of fishing the northwest Atlantic for vessels to bring small boats with them, but only for inshore fishing. Now the French began to launch them on the offshore banks, which allowed the use of more longlines.

The French vessels known as "Terra-Neuvas" used fairly large shallops, launching one from each side of the vessel. It was dangerous work. Some vessels made the shallop fishermen pay out a lifeline that kept them attached to the ship. Others failed to do so; some fishermen got lost on the water and died. The Minister of Marine ordered the fleet to use the lifelines; but they interfered with mobility, and the fish guts near the ship attracted dogfish that kept away the cod. Soon the shallops habitually fished with no lifeline to the vessel. The French also began replacing iron hooks with new mass-produced steel ones.[36]

By 1820, old French fishermen as well as English fishermen on the Grand Banks were complaining that longlines, forbidden by the English government, were destroying the species. Before, large mother codfish had escaped the handline; the longline, set closer to the bottom where they stayed, captured them. "Full of eggs, they yielded to the greedy fisherman who had no care for the future." Some French voices also protested that the longline lowered the quality of the fish, because of the longer time on the hook. But the longline gave bigger catches, and it spread irresistibly.[37]

Longlines spread to New England in the 1840's and to the Maritimes in the 1850's, if not earlier.[38]

The Maritimes and Quebec hit their stride

Although the Americans dominated the northwest Atlantic vessel fishery, with the French also strong, the Maritimes and Quebec were becoming powerful, with their own vessel fleet and thousands of small boats. The Anglo-American and Anglo-French wars gave them the chance to expand, which they seized. The end of the wars brought a setback. A short depression struck the B.N.A. coast, especially severe in Newfoundland. The Americans regained fishing privileges by the 1818 convention. Then the provinces lost markets in Europe, where French and Norwegian competition was resurging. The growing competition in Europe turned B.N.A. producers more to the markets of North America, the West Indies, and "the Brazils," the newly independent states in South America. Here, too, they faced renewed competition from New Englanders.

Still, the Maritime and Quebec fleet kept expanding. Nova Scotia and southern New Brunswick's Bay of Fundy shore shared many of New England's advantages: a good position for trading, no ice, a long fishing season, and a good mix of species. They had greater fish resources than New England. And they had received an influx of pre-Loyalist and Loyalist settlers from that region, bringing with them traditions of self-government and commerce.

Strong little fishing towns had sprung up. Yarmouth in 1828 had 65 vessels averaging 46 tons in coastal fishing and trading, with some 20 of them trading to the West Indies. Barrington had 69 vessels and 62 boats. Digby, Chester, Port Medway, Liverpool, Sydney, Pictou, and many other growing ports had vessels working from the West Indies to Labrador. Halifax had six ships, 67 brigs, and 77 schooners engaged in coasting, fishing, and the West Indies, Brazil, and European trade. The communities had many smaller craft doing day fishing.

In particular, Lunenburg by 1828 had more than a hundred vessels, around 20 of them trading with the West Indies, and was pulling ahead of other towns. Lunenburgers more than most others used joint-stock financing of boats: townspeople would back up promising skippers by buying into the 64 shares of a schooner. The joint-stock system helped Lunenburg mount a large fleet for the bank fishery.

Cape Breton in 1828 had 340 registered vessels, averaging 50 tons, probably more than 1,000 smaller craft, and a number of trading establishments besides the Jersey houses. By 1843, it was said that the Canso and Cape Breton areas had about 5,000 fishermen, with more than 120 shallops and 1,700 boats, and that as many might be employed in the rest of the province. That made a total of roughly 10,000 fishermen in Nova Scotia, with perhaps 3,400 boats.[39]

Most Maritimers and Quebecers fished nearby grounds; some went further afield. Nova Scotia and New Brunswick, chiefly the former, in 1830 sent 100–200 vessels and 1,200 men to the Labrador. In 1831, some 27 schooners fitted out in the Magdalens, with 10 of them fishing the Labrador.[40] In 1840, Nova Scotia exported about 327,000 quintals of dried fish, a considerable amount, probably equivalent to about 81,000 tonnes live weight. The province also produced 71,600 barrels of green fish and 27,750 boxes of smoked fish. The West Indies took well over half of the dried fish, and close to half of the green and smoked fish. Exports went also to other B.N.A. colonies, the United States, the Brazils, Britain, Europe, and Africa.[41]

In Prince Edward Island, the fishery had gotten off to a slower start. An 1803 report noted about 70 vessels on the island, mostly schooners owned by French settlers on the north shore. But fishing was an irregular occupation. The vessels also served for trading farm produce to Halifax and Newfoundland. American vessels would continue to dominate the fishery around P.E.I. for decades to come.[42]

Fishermen often live poorly

In the Nova Scotia Assembly in 1831, it was stated that "the very existence of Trade in these Northern Colonies depends upon the prosperity of the fisheries, which are the principal support of the Trade to the West Indies: we could not supply the Islands with Timber, and numerous other articles, if our Fisheries failed, as that staple article affects directly or indirectly every other branch of Commerce from these Atlantic Colonies." An earlier petition made the same point: "when the fisheries and the attendant Commerce flourishes, the effects circulate beneficially through every branch of industry, and the farmer, the woodsman, and ... the mechanic all receive there-from an immediate and general impulse."[43] In Lunenburg, cod traditionally meant so much that a church steeple there still sports a weathervane in the form of a codfish.

But growth was not always impressive, nor prosperity universal. Small open boats predominated by far. Where vessels existed, they tended to be smaller than U.S. schooners (except perhaps at Lunenburg). There were many complaints of the Maritimes lagging behind the United States.

Because of their resources and location, Nova Scotia and the Bay of Fundy offered the best chance for fishermen to climb into a better life. But a typical report of the 1830's found Grand Manan Island, on the New Brunswick side of the Bay of Fundy, to be behindhand. The people would fish, plant a few potatoes, and do well at neither, said the report; they needed to copy the United States and use nets instead of hooks for mackerel.[44] And in Nova Scotia it was said that fishermen, "by dividing their time between Coasting, Farming, and Fishing ... fail in producing any good results."[45]

In Nova Scotia as in Newfoundland, it was common for fishermen to cut their own timber and build their own craft. Thomas Chandler Haliburton wrote that "the Nova Scotian ... is often found superintending the cultivation of a farm and building a vessel at the same time; and is not only able to catch and cure a cargo of fish but to find his way with it to the West Indies or the Mediterranean; he is a man of all work but expert in none."[46] Fisherman-owners, said to be mostly poor, would get their outfitting on credit (at high rates). Crewmen at the time generally got wages, contrasting with the American practice of dividing the catch into shares.[47]

A member of the Nova Scotia Assembly, from Barrington, in the relatively strong fishing area of southwest Nova Scotia, in 1827 said, "I have never yet known among my constituents one solitary instance of a man getting beforehand by fishing and fishing only; those who own vessels or parts of vessels did not earn them by fishing; they earned them in better times, by sailing coastwise, carrying plaster of paris, etc...." The fishermen, he continued,

are the main staff and support of the commerce of their country, they are the greatest source of revenue; from their labours originate the principal article of exportation; their hard earnings have helped to enrich many of those who are engaged in commercial pursuits, and have served to aggrandize their country. But they themselves, although they compose a large proportion of the population are literally in a state of bondage.... Their unprofitable callings have rendered them destitute of the means of improvement, and doomed them to perpetual servitude, their education and morals being almost totally neglected."[48]

Jersey houses dominate the Gulf

Despite individual hardships, the marine economy was on the rise throughout the Maritimes. In the Gulf of St. Lawrence, with a more seasonal fishery further from markets, a smaller number of larger companies were more dominant. Channel Island firms such as the Robins (with locations including Paspébiac, Percé, Grand River, and Newport) ran a truck system that put fishermen under semipermanent obligation, with poverty frequent. Here as elsewhere, fishermen showed versatility in farming and fishing. But they received only small lots of ground for farming, which forced them more into fishing. They got paid half in goods, half in cash, which they could spend only at the company store. If the fishermen failed to pay debts, the owners might press them onto ships. But a certain level of debt to the company was part of life; fishing families never really expected to get clear of debt.[49] Schools were ruled out. "If they were educated," wrote Philippe Robin, "would they be any cleverer as fishermen?"[50] Complaints arose of bondage.[51]

It is worth noting, however, that the companies themselves were often close to the margin. Many Channel Island firms that located in the Gulf soon went out of business.[52] The Channel Island companies at first owned most of the boats. But fishermen gradually began buying their own. By the end of the 19th century, fisherman ownership dominated the fleet.[53]

Maritimers produce more herring, mackerel

As the Maritimes' population rose and transportation improved, smaller fisheries turned into larger-scale commerce. Commercial herring and mackerel fisheries had begun with the pre-Loyalists and Loyalists in the second half of the 18th century. Beach-seining was a typical method, as were gillnetting and hook and line. In Nova Scotia, Digby production rose from 630 barrels in 1824 to some 5,600 in 1826. Mackerel exports from Halifax quadrupled from about 19,000 barrels in 1839 to 83,000 barrels in 1846. Exports of pickled fish, which often meant herring and mackerel, rose from about 61,000 barrels to 136,000 barrels.[55] In New Brunswick, however, a government report complained at mid-century that no mackerel fishery existed in the province.[56]

Though herring occur in most parts of the Atlantic coast, the Bay of Fundy is the most consistent producer. Early in the 1800's, a common fishing method at Grand Manan Island in the Bay of Fundy was torching. Fishermen working at night would light a torch in the bow, and with a dipnet scooped up the herring that followed the light.[57] Torching remained common at Grand Manan and Campobello until the middle of the 20th century.

There was some regulation of the important fish-

Perley's portrait of the Jersey houses

The New Brunswick government appointed the lawyer and naturalist Moses H. Perley in 1848–1849 to report on the province's fish and fishery. Perley's perceptive report described the breadth and sway of the Jersey houses. He pointed out the strong presence of British fishermen on the water, who still came over to work the Gulf along with local fishermen, and of foreign goods in the shops, reflecting contemporary trade patterns.

- The Jersey merchants ... prosecute these fisheries with great zeal and assiduity, and, as it is believed, with much profit. ... They employ upwards of one hundred vessels ... besides the smaller craft required upon the coast. Two of the leading Jersey firms, Messrs. Robin and Co. and Nicolle Brothers, are supposed respectively to afford employment, directly or indirectly, to nearly one thousand persons.

- On the beach at Paspebiac, is situate the depot of the wealthy and well known firm of Charles Robin and Co., of Jersey... . Every spring, a whole fleet of ships and brigantines belonging to the firm, arrive at Paspebiac from Jersey, with double crews, and all the necessary stores for the season. These vessels are moored in front of the beach, their sails are unbent and stored, their topmasts and yards are struck and housed. The whole of the vessels are placed in charge of one master and crew, who take care of them during the summer, and issue the salt, with which they are ballasted, as it is required. The rest of the masters and crews are dispatched in boats and shallops to various parts of the Bay to fish, and collect fish from those who deal with the firm. When the fishing season is over, these vessels depart with cargoes for the West Indies and Brazil, but more frequently to the Mediterranean—to the Ports of Messina and Naples.

- The 'fishing rooms' at Miscou are shut up in the winter season, and left in charge of one of the above residents, who is called the 'room keeper.' The Jersey men employed here during the summer, either return to Jersey for the winter, or go to the Mediterranean in the vessels which take the dried fish to the markets there, returning to their posts in the spring. They are completely birds of passage, having no tie in this Province, or any interest in its general prosperity.

- [In the Shippagan fishery of Wm. Fruing & Co., one of the Jersey houses:]

... there were sixty boats engaged in fishing, averaging two men and a boy to each boat. ... Nearly all the fishermen at this establishment were French settlers, who had small farms or patches of land, somewhere in the vicinity, which they cultivated. ... Those who are too poor to own boats hire them of the firm for the season....

The fishermen are allowed for a quintal of cod ... ten shillings, and for ling and haddock, five shillings,—the amount payable in goods at the store of the firm, on Point Amacque, where a large quantity of foreign goods is kept, over every variety. Here were found Jersey hose and stockings—Irish butter—Cuba molasses—Naples biscuit, of half a pound each—Brazilian sugar—Sicilian lemons—Neapolitan brandy—American tobacco—with English, Dutch, and German goods,—but nothing of Colonial produce or manufacture, except Canadian pork and flour.

- Some of the residents at Shippagan, who are in more independent circumstances, prosecute the fisheries in connection with their farming, curing the fish themselves, and disposing of them at the close of the season to the Jersey merchants, or to others, as they see fit.[54]

ery at Grand Manan, also prosecuted by Nova Scotian and American vessels. The visiting Moses Perley was told that vessels had been restricted to 30 fathoms of net; boats, to 15 fathoms. But policing could be a problem. At times there were a hundred vessels fishing, with a government boat trying to keep law and order. "Nets were continually destroyed or stolen, especially during dark and windy nights.... It was said, that boats with old scythes attached to their bottoms, had been rowed swiftly among the nets, by which great damage had been done."

By 1850, weirs for herring and mackerel were common on the New Brunswick side of the Bay of Fundy. Aboriginal peoples had used stone or brush weirs; white fishermen apparently adapted them. Nova Scotia fishermen used the brush weir at least from the later 1700's. Weirs seem to have become

Curing fish at a merchant establishment in the Gulf later in the century. (Photo by Quebec photographer Jules-Ernest Livernois, 1851-1933. Library and Archives Canada, PA-23872)

more common around 1820. On the other side of the Bay of Fundy, at Campobello, New Brunswick, and Lubec, Maine, fishermen began building weirs, which gradually spread throughout the area. Weirmen drove large wooden stakes into the bottom, fastened poles atop the stakes, and put twine on the poles to form a semicircular trap. Fish schooling close to shore followed a leader fence into the weir; fishermen then closed the weir's mouth with a "shutoff" net. When the purse-seine became common, weirmen began using small purse-seines to take up fish within the weir. Weirs became a mainstay of the southwest New Brunswick herring fishery.[58]

Lobster fishery starts up

One of the greatest commercial fisheries was just starting up, thanks to new technology. The first person to eat a lobster must have had great nerve, to hope for anything good behind the claws, spidery legs, and bulging eyes. But people gradually learned. Aboriginal people captured lobsters,[59] and the French pioneers at Port Royal in the early 1600's caught them by hand. Maritime fishermen in the 1700's gaffed, hooked, and speared lobsters for small local markets. Settlers at the Bay of Chaleur used thousands for fertilizer.

Early in the 1800's a commercial fishery for lobster started in Massachusetts and spread east, reaching Maine by the 1840's and the Maritimes by the 1850's and 1860's. Fishermen worked from row or sail boats, using a hoop net with bait in the centre. Fishermen might attach 20 or more nets to a cable, forming a trawl.

At first, markets remained local and prices low. But that changed when lobsters became keepable and transportable. France in the early 1800's developed the preserving of food in containers, initially glass jars. The Underwood company of Massachusetts became a leader in North America, packing lobster and salmon in the 1820's, at first using boiling water and glass jars. Others adopted tin canisters.[60]

In 1839, at Saint John, New Brunswick, Tristan Halliday canned lobster and salmon, the first canned salmon in North America. After seeing Halliday's operation at Saint John, Upham Treat, an entrepreneur near Eastport, Maine, began canning lobster, salmon, and mackerel. Later, Treat put up codfish, beef, and mutton as well as lobster. Charles Mitchell, a Scot, canned salmon and meats at Halifax in 1840, and later canned fish in the U.S., helping the practice to spread. Around 1845, a cannery started up at Portage Island on New Brunswick's eastern shore, and later in the 1840's others operated at Kouchibouguac, N.B. and Yarmouth, N.S.[61]

Lobsters were abundant; a Prince Edward Island account in 1839 noted they were so plentiful that older settlers despised them. "They should never be permitted to appear at dinner, and should not be eaten for breakfast or supper above once a week."[62] But consumers elsewhere wanted them; canneries would soon multiply faster than lobsters. For most of the century, canning remained a laborious by-hand process involving cutting strips of metal and soldering the sides, bottom, and top.[63]

Brush weir of an early type. (From Goode, 1887, courtesy of Bill McMullon)

Governments get more active

Partly from natural vigour, partly from impatience at British controls, the B.N.A. colonies were gradually taking more control of their own affairs. Following the practice of England and New England, every province now had an elected legislature, starting with Nova Scotia in 1758. But British governors and their associates still controlled an over-large share of legislation, finances, and official appointments. Governors and assemblies frequently got into disputes. The elected legislatures themselves were not necessarily very representative; in Nova Scotia, for example, Halifax merchants controlled many activities.

Reform movements campaigned to replace the ruling cliques and councils. Nova Scotia and the Canadas gained Responsible Government in 1848, Prince Edward Island in 1851, New Brunswick in 1854, and Newfoundland in 1855. This was a turning point for democracy; colonial cabinets now held power only with the majority support of the elected assembly.

During the evolution to Responsible Government, fishery interests made representations to elected representatives and officials, whoever might serve them best. Colonial governments, with occasional help from London, addressed the fisheries mainly through trade regulations, subsidies, and enforcement against foreign fishing.

The British mercantile system still applied. Complex regulations gave Britain and her colonies a preferential trading position with each other, and made it difficult for the colonies to trade outside the British network. But North Americans had never followed the system all that closely, and now it was bursting at the seams. B.N.A. interests pushed for trading freedom for themselves and for tariff and such restrictions against others, as it suited them. As time went by, Britain herself wanted more freedom from the cumbersome system and its incessant colonial demands.

Sea-fishery conservation got almost no attention, despite some reports of inshore depletion. The authorities paid more attention to conservation in the inland fisheries, but laws were often ineffective.

Maritime governments apply subsidies

Nova Scotia had already made use of subsidies before the War of 1812. During the post-war depression, the province gave bounties on cod and "scale-fish" (a common term for groundfish other than cod, such as haddock and pollock). Then, when the British government in 1823 allowed American trade with the British West Indies, the colonies bemoaned the change. They also complained about the duties they had to pay on fishing equipment they imported for their own vessels.

Nova Scotia provided more assistance. In 1824, bounties went onto exports of "merchantable" fish (that is, of good tradeable quality) to Europe, Africa, Mexico, and South America. In 1828–1830, additional bounties went onto cod exports, widening the rules to include fish shipped in the vessels of other nations. The colony made similar attempts to aid the mackerel and other fisheries, and also encouraged shipping by lowering some duties. Bounties helped draw several vessels into the whale fishery.[64]

In 1828, Nova Scotia began encouraging the industry to organize its own improvements. A new society for encouragement of the fisheries offered premiums for fish taken on the banks, on the Labrador, and in the Gulf of St. Lawrence, as well as for vessels landing the most "merchantable" fish in Halifax.

New Brunswick made less use of bounties, but in 1824 and 1825 paid vessels 20 shillings per ton. As for business provisions, New Brunswick had the advantage of good supplies of salt, arising from her extensive timber trade. Nova Scotia fishermen often dealt with Saint John for salt.[65]

In 1825, the Prince Edward Island legislature passed a Fisheries Act, setting conditions for hiring and employment of fishermen, establishing an April 15 to November 1 season, and providing some protection for oyster beds. Another act in 1829 provided bounties on vessel tonnage and cod exports. The fleet was still small: according to one report, only 11 vessels and 38 men fishing commercially. Bounties continued intermittently, to no great effect.[66]

Governments try more quality, conservation rules

Apart from providing bounties, colonial governments in the 1791–1848 period gave somewhat more attention to regulating the fishery itself, with scattered regulations for quality and conservation. In Nova Scotia, Annapolis County apparently appointed fishery officers annually as early as 1772. These were nominated by grand jury and appointed by Sessions Court. In 1797, seven such appointments took place in the province, for a culler of fish, an inspector of smoked herrings, an inspector of pickled fish, a gauger, and three overseers of the fishery.[67] In 1827, Nova Scotia passed regulations about the making of barrels, along with other pickled-fish laws and amendments; and in 1828, it set up an inspection act, amended in 1829, governing pickled fish.

Such measures apparently had too little effect; complaints about quality continued. It was said that the use of beach-seines at Canso produced poorer quality mackerel than hook-and-line fishing, and that the rotten fish on the beach were driving good fish away. Speakers in the assembly in 1829 wanted stricter enforcement, and talked also about problems of administration. It was said that Nova

Scotia alewives had "almost entirely lost their repute" and were displaced by Scotch herring. Bad fish was being passed off in the West Indies as the "nauseous food of the forlorn African slaves." An inspection law "would be an act of humanity."

In 1833, Nova Scotia passed a bill providing for the appointment of fishery inspectors for all districts.[68] To judge from later reports, these were part-time positions and none too efficient. Prince Edward Island in 1829 passed its first quality rules, regulating the size of fish barrels and enforcing inspection of exports.[69]

In 1845, New Brunswick passed stringent laws for salmon conservation. These went more or less unenforced.[70] As noted earlier, New Brunswick had also restricted the amount of netting to protect the Grand Manan herring fishery.

In 1823, Quebec introduced regulations governing fish quality, especially for pickled fish—salmon, herring, shad, and sturgeon—and appointed inspectors at Montreal and Quebec. In 1824, the province passed conservation regulations, especially for salmon. These actions seem to have had little effect. Concerns were rising about the fishery, but neither the Maritimes nor Quebec was taking firm hold.

British seize American vessels

Meanwhile, problems offshore led to additional government action. In the cod fisheries, the large, fast American vessels sometimes ran down Nova Scotians. Some captains strengthened their bowsprits for the purpose, and carried guns. New England vessels attracted large numbers of Maritimers and Newfoundlanders to migrate and fish with them; Nova Scotian masters of U.S. vessels ("white-washed Yankees") acquired the worst reputation.[71]

Some Nova Scotians wanted bigger banking vessels, more capital investment, and a more dedicated fishery, rather than fishermen dividing their time between coasting, farming, and fishing. To invest, one company told the Assembly, capitalists had to be sure that the inshore fisheries within treaty limits belonged to British subjects.[72] They needed enforcement.

In 1835, Nova Scotian authorities seized four U.S. vessels. In 1836, the province passed the Hovering Act, which let revenue officers board vessels coming within the three-mile limit. In 1838, the imperial authorities decided to station a small armed vessel ("revenue schooner") on the Nova Scotia coast and another at Prince Edward Island. In 1840, two U.S. vessels were seized for purchasing bait. New regulations and increased use of revenue schooners appeared to lessen the American fishery at Nova Scotia.[73]

The Americans argued constantly with the British about whether the three-mile limit should start from the coast or from headland-to-headland closing lines. The 1843 seizure of the schooner *Washington* in the Bay of Fundy, and another seizure within a headland-to-headland closing line at Cape Breton, became a stormy issue. Great Britain ordered that U.S. fishermen be allowed into the Bay of Fundy, the only bay conceded to be open.[74] Later, in 1853, when Britain and the United States were trying to settle some disputes in a more friendly atmosphere, an umpire ruled that the Bay of Fundy was not a bay as included in the 1818 treaty, because of its great size.[75]

Nova Scotia's sister provinces acted more slowly. Prince Edward Island passed a hovering act in 1843, New Brunswick in 1853. Nova Scotian authorities or British patrols at Nova Scotia continued to make the most seizures.[76] All told, from 1818 to 1851, the British colonies seized 51 vessels, destroying 25 of them. In the same period, the Americans made at least 16,000 voyages to the Maritimes. No doubt many sneaked inside the three-mile limit, but the more aggressive approach by the provinces kept the Americans on better behaviour than would otherwise have been the case.[77]

Colonial trade system collapses

Meanwhile, in the now major shipbuilding trade, the centre of gravity was shifting to larger vessels at the bigger yards, including those of Quebec and the Great Lakes. In the Maritimes, the building of fishing vessels became more of a cottage industry, at small boatyards all along the coast.

Greater changes were coming. The British colonial trade system—the Navigation Laws, the tariffs, and the whole attempt at a self-reinforcing imperial network—was under strain both at home and in the colonies. Britain was leading the Industrial Revolution, her factories pouring out goods that needed markets. Manufacturers and politicians envisaged the benefits of free trade, whereby they could simplify the system, receive raw materials without restriction, enjoy cheaper food imports, increase industrial employment, and send out finished goods to a grateful world. At the same time, the colonies themselves were trying, in an ambivalent way, to break out of the old system. Many Britons began thinking of the colonies as a burden; it might help everyone if they were more separate.

The complex trade laws were loosening up. In 1826, Britain had allowed more free ports for transhipment to open up in the Maritimes. Other changes allowed foreign vessels to supply the colonies directly, and also let colonial vessels trade more freely with non-British areas, including Brazil and other South American countries. An important measure in 1830 gave the B.N.A. colonies power to impose their own protective tariffs.

A further series of reforms, notably in 1846 and 1849, did away with the remainder of the old colonial system. Now foreign ships could come and go freely. The colonies could trade as they wished, repeal preferences on British goods, and use their tariffs to restrict competition or gain revenue. The old mercantile system was swept away.[78]

With more freedom came less security. The shock of freer trade coincided with the depression of the 1840's. Maritime governments began thinking less of the marine economy and more of railways, manufacturing, and the continent.[79] Sentiment grew for "Reciprocity" with the United States—a mutual lowering of duties. The Maritimes wanted free trade just when the Americans wanted more fishing privileges in the British colonies. Both would get their wish in the 1850's.

Newfoundland: most fish, least prosperity

Like the Maritimes and Quebec, Newfoundland grew stronger in the 1791–1848 period, but with more difficulty and less progress.

In the 1790's, Newfoundlanders were gaining ground in the fishery, while British and French competitors declined. British conflicts with the United States meant that less money and fewer emigrating Newfoundlanders flowed towards the United States. As well, British restrictions at first held back New England's trade with the British West Indies, opening new opportunities for Newfoundlanders.

Newfoundland's population rose from about 10,000 in 1776 to about 122,600 in 1851. But this was still small, compared with more than half a million in the Maritimes and nearly 900,000 in Lower Canada.[80] Newfoundland faced harder conditions than other provinces. Trees were smaller and slower growing; farmland was scarce. Manufacturing, slowly increasing in the Maritimes to serve the colonial population, was uncommon in Newfoundland. Some new fisheries were developing, notably for seals; but the colony was largely captive to cod, and ordinary fishermen captive to the merchants.

Newfoundland had the most codfish, the least prosperity, and the least power. In 1791, the colony still had no legislature and no power to run its own fishery or other affairs. Contemporary accounts often noted poverty and hardship. In 1799, the island's governor wrote that

> ... unless these poor wretches emigrate they must starve, for how can it be otherwise whilst the Merchant has the power of setting his own price on the supplies issued to the Fishermen, and on the Fish which these men catch for them. Thus we see a set of unfortunate beings working like Slaves, and even hazarding their lives, when at the expiration of their term (however successful their exertions) they find themselves not only without gain, but so deeply indebted, as to force them to emigrate, or drive them to despair.[81]

But Newfoundland with its huge codfish resource was developing a fleet and a mind of its own. As for the Maritimes, Britain's wars with France and United States speeded growth. By 1795, the Newfoundland saltfish industry employed about 7,100 resident men. "By 1805, it appears that the centre of the Newfoundland fishery was shifting from the West Country to St. John's."[82] By the early 1820's, resident fishermen no longer relied on migratory servants; they were numerous enough to take the catch by themselves.[83] Residents had developed a small bank fishery, as well as a large fishery closer to shore.

Seal fishery employs thousands

Much of Newfoundland's new strength and self-assertion came from sealing. The Anglo-French wars beginning in the 1790's boosted the seal fishery. With less French presence in their way, Newfoundlanders pushed northwards. By the end of the century they were moving offshore, using sailing vessels to go to the pupping grounds on late-winter ice floes.

The cod fishery by Newfoundland-based vessels provided ships; the seal fishery in turn aided growth of the bank fleet. In 1800, about 50 vessels went out for seals. Merchants in St. John's and Conception Bay would send out vessels of up to 75 tons, with between 10 and 20 men, heading north around mid-March to meet the ice pack. Gunners shot the larger seals; men with clubs killed the others, dragging

Killing seals in Newfoundland. (From Prowse's *History of Newfoundland*)

them back to the vessels for skinning. A successful vessel could return in about five weeks, unload, and make a second trip. After the seals were landed the fat was put into puncheons and left to melt in the sun, then put into vats where water was drained off from the bottom and oil taken from the top. With the oil extracted, the blubber or remainder was boiled to produce common seal oil.[84] A seal would yield 10–15 gallons of oil.[85]

As Europe grew, demand for seal oil increased. By 1799, St. John's fishermen were capturing some 80,000 seals, Conception Bay people about half that number. In 1804, some 1,600 men took 156,000

Sealing song

The seal fishery meant money at the end of hard winters, and adventure on the ice. As they did to much of life, Newfoundlanders applied music to the seal fishery. Here are excerpts from one song:

The Block House Flag is up today to welcome home the stranger
And Stewart's House is looking out for Barbour in the Ranger
But Job's are wishing Blandford first who never missed the patches
He struck them on the twenty-third and filled her to the hatches.

The first of the Fleet is off Torbay,
All with their colours flying:
And Girls are busy starching shirts
And pans of beefsteaks frying.

Though some may sing of lords or kings,
Brave heroes in each battle,
Our boys for fat, would gaff and bat,
And make the whitecoats rattle.

They's kill their foe at every blow
(Was Waterloo more grander?)
To face, who could, an old dog hood
Like a plucky Newfoundlander?[90]

The town of St. John's, Newfoundland before the fire of 1846. (From Prowse's *History of Newfoundland*)

seals. By 1805, 131 ships went out to the dangerous fishery; 25 were lost in the ice. By 1834, St. John's alone had 125 ships, with 3,000 men sealing. According to Prowse, the fishery changed social habits for some, getting them working in late winter rather than drinking and dancing.[86] Pelt production rose from an average 197,000 in 1816–1820 to 527,000 in 1841–1845. By mid-century, the seal fishery might employ more than 10,000 men and take more than half a million seals.[87]

For Newfoundlanders, the seal fishery was a historic achievement. It was the first fishery they created, the much-needed complement to the cod fishery, and the chief spur to new growth. The departure of vessels for the fishery became a great occasion, with toasts of "bloody decks." One writer in the 1840's observed that "[t]he interest of every individual, from the richest to the poorest, is interwoven with [the seal fishery], and the prosecution of the voyage causes more anxiety, excitement, and solicitude, than any other business in Newfoundland, or probably in the world."[88]

Nova Scotia's participation in the seal fishery remained small. In the years 1827–1829, a few Nova Scotia vessels took seals in Newfoundland, but with disappointing results. About 20 small Cape Breton vessels caught seals in the Gulf.[89]

Western boats and Newfoundland jacks

Small boats dominated by far in Newfoundland. But the seal fishery helped boost the vessel fleet. The first decked vessels for the seal fishery were 40–50 feet long, with 14–15-foot beams. According to Prowse, "the schooner rig came into vogue about the time of the commencement of the ship seal fishery. In the early accounts of this industry a distinction is made between the fishery as prosecuted in shallops and in schooners. The only novelty introduced about 1798 was the use of larger vessels in this business."[91]

Different types of schooner picked up their own names. The smaller Newfoundland jack was "a bluff, two-masted decked vessel, schooner-rigged and varying from 5 to 20 tons ... used for various fisheries purposes." These vessels might be 25–30 feet long. The larger western boat or cape boat (used west of Cape Race on the south coast) ranged 15–30 tons and might be 40–50 feet in keel length. Jacks and western boats stayed closer inshore than the larger schooners.[92]

Labrador fishery starts to grow

Annexed to Quebec in 1774, Labrador and Anticosti Island (but not the Magdalen Islands) came back to Newfoundland in 1809. Quebecers in the salmon and seal fisheries protested. In 1825 the Labrador coast from Blanc Sablon west, including Anticosti, returned to Lower Canada (Quebec). Coastal Labrador now had its final political border. As a geographical expression, "Labrador" at first continued to include much of the Quebec North Shore, but as time passed it began more clearly to signify the area north of Blanc Sablon.

Some fishing had taken place below Sandwich Bay (the area of present-day Cartwright) from the 1750's.[93] Over time, more vessels moved northward, up to and around the easternmost point of Cape Charles. Fishermen from all around began vying for Labrador cod.

Peace brings challenges

The fishery resources in Newfoundland continued to amaze travellers. Anspach about 1818 described Conception Bay at night during the capelin-scull, with whales and cod leaping, capelin everywhere, the whole surface of the bay covered, and women all along the shore with barrows and baskets, loading fish, while fishermen caught them with cast-nets for bait.

Fish prices had soared during Britain's wars with America and France. The loss of American market paled in comparison with gains in Europe. Newfoundland's cod exports nearly doubled, from about 660,000 quintals in 1804 to more than a million in 1815 and 1816. In the latter year, more than half went to Spain, Portugal, and Italy, followed by the West Indies, Great Britain, and British North America.[94]

But the wars ended by 1815. Fishing competition from the French and Americans, as well as market competition from those countries, the Norwegians, and others, brought a drop in demand. A harsh depression followed, with many bankruptcies. Emigration to the continent increased. The new merchant class weathered this storm, and growth gradually restarted. Meanwhile, another market was coming on stream. Exports to Brazil rose from about 2,000 quintals in 1814 to 64,000 quintals in 1825.[95]

Sealing, fishing, and shipping made the shipyards grow. In 1804, Newfoundland was said to have 150 schooners.[96] By 1830, she had 18 banking vessels, and fishing along the shore were 301 "island vessels" and 3,797 boats.[97]

From the 1830's into the 20th century, Newfoundland's schooner fleet grew to number many hundreds. The vessels worked mainly along the huge coastline of Newfoundland and Labrador, drying fish as they caught them, rather than fishing the offshore banks. Newfoundland's fishing schooners provided a base for trading-schooner development, though never on the scale of the Maritimes, which had more people and goods. Around the 1840's, according to Prowse, Conception Bay merchants started buying "slop-built" vessels from the provinces.

Labrador fishery becomes huge

Much of the growth stemmed from Labrador, where the Newfoundland fleet had lots of company. In 1820, some 5,800 American fishermen worked in the region. Vessels from Lunenburg, N.S. in the 1820's found it their most vital fishery.[98] In the 1830's, Nova Scotia and New Brunswick would send 100–200 vessels to the Labrador. In 1831, of 27 vessels fitted out in the Magdalens, 10 went to the Labrador.

The English overseas fleet, though fading, was still marginally in the picture. And even as it dwindled and disappeared, English companies maintained enterprises in the Gulf, Newfoundland, and Labrador.[99] It was difficult, in the case of many larger outfits, to define them as belonging either to England or to Newfoundland.

During the Napoleonic Wars, Newfoundlanders had made use of the abandoned French Shore in northeast and western Newfoundland. After peace in 1815, the French reoccupied their shore, forcing many Newfoundlanders north to the Labrador coast. From 1814 to the late 1820's, their fishery at Labrador multiplied sixfold. In 1825, some 200 vessels came to Labrador from Conception Bay, and 60–70 from St. John's. About 5,000 Newfoundlanders took part in the fishery.[100] Besides cod, they might take smaller quantities of seals and salmon.

Many Newfoundland schooners would first go to the seal fishery from March to May, then head for Labrador until October. The fishery at least in the early stages used mainly seines at Labrador, since

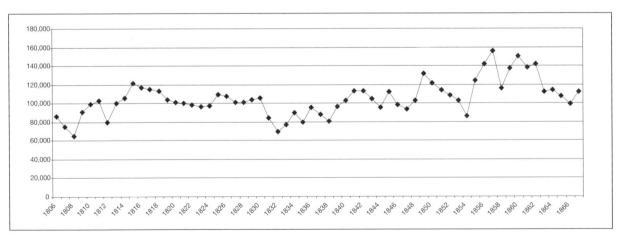

Exports of dried cod from Newfoundland, 1806-67 ('000 lbs).

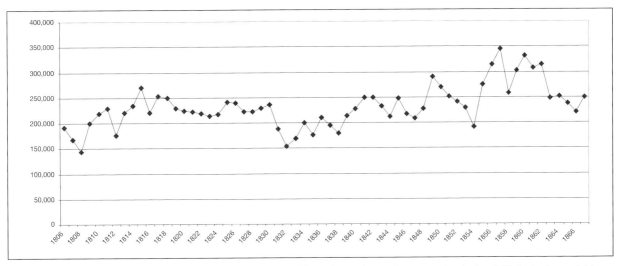

Cod landings in Newfoundland, 1806-67, in tonnes, calculated from dry-cod exports, conversion factor 4.88 dry to live weight.

codfish were plentiful near the beach. "Floaters" moved along the coast, often following north the capelin that they used for bait. "Stationers" brought small boats with them and fished from one place, sometimes for a fishing establishment on the nearby shore. In 1829, English owners in Dartmouth had six or seven establishments at Labrador, and Jersey houses had four or five. A small resident population of "liviers" gradually developed.

Although the Newfoundland fishery in Labrador was growing, the competition was still strong. Newfoundland merchants complained at mid-century about the immense numbers of vessels from the United States and the neighbouring provinces, and about their monopoly of trade with Labrador residents.[101]

"Fishocracy" puts fishermen last

The Newfoundland merchant class, although eventually competing with the West Country interests, had stemmed from them in the first place. Control of the larger St. John's firms still generally rested in England. Owners of merchant firms sometimes retired back to the mother country. Of major merchant firms in Newfoundland, the Sheas, Crosbies, and Harveys operated from at least 1770. Baine Johnston operated from 1780. Job Brothers originated in Devonshire, the West Country, in the 1700's. Bowring's operated from 1811.[102] Jersey houses were strong on the south coast and along the Labrador.

In the 1830's, the Newfoundland "fishocracy," as one contemporary account called it, held in first place the main merchants, high officials, and some lawyers and doctors; second, the small merchants, important shopkeepers, secondary officials, and more lawyers and doctors; third, the grocers, master

mechanics, and schooner holders; and then, at the bottom of the hold, the fishermen.[103] But recent studies suggest that the better-off fishermen with bigger vessels could enjoy a higher status.[104]

Merchants owned some vessels; smaller-scale planters owned and operated others. With their own boats and the help of their families and fishing "servants," planters could "make" fish more cheaply than merchants or other boat-keepers using wage workers. The rest of the people in the fishery, the majority, had small open boats of their own or worked as fishing servants. Illiteracy was widespread.

Planters and smaller-scale fishermen continued suffering from debt to the major merchants. Contemporary accounts charged that the merchants met together yearly to set the price of fish, and to adjust the price of provisions so that the planters could barely get by. Merchants in some areas issued their own money, good only at their own stores.[105]

Prominent among the wage fishermen who hired out to merchants or planters were Irishmen. Some came over under bond, which they worked to pay off. To quote one account: "The slavery of the Newfoundland fishermen, thus commenced upon their first entering the country, is perpetuated by a system of the most flagrant and shameful extortion.... The prices are so enormous that the original debt due for the passage money of the emigrants instead of being diminished by the hardest and most faithful servitude, continues rapidly to increase."[106]

Herring, mackerel, lobsters gain attention

Although cod and seals dominated, different fisheries were beginning to add variety. Newfoundland exported more than 10,000 barrels of herring most years in the 1840's, and almost 21,000 barrels in 1845. Cured and barrelled herring were however a minor part of the catch, which mostly went unrecorded.[107] Most herring went for bait, sold to the great fleets from elsewhere, especially after the French introduced the longline. French vessels bought bait on the south coast, and sometimes Newfoundland fishermen carried bait to St. Pierre for sale.[108]

Mackerel are at the edge of their range in Newfoundland. But they were plentiful at times, so much so that they became a nuisance to fishermen and were used as manure. Small quantities got exported from 1814 into the 1830's. When mackerel then became scarce, it was said that they had been "cursed off the coast."[109]

Newfoundland first exported lobster in 1856, possibly live, for the first canning was said to take place only in 1858, on the west coast near the Bay of Islands.[110]

"Five and twenty years a whaler"

The Newfoundland folk song, Jack Was Every Inch a Sailor, "five and twenty years a whaler," reflects the presence of an industry that, while never large, was stronger than elsewhere on the B.N.A. coast.

New Englanders had earlier fished whales off the island's southwest coast. After conflicts between England and America interrupted this fishery in 1807, a Newfoundland-owned operation started up. A bounty helped the whale fishery, and it remained important until the middle of the century. Harpooners would row out in small boats from shore. One whaling station lasted from 1850 until 1890, probably catching mostly small whales.[111]

Some whaling took place in the other B.N.A. colonies. A Province of Canada report for 1858 noted, "Few persons are aware of the value of the whale-fishery of our Gulf and River." The Gaspé fishery that year produced 1,624 barrels of oil.[112]

Government enacts bounties

British laws had held back settlement and civic institutions in Newfoundland. But in 1813, after years of pressure from residents, Great Britain began to authorize the sale of land. The British still wanted to discourage settlement, but imperial authorities had to recognize the growing strength of residents. As government slowly developed, imperial and then local authorities often contended with fishery issues and requests for help.

The earlier cod bounties created under Palliser's prompting to help British-based vessels fishing at Newfoundland ended in 1803. But according to H.A. Innis, in 1806 and 1807 an imperial bounty of two shillings a quintal applied in Newfoundland as in Nova Scotia. The drastic losses in the post-1815 slump brought another cod bounty of three shillings a quintal for cod put up in Newfoundland.[113]

After 1820, Great Britain allowed foreign vessels to supply Newfoundland fishing stations with foreign goods and produce. The loosening of restrictions and breaking of the monopoly of British shipping strengthened local traders. Some larger British companies gave way to smaller Newfoundland ones, often operating small schooners that travelled the coast peddling goods and buying fish, "stepping in between the merchant and his planter, and buying the fish from under his nose as it were."[114]

From 1824, new laws gave Newfoundland an elected assembly with control over some finances. By 1833, there was full representative government with a house of assembly, although the new system had growing pains, and some merchants opposed it.

The growth of competition from the French and from New England increased the demand for political power. The Newfoundlanders wanted more fishery control, more conservation measures, and more laws controlling the sale of bait to non-Newfoundlanders. Outside the fishery, more government machinery came into place to help agriculture, industry, and trade.[115]

Colonies battle over bait

All over the coast of British North America, the growth of the herring, mackerel, capelin, and even the clam fishery owed much to local, French, and American purchases of bait, especially as longlines multiplied. In Newfoundland, the demand created heavy herring sales from the south coast. Arguments raged about the rights and wrongs of selling bait to their competitors, particularly the bounty-fed French fleet from overseas and St. Pierre.[116]

The colony made various attempts to restrict the sale of bait. In 1845, Newfoundland passed a tax on bait fish for export. In 1846–1847, a British cruiser helped enforce the new rule, while also guarding against smuggling and other encroachments. But there was little visible effect on the French fishery, as Newfoundlanders smuggled bait to French vessels or to St. Pierre and Miquelon.

The same bait tax affected the United States and the Maritimes. Nova Scotia protested, and in 1849, new Newfoundland legislation exempted the British colonies. Sales of herring from Fortune Bay and St. George's Bay immediately resurged, as fish went to Halifax and the United States for food and bait.

Taken altogether, the Newfoundland bait tax had almost no effect. But it signalled Newfoundland's interest in controlling the local resource for its own purposes.[117]

Beach seining for capelin to be used as cod bait, at a French fishing station in Newfoundland, 1844. (Library and Archives Canada, C-8857)

On the Pacific

Hudson's Bay Company exports Pacific salmon

By the time the first Spanish sailors saw the coast of British Columbia in 1774, the European fishery on the Atlantic was nearly three centuries old. The white man's fishery was late starting in British Columbia, and remained slow in the 1791–1848 period.

By the end of the 1700's, New England whalers were already ranging the coast. Early in the 1800's, the North West Company and Hudson's Bay Company explored parts of the region and set up fur-trading posts. Trapping rapidly depleted the sea-otter population.[118] In 1827, the H.B.C. (early settlers in British Columbia said the initials meant "Here before Christ") set up Fort Langley, 33 miles up the Fraser. In 1829, the post began curing and exporting salmon. Native men caught the salmon; women cleaned, cured, and packed them.

The Native fishery, already taking large quantities for thousands of years, had cured salmon without salt, letting sun and air do the work. The H.B.C. added salt. By 1835, Fort Langley exported more than 3,000 barrels of pickled salmon, chiefly to Hawaii, as food for whaler crews wintering there. In 1851, the company set up another fishery on San Juan Island, which lasted until those islands passed to the United States, in 1872.[119] But so far, the non-Native B.C. commercial fishery was trifling, compared with the Atlantic.

On the Atlantic: How good was the Golden Age?

The Atlantic coast by 1848 had a large and growing fleet. The Maritime provinces stood fourth in magnitude in the register of the world's ocean shipping. And the B.N.A. fishery including Newfoundland was overtaking American production.

Nova Scotia led in population and trade, her vessels sailing the seas of the world. The other Atlantic provinces shared in the marine economy, though in different proportions, with New Brunswick paying more attention than Nova Scotia to the forests and shipbuilding, Prince Edward Island looking more to agriculture, and Newfoundland most dependent on fishing.

Looked at as part of the shipping–trading–shipbuilding complex, the fishery was vigorous. Looked at by itself, it showed weaknesses. Testimony was frequent about the poorer situation of B.N.A. fishermen compared with those of the United States. Many fishermen migrated from Newfoundland to the Maritimes, or from the entire B.N.A. coast to New England, in a chain of deprivation.

There may be accounts of rich fishermen in the first half of the 19th century, the Golden Age of the Maritimes. The writer has come across none. Who then built the big old houses that adorn so many Atlantic towns? Evidently, the owners of larger enterprises, no doubt some in fishing, but probably mostly in lumbering, shipbuilding, shipping, trading, and other businesses. Ordinary fishermen had little chance at education, money, and power.

Were fishermen any worse off than other ordinary citizens? It seems so, on the average. Contemporary accounts frequently mentioned their poor situation. And working on the water isolated them somewhat. They were less inclined than farmers to study, plan, or organize for their mutual benefit. One suspects that then as later, a few did well and a larger number adequately, while most suffered from low incomes and instability.

Yet it is hard to pass an unrelieved dark judgement on the B.N.A. fishermen's life. For all the talk of serfdom in the fishery then and later, fishermen generally preserved a spirit of independence and self-reliance. They usually owned their houses, and on the Atlantic coast in general seem to have owned most of the small boats since early on. Sometimes it is hard to know if a fisherman is more like a peasant or a king. As for the large merchant firms: if they were harsh creditors to much of the coast, they were also fountainheads of commerce.

Fishery management mainly trade management

As before, colonial laws and actions affecting fisheries in 1791–1848 mainly stemmed from the desire to prosper in trade. That desire brought the bounties, the tariffs, the negotiations over fishing rights and privileges, and in Newfoundland, the attempts to control the bait trade. Quality regulations stemmed from the same source.

There was little apparent concern about the welfare of individual fishermen, who for the most part had little education, money, or influence. Depending on their status as property holders, they might or might not be able to vote for members of the elected assemblies, which in turn had less power than the governing councils. These were essentially oligarchies of appointed officials, drawn from merchant, legal, and official sources, and serving sometimes for decades.

There was still almost no attempt to understand the behaviour and biology of fish, or to protect them. Although local authorities appointed local overseers, there were few laws, inconsistent application, and little apparent effect. Outside of salmon regulations, what is now called resource management was generally absent.

Some long-term tendencies had now appeared:

- a concern with fish quality, without the will to tackle it fully;

- a similar lack of thoroughness in conservation; and

- more of a tendency in Newfoundland than in the Maritimes and Quebec to make powerful use of regulations, as in the attempt to control the bait trade.

Occasional half-hearted attempts to encourage improvements took place, as in Nova Scotia's 1828 premiums to encourage better products. But mainly, government's approach was fragmented, as indeed was the industry, scattered along the coast and with no central voice.

Still, governments were taking a more active role than in earlier periods. Bounties, tariffs, and the like must have had at least some influence, or they would not have stimulated lobbying and protests. Judging by contemporary comments, it is safe to say that subsidies, especially those by France, kept boats fishing that otherwise would have left the industry.

CHAPTER 5.
1848–1867: Fishery problems accumulate; strong laws emerge

A s Responsible Government came into place from 1848 on, the B.N.A. colonies were still growing fast. Back in 1820, the Maritimes and Newfoundland had numbered some 200,000 people, mostly Loyalists and immigrants from the Celtic fringe of Great Britain; by 1870, their population had multiplied to some 900,000, now mostly native-born.[1] The Province of Canada was growing even faster. In 1871, shortly after Confederation, Quebec and Ontario had three times the population of the Maritimes and Newfoundland.

On the coast, the great majority of people still lived in rural areas. The fishery, along with shipping, trading, and shipbuilding in many areas, remained vital to the economy. The cod fishery was at a high level. In the 1850's, dried cod exports from Newfoundland often surpassed a million quintals,[2] which (according to present-day conversion tables) would equal nearly 250,000 tonnes live weight. The French bank fishery was said at that point to take another 1.2 million quintals, which if true would bring the total round weight catch of cod to half a million tonnes, without even counting the American and Canadian cod catches.

In 1866, some 20,000 fishermen, men and boys, in the Maritimes and the Province of Canada produced a value of $4–$5 million. By 1870, both Canada and Newfoundland were producing a greater value of cold-water sea fish than New England (the figures were $7 million for Canada and almost $7.3 million for Newfoundland, versus $5.3 million for the United States). They were outfishing the Americans, even though the New Englanders still operated the biggest vessel fleet in tonnage. In 1876, New England had "a total of 80,000 tons of shipping employed in the pursuit of the Cod and the Mackerel alone.... It is very considerably more than the total tonnage [61,551 tons] employed by the Newfoundlanders in the prosecution of their enormous fisheries, and it represents nearly double the amount of tonnage [44,881 tons] employed in the fishing industry of the Dominion of Canada." In cod, Canada and especially Newfoundland were now out-producing Scotland (by far), France, and Norway. Canada and Newfoundland combined were beginning to rival Scottish and Norwegian output of herring.[3]

In the 1848–1867 period, the fishery helped bring about the Reciprocity (free-trade) treaty with the United States. It also figured in the lead-up to Confederation. Yet, in the Maritimes and Quebec, the industry began sliding from centre stage in this period. Even as the Atlantic economy of shipping, fishing, and trading reached its peak, politicians and entrepreneurs were looking landward.

Settlers were pouring into Central Canada and the American midwest. Great inland cities were growing; railways were stretching across the land. New processes were enabling mass production of steel and oil. A modern economy was basing itself mainly on land transport, land resources, and a continental population. The tide was turning away from the marine economy. The sea fishery was starting to look like a poor cousin.

The fishery's labour force was generally undereducated and unorganized, and usually called little attention to itself. Men worked out of sight on the water, shipped out their fish to foreign countries, and tended to have less involvement in community affairs than those on shore. The fishery was almost a permanent frontier, little understood by landsmen, and lacking the will or capacity for self-regulation.

Governments continued to tinker with river regulations, but largely left the sea fisheries alone, except for the efforts to keep out Americans. As it happened, the most far-reaching fisheries legislation came from the inland colonies with the smallest fisheries. The Province of Canada enacted laws that would pass over into the post-Confederation Fisheries Act, making it one of the strongest such laws in the world.

Dealing with the Americans: from resistance to Reciprocity

The New England fleet that helped build the United States also pushed the Maritimes towards union, or at least co-operation. The motive was fisheries self-defence.

The fast-growing B.N.A. fishery depended heavily on open boats or small decked vessels fishing close to shore.[4] The Americans had big, fast schooners, which often disregarded the three-mile limit. Moses Perley reported at mid-century that "from seven to eight hundred sail of American fishing vessels enter the Gulf of St. Lawrence annually; and scattering over the whole of its wide extent, with little heed of the limits to which they are restricted by Treaty, pursue their business unmolested, and but rarely leave their stations without full and valuable fares."[5]

Conflicts sometimes occurred. At Grand Manan Island, just off New Brunswick's Passamaquoddy Bay, rioting erupted early in the century, as residents tried to keep Americans outside the three-mile limit.[6] As Moses Perley was visiting the island at mid-century, a schooner from Lubec, Maine, anchored on the hake-ground. "The fishermen ... gathered around the schooner in their boats, and desired the skipper to move off; on his refusal, they pulled towards the shore to bring off an additional force with fire arms, when the skipper lifted his anchor and made sail."[7]

Relations could also be friendly. Inshore from Grand Manan, in Head Harbour passage off Campobello Island, Perley found that "the fishing boats from Eastport, and other places within the limits of the United States, fish equally, and mingle freely with the British boats on their fishing grounds, near West Isles, where the fish are most numerous.... It is a very gay scene on a fine day, to mingle with some two or three hundred boats fishing in the big eddy, lying so closely together as to leave little more than space between to pull up the fish."[8]

But elsewhere, the large U.S. vessels raised alarm. Fishery frictions brought about concerted action among the provinces well before Confederation. In 1851, at a meeting in Toronto, the Province of Canada agreed to provide a steamer or two or more sailing vessels to police the Gulf and the Labrador. Nova Scotia agreed to continue operating at least the two vessels she had. New Brunswick said she would supply at least one vessel for the Bay of Fundy, the only bay conceded to be open (outside three miles) to the Americans. In 1852, Prince Edward Island also supplied a vessel.[9]

Most important, Great Britain announced in 1852 that she would station a force of small sailing vessels and steamers to enforce the 1818 Convention.[10] As more vessels took up guard, New Brunswick seized an American vessel at Grand Manan,[11] and a Nova Scotian armed cutter, the *Telegraph*, seized an American mackerel schooner off P.E.I. A Gloucester newspaper reported in 1852 that "the commanders of British cruisers are in the habit of disguising their vessels as fishermen, so as to decoy the American vessels within their reach, when they become fishers of men and of prize-money. ... Some of the tricks resorted to by some of the provincial officers would disgrace any sailor."

Patrols disrupted Americans fishing in the Bay of Chaleur and within three miles generally, which cut the New England mackerel catch in half. To make matters worse for the Americans, an October gale in 1852 sank 21vessels off Souris, P.E.I.,[12] this after the even worse "Yankee Gale" of the previous year, which sank scores of American and some B.NA. vessels.

While the Americans wanted more fish, even reducing their tariff in 1848,[13] the Canadians wanted more U.S. market for fish and other goods. The Canadian actions against U.S. vessels created a climate of tension, mixed with opportunity.

Reciprocity brightens the picture

Meanwhile, in the late 1830's, Great Britain had freed up trade, ending the elaborate colonial trading system and Navigation Laws. This shook commercial interests in the Province of Canada, driving merchants to seek a new arrangement with the United States. The coastal colonies that had fought to protect their fisheries from the Americans now saw opportunity in opening them. In J.B. Brebner's words:

> [f]ree entry of Nova Scotian fish, coal, ships, and timber to the American market was worth considering, particularly if there was any danger that Great Britain might insist on settling the fisheries controversy without deferring to Nova Scotia. New Brunswick was naturally less concerned about the fisheries than about a market for her lumber and ships. Prince Edward Island, whose agricultural wealth made her a natural supply point (and smuggling centre) for American fishermen, was quite willing to exchange the inshore fishery for free trade. Newfoundland, at first apathetic, woke up in time to see the advantage of the American market for her fish.[14]

At first the United States was only lukewarm to the idea of freer reciprocal trade. Britain's naval moves against American fishermen heightened U.S. interest in an agreement. Great Britain herself, eager to avoid further fishery conflicts with the U.S., supported Reciprocity, as did her colonial governors. The ultimate outcome was the Reciprocity Treaty of 1854–1866, closely associated with the fishery.

With proclamation of the treaty on September 11, 1854, each side could fish in the other's territorial waters (the Canadians as far south as 36 degrees north, near Cape Hatteras). Besides fishing inside three miles, American fishermen could buy bait and supplies in B.NA. ports, and cure fish there, just like the local fishermen. Neither side could fish salmon, shad, or any river species in the other country. Duties came off fish, flour, grain, lumber, coal, and other natural products; these Canadian goods could enter the American market unrestricted. Three commissioners, including Joseph Howe, Nova Scotia's great advocate of Responsible Government, were to administer the treaty.[15]

Results were significant. The U.S. fishery remained strong in Canadian waters until the U.S. Civil War of 1861–1865. But the treaty also stimulated Canadian fisheries. Mackerel had been an American fishery. Now, with tariffs gone, Canadians began fishing mackerel all along the coast, using mainly handlines.[16]

Fishery exports shot up and the general economy improved in the Maritimes, though partly from other factors, such as the boom in railway building. Trade increased year by year with the United States, especially exports of fish, lumber, and coal. So many U.S. vessels spent money for supplies at the Strait of Canso, P.E.I., and elsewhere that some people feared the American economic presence would bring annexation.[17]

In Prince Edward Island, the legislature supplemented the Reciprocity treaty by allowing Americans to buy land along the shore. American schooners that previously had sailed home with their catch now could land it locally and go back fishing. An influx of American-controlled companies boosted the fishery. By 1861, there were 89 American companies operating. Exports of cod doubled from 8,500 quintals in 1854 to 19,000 in 1859, a peak year. Pickled fish exports went from fewer than 4,000 barrels in 1854 to more than 17,500 in 1863. Agriculture exports increased even more rapidly.[18]

Reciprocity left a mark in the minds of Atlantic coast people. It coincided with the summit of the fishing—trading—wooden boatbuilding marine economy. A number of factors, such as the replacement of wooden ships by steel, weakened that economy later in the century. But many people saw the main problem as the loss of Reciprocity. For generations, the feeling lingered among some Maritimers that had free trade remained, the region might have avoided its hundred-year decline relative to Central Canada.

Fresh-fish trade grows

Meanwhile, the fish trade was changing. American demand for fresh fish rose as the population grew and railways spread. In 1858, the first fish shipments went in ice from Boston to New York. Mechanical refrigeration appeared in the 1860's, but took years to become common in the fishery. Most ice was cut from frozen ponds and kept in icehouses. The trade in fresh fish from inshore grounds and nearby banks began displacing some of New England's trade in cured fish.[19] As fishing pressure increased on nearby grounds, bigger and faster American schooners turned more to grounds off Nova Scotia.

The fishery for halibut, a tasty fish which keeps well, exemplified the pattern of search and deplete. Sold fresh from nearby waters, halibut were also salted. By 1850, fishing vessels had exhausted inshore stocks, and were chasing halibut on Georges and other banks. In the 1860's and 1870's, Gloucester fishermen fished down the halibut along the Canadian Labrador. Around 1865, the Americans started fishing halibut systematically on the Grand Banks. Vessels salting halibut fished as far away as Greenland. Their increasing depletion forced many fishermen to switch to cod on Georges and the Grand Banks.[20]

The hunt for halibut would continue, making them scarce off Atlantic Canada and barely present off New England, where at the end of the 20th century there was no longer a commercial fishery targeting halibut directly.

Mackerel and herring fisheries increase

Americans also led the growing fisheries for mackerel and herring. In 1851, it was said that 1,000–1,200 American vessels went to the Gulf of St. Lawrence for mackerel.[21] Perley described their fishery thus:

> The mode of fishing pursued by the American mackerel fishers who frequent the Gulf, is that with the line, called "trailing." When a "schull" is met with, the vessel, generally of 60 or 80 tons burthen, is put under easy sail, a smart breeze (thence called a mackerel breeze) being considered most favourable. ... If no fish are in sight, the American mackerel fisher on reaching some old resort, furls all the sails of his vessel, ... and commences throwing over bait, to attract the fish to the surface of the water. The bait is usually small mackerel, or salted herrings, cut in pieces by a machine, called a "bait-mill".

Fishermen would also grind up clams for bait. Having attracted the mackerel, they caught them by handline. At Canso they also used beach-seines and driftnets.[22] Use of the purse-seine increased, starting with the part of the fleet that worked around Maine and Massachusetts.[23]

Frozen fish, fertilizer enter commerce

By 1855, New Englanders were also selling naturally frozen herring from northern grounds for bait or food—probably the first commercial frozen-food use in North America. The first trips came from Newfoundland. Later, fishermen got frozen herring also at Nova Scotia and New Brunswick. Frozen herring sharply boosted the B.N.A.–U.S.A. herring trade.[24]

By mid-century, Americans in eastern Maine had small "works" for making oil from herring, using the refuse for fertilizer, or "guano." Upham Treat, mentioned earlier, set up at Treat's Island, on the border between Campobello Island, New Brunswick, and Eastport and Lubec, Maine. Along with other products, Treat produced smoked herring, oil, and guano, and he also experimented with a guano from rockweed.[25]

What were later called reduction fisheries spread from Maine into the Maritimes. Machinery cooked, pressed, and then dried the resulting particles of fish or fish offal, producing oil and fertilizer (and later, fish-meal for animal food). The low-value product lent itself to high volumes, so fishermen pursued abundant, closely schooling pelagic species such as herring.

B.N.A. colonies build up mackerel, herring trade

As the purse-seine, which needed no bait, made offshore fishing easier, the Americans had less need to enter the Canadian three-mile zone for mackerel.[26] They could buy it, for bait or food, from the growing Maritime supply. By the 1860's, Canadian exports of pickled mackerel to the U.S. often ran above 60,000 barrels and $400,000.

A varied trade in herring also built up, with bait an important part.[27] The Bay of Fundy and eastern Maine were a hot spot for herring production, by weirs, gillnets, and torching. Vessels from Passamaquoddy Bay, including Eastport and Lubec on the American side and Campobello on the Canadian, filled seasonal gaps in supply by seeking herring elsewhere.[28] In 1850, for example, vessels from Campobello went after herring at Newfoundland, the Magdalens, and St. George's Bay and the Tusket Islands in Nova Scotia. In 1867, Campobello vessels were still prosecuting the Newfoundland herring fishery.[29] Nova Scotia vessels also went far afield, trading vegetables and other goods for Newfoundland herring.

Bay of Fundy fishermen prosecuted a winter fishery from the U.S. border to Point Lepreau in New Brunswick, selling frozen as well as fresh herring to American buyers, and gradually supplanting the Newfoundland frozen herring trade. Elsewhere as well, businesses were smoking and pickling more herring. The production of cured herring in the Maritimes and Quebec built up to more than half a million barrels a year by 1869. By the 1870's, New Brunswick typically took 43 per cent, Nova Scotia 32 per cent, Quebec 21 per cent, and Prince Edward Island 4 per cent of the catch. With Newfoundland added in, the B.N.A. catch may have equalled a million barrels of herring yearly by the 1870's.[30]

Maritime fishery: large but lagging

For the coastal economy in general, some progress was taking place in manufacturing, agriculture, and other employment, with timber and shipbuilding especially important in New Brunswick. But the provinces depended heavily on the marine and fishery trade. In Nova Scotia, nearly one-half the exports were fish.[31]

As noted in the last chapter, the New Brunswick government in 1848–1849 engaged Moses Perley to inquire into the province's fisheries. His reports, published in 1852, gave a thorough picture of New Brunswick fisheries, with information on neighbouring provinces as well. Perley reported underdevelopment in some fisheries, depletion in others, and shortcomings in conservation and in quality. He described the poverty of many fishermen, especially on the Gulf of St. Lawrence, and the lack of education and organization in the industry.

In the Gulf, Perley said, the American fleet dwarfed the local one. "The deep-sea fishery for cod is not prosecuted to any great extent in the Gulf by the people of New Brunswick. A few schooners proceed from the fishing stations in the County of Gloucester ... to the Bradelle Bank, about fifty miles from Miscou. ... [At the same bank] in the summer of 1839, H.M.S. Champion ... passed through a fleet of 600 to 700 sail of American fishing schooners, all engaged in cod fishing."

On the B.N.A. side, the large "Jersey houses," companies like that of the Robin family, still dominated in the Gulf. Elsewhere, there appeared more of a mix of individuals and small companies. Indeed, the Gulf itself would develop more variety in the years before Confederation, with new enterprises setting up beside the Robins and the Janvrins.[32] Perley found many things behindhand in the Gulf. He painted a more positive picture of the fisheries at Campobello, Grand Manan, and elsewhere in the Bay of Fundy; but there, too, he pointed out problems, along with opportunities for progress.

Perley suggested that New Brunswickers could well adopt the longline (bultow) method of cod fishing, which the French had introduced and the English were now adopting at Newfoundland.[33] Trawls also offered opportunity. "This net is greatly used in the fisheries of the British Channel, where it is called the ground-net, drag-net, trawl, or trammel," Perley wrote. "It is a triangular net, with a mouth from twenty to thirty feet wide, and one foot high; this is so suspended from, and drawn after the fishing smack, as to scrape along the ground, and capture whatever swims within a few inches of the bottom.... By the convention between England and France, relative to the channel fisheries, it is stipulated that no trawl-net shall be used of which the meshes are less than an inch and three quarters from knot to knot." Little was known of the flatfish in the Gulf, since nobody used trawlnets, the best means for taking them. Perley was making an early call for a method that became popular in Canada only in the next century, when it would cause endless controversy.

New Brunswick should also, he said, make more use of the railway in marketing fish. "The fresh salmon, packed in ice, which were sent last season from Saint John to Boston by the Steamers, owing to the facilities of transport in the United States, in three days after they left Saint John, appeared at table, in prime condition, at Albany, Buffalo, Niagara Falls, New York, and Philadelphia. ... Aided by railways, the fisheries of the Gulf of Saint Lawrence, now of too little importance, and such limited value, would take rank as one of the highest privileges of New Brunswick—its unfailing source of wealth forever hereafter."[34]

Early days in the New England lobster fishery. (From Goode, 1887)

Lobster fishery builds up

Perley visited the entrepreneurial Upham Treat's establishment, on the American side of Passamaquoddy Bay. Here Canadian boats would bring salmon from Saint John on ice, and American fishermen would bring live lobsters in "smacks, with wells," from the westward. "When too many arrive at one time, they are placed in the tide, in a sort of crib, or pound, enclosed with high palings, and there fed until they can be boiled and packed."[35] This was an early example of the lobster pound, which became a common sight in Passamaquoddy Bay and elsewhere in the Bay of Fundy.

Lobsters were cheap; Treat paid $5 per hundred, or five cents each. "Lobsters are found everywhere on the coast," Perley reported, "and in the Bay of Chaleur, in such extraordinary numbers, that they are used by thousands to manure the land. ... Every potato field [near Shippagan and Caraquet] is strewn with lobster shells, each potato hill being furnished with two, and perhaps three, lobsters."

Canneries were starting to appear. "Within a few years, one establishment has been set up on Portage Island, at the mouth of the Miramichi River, and another at the mouth of the Kouchibouguac River, for putting up lobsters, in tin cases, hermetically sealed, for exportation. ... The preservation of lobsters, in this manner, need only be restricted by the demand, for the supply is almost unlimited." The lobsters were taken with small hoopnets. At Grand Manan, in the Bay of Fundy, one could take lobsters "with a gaff, in almost any quantity."[36]

Under-exploited species abound

Under-exploited species, as they came to be called in the 20th century, loomed large in Perley's thoughts. Crab of all sizes were available in the Gulf, and shrimp in endless quantities. "At times, the waters of the Straits of Northumberland appear as if thickened with masses of shrimps moving about...." In the event, crab and shrimp fisheries would only develop more than a century later.

Among the finfish, Perley reported, "the common flounder is found in such abundance in the Gulf, that it is used largely for manuring land." As for herring, "The magnificent and unlimited herring fishery of the Gulf of Saint Lawrence and Bay of Chaleur, barely furnishes a sufficient quantity for export to prevent herrings being altogether omitted from the returns. Of all the fisheries of the Gulf of Saint Lawrence, none could be increased to a greater extent...."[37]

Yet the herring population gave Perley some cause for concern. In the Bay of Chaleur, Perley reported, herring had fallen off greatly in number. Just outside Bathurst Harbour,

> there was a beach where the herrings were accustomed to deposit their spawn in immense quantities, and the place was thence called 'Herring Spawn Beach.' [A local observer] has seen the spawn thrown up on this beach by the surf in long thick rolls, or masses, which were carted away by the neighbouring farmers, and used as top-dressing for their fields! As a matter of course, this shameful proceeding destroyed the herring fishing at that place completely, and injured it all along the coast."[38]

At spots in the Bay of Fundy, the sticky herring spawn would fill the water so thickly that "it was said to be no unusual circumstance for the net-rope (9 thread ratline) to be found in the morning as thick as a man's arm with the spawn, while a vessel's cable would be increased to the size of a five gallon keg." But at Grand Manan, he reported, "the enormous destruction of herrings, and their spawn, at the Southern Head of Grand Manan, is an evil which demands immediate remedy."[39]

Despite repeated cries of alarm and some attempts at regulation, the latter fishery at the southern end of Grand Manan would later shrink and virtually disappear. Meanwhile, reduction plants were adding pressure on herring. By 1867, the government was hearing complaints that reduction plants took too many fish.[40]

Report points to depletion

Perley's report noted various declines. "The quantities of salmon in the ... Restigouche and Miramichi, at the first settlement of the country, were perfectly prodigious; although many are yet taken annually, the supply diminishes from year to year. And this is not surprising when it is considered, that many of the streams formerly frequented by salmon, are now completely shut against them, by mill dams without 'fishways'...."

The mill-dams provided water power to saw the logs that fed the boat-building, house-building, and timber trades. But this power came at the cost of blocking and polluting streams. "[T]he injuries arising from saw-dust, and mill-rubbish, being cast into rivers and harbours; and the wholesale destruction of salmon on their spawning beds far up the rivers, have all been pointed out in this Report."[41] By the time of Confederation, salmon exports from New Brunswick had begun to decline.[42]

People in eastern New Brunswick had begun fishing oysters. But "from the manner in which the oyster fishery of the Gulf Shore is now being conducted," Perley reported, "all the oysters of good quality will, in a few years, be quite destroyed." The following decades would see many more lamentations about oysters.

As for trout, "the destruction of these beautiful fish takes place by wholesale, upon many rivers in the northern part of the Province...." And "the gaspereaux fishery has been considered of so much importance, that various Acts of Assembly have, from time to time, been passed for its regulation and protection. But these laws have either been neglected, or not properly enforced, and this fishery is rapidly declining."

In the saltwater fisheries, "the Bay of Chaleur cod are more prized in the markets of the Mediterranean, and will, at all times, sell there more readily, and at higher prices, than any other. ... There has been great complaint of late years, in the upper part of the Bay of Chaleur, of the falling off in the cod fishery.... [O]n the Gaspé, the fishing establishments are deserted, and going to ruin."[43]

Depletion: true or false?

Perley gave no long-term and comprehensive statistics, but he was nobody's fool. One has to believe "depletion" was taking place in various fisheries at mid-century, both inshore and on the banks. But how serious was it?

Many reports must have only reflected the dropping back of fish stocks to a lower level of abundance, where they could still produce healthy volumes. An unfished fishery produces nothing; any fishing must cause some depletion. From a modern perspective, there exists an optimum fishing level (although it varies with environmental conditions) that safeguards the stock while producing good catches. Beyond that level, a fishery can suffer from "yield overfishing," which reduces potential harvests. More intense fishing can produce the worst effect: "recruitment overfishing," which interferes with the stock's ability to reproduce itself.

In Perley's time, the most visible damage was taking place in freshwater fisheries. Mill-dams generally meant death to salmon and other species. But even in the sea fisheries, the depletion of halibut and the difficulties of the Grand Manan spawn-herring fishery provided examples of serious overfishing.

Quality problems continue

"[T]he laws which exist for regulating the inspection of fish, are everywhere treated as a nullity," Perley reported. "[A]ll the fish taken in the Bay of Fundy, on the New Brunswick side, are very badly cured, whether pickled, dried, or smoked; and there is besides, great deficiency in the weight of barrels of pickled fish. In fact, no reliance whatever can be placed upon the inspection, or the weight of fish...." This failing "is such as to prevent those fish obtaining the best prices, and prohibits their being sent to distant foreign markets...."[44]

Perley described fish-handling in the hake fishery at Grand Manan:

[T]he fishermen set up pieces of board upon the open beach in a temporary manner, on which the fish were split; they could not be said to be cleaned, as no water was used in the operation. [The fish after being gutted and split] in a clumsy manner, with uncommonly bad knives, were thrown down upon the gravel; thence they were carried off on handbarrows, upon which they were tossed in a heap, three or four at a time, with pitchforks. From the barrows the fish were pitchforked into the scale to be weighed; from the scale they were again pitchforked upon the barrows; and being carried off to the pickling casks, were once more pitchforked into the pickle; by this time the fish were perforated in all directions, and looked little better than a mass of blood and dirt.[45]

Also at Grand Manan, "the writer met about sixty fishermen, and explained to them the imperfections in their cure of herrings.... The fishermen were told, that besides foreign markets which might be opened under a better system of cure and inspection, there was in Canada an extensive demand for well cured fish, as also in the Western States bordering on the great lakes." But Perley met skepticism. "To this it was replied by the fishermen, that unless the system was general, it was useless for any one person to cure his fish better than his neighbour, as he would obtain no better prices, all the fish from each locality being classed together, and bearing one [low] price."[46]

The fishermen had put their finger on a problem that would last for generations and in some areas still persists. Pitchforking, for example, remained common until the latter 20th century. Why should the fisherman produce better quality if the local market gave him no reward?

Some fishermen live in "bondage"

Especially on the Gulf shore, Perley found poverty and lack of power:

All the settlers at Point Miscou complained bitterly of their poverty, and state of bondage. They said they were completely in the hands of the Jersey merchants, to whom they were indebted, and who dictated their own prices and terms of dealing. They appeared to feel very much the want of a school; and they stated the surprising fact, that they had never been visited by priest or clergyman of any denomination. The children are growing up unbaptised, and in total ignorance....

[T]here are like bodies of fishermen at other localities in the northern part of the Province, who are held in nearly the same state of poverty and bondage. The more favored inhabitants of New Brunswick ... will no doubt be surprised to learn, that there are any of their fellow subjects, dwelling in the same colony, who are even in a worse position than southern slaves....[47]

Newfoundlanders lead Gulf sealing

Perley reported that in the Gulf of St. Lawrence, "sealing is carried on very extensively from Newfoundland, in schooners of about eighty tons burthen, with crews of thirty men. It is attended with fearful dangers; yet the hardy seal hunter of Newfoundland eagerly courts the perilous adventure."

The other colonies took a smaller part in marine-mammal fisheries. New Brunswick vessels at mid-century did some sealing and whaling in the Gulf, commonly taking humpbacks.[48] In 1859, ten whaling vessels fitted out at Gaspé.[49] And Cape Bretoners, encouraged by a small Nova Scotia bounty, had begun taking some seals on the "seal meadows" of floating ice.

Nova Scotia leads Maritime sea fishery

Nova Scotia had a far stronger vessel fishery than New Brunswick. Perley quoted a Nova Scotian:

[O]ur 'Bankers' are generally of small size, from 20 to 50 tons, neither so well constructed, fitted or found, as those of the Americans. Our vessels go to sea, from the 1st of April, to the 1st of May. They continue cod-fishing, on the various banks, between Cape Sable and Cape Canso, until about the 10th of June. ...

In June, our 'Bankers' proceed to Cape Breton, the Gulf of St. Lawrence, or the Labrador, whence they return with cargoes of cod, seal-skins, &c. Many reach home about the last of August, and commence the catch of dog-fish, which are valuable on account of the oil their livers yield.

The fishing for dog-fish having slackened, our vessels are next engaged in taking herrings and mackerel, continuing to fish for the latter until late in November. During some seasons, this is done with nets and seines; but the quantity taken in the seines is sometimes very large, and then the cure is not so good....

The second branch, the shore or boat fishery, is carried on to a greater or less extent, along our whole coast. Whale-boats manned by 2 to 4 men, and large sail boats, undecked, are used.[50]

According to other sources, the Nova Scotia fishery employed more than 10,000 men in 1851, rising rapidly to more than 20,000 by 1874. The province in 1851 had 812 decked vessels. In 1857, there were nearly 150 Nova Scotian schooners at Blanc Sablon alone on the Labrador. Lunenburg was a force in the large-schooner fleet. In Nova Scotia's local, smaller-boat fleet, the number of boats rose from 5,161 in 1851 to 7,670 in 1869, as shown in Table 5-1, and would double by the end of the century.

Table 5-1. Nova Scotia fleet, selected years.			
	Vessels	Boats	Men
1851	812	5,161	10,394
1861	900	8,816	14,332
1869	635	7,670	17,557

In groundfish, Nova Scotia by the late 1860's and early 1870's typically took 80 per cent of the haddock, hake, and pollock catch, and 54 per cent of the cod. Quebec was next with 38 per cent, then New Brunswick with 6 per cent, and Prince Edward Island with only 2 per cent of the cod. Although lobster would become P.E.I.'s specialty, as late as 1873 the island had only two lobster canneries.

In pelagic fisheries, Nova Scotia also took 80 per cent of the mackerel. Only in herring did it trail New Brunswick. (At some point, New Brunswickers picked up the nickname of "herring chokers," a term sometimes shared with other Maritimers and with Scandinavians; Nova Scotians became "Bluenosers.")[51]

Maritimes enact more fishery laws

Perley's report described much that was vigorous, more that was worrisome. No colony at mid-century had a thorough system to run the fisheries. There were scattered efforts. New Brunswick and Nova Scotia each had a fisheries committee of the legislature. Prince Edward Island and Newfoundland had no specific fisheries administration. In Upper and Lower Canada before 1867, the Crown Lands administration ran the fisheries. But everywhere, the mechanics of regulation were weak.[52]

In New Brunswick, several of Perley's recommendations took effect in the New Brunswick Fishery Acts of 1851 and 1852. For conservation, the laws spelled out closed seasons for salmon and for the Grand Manan herring fishery. Every dam to be built had to have a fishway; and no slabs, edgings, or other mill rubbish—sawdust excepted—could be put into the rivers.

Perley had reported enforcement problems, quoting a complaint from the Miramichi area that while white settlers netted, seined, dragged, and speared salmon everywhere, overseers did nothing. This was natural because they received no pay, and if they prosecuted an offender they might have to pay the costs from their own pockets. Perley recommended what would now be called user fees. "A moderate assessment upon all salmon nets in use, should be levied, and applied to the payment of the overseers of the fisheries for their services." Similarly, rents arising from "fishing rooms" should go towards fisheries improvements.[53]

The 1851 law authorized the Lieutenant Governor in Council to make regulations and appoint wardens. But the system soon became laughable. One warden learned about his appointment only two years after the fact, when he saw his name in the provincial Gazette; during the two years, he had received no pay.[54]

New Brunswick experiments with leases

Perley praised the idea of leases. The thinking was the same as among supporters of limited-entry licences and individual quotas a century and a half later: if someone had direct responsibility for the well-being of fish and could profit from their abundance, then self-interest would impel them to do their best for conservation.

The oyster fishery prompted thoughts of both leases and aquaculture (to use the modern term). The fishery could be improved "by judicious regulations and restrictions, as by encouraging the formation of artificial beds, or 'layings', in favourable situations," Perley wrote. "Several persons on the coast intimated to the writer, their desire to form new and extensive beds in the sea water, by removing oysters from the mixed water of the estuaries, where they are now almost worthless, if they could obtain an exclusive right to such beds when formed, and the necessary enactments to prevent their being plundered."[55]

Proprietors of land along salmon rivers already owned the neighbouring fisheries. Perley recommended granting leases for crown land. "The fisheries belonging to the Crown, in the rivers whose banks are ungranted, should be leased, on condition that each lessee should fish only at the proper season, and protect the river at all other times." An example was the Nepisiguit River, flowing through northern New Brunswick into the Bay of Chaleur: "[T]his fishery might be leased to some responsible person, who should be allowed to fish the river, during the proper season only, and bound to protect it at other times, which would then, in all probability, be done effectively."[56] Perley also recommended that the government lay out and lease fishing stations (fishing rooms) on provincially owned seashores of the Gulf of St. Lawrence. The legislation following Perley's report allowed the Lieutenant Governor to grant leases or licences of occupation for fishing stations on ungranted (i.e., crown-owned) shores, beaches, and islands of the province, for up to five years.[57]

Perley was an early advocate of sport fishing in New Brunswick, and his suggestion of leases found use in that domain. In 1863, the province began leasing specific waters for angling; the Dominion government took over that role in 1867.[58]

While advocating regularized leases, Perley took issue with the old Saint John Harbour lottery, where the city would grant fishing rights by lottery every year, and fishermen would then buy the lots from the winners. The lots should instead be auctioned off.[59] It appears that the civic corporation began holding the lottery for fishermen only, then renting them the berths, and realizing $2,500 a year.[60]

Educating and organizing fishermen

Perley's report recommended a form of trade school for fishermen. "The establishment of a few superior schools at Grand Manan, Campo Bello, and West Isles, and probably in some other locations, where the young fishermen should be taught book-keeping, navigation, some knowledge of astronomy, and such other branches of learning as might be useful in their calling, would be one of the greatest boons that could be conferred...."

No schooling emerged from this recommendation. But provincial governments in the 1850's did try to get fishermen better organized, and to encourage better practices. In 1851, New Brunswick funded Fishery Societies, the government providing three times the money subscribed by members. The societies were to use the funds for improving the fishery, and in particular for "premiums" (i.e., prize money) on best-quality fish. Fishery Societies started up at Grand Manan, Deer, and Campobello islands in the Bay of Fundy. At Campobello, an organizer deemed these societies "the commencement of a new era for fishermen...."[61] The Campobello Fish Fair, originating from the Society, lasted well into the 20th century.[62] The other New Brunswick societies left little trace.

Nova Scotia, P.E.I. make weak attempts at management

Nova Scotia in 1853 created the Provincial Association for the Protection of the Inland Fisheries and Game of Nova Scotia. This group worked to restore some rivers. By 1867, however, it was languishing. Also in 1853, declines in Nova Scotia rivers prompted an act that established wardens in every county. This brought no benefits of any consequence. One Captain Charnley promoted a more elaborate scheme to frame a complete set of fishery regulations. This fell through, but it shows that people were getting concerned.[63]

Fishing methods drew more governmental attention. In the sea fishery, Lunenburg fishermen had begun experimenting with longlines in the 1850's.[64] Some Nova Scotia and New Brunswick fishermen protested that this method caught the larger mother fish and injured the spawning stock. In the Nova Scotia Assembly, the longline drew criticism as "one of the evils produced by [the French] bounty system—the natural offspring of a vile parent." The French government provided an "enormous bounty of ten francs ... for every quintal of fish caught by their fishermen." It was said that the Grand Bank was seriously injured and that Banquereau bank, where fishing had only recently begun, had already suffered complete ruin from longline fishing. Bay of Fundy fishermen also complained about the longline.[65]

Around 1862, Britain on Nova Scotia's behalf relayed a request that France take steps to prevent longlines from depleting the banks. The French politely told Nova Scotia to mind its own business: it was preferable to let each country run its own fishery.[66] In the Maritimes, longlines continued to spread.

Nova Scotia took other occasional actions. In 1851, the colony set up a bounty for those engaged in the hook-and-line mackerel fishery for three months or more.[67] The same year, Nova Scotia also enacted a general law on regulation and inspection of certain products including fish. Every county was to have a chief inspector of fish who could appoint deputies. The law established grades ("qualities") for mackerel, salmon, some other pickled fish, and smoked herrings. Products destined for export had to be inspected, on penalty of a five-shelling fine for every cask.[68] Prince Edward Island paid less attention to the fisheries. In 1851, however, the island government set up some bounties for the fishery.

All told, during the pre-Confederation years and especially during Reciprocity, no Maritimes government took firm hold of the fisheries. Although the industry was mixed up with trade regulation, subsidies, and sovereignty, few regulations applied to the fishery operations themselves, except for sporadic regulations on inspection, and the odd one on mesh size or fishing places on the shore.

Newfoundland: cod, seals, and steel

By the start of the 1848–1867 period, the Newfoundland cod fleet was a major force. From the middle to the end of the century, cod exports typically ranged between a million and 1.4 million quintals. Vessels mainly prosecuted the Labrador fishery, the term "Labrador" still including part of the Gulf's North Shore as well as more northern areas.[69] Although Americans were still strong in the Labrador vessel fishery, and Maritimers and Quebecers fished there as well, Newfoundlanders had a strong presence. Compared with European countries, Newfoundland was well ahead in cod production, as shown in Table 5-2.[70]

On the Labrador, one man could in season catch as much as two tons a day. Wives and families helped in curing the fish. Cod-liver oil was still extracted,[71] and would become a health product. Fishermen were making more use of large-scale methods—cod-seines and longlines—which contributed to quality problems.[72]

The Newfoundland vessel fishery moved gradually northward on the Labrador, especially after cod catches in northern Newfoundland suffered a downturn. Vessels were now fishing north of Cartwright and Sandwich Bay, and in the 1860's going north of Cape Harrison. By 1876, some 400 craft were fishing north of Cape Harrison. Most carried eight men, three fishing boats, and one shore boat.[73]

Relatively few Newfoundland vessels fished the offshore banks. Smaller boats fished local waters wherever there were settlers—that is, along most of the east and south coast, and increasingly on the west coast. By 1857, there were about 3,000 Newfoundland settlers on the French Shore. That same year, Newfoundland as a whole had about 38,600 males catching and curing fish; they accounted for 90.4 per cent of all persons occupied. The fishery was the economy.

Sealers move to steel and steam

The seal fishery was now taking enormous yields—sometimes more than half a million pelts in the early 1850's, though catches soon dropped back to below 400,000 for the remainder of the period. In 1857, at the peak of Newfoundland participation, 370 ships and 13,600 men took part. Vessels had grown to the 100-ton range, with crews of more than 30 men.

The seal fishery had helped to build up the schooner fleet, and indeed the colony itself. The Newfoundland historian Shannon Ryan has said that if cod brought the settlers, seals allowed them to stay there.

The bulk of seals were on the ice-pans of the "Front," on the broad Atlantic off northeast Newfoundland and southern Labrador. Steam-powered vessels, now common on the oceans, offered sealing enterprises more carrying capacity and strength around the ice. In 1863, the Grieves firm purchased the *Wolf*, and Baine Johnston the *Bloodhound*. Other leading companies followed. The large, new Scottish-built vessels became known as the "wooden walls" of the seal fishery.

The new technology wrought a transformation. First, it accelerated the centralization of the seal fishery, and the saltfish trade with it, under the control of St. John's merchant firms that could afford steamers. By the middle and late 1860's, St. John's

Table 5-2.
Cod production by selected countries and years (as presented by H.Y. Hind, 1877).

YEARS	SCOTLAND		NORWAY	NEWFOUNDLAND	CANADA	FRANCE
	Cod, Ling and Hake Cured	Cod, Ling and Hake Exported	Cod, Ling, Hake, Pollock, Haddock Exported	Codfish Exported	Codfish Catch	
	Cwt.	Cwt.	Quintals	Quintals	Quintals	
1846 to 1850	90,486	25,832	537,450	980,336		About 500,000 quintals a year from the northwest Atlantic
1851 to 1855	104,780	21,499	605,737	953,858		
1856 to 1860	108,968	32,847	666,076	1,220,154		
1861 to 1865	108,658	41,011	751,382	1,056,551		
1866 to 1870	126,032	50,925		1,130,176 [1]		
1871 to 1875	151,375	64,159		1,333,009	785,426	

1. Mean of the years 1867-69-70.

Return of the *Hunter*

The great chronicler of Newfoundland, D.W. Prowse, left this account of a sealing schooner's return:

[P]lanters, with their families and household belongings, including their dogs and goats, used annually to transport themselves to the Labrador for the fishing season. Until within the last few years all this immense traffic was carried on by sailing vessels, mostly small schooners; there was overcrowding, and a great want of proper accommodation for the women ....

I have a lively remembrance of one episode in this Labrador emigration. All the first three weeks of November 1868 there was terrible anxiety in Brigus about William John Rabbit's brig, Hunter; she was known to be an old vessel, and was much over due . ... One morning I was awakened by an unusual noise and disturbance in the settlement; it was just daylight. As I peered through the window I saw a bare-headed fellow, with his garments loose about him, ... rushing through the streets, shouting at the top of his voice, The Hunter is coming! The Hunter is coming!

The shout electrified Brigus; women and men half dressed, wild with intense excitement, rushed out to see her and in a short time the battered old brig was seen slowly coming round the point; her sails hung about her loose and ragged, her old gear was weather-beaten and dilapidated, her stumpy spars bore only her topsails; but what joy! what intoxication of delight! that clumsy old vessel brought to Brigus on that fine sunny November morning.

With many others I went on board the Hunter, and as long as I live I shall never forget the scene—the women and children, goats, pigs, and dogs, crowded in her hold. After seeing and smelling, I believe I can now form an idea of the horrors of the middle passage, and the odours and sufferings of the chained negroes in the slaver's hold.[74]

dominated by far. Outport firms, some still owned in England, were fading out.

Second, every new steamer displaced many sailing vessels, changing the make-up of the fleet. (The steamers themselves carried square-rigged sails, but below they had the big, powerful engines.) In 1853, St. John's and Conception Bay had nearly 300 sealing vessels; in 1862, they still had 187. By contrast, a report in 1874 said that "the seal fishery this year will be chiefly prosecuted by steamers—no less than 23 being about to leave or have left, St. John's on that voyage. The few sailing vessels of the sealing fleet left port last week...." By 1881, there were only some 15 sailing vessels going to the ice, mainly from Harbour Grace on Conception Bay. Some small cod-fishing schooners continued to take seals inshore, especially on the southwest coast of Newfoundland; but the main sailing fleet disappeared from the seal fishery.

Even before the steamers, observers were beginning to worry about the seal resource. Some suggested preventing the wasteful "panning" of seals: sealers would kill and leave seals on the ice, and later be unable to recover them. Vessels also began to make two trips, killing old, breeding seals when they ran out of young. The government made inconclusive attempts at legislation. Nothing concrete would happen until the 1870's, when legislation set seasons and penalized the killing of "cat" seals below a certain weight. Subsidies would come in their turn.

As steamers took over, contemporary accounts lamented the loss of employment and waterfront life. An 1871 report noted that

[t]he employment of steamers, advantageous as it is in some respects, will not only have the effect of apportioning the greater part of the voyage to the capitalist, but will also throw out of employment half the men formerly engaged in sailing craft. We take, for instance, either of the largest trips of last spring—say 28,000 seals, this would give a fair paying voyage to nine sailing vessels carrying four hundred men; while the steamer did the work with about 180 men. Is this an argument against steam? Certainly not. ... Steam has shown what it can do in Sealing as well as in its application to other industries.... But the facts still warn us that in those prosperous results which we thankfully acknowledge, the conditions of abiding good to the general population are wanting. ... The point inevitably and most forcibly suggested, is the necessity of new employments for those who will be cast adrift by the abandonment of sailing vessels....[75]

The writer had identified a puzzle that would bedevil fisheries management in many future conflicts. When new methods appeared in the land-based economy, free enterprise, capitalist competition, and the invisible hand of the market generally got the benefit of the doubt, and rightly so. If new technology threw hundreds out of work, still it seemed that something else would turn up, productivity overall would advance, and human happiness would increase. But the fishery could behave differently. The resource might shrink, and real productivity decrease. (In Newfoundland, as the steamers decimated sailing-vessel competition, pelt production shrank.) Alternative employment might be non-existent. And communities might fade or die.

Smaller boats might in some cases return more on investment, but even so be unable to compete with larger vessels, which could rake up the fish first, producing more per man but at a higher cost. An operator with an older, paid-up craft sometimes felt that to have any chance at all to compete, he must borrow money and invest in a bigger boat or better gear; otherwise he was dead in the water. Thus, the race for the fish could confound economic rationality and derail long-term planning. True business cost-effectiveness often went unreckoned, let alone the less tangible costs to the resource and the community.

If economic factors were often murky, still the changes were clear to those who lost out. Every panic in a declining fishery would find its politician, to press for special measures to preserve jobs, a community, or a "way of life."

Scattered attempts at organization

Of the B.N.A. colonies, Newfoundland had the least-diversified economy, the most dependence on the fishery, and the most homogeneous fishery, with seals and cod employing practically everybody. Those circumstances fostered more labour activism than in the Maritimes. In 1832, sealers went on strike in Harbour Grace and Carbonear, to pressure the merchants for payments in cash rather than in kind or "truck." When results were unsatisfactory, "a large body of men boarded [the Ridley firm's] vessel Perseverance with saws, axes and guns and, forcing the officers sleeping below to stay where they were, caused considerable damage." The sealers won their point.

Other protests in the following years brought further gains. In 1845, a captain Henry Supple organized major meetings on behalf of the sealers. At a "Monster Meeting" on Conception Bay, one account reported, "the procession, six deep, then advanced into the town of Brigus and halted on the Pond. The number it is said amounted to fully 3,000 able-bodied men!" In 1853, an account of a sealers' protest lamented its "brutal violence and intimidation." Among shoremen, by 1855 a Seal-Skinners' Union had formed.[76]

After the 1850's, apart from occasional refusals to work, labour peace generally prevailed—"a sad commentary on the depressed state of the industry rather than a happy reflection of improved conditions."[77]

Outside the seal fishery, the first fishermen's organization with any continuity took quite a different shape. In 1862, the Heart's Content Fishermen's Society came into being. From this followed, in 1873, the Society of United Fishermen (S.U.F.), which spread to many communities and lasted for many years. S.U.F. halls still stand in Newfoundland. Did the S.U.F. fight for better prices, laws, or living conditions? On the contrary, it was a Protestant lodge, with robes, parades, and all the trappings. True to the religiously divided nature of Newfoundland, the Catholic side in 1871 answered with its own fishermen's lodge, the Star of the Sea.[78]

In 1863, a Newfoundland "Fishermen's Society" protested to the government against the use of capelin for fertilizer; against bultows; against herring-seines; against the cod-seine; against cod-nets; against the jigger; and against the taking of herring in spawning season.[79] Then as later, fishermen would unite more readily to complain to government than for any other reason. But as in the Maritimes, these early organizations, apart from the S.U.F., left little trace.

Fishery issues hasten Responsible Government

In the 1860's, with Lower Canada offering new competition on the Labrador coast, and the Americans a constant presence, Newfoundland tried to protect its own interests. To discourage the foreign fishermen–traders, the colony applied duties on imported goods.[80]

Meanwhile, Norway was offering more competition in European markets.[81] So was France, which was catching and drying fish on Newfoundland's west and northeast coast. Newfoundland authorities worried about the French straying out of bounds. In 1853, Newfoundland asked Great Britain to send a war steamer in winter and spring, to control the French fishery. Great Britain suggested that Newfoundland, like Nova Scotia, make use of its own schooners. That cost money, and Newfoundland's attempts to organize funding for fishery patrols helped the move towards Responsible Government.[82]

The seal fishery also hastened the emergence of local government, by strengthening the economy and helping to wean Newfoundland from British mercantile interests. Indeed, Shannon Ryan asserts that "Newfoundland would not have acquired representative government if the cod fishery had remained the sole industry."[83]

Representative government had come into place in 1832; the push continued in following years for full Responsible Government, with the executive responsible to the elected members. One speaker told the Assembly in 1854:

It is no uncommon occurrence in Newfoundland for a planter to fell, and bring out of the forest, timber and other materials necessary to construct a vessel, to build her from keel to topmast, and afterwards to take charge of and navigate her in prosecuting the trade of the colony. Surely then such men are not to be supposed devoid of that intellectuality which would qualify them to become the recipients of a system of constitutional rule under the enjoyment of which they observe their sister colonies thriving.[84]

After a parliamentary commotion partly involving fishery issues, Newfoundland achieved Responsible Government and full colonial status in 1855.

Newfoundland's "Magna Carta"

Then in 1857, a major issue struck to the saltwater soul of Newfoundland. Great Britain agreed that France's rights entitled her to exclusive use of sections of the coast. This would effectively bar Newfoundlanders from parts of the island. The Newfoundland government objected, contending that the British agreement went against the principle of "local assent." The colony had joined the Reciprocity treaty with the United States only after receiving the assent of the local legislature; the same principle should apply to the French agreement.

Beset by Newfoundland protests, which were supported by the other colonies, Great Britain withdrew from the agreement with France. The defeat of the convention strengthened Newfoundland's control over her natural resources, and became known as Newfoundland's Magna Carta.

Even so, Great Britain still tried to keep the colony on a short leash. When Newfoundland wanted to control the land along the French Shore, London resisted the idea. But by the 1880's, Britain conceded that Newfoundland had territorial jurisdiction over the French Shore.[85] Magistrates and police were appointed, and members of the legislative assembly were elected. France retained her shore-fishing rights, but the population of Newfoundlanders increased.

The French fleet was still sizeable, though varying with fishery conditions, and heavily subsidized. In 1860, 123 ships with 3,900 men fished the banks, and 105 ships with 6,200 men fished the northeast coast. In 1867, 137 ships fished the banks, 34 the northeast coast. That same year, the French island of St. Pierre operated some 95 sloops and small schooners fishing the banks and Gulf, and 350 smaller craft, employing in total some 1,500 men.[86]

Bait trade grows

As longlines and sea-launched dories put millions more hooks in the water, foreign demand for Newfoundland bait kept growing. In 1855, for example, 40 schooners, 684 boats, and nearly 2,500 men on Newfoundland's south coast were catching bait to sell to the French. In Fortune Bay, the bait trade

may have matched the cod trade at times. In 1864, Newfoundlanders exported no less than 74,000 barrels of herring and 40,000 hogsheads of capelin to St. Pierre for bait. Supply and demand fluctuated so greatly that prices ranged from 1 to 30 francs a barrel.[87]

New Englanders also wanted bait, especially after the 1854 Reciprocity agreement favoured more American fishing in B.N.A. waters. Besides fishing bait on their own, New Englanders as noted earlier bought naturally frozen winter herring from Newfoundlanders, carrying it back as bait for the Georges Bank fishery in February, and also for sale as food. Americans in the 1870's encouraged the building of icehouses in Newfoundland to store bait, and to provide ice to U.S. vessels for storing their fish on board.

Newfoundlanders also increased their export trade in food herring, though it remained small compared with the bait trade: some 42,600 barrels cured and exported in 1874, out of more than 270,000 barrels caught altogether.[88]

As more foreign vessels sought to catch or buy herring in Newfoundland, especially at Fortune Bay on the south coast and the area between Blanc Sablon and Indian Tickle on the Labrador, complaints began to rise about damage from the seine fishery.[89]

Newfoundland sets more rules

After the 1857 "Magna Carta," the Newfoundland government became more active in fisheries management, holding hearings and passing regulations. In 1858, the colony set certain mesh-size regulations, and also prohibited bait exports from April to August, to hinder foreign fishing.

Newfoundland in 1862 restricted the use of purse-seines by its fishermen, and set closed seasons. These were presumably conservation measures coupled with a concern to slow down the flow of bait to foreign vessels.

When Reciprocity ended in 1866, Newfoundland stopped American fishing within three miles, except for the Treaty Shore on the southwest and west coasts. The American vessels turned more to buying bait, rather than fishing it.[90] Newfoundland believed it had a weapon in the control of bait, and would make use of it in coming decades.

Canning spreads to the Pacific

At the other end of the future Canada, the Pacific fishery by white men at the outset of the 1848–1867 period was still tiny, as was the colony itself. In 1849, the Hudson's Bay Company (H.B.C.) took over Vancouver Island for purposes of colonization, and in 1856 James Douglas became governor. At mid-century there were only a handful of white fur-trading posts on the coast and up the rivers in the inte-

Native people in British Columbia had a widespread fishery. This picture from about 1870 shows houses for dried fish – salmon caches. (Library and Archives Canada, C-24288)

rior. But growth would now be rapid, triggered by a gold rush on the Fraser River.

Fishing at first remained confined to the Hudson's Bay Company's exports of cured fish. In 1858, the beginnings of the gold rush brought clashes between Native people and miners interfering with their fishing sites.[91] The H.B.C. that same year ceased exporting Fraser salmon. But in 1863, a new salmon-curing plant started up at Beechy Bay, and in 1864, another on the Fraser.

By then, new technology was taking a hand in salmon production. In 1864, canning of Pacific salmon began on the Sacramento River in California. Experimental canning took place in British Columbia in 1867. By that time, the gold rush had brought thousands of new settlers to both California and British Columbia, increasing the demand for preserved food.[92] A great industry was about to take off.

The United Canadas: more laws, stronger administration

By 1848, the Province of Canada was already by far the most populous part of British North America, and it had an important sea fishery. The Gaspé and the North Shore of the Gulf produced more than $1 million by 1864, with cod counting for nearly two-thirds, followed by herring, seal oil, cod-liver oil, and

Indians catching salmon on the Fraser River, ca. 1889. (Detail, Library and Archives Canada, PA-148734)

salmon. The fleet had 157 vessels and about 2,600 undecked boats, almost all cod handliners.[93]

In Upper Canada, the future Ontario, people were fishing the Great Lakes to feed the rising population. Early settlers had found great abundance of whitefish, trout, sturgeon, and other species. Lake Ontario's landlocked Atlantic salmon sometimes

72

swarmed so thickly that one could catch them with a shovel, or by hand.[94]

On Lake Erie, commercial fishing began on the American side around 1795. As new settlers poured into Upper Canada, a beach-seine fishery for white-fish developed as early as 1807 on Lake Ontario. Beach-seining spread by 1815 to the Ontario shore of Lake Erie. The year 1815 saw seine sets on Lake Erie. By mid-century, fishermen were using gillnets and also the pound-net: a fixed enclosure of nets on stakes near the shore. It kept the fish alive, an advantage for marketing. As elsewhere, beach-seines would gradually fade away, as the original abundance got fished down. The first steam fishing tug came to Lake Huron in 1860.[95]

The two Canadas had a small fishery compared with the Maritimes. Yet the Province of Canada would set up the strongest laws and administration in the pre-Confederation period. Early regulations were modest and scattered, as in the Maritimes. An 1807 law in Upper Canada set rules on fishing gear for Lake Ontario, and banned nets from the mouths of salmon streams. Closed season, more rules on gear, and requirements for fishways followed.[96]

The big advance came in the 1850's. An act in 1853 aimed to develop the sea fisheries, and made clear that all British subjects could fish on the Labrador coast (including much of the present-day North Shore), where jurisdiction had been somewhat confused. The 1853 act stated that fishermen from Canada "have been of late years by strong hand prevented from making [fish] on the coasts thereof and islands contiguous thereto."

Further regulations in 1856 aided the build-up of the sea fishery. Fishermen from Gaspé and the Bay of Chaleur settled in such places as Natashquan and Sept-Îles; in the latter village, six new cod-fishing operations started up in 1857. By 1876, it was said that 17 Gaspé firms had 30 establishments on the North Shore. From Godbout at the mouth of the St. Lawrence to Blanc Sablon, they employed an estimated 1,225 people and 300 fishing vessels.[97] To the east of Harrington Harbour, many new settlers came from Newfoundland; towns along this coast still have a Newfoundland aspect and accent.

Fortin, *La Canadienne* police the Gulf

Meanwhile, Pierre Fortin commanded *La Canadienne*, the Canadas' patrol schooner. Year after year from 1852 to 1865, Captain Fortin covered the Gulf of St. Lawrence, settling disputes, showing the flag to American vessels, talking up fisheries improvements to the government and to the fishermen, and presenting reports.

Fortin had a hand in many activities. For example, his successor, Théophile Tetu, reported in 1867 on the results of Fortin's oyster planting. Tetu reported in the same year how the abolishing of spears and weirs, and the reducing of nets, had brought an increase in the abundance of salmon. "The system is working."[98]

Strong administrative system takes shape

In 1857, the Province of Canada enacted a Fishery Act, which was strengthened in 1858. This legislation set up the strongest fishery administration of any B.N.A. province, encouraged artificial propagation of fish (i.e., early aquaculture), forbade certain forms of pollution, and provided for controlling the fishery through licences and leases.

On the administrative side, fishery powers previously vested in municipalities now went to the Governor in Council. The 1858 legislation made the Commissioner of Crown Lands responsible for carrying out fisheries regulations. The government now could appoint a superintendent of fisheries for Upper Canada and one for Lower Canada. For Upper Canada, the first superintendent of fisheries was John McCuaig, who had one officer below him. By 1866, Upper Canada had 18 fishery overseers, along with a number of guardians appointed seasonally to protect spawning grounds. Richard Nettle became the first superintendent for Lower Canada, in charge of ten overseers. A select committee of the Legislative Assembly was to report on the operation of the Fishery Act.[99]

Leases, licences come into effect

Licences and leases, which would become major tools of fishery management, also got their blessing in the 1858 act. The idea had earlier roots. In the early days of the North American fishery, the English had granted exclusive fishing privileges or rights in river fisheries; the French, in both river and sea fisheries. After the fall of New France, Governor Palliser overturned exclusive privileges for private parties on the Labrador; and they seem to have vanished from the sea fisheries.

But river rights persisted. It must have seemed natural that the fishery alongside a piece of land would go with the piece of land. Salmon "proprietorships" were and still are common in New Brunswick. On various rivers, major examples being the Miramichi and Restigouche, someone buying a riverside property will typically purchase the fishing right along with it, even though the fish themselves are a public trust, and government still controls the manner of the fishery.

In the Province of Canada, the government in 1845 refused to issue a new lease for a fishery on the River St. Clair, "the Atty. General having reported that the right to fish in the Sea &c. is a public right." Still, the government had "leased or issued licences of occupation to crown lands that fronted on desirable fishing sites. In the case of a seine fishery such a licence gave effective control of the fishery to the licencee."[100] In 1852, after several parties at

Fortin's *La Canadienne*. Sketch by marine artist F.R. Berchem, based on a contemporary woodcut. (Courtesy of Bernard Collin, Canadian Coast Guard)

Tadoussac applied for a lease of the exclusive privilege of fishing "porpoise" (presumably beluga whales), it was decided to offer the lease to public competition. A similar competition took place at Malbaie. That shows the idea of leases already present (as in New Brunswick, where Moses Perley had promoted leases). In Lower Canada, Captain Pierre Fortin in 1856 suggested licences for the Gulf of St. Lawrence fishery. These were, however, seen as contrary to the law of 1853, which had spelled out that all British subjects could fish on the Labrador.

The 1857 act noted that "[e]very subject of Her Majesty who shall be in peaceful possession of any fishing station for Salmon or Seals ... shall be deemed the owner thereof...." However, it made no direct mention of leases or licences. But a fisheries official appointed under that law would promote them.

In September 1857, Superintendent Richard Nettle of Lower Canada reported destructive fishing on salmon rivers. Indians were spearing salmon by torchlight, and "vessels of all descriptions, and from various places, are fishing in every bay and river along the shore," both netting and spearing. "[I]t is impossible to ascertain the quantity of fish taken on the north shore this year. ... [T]he many hundreds of nets that have been placed in the rivers and bays ... together with the vile practice of spearing, has almost totally destroyed them." Nettle concluded: "I have now the honour to suggest what I conceive to be the only effective remedy to prevent the utter destruction of the salmon fisheries of the St. Lawrence; and would beg to recommend:

"... That the salmon fisheries of the St. Lawrence and its tributaries (with the bays included), be leased by *public competition and tender*."

The revenues, Nettle added, would pay for effective protection by vessels. Their good effects would be seen "in affording a guaranty to the well disposed fishermen, and by being a terror to the lawless."[101]

The 1858 Fishery Act, assented to April 16, stated that "[t]he Governor in Council may grant special fishing leases and licences on lands belonging to the Crown, for any term not exceeding nine years, and may make all and every such regulation or regulations as may be found necessary or expedient for the better management and regulation of the Fisheries of the Province."

Nettle, Whitcher promote licences, leases

It appears likely that Richard Nettle, Superintendent of Fisheries for Lower Canada, was the chief inspirer of the licensing power. Reporting on his visits to rivers and bays, Nettle made frequent reference to the benefits of licensing, as in this instance:

> As an instance of the necessity of regulating the nets, the following will suffice: Mr. Joseph Eden, of Gaspé Basin, had a fishery opposite his property, which was very productive. His neighbour, seeing his success, ran out a net a few hundred yards in advance of his, and nearer the mouth of the river. The consequence was the entire destruction of Mr. Eden's fishery, and the establishment of his neighbour's.

> The process is similar on all the rivers, and calls for immediate action. Proper salmon-fishing stations should be established, a licence granted, and a nominal rent charged.

W. F. Whitcher (Library and Archives Canada, PA-175336)

Nettle noted overseas examples of leasing: the salmon fisheries of the River Tay had been leased for the "incredible sum of £18,500 per annum."

Another Crown Lands employee, William F. Whitcher, had started in 1848 as a clerk, and was rising through the ranks. In 1858, he was reporting to Nettle. This account by Whitcher reflected attitudes of the time towards Native people.

> I succeeded by warnings and personal vigilance in deterring several Indians and others from spearing salmon within reach of the Saguenay district.

> The major part of these abuses occurs through ignorance and misbelief, rather than from wilful or perverse offending. Exceptions, of course, are there always found, but even they are somewhat mitigated by the prejudice of habit and the blindness of that stubborn determination which characterizes almost all of those inured to half-savage, rude and secluded life. Moreover, such peculiarities oftener yield to than successfully withstand a firm purpose, patiently explained, and administered in a considerate spirit, at once cautious, prudent and imperative.

Whitcher in 1858 summed up the rationale of leases and licences, in terms much like those of some late-20th century economists, though with more eloquence. He railed against the illusory easy money of fishing and the dissipation of returns among too many "suicidal occupants," with "proceeds appreciable to none," while the resource shrank.

> There is ... one other subject towards which it is desirable to direct your earnest attention. I mean the speedy leasing of all the superior salmon fisheries upon the Lower St. Lawrence and its tributaries, and bringing the numerous inferior coastwise stations under control of a petty license system.

Indiscriminate free fishing here is productive of many social evils,—it destroys also these valuable fisheries,—and of positive hardships it produces a plentiful crop. The custom affords facilities and abounds with temptations to lead dissolute and lazy lives. I could point out frequent examples of able-bodied men having thereby lapsed into an improvident and idle existence. Individuals who might earn for themselves and families the comforts and competence which reward industrious perseverance in agricultural pursuits (despite all rigors of climate and inferiority of soil) now while away the precious seasons in half-starved and pseudo-savage indolence. Enticed by habit, or tempted by (too often illusory) hopes of speedier gain, many forsake their farms and waste their little labor on a precarious fishery, to properly work which they have neither means nor energy.

When winter arrives they are reduced to want, and leaving their shivering families to brave out impending starvation, some betake themselves to the companionship of Indian hunters, and the mingled excitement, toil and idleness of the trapper's winter campaign. Doubtless to prescribe suitable fishing locations, and in return for the protection and regulation extended to the holders exact a small rental, would have the effect of weeding out these suicidal occupants, and throw into the hands of such of their neighbours as can afford to harvest it, a remunerative extent of water limit. Having therein exclusive privileges these could even invest sufficient capital (commensurate with their scanty means) for the purpose of deriving a beneficial return. Whereas, at present the selfsame grounds at each returning season become so numerously occupied as to make the proceeds appreciable to none, whilst at the same time the source of supply is fast dwindling away. ...

The leasing of streams and licensing smaller fisheries, should be so applied as to afford in many respects at least an incidental protection to the salmon and sea-trout fisheries generally.[102]

Whitcher would take charge of fisheries for Upper and Lower Canada and become Commissioner of Fisheries after Confederation. In that position he would continue to promote licences and leases.

Using licences and leases

In an 1858 report, the Commissioner of Crown Lands noted:

In so far as regards the chief commercial fisheries upon the waters of Upper Canada, a system of leasing all vacant public lands still belonging to the Crown and accessory to carrying on the fishing business, has been already adopted....

With respect to the Salmon fisheries of Lower Canada, it was deemed advisable to expose presently to public competition various valuable net-fishings at the mouths of certain well known Salmon Rivers tributary to the Lower St. Lawrence. Tenders (due 15th March) have been invited for five years' lease of those streams....

The report continued that the freshwater fisheries' "regulation and conservance go hand in hand with the principle of their economic development."[103]

Various government measures reflected the new power. In 1858, an Order-in-Council specified that in Lower Canada anyone wishing to fish salmon and sea trout needed a licence. That same year, the Commissioner of Crown Lands was authorized to advertise certain fisheries in Europe and America,

evidently to invite bids. A notice from the Commissioner of Crown Lands specified in January 1859 that the Superintendent of Fisheries for Lower Canada was empowered to grant season licences (May 1– July 30) for fishing stations for salmon and sea trout on Crown Lands bordering the St. Lawrence and its tributaries, at "discretionary rentals."[104]

Now leases and licences were clearly in the law, at least for crown lands, together with the broad power to make "all and every such regulation" towards "better management and regulation." But the new powers were already raising complex questions that would recur in future. Who deserved and who would get access to the fish, and by what means?

John McCuaig, Superintendent of Fisheries for Upper Canada, made inquiries among fishermen about the best ways to proceed with leases and charges. The most money would come from opening leases to speculators, he said:

But the questions recur;—would this method be considerate towards the past and present occupants? And would it be the fittest mode of preserving control over the Lake Fisheries, and stimulating their development as a natural supply and commity of trade? It must nevertheless be borne in mind by those concerned that the "vested interests" which exist after a lengthy enjoyment of the profits of free fishery have obtained by sufferance only, and might be said to have already amply indemnified themselves.

McCuaig recommended leasing vacant stations, on land and water, to the first applicant at valuation, or exposing them to public sale. Where there were adverse claims, the location or fishery privilege should go to the highest bidder. As for charges for fishing elsewhere, "some would prefer a licence fee on each fishing boat; others mention the exaction of a toll upon each fisherman's take per barrel." McCuaig recommended instead charging by the length of net.[105]

The imposition of licences in Upper Canada ran headlong into opposition from fishermen who thought they already owned fisheries through occupation. A.B. McCullough, the authority on the Great Lakes fishery, writes that "[a] man appointed to report on violations of the Fishery Act at Burlington Beach was severely beaten, and in 1863 Superintendent William Gibbard disappeared while investigating fishing violations at Manitoulin Island. He was presumed to have been murdered."[106]

In Lower Canada various leases came into place, with various difficulties. For example, in 1860, holders of privileges on the Escoumins, Moisie, and Godbout rivers in Lower Canada applied for remission of their rents for that year. A select committee of the Legislative Assembly in 1864 noted that leasing of salmon rivers had been done precipitately in 1859, and it recommended compensation to fishermen dispossessed thereby.

Still, the new system was firmly in place. Self-interest, it was thought, would give lease-holders both profit and the incentive to conserve the resource.[107]

Salmon culture heightens interest in leases

The new interest in leases went hand in hand with new hopes for salmon culture. In 1857, Richard Nettle raised trout and salmon artificially near Quebec City, the first person to do so in Canada.[108] Although it sounds magical, salmon breeding rests upon a simple basis: one squeezes milt from the male and eggs from the female, and mixes them in a bucket. With proper handling, one can achieve a far higher survival rate of the fertilized eggs than in the wild.

The 1857 Fishery Act mentioned salmon culture for the first time. "For the purpose of encouraging and affording information with respect to the production of salmon and other fish, an apparatus for the artificial propagation of fish shall be kept in the department of the Commissioner of Crown Lands." The "apparatus," including spawning boxes and a pond, reflected Nettle's work.

Nettle was in contact with fish breeders in Ireland for information and for supplies of salmon ova. His December 1857 report mentioned "several parties ... who are anxiously awaiting the action of the Government as regards the leasing [of] the several salmon rivers within the Lower Province, and who would, where it was necessary, immediately commence the breeding of salmon on a very large scale."[109]

Fish culture and leases were entwined. In Lower Canada in 1857, a Mr. Boswell of Quebec bought the seigniory of Jacques Cartier, with "the old French rights," in order to restore the Jacques Cartier River by salmon culture; he abandoned the project when it became clear the government could afford no protection to his salmon.[110] In December 1859, certain fishery rights were granted on lakes Megantic, Louisa, and Aylmer to a Mr. De Courtenay "with a view to initiate a system of propagation &c." In Upper Canada, Samuel Wilmot in 1865 began culturing salmon. In 1866, an Order-in-Council set aside Wilmot's Creek, near Newcastle, Ontario, for natural and artificial breeding of salmon. Wilmot in 1868 became a fishery officer, and spent a long career in fish culture.[111]

From these small beginnings, fish culture would become an overwhelming trend later in the century. Although associated at the outset with the idea of private leases, it would in the end take place under government auspices.

New laws have broad reach

The Province of Canada legislators spread regulations widely. The 1857 act brought in what we would now call habitat-related provisions. Owners of dams or slides on any salmon river had to provide a fishway, between June 1 and October 20, of such form and dimensions as determined by the Governor in Council. And anyone throwing ballast overboard in a river, harbour, or roadstead where sea fishing took place, or throwing fish offal into a river or within three miles of the coast of the mainland or any island, could incur a fine not exceeding 20 pounds.

The ambitious act also set closed seasons (August 1–March 1 in Lower Canada, September 1–March 10 in Upper Canada) when no one could catch salmon except with a rod and line. It outlawed using torches for salmon in Lower Canada, and torches and spears for salmon, maskinongé, speckled trout, or bass in Upper Canada. It set seasons for other species, and specified that no one could construct a fish pound in any river or brook.

Other provisions concerned fishermen and their employers. No one could seize any boat or gear necessary to the subsistence or fishing operations of a fisherman during the season May 1–November 1, except for the recovery of penalties or fines imposed under the act. This appears to have protected fish-ermen from creditors for the duration of the season. Another rule benefitted employers, setting penalties for anyone breaking a written agreement to fish. As for enforcement, half the fines and forfeitures paid by anyone under the act would go the Crown, the other half to the complainant. (Similar provisions often occurred in the pre-Confederation fishery statutes of other provinces.)

The 1858 act required a permit to take oysters, set mesh-size limits for cod-seines and salmon nets, and made other conservation provisions. A later regulation forbade fishing cod or halibut by long-lines (set-lines or bultows) within three miles of the Magdalens; this reflected the typical resistance to the new gear.

The 1858 act also addressed development, mak-ing bounties available for vessels in the seal, cod, mackerel, herring, and whale fisheries ($3 per vessel ton for three months' consecutive fishing, more for longer periods) The crew should get one-third of the bounty, the owner the rest.

Another act in 1859 gave the Governor in Council authority to appoint inspectors, who would inspect fish and oil voluntarily submitted to them and would certify the products that met government standards. The idea was that the manufacturer would want to submit his fish, since the government stamp of approval could help him sell them. The only penal-ties in the act would bear against the Inspector him-self for misperformance, or against anyone interfer-ing with the stamps or brands that he put upon bar-rels or boxes.[112]

1865 act combines strong powers

In 1865, a new fisheries act for the United Canadas integrated licences and leases, broad man-agement powers, and another major element: the power against "deleterious substances." In a mem-orandum to Alex Campbell, Commissioner of Crown Lands, W.F. Whitcher explained that the act would make clear the power of the Crown and provide for local overseers. It would also provide for granting of leases, and for shorter and more flexible closed sea-sons. And it would deal with fishways and types of fishing gear.

Legislators took a lively interest. Debate touched on the fishing rights of seigneurs in Lower Canada; Judge Dorion's decision that a private person can own a fishing right; the aim of the new law only to regulate, not to own, private fishing enterprises; the shortcomings of previous bills that provided only a short lease period, thus attracting few investors; and the Crown ownership of all fishing waters in Upper Canada. Discussion also ranged over the depletion of fisheries, the pollution of rivers by sawdust, and the use of fishways and closed seasons. Some men-tioned the possibility of abolishing brush weirs and

fixed fishing gear, limiting trout gillnets to five miles offshore, restricting seines and bultows, and so on.

Eventually came the act, assented to September 18, 1865. Regarding licences, Section 3 said that

[t]he Commissioner of Crown Lands may, where the exclusive right of fishing does not already exist by law in favour of private persons, issue fishing leases and licences for fisheries and fish-ing wheresoever situated or carried on, and grant licences of occupation for public lands in connec-tion with fisheries; but leases or licences for any term exceeding nine years shall be issued only under authority of an order of the Governor General in Council.

The earlier act in 1858 had called for leases and licences only on Crown lands; this 1865 act appeared to allow leases and licences anywhere.

On April 16, 1867, shortly before Confederation, two more sweeping rules came into effect. In Lower Canada, there would be no netting of salmon with-out a lease or licence; in Upper Canada, there would be no netting of anything without a lease or licence.[113]

The strong powers of the Province of Canada's fishery legislation would become Canadian law after Confederation in 1867. Licences and leases would be prominent, but their application rarely as thor-ough and effective as early administrators hoped.

Law prohibits "deleterious substances"

Today the federal Fisheries Act still stands as Canada's strongest environmental legislation. This power as well derives from laws in the Province of Canada.

Various acts in the 1830's and 1840's had pro-hibited the dumping of mill wastes in navigable streams. After the 1857 legislation forbade the dumping of ballast and fish waste, that of 1858 included provisions against throwing lime, chemi-cals, or drugs into waters frequented by certain species of fish.

The 1865 act, perhaps influenced by contempo-rary legislation in Britain, broadened the terminolo-gy. Section 18 of the 1865 act said that

[w]hoever throws overboard ballast, coal ashes, stones, or other prejudicial or deleterious sub-stances, in any river, harbour, or roadstead, or any water where fishing is carried on, or throws overboard or lets fall upon any fishing bank or ground, or leaves, or deposits, or causes to be thrown, left, or deposited upon the shore, beach, or bank of any water, or upon the beach between high and low water mark, inside of any tidal estuary, or within two hundred yards of the mouth of any salmon river, remains or offals of fish, or of marine animals, or leaves decayed or

decaying fish in any net or other fishing apparatus, shall incur for any such offence a fine not exceeding one hundred dollars, or imprisonment for not more than two months....

The "deleterious substance" phrase remains in today's Fisheries Act, as a cornerstone of environmental protection.

Wide powers include "better management"

Along with the key provisions on licensing and pollution, the 1865 act provided for the making of "all and every such Regulation" as might be needed for "better management and regulation of the sea-coast and inland fisheries." This too passed over into the post-Confederation Fisheries Act.

Thus, pre-Confederation legislators created a powerful set of tools, reaching well beyond conservation into fish quality, development, pollution control, business dealings, licences and leases, and "all and every such regulation" as was needed. In the Dominion of Canada, future ministers would often use the act for economic and social purposes.

The fisheries and Confederation

In the early 1860's, the Dominion of Canada was no more than a possibility. No one knew if colonies and political parties would put aside their differences to form a new country. But the international context would help move the B.N.A. colonies in that direction. Changes in the fishery played their part.

Dories, longlines intensify fishery

Back in 1848, New England bankers were still fishing cod by handline from the deck of the schooner. Inshore fishermen often fished from dories, flat-bottomed, flare-sided rowboats which seem to have developed before 1800 in Massachusetts. Anyone who has ever been in a dory knows what a wonderful craft it is—tippy yet hard to capsize.

The French were already launching small boats from fishing ships on the banks. Then someone in New England realized that dories with removable thwarts could easily stack inside one another on the deck of a schooner. By 1855, schooners carried 13-foot dories to the offshore banks, launching them at sea to increase the range of handline fishermen. By 1860, they used dories for longlining, as that technology spread from the French to other fleets. The Americans found that the longline caught far more of the older, larger cod, many weighing nearly a hundred pounds, although "the large fish were nearly all caught up in time...."[114]

Schooners gradually increased in size from 45 tons to a typical 75 tons, "clipper" size, by 1885.[115] Longlines multiplied the number of steel hooks, now mass-produced; dories multiplied the number of longlines; large schooners could carry more dories—by 1860, the ingredients had come together for a great increase in fishing power.

Dory fishing on the Grand Banks. (Thomas Wesley McLean drawing. Library and Archives Canada, C-69716)

War delays American expansion

Civil War tore at the United States in 1861–1865, setting back the American fishery, and injuring American–British relations. The Americans felt with some justification that Britain aided Confederate naval actions. With the war over, anti-British feeling surfaced. In 1866, the United States pulled out of the Reciprocity Treaty; this meant no more American fishing within three miles of most of the B.N.A. coast. As well, Washington repealed the old bounty, in effect since 1813, which had given a certain amount of money per vessel ton to the owner and to the fishermen. The Americans still had shore privileges in Newfoundland, Labrador, and the Magdalens under the 1818 Convention. But Newfoundland in 1862 had begun imposing strict duties on goods and spirits traded by foreign vessels; this impeded fishing and buying of bait.[116]

Under these blows, the fleet's tonnage dropped, by one report, an amazing 70 per cent from its 1862 level.[117] In the Gulf of St. Lawrence, part of the decline came as mackerel fishermen realized that the new purse-seine, though superior to the mackerel jig, got better catches elsewhere than in the Gulf. At Labrador, the American cod fishery dropped off sharply, not just because of Newfoundland's attitude and the loss of reciprocal fishing, but also because of new opportunites offshore.[118]

Americans shift to the banks

From about 1870, American fishermen with their bigger, faster schooners turned mainly to the larger fish of the banks, everywhere from Georges to the Grand Banks, where they had more room for their longlines.[119] A significant American fishery remained along the coast, especially for bait. But the remainder of the century was chiefly the heyday of the offshore banks fishery, by dory schooners using longlines and salting down the fish for drying ashore.

Fishing had already been hard and adventurous. Launching men in dories, in the waves and fog of the offshore banks, made it even rougher. Typified in Kipling's Captains Courageous, the Americans were tough birds, drivers and achievers, tossing on the water far from home. Their fierce struggle showed up in the losses at sea. In the years 1831–1875, Gloucester alone lost 333 vessels and 1,590 men, many off the Maritimes and Newfoundland (whence a good many had migrated to New England in the first place).[120] The Canadian fishing fleet also lost vessels: 105 off British North America from 1868 to 1876, compared with 62 U.S. vessels on those same shores during that period.[121]

Meanwhile, the dory system spread to other fleets. The French first kept to their larger shallops. But by 1872, local schooners at St. Pierre and Miquelon began using American-style dories.

Vessels from metropolitan France still made some use of the heavier shallops, but by the end of the 1870's abandoned them for the dory.[122] The Portuguese had resumed fishing on the Grand Banks in the 1830's, and they too would take up the dory.[123]

End of Reciprocity rocks Canadian fishery

In Canada, the 1866 collapse of Reciprocity caused turmoil. American duties against imports of Canadian fish became a major obstacle. An attempt to renew Reciprocity saw the United States grudgingly offer to admit free of duty only a short list of items such as burrstones, grindstones, and rags. They also offered to repeal bounties to U.S. fishermen (which they soon did anyway); duties against Canadian fish would, however, remain.[124] The negotiations came to nothing.

In the Maritimes, some operators sold their vessels and reverted to boats. Nova Scotia fishermen faced a worsening situation, especially with an accompanying crisis in the river fisheries, particularly for salmon. The province provided some relief funds as emergency aid for fishermen.[125]

Americans still wanted access to Canadian fish, and sometimes breached the three-mile limit. A Gloucester newspaper reported in 1866 that despite the loss of shore-fishing privileges, "from 30 to 40 sail of vessels will be added to the fleet, and although the business will be attended with considerable risk, yet our fishermen are not scared at trifles; they will keep a sharp lookout for English cruisers and get good trips in spite of them."

American threat spurs interest in Confederation

The United States was feeling truculent, resenting the way Britain and British North America had accommodated the Confederacy. Manifest Destiny, the recurrent American ambition to absorb Canada, had become a powerful impulse. Many of the political leaders in Britain would have been happy to see Canada become part of the United States. Old-country feuds also entered the picture; Irish-American members of the Fenian movement sought to damage Britain by mounting raids to take over the northern provinces.[126]

Wary of the Americans, the B.N.A. colonies also perhaps harboured a certain resentment, or need of dignity, in relation to the Mother Country. Sir Charles Tupper, premier of Nova Scotia, said in 1860 that "[a]t present we are without name or nationality.... What is a British-American but a man regarded as a mere dependent on an Empire which, however great and glorious, does not recognize him as entitled to any voice in her Senate, or possessing any interests worthy of Imperial regard."[127]

Shining above all else in the 1860's was the promise of economic growth and Canada's expansion from sea to sea. Few talked about growth in the sea fishery. More attention went to shipping, since the Atlantic coast would be the junction point of a great trade, with high prospects in railways, manufacturing, government services, and the like. Maritime politicians looked for progress through integration with the larger Canadian economy. Nova Scotia's possession of coal mines and other advantages would make her "the great emporium for manufacturers in British America."[128]

Tupper tries to bar out Americans

Still, the fishery was a great industry which needed protection from the Americans. Denying them fish might induce them to renew Reciprocity, and if not, would at least keep them out of the way. Nova Scotia purchased a small ship, and in March 1866 announced its intention to seize and forfeit American vessels within three miles.[129]

Premier Tupper's muscle-flexing was at odds with imperial Britain, now anxious to accommodate the victorious American union. London wanted American fishing privileges restored. Politicians in Upper and Lower Canada (the latter with its own fishery interests in Gaspé and the North Shore) followed Britain's lead. They were reluctantly willing to let American vessels in, provided they paid a licence fee; and they had their reasons.

Province of Canada exercises diplomacy

In an 1866 letter, the Province of Canada authorities reassured Great Britain that neither old Canada (the united legislature of Upper and Lower Canada) nor the incipient new Canada wanted conflict with the United States. At the same time, the fishery held great importance for new Canada. Post-Reciprocity, American tariffs had made it necessary to seek other markets. Control of the supply of fish could open new channels of trade.

Now Britain was suggesting that Canada restore American fishing privileges, in hopes of getting Reciprocity renewed. Such a policy would be wrong, the letter said, and if pursued would eventually bring evil consequences, with even less likelihood of Reciprocity. The United States would see the granting of privileges as weakness; this would bring more danger of collision, "till neither country could recede with honour." Even so, the letter said, to avoid conflict the Province of Canada was inviting the Maritimes to unite with Canada in issuing licences for a moderate fee for this year only.[130]

This moderate proposal found British approval. In June 1866, the Americans were allowed to fish within three miles, and were supposed to pay a modest fee of 50 cents per vessel ton.

In Nova Scotia, Tupper still opposed the whole idea; offering any privileges to the U.S. would hurt the chances for a renewal of Reciprocity. He spoke of English giveaways of the fisheries. At one point in the heated atmosphere, Tupper said that the fisheries question could block Confederation.[131]

But Tupper also spoke of Nova Scotia's weakness without Canada. He told the Nova Scotia legislature in March, 1867, that the fishing licences had been "a compromise suggested to the British government by Canada. The Canadians were ready to license the fisheries, and standing as we do today, we are at the mercy of Canada. If Canada falls, we must fall. We have no status by ourselves, we have no standing in relation to the Empire apart from Canada. ... It is well known that the voice of Canada has always been supreme, although we have the largest interest in the fisheries."[132]

So the Americans got their licences, but Canadians got little in return. Revenues from licence sales were supposed to help provide money for a protection fleet. But of an estimated 800 American vessels taking more than $4 million worth of fish in 1866, only 451 paid the fees. Revenue came to only $13,000. Meanwhile, Canadians paid $220,000 duty on fish sent to the United States.

Tupper did act unilaterally, just before Confederation, to double the licence fee charged by Nova Scotia, without informing any of the other colonies or England. This move had little effect. Confederation would come into place, in July of 1867, during a strained atmosphere in the fisheries.[133]

Why federal jurisdiction?

As late as the pre-Confederation Quebec Conference in 1866, the leaders of Nova Scotia, New Brunswick, and the Canadas still agreed that both federal and provincial governments would have jurisdiction over fisheries. Later, however, when the London Conference made the final listing of who would control what, the federal government alone got the charge of "sea coast and inland fisheries." Why?

International pressures probably affected the decision. Historian Kenneth Pryke, while pointing to a general centralizing tendency at the London Conference, adds that federal fishery jurisdiction "was obviously caused by the dispute of the previous spring over fishing licences...." And H.A. Innis suggests that the federal power over fisheries reflected the importance of Confederation as a means of resisting New England.[134]

The Maritimes had frequently turned to Great Britain for help against the Americans; but now London was growing more reluctant, and wanted accommodation with the United States. Therefore, it would indeed seem natural for the coastal provinces to look to the new Dominion government for enforcement strength. Nova Scotia's Tupper may have thought he could persuade the future Canada to take a firm stand; indeed, that was to happen after Confederation.

Apart from enforcement strength, other factors may have been at play. One can speculate that the fisheries already had enough problems that provincial politicians at the London Conference would just as soon hand them over to the federal side. The Conference agreed that the federal government could for the benefit of Canada undertake works that would normally fall under provincial jurisdiction.[135] It is worth noting that shortly after Confederation, finance Minister Francis Hincks would say that "the fisheries are a mere expense...." To drift further into speculation, the Province of Canada was passing strong new rules for fishery administration; perhaps Canadian delegates believed in a strong government role, and thought the federal side best equipped for it.

Some observers today would suggest another reason for federal jurisdiction: that provincial interests compete for fish, and the federal government needs jurisdiction to arbitrate. But pre-Confederation, when the resource was still relatively strong, and people could fish where they liked and take all they wanted, that reasoning would have been weaker.

Whatever the reason, Confederation made federal jurisdiction over fisheries strong and clear. Fisheries thus reflected Sir John A. Macdonald's desire for a strong central government.

Sea fishery on the sidelines

By the time of Confederation, the fishery had become, as it would remain, an industry of contradictions. The historical foundation of the Atlantic coast, it was still all-important in many areas. In Newfoundland, separate from Canada until 1949, the fishery remained central. But in Canada, despite its great regional importance, the fishery was moving to the edge of the national economy. In the new Confederation, most attention would go towards manufacturing, metropolitan trades and services, railways, and the interior of the continent.

The fishery still had vigour. Accounts of the period on the Atlantic coast often portray bustling fishery centres full of wharves and "forests of masts." And it still offered potential for expansion, as shown by the spread of the longline fishery on the banks, and by the development of herring, lobster, and on the Pacific, salmon fisheries.

Yet the fishery was only part of the story. The bustling Atlantic ports drew much of their strength from the shipping and carrying trades. And if there was a Golden Age of the Maritimes, there was no Golden Age of the fishery, at least in terms of cash income. Few if any accounts of the period associate fishing, by itself, with wealth. Rather, they portray

hard labour and risk, with wrecks and drownings frequent. By the time of Confederation, the fishery, with its many reports of poverty and distress, seemed to present more problems than promise.

Why were Canadians more regulatory?

Fishing pressure, local problems, and consequent regulations appeared earlier in New England than in the B.N.A. colonies. Yet the Province and then the Dominion of Canada would turn out to be more regulatory. Why was this so?

It is often noted that "peace, order, and good government" stamped the northern colonies, rather than "life, liberty, and the pursuit of happiness." In much of Canada, the governmental authorities came first and then brought the people. Private industry had less momentum and sway. But if that is true in general, the B.N.A. colonies still varied. The most far-reaching fishery legislation came not from the Maritimes but from the Province of Canada, largely inland.

Perhaps it had to do with the atmosphere of growth and progress, with immigrants pouring in, industries developing, and people willing to tackle issues. But for licensing in particular, the assertive legislation may have stemmed as much from specific circumstances. Fishing for salmon, whitefish, and other species took place largely from fishing stations on the land. Leases and controls for land were familiar, and were easily extended to adjacent waters.

Finally, the character of early legislators and officials, energetic men like Richard Nettle and W.F. Whitcher, perhaps helped shaped the regulatory frame of mind at the outset. In any case, the Province of Canada would bequeath to the new Dominion a strong and durable Fisheries Act.

Regulation more wide than deep

Although pre-Confederation B.N.A. authorities already had somewhat of a regulatory bent, particularly in the Province of Canada, one can overstate its influence. The varied laws in the different colonies were touching many matters: sovereignty, conservation, fish quality, trade, and development.

But for the Maritimes in particular, regulation most often worked superficially. On fish quality, for instance, governments occasionally noted problems and appointed overseers, but never made a major push. Conservation measures were weak. Reports presented just after Confederation painted previous management as generally a failure. It had gone wide rather than deep, making little attempt to get to the heart of matters.

Management ideas in British North America reached deeper when Perley and Whitcher dreamt of using licences and leases systematically, to put private enterprise to work for the benefit of common property. But they were thinking mainly of salmon. Indeed, the bulk of regulation and regulatory systems so far pertained to the river and nearshore fisheries—tiny in comparison to the great fishery for cod and other groundfish, which went largely unregulated.

The Atlantic sea fishery was an old and individualistic industry, with still-abundant resources, essentially a going concern despite its problems. A speaker in the Newfoundland House of Assembly noted in 1864 that "there appears to be no decline in the cod fishery ... but ... a great increase has taken place in the population, and, consequently the produce has to support a much larger number of fishermen and their families; and should the population continue to increase, and trust to the fishery for their subsistence, the natural results must inevitably follow."[136] But such voices were few. The groundfish fishery went on as before, slipping a little in catches here or in quality there, but never inspiring much public thought, let alone thoroughgoing reform.

The fisherman and the farmer

By now farmers enjoyed more attention. By mid-century they were a political force in the Maritimes. The farmers' advance would continue over the next century, with educational efforts (Nova Scotia got an agricultural college in 1885), extension workers, representative organizations, co-operatives, credit unions, marketing boards, crop insurance, and stabilization plans. Fishermen had few parallel efforts until after the Second World War.

To some degree this was natural. Even in the Maritimes there were more farmers, whether subsistence or commercial. By the 1880's, Nova Scotia had more than twice as many farmers as fishermen (roughly 59,000 compared with 24,000); New Brunswick had four times as many (50,000 against 12,000); and P.E.I. five times as many (21,000 to 4,000). Canada as a whole had more than ten times as many farmers (723,000) as fishermen (65,600).

But there were other reasons why fishermen were less organized and got less attention. They worked in a different world, out on the water. They were unseen, undereducated, scattered by geography, and often individualistic by temperament. People set up courses and wrote books and articles about agriculture. The fishery stimulated far less thought. Planting and tending fields demanded planning; the fish came by themselves. Fishermen were part of a strange, almost a wild occupation, hunting animals in the sea.

While seldom rich, farmers at least had the power of private property and political presence. Fishermen had neither in the full sense, although they often won political attention by ad hoc efforts. One can view the Western world as built on three pillars: private enterprise; the vote; and the underlying, less tangible infrastructure of education, information, and constructive attitudes, including the sharing of responsibility. The common-property fishery was already weak on all three.

But the fishery still had its saving graces. Life was local and direct, with a certain pride and freedom, especially if you owned your own boat. The challenges of the Atlantic fishery produced many fine, self-reliant families, who made decent or at least tolerable livings. They created communities where the conversations, humour, weather, beauty, romances, fights, and business dealings could absorb one like a great play that obliged you to participate. One might be forced to outmigrate, but when you grow up in a fishing community, your soul never leaves it.

PART 2: 1867–1914
BRINGING LAW AND ORDER TO THE FISHERY
CHAPTER 6.
National and international events, 1867–1914

After Canada's Confederation in 1867, as coal, steel, and engines extended their reach, the marine economy of wood, wind, and water started to fade. Railways and roads subsidized by governments, including the Intercolonial Railway from Halifax to Quebec, completed in 1876, were shifting business from schooners to the shore. On the sea, steamer routes, often subsidized, cut into the wooden-vessel trade. British timber preferences had ended, and in any case, the timber industry was moving on to new forests of the interior.

Purse-seines became more of a factor, as in the Bay of Fundy herring fishery. (From Goode, 1887)

Even the coastal provinces wanted to be part of the new continental industrialism, especially as an economic downturn afflicted much of the Western world in the last quarter of the century. The east coast was only a fringe player in the new economy. Boatbuilding and the shipping trade were in decline. For many people, the fishery now seemed their only chance. But it would prove unable to support, by itself, the old marine prosperity.

Before Confederation, research and management for the sea fisheries had remained almost non-existent. Now the rapid development of lobster canning on the Atlantic and salmon canning on the Pacific would change the picture. Plants were going up in any empty cove. It took only two decades for the lobster- and salmon-canning industries to go from zero to practically maximum production, bringing threats of overcrowding and catch decline.

Alarmed by reports of overfished lobster or declining shad or disappearing sturgeon, the new Department of Marine and Fisheries passed regulations by the score. The rules became more and more detailed, sometimes taking a social or economic twist. Rather than following the laissez-faire approach typical in the neighbouring United States, the Canadians tried to watch and regulate everything in sight. And fishery clashes with the United States helped define Canadian sovereignty.

The first Minister, Peter Mitchell, and Commissioner of Fisheries, W.F. Whitcher, instilled an active approach, which lasted. Department officials and royal commissioners wrote new rules for scores of fisheries.

While the fisheries service worked hard, the thoroughness of applying rules varied. For example, attempts at limiting the number of licences ran into trouble. Still, Mitchell, Whitcher, and Whitcher's successor, E.E. Prince, rarely hesitated to tackle an issue. Managers of the day wrote the regulations that dominated fishery management until after the Second World War and continue to mark the Canadian management approach.

First Minister finds bad conditions

The years from 1867 to about 1880 saw the new Dominion set up strong fisheries legislation and a national administration. The Department of Marine and Fisheries came into being on July 1, 1867. Peter Mitchell, a native of Newcastle, N.B., became Minister in Sir John A. Macdonald's government. A former premier of New Brunswick, Mitchell had helped bring the province into Confederation.

For the provinces, earlier fisheries legislation stayed in effect for the time being. Mitchell commissioned reports on the fisheries of Nova Scotia and New Brunswick, the other partners in the new dominion. Painting a picture of mismanagement and distress, the reports gave Mitchell reason to put forward a new fisheries act like that of the old Canadas.

Thomas Knight's *Report on the Fisheries of Nova Scotia*, commissioned by Mitchell, said that the fisheries in that province employed at least one-fifth of the adult male population. Nova Scotians took part in the bank, Labrador, Gaspé, and Gulf fisheries. But locally, fish had become fewer. Weirs and bultows had had bad effects. Lack of bait had become a problem. Rivers needed fishways. Salmon, gaspereau (alewife), shad, and sea trout faced extermination. And the province needed fishery societies. Knight noted the superiority of the previous Canadian fisheries act, still in effect in Ontario and Quebec. W.H. Johnston also reported on Nova Scotia fisheries, listing many causes of damage. The province had fish inspection laws by county; it needed a central system.

Hon. Peter Mitchell, Senator, Minister of Marine and Fisheries, in July, 1869. (Library and Archives Canada, PA-25313, detail)

Mitchell was told that the loss of Reciprocity had harmed Nova Scotian fisheries; Nova Scotians opposed licensing foreign vessels unless it meant a return to Reciprocity. With many foreign vessels on the scene, more than 1,000 in the Gulf mackerel fleet, and bait supplies down, local fishermen remained poor. The inshore net fishery appeared insufficient. The fishery needed to combine inshore and offshore; but to get into the offshore fishery, fishermen needed government capital aid, or bounties.[1]

Reporting from New Brunswick, W.H. Venning noted that fish is "an absolute wealth, needing neither time nor labour." But resource problems were many. Fishermen disobeyed New Brunswick's laws, such as the one prohibiting driftnetting in harbours and rivers. Wardens and overseers had proved useless. At Grand Manan, weirs were destroying the herring fishery, the spawning grounds needed more protection, and fishmeal plants were taking great quantities of small herring. Destruction was widespread.

As for fish products, New Brunswick's inspection laws remained unenforced. Venning suggested that New Brunswick needed a fisheries act like that of the Canadas, to avoid a U.S.-style depopulation of rivers "by practices which all sensible men deplore."[2]

Thus, well-documented reports from the two Maritime members of Confederation pointed out the many troubles of the fishery and suggested a law like the 1865 act of the Province of Canada.

Administration comes into place

Meanwhile, the official apparatus was taking shape. William Smith, a native of Scotland, became deputy minister of Marine and Fisheries on November 11, 1867. Although "not a popular man," Smith gained a reputation for independence from politics and for frugality in his administration.

In Smith's early years, the entire headquarters staff for Marine and Fisheries numbered only 25 or so, with perhaps 1,200 people in the field. Probably fewer than half of these worked in Fisheries. The other duties included lights and lighthouses, buoys, pilotage, ports and harbours, wharves and piers, government vessels, and much more.[3]

W.F. Whitcher of the Province of Canada administration became the chief fisheries official for the new Dominion. Deputy minister Smith's name rarely appears in the annual reports on fisheries, although he must have exerted some influence. Instead, Whitcher's name appears next to Mitchell's, and on circulars and notices to the fishing industry. He seems to have kept direct charge of Ontario and Quebec fisheries, while Nova Scotia and New Brunswick would get their own inspectors. By 1873, Whitcher was signing annual reports as "Commissioner of Fisheries."

Whitcher had already dealt with fundamental fisheries legislation for the Province of Canada. Now he would be present at the creation of the federal Fisheries Act. Whitcher "was regarded as an able authority and a courageous administrator."[4] Under Mitchell, he set up the federal fisheries service. He pushed for compulsory inspection of fish products. He imparted the idea of licensing and leasing fisheries. He backed the establishing of an extensive network of hatcheries and fishways. He also helped to win, in 1884, the setting-up of Fisheries as a separate department for a number of years.

Fisheries Act copies Province of Canada law

In the 1868 annual report of the Department of Marine and Fisheries, Minister Peter Mitchell wrote that at Confederation the fisheries of the Maritimes were poorly managed and nearly exhausted. Since the earlier laws of the Province of Canada were working, he had made inquiries through Whitcher, and found he should extend the same system to the whole new Dominion. Existing Nova Scotia and New Brunswick laws could also remain in force, possibly to change later.[5] For the time being, Nova Scotia kept its power to appoint overseers of the fishery.

The Fisheries Act ("An Act for the Regulation of Fishing and Protection of Fisheries") received royal assent on May 22, 1868. Elements included the power to make "all and every" regulation for "better management and regulation" of the sea-coast and inland fisheries; the power over licences and leases; the prohibition of deleterious substances; provisions for fish culture, fishways, protection of young, and fish sanctuaries; restrictions on size or use of some types of gear; closed seasons; and many other specific regulations, mainly for salmon and inland fisheries. The only regulation for the cod fishery set minimum sizes for seines. Other rules affected the oyster fishery and the mode of taking seals and whales. Some amendments to the act took place in 1886 and 1906.[6]

Using the Fisheries Act

The federal Fisheries Act has stood up well; its original provisions, especially on licensing and pollution, contained the strength to meet most requirements of present-day management. The act controls some matters directly, and for others provides power to make specific regulations. These latter multiplied over the years. To make or change a regulation, the department through the minister must seek an Order-in-Council. The Privy Council (in theory, all members of cabinet; in practice, a committee) endorses regulations that ministers put forward.

Today, as management issues arise, ministers and departmental officials can sometimes set policy deriving from basic provisions of the Fisheries Act itself. At other times, they prefer to spell out regulations. The latter course provides concrete, visible rules, but those very rules sometimes seem to tie the department's own hands.

If a fishery officer or manager faces an emergency conservation problem, the act provides some authority to vary the regulations. In the case of a more complex problem, department officials may try to get voluntary compliance from the industry, pending passage of a new regulation. In the 1980's and 1990's, fishery managers more and more wrote in fishing rules as part of the licence-holder's "conditions of licensing."

Fisheries Protection Service fosters sovereignty

The Macdonald government coupled the Fisheries Act with "An Act Respecting Fishing by Foreign Vessels." This closely resembled Nova Scotia's Hovering Act of 1836. Foreign vessels needed a licence to fish in Canadian waters; for offences, Canadian authorities could seize and forfeit vessels, equipment, and catch. This law got strengthened in May 1870.[7]

Like the Fisheries Act, the act respecting foreign vessels was to serve well. Its present-day successor is the Coastal Fisheries Protection Act.

Americans resist licensing rules

The act respecting foreign vessels came during a tense period. The United States, despite its increased bank fishery, still wanted access to fish on Canada's coast. Some Americans wanted Canada, period. "Manifest Destiny" was in the air. Irish-Americans were agitating for an American takeover of Canada. Meanwhile, the U.S. government was seeking compensation for British co-operation with the *Alabama* and other Confederate vessels. The Alabama Claims were an emotional point of national honour, such that the influential Senator Sumner suggested at one point that Britain abandon North America entirely. British leaders such as Gladstone half expected, and some perhaps hoped, that Canada would become part of the United States.[8]

After Reciprocity ended in 1866, the British had prevailed on British North America to let U.S. vessels fish within three miles of the shore, provided they paid a licence fee of 50 cents a ton. Tupper in Nova Scotia had raised the fee to $1, and the other provinces had followed suit. After Confederation, with the United Kingdom still controlling foreign relations, Mitchell and Tupper successfully pressed the British for another increase, to $2 per ton.[9] Tupper, now in Macdonald's cabinet, thought this move would move the Americans towards a restoration of Reciprocity.

The more immediate result was an increase in scofflaw behaviour. Fewer than 300 American vessels paid the fee in 1867; only 68 did so in 1868. In 1869, only 12 of 162 vessels stopped for examination possessed licences. On Passamaquoddy Bay, U.S. vessels sometimes registered on the Canadian side, copying British numbers.[10] In the words of historian J.B. Brebner, "The Americans simply refused to admit that the Canadians would dare to exclude them, and the British Navy, which had the task of policing the shores, was much too tender and anxious to avoid trouble to disillusion them."[11]

Mitchell, Macdonald take on Americans

Macdonald's government acted forcefully to gain what it could from the situation. Britain was no longer the reliable guardian of yore. In the spring of 1868, Mitchell set up a new marine police to guard the fisheries and borders. He ordered six new vessels, at the price of nearly a million dollars.[12] These vessels, Canada's first armed marine police, would be of great import in what followed.

As part of their soft approach, the British in 1866 had ordered the navy to let the Americans alone unless they were fishing within three miles of shore, or within bays and harbours less than ten miles wide at the entrance. Even then, American vessels were to get three warnings before having to either purchase a licence or leave. Mitchell wanted tougher enforcement. In 1869, he got the British to supplement their vessel patrols by stationing boat crews from men-of-war along the coast. But Britain rejected requests for a strict and thorough policy.

Canada now had vessels of her own on the grounds, including at times the chief ship in Dominion service, the *Druid.* When the Lords of the Admiralty tried to

Mitchell's fleet versus the Americans, as seen by the *Canadian Illustrated News*, March 12, 1870. (Library and Archives Canada, C-48733)

place them under control of the British naval fleet, Mitchell successfully resisted. In 1868, he won British permission to reduce the number of warnings to U.S. vessels from three to one. Yet the number of licences bought kept decreasing. The British were lenient, the Americans truculent, some carrying rifles. At the same time, Prince Edward Island was undermining Canada.[13]

Prince Edward Island helps American fleet

Prince Edward Island's fishery had grown since mid-century. In 1851, the island had set up a bounty system to promote the fishery. This and the prosperous trade of Reciprocity had increased the island's vessel fleet. When Reciprocity ended free trade in fish, the vessel fleet declined. The boat fleet loomed large in the P.E.I. fishery.

Now, with Canada trying to keep Americans out of her inshore fisheries, P.E.I. opened her shores. Thus the island gained trade and work on U.S. vessels. P.E.I. took over the great Canso trade in bait, barrels, other supplies, and transhipment; it became an American fishing base just off the shores of Canada. When Canada objected, Great Britain eventually backed Prince Edward Island. Even outside P.E.I., considerable sympathy existed for the American vessels, which were often manned by Canadians, especially from the Canso area.[14]

Canada closes three-mile zone, confiscates American vessels

While pressing the British for tougher enforcement, Mitchell made overtures to the Americans. He declared in 1869 that the two countries could make a satisfactory fisheries agreement if the United States established beneficial trade relations. But discussions that year came to nothing; the Grant administration rejected Canadian proposals for renewed Reciprocity. Thus the 1860's ended with Americans making free in Canadian waters, and refusing better trade relations. Mitchell, aggressive by nature, was growing impatient, and so was the Macdonald government.

On January 9, 1870, Canada discontinued entirely the licensing of American vessels, and excluded them from fishing within three miles. Mitchell deployed his new patrol fleet, issuing tough instructions to the captains and crews. The police vessels so resembled American fishing schooners that they could approach without raising alarm. Officers boarded some 400 U.S. vessels; the Canadian government seized and condemned 15 of them.

The Canadian actions greatly perturbed London, and forced Britain's hand. The U.K. authorities sent naval reinforcements to help the Canadians, but insisted on temperate behaviour towards the Americans. Meanwhile, Mitchell further annoyed the Colonial Office by questioning whether the British really wanted to catch Americans. Now that he had the British government's alarmed attention, Mitchell encouraged the

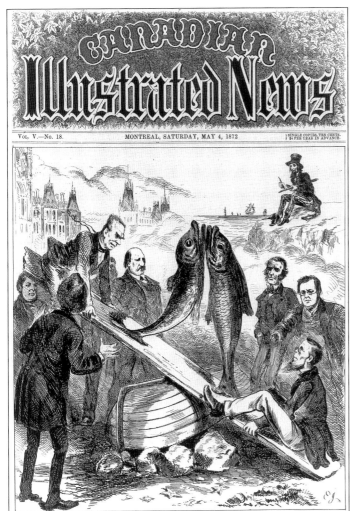

Vol. V.—No. 18.   MONTREAL, SATURDAY, MAY 4, 1872   SINGLE COPIES, TEN CENTS. $4 PER YEAR IN ADVANCE.

SKETCHES FROM THE CAPITAL.—A GAME OF SEE-SAW.—BY OUR OWN ARTIST.

The *Canadian Illustrated News*, May 4, 1872, on the Canada-U.S. fisheries negotiations. (Library and Archives Canada, C-58593)

Canada wins recognition, fisheries free trade

Macdonald, with strong support from Mitchell, Tupper, and others in the cabinet, pressed his fellow British commissioners for satisfaction for Canada. Francis Hincks, Minister of Finance, wrote to Macdonald in Washington that "[w]e have no object in refusing [the Americans the fisheries and use of the St. Lawrence River], on the contrary the fisheries are a mere expense. Our equivalents that should be pressed are full reciprocal trade—If we yield on this England must compensate us. But we cant yield the fisheries without at least free importation of our fish and free or low duty coal lumber and salt, particularly the first."[16]

Macdonald took the toughest line he could, telling the Colonial Office through Tupper that "Canada considers inshore fisheries her property and that they cannot be sold without her consent." If fisheries provided training for naval seamen, he asked, why encourage the United States in getting the same kind of training?[17] But Britain preferred appeasing the Americans. Macdonald commented that the other British commissioners "seem to have only one thing on their minds; that is, to go home to England with a treaty in their pockets, settling everything, no matter at what cost to Canada."[18] He noted that "[t]he American Commissioners have found our English friends so squeezable in nature, that their audacity has grown beyond all bounds."[19]

Canada never got the full free trade she wanted. Great Britain forced a settlement, reflected in the 1871 Treaty of Washington, involving ten years' free fishing in each other's waters, starting in 1873; free trade in fish and fish oil (except inland fish and fish packed in oil); and a sum of money to Canada and Newfoundland, to compensate for the greater value of the fishing privileges they were offering. The exact sum of money, America's *Alabama* claims, and the Pacific coast boundary dispute about the San Juan Islands all went to arbitration. Meanwhile, American schooners flooded back into Canadian waters, many still chasing mackerel in the Gulf.

Joseph Howe, Nova Scotia's old crusader against Confederation, said that England had bought peace at the sacrifice of Canadian interests. Macdonald remarked that Howe's speech on the matter was "more untimely than untrue," and he wrote to Tupper, "My first impulse was to hand in my resignation."[20]

Still, considering the British and American forces arrayed against Canada, the young Dominion had made good headway for its fish trade and its international standing. Mitchell, Macdonald, and the cabinet had put the national strength on the line. The first real test of Canadian sovereignty, the fisheries dispute

Canadian government to propose that Britain and the United States create a Joint High Commission, with the Dominion represented on it.

Meanwhile, the Canadian seizures had provoked American anger, helping bring annexationism to a peak. President Grant stated in December 1870 that the Dominion of Canada, a "semi-independent but irresponsible agent has exercised its delegated powers in an unfriendly way."

While some voices called for reprisals, others called for peace. America needed good relations with Britain, partly for financial reasons; and Britain wanted good relations with America, partly because problems were looming in Europe. The result was the announcement in February 1871 of the Joint High Commission that Canada had wanted. Prime Minister Macdonald became a member.[15]

strengthened it. Even though Britain had the main say in negotiating the Treaty of Washington, it was Canada, including Mitchell and his marine police, that had brought the Americans and British to the negotiating point. The Treaty of Washington in its way marked U.S. acceptance of transcontinental Canada. It began a lasting tradition of peaceful relations and mutual respect between Canada, the United States, and Britain.[21]

Government turns to tariffs and "National Policy"

Although fisheries disputes had precipitated the Treaty of Washington, that same treaty helped remove the fishery from the centre of the diplomatic stage. In future years, the fishery could still provoke international tensions. Except in Newfoundland, however, it was never again such a major and continuous factor in high policy.

After the Treaty of Washington brought only fisheries free trade, Alexander Mackenzie's Liberal government, taking power in 1873, tried to restore more general free trade, and failed. When Sir John A. Macdonald returned to power in 1878, the government went the other way, with new tariffs on manufactured goods. The tariffs increased costs to Maritime fishermen and farmers, and impeded trading. High tariffs became part of the "National Policy," which included completion of the Canadian Pacific Railway to open the west, and subsidization of fast steamer service to Europe and Asia, to help exports.

Halifax Award provides bounties

Pursuant to the treaty, a Fisheries Commission met at Halifax in 1877 to settle the payment to Canada for fishing privileges. Mitchell's successor as Minister, Albert Smith, took a strong role that earned him a knighthood. The "Halifax Award" payment took place in 1879: Newfoundland got about $1 million, Canada about $4.5 million.[22]

Maritime members of Parliament lobbied the Canadian government to direct the money to the coast. In 1882, Parliament resolved that the government would grant $150,000 annually for fisheries development. This led to the Deep Sea Fisheries Act of May 17, 1882, authorizing "an Annual Grant for the Development of the Sea Fisheries and Encouraging of the Building of Fishing Vessels." Small payments went directly to fishermen. These bounties would last for many years, becoming less significant over time. (The writer's father, a fisherman all his life, claimed for the bounty in only one year, 1937, and got enough to buy a pair of boots.) The annual grant changed in 1891 to $160,000, and there it stayed until the 1960's, when the government decided to put the money into general programs for fishermen. Total bounty payments from 1882 to 1967–1968 amounted to about $13.7 million.

Prince Edward Island joins Canada

Mitchell's immediate successors made no great changes. Albert J. Smith served as Marine and Fisheries Minister from 1873 to 1878, in the Liberal government of Alexander Mackenzie. (As premier of New Brunswick, Smith had almost prevented Confederation, which he had regarded as a devious scheme from the "oily brains of Canadian politicians.")[23]

While Smith was Minister, Prince Edward Island joined Confederation. During the Canada–United States fisheries dispute, P.E.I. had prospered by aiding the American fleet. But the renewal of fisheries reciprocity in 1871 had put Canada and P.E.I. back on an equal footing. Meanwhile, the island had got railway fever, building a line from one end of the colony to the other, and also building up a large debt. In 1873, P.E.I. took shelter in Confederation, and Canada took over her debts.

On October 7, 1875, the Fisheries Act was extended to the new province. When Sir John A. Macdonald's Conservatives returned to power in 1878, James C. Pope, the shipowner and former premier who had brought Prince Edward Island into Confederation, became Minister of Marine and Fisheries until 1882.

The spread of regulation

Staff grows to 600

The early years of the fisheries service, from Confederation to about 1880, set the mould in several respects: the creation and staffing of the branch; the passing of a powerful Fisheries Act and legislation to control foreign vessels; the use of licences and leases; the operation of hatcheries; and the setting of general conservation regulations, such as the prohibition of explosives and, for some fisheries, a weekly closed time. The following decades, from about 1880 to 1914, would see hundreds of specific fishery regulations take hold across the country, many authored by roaming royal commissions.

Under successive ministers and Whitcher, the field staff became substantial. Some impetus came from a Dominion-wide survey by a House of Commons Select Committee on Fisheries and Navigation, in 1868 and 1869, which detailed observations and complaints from overseers, fishermen, and others familiar with fisheries in their area.[24] Already by 1871, there were more than 90 overseers (who had *ex officio* power as magistrates), usually earning $100 or more, and 160 wardens, usually earning $25 or so.

Headquarters was small; even in 1880 the "Establishment Staff" of the entire Department of Marine and Fisheries in Ottawa, including Minister J.C. Pope, numbered only 25. Pope earned $7,000; deputy minister William Smith, $3,200; Commissioner of Fisheries W.F. Whitcher, $2,400. By that same year, the Outside Service (that is, outside Ottawa) of the fisheries branch had grown to 594 fishery officers, including inspectors, overseers, and wardens. These included only two in British Columbia (Inspector A.C. Anderson, salary $600 yearly, and Overseer George Pittendreigh, salary $500). Fifteen of the outside staff worked in the ten hatcheries. Fish culture took more than $29,000 of fisheries spending, which totalled more than $86,000. Nova Scotia had the most officials, 240 of them.

The inspectors and below them, the overseers, were the main figures. These were the days before civil service reform and the merit principle; appointments were mainly political. For wardens especially, work was part-time. Their main job might be farming or something else. Local officers at first received no pay in winter, which contributed to a large turnover.[25] (Names of the positions would change over the coming years, but generally a warden, guardian, or patrolman served seasonally, as hired by an officer of higher status. Some were hired with their boats, to patrol specific areas.)

Until Mitchell put his marine police on the water, the department had had only one patrol vessel. Pierre Fortin, who had for many years kept law and order in the Gulf of St. Lawrence, in 1867 gave over command of *La Canadienne* to Théophile Tétu. Captain Tétu immediately got involved in similar work: for example, pulling down scaffoldings from river falls where Indians with torches speared salmon, and holding inquests on criminal matters. Tétu put forward an idea that people involved in the fisheries have often had to relearn; it was untrue, he said, that the fish in the Gulf of St. Lawrence were declining; rather, the number of boats had increased, leaving fewer fish for each. Tétu unfortunately died at 34, during the 1868 voyage, but the Gulf patrol continued.

Mitchell boosted the protection fleet right after Confederation. But after the fisheries dispute with the Americans ended in the early 1870's, his marine police force faded from the scene. By 1880, there was again only one vessel in the Fisheries Protection Service, still for the Gulf and lower St. Lawrence.[26]

Fear of depletion brings new regulations

As the staff grew, so did regulations for conservation, which appear often to have come from the bottom up. The reports from Whitcher's fishery officers remain impressive today. Clearly written, they show closeness to the fisheries, awareness of their complexities, and a concern with concrete matters and the lives of people. The early officers watched, listened, and made suggestions, just as royal commissions later did in a more organized way. If departmental and ministerial judgement found the action desirable, another regulation entered the books, supplementing the local rules inherited from the pre-Confederation provinces.

The field staff had many causes of concern. In the department's first annual report after Confederation, a New Brunswick official reported on the "total disregard" of fishery laws. Another officer wrote that "no country

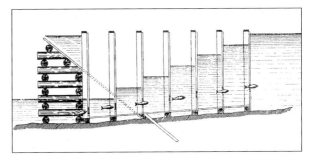

The department built many fishways to offset mill-dams. This fishway design appeared in the 1891 Annual Report.

had so many rivers for fish as Nova Scotia, and none were so destroyed."[27] By now concern about depletion was spreading, especially in freshwater fisheries.

For river species, mill-dams and sawdust remained a problem. New regulations forbade putting sawdust into navigable streams. In addition, pulp and paper mills were spreading in Canada after 1864. Whitcher in the 1870's noted the poor enforcement of laws affecting mill offal. "The general enforcement of these statutes is rendered almost impossible by the persistent indifference and active antagonism of the manufacturing interest." Circulars from Whitcher in 1875 and 1876 advised his Quebec officers to be discreet in enforcing the Sunday close on salmon fishing, so long as rivers were getting enough escapement; but they were to be harsh on mill rubbish. Another 1876 circular advised fishery officers in Nova Scotia and New Brunswick that there would be no more keeping of half the fine assessed against anyone. Thus ended a practice that had appeared in provincial regulations long before Confederation.

Various regulations of course applied to the salmon fishery. And a landings tax came into effect on July 22, 1875. The tax on salmon caught by net was to be 40 cents per 200 pounds; on bass, 20 cents per 200 pounds.[28]

Sea fisheries get less regulation

The fisheries service was slower to regulate the marine fisheries than the freshwater fisheries, where one could see the damage. But there were signs of concern. Whitcher wrote in 1874 that "the department has always avoided placing any restrictions on the pursuit of the deep-sea fisheries. At the same time it may be necessary to regulate participation in them by such means as shall obviate collisions and mutual hindrance." Thus, rather than abolishing cod-seines on the Labrador coast, one should separate the beach-seines from the hook-and-line boats. In 1875, Whitcher sent a circular to fishery officers, asking about the effect of bultows (that is, longlines, also known as trawls). The answers must have reassured him, since no regulation came into effect.

The Commissioner of Fisheries also expressed concerns in the 1870's about practices in the oyster and lobster fisheries. And in the seal fishery, the new fishing power of steamers, some from overseas, brought danger. Whitcher wrote in the 1874 annual report that

> [t]he inevitable fate attending excessive pursuit of the fauna of field forest and flood, threatens speedy extinction of seals in the Gulf of St. Lawrence. While seal hunting on the ice was carried on from sailing vessels and by shore-nets, the vicissitudes of the pursuit afforded some natural protection to this animal, and its numbers kept up a flagging pace with the legitimate annual destruction. But the recent employment of steamers has overcome many former difficulties, and enables the sealers to pursue their prey with indiscriminate slaughter. ... There were at one time last season engaged in this destructive business, on the Arctic seal grounds, nearly forty steamers and as many sailing crafts from various European ports; and so great was the havoc committed that it has excited universal apprehension. About the same time extensive operations by American steamers in the Gulf of St. Lawrence also attracted attention.

Whitcher commented that he could not recommend restricting Canadian sealers until, through mechanisms of the Treaty of Washington, the American sealers also faced controls.

Whitcher promotes licences and leases

Behind the specific regulations, Whitcher was promoting a wider scheme. In 1873, a number of Nova Scotians petitioned Mitchell, the Minister, about the decline of the river fisheries, the scarcity of wardens, and other problems. Whitcher responded to Mitchell that yes, there were few wardens. More serious, however, was the failure to apply the Ontario and Quebec leasing system to Nova Scotia and New Brunswick, since giving out leases and licenses created less fuss and brought more revenue.

Whitcher noted in the annual report that

> [t]he angling divisions of several salmon rivers on the St. Lawrence are now vacant, and others will be disposable in the course of next spring. These privileges it is proposed to advertise, and to invite offers to rent the same. When occupied by sportsmen the rivers receive increased protection; and besides contributing to the fishery funds they also become subject to local guardianship at private cost, and in that respect cease to be a charge on the public revenue.

Elsewhere, he dwelt on the removal of salmon nets from the Restigouche and Moisie rivers, and how it had increased the fish.

> At both places it is now clearly proved that immoderate netting is a serious hindrance to the restoration of the salmon fishery, and a positive disadvantage to the fishermen themselves. It also is quite as clearly established that a moderate quantity of nets, judiciously situated, render at once a far more profitable return to the owners and admit of maintaining a permanent stock of mature salmon. This fact has a peculiar bearing on the regulation of the salmon fishery. The occupancy of salmon stands under formal titles enables the occupiers to economize both their own capital and labour and the public property in salmon. Where the fishery is carried on in a desultory and improvident manner, under such incitements to excess as are created by contentious rivalry and the prospect of mere temporary gain, it is extremely difficult to control fishing operations within reasonable bounds. But, on the other hand, where occupants can rely on the permanence of their holdings, and enjoy in successive years the benefit of their own moderation in each preceding season, the department finds very little difficulty in controlling the pursuit.

Whitcher added that "it is not easy to convince fishermen how much cheaper and more profitable it is in their own interest to conform to the same principles on which legal protection is founded and the departmental regulations are enforced. Nothing short of the plainest examples appear to be sufficient to attract their earnest attention." But examples like the increase in Moisie River salmon should do it. (Like many of his successors, Whitcher expressed surprise that fishermen failed to realize what the department was doing for them, without, however, undertaking any educational campaigns to get them on side.)

Licences and leases in the fishing business failed to work out as Whitcher wanted. The new leases in the New Brunswick salmon fishery would precipitate a court case that weakened federal jurisdiction in freshwater fisheries. Apart from certain river and shore fisheries, leases never become common on the Atlantic, and licences, where they existed on that coast, remained mainly a formality for a century.

On the Pacific though, licences and leases soon became important in fishery management. And in the 1960's and 1970's, licences would reassert themselves on the Atlantic, as "limited entry" became the fundamental tool of fishery management.

Whitcher's rationale for licences and leases

In the 1873 annual report, Whitcher gave a long exposition on licences and leases, worth reproducing at length because it addresses a central issue in fisheries management.

It is respectfully suggested that the system of leasing and licensing fishery privileges under the Fisheries Act, already introduced in the Provinces of Nova Scotia and New Brunswick, be now further extended conformably with the practice existing in the Provinces of Ontario and Quebec.

In these Provinces the system has been brought gradually into operation since the year 1856. It is confined almost exclusively to salmon and sea-trout fishery in Quebec, and to white fish and salmon trout fishery in Ontario. There is still open a large field for its extension, without encroaching on the deep sea fisheries for cod, halibut, mackerel, herring and other scale fishes. At the date of Confederation a similar principle existed in Nova Scotia and New Brunswick, but was limited in its application to very few instances. The Provincial Government in Nova Scotia had issued one lease of oyster beds; and the Government of New Brunswick had granted one lease of salmon fishery, at nominal rents. Besides these dues on leases a small tax on salmon nets was payable to the municipal authorities; and under an Imperial grant of fishery rights in St. John Harbour, the civic corporation rented fishing berths to the local fishermen by lottery, realizing about $2,500 per annum. Also fishery rents of $598.78 per annum were paid by the salmon fishers on the Naval Reserve at Portage Island, N.B., under the title of fishing "lots" from the Admiralty, which rents were applied to local purposes. Since Confederation some special licenses for trapnets were issued in Nova Scotia, and in New Brunswick several season licenses for salmon fishing with nets, and a few leases for salmon angling have been granted.

The Fisheries Act evidently contemplates the system of granting titles for fishing privileges as a basis of administration. Certain of its provisions are predicated on the supposition that leasing and licensing would become general, providing always for necessary exceptions as to legal titles, prior occupancy and preferential claims.

It is unnecessary, after several years of its beneficial operation, even though but partially carried out, to explain at length its advantages. Primarily, it systematizes the fishing business, and it also induces private expenditure both in guarding and improving the streams, which outlay would otherwise require to be defrayed from public funds. Secondarily, it promotes investment of capital, and gives permanence and security to fishing industries, enhancing the value of fishing privileges to both individual fishermen and the public, which hitherto had but a fitful existence and were fast becoming altogether unproductive. Revenue is only an incident and not a main object.

There were reasons of state for not superseding the Provincial Fishery Laws in Nova Scotia and New Brunswick by Dominion legislation when the Maritime Provinces were confederated. Like reasons have since prevented anything further being done beyond merely introducing the leasing and licensing principle into those provinces in a few instances where precedents had been set by the Provincial Governments. This department essayed on two occasions to advance another step, but made no progress....

Legislation is not required; no assimilation of laws is requisite. All that is necessary is, by departmental action, to proceed with leasing and licensing fishery stations in those provinces just as has been done in Ontario and Quebec. But, as the matter has been considered in the light of a "policy," it may be deemed advisable to confirm the proposed action by an Order in Council, in the form of a Fishery Regulation, prohibiting such kinds of fishing as it is intended to lease or license, except under authority of leases or licenses. This is the same course as was pursued for Ontario and Quebec.

It may be advisable to act first on the numerous applications which are fyled, and in other instances where no adverse circumstances of conflicting demands exist. Attention should be directed to carrying out this system with every regard for the obvious desirability of enlisting the sympathies of the public and promoting the truest interests of the fishermen. There should be a thorough examination into each case; and the greatest possible care and precautions should be observed in order to avoid doing violence to the prejudices, or injury to the position and interests of persons affected thereby. Scrupulous regard will require to be paid to priority of occupation and recognized user. A careful distinction must be observed between the deep-sea and inland and the estuary and river fishings. These latter should alone, in my humble opinion, be subject (for the present at least) to the system of occupation under lease or license.

The undersigned considers it undesirable to anticipate the production of direct revenue from fishery rentals, the rates of which are for the most part nominal. Any system of regulation and economic use of fishing privileges under titles may be more profitable adapted as an auxiliary to protection of inland fisheries, and to enhance their productive value. It is not improbable, however, that in due course of time sufficient funds may be derived to render the service self-sustaining.

In the 1876 annual report, Whitcher brought forward further arguments: "Besides securing fishermen in the exclusive enjoyment of certain fishing privileges and obviating all disputes, the plan of leasing or licensing enables us to dispense with the numberless and cumbrous regulations which at present exist, as conditions could be embodied in the leases or licenses equivalent to prohibitory on directive regulations." And rather than interfering with holders of fishing stations, such defined privileges would render permanent the occupations that were now temporary and questionable.

Except that they are better written, Whitcher's words could have appeared in any number of studies in the 1970's and 1980's, when many people rediscovered the concepts of limited entry, quasi-property rights, and "resource rent" or "cost-recovery."

Simple regulations represent deep powers

As of 1886, one could still summarize the Dominion's main fishery laws on a single page. Overtopping various local regulations, they set closed seasons and weekly closed times, forbade net fishing in "public waters" (evidently crown land) except under lease or licence, controlled net sizes and barriers and the use of explosives or poisons, and provided for fish-passages at mill-dams.

If the list shown (on next page) seems simple, still the powers ran deep. The fisheries branch already used almost every method of regulation that would appear later. It could control who would fish, how, when, where, and for what. The department already granted licences for both common-property and quasi-private fisheries. It already applied what we now call the user-pay principle, with a landings tax on salmon. And the law protected Canadian waterways from blockages and deleterious substances. That being said, regulation was tempered by caution, as in Whitcher's early reluctance to intervene in the deep-sea fishery.

The 1878 Annual Report carried illustrations of the Dominion Hatchery at Newcastle, Ontario, Wilmot's original site.

Wilmot promotes hatcheries

In the 1867–1914 period, fish culture loomed large in departmental thinking. As noted earlier, leases and licences in the Province of Canada had been partly linked to the idea of fish culture. Now hatcheries became prominent, and this new effort owed much to one man's work.

After Richard Nettle's early experiments in the 1850's, Samuel Wilmot took an interest in the Ontario salmon at Wilmot's Creek, near Newcastle, Ontario. (Ontario salmon looked identical to Atlantic salmon but apparently spent their lives in Lake Ontario.) The fall spawning runs up the stream, past Wilmot's farmhouse, were declining. In the early 1860's he built troughs in his basement to raise salmon. First he collected eggs from the stream; then he switched to stripping eggs and milt from mature salmon and mixing them in a pail. This artificial method produced a much

THE FISHERY LAWS OF THE DOMINION.
TABLE OF CLOSE SEASONS ON 1st JANUARY, 1886.

Kinds of Fish.	Ontario.	Quebec.	Nova Scotia.	New Brunswick.	P. E. Island.
Salmon (net fishing).........................	....	Aug. 1 to May 1.	Aug. 15 to March 1.	Aug. 15 to March 1.	....
do (angling)	....	Sept. 1 to May 1.	Sept. 15 to Feb. 1.	Sept. 15 to Feb. 1.	....
do do Ristigouche River..	....	Aug. 15 to May 1.	....	Aug. 15 to May 1.	....
Speckled Trout (*Salmo Fontinalis*) ...	Sept. 15 to May 1.	Oct. 1 to Jan. 1.	....	....	Oct. 1 to Dec. 1.
Large Grey Trout, Lunge and Win-ninish.		Oct. 15 to Dec. 1.	....	....	....
Pickerel (Doré)	April 15 to May 15.	April 15 to May 15.	....	....	....
Bass and Maskinongé	April 15 to June 15.	April 15 to June 15.	....	....	....
Whitefish and Salmon Trout...............	Nov. 1 to Nov. 30.	....	....	....	....
Whitefish	....	Nov. 10 to Dec. 1.	....	....	....
Sea Bass	....	....	....	March 1 to Oct. 1.	....
Smelts	....	....	April 15 to May 15.	April 15 to May 15.	....
		Bag net fishing prohibited, except under license.			
Lobsters	....	Aug. 20 to April 20.	Aug. 1 to April 1. (West coast) Aug. 20 to April 20. (North coast)	Aug. 1 to April 1. (South coast) Aug. 20 to April 20. (North coast)	Aug. 20 to April 20.
Sturgeon	....	....	....	Aug. 31 to May 1.	....
Oysters	....	June 1 to Sept. 15.	June 1 to Sept. 15.	June 1 to Sept. 15.	June 1 to Sept. 15.

NOTE.—The fishery laws only partially extended to British Columbia and Manitoba. Close seasons in the latter province are : Whitefish, from 20th October to 1st November; and speckled trout, from 1st October to 1st January.

SYNOPSIS OF FISHERY LAWS.

Net fishing of any kind is prohibited in public waters, except under leases or licenses.

The size of nets is regulated so as to prevent the killing of young fish. Nets cannot be set or seines used so as to bar channels or bays.

A general weekly close time is provided in addition to special close seasons.

The use of explosive or poisonous substances for taking fish is illegal.

Mill-dams must be provided with efficient fish-passes. Models or drawings will be furnished by the Department on application.

The above enactments and close seasons are supplemented in special cases, under authority of the Fisheries Act, by a total prohibition of fishing for stated periods.

Fishery laws of the Dominion, from 1886 Annual Report.

higher proportion of fertilized eggs. Wilmot became convinced that hatcheries would save the salmon.

He failed to get permission to ranch salmon, that is, to release them to grow and then to be the exclusive harvester for a portion of the Lake Ontario shore. The government of Upper Canada, however, arranged with him to operate a hatchery on Wilmot Creek. With spawn taken in 1866, Wilmot produced nearly 15,000 fry. The government expanded his hatchery, which over the next half-century, it would produce millions of fish of various species.

In New Brunswick, an early experiment with a private hatchery on the Miramichi ran into complications, partly stemming from American involvement. The department revoked the licence.[29] But the department itself, adopting a proposal by Wilmot, was by 1873 building hatcheries on the Restigouche and Miramichi and at Gaspé. In 1875, special P.E.I. regulations set aside the Midgell, Morell, Dunk, and Winter rivers for natural and artificial propagation of salmon.

In 1876, the department appointed Wilmot Superintendent of Fish Culture for Canada, which post he held until 1895. Hatcheries came into being for many species, but with the emphasis on salmon. The department would invest large amounts of money and manpower in this effort, and for decades hatchery officials would write glowing, self-hypnotized reports about progress in fish culture. But measurable results were few. In New Brunswick, salmon catches peaked in the 1870's, perhaps because of additional fishing effort, then declined until the 1930's.[30] Some rivers, despite hatcheries, would lose all their salmon, as had already happened for many New England streams.

Environmental changes far outweighed any benefits from the hatcheries. Deforestation shrank some streams and subjected them to greater extremes of temperature. Loss of food from the riverbanks, silting, lower water levels, dams, and fishing pressure all conspired to weaken the salmon.

Although Whitcher encouraged hatcheries to compensate for other losses, he saw their limits. In 1874, he wrote that

> while it is true that fifty or sixty years ago, almost all the considerable streams in Ontario, Nova Scotia, New Brunswick, and parts of Quebec were resorted to by anadromous fishes, it is also true that the conditions of many of them have undergone a total change. The forest has been cleared along their banks and thinned out to such an extent even to their head waters, that the snows of winter and the rains of summer are much more rapidly evaporated, and what were once full streams flowing through virgin forests, are now, in the hot season, mere rivulets meandering through meadows and cultivated fields. The once secluded spawning beds are now crossed and recrossed by herds of grazing cattle, and often for miles but a mere thread of water trickles over the bars and gravel beds. While the settlement of the country has produced these changes in our rivers, the erection of mills and dams on most of them, and the prosecution of lumbering operations on all of them, have worked still greater changes. Most of these dams were erected many years ago, before any laws were enacted for the preservation of fish, and the consequence is, that a very large number of the smaller rivers have been deserted by their finny denizens, and it is very doubtful whether, under these altered circumstances, they can ever be restored, even were the costly experiment of restocking them by artificial culture tried.[31]

Hatcheries were indeed tried, but hopes of them restoring abundance would gradually fade.

Canada's fishery in 1880

The fisheries branch was dealing with a sizeable industry, worth a reported $14.5 million in 1880, and perhaps more (the fishery was often thought to be undervalued because of poor statistics). Canning was bringing a new industrialism to the fishery, with salmon factories starting up on the Pacific and sardine factories and lobster canneries on the Atlantic. Although the Canadian sea fishery was bigger than ever, growth in cod was levelling off.

Nova Scotia's fishery in 1880 had a product value of $6.3 million. Cod was most valuable at nearly $2.5 million; barrelled mackerel ($1.3 million) came second, followed by canned lobsters ($612,000), barrelled herring, haddock, fish oil, hake, pollock, and other products. The province employed 731 vessels with 6,748 men and 11,210 boats with 22,798 men. The main trade was, of course, in salted, dried fish; some was also wet-salted or smoked. The industry also sold fresh salmon in ice as well as smoked, canned, and barrelled; and there was a small fresh fish trade at Digby and Halifax.

New Brunswick's fishery was worth more than $2.7 million. The most valuable product was canned lobsters ($710,000), followed by barrelled herring, salt cod, hake, barrelled mackerel, fresh salmon in ice, smoked herring, sardines, and other products. The New Brunswick industry employed 220 vessels with 1,175 men and 4,219 boats with 7,391 men.

Prince Edward Island's fishery yielded more than $1.6 million. Canned lobsters were even with New Brunswick's at $710,000, followed by barrelled mackerel, cod, barrelled herring, barrelled oysters ($61,000), and other products. The P.E.I. fleet had 32 vessels with 161 men and 1,383 boats with 3,864 men.

Quebec's fishery in 1880 was worth more than $2.6 million, almost equal to New Brunswick's. Cod came to about $1.5 million; lobster, to $76,400; herring, to $73,800; sealskins, to $25,600; and whale oil, to $5,400. The fleet had 166 vessels with 843 sailors and 3,398 fishing boats with 10,692 fishermen and shoremen.

Ontario's fisheries in 1880 yielded products worth about $445,000. Whitefish were most valuable, followed by trout, herrings, pickerel, and sturgeon. (Overfishing would soon deplete most of these valuable species, with others coming to the front.) The Ontario fishery used 18 vessels with 54 men and 865 boats with 2,076 men.

British Columbia's fisheries in 1880 were worth more than $713,000, less than five per cent of the Canadian marketed value. Canned salmon led at $401,000, followed by fur sealskins ($163,000), dogfish–seal–porpoise, and various other products including small amounts of halibut (fresh) and herring. The B.C. fishery used 4 steamers, 10 schooners, and 317 boats, along with 93 cedar canoes in the sealing fleet.

All told, Canada's fishery in 1880 employed 8,757 men in 1,181 vessels and 52,577 men in 25,266 boats, for a total of 60,657 men and 25,266 fishing craft. Departmental reports still gave no figures for the many thousands who worked onshore in the fishing industry, or for the freshwater fisheries on the Prairies. Although Canada's vessel fleet had grown to nearly 1,200, it still lagged behind that of New England. As of 1886, that region employed about 18,000 men in 1,956 vessels, many of which still fished off Canada. There were 1,530 vessels in "food fish," 215 in shellfish and lobster, 177 in whales and seals, and 34 in menhaden.

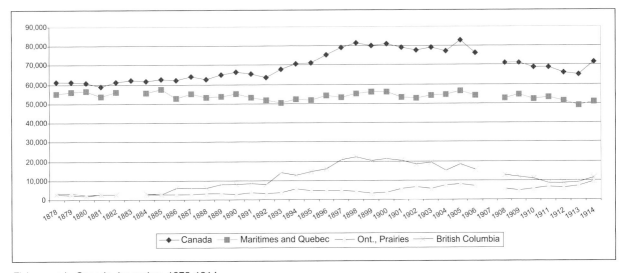

Fishermen in Canada, by region, 1878-1914.

Foreign frictions return

United States ends fisheries free trade

In the disputes and negotiations leading to the 1871 Treaty of Washington, Canada had won only fisheries rather than general free trade. In the 1880's even that vanished, because of anti-Canadian resentment in New England. American fishermen complained about "unfair" Canadian competition under fisheries reci-

procity. They also got worked up about the Halifax Award of 1877. New Englanders thought $5.5 million an outrageously high price for fishing privileges, on top of which, the money was subsidizing their competitors.

At the same time, the American interest in Canadian and Newfoundland nearshore fish was dwindling. The number of mackerel vessels operating in the Gulf of St. Lawrence fell from 254 in 1873 to only 1 in 1882. American groundfish fishermen were continuing to abandon shore-drying, turning to salt and fresh fishing on the banks.[32]

In 1883, a joint resolution of the Senate and the House of Representatives gave notice that the fishing articles of the Treaty of Washington would end, effective 1885.[33] In 1887, Spencer Baird, the U.S. Commissioner of Fish and Fisheries, summed up the reasons for New England's loss of interest in B.N.A. waters. In essence, although a large American fleet still came north, there were now fewer Canadian fish within the three-mile limit and less American need of them, either for food or for bait.

The halibut fishery near the provinces had dwindled. Since 1875, American fishermen had been finding halibut in deep water only. This in turn lessened the need for bait from B.N.A. inshore waters. In the offshore halibut fishery, vessels needed only a little bait, one or two days' worth; they then used "refuse fish" or small halibut for bait.

Offshore cod vessels from Gloucester, using trawls (that is, longlines) continued to use fresh bait from the provinces. Many of the Gloucester crews were Canadians, and they urged the owners to get fresh bait in Canada, largely because they had relatives there. But other cod vessels tended to use handlines and salt bait, which they could bring from home. Inshore in New England, the catch by gillnets, which needed no bait, was increasing. This further cut down the need for frozen bait herring from New Brunswick, Nova Scotia, and Newfoundland.

About all the Americans now wanted within the three-mile limit, Baird said, was mackerel. But there again, they now had bigger vessels and purse-seines with which to fish their own mackerel offshore. Most American fishermen fished no closer than 25–30 miles to the Canadian shore, and they were most likely to be 100–200 miles from shore. Recently, they had been catching more mackerel off the United States itself.

The 1818 Convention still permitted U.S. vessels to fish freely along the Treaty Shore, including the Magdalens, part of Newfoundland, the Quebec North Shore, and Labrador. But in these places, Baird said, the fishery had dwindled to a low level.

Bait from Canada now had no vital importance, especially since the agitation about Canadian bait had spurred the home-caught bait fishery in Maine and Massachusetts to increase. Writing two years after the fact, Baird concluded that abrogating the fishery provisions of the Treaty of Washington had had little effect on U.S. fisheries.[34]

Salt fish slows down

But the loss of the treaty did affect Canadian fishermen. The Americans imposed heavy duties on fish products. Market trends themselves—the growing use of fresh fish and meat—hurt the trade in dried cod and other salted products. As America began producing sugar beets, the prosperity of West Indies planters declined, further weakening the market for B.N.A. saltfish. The accompanying decline of wooden shipbuilding and shipping made it harder to earn a living by combining fishing with other trades.

The American abrogation helped weaken the coastal economy relative to the rest of Canada. The fishery remained a great employer. With its manifold linkages to boatbuilders, fish dealers, equipment makers, and local tradesmen of all sorts, it supported most of the coast. But as the linked marine economy declined, the fishery was taking on more clearly a role that became traditional: the catch basin, the employer of last resort.

American tensions return

The end of fisheries reciprocity in 1885 would reintroduce a degree of international tension on the Atlantic. In the 1880's and 1890's, Canada also faced foreign challenges on the Great Lakes and the Pacific and Arctic coasts, with the Department of Marine and Fisheries on the front lines.

Sovereign matters involved ministers. With Macdonald's Conservatives still in power, A.W. MacLelan of Nova Scotia served as Minister of Marine and Fisheries from 1882 to 1885. George E. Foster of New Brunswick took over for the years 1885–1888. Foster entered politics mainly by chance. Prime Minister Sir John A. Macdonald happened to pass by when Foster was giving a public lecture on temperance. Though no abstainer himself, Macdonald found Foster's speaking impressive, and ordered, "Get him into politics."[35] Foster entered Parliament a few months later, and soon became Minister. It was the start of a distinguished political career, including many years as Minister of Finance. At the outset, Foster had his hands full with fisheries.

The patrol fleet had dwindled with the Treaty of Washington to only one vessel in 1880, and that for the Gulf and Lower St. Lawrence, the sailing vessel *La Canadienne*. A 154-foot cruiser of the same name took over in 1881, under William Wakeham.

As fisheries reciprocity ended in 1885, in the words of the 1886 annual report, "no other course was then left the Canadian government but to adopt measures for the protection of its rights." Canada re-imposed strict enforcement against American fishing vessels. Fisheries added eight other vessels, bringing the fleet to nine, in a new Fisheries Protection Service.[36]

A large U.S. fleet still came north. Although they no longer had privileges inside three miles in the Maritimes and Gaspé, nothing stopped them fishing on the Treaty Shore at the Magdalens, Newfoundland, and Labrador. Off the Maritimes, they could fish just outside the three-mile zone, and some would try to slip inside it.

New patrol fleet seizes U.S. vessels

Outside the Treaty Shore, by the terms of the 1818 Convention, Americans could enter Canadian waters only for shelter, wood, and water. There would be no fishing inside the zone, no transhipping crews or fish, and no buying of bait. To enforce the rules, minister George Foster told his fishery patrols that they had "full authority." While displaying a conciliatory approach, "you will accost every foreign fishing vessel." In 1886, the protection service made some 700 boardings, and in 1887, made more than 1,300 boardings. The Canadians seized several U.S. vessels.[37]

Americans raised a hue and cry. Politicians took up the cause against Canadian "brutality" and "inhumanity." The Canadian government had even penalized its own fishermen for supplying U.S. vessels in waters outside the three-mile limit. What if American fishermen wanted to enter Canadian ports just to buy a newspaper? What if they wanted to bury their dead?[38]

If American protests sounded exaggerated, still some Canadian officers may have overdone the enforcement. After a U.S. vessel bought food in Prince Edward Island, Canadian patrol officers reportedly gave an emetic to an American sailor to make him disgorge his illegal meal.[39]

The United States retaliated with a "Non-Intercourse Act," allowing the government to bar Canadian vessels from U.S. ports and to bar Canadian fish or anything else from U.S. markets. But Canadians had at least some sympathy in the United States. The Boston Fish Bureau, apparently representing importers, in 1886 asked for renewed Reciprocity. The Bureau said that the complaints by U.S. vessel owners that Canadians were hurting them were a pretence; it was Canadians who manned the U.S. vessels anyway.

The question of Canadians in the U.S. fleet came up several times. The high out-migration from the Maritimes to New England included many mariners, such as the famous schooner-builder Donald MacKay, and Joshua Slocum, the first man to sail alone around the world. Some said that U.S. operators brought in Canadian fishermen because they could get them at lower wages. But it appears that Americans manned most of the fleet. An 1887 report said that of 14,240 fishermen in the U.S. North Atlantic fleet, 78 per cent were American.[40]

"Modus vivendi" calms waters

U.S. and Canadian negotiators tried in 1888 to replace the Treaty of Washington with a modified arrangement. Massachusetts protests defeated it.[41] That same year, however, the two sides and Great Britain arrived at a more limited "modus vivendi." Though still barred from fishing within three miles (except on the Treaty Shore), American vessels could now come into Canadian or Newfoundland ports for supplies, repairs, or transhipment of fish or crews, on payment of a yearly licence fee.

The modus vivendi came into force pending negotiations for a new treaty. But the two sides never concluded the new treaty.[42] The modus vivendi of paying licence fees for port privileges was to last until the

1920's. The arrangement calmed the seas. By 1890, there was only one seizure, of the schooner *Davy Crockett*, for fishing from dories within the three-mile limit in the Gulf.

Despite its new port privileges, the U.S. fleet was making less use of Canadian waters, instead sticking closer to home. At LaHave Bank, off southwest Nova Scotia, New Englanders developed an important fishery for fresh haddock. In general, however, the New England fishery on Nova Scotia's banks and the Grand Banks was on the downswing, declining from 1880's levels of 300 vessels and up to only 60 vessels in 1910.[43]

Canadians had hoped the modus vivendi would bring better trade relations. But the United States in the 1890's increased its general tariffs, hurting the Canadian economy. And the U.S. takeover of Puerto Rico in 1898 cost the Canadians a duty-free market.

Pacific fur seals bring new dispute

By the time of the 1888 modus vivendi on the Atlantic, another conflict had erupted on the Pacific, far north in the Bering Sea, when Canadians encroached on what the Americans considered their own resource.

After the 1867 Alaska Purchase, Americans had begun a land-based seal harvest at the Pribilof Island rookeries. This was the United States's first big benefit from the purchase. In the 1880's, American vessels from San Francisco and Canadian ones from Victoria began cutting into the crop, taking seals from the same Pribilof-bred stock but using schooners in the open water. By 1886, 20 Canadian schooners employed 79 regular seamen and 380 Indian hunters. The vessels launched the Native people in canoes for daredevil ocean chases. As fishery Overseer Alexander C. Anderson described it in the 1880 annual report,

> It is only with the aid of the Indians of the West coast, expert in the management of canoes and habituated to this chase, that success is at present found to be attainable. These are hired upon shares, receiving one-third, I believe, of the produce of their chase or the equivalent in cash. Small schooners are equipped, on board of which the hunters are received, with their canoes, one of which is required for every two hunters. Lying-to off the banks the canoes, weather permitting, are speedily launched, and the seals, while sleeping on the surface, are cautiously approached. The spear only is employed, the head of which disengages itself from the shaft as soon as the prey is struck. To this barbed head a line is attached. If not killed outright the wounded victim is said to attack its pursuers with much ferocity, but a blow on the head from a club kept constantly in readiness soon terminates the unequal conflict. Of course accidents occasionally occur, and the whole scene is described as being very exciting to those who,

Part of Canadian sealing fleet laid up for the winter, Victoria harbour, B.C., 1891. (Library and Archives Canada, PA-110161, detail)

from the deck of the attending schooner, watch the progress of the chase.

Although their own countrymen also did pelagic sealing, Americans blamed Canadian sealers for a decline in their harvest. In 1886, American "revenue cutters" seized three Canadian sealing vessels on the high seas. British protests secured their release. The Americans pressed Great Britain and other countries to agree to an international closed season. Canadians protested that the proposed limitations would destroy the main part of their season, and might imply exclusive American rights in the Bering Sea.

The Americans seized more Canadian vessels in following years, while letting their own pelagic (that is, open-sea) sealers keep operating. U.S. authorities claimed sovereign control of a large part of the Bering Sea and north Pacific as part of the Alaska Territory and forbade the killing of fur-bearing animals without permission. Canadian authorities seized American sealers in retaliation.

Meanwhile the Conservatives were changing ministers. Charles Hibbert Tupper replaced Foster in 1888 and remained Minister of Marine and Fisheries until 1894. (He was the son of Charles Tupper, the Nova Scotia premier who had fought earlier fisheries battles and who in 1896 would serve briefly as Prime Minister.) The new minister took a firm line in the seal dispute.

In 1891, a British–American agreement closed the high-seas fishery to the Canadians, pending arbitra-tion. British and American cruisers expelled 41 vessels (averaging about 65 tons) from the Bering Sea. But in 1893, an international tribunal upheld the Canadian right to hunt seals in international waters, subject to some conservation restrictions. Congress delayed but finally paid $473,000 (U.S.) in compensation to Canadians for previous interference. A minority of the tribunal, however, supported the idea of American property rights in the seals on the high seas, since they were conceived, born, and reared on American soil.

The Americans ended their own pelagic sealing in 1897, pursuing only a land-based hunt. But Canadians kept sealing, despite American opposition and harassment. For example, the Victoria Company sent 16 vessels north in 1908. A tortuous dispute involving Canada, the United States, and Great Britain dragged on. Meanwhile, with Russia and Japan also in the fishery, seal stocks showed the strain of over-exploitation.

In 1911, an international conference, with W.A. Found, a top official of Marine and Fisheries advising the British plenipotentiaries, buried the lingering dispute. It was agreed that the United States alone would conduct the fur seal hunt, on land at the Pribilof Islands. The other countries involved—Canada, Japan, and Russia—would receive a share of the proceeds. Canada got 15 per cent, and Canadian fishermen got compensation for previous losses. This arrangement brought peace.[44]

100

The *Petrel*. (Courtesy of Bernard Collin, Canadian Coast Guard)

Gunfire on the Great Lakes

Meanwhile, the Department of Marine and Fisheries from 1888 patrolled the Great Lakes by ship.[45] Americans were a multi-element threat: fishing on the Canadian side of the border; and also controlling much of the fishery by Canadians, especially after American tariffs from 1890 virtually excluded Canadian fish dealers from the U.S. market.

Charles Tupper and his 1894 successor John Costigan stepped up enforcement. On the Great Lakes as elsewhere, Dominion steamers combined fishery patrols with customs and other duties. In the mid-1890's, American barges pulled by tugs were dumping garbage from Detroit on the Canadian side. In 1895, Captain Edward Dunn of the *Petrel* fired a shot across the bow of a garbage tug. The U.S. offenders received fines in Canadian courts. An American inquiry later vindicated the Canadian action. Another time, Dunn arrested a large party of wealthy Americans on a sport-fishing excursion, causing consternation mixed with amusement on the American side.[46]

In 1903, patrolling against American poachers, the *Petrel* under Captain Dunn fired rifle shots against an American vessel, the *Silver Spray*, which escaped. In the same episode, the *Petrel* seized three other tugs and confiscated hundreds of nets. In 1905, the new patrol vessel *Vigilant*, under Dunn, accidentally capsized the American vessel *Grace M.* Two fishermen drowned; Dunn rescued three others.[47]

Fishery frictions run coast to coast

Apart from the bigger Canada–U.S. disputes, lesser frictions cropped up in the late 19th and early 20th centuries. Often they involved Canadians pressing for conservation in international waters while the Americans took a freer approach.

As early as 1872, Minister Peter Mitchell called for international cooperation in conservation; otherwise,

U.S. fishing, as at border areas in Lake Erie and Lake Huron, could override the effect of Canadian regulations. Two years later, Canada proposed a reciprocity treaty that would among other things deal with Great Lakes fisheries. This idea went nowhere.[48] Problems continued here and elsewhere. American trapnets on the Great Lakes became more and more common. In 1891, the department's annual report complained of U.S. overfishing, and remarked on the superiority of Canadian regulations.

A prophetic sentence in that report noted that the jurisdiction of individual states over U.S. fisheries made it difficult to get any regulations. This would happen over and over. Proposed conservation agreements would threaten some local concern, and since the U.S. had no strong national fisheries agency, local interests usually prevailed with American politicians.

The Canadians in the early 1890's wanted U.S. action to reduce pollution of the upper St. John River, where American mills outnumbered Canadian by ten to one; to restrict pound-nets in the Great Lakes (where Canadian officials said their own conservation measures let Canadian fishermen catch as much as the Americans with only one-quarter as many nets); to control the fishery for Fraser River salmon; and to control purse-seining for mackerel. In 1892, a Canada–U.S. conference on fisheries took place, with U.S. delegates expressing admiration for Canadian regulations. A joint commission in 1893, consisting of William Wakeham for Canada and Richard Rathbun for the United States, began examining fisheries and pollution in boundary waters. In 1897, the commission published an extensive report on "the Preservation of the Fisheries in Waters Contiguous to the United States and Canada" and suggested corrective measures. The Canadian department hailed this report. The U.S. Congress refused to approve it.[49]

On the Pacific, more American canneries were now taking Fraser River fish. American fishermen intercepted homing salmon as they passed through American waters on their way to spawn in the Fraser system. As a 1905–1907 federal royal commission noted, the American expansion had put Fraser stocks under a strain.

While Canada had disallowed the purse-seine in British Columbia, the U.S. permitted fishermen to use trapnets, purse-seines, and even fishwheels, a marvellously efficient method, with the river current turning a wheel fitted with scoops that took up the fish. When the Americans began to take the lion's share of salmon from Canada's Fraser River, the fisheries service in 1904 decided to let Canadians equalize their chances by using trapnets, drag-seines, and purse-seines.[50] Thus the Americans, in a way, fostered use of a powerful method that might otherwise have stayed under ban.

In 1908, the British, Canadian, and American governments negotiated a proposed treaty governing fisheries in shared boundary waters coast to coast, from the Fraser to Passamaquoddy Bay. The tentative agreement called for an International Fisheries Commission to draft regulations. But the treaty never went into effect, largely because of resistance from the U.S. industry.[51] Other attempts to negotiate treaties on Fraser fishing, in 1919, 1921, and 1930, would fail, before final success.

Canada and the United States made more progress with water than with fish. In 1909, the Boundary Waters Treaty with the United States set up the International Joint Commission (I.J.C.), with three commissioners from each side, to deal with boundary waters' protection, apportionment, and development (for example, through hydro power). The I.J.C. began work in 1912. It has reported and made recommendations on scores of issues, such as diversions of waters. It is generally considered to work well.

Fisheries Protection Service lays keel for navy

In the run-up to the 1871 Treaty of Washington and again in the 1890's, the fisheries patrol service had protected sovereignty. In the early 1900's, the patrol service became the nucleus of the Canadian navy.

Ministers changed in the meantime. A New Brunswicker, John Costigan, known as a spokesman for Irish Roman Catholics in Canada, served in 1894–1896. In Wilfrid Laurier's new Liberal government, Louis H. Davies of Prince Edward Island held the Marine and Fisheries portfolio for several years, until his appointment to the Supreme Court. James Sutherland of Ontario in 1902 took over the portfolio for ten months. A Quebec representative, J. Raymond Prefontaine, became Minister from November 1902 until his death in 1905. Prefontaine was the first Minister to visit British Columbia.

Laurier himself took over as Minister for two months, then, in February 1906, appointed his protégé Louis Philippe Brodeur, who served until 1911. During Brodeur's tenure, Parliament in 1908 set up the Standing Committee on Marine and Fisheries. Brodeur would be a key figure in creating the Canadian navy.

While Canadian national consciousness was growing, so was Great Britain's feeling of imperial destiny. Little Englandism had waned; British pride in the Empire was waxing strong. Competing with the French and German empires, the British renewed their interest in sea power. The idea arose that Canada and the other colonies should give money to keep the Royal Navy supreme at sea.

In Canada, opinions were mixed. The British government, the many imperialists in Canada, and the Conservative opposition all wanted Prime Minister Laurier to help fund the Royal Navy. Others began to think Canada could launch a homegrown navy for protection on the coasts and Great Lakes. The fisheries patrol fleet provided evidence.

Back in 1886 when the Fisheries Protection Service had expanded, the government had pressed existing Canadian vessels into service. In the 1890's, the fleet gained force as the government built and bought five armed vessels, all over 100 feet in length. In 1900, the protection service still had only nine vessels, but they were better ones. From 1894, the crews had uniforms.[52]

Concern remained about possible American encroachments, in light of the continued American presence in the Arctic and at Newfoundland, occasional Great Lakes incidents, and differences over the Alaska boundary. A controversial tribunal in 1903 awarded the United States its present territory in the Alaska panhandle, running down the northwest coast of British Columbia. Part of the judgement set the so-called A–B Line, running between two American points of land at the southern end of the panhandle. (Fishery disputes related to American claims to a territorial sea running off from the A–B Line have occurred off and on since then.)

Patrol fleet turns military

After a 1902 colonial conference on defence, the Fisheries Protection Service took on more naval trappings. That same year, Marine and Fisheries became responsible for Arctic sovereignty. In 1904, the department acquired three more armed vessels, including the 176-foot *Vigilant*, considered to be the first modern warship built in Canada, and the 200-foot *Canada*. With 73 officers and crew, the *Canada* operated as a man-of-war and began training a naval militia. By 1909–1910, the protection fleet had 13 vessels and 255 men. Some of the other cruisers were small warships with ram bows and cannons.

The department in 1904 under Minister Prefontaine had drafted a Naval Militia Bill that would have created a larger force, up to 800 officers and men, for Marine and Fisheries. This never went forward, Laurier preferring to avoid controversy at the time. Meanwhile, Laurier's government discouraged the idea of Canadian contributions to the imperial navy. In 1907, at another colonial conference on defence, Marine and Fisheries Minister Brodeur told British and colonial representa-

The 176-foot *Vigilant*, built at Polson Iron Works, Toronto.

Crew of the Fisheries Protection Service cruiser *Canada*. (Library and Archives Canada, PA-123950)

tives that Canada had already taken a reasonable share of naval expenditures through the Fisheries Protection Service and other works.

That same year, a much-noted around-the-world cruise by 16 American battleships helped reawaken Canadian concerns about naval weakness. The department in 1908 reorganized itself on more naval lines. Charles E. Kingsmill, a Canadian captain in the Royal Navy, left that post to become Director of Canada's Marine Service. George Desbarats, former director of the government shipyard at Sorel, Quebec, became deputy minister. Parliament in 1909 approved a resolution that supported the build-up of a naval militia under Marine and Fisheries.

Meanwhile, in Britain concern was growing about the increasing strength of the German navy, as reflected in the "Dreadnought Crisis" of 1909. In Canada, this both heightened the consciousness and confused the perception of the naval issue. The Dominion was already strengthening its fisheries patrol fleet, with an eye to its use as a naval militia. Now some voices clamoured for a more forceful and visible step, whether a financial contribution to the Royal Navy or a bigger Canadian navy that was more than a militia and paid more attention to imperial needs, rather than just Canada's coastal defence.

After weighing the factors, Laurier's government finally decided on a more visible Canadian navy. It was Brodeur, Minister of Marine and Fisheries, who introduced the Naval Services Act in Parliament, establishing the Royal Canadian Navy as of May 4, 1910. On June 3, 1910, Brodeur became the first Minister of the Naval Service, while retaining Marine and Fisheries. Deputy minister George J. Desbarats also did double duty until, in 1911, Alexander Johnston replaced him at Marine and Fisheries, and Charles Kingsmill, the department's Director of the Marine Service, became Director of the navy.

In 1911, Brodeur left his two portfolios for the Supreme Court of Canada. Rodolphe Lemieux, another Quebec representative, served only two months before the Laurier government lost power in October, defeated by its renewed attempt at reciprocity with the United States and by its creation of the navy. French Canadian nationalists and federal Conservatives had joined in protesting Laurier's naval policy—the former, for doing too much; the latter, for doing too little.

In Robert Borden's new Conservative government, Sir John D. Hazen, formerly premier of New Brunswick, became Minister of Marine and Fisheries and, ex officio, Minister of the Naval Service. The navy survived the Borden government's initial lack of sympathy, but entered the First World War with only two ships of its own.[53]

Marine and Fisheries helps secure Canadian Arctic

The Department of Marine and Fisheries also asserted Canada's claims in the Arctic. These were by no means certain; other nations had mounted a stronger presence in the north.

American whalers were fishing the Arctic, both east and west. As steam engines became common after

1850, whalers also gained efficiency from the shoulder gun, firing timed charges. Then, by 1870, came the bow-mounted harpoon gun, firing explosive missiles. Demand was still good; although kerosene and petroleum had now replaced whaleoil lamps, the animals still supplied lubricating oil, baleen for corset stays, and other materials. Whalers pushed further into the eastern Arctic. In 1860, New England whalers opened the northwest section of Hudson Bay. In 1874, the United States asked permission for whaling and mining at Baffin Island. The perturbation over this request helped bring about Britain's 1880 transfer of her Arctic claims of sovereignty to Canada.

Gordon, Wakeham, Bernier lead northern missions

The transfer needed backing by a Canadian presence in the north. The Canadian government sent expeditions north in 1884, 1885, and 1886. Commanding the expeditions was Andrew Robertson Gordon, formerly a lieutenant in the Royal Navy, now an officer of the fisheries patrol service. For the first expedition Gordon chartered a sealing steamer from Newfoundland, the *Neptune*. After scouting around, Gordon recommended Churchill, a post of the Hudson's Bay Company (H.B.C.), as a railway terminal (which it became decades later, in 1931).

Gordon's expeditions were only a start, and the foreign presence in the Arctic was still strong. On the west, American whalers were penetrating further; from 1880 to about 1914, they operated in the Beaufort Sea. Meanwhile, whalers from overseas persisted in the eastern Arctic. As well, Norwegian and American explorers raised the foreign profile in the north.

Captain Joseph-Elzear Bernier (centre) and his crew at Winter Harbour, Melville Island, N.W.T., July 1, 1909. (Library and Archives Canada, C-1198)

The Canadian government undertook several Arctic expeditions at the turn of the century. These included work in Hudson Bay and Labrador by the *Diana* in 1897, under William Wakeham of the fisheries patrol service. On Cumberland Sound, at a Scottish whalers' depot, Wakeham hoisted the Union Jack and proclaimed Baffin's Land and everything adjacent as being now and always under British sovereignty.

In 1902, Marine and Fisheries took charge of Arctic sovereignty. In 1903, the Laurier government began sending North-West Mounted Police to Herschel Island off the Yukon coast and to Hudson Bay. They were to show the flag and keep law and order, sometimes a problem with drunken American whalers recruited by crimps. The *Neptune*, the same vessel used by Captain Gordon in the 1880's, transported Mounties to the north.

In 1906, Canada amended the Fisheries Act to proclaim Hudson Bay wholly territorial waters of Canada. Between 1904 and 1911, the government mounted several cruises by the vessel *Arctic*, under the mariner Joseph-Elzear Bernier. (As a young man, Bernier had served under Pierre Fortin on *La Canadienne*.) Conducting explorations and collecting customs dues, Captain Bernier reported back to the Minister of Marine and Fisheries as a Fishery Officer. Bernier noted after his 1906–1907 cruise that the number of American and Scottish whalers in Baffin and Hudson Bay, as high as 600–630 in earlier years, had declined to about 50, and these, during his visit, were finding no whales.

Bernier proclaimed sovereignty at various sites, and in 1909 unveiled a plaque on Melville Island officially claiming the Arctic islands for Canada. In 1910, he began issuing licences to the foreign whalers, thus affirming Canadian sovereignty. By the end of the decade Canada's claim to the Arctic lands appeared fairly secure. Further developments in the 1920's nailed it down.

Meanwhile, the bowhead whales in the Arctic were dwindling, and other materials were replacing baleen.

Harpoon gun on a British Columbia whaling steamer. (Library and Archives Canada, PA-40995, detail)

By the First World War, Arctic whaling had all but ceased.

Fisheries department wins, loses "separate" status

While the Department of Marine and Fisheries was fortifying Canada's national stature, the fisheries service was trying to assert its own importance. Back in 1881, when James Pope was Minister, Whitcher had reported that the fisheries branch employed nearly 700 "highly-intelligent" officers, overseeing a $16 million industry that exported 40 per cent of production. He made clear his displeasure about insufficient respect given to the fisheries side of the department in the annual report. The Fisheries Statements, he complained, were "sandwiched in between 'the report of the director of the time ball at St. John, N.B.,' and sub-reports on 'sundry marine hospitals'." Whitcher went on to praise his fishery officers, point out the importance of the fishing industry, and imply criticism of deputy minister William Smith.

Whatever the influence of Whitcher's outburst, in 1884 the government briefly divided Marine and Fisheries into two departments, each with its own deputy minister, but both reporting to the same Minister (MacLelan, then Foster, then Tupper). Annual reports carried the name of the Department of Fisheries. (It was under Fisheries, not Marine and Fisheries, that Canada in the mid-1880's built up the Fisheries Protection Service.) John Tilton became deputy minister of fisheries until he retired in 1891. The department reunited, and William Smith in 1892 again became deputy minister of Marine and Fisheries until he retired in 1896. F. Gourdeau then took over until 1909, followed by G.J. Desbarats.

Meanwhile, Whitcher was gone, having last signed the annual report as Commissioner of Fisheries in 1882. That title dropped out of sight during the years when Fisheries was a separate department, then returned in 1893.

Prince, Gordon become prominent

That year, the department recruited Edward E. Prince, then teaching zoology at St. Mungo's College, Glasgow, Scotland, to serve as Commissioner of Fisheries. When Samuel Wilmot retired in 1895, Prince also took over fish culture.

Prince became the dominant figure in fisheries management until the First World War. He took an interest in every aspect of the fisheries. He wrote paper after paper on fisheries management and biology. And he chaired many of the royal commissions that set the fishery regulations that still prevail.

Although he also took on the title of General Inspector, Prince never occupied himself greatly with day-to-day management. He recognized the need of licence limitation in the lobster fishery, but was never able to carry it through. He recognized underdevelop-

Professor Prince (left), ex-Chief Justice McGuire, and Dr. Euston Sisley at Big Quill Lake for the 1910 Alberta and Saskatchewan Fisheries Commission.

ment in the fisheries, but again, the department's work was less than thorough.

In 1909, Prince moved entirely out of executive duties to international and commission work. One gets the impression that in his later career Prince almost retreated to his study, leaving the day-to-day work to W.A. Found and others. Found, who joined the fisheries service in 1898 as a secretary, by 1911 took the newly designated post of Superintendent of Fisheries, and he would dominate after the war.

As for Prince, some of his successors tended to describe "the genial professor" in slightly irreverent tones. A.G. Huntsman, the scientist who most influenced fisheries research in the first part of the 20th century, downplayed Prince's abilities as a scientist, though noting his abilities as a naturalist. Still, Prince remains a fundamental figure in both management and science.

Another official shone, though more briefly, in the last part of the century: Andrew Robertson Gordon, the fisheries patrol officer who led Hudson Bay expeditions in the 1880's. Gordon continued in the patrol service, and became Commander of the Fishery Protection Fleet from 1891 until his death in 1893.[54] His reports displayed keen intelligence and vision.

In 1889, Gordon reported on the practicalities of setting up a tide and current survey. His work helped bring about annual tide tables and the 1893 creation of the Tidal and Current Service.[55] Following this and other early efforts, notably the Georgian Bay survey and subsequent work, the Canadian Hydrographic Service started officially in 1904.

Like Whitcher, Gordon was an advocate of licensing. He recommended after his third Hudson Bay expedition that the government begin issuing licences for foreign whalers. He intimated his objections that only the Hudson's Bay Company and American fishermen should be getting the dollars from northern resources. Gordon also supported renting out salmon rivers.[56]

As for salmon conservation, Gordon took what we now call an "ecosystem approach." He wrote that New

England's destruction of her anadromous resources was now complete. But "you cannot injure or destroy one fishery without affecting another." The shore fed the sea, said Gordon, but New England had already destroyed its shores, and Canada was starting to do the same. Woodcutting was making rivers unstable. The rivers needed timber around them, needed fishways, and needed the absence of pollution.

Gordon maintained that there should be no more taking of anadromous fish for bait; rather, there should be icehouses and refrigerators to preserve herring and other sea species. Some years later, the department started a bait-freezer program. Gordon like others quarrelled with the use of purse-seines, giving reasons in the annual report of 1890 why they should be banned. He perhaps influenced departmental thinking, for a ban followed in 1891.

The patrol captain suggested a Fisheries Loan Board, to help Canadians compete with the subsidized French and the highly capitalized Americans. Half a century later, provincial governments began setting up such loan boards. Gordon also recommended recording catches on statistical charts, ruled off into squares. Only in the 1930's and 1940's did the Fisheries Research Board catch up with this idea.

Finally, Gordon successfully pushed to create a Fisheries Intelligence Bureau, which started up in 1889 and lasted for many years. The Bureau collected information and reported to the industry on the location of fish schools off the coast, supplies of bait, and so on.[57]

Fears of depletion rise

By the 1880's, even before Prince and Gordon arrived, overfishing and depletion worried the department as never before. The lobster catch had peaked in the 1880's; now signs of decline were obvious. Newfoundland cod had dropped off enough to cause alarm in Canada. There were reports—some backed by statistics, some not—of decline in Great Lakes salmon and other lake species, whales, walrus, mackerel, stur-

Sturgeon of 1,020 pounds, New Westminster, B.C.
(Library and Archives Canada, PA-40976)

geon, and various inshore species. Commercial fisheries for some species had only recently developed, and already they seemed to be wearing out.[58]

Statistics were still poor. Abundance was declining by an unclear amount. Later, in the 20th century, biologists would sometimes scoff at old tales of fish so thick "you could walk across the rivers." Yet some of the old reports by observers did show prodigious amounts of fish, sometimes of enormous size, such as the 1,000-pound sturgeon that a B.C. Indian caught in 1880. The widespread complaints suggest that some inshore and river fisheries were indeed dropping sharply.

One sign of the times was the gradual falling-off in the use of beach-seines, which depend on fish coming right to the shore. The department itself placed beach-seines under strict regulation in many areas, in an attempt to conserve what fish remained near the beach. Rules were specific, as in a regulation requiring that anyone using a beach-seine at Peggy's Cove, Nova Scotia, had to live within five miles of the village.

Annual reports of the department were full of worries. For example, the Maritimes shad fishery had greatly declined. The inshore fishery at Grand Manan, New Brunswick, had deteriorated; dogfish got part of the blame. The oyster fishery was always a source of scandalized complaints in reports of the period.

As resource and industry problems became more prominent, the government groped its way into a three-fold response: more fish culture, more science, but above all, more regulations, set by a host of royal commissions.

Fish culture reaches a peak

Work continued on fishways,[59] but hatcheries caused more excitement. Samuel Wilmot's early experiments had gotten wide attention as a means to preserve and increase Great Lakes salmon. By the 1890's, hatchery enthusiasm prevailed on the Pacific and across Canada. Dozens of hatcheries dotted coastal and inland waters.

Besides salmon, which got the most attention, the hatcheries produced shad, whitefish, pickerel, trout, and other species. Canada's 1891 lobster hatchery at Bayview, New Brunswick, on the Northumberland Strait followed an earlier lobster hatchery at Newfoundland, where hatcheries were also popular. Among other efforts, that colony set up a cod hatchery at Dildo, in futile hopes of reproduction.

Little evaluation of hatcheries took place. Fish culturists assumed that if they were producing great numbers of fry, it must be a help. Americans were doing the same thing. Spencer Baird, the renowned U.S. Commissioner of Fish and Fisheries, for years used the survey vessel Fish Hawk to distribute millions of fry of shad, lobster, and other species at sea. Acclaim was widespread; meanwhile, research languished.

Year after year in Canada, Samuel Wilmot, in charge of fish culture for the Dominion, reported on the great

Clockwise from top left: salmon hatchery at Harrison Lake, B.C.; lobster hatchery at Canso, N.S.; inside a fish hatchery at Point Edward, Ontario; dumping fry from the Point Edward hatchery into a lake, April, 1911. (Library and Archives Canada, PA-11609, PA-20724; PA-60797; PA-133228)

progress that hatcheries were making. British Columbia got its first salmon hatchery, opposite New Westminster on the Fraser, in 1884. By that time, Quebec had four hatcheries, Ontario, New Brunswick, and Nova Scotia two each, and Prince Edward Island one.

The numbers kept growing. By 1910, British Columbia had eight hatcheries, Quebec eight, Ontario five, New Brunswick five, Nova Scotia five, and P.E.I. three. Hatcheries were now planting more than a million fry per year. But some officials were beginning to doubt their value. One B.C. fisheries inspector noted that for enhancing salmon, clearing streams was 100 per cent better. In Newfoundland as well, doubts were cropping up. In Atlantic Canada, lobster hatcheries would close in 1917, though hatcheries remained for other species.

Fish transplants become popular

Like hatcheries, the transplanting of fish seemed to allow the creation of new abundance from almost nothing. Fishery Inspector Thomas Mowat tried to transplant lobsters to British Columbia in the 1880's. New attempts took place in 1896 and 1905. Other experiments in the 1890's and early in the 20th century put whitefish into B.C. lakes and Atlantic salmon and oys-

ters into B.C. waters, with some survival in the latter case. On the Atlantic, the fisheries service experimented with keeping egg-bearing female lobsters in pounds and returning them to sea in the closed season. It also tried transplanting black bass to B.C. and western Ontario.

Prince noted in 1898 that more scientific knowledge was needed; lots of amateurs were now transplanting fish, and could do more harm than good. Carp soon provided a case in point. In 1896, some carp escaped from a stocked pond into the Great Lakes system (others may have got there from different sources) and spread in the Great Lakes and Manitoba, where they damaged not only native fish but also birds, by their effect on aquatic plants. A "coarse fish," carp provide only a trifling commercial fishery. Mainly they are undesirable. Control measures have had little success. In short, a chance introduction of a new species brought a long struggle to undo the damage.[60]

Biological stations begin

Until the late 19th century, Canadian fishery management still had little connection with science. Overseas, fisheries biology was advancing. In England, the Marine Biological Association was formed in 1884, and a fisheries laboratory took shape in Plymouth four

107

The first floating laboratory, at St. Andrews, N.B.
(Photo courtesy of Bill McMullon)

years later. There, E.W.L. Holt did pioneering work on biology, demonstrated declines in abundance, and proposed some of the earliest modern regulations. The Dane C.G.J. Peterson was also beginning fundamental work on population dynamics and the need to control fishing.

In Canada, Whitcher had earlier praised the fisheries work of the Natural History Society of Montreal. In the 1890's, Prince and others spread the idea of a biological station. The Royal Society of Canada took up the cause. In 1898, the Minister of Marine and Fisheries, Sir L.H. Davies, approved a proposal by the Royal Society, and Parliament in June authorized $7,000 to pay for construction and one year's operation of a floating laboratory based at St. Andrews, N.B.

Built in 1899, the floating laboratory spent time at her home base in St. Andrews, at Canso, N.S., at Malpeque, P.E.I. (working on oysters), and at Gaspé. She sprang a leak in 1907 while being towed up the St. Lawrence River, and in 1908 disappeared from the records. In 1908, the fisheries service built a shore laboratory at St. Andrews and another at Nanaimo, on Vancouver Island.[61]

Although Ontario had by now strongly asserted itself in fishery matters, the federal government, pushed by Prince, in 1901 set up a biological station at Go Home Bay on Georgian Bay. This station continued until 1913.[62]

The early biological stations kept a certain distance from the department and the fishing industry. As suggested by the Royal Society, the first St. Andrews-based station was administered by a special Board of Management. The board included Prince, as director, and several university professors. The station was to work closely with Canadian universities; their investigators would do as much of the scientific work as was practical.

Science separates from management

In the early years, the biological stations still got their budget from the department. But when the scientists requested money to buy some German and French publications, it seems that it was W.A. Found,

the new strongman of the department, who refused. The board got angry, went to the politicians, and in 1912 got a separate mandate and budget from Parliament, as the Biological Board of Canada.[63] Although Prince, who had chaired the earlier board since 1900, continued to head the new one until 1921, the university-based scientists kept an arm's-length relation with the department.

In *The Aquatic Explorers*, Ken Johnstone has chronicled the board's many achievements in science. But the split held the board back from experimenting in management, and it discouraged the department from asking scientists to look at industry questions. The main connection was Prince; but the tenuous department–board relation would weaken when he left. While the fisheries service did its rough and ready experiments in management, university professors and their students continued basic research on biology, migrations, and so on. The idea was always that the research would ultimately benefit the resource and the industry. But it took decades before the industry or the department gained much influence on the workings of the board.

Eminent among the early scientists was A.P. Knight of Queen's University. In 1901, the board began the scientific series *Contributions to Canadian Biology*, which in 1925 changed its title to *Contributions to Canadian Biology and Fisheries* and in 1934 to *Journal of the Biological Board of Canada*.

In another form of organized knowledge, the fisheries service itself, at the end of the 19th century and in the early years of the 20th, operated a small fisheries museum in Ottawa.

Officials, royal commissions spread regulations

While fish culturists and early scientists made their efforts, the greatest force for fishery conservation was the regulations spreading throughout the freshwater fisheries and into some sea fisheries. Many emanated from local complaints and the suggestions of fishery officials. (The corps got a boost in 1893, when local fishery officers began getting half-pay through the winter, even though rivers were frozen and inland fisheries halted.)[64] Their conservation-mindedness showed when fishery inspectors from across the country met in Ottawa in 1891. Apart from their recommendations to the royal commission then studying B.C. fisheries, they expressed these views:

- purse-seines should be banned;

- spear fishing should be banned;

- the lobster fishery needed closed areas;

- gillnets were just as dangerous as pound-nets (then a subject of horror);

- there should be new closed times for shad, trout, and other fisheries;

- there should be a seasonal closure (June 1–September 1) of mackerel netting during the day;

- there should be no trawls (that is, longlines) or bultows within two miles of shore during the night on bays;

- there should in some fisheries (e.g., the herring spawning grounds at Grand Manan, N.B.) be spawning "sanctuaries"; and

- fishermen's buoys should be marked.[65]

Most of these suggestions became regulations. Rule-making continued apace. For example, in 1894, the ban on rockets and explosives in certain fisheries got extended to cover all species.[66]

Beyond the fishery officers, the chief wellspring of rules was that Canadian favourite, the royal commission. In response to fishery declines, from 1890 to about 1920 the department set up dozens of royal commissions, including at least nine on B.C. fisheries alone. A handful of such experts as were available, often including E.E. Prince, would talk to people in the fishery concerned and draw up regulations, generally on fish size, gear, or seasons, for specific fisheries. The royal commissions of this period set the general shape of fisheries management that lasted to the 1960's.

This semi-grassroots approach had its strengths then as today. At commission hearings, local people get their say. If positions have hardened between the industry and bureaucracy, the commissioners can often override set attitudes to bring in a new approach.

But this approach also opens the possibility of a hodgepodge of locally suggested regulations. This was even more the case at the turn of the century, when real fisheries knowledge was still lacking. Fishermen and processors advanced many reasons for depletion; and often, each believed they had the sole answer. Prince noted in 1898 that experienced fishermen in the Bay of Fundy had advanced 16 reasons for fluctuations in the herring industry.

Everywhere and always, fishermen tended to associate depletion with rival types of gear. For example, in Prince's time, Bay of Fundy gillnet fishermen blamed weirs for overfishing herring. Among other fishery villains cited in the 1890's: longlines dropped dead fish to rot and destroy grounds; purse-seines had been harmful; the pound net was deadly; and so on.

Prince had an eye for basic matters. He wrote in 1898 that when the Newfoundland bank fisheries declined from 330 vessels in 1889 to 58 in 1894, great alarm followed, and various parties advanced 59 separate reasons. The real cause, Prince suspected, was Newfoundland's backwardness in fishing methods and marketing (although, as Shannon Ryan points out, the decline in government subsidies for the banking fleet was a major factor).[67]

If Prince was conscious of complexities and conflicting explanations, he was also alarmed about depletion. For example, he noted in 1898 that overfishing had damaged shad, lobsters, oysters, menhaden, and the Great Lakes in general. And Prince himself was probably the greatest regulation-writer of all.

What reasoning lay behind the many new rules? Prince noted in 1902 that

> [f]our main interests have been prominent in the forming of fishery regulations generally. These are: First, the interests of the fish. If there were no fish there would be no fisherman and no fishing industries. ... Second, the interests of the fishermen as an industrial community. ... Third, the interests of the state as a whole. ... The public interest ... may not always coincide with the first or second interest described above, indeed they may come into serious collision, and many authorities might be quoted to show that the public interest should be paramount. ... Fourth, international interests, ... which have often reached a stage so crucial and perilous as to over-ride the interests of the fish, the fishermen, and the nation.[68]

Prince had pinpointed the problem of conflicting objectives, to which the department in future years, despite many attempts, could never find a formulaic solution.[69] Neither would government attempt, until nearly a century later, to apply education and democracy in such a way that the people affected would set the priorities as circumstances changed. Instead, regulation would generally continue as it began: when complaints emerged, the political and official seismographs would note it, then civil servants and ministers would consult in a rough and ready way, make whatever decision seemed best at the time, and hang on until the next problem surfaced.

Inspection law gets stronger

In the 1870's, Whitcher had tried and failed to get compulsory inspection of fish products. Under his scheme, fishery officers were to inspect fish, as well as carry out their other duties, and introduce "a light scale of fees" for inspection, which would supplement "their nominal pay in the protective service." Although Parliament passed a general Inspection Act, Whitcher's plan never came into full force.

Problems continued, notably in the curing of herring. In 1890, deputy minister Tilton wrote a long justification, with many quotes from interested parties, on the need for an inspection act. The same year saw the new law come into effect.[70]

The Inspection Act remained voluntary. Producers could, if they wished, request the department to view their product. As before Confederation in some instances, it was thought that shippers would want to get the government stamp of approval on their barrels, as an aid to marketing. This supposed self-interest turned out to be a weak lever; many shippers continued exporting fish approved only by themselves.

Government subsidizes fresh-fish transport

Besides regulation and resource enhancement, the department began looking more towards development.

The growth of the American fresh-fish fishery was obvious. And by the early 1870's, Whitcher was noting the increase in the Canadian fresh trade, which gradually grew stronger. Fresh salmon packed in ice went regularly by steamer from Saint John to Boston, and by 1879 it was being sent to Great Britain. By 1878, steamer connections also enabled live-lobster shipments from Yarmouth to Boston. Oysters, mackerel, herring, groundfish, and other salt- and freshwater species all formed part of the small fresh-fish trade. By 1885, Digby, Yarmouth, and Lunenburg were all producing some fresh fish in winter, when it would keep easier.

Train transport aided the fledgling industry. Mixtures of salt and ice kept the fish cool in boxes and storehouses. Fish-handlers used natural ice cut from ponds and kept in icehouses, or they used the new artificial ice. Cold storage and freezing became more common at the end of the 19th century. Household refrigeration developed more slowly, becoming important only in the later 1930's.

Maritimers still stuck mostly to dried, wet-salted, and pickled fish. Parliament in 1907 passed an act authorizing payment of 30 per cent of the cost of building cold-storage warehouses. That same year, to encourage the fresh-fish trade, the Canadian government decided to subsidize transport of fresh fish to market, in chilled rail cars, from both the east and the west coasts. (The cars held ice and salt in chests at each end to cool the air; the chests were replenished at stops along the line.) The one-third subsidy on shipments of less than carload lots, by express, did appear to help the fresh-fish trade. It was the first major attempt to aid marketing.

CHAPTER 7.
Freshwater fisheries, 1867–1914

The inland fisheries of the old Province of Canada had produced the main elements of the Fisheries Act, including the licensing power. Freshwater conservation problems generated much of the fisheries service's early work. But while those problems continued in the 1867–1914 period, the federal government by the turn of the century lost much of its control of inland fisheries.

Great Lakes fisheries face depletion, foreign control

In the Great Lakes area, the Department of Marine and Fisheries continued to make use of licences and leases. "Department officials ... reviewed applications for licences and approved or denied them, on the basis of the number considered safe for the fish population and an assessment of the application's worth." Fishermen could not transfer the licence to someone else. The department often refused to license previous offenders. Having fished in a place for a long period was no guarantee of a licence.

Enforcement posed a major challenge; some fishermen hid out and fished as they liked. By the 1890's, the fisheries branch had 33 overseers around the lakes and 90 paid officials. As elsewhere, overseers made no great money, and usually had to do other work as well. Members of Parliament influenced the appointments. Still, many were excellent officials.[1]

While commencing a long, rearguard action to protect salmon on the Atlantic coast, the fisheries service was already losing the battle in Ontario. Despite Whitcher's system of leases and licences, which applied in some locations, and despite the work of Wilmot and other fish culturists, Lake Ontario salmon declined rapidly after about 1879. Industry, urbanization, changes in water level, and fishing pressure all contributed. By about 1890, salmon had vanished from

Lake Ontario, including those from Wilmot Creek, where salmon had once been so plentiful that women could seine them with their flannel petticoats.[2] But there were still many other species, such as whitefish and pickerel. With many fishermen living close to market, lakes Erie and Ontario led Canada in developing a strong fresh-fish industry. Train and steamship service from the 1850's boosted the fresh trade. Smaller operations first dominated the fishery. But as A.B. McCullough has outlined, better technology in the last quarter of the century—steam tugs, more nets—needed bigger money. Several larger firms emerged, some passing into American hands.

Indeed, Americans generally dominated the 19th-century fishery on both sides of the border. American vessels crossed the marine boundary, and American dealers controlled many operations on the Canadian side, despite licensing regulations. The Chicago-based Booth Packing Company dominated the Lake Superior industry by the turn of the century, and became a major force on the upper lakes and Lake Ontario.

The smaller Canadian independents fared best on Lake Erie, where they were close to both fish and market. Canadians gained more control in the early 20th century. As elsewhere, Native fishermen generally lost out.[3]

By the early 20th century, Ontario fishermen regularly used motor-powered gillnet boats. They had begun to use extra-deep gillnets, or "bullnets," to catch lake

Indians fishing in rapids at Sault Ste-Marie, Ontario, ca. 1885. (Library and Archives Canada, PA-164372)

Fresh fish in boxes on the Bay of Quinte in Ontario. Until the 1980's, boxing fish was uncommon on most parts of the Atlantic coast. (Library and Archives Canada, NL-13105)

Fishing fleet in Goderich, Ontario, 1884. Steam tugs in the late 1800's often towed sailing craft to port. (Library and Archives Canada, NL-13100)

herring. Lake Erie was always the most important lake, sometimes equalling the production of the other Great Lakes combined.

In the last quarter of the 19th century, concern about freshwater species led to great alarm about the pound-net. The annual report of 1891 lamented that "Fishing from morning till night and from night till morning, in season and out of season, and all through every season, for all kinds or sizes of fish, it abates not its ravages for any cause but exhaustion." The report noted that the pound-net had depleted many fisheries in the northeastern United States and Great Lakes. Yet Canada had to permit some use of it, so that her fishermen could compete with American fishermen.

To mitigate the effects of the fearsome pound-net, the department experimented with escapes for smaller fish, restricted the number of pound-nets per fisher-man, and set closed times for pound-nets. Despite all the alarm, pound-nets lingered on; in the late 20th century, some still existed on the Great Lakes.

In 1893–1894, a commission on the Ontario fisheries brought about new closed seasons. It also recommended changing mesh size back to 5 inches from 4 inches, recommended regulations for pound-nets and their meshes, proposed spawning sanctuaries, and waxed eloquent on the causes of depletion, including sawlog damage.

The department faced many obstacles in enforcement and compliance, stemming not only from illegal fishing by Canadians and Americans, but sometimes from the political side. To quote A.B. McCullough:

> The local fishermen [on the Detroit river] would apply to the Commissioner of Fisheries, E.E. Prince, for an extension of the fishing season. Prince would refuse, the fishermen would apply to their members of Parliament, the members would approach the minister, and in most years the minister would overrule Prince and grant an extension of the season.[4]

The federal department after Confederation had never strictly limited the numbers of fishermen on the Great Lakes; rather, it had made use of exclusive leases, mainly for salmon, until the salmon disappeared. In the 1890's, however, the department responded to catch declines in the western end of Lake Erie by restricting the number of licences in certain areas. The overall number dropped.[5]

Provinces take control of some freshwater fisheries

Across eastern Canada, the river fisheries were already an old source of worries. Species of concern included sturgeon and bass; the department in 1892 forbade catching the latter on the St. John River, to save them from "utter ruin."[6] The department also passed regulations restricting hoop nets, often used in rivers. The sturgeon fishery was falling off in the 1890's. Fishery officers protested the use of longlines for sturgeon. Over the next couple of decades, the sturgeon fishery declined to near insignificance.

Salmon was the river species of prime interest, and a frequent source of federal regulations. But salmon also prompted the first major weakening of federal control.

The British North America Act had given the Dominion clear power over "Sea Coast and Inland Fisheries," and the department had done as it pleased with all fisheries. But the provinces were beginning to assert their powers in many spheres. By the 1880's, Sir John A. Macdonald no longer dared to blithely disallow provincial laws. In following decades, a long series of legal cases, judged by the Judicial Committee of the Privy Council in Britain, chipped away at the power of the central government. Judgements tended to stress provincial powers over property and civil rights.

The federal fisheries powers were an early target. The first big crack in federal jurisdiction came in the case of *The Queen versus Robertson* in 1882, involving salmon-stream leases in New Brunswick. Commissioner of Fisheries W.F. Whitcher in the 1860's and 1870's had set great store by salmon leases, believing that private interest and the profit motive would foster conservation. But leases had failed to head off the demise of Great Lakes salmon, suffering from environmental changes; and now they would suffer a legal blow.

In The *Queen versus Robertson*, Canada's Supreme Court decided that although the Dominion government could legislate in regard to all fisheries, it had no power to interfere with, control, or grant exclusive fishery leases in any non-navigable rivers. The public right of fishing extended only to tidal waters. In the non-tidal waters, it made no difference whether the riverbed or soil belonged to the Crown in right of a province, or to a private owner holding a title from the Crown; still, the federal government had no power of leasing.[7]

Whitcher wrote to holders of federal leases in New Brunswick, telling them the situation had changed. The province in 1883 began auctioning off angling leases, to sport-fishing interests, on long stretches of salmon rivers running through crown land.

Leases continued elsewhere, in some inland and estuarial fisheries. Licences and leases provided order and gave the department a deterrent effect, through the threat of cancellation.

Yet, the ideology of licensing and leasing had lost some steam. Whitcher's theories about the efficacy of private enterprise in both business and conservation had been largely linked with leases that gave a fair degree of control to a single owner. Now, fewer departmental statements touted the value of leases for conservation and private enterprise. It would seem that the demise of Lake Ontario salmon, the loss of control in New Brunswick, and, perhaps, Whitcher's retirement in the 1880's all weakened the idea.

On the Great Lakes, the department continued its licensing strictness, sometimes cancelling licences.[8] But provinces were flexing their muscles in various spheres, including the fisheries. By 1891, various provinces had passed relevant acts. They seem to have wanted more fishery control partly for its own sake, partly because they hoped to gain revenue through licence fees. In 1892, Quebec questioned federal rights over inland fisheries, in particular a lease granted on the Richelieu River. In 1894, Ontario passed a code of fishery regulations. In 1897, British Columbia did the same.

The question of who controlled what caused another referral to the courts. This brought about an 1898 judgement by the Judicial Committee of the Imperial Privy Council that split authority over Canada's fisheries. The judgement confirmed that the federal government had exclusive competence to enact fishery regulations and restrictions and had the right to impose "a tax by way of licence as a condition of the right to fish."

But, the provinces had all proprietary rights in respect of fisheries that they held before Confederation, when ownership of river banks often went together with fishing rights. The provincial government could also tax provincial fisheries in addition to any tax imposed by the Dominion Parliament.

After the 1898 judgement, the government of Ontario took over licensing, leasing, protection of fisheries, and later, between 1913 and 1926, the federal hatcheries in that province. This was a long and quarrelsome process, as the two governments jockeyed over various elements of fishery control, but it ended with a general federal withdrawal. The Dominion closed its Great Lakes biological station after 1913. Federal fisheries patrols on the Great Lakes ended in 1922.[9]

Meanwhile, Quebec also took over management of inland fisheries, and between 1913 and 1920, federal hatcheries. A delegation of sea-fishery management to that province would follow in the 1920's. Other provinces continued as before, except that British Columbia began asserting more power over inland fisheries.

The federal department kept its power to license fishermen in tidal waters. It also kept control of coastal processing plants, licensing lobster and salmon canneries. As well, the 1898 judgement also said that the federal government could tax any provincially-issued licences. It confirmed that the federal government owned public harbours and their fisheries. And a further decision in 1899 said that Canada could regulate

113

on the fisheries in such a way as to affect provincial rights of ownership, even though the province retained ownership.[10]

The jurisdictional decisions and subsequent arrangements left a complex snarl. In non-tidal fisheries, provinces controlled proprietary rights but the federal government had management jurisdiction. Over time, the two levels of government worked out practical arrangements. In general terms, where a province administered the fishery, it would recommend regulations to the federal government. The federal government normally enacted them as recommended, and the province enforced them. For salmon, however, the federal side controlled salt-water licensing, controlled openings and other conservation measures for almost all waters, and did most enforcement. For freshwater trout and such, the provinces generally took the main hand in management, but federal officials might play a role. The provinces generally issued freshwater licences.[11]

In New Brunswick the situation is particularly complicated. *The Queen versus Robertson* had derived from the "riparian rights" of property owners to fish. Those ownership rights remain today, notably on the Miramichi and Restigouche, the greatest salmon rivers, where private ownership and crown leases leave little fishing space for ordinary citizens, and to a lesser extent on the St. John and Nepisiguit rivers. Private deeds mention the fishery. In addition, the province leases fishing access on crown lands.

The situation is less pronounced in other provinces, some of which have withdrawn private rights, as Nova Scotia did in 1911. In all cases, the federal government still has ultimate jurisdiction over the time and manner of fishing, the fish being a public trust.[12]

Ontario fisheries keep declining

On the Great Lakes, provincial management brought little good news. Ontario took a more lax approach to fisheries management, more like that on the American side.[13] The province's fisheries kept dropping off. "No words can exaggerate the former plenitude" of the Georgian Bay fisheries, said the federal department's annual report of 1908. Now, whitefish had declined; less desirable coarse fish had increased. The provincial and federal authorities had restricted both trapnets and seines. But nothing availed. The Great Lakes were well launched into a pattern of overfishing desirable species, to be replaced by less valuable fish.

Sturgeon, once considered a nuisance and used for fertilizer, had become scarce by 1900. Dams hastened the decline of this river-spawning species. Lake trout also became scarce by the turn of the century. Whitefish catches dropped, with overfishing and environmental degradation the likely culprits. As whitefish declined in Lake Erie, herring took over—an early example of a less valuable species moving into a vacant ecological niche. Even if dollar value dropped, total production could remain fairly stable.

Nets on the shore full of fish, 1907. (Library and Archives Canada, PA-60883)

But catches of lake herring themselves crashed on Lake Erie in the 1920's, probably because of a too-intensive fishery.[14] Meanwhile, sauger commenced a slow decline after 1916, again apparently because of overfishing and environmental losses. Blue pike landings began to fluctuate extensively by 1915, probably because of heavy fishing. As other species declined, walleye and yellow perch began taking on more and more importance.

To top matters off, the sea lamprey, a destructive predator long present in Lake Ontario, in the 1920's wreaked new damage in that lake. It moved through Lake Erie into Lake Huron in the 1930's, harming fish stocks in the latter lake. Only after the Second World War would federal and provincial authorities find a lampricide that could be used without harming other stocks.[15] Other nuisance fish—carp, smelt, and alewives—spread in the Great Lakes, in some instances because of unwise plantings.

By the end of the First World War, the pattern was set for Great Lakes fisheries of the future. J.H. Leach and S.J. Nepszy of the Ontario Ministry of Natural Resources in the 1970's summed up changes in the Great Lakes fishery thus:

> The dominant and most consistent cultural stress acting on the [fish] community has been the commercial fishery. All species important to man have been affected. Throughout its history the commercial fishery of Lake Erie has been inadequately regulated... . The progressive exploitation of stocks in a multi-species fishery has been described by Regier and Loftus ... as the 'fishing-up' sequence. Basically, this process is described as the gradual shift in fishing effort from higher-valued to lower-valued species as the preferred stocks pass their peaks and decline in abundance. It encompasses the ability of fish

ermen to increase fishing efficiencies through improved gear, more and larger vessels and a growing knowledge of the behaviour of fish. This process began in Lake Erie in the 1800's and is still continuing.[16]

Manitoba fishery builds up

Even after the jurisdictional decision of 1898, the federal government still controlled the prairie fishery. In Manitoba, by the early 1880's, a commercial fishing station had begun to supply Winnipeg. After the railway reached Winnipeg in 1885, the city began growing in a headlong rush.

A gillnet fishery developed on Lake Winnipeg and soon began exporting. Important species included whitefish, pickerel, and pike. The industry also produced caviar and gold-eyes, along with coarse fish of less value. The prairie fishery grew from a value of $30,600 in 1876 to $745,500 in 1896. White entrepreneurs often used Native fishermen for labour. American companies at first controlled much of the fishery, although this lessened after 1905.

The Fisheries Act came into effect for Manitoba on October 1, 1880. In 1884, the department appointed an inspector at Winnipeg and an overseer for the Qu'Appelle district. After Wilmot made a report on Lake Winnipeg and Manitoba, the department in 1891 set up a proper organization and passed initial regulations for Manitoba and the Northwest Territories.

As in British Columbia, conflicts emerged between the Indians and the new fishery managers. There were reports of some Indians defying the closed seasons and selling illegally caught fish to traders.[17]

Fishing boats at mouth of Little Saskatchewan River, Manitoba, 1890. Note steam tug behind the sailboats. (Library and Archives Canada, PA-52766)

Quotas come to prairie fishery

As whitefish got scarcer, fears grew of over-exploiting the Lake Winnipeg fishery. The confined prairie lakes, like the river systems, made it obvious that instead of "lots more fish in the sea," there were only a limited number. On the Pacific, this visible vulnerability would lead to licence limitation. On the prairies it led to quotas.

In 1892, the department restricted fishing to the northern part of the lake and ordered that no company could fish more than 20,000 yards of net. This in effect limited each company to less than 20 boats. The department was limiting the amount of effort per company, without limiting the number of companies.

In 1894, on April 14, an Order-in-Council prohibited tugs—that is, vessels with motor power, often used for towing sailboats or skiffs to the dock—except in shipping. New rules were again fighting new technology, only to give way over time.

A commission of enquiry noted the depletion in Lake Winnipeg. Prince and the other commissioners drew up rules to

- abolish the steam-tug licence belonging to the Commercial Company, "the object being to remove all control by commercial companies or combines, and to place the fisheries, as far as possible, in the hands of the *bona fide* fishermen." Provision would be made for other tugs to tow the smaller boats of ordinary fishermen;

- delimit the area to be fished in summer;
- gradually increase mesh size in the whitefish fishery, to raise the age of capture;

- reduce the amount of nets and gear to be used by tugs (regulations already confined fishermen to baited hooks and gillnets only); and

- limit the total annual summer catch of whitefish.

There followed a whitefish catch quota of 2.5 million pounds in Lake Winnipeg. Lake quotas were to continue in Manitoba; Ontario authorities also used them for whitefish. This was probably the first systematic use of quotas in Canadian fisheries. Also notable was the concern to protect "bona fide" independent fishermen, a policy set on other than conservation grounds.[18]

All this was highly active regulation, with licences prominent. The royal commission noted that for 20 years, the fisheries service had issued two kinds of licences: a regular commercial one and a "domestic" one allowing a limited fishery. Problems had arisen with domestic licensees selling fish com-

Species	1895	1900	1905	1909
Whitefish	4,270,319	5,872,400	8,005,000	4,662,100
Pickerel	931,190	2,275,100	6,900,000	5,750,400
Pike	689,395	444,300	3,790,000	3,067,100
Sturgeon	104,240	981,500	600,000	94,300
Tullibee	278,800	204,200	2,074,000	834,200
Catfish	79,724	184,400	500,000	87,200

Table 7-1. Manitoba production (in pounds), selected species and years. [20]

mercially. The commission recommended doing away with the domestic licence, letting settlers and Indians fish up to 100 yards of net, and reforming the cumbersome system that saw licence applications sent to Ottawa.

The commissioners also worried about the marketing system, and made complaints that stayed common in the prairie fishery for decades to come. American buyers were exercising undue powers and using shady practices in the marketplace. The commissioners suggested measures to ensure fair supplies to Canadian retailers.

About another recurring problem, the commissioners had this to say:

> Requests for extensions of fishing times are sent to Ottawa and urged with great force, owing, in some cases, to a serious shortage in the season's catch of fish. ... To this commission it appears strange that, during a season in which the fish appear ... to be especially scarce, requests should be made for an increased destruction of them. ... In our opinion, when the fish ... appear to be scarce, that is precisely the time that they should be conserved. [19]

Prince and his colleagues had pinpointed a problem that would keep recurring. Measures that to regulators meant conservation, to fishermen meant desperation, even though conservation, if properly carried through, would best suit their long-term goals.

Overall production doubled from 10.7 million pounds in 1900 to 20.5 million pounds in 1909. Table 7-1 shows production for major species in Manitoba.

Federal authorities often ignore freshwater lessons

Since this book deals mainly with fisheries management by federal authorities, it will mostly omit the history of Ontario fisheries management from now on, and will say little about prairie-fisheries management after 1930, when the provincial governments took over management. But it is worth noting that, as with the demise of Ontario salmon and the ecological shifts in the Great Lakes, freshwater fisheries have often shown how easily fishermen and society in general can damage a fishery. Fisheries in Canada's rivers, the Great Lakes, and the Prairie provinces have been pioneers of problems.

Freshwater managers also tinkered with some solutions, such as overall catch quotas and individual quotas, well in advance of the federal side. Whatever lessons freshwater fisheries have had to teach, the much bigger sea fisheries have usually ignored.

CHAPTER 8.
On the Atlantic, 1867–1914

Although Newfoundland was still a separate country, she shared some conditions with the Maritimes and Quebec. The following will touch on some coast-wide elements, before dealing more specifically with the Maritimes and Quebec, and later with Newfoundland.

Baiting longlines in the schooner fleet, 1910. (Library and Archives Canada, C-37550)

Atlantic economy falls behind

In the 1867–1914 period, while Canada was taking a bigger place in the world, the Atlantic coast was taking a smaller place in Canada. At the end of the 1870's, the Maritimes still had the third or fourth biggest fleet in the world. Some growth was continuing on the coast; industrialism boosted the coal and steel industry in Nova Scotia. But the main Canadian advances took place in Quebec and Ontario. It seemed easier for steel to move to central Canada than for industry to move to the Maritimes. Newfoundland shared the Maritime and Quebec weaknesses, but with fewer offsetting opportunities.

Economic conditions were changing. An international depression beginning in 1873 affected various sectors, including New Brunswick's timber trade. The "National Policy" after 1878, with its high tariffs to protect manufacturing and its encouragement of western settlement, also subsidized both railroads and steamships. The steamer service may have helped

exports, but did nothing for Maritime builders and operators of wooden boats. They lost further ground in both the deep-sea and the coastal trade. Schooners were prey to fire and storm and shipwreck. (For example, the Newfoundland seal fishery from 1810 to 1870 lost an estimated 400 vessels and 1,000 men.) As safer, more reliable steamship service to the West Indies, Brazil, and other overseas destinations cut into the wooden-vessel trade, some vessel owners became more interested in pursuing investment opportunities on shore. Many shipyards closed.

Some shipping and trading interests opposed Confederation, and later generations of Maritimers sometimes blamed the new tariffs for weakening the coastal economy. Yet, at the time, tariffs also got strong support in parts of the Maritimes. Some Maritime capitalists built manufacturing companies, which were often successful. The steamers displacing wooden boats, the railroads, and the pull of the continental economy probably did more than the tariffs to leave the Maritimes in relative backwardness. In

Newfoundland, then separate from Canada and its tariffs, the marine economy also declined.

The last major pillar of the old coastal economy, the fishery, would still see growth in some areas, but could not by itself offset the gradual decline in other marine industries or provide prosperous work for all. Coastal population was still increasing: in the Maritimes, from fewer than 800,000 in 1871 to more than 900,000 in 1911; in Newfoundland, from about 197,000 in 1884 to about 243,000 in 1911. But Ontario and Quebec were now far bigger, and growing faster.

Report warns of overfishing

Although the fisheries service in its early years worried mainly about freshwater conservation, a remarkable report in 1877 pointed to future problems in the sea fisheries.

To help prepare Canada's case for the Halifax Award, the department had the Canadian naturalist and geologist Henry Youle Hind prepare a report on the fisheries and their value. Hind looked at Newfoundland's as well as Atlantic Canada's fisheries. Two-thirds of a century later, the famous biologist A.G. Hunstman called Hind's report "a noteworthy but little known presentation of available oceanographic knowledge of Canadian Atlantic waters on the causation of the fisheries."[1]

Hind argued that overfishing was becoming common in the United States and Canada. If the fishery was, as some said, inexhaustible, why had menhaden disappeared from eastern Maine and the Bay of Fundy? Why had the cod fishery declined so badly in New England? Gloucester had increased its deep-sea fishery not only because railways and ice had encouraged the bank fishery, but also because inshore fish had disappeared. In the Dominion the problem had a special nature: much of the British and foreign fishery was pursuing bait fish, removing the food of more valuable species.

Hind's report noted that the inshore fishery remained the bulwark of the fishery in both countries. The American deep-sea fishery provided only one-fifth of the U.S. fishing income of $43 million. Although the United States had fewer resources of its own, its roving cod and mackerel fleet had nearly twice the tonnage of the entire Canadian fleet. Since the Treaty of Washington, the Americans again could fish anywhere. U.S. vessels pursued Georges Bank cod, using northern herring for bait, then mackerel off New England and Canada, then cod and other species that came close inshore as the water warmed.

Hind stressed the great importance of Canadian and Newfoundland bait for the U.S. fleet (although, as noted earlier, that demand was lessening even as he wrote). It was Newfoundland herring that had made possible the Georges Bank winter fishery. The French fleet also depended on shore-caught bait. But the bait fishery as now pursued was "suicidal."

In the Gulf of St. Lawrence, the bait fishery was the primary American fishery. American vessels also used herring from the Bay of Fundy. They now made more use of purse-seines, with their great catching power. And they bought huge quantities. They were reducing Nova Scotians and Newfoundlanders to catchers of bait, and were ruining the fishery. To let U.S. vessels continue fishing in the Canadian zone would further diminish the supply of fish, and would ultimately diminish the deep-sea fishery, because the shore affected the sea. Many areas had already lost fish; others were likely to do so.

Canada's fishery had suffered from underestimation of its value, Hind maintained. While Nova Scotia vessels remained rare on the banks, the United States had many bankers at the Grand, Georges, and Le Have banks; and they were bigger, ranging from 70 to 120 tons. Even so, the U.S. sea fishery off British North America equalled only a bit over half the combined fishery of Canada and Newfoundland. And Canada's commercial marine fleet, mainly fishing vessels, fell behind only those of the United Kingdom, the United States, Norway, and Italy.

British North America simply had more fish than most areas of the world. Places such as Canso Gut, the south shore of Newfoundland, and the Strait of Belle Isle got constant food and ova from elsewhere and could better withstand heavy fishing. But other areas could rapidly become depleted.

At Newfoundland, mackerel had once been plentiful, but then declined. Cod had declined in Chaleur Bay and the southern Gulf, because people had taken so much herring and capelin for bait and for manuring the land. Newfoundland had passed laws against using capelin as manure. Also destructive were bultows (longlines) set near bottom, where they took the mother (roe) fish. Hind noted increased exports of roe in recent years. Seines and traps also damaged the resource.

Hind's main message was the value of Canada's resource and of what Americans took from it, and the associated dangers. He also discoursed on science and other fishery matters. For example, he outlined the arrival dates of cod at different places on the Labrador, and defined the seasons of other fisheries. He noted that fishermen, then as now, would resist co-operating to advance fisheries science, for fear it might bring further regulations. Hind noted in passing the Norway–United Kingdom agreement on methods of killing harp seals, both for reasons of humaneness and for manufacturing purposes.[2]

The Maritimes and Quebec

Engines change the industry

For the Maritimes and Quebec specifically, the 1867–1914 period saw important growth in the lobster

Drying fish at Digby, N.S. (Library and Archives Canada, PA-20733)

and herring fisheries, where factories sprouted in a new industrialism. Starting in the United States, cold-storage plants appeared for fish and bait. And the mechanical "reduction" of fish and offal into fertilizer spread eastward.

The main fleet still consisted of many thousands of rowboats and open sailboats, operated by one man or a small crew, and hundreds of schooners, often family-owned. As lobster and herring picked up, more fishermen could switch between species, and buyers, during the year. Many mixed the fishery with other work, whether subsistence farming, work in the woods, or other employment they might pick up at home. They might also make seasonal migrations to the city.[3] Steam power moved only slowly from shore to sea, with a few trawlers appearing at the turn of the century. The bigger technological change at sea was the gasoline engine, in the early years of the century.

By 1910–1911, some 5,000 boats in the Dominion used gas engines; by 1917, some 15,000. In the "make-and-break" engines, a magneto turned by the engine itself produced the spark. Well-known Maritime-made engines included the Acadia and Stewart Imperial brands. The gas engine opened the small fisherman's way towards something better than rowboat or sloop. He could cover greater distances with more reliability. On Cape Sable Island, Nova Scotia, the "Cape-Island" style boat, with a wheelhouse and cuddy forward, appeared early in the 1900's. Larger gas boats began fishing the nearer Atlantic banks by 1918.[4]

Groundfish trade shows little growth

Meanwhile, the foundation fishery of groundfish saw mixed results. Gillnets were becoming more common, as were cod-traps in the northern Gulf. In southern areas, the fresh-fish trade began to grow. More off-shore schooners and eventually a few trawlers appeared.

But the old trade in dried salted cod showed problems as often as progress. Production of dried salted cod reached a peak in 1886, then fell. Innis lists the reasons as the decline of wooden sailing vessels, with disastrous results for many ports; weakening markets, in countries dependent on cane-sugar, because of competition by beet-sugar; an increase in tariffs in importing countries; competition from meat products; and competition from the fresh-fish trade.[5]

The big merchant firms of the Gulf were changing their role in the last part of the century. Cod markets were soft. Prices in 1880 dropped to their lowest levels since 1830. In 1886, the two Robin firms in the Gulf ran into trouble, as did the Le Boutillier company. Competing firms were more and more a threat. The Robin interests soon went through reorganization and a merger. They turned more towards retail business, although retaining cod production.

Although the Robin interests no longer dominated the Gulf as in the old days, the company remained active. In 1910, Robin Collas sold its Canso plant to the Maritime Fish Company; but it bought the Lunenburg plant of the Atlantic Fisheries company plus two other saltfish companies, forming Robin, Jones and Whitman. After the war, however, its salt-fish operations declined.[6]

Lunenburg leads bank fishery

Nova Scotia after Confederation developed a strong fishery on offshore banks including the Grand Banks. The reasons were various. To some degree, Newfoundland edged its rivals offshore, by making it harder for Maritimers and New Englanders to get bait from Newfoundland shores or to trade with Newfoundland and Labrador residents. At the same time, there were difficulties with inshore fisheries. The Labrador fishery had suffered bad fluctuations. In Nova Scotia's own shore fishery, fish had become scarcer. Mackerel, under heavy U.S. fishing pressure, had become an uncertain fishery.

Nova Scotia vessel operators had to rethink operations. Lunenburg in particular followed the American lead in turning to the offshore banks around 1873. At the same time, the Lunenburg fleet adopted Gloucester-style flat-bottomed dories, and turned to longlining. The use of baited trawls at first caused protests, and fishermen had some difficulties finding bait. While Americans were more and more selling fish fresh, Lunenburgers mostly salted their fish, especially since the American move towards fresh gave Canadians more room to sell saltfish in the West Indies market.[7]

Lunenburgers moved ahead in the offshore, larger-vessel fishery, with schooners of 80–100 tons. Why did their small town take the lead over such nearby ports as Liverpool and Lockeport, equally well situated, and indeed over Newfoundland towns almost bordering the Grand Banks? No major factors suggest themselves, except the Lunenburgers' habit of hard work, combined with a willingness to invest in local fishing operations. Joint stock companies, often with fishermen prominent in them, owned schooners.

Lunenburg's fleet by 1910 grew to more than 140 vessels. Many obstacles, including the U.S. consumers' switch to fresh fish, the loss of some Latin American markets, and stronger competition from other countries with trawlers, worked against the Lunenburg fleet. But Scott Balcom, the authority on this fleet, notes that its efficiency helped stave off the consequences until after the First World War. Even then, Lunenburg entrepreneurs survived the Great Depression to build up the single strongest fleet and biggest fishing company in Atlantic Canada.[8]

The first trawlers appear

As the American fresh-fish industry grew in the late 19th and early 20th centuries, some Nova Scotians also began shipping fresh fish to market. A few trawlers began showing up after the turn of the century for the fresh-fish trade.

Towed trawl nets (not to be confused with longline "trawls") had become common in Europe in the 19th century. Fishermen had first used a fixed beam to keep the net's mouth open as it moved along the bottom. Then beam trawls gave way to otter trawls. These had two wooden "doors" that met the water in such a way as to hold the net open. In 1876, steam-powered trawlers appeared in France.

The year 1897 saw the first use of a steam trawler in Nova Scotia.[9] A number began working from Canso around 1910. As of 1917, there were four steam trawlers on the Atlantic and one on the Pacific. The Atlantic trawlers were said to have "immensely stimulated the trade in fresh fish, by the regularity with which they land supplies." The proportion of cod, haddock, hake, and pollock being dried declined; increasing amounts went fresh, frozen, wet-salted, smoked, or canned. Cold storage to hold fish chilled or frozen became more common by the war.[10]

Still, the great majority of vessels in the "offshore" Atlantic fishery were sailing and, later, sail-plus-engine

The *Rayon d'Or*, an early trawler. (Maritime Museum of the Atlantic, MP15.87.23)

craft using longlines. Often, family-run operations owned anywhere from one to a couple of dozen schooners.

Department uses bait-freezers to organize fishermen

In the later 19th century, "cold-storage" plants using the new mechanical–electrical methods of refrigeration were appearing here and there, to keep fish chilled or frozen. By 1910, departmental official R.N. Venning was making the grandly wrong prediction that lobster boiled in the shell and then frozen would displace the still-young live lobster trade. But freezing and cold storage for the fish trade never took strong hold until after further technical advances, from the 1920's onward. Meanwhile, the federal department sponsored freezers for bait, to help fishermen in both their work and their organizations.

Around 1891 the department began issuing bait bulletins, saying where fresh bait might be had, and giving out information on constructing freezers that used a brine of salt and ice. After an 1898 royal commission under E.E. Prince on the lobster fishery, the department got legislation that let it provide bait-freezers, with bonus money for associations that maintained them. By 1908, 45 such freezers had come into operation.

Before Confederation, New Brunswick and to some degree Nova Scotia had sponsored fishery societies to encourage those in the industry to get better organized and follow good practices. The bait-freezer experiment was the first notable post-Confederation attempt in the same direction. There were 62 bait-freezers by 1909. But the program would fade away in future years.

Groundfish goes mostly unregulated

Most regulations still concerned the river and shore fisheries. For the greatest fishery of all, groundfish, the fisheries service made few specific rules and sponsored no major royal commissions. The enormous, pre-existing saltfish trade simply underlay everything else.

For the cod fishery, a 1911 summary of regulations specified that there would be no seining cod within half a mile of where hook-and-line fishing was going on. Cod-seines had to have at least 4-inch mesh in the arms and 3-inch mesh in the bottom. In the Gulf of St. Lawrence, nearshore trapnets required licences; regulations affected their spacing and mesh-size, and jiggers were prohibited. These were coastal regulations; the deep-sea fishery went unrestricted.[11]

Scattered attempts at development took place. In 1891, the department subsidized one Cathcart Thompson of Halifax to experiment with drying fish by absorbent pads. This marked the beginning of a long series of attempts to find a way to dry cod inside, thus to overcome the rain and fog that often interfered with outdoor drying. In 1905, the fisheries service put up a cod-drying plant at Souris, P.E.I., and leased it out in

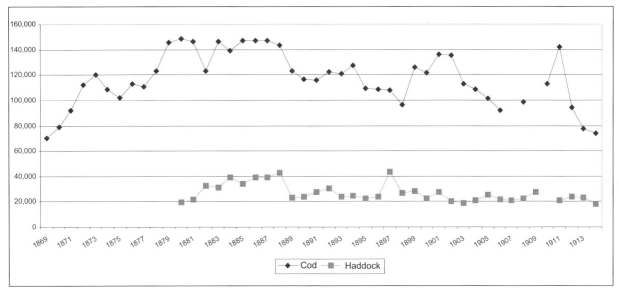

Cod and haddock landings (tonnes) in the Maritimes and Quebec, 1869-1914.

1909. And as noted earlier, the government from 1907 subsidized transport of fresh fish.

Pelagics: canning technology changes herring fishery

The Maritimes already produced salted and smoked herring. Perley's mid-century report had described a substantial fishery, especially in the Bay of Fundy. And Americans in particular had prosecuted a mackerel fishery in the Gulf. Now, canning technology would create a sardine-herring industry in the Bay of Fundy. Meanwhile, steam-powered purse-seines would lead to fears of mackerel depletion and a ban on purse-seining.

At the outset of the 1867–1914 period, the bait trade was of course still strong. Americans would both catch

Eastport ladies packing sardines. (From Goode, 1887)

and buy herring off Canada. As before, New Brunswick fishermen along the Bay of Fundy shore in winter sold frozen herring to American buyers.[12] In 1874–1875, U.S. interests were sending 30-odd schooners to New Brunswick (and 20-odd to Newfoundland) to buy herring for bait. They would then make early trips to Georges, and later fish the Grand Bank and the Western Banks off Nova Scotia.[13]

Meanwhile, Americans were applying the canning process with mixed success to many species and products: eels, menhaden, smelt, smoked sturgeon, halibut (on the Pacific coast), mackerel, scallops, smoked trout, smoked pike, smoked carp, caviar, finnan haddie, codfish balls, and chowders. Eastern Maine and southern New Brunswick had not only full-grown herring but an abundance of small ones, especially at Passamaquoddy Bay.

The idea of using small herring as "sardines" derived from Europe, especially France. After an earlier attempt failed, New York businessmen set up the Eagle Preserved Fish Company at Eastport in 1875. This company and the many that followed often used French labels to attract customers. By 1886, 45 sardine plants in eastern Maine employed 4,315 people. The number kept growing; by 1900, Maine had about 75 sardine factories, 51 of them in the Eastport–Lubec area.

Processing took a lot of work: salting the fish; spreading them on flakes (wood or wire frames); drying them outside by sun or inside by artificial heat; frying them in oil or steaming them; making the cans; filling the cans with sardines, oil, spices, and so on; soldering the tops on the cans (some men could solder a thousand cans a day); "water-bathing" the cans in boiling water; making a small hole to vent the superheated air from the can; resealing the can with a drop of solder;

121

and cleaning and boxing the cans. Around the turn of the century, factories got machinery to make and seal cans. Sardine processing in the 20th century became simpler and more mechanized, but still required, at least until recently, individual packing of every tiny fish.[14]

Herring-weir fishery builds up

Eastport and Lubec factories got their herring from the rapidly multiplying brush weirs, mostly on the Canadian side. Weir fishermen took out the fish with seines, either hauling them onto the nearby beach or making a circle inside the weir and brailing out the fish with dipnets. Small open sailboats, 18–30 feet long, took the herring to the factories. By the 1880's, some plants used steamers to tow the boats in.

The Canadian government soon made weir-builders obtain a licence, for $5. A new weir could be no closer than 600 feet to another, and was not to interfere with the other's fishing. The local fishery officer inspected the location and gave his recommendation to the head inspector for the province, who issued the licence. Once a person got a licence, he received preference, and could keep the licence even if he failed to build or maintain his weir. If someone else wanted to take over a licensed but unused site, he had to get a statement from the licence-holder that he had no objection. Some fishermen purchased a number of privileges and sold them to their neighbours. It was an early form of licence limitation.

After 1880, fishermen on the Fundy shore also began to use drag-seines along the shore outside weirs. Soon 50 or 60 seining crews, using large boats, were at work. Then a cannery proprietor introduced the purse-seine to the area, for seining in open water. A small sailboat successfully used a 250-fathom purse-seine.

Weir fishermen protested that because of the inrush of drag-seines, they were losing the benefits of the weir privilege for which they paid. The state of Maine had by this time banned the seining of herring. The Canadian government decided that the weir fishery was an established industry that deserved protection. In August 1886, the government banned both drag-seines and purse-seines in this fishery. A more general ban on purse-seines was to follow.

Meanwhile, the weir fishery kept growing, to supply the new factories. By 1893, Passamaquoddy Bay all told had 240 weirs. American tariffs on canned sardines held back the building of factories on the Canadian side.

In the 1880's, Canadians got a small sardine factory going at St. Andrews, near the U.S. border. It was at nearby Blacks Harbour, though, that the industry planted its feet firmly. Connors Brothers started up in 1885; in 1895, the large Connors plant was still the only one operating in Canada.[15] Other plants followed, although, over time, many would come under Connors's control. The sardine industry would become a major employer in southwest New Brunswick.

Purse-seining herring inside a weir. (From Goode, 1887)

Fisheries service tries to improve large-herring fishery

Meanwhile, the fisheries branch made some attempts to further develop the fishery for larger herring, then produced smoked or pickled in various parts of the Maritimes. In 1889, the department commissioned a report on the herring fisheries of the United Kingdom and Holland. The committee involved heard the same message from Canadian fishermen that Moses Perley's inquiry had heard in the Bay of Fundy four decades earlier: better quality required a price differential. But the final report suggested everything else but that.

In the early 1900's, E.E. Prince brought over J.J. Cowie from Britain, along with half a dozen "Scottish herring girls," to demonstrate better curing methods. Cowie, who would become a department stalwart, also demonstrated "drifting" gillnets at sea in the British method. Neither effort had great impact.

The herring fishery was already substantial. In 1890, Captain Gordon of the Fisheries Protection Service had reported that the herring and mackerel fishery on Canada's Atlantic coast probably used in total 1,110 miles of nets. By 1914, the herring fishery in the Maritimes and Quebec was landing 90,000

"Scottish herring girls" in Canada.

tonnes or more a year, a bigger yield than the cod fishery, although yielding a lot less value.

Among pelagic and estuarial species, the Maritimes smelt fishery also grew rapidly in the 1880's. The department in the 1880's extended its smelt regulations from New Brunswick to Nova Scotia, with special licences for bag-nets.

Purse-seines inspire fear

American fishermen by the 1870's, in their own and more northern waters, often used large beach-seines and purse-seines, set and hauled with steam power. From May to August on Newfoundland's northeast coast, American vessels in the shore fishery used seines that ran 333–336 metres in length and up to 28 metres deep; they could take 30,000 codfish in one haul. They also used purse-seines, notably for mackerel. Around 1888, steam-powered vessels followed schooners into purse-seining, using winches and large nets of 190–225 fathoms' length. Seines and purse-seines caused worries in both the U.S. and Canada.[16]

The Canadian government began cutting away at the use of seines in the sardine fishery, as already noted, and elsewhere. As early as 1876, an Order-in-Council forbade any cod fishing with purse-seines less than one-half mile from fishermen using hooks and lines. An 1876 Order-in-Council forbade the use of shad nets (apparently meaning beach-seines) greater than 250 fathoms' length in Albert and Westmorland counties, New Brunswick. An 1877 Order-in-Council forbade using gaspereau- and bass-seines in parts of New Brunswick and using seines to catch smelt anywhere in Canada.[17]

Mackerel scarcity in the Gulf of St. Lawrence, beginning in the late 1880's, helped to focus concerns about seines. The Americans had made use of seines; inshore abundance had dropped. Department officials and fishermen blamed the purse-seine for mackerel depletion. The department set various mackerel-protection regulations, including in 1892 a prohibition against leaving mackerel, herring, or gaspereau nets in the water during the day, and against tying together a fleet of gillnets longer than 60 fathoms. The chief regulation, however, involved the purse-seine.[18]

The Canadian government, having already banned purse-seines in the sardine fishery, approached the United States for an international ban on the gear. Otherwise, if American fishermen kept using purse-seines off Canada, it would be difficult to ask Canadian fishermen to give them up. When no action came from the U.S. side, the Canadian government banned purse-seines anyway, on August 28, 1891. Anyone using a purse-seine would face a penalty of $50–$500, and the government could also confiscate his vessel, boat, "and apparatus" used with the seine. The ban would last on the Atlantic till the 1930's.

When the longline and the cod-trap had first appeared in the groundfish fishery, they too had caused complaints, yet made their way into common use. But the purse-seine for pelagics, able to take greater volumes in a single haul, suffered damnation for several decades.

Shad fishery dwindles

Meanwhile, another herring-like species fell on hard times. Prince chaired the Dominion Shad Fishery Commission of 1908–1910, which consisted in large part of laments for a vanishing fishery. The report noted that back in the late 1860's, probably 60 farmers had fished shad at the head of the Bay of Fundy. The fishery had expanded; by the mid-1870's, about 40 boats fished at the head of the bay, another 130–140 at Pré d'en Haut, and more than 90 others from Port à Pique to Economy Point. But now, the commission said, the last-named area had only 10–12 boats. The whole fishery seemed a story of abundance followed by depletion.

This fishery never regained anything like its former importance. Today it is thought that much of the decline stemmed from dams and pollution in American rivers where it seems most shad originate. At the time, however, fish weirs got much of the blame for shad depletion.[19] Shad weirs at the head of the Bay of Fundy gradually declined in any case as the abundance of fish dropped.

Shellfish: lobster fishery spawns 600 factories

As early as 1873, six years after Confederation, the Commissioner of Fisheries, W.F. Whitcher, warned about overfishing lobster:

> This fishery has but lately assumed commercial importance, and is prosecuted chiefly on the coasts of Nova Scotia and New Brunswick. In the former Province about forty, and in the latter Province about twenty-four, factories are now in operation for the preparation and canning of lobsters. ... [The rapidly increasing volume and

Tending lobster traps at The Ovens, Lunenburg, N.S., 1879.
(Library and Archives Canada, PA-51113)

value] point to the necessity for economising and perpetuating the natural supply. It seems that excessive fishing has exhausted the lobster fishery along the north-eastern coast of the United States; and that the enterprise which was embarked in the same has now been transferred to Canada. Such being the case, if the same indiscriminate fishing should be practiced on our coasts, similar results might occur. Doubtless, for a short time all persons interested would prosper, and the country may appear to benefit by the rapid and extensive development of this resource; but a period of reaction must necessarily ensue, commencing sooner or later in an enfeebled or exhausted condition of the fishery. ... There is nothing easier than to exhaust a shellfish fishery, and nothing harder than to revive it. The oyster fishery of the country should serve us as a warning example.[20]

The Atlantic lobster fishery was expanding explosively as Canadians followed the American lead. From 1873 to 1883, the number of lobster factories in the Maritimes went from 60-odd to nearly 600, including about 100 in P.E.I. In Quebec, lobster canneries increased from 11 in 1877 to 99 in 1889.[21] Most operations were small and labour intensive, with workers soldering the tins by hand. Some exports of live lobster to the U.S. began.

Fishing regulations soon followed. The first rules in 1873 forbade the taking of spawn-carrying lobsters, lobsters weighing less than $1^{1/2}$ pounds, and soft-shelled lobsters (newly moulted and less marketable). Whitcher consulted his fishery officers by letter about closed seasons, which came into effect in 1874. Besides a closed season everywhere in the Maritimes during July and August, the new rules set a nine-inch size limit on overall length. The Maritimes industry made vigorous objections.

New regulations in 1877 and 1879 set different closed seasons for different parts of the coast, and made them longer. The idea was partly to take lobsters when they were in the best condition, partly to reduce exploitation. Gordon DeWolf's 1974 study on the economic effects of lobster regulations points out that "an important effect of the long closed season was to make lobster fishing a part-time activity. There were complaints from fishermen who depended upon sales for their winter supplies ... and from cannery operators because the season was too short."

For fishermen, lobsters became not only a supplement to groundfish and other fisheries, but in some cases their mainstay fishery. With new factories setting up, fishermen might more easily get out from under the thumb of previously dominant merchants. In the Gulf of St. Lawrence, lobsters helped break the hold of the old Jersey houses. At the turn of the century, the gas engine would further increase the small-scale fisherman's feeling of independence.

Regulations put no control on the ever-rising number of fishermen and traps. Still, in the 1880's the fisheries service gave at least some consideration to licences and leases. Gordon DeWolf notes that

[a]mong cannery operators, disputes often arose regarding property rights. A cannery might control a 2-to 4-mile frontage and, if successful, might attract other canneries to the same region. Established canneries complained of unfair competition and pressed for a leasing system that would give them defined property rights. Fisheries inspectors also pressed for a leasing system to stabilize the industry. ... They argued that canneries would be much more concerned with protecting lobsters if they were guaranteed future property rights. The federal government favoured free competition and argued that exclusive fishing rights would create a monopoly and take away bargaining power of fishermen.[22]

As the early rules went through various adjustments, a new market was developing for live lobster. Fishermen already knew about lobsters' ability to survive on shore. Early in the 19th century some Nova Scotians had sent several barrels of live lobster on a sailing vessel to King George III. Now, as the commercial fishery developed, shipments of live lobster to New England by sea began in the late 1870's. In 1891, the department helped a Captain McGray in an attempt to ship live lobsters to England.

Steep catch decline brings new regulations

By the mid-1880's, it seemed clear that lobsters were overfished and something must be done. An 1887 royal commission tried among other things to protect lobsters during their spawning period. Their recommendations led in 1887 to new closed seasons: west of Canso, July 1 through December 31, and elsewhere, July 15 through December 31.

Early in the 1890's, Samuel Wilmot and E.E. Prince suggested stricter licensing in the lobster fishery, including restrictions on the number of canneries.[23] (From 1893, canneries paid licence fees proportional to the volume canned: $2 per 100 cases.) There was little result. The value of lobster kept climbing. By the end of the century canned lobster would rival salt cod in value. But the average size of lobsters caught had dropped from the two- to four-pound range to a pound or less.

By then, a series of reports and regulations had culminated in an 1898 royal commission under Prince. This report noted that "the failure of the mackerel, cod and other fisheries, has had a great deal to do with compelling a large number of fishermen to take up lobster fishing with the result that the fishery has become practically the staple industry along large portions of the coast."

Although U.S. interests had pioneered the lobster canning industry, the report said, still it would be better if Canadian factories belonged to Canadians. The department should cease issuing licences to U.S. owners. The canneries should be spaced out at equal distances along the coast. Some might well ease the tight situation by diversifying into the canning of fruit. There should be temporary lobster reserves, closed to fishing. And those in the industry needed more information and education about it.

Nothing significant seems to have emerged from most of these recommendations. The royal commission did, however, result in the creation of six lobster districts, with longer fall and winter seasons in the south, shifting as the year advanced to shorter summer seasons in the north. Regulations also created size limits, larger in the south and smaller in the "canner" districts of eastern Nova Scotia, Cape Breton, and the Gulf.

Prince's 1898 royal commission had noted that it was "impossible to state the effect of past regulations." But for the most part, his commission only elaborated on previous regulations, making them comprehensive for the Maritimes. Prince's regulations, with periodic readjustments of seasons, district, and size limits, remained the basis of lobster management until the 1960's.[24]

Despite the new regulations, Canadian landings kept dropping, from about 45,000 tonnes in 1890 to about 30,000 by 1909. The number of canneries rose from 578 in 1896 to 760 in 1900, then fell back to 593 in 1909–1910.

Fishermen were taking younger and younger lobsters. After about 1900, the gas engine became popular. With it, the fisherman who had fished maybe 75–90 traps from a dory or sailboat could fish 250–300.

Wakeham's report loosens regulations

Given the alarming decline of lobster, the department tried to tighten up on lobster licensing. This action brought complaints and demands for a new inquiry, duly carried out in 1909 by the redoubtable Commander William Wakeham, officer in charge of the Gulf fisheries division. Wakeham was less alarmed than his predecessors about overfishing. Yes, the size of lobsters had gone down; but the canned pack had stayed fairly stable because fishing effort had increased, with more boats, better traps, and motorized boats working more ground. Although one couldn't drain the fishery forever, still the lobster showed "wonderful vitality" against destruction.

In today's parlance, one would call Wakeham a deregulator. He recommended open entry for fishermen, with no licence necessary; and open entry for canneries, though without allowing too many canneries for any one person. Strikingly, he recommended no size limit except in the Bay of Fundy, which served the live-lobster trade. But, he wanted stricter enforcement.

New regulations in 1910 applied size limits of $4^{1/2}$ inches' carapace length in Charlotte and Saint John counties, New Brunswick, and a 9-inch overall size limit for the rest of the Bay of Fundy. Elsewhere, size limits vanished. Following Wakeham's recommendation, however, regulations for the rest of the Maritimes set a minimum lath spacing, to let small lobsters crawl out of the trap. This regulation got rescinded in 1914.

The department asked factories to report the number of fishermen, traps, workers, and so on that they used. The fisheries service expressed the intention to favour fishermen's co-operatives. But regulation remained basically the same, based on seasons, protection for egg-carrying lobsters, and in the Bay of Fundy, size as well.[25]

The whole affair ended for many decades any serious attempt to limit lobster licences. Wakeham's dismissal of depletion as exaggerated and his questioning of size limits may also have contributed to a certain weakening of enforcement, never that strong in the early fishery.

Prince fails to get limited entry

The lobster fishery continued to fall off. Around 1900, catches ranged around 32,000 tonnes; by 1920, less than 20,000 tonnes. The number of Maritimes canneries kept dropping, to 512 in 1920.

In 1912–1913, Prince headed another royal commission, on Atlantic shellfish. Prince noted that fishery patrols were inefficient and needed a host of improvements, including motorboats.

The commission circled back to the old question of licence limitation. The Maritimes lobster fishery in 1913 had 25,000 fishermen. Despite industry resistance, said the commissioners, the department should resume the licensing strictness of its earliest years and limit entry into the industry. Prince wrote, "So long as the taking of lobsters on Canadian shores is a free fishery, so long will it be difficult to carry out the preservative measures that are desirable." The lack of such measures would exhaust the fishery, he said. But it was to take more than half a century before the department limited entry.

Another major recommendation also got stalled. The royal commissioners noted that lobster fishermen might well be licensed to fish in specific areas only. Such a regulation would come to pass, but only decades later. At the same time, the commissioners dismissed the suggestion of limiting boats to 300 traps each, saying it would never work. In the 1960's and 1970's, however, trap limits too would become part of lobster management.

The royal commission noted that tagging of lobster had proved them non-migratory, a conclusion that has stood up fairly well over time (although more recent studies have shown migrations of more than 100 kilometres in some areas, and larval transport may cause events in one area to affect another).

Prince's royal commission of 1912–1913 resulted in no great changes. The commotion and concern about the Maritimes lobster fishery subsided to a lower level, without vanishing. Policing of lobster regulations continued to occupy many of the fishery officers and guardians. Seizures of boats for illegal fishing took place from time to time.

Scallop fishery starts up; oyster fishery slows down

In the 1890's a small fishery for scallops started up on New Brunswick's Bay of Fundy coast, around L'Etang (near Blacks Harbour). Only around 1920 would the better known Digby, N.S. scallop fishery begin.[26] Far more attention went to the oyster fishery, which was visible and close to shore.

The fishery was already causing concerns in the early 1870's. A closed season of June 1–September 15 began in 1875. In 1887, the government appointed a commission to "enquire into and report upon the lobster and oyster fisheries of Canada." An expert hired from the United Kingdom, Thomas Kemp, damned the excesses of overfishing, cupidity, and stupidity that had harmed the fishery; he suggested size, season, and licensing regulations. Deputy minister of Fisheries John Tilton warned of the threat of extinction. In 1891, the department began licensing oyster fishermen (without limiting their numbers), and granting leases. It also circulated a petition to the public to help strengthen its hand in oyster regulations. In 1893, the government passed further regulations dealing with sizes and seasons. All this failed to bring a cure. In 1898, the distinguished Prof. W.F. Ganong noted that the fishery's fate must be either vigorous government interference or a slow death.

Production peaked in the 1880's and early 1890's, a high point of 64,600 barrels coming in 1882. But then, from 1902 through 1909, production never reached 40,000 barrels, in spite of increasing demand. And as Prince noted in the 1912–1913 royal commission on shellfish, not one person made a living from oysters.

Observers blamed the depletion on free fishing and a general lack of care. To quote the *Commission of Conservation Canada, 1911*: "In early years, oysters were actually burned in order to obtain the lime contained in the shells. Ice fishing, which was not prohibited by law until the past decade, was another prolific cause of waste. The oysters were raked up from the bottom through a hole in the ice, the large ones sorted out for market and the small ones left on the ice to perish with the cold." Before a law was passed, similar destruction took place in the summer fishery. And farmers damaged the beds by using power diggers to get shells, which were valuable as fertilizer.

Laws by now had restrained the power diggers; set a Dominion closed season; prohibited ice, night, and Sunday fishing; set a size limit; and made licences mandatory. Officials saw equal hope in leases, which might stimulate private efforts at conservation.

Before the 1898 judgement on jurisdiction, the Dominion government had granted leases for considerable areas. The court judgement created a deadlock over who would issue licences, since the Dominion government controlled fishing and the provinces claimed authority over the oyster grounds along the shore. Few leases still existed in 1910.

That year, to break the deadlock, the department got a law passed amending the Fisheries Act to allow agreements whereby the provinces would administer leases. Agreements came into force with Prince Edward Island, Nova Scotia, New Brunswick, and British Columbia. But the provinces refused to take over administration of the public beds as well. The federal–provincial jockeying meant that oyster culture would see little progress for years to come.

Meanwhile, Ernest Kemp kept writing long reports on the proper way to breed oysters and regulate the fishery. The fisheries service paid him to get a steam tug and use it to seed oysters and clean oyster beds.[27]

Dogfish foil development

By the early 1900's, the department had made various attempts at research and development, including Kemp's work on oysters, some early research on mackerel, exploratory fishing on the Pacific coast, and the great push on fish culture by Wilmot and others. Other attempts at development mostly involved underutilized species, and mostly petered out.

The quintessential underutilized species is dogfish. They are a nuisance, preying on other fish and destroying lines and nets. So why not fish them, to make some money and help conservation?

A special report by Prince in 1903 considered the dogfish question. He noted suggestions of dynamiting them, or inoculating them with fatal diseases. In 1904, the government decided against a bounty, and instead experimented with a "reduction" fishery, to grind them down for fish oil and fertilizer. The government put up a reduction plant at Canso, N.S. The government

Dogfish reduction works, Canso, N.S. (Library and Archives Canada, PA-20728)

bought the nuisance fish and produced fertilizer or fish-meal from them. In the next few years, two more plants came into being.

But in the next decade the plants faded away. Dogfish turned out to be harder to catch than people had thought, they were hard to handle, and the whole venture was uneconomic. This was to be the first of many unsuccessful attempts to deal with dogfish.

In another report in 1907, Prince urged attention to other underutilized species, such as catfish, sturgeon, skate, rock eel, roe in general, abalone, mussels, kelp, walruses, and whales. He was careful to note that market and enough volume to make a living were essential.

Following years brought little development for most of these species. A small fishery continued for eels. But sturgeon soon became scarce. Whales in Newfoundland were heading into a short boom and bust. Walruses remained underutilized because they were extremely scarce, never having recovered from the early fishery.[28]

After the Second World War, more "underutilized" species would take off, including, from Prince's list, roe, mussels, seaweeds, abalone (in British Columbia), and whales. But even then, development of fisheries was a story of fits, starts, and many failures. Prince was right to point to volume and especially to the market as key elements.

Salmon get close regulation

Maritime and Quebec salmon were already a major concern from decades past. They supported a relatively small commercial and a growing recreational fishery. By late in the century, "American capitalists" were entering the sport fishery in strength. Anglers soon learned how to lobby. The department's annual report of 1891 noted the Restigouche salmon club's various grievances.

Despite provincial control of private fishing rights, the fisheries service still set the conservation rules for all salmon fisheries. By 1910, the service was regulating salmon every which way, including closed seasons for net and fly fishing, size limits (three pounds), mesh size limits, spacing of nets, and no fishing within 200 yards of spawning streams. There were also specific regulations by province, including in New Brunswick a net licence fee of three cents per fathom. Officials allowed Native persons to fish for food.

A long struggle was now well under way to protect a species that became, for many people, almost the object of veneration. The beauty and vitality of Atlantic salmon, their long migrations past myriad threats to their spawning grounds, and their strength combined with vulnerability seized the imagination of school children, anglers, conservationists, and fishery officials themselves. Scientists and managers would labour for generations on behalf of Atlantic salmon, in a contest that still continues.

Stream thick with fish, ca. 1878-1883. (Library and Archives Canada, PA-214194)

Native fishermen take porpoises

Native people on the Atlantic generally took no prominent part in the commercial fisheries. But they were the chief figures in one lesser known fishery: on both sides of the Bay of Fundy, Mi'kmaq and Passamaquoddy hunters took porpoises, which yielded a superior lubricating and fuel oil. In southern New Brunswick's Passamaquoddy Bay, according to an 1880 American magazine article, "porpoise-shooting affords to the Indians of the Passamaquoddy tribe their principal means of support." The Native hunters would shoot and then spear the porpoise, hauling it aboard a frail craft, even in rough weather. The continued growth of petroleum-based oils ended the porpoise fishery.[29]

Regulation: more active than thorough

Besides the major royal commissions, other inquiries took place. For example, in 1903, a royal commission looked at the sardine fishery in the Bay of Fundy. In 1903–1904, the fisheries of New Brunswick in general and Gloucester County in particular came under investigation. In 1906, John F. Calder investigated U.S. ownership of fishing licences on the Canadian side of Passamaquoddy Bay, N.B. Another royal commission, starting in 1908, followed the decline in the Bay of Fundy shad fishery. And an Order-in-Council of September 12, 1907, set a host of regulations for fisheries across the Dominion.[30]

The fisheries branch was intervening more deeply, trying to restrain some fisheries, such as lobster, and to build up others, such as dogfish. It was building hatcheries, providing bait-freezers to fishermen, and subsidizing fresh-fish transport to help the trade grow. (The Dominion government also provided many public wharves in fishing ports.) A statistical system recorded production, science was growing, and administrators like Prince were thinking about every aspect of management.

But in few instances was management thorough and far-reaching. Licensing and limited entry never came into full play. Nor was there, then or later, much evaluation of the effects of regulations. Instead, officials would watch developments and react, often through royal commissions, in whatever way seemed sensible and acceptable at the time.

More growth than true progress

The 1867–1914 period saw great growth in the Maritimes and Quebec fisheries as groundfish peaked (and then declined), lobster canneries multiplied, herring landings increased, and gas engines changed operations on land and sea. But that growth came more by expansion than by improvement.

Maritime fishermen in the 1880–1910 period generally lacked organization, and held no great place in Canadian politics. Although information is scarce on incomes, the ceaseless out-migration suggests that they were generally low. Captain Gordon of the Fisheries Protection Service wrote in 1890 that the crewman on a high-line vessel might earn $150; on a "high average" vessel, fishing groundfish, mackerel, and herring, more like $88 for the season. He would get one-quarter of whatever he caught by hook and every fifth barrel that he caught by net. The owner on an ordinary vessel would get about $280.[31]

Generally speaking, of all Atlantic fishermen, those in the southern Maritimes did best. Always allowing for exceptions, the farther north and east the fisherman lived, the less substantial his boat and income were likely to be. Although the situation varied, fishermen mostly lacked money and power, and were now part of a trailing, rather than a leading, element in the economy.

Cowie laments lack of progress

J.J. Cowie, Prince's recruit who served the department for decades, published an article, "The Non-progression of the Atlantic Fisheries of Canada," in the department's 1909–1910 annual report. The concentration on saltfish, rather than fresh fish as in the United States, had held back development. Only recently, Cowie wrote, had anyone put smoked fillets onto the Halifax market. Retailing was poor. Despite the new subsidy, transportation remained a problem. Canadian fish products needed promotion. He advised bringing in an inspection program and revamping the inadequate Fisheries Intelligence Bureau.

Cowie's vigorous critique was the department's most open commentary to date on the slowness and slackness of the Atlantic industry. In the decades following, similar litanies recurred over and over. Different attempts at development got under way but met no great success, as though some essential element was missing.

Back in the mid-19th century, the heyday of a vigorous wood–wind–water economy, the Maritimes had been stronger. In the post-Confederation period, could business and political leaders have kept the Atlantic coast a trade entrepot where boatbuilding, fishing, and transport continued to reinforce each other? That alluring vision is probably impossible. Of the old pillars of the marine economy—lumber, shipping, trading, and fishing—in the 20th century only fishing retained anything like its old dimensions. Steam and steel put the others out of contention. The fishing industry by itself could hardly restore the general vigour of the old multi-element, self-reinforcing marine economy.

CHAPTER 9.
Newfoundland, 1867–1914

Weak economy, weak fishery

With a growing population—up fivefold during the century—but no growth in the fishery, Newfoundlanders in the last quarter of the 19th century felt hard-pressed. The fishery was still life and death to the country, with alternative employment scarce. But, as in the Maritimes, Newfoundland's leaders were putting hopes elsewhere than the fishery. The government commenced railway building in 1880, while the Macdonald government was setting up the completion of the Canadian Pacific Railway (C.P.R.). Five years later, when the C.P.R. reached British Columbia, the Newfoundland railway from St. John's had reached only the nearby port of Harbour Grace.

Big cod fish from trap at Battle Harbour, Labrador. The larger fish measured 5 feet, 5 inches, and weighed 60 pounds. From R.E. Holloway's 1910 book, *Through Newfoundland with the Camera*. (Library and Archives Canada, C-76178)

The colony went deeply into debt to build its railway. Fire destroyed most of St. John's in 1892. In 1894, the banking system collapsed. Canadian banks moved in. For a period, there was interest in Confederation. But when Canada objected to assuming Newfoundland's entire $16 million debt, discussions broke off. In 1895, Britain had to help out with funding.

Newfoundland was going through the same convulsions as the Maritimes after the decline of the wood–wind–water economy, only worse. Despite her legendary abundance of cod, Newfoundland had a less varied, more seasonal fishery than the Maritimes, and even less linkage with the new railway–agriculture–manufacturing economy of the continent. The colony struggled on, gradually completing the railway, encouraging foreign investment, promoting mining projects, and so on, with an optimism that makes Maritimers look like morose existentialists. But building up other industries was to take a long time.

International events:
Newfoundland resists foreign fleets

At the outset of the 1867–1914 period, Newfoundland was still trying to assert her power in the sea fisheries. As Newfoundland built up her fishery on the Labrador, the government moved against other traders in the area, including U.S. schooners. Duties on foreign goods came into effect. The old fishing-ship firms, now mainly or only traders, finally died out completely, the last of them disappearing in the general depression that began in 1873. As both Newfoundland and Lower Canada expanded their fisheries on the Labrador, Nova Scotia's position suffered.[1]

When restored fisheries Reciprocity reopened all of Newfoundland's coast in 1871, some American vessels began seining their own bait rather than buying herring from Newfoundlanders. They would sometimes bar off herring in coves until winter came, when they could transport frozen herring south, for food and bait.

Newfoundland had already restricted use of the purse-seine by its own fishermen. In 1876, Newfoundland passed a law prohibiting seining between October 20 and April 25, the season when the Americans sought frozen herring. Nor could fishermen bar herring off in coves.

Fortune Bay incident riles Americans

On January 6, 1878, occurred what became known to governments as "the Fortune Bay incident." As Americans were seining at Long Harbour, Fortune Bay, a Newfoundland crowd drove them away and destroyed several nets.[2]

Now rose a question left unresolved by the 1818 Convention—that is, whether Canada and Newfoundland could regulate the Americans when fishing close to shore. The British government maintained that the Treaty of Washington and earlier treaties had admitted the Americans to a regulated fishery, not a free one; the Americans needed to comply with local rules. The Americans contended that the British side was whittling down their rights under the treaty, and that in any case, the American fishermen deserved compensation for the violence used against them.

Further incidents followed, at Conception Bay in Newfoundland and in Cape Breton. The resistance cut into American fishing; the U.S. Commissioner of Fisheries, Spencer Baird, noted in 1887 that because of northern opposition to catching of bait, Americans had in large part abandoned the practice, especially after the Fortune Bay incident. They now mostly bought herring, or they hired provincial fishermen to catch it for them.

If the Fortune Bay incident helped push the Americans out of the herring fishery, it also helped push them out of fisheries Reciprocity. Adding to the discontent over the Halifax Award, the Fortune Bay incident helped prompt the Americans to abrogate the fisheries provisions of the Treaty of Washington in 1885. As for Canada, the U.S.–U.K. modus vivendi let Americans use Newfoundland ports from 1880 on, so long as they paid a licence fee.[3]

Laws restrict sale of bait

When fisheries Reciprocity ended, Newfoundland lost some of her small American market. She now depended more on Europe. But markets there were softening too, as France and Norway increased their competition.

Once again, Newfoundland politicians saw the control of bait as a lever of power. Foreigners, especially the heavily subsidized French, were using bait bought in Newfoundland to catch cod. Why not halt the sale of bait to foreigners?

In 1886 and 1887, legislation restricted the sale of bait to foreigners, exempting Canadians. Later, Newfoundlanders accused Canadian fishermen of weakening the Bait Acts by selling bait to the French at St. Pierre and Miquelon, including herring from Cape Breton, the Magdalens, and Newfoundland itself. Meanwhile, the French, with less Newfoundland bait, were turning to salt bait, and also using bait that they fished on the banks. How much influence the Bait Acts had in discouraging the French fishery is unclear.

In 1890, a modus vivendi let the French buy licences for port privileges, much like Canada's and Newfoundland's arrangement with the United States. This eased the situation, though tensions remained.[4]

Canada torpedoes Newfoundland Reciprocity

The new bait legislation got reaction from Americans. Unable even to buy bait from Newfoundlanders after the Bait Act cut off purchases, they had to fish bait themselves on the Treaty Shore, or bring it from elsewhere. This inconvenienced the still-important, though declining, American fishery at Newfoundland. In 1890, Robert Bond, a minister in the Newfoundland government, negotiated with U.S. Secretary of State James Blaine a limited reciprocity treaty that would, among other things, let the Americans fish or buy bait, and let Newfoundland export fish duty-free to the U.S.

Canada, fearing American expansionism, made various objections. Why should the U.S. benefit more from Newfoundland fisheries than Newfoundland's fellow colony? Any reciprocity arrangement should benefit all British North America. After a diplomatic fuss, Great Britain blocked the "Bond–Blaine Treaty." Newfoundlanders saw this as a major injury; generations later, some still resented it.

Stung by Canada, Newfoundland again restricted the sale of bait, except to Americans, who could again buy it. Canada retaliated with duties on Newfoundland fish; and Newfoundland, with duties on Canadian flour. In 1892, the tariff contest relaxed. Canadians could again buy licences to buy bait.[5] Meanwhile,

Americans bought large amounts. By 1898, in Placentia Bay, two to three thousand Newfoundlanders would sell herring to the foreign vessels, with hundreds more fishing at Bay St. George.

In 1902, Bond, now Prime Minister, tried again with the "Bond–Hay Treaty," negotiated with U.S. Secretary of State Hay. This time the U.S. Senate, in 1904, refused approval.

Newfoundland wins right to regulate foreign fishing

The colony returned to the charge against the Americans. On the Treaty Shore of the southwest and west coast of the island, the Americans still claimed the right to fish on their own terms, while Newfoundland claimed the right to regulate their fishery. A prime issue was whether Americans could use purse-seines, outlawed for Newfoundlanders.

After the U.S. Senate rejected the Bond–Hay Treaty, Newfoundland again restricted the sale of bait to Americans. The colony's Foreign Vessels Fishing Act of 1905 provided for forfeiture of U.S. vessels using Newfoundland bait or crews. But restricting sales of bait was often difficult. Fishermen of the south and west coasts, far from St. John's, had their own profits to think of. Some smuggling continued. Also the Americans caught their own bait when they could.

Still, the restrictions held enough strength to irritate the Americans, who took up the issue with Great Britain. The whole fuss ended in another modus vivendi in 1906, which let the Americans use purse-seines under certain restrictions on the Treaty Shore to catch bait and let them employ Newfoundlanders to fish with them outside three miles.

The modus vivendi inspired new protests in Newfoundland. Great Britain and the United States finally submitted the whole question of American fishing to the Hague Tribunal, whose judgement in 1910 affirmed in large part Newfoundland's rights to regulate U.S. fishing and control the sale of bait. Total sovereignty over the territorial waters along the Treaty Shore rested in Great Britain and its possessions, that is, Newfoundland. But Newfoundland's regulation must be reasonable, and treat local and American fishermen equally. Newfoundland still had to satisfy the Convention of 1818, which let the Americans use the Treaty Shore.

The judgement also settled some contentious related questions. For example, the American vessels had a right to hire Newfoundland or Canadian crews, but Newfoundland or Canada could restrict their own citizens from such employment (rather than forbidding the Americans to hire them). And the three-mile limit should follow the sinuosities of the coast, except for straight lines drawn across the mouths of bays.

In any case, the Americans were now less interested in drying fish on the Treaty Shore, or even in fishing bait for themselves. Their fleet was shrinking in northern waters. After the Hague decision, fishery relations between the two countries proceeded in a more peaceful, less disputatious manner.[6]

Newfoundland gains control of French Shore

The lobster fishery caused another complicated clash. Newfoundland settlers began developing a lobster fishery around 1880. So did French operators, and both nationalities wanted to use the French Shore in western Newfoundland. Although the French had no rights of settlement, they claimed exclusive fishing rights on the French Shore. They now contended that the same rights that let them dry cod should let them can lobster.

Newfoundlanders were pressing westward. Great Britain at first resisted Newfoundland's assertiveness, but by 1881 conceded that the colony had territorial jurisdiction over the French Shore.[7] As Newfoundlanders began setting up lobster factories on the French Shore, the French protested. In 1887, a French warship destroyed some property. Charges and countercharges followed, Newfoundlanders protesting against the French lobster factories, and the French complaining about Newfoundlanders encroaching with cod-traps.

In 1889, Newfoundland outlawed cod-traps on the French Shore. A modus vivendi gave each lobster packer, French or British (that is, Newfoundlanders), a specified strip of coast, under the command of French or British commodores. British lobster factories increased to 59 by 1887. The French had fewer: in 1894, 14 lobster stations, with 15 vessels and 649 men. The rule in the 1890's was no new factories without mutual consent. The lobster situation remained an uneasy stalemate.[8]

Newfoundland wins more control

By 1904, Britain and France were moving into the "Entente Cordiale." As part of the settlement of differences, they agreed that France would abandon her claims to exclusive use of the French Shore, which ran

Fishing vessels bound north for Labrador. (Library and Archives Canada, C-19130)

up the whole west coast and back down the northeast coast as far as Cape St. John.

This opened territory in a clear manner for fishing. The French lobster factories vanished; Newfoundlanders took over. France gave up the right to land and dry fish. Britain made a monetary settlement with France, but also agreed that France would have rights to an equal summer fishery. French fishing rights thus continued, though modified.

The Newfoundland authorities continued to worry about smuggling of bait to St. Pierre and Miquelon, and railed against Canada for allowing bait sales from the Maritimes to the French. But in following years the French Shore problem faded. The number of vessels coming to Newfoundland's shores from France and St. Pierre dropped. Still, St. Pierre's own locally based fishery continued; and France's wet-salted fishery on the Grand Banks, which had 226 vessels in 1904, continued in a fairly steady fashion until the Second World War.[9]

The fishery, 1867–1914: large Newfoundland fleet produces little growth

In the 1870's, Newfoundland possessed a fleet of similar size to those of the Maritimes and Quebec, although with fewer banking vessels. The number of men fishing, more than 50,000 in Newfoundland, was also broadly similar. But Newfoundland had less room for growth. The cod fishery, which had employed large numbers of small open boats for centuries, in the last part of the century reached a peak and dropped off. The same pattern prevailed in the Maritimes and Quebec, but they could divert more energy to other fisheries. Their herring and lobster resources, and others to be developed later, exceeded those of

---

Newfoundland's fishery in the 1870's

H.Y. Hind's 1877 report, referred to above, showed a major Newfoundland fishery. The colony was generally ahead of Norway in producing cod. And in North America, "the mean annual yield of the Sea Fisheries of the United States,—the greater portion of the catch being made in waters off British American coast lines,—is not much more than half of the combined catch of the Dominion and Newfoundland."

Hind's report described a large fleet fishing mostly along the shore.

In Newfoundland the DEEP SEA FISHERY, as distinguished from the fishery pursued in coastal waters, or within three Marine miles from the shore, has scarcely a separate existence. The vessels which are enumerated in the census are used chiefly for the purpose of sailing from one Coastal Fishery Station to another on the Island of Newfoundland, or for the Labrador Fishery. The total number of boats employed in the Shore Fishery was 18,611 in 1874 and 14,755 in 1869, and the number of persons engaged in catching and curing fish 45,854 in 1874 and 37,259 in 1869. The number of vessels was 1,197 during 1874, with a tonnage of 61,551 tons, manned by 8391 fishermen sailors. Sealers are included in the enumeration. ... The appliances used in these Fisheries indicate, to a certain extent, their character. Where decked vessels are necessarily employed beyond the limit of three Marine miles from the shore, it is essentially a DEEP SEA FISHERY. Where open boats only are used, it is in general a COAST FISHERY, although, as in the case of Newfoundland, the depth of water near to the coast line may vary from 10 to 100 fathoms and more. Where the fishery is pursued from the shore, but with the use of open boats, as in the taking of Mackerel, Herring, and especially Caplin, Smelt and Launce, it is a STRAND FISHERY. Both the Deep Sea Fishery and the Coast Fishery are dependant in a very large measure on the Strand Fishery for Bait. The character of the Newfoundland fishery is further indicated by the large number of Fishing Rooms, in actual use. In 1874, these amounted to 8902 in number, in 1869 to 7,444.

The Maritimes and Quebec shore fishery in 1876 had 20,241 boats, slightly more than Newfoundland's. The vessel fleet numbered 1,379, also slightly more than Newfoundland's. But Maritime and Quebec vessels tended to be smaller, an average 32.5 tons. Of the total of 21,620 craft, about 1 in 15 was a decked vessel, the same proportion as in Newfoundland. Broadly speaking, the two fleets were equal.

Employment was also similar in the two regions. In 1874, Newfoundland's shore fishery had 45,854 persons catching and curing fish; the vessel fleet had another 8,394 "fishermen sailors." Fishing employment including sealers totalled 54,248. The Maritimes and Quebec in 1876 had 40,023 men in the shore fishery and 9,097 in the vessel fishery; altogether, 49,120 persons catching and curing fish.

Cod was of course Newfoundland's biggest fishery, by far. In 1874, dried cod was worth $6.1 million; seal oil, $610,000; pickled herring, $569,000; sealskins, $518,000; cod oil, $470,000; and salmon, $118,000.

Spreading codfish to dry, with schooners in background. (Library and Archives Canada, C-74893)

Newfoundland, and could offset the decline of employment in the groundfish fishery.

Subsidies encourage bank fleet

In the 1870's and 1880's, Newfoundland's cod production was higher than ever. But fishery growth was falling behind population growth: as the number of Newfoundlanders more than doubled between 1845 and 1891, cod exports rose only 25 per cent. In 1880, the Joint Committee of Council and House of Assembly was saying that the fisheries offered little hope of increase.

Meanwhile, the adoption of steamers had lessened employment at the seal hunt. Seal production itself declined after the 1860's. The historian Prowse wrote in the 1890's that "[p]olitics and steam have done more than any other cause to ruin the middle class, the well-to-do dealers that once abounded in the outports." Steel sealing vessels weakened the wooden-boatbuilding industry around the coast, as did the steel steamers setting up new routes in the 1880's to the West Indies and Brazil.

In the 1870's and 1880's, the government provided subsidies to assist the fishing and shipbuilding industries. The subsidies helped build up the bank fishery, especially from the south coast. The number of banking vessels rose to 279 in 1890. Although the numbers soon dropped back as the subsidies disappeared, the bank fishery remained stronger than before. In 1901, the bank fleet numbered 118 vessels, with Grand Bank the main port.

(The south coast would remain dominant on the banks. The 1937 Commission of Enquiry on Newfoundland fisheries noted that only Fortune Bay and Placentia Bay had banking fleets, "those areas being nearest to the Grand Banks, where conditions are most suitable. The largest vessels go as far as Labrador and Straits of Belle Isle to fish during the months of September and October, and in recent years, some have gone as far North as Greenland."[10])

Meanwhile, with American and some Canadian schooners turning offshore to the Grand and southern banks, Newfoundland vessels were becoming more dominant on the Labrador. The Labrador floater fishery in the decade 1900–1910 reached its high point; in one season, more than 1,400 vessels took part. The schooner fleet made extensive use of superannuated Nova Scotia vessels.

Cod-traps multiply

Although earlier decades had seen precursors, the Newfoundland cod-trap came into being only around 1871. It was the work of W.H. Whiteley, a planter at Bonne Esperance on the lower North Shore of Quebec. Whiteley was also, from 1867 to 1897, Canada's Fishery Overseer for the Bonne Esperance Division (about 60 miles of coast).

In his 1876 licence application to the Canadian authorities, Whiteley called his new device a "pound net." In the cod-trap, as in many other fish traps, a leader net running out from shore diverts the fish into the mouth of an enclosure. The cod-trap, in one

133

Hauling a cod-trap at Indian Harbour, Labrador.
(Photo by R.E. Holloway. Library and Archives Canada,
C-46721)

mated 400 vessels on the coast, mostly from Newfoundland.

New cod fishing regulations for the Gulf of St. Lawrence required that each trap have a licence. Residents got prior claim to location. Cod-traps were not to interfere with salmon fishing. There must be at least 250 yards between traps, and there could be only one trap per vessel. Each cod leader had to extend from shore. A minimum mesh size applied, and a fee of 50 cents for each fathom of leader. Bultows or gill-nets were prohibited within three miles of any island; and jiggers were prohibited.[11]

The cod-trap would find its main use in Newfoundland, where it would become, in many areas, the dominant gear.

description, "resembles a room with twine walls, a floor and a door."

The cod-trap spread rapidly, to the point that Whiteley himself began to consider it a menace. On Quebec's Labrador coast, shore fishermen soon were complaining about schooners using traps. In 1894, Whiteley recommended regulations to control the esti-

Higher volumes bring lower quality

In comparison with hooks on cod-seines in the shore fishery, the cod-trap produced bigger volumes, and extended the season. As steamers made it easier to get passage to Labrador, the number of fishing stations increased. The merchants' control over operations,

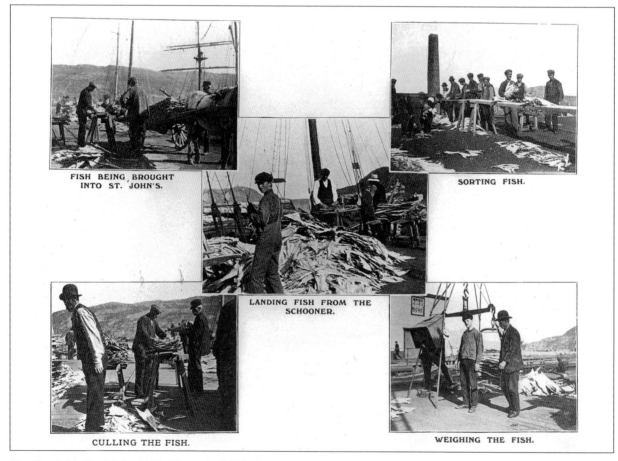

FISH BEING BROUGHT INTO ST. JOHN'S.

SORTING FISH.

LANDING FISH FROM THE SCHOONER.

CULLING THE FISH.

WEIGHING THE FISH.

Handling fish in St. John's, 1910. (Photos by R.E. Holloway. Library and Archives Canada, C-76173)

including production quality, decreased. Part of the reason was that cod-traps produced bigger catches of smaller, younger fish, with less time for curing. More and more, fishermen sold their salt cod "tal qual"—a single price, whatever its quality.

Meanwhile, markets weakened as some customers faced harder times. The United States was turning more towards fresh fish rather than salt. Some countries such as Puerto Rico and Cuba, under American tutelage, were raising tariffs. Only Brazil provided a growing market in the latter part of the century. In Europe, France and Norway were strong competitors, aided by trawlers.

Newfoundland's marketing system was weak. Unable to hold any kind of price, exporters large and small raced each other to the overseas markets to unload their fish. The first shipments to arrive in Spain got the best price; this prompted merchants, seeking large early cargoes, to buy much of the fish from Labrador ungraded (tal qual), for one price. Selling on consignment became more and more common. Weak marketing and a weak industrial structure led to low and unstable prices to the fishermen.

As the century drew to a close, some of Newfoundland's old, major merchant firms retrenched into wholesale supply and merchandising. Smaller enterprises, ranging from local merchants with several vessels to single families with one small sailboat, competed with the old-time merchant operations. But they failed to match the product quality.[12]

Life around the bay

Not everyone fished the Labrador. St. Mary's Bay on the southern Avalon Peninsula was illustrative, with a mixture of small and larger craft, jacks and western boats that mostly fished in the bay or near the mouth, and some larger vessels that went to the Grand Banks. Local merchants had stores and small fleets, and could

Women tending fish flakes in Newfoundland, 1903.
(Library and Archives Canada, PA-124429)

Fishing boats outside St. John's harbour, ca.1909.
(Library and Archives Canada, C-37556)

advance limited credit. For substantial credit, boat-owners dealt with merchants in St. John's.

Boat-owners would finish up the year and take their fish, maybe 30 or 50 or 100 quintals, to St. John's on their own vessel, or send it aboard somebody else's, to firms such as Job Brothers or Bowring's. They would also send or take a list of provisions—everything they needed for the next year—which would come back by the same boat. Very little money might be left after buying supplies. By "the hungry month of March," money would be scarce indeed. Planters would sometimes employ hard-up people who would work just for food and maybe a few articles of clothing.[13]

Then life would pick up as the new season approached, and fishermen prepared boats and nets. The Newfoundland politician Peter Cashin wrote that in the 19th century

> [m]otor cars or trucks did not exist. Motor engines for boats were not even dreamed of, and everything depended for speed on a fair wind for the coaster, two pairs of oars for a dory and four eighteen foot spruce oars and a sculling oar for a trap skiff.

> ... Great activity prevailed during the months of April, May and June. Hundreds, yes, thousands of fishermen planters and fishermen came to St. John's to procure supplies from the various merchants in order to prosecute the cod fishery. The small harbour of St. John's was literally crowded with vessels, schooners, western boats and jacks, from nearly every part of the country.[14]

Around all the bays, most dealings were in truck. Local merchants who sold household and other goods also bought fish and would deduct the value of a man's fish from the amount he owed for goods. In some years, a fisherman might never see cash. The diet was largely fish. When chickens were imported to the northeast coast, some people unused to the creatures preferred to keep them as pets.

The fishery at the turn of the century relied on handlines, longlines, cod-traps, gillnets (very few), and cod-seines (fast declining). After the Bowring firm's unsuccessful trawling experiment in 1899, nobody else used trawlers until the mid-1920's. Gas engines appeared in Newfoundland by 1910. By 1914, an estimated 4,000 Newfoundland fishermen used gas engines, bought mainly from Canada.[15]

Death at sea remained common. In 1885, a gale wrecked 70–80 vessels on the Labrador coast, killing about 70 men, women, and children. In 1902, a more ordinary year, 17 people lost their lives in the bank fishery; 8, in the shore and Labrador fishery (the Gloucester fleet lost even more).

People might fish the whole season at home, or travel long distances. Stationers might move to their location in spring by motorboat, schooner, or coastal steamer, frequently with their families. Floaters moved their schooners from place to place along the coast. Both stationers and floaters returned to shore daily. The Labrador floater fishery in the decades 1890–1910 reached its high point; in one season, more than 1,400 vessels took part.

Local management only beginning

Although vigorous in international matters, Newfoundland was only beginning to address conservation itself. Over the years, the oldest colony had built up fewer local fishery regulations than the Maritimes. There was less river fishing, where local problems spawned local regulations. Indeed, local government barely existed; and a colony that depended totally on fishing still in the 1880's had no fisheries department.

In that decade, however, the colony passed some conservation regulations such as mesh size restrictions.[16] Then, in 1890, the government set up the Newfoundland Fisheries Commission and hired a Norwegian expert, Adolph Neilsen, to take charge of matters. In 1898, the Commission became the Department of Marine and Fisheries. Around 1902, the authorities created a Newfoundland Fisheries Board, attached to the department, which acted as an advisory council.[17]

There were some regulations such as those prohibiting cod-traps in certain areas, to avoid gear conflicts. But the huge groundfish fishery saw few restrictions. More attention went to encouragement. Newfoundland began subsidizing the refrigerated (by ice and salt) storage of bait, the first such operation starting in 1893 at Burin. Fishermen failed to keep up the operations.

In other development work, the department distributed directions for curing and packing herring and for making cod-liver oil. A new Norwegian method of making that oil had brought some additional value to the fishery. The government also tried to encourage better quality in making salt cod.

The industry itself was adopting some new approaches. The cod-trap was rapidly replacing cod-seines. By 1870, the Job Brothers concern in Newfoundland was producing fertilizer at Bay Bulls, Catalina, and Lance-au-Loup. The company built two small steamers to collect material.[18]

Salmon, herring fisheries get more regulation

The new Newfoundland fisheries department concerned itself, like the early Canadian department,

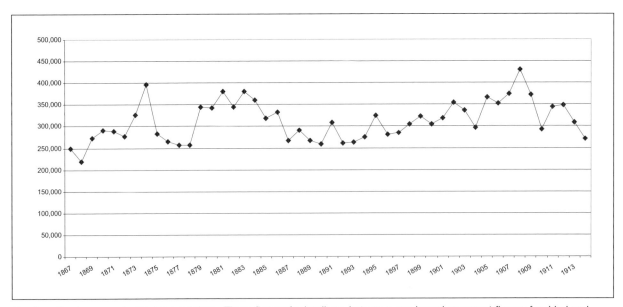

Newfoundland was producing large catches. These figures for landings, in tonnes, are based on export figures for dried cod, with a conversion factor of 4.88.

largely with river and inshore conservation: prevention of illegal netting, provision of fishways, and such. Nets were ruining the Gander River salmon, officials noted in 1902. Instead of trying to make up for it with expensive hatcheries, they said, one might do better to assist nature by letting the fish run freely. The department installed the first fish ladders in 1904, decades behind Canada.[19]

Around 1900, Newfoundland interests began exporting fresh salmon to the United States, as Maritimers had done for some decades.[20] On the Labrador, Newfoundland schoonermen would simply break ice off the icebergs to pack them. Fresh salmon gradually replaced the old trade in pickled salmon. The department restricted the use of nets in salmon rivers.

Newfoundlanders were already salting herring for export. The Newfoundland government encouraged the use of herring for food, importing Scotch packers in 1898–1899 to demonstrate the cure of herring. Some herring were packed following the superior Scotch cure; but as the governor of the day noted, the "industry does not seem to increase." Then as now, the Newfoundland herring business was an up-and-down affair. At Fortune, on the south coast, it was said that gurry had ruined the grounds. And early in the century, the herring fishery suffered a downturn at Labrador.

Americans and Canadians continued to buy frozen herring to take back to their own shores, for bait or food. To improve returns, the government set a minimum price of $1.25 per barrel. This was probably the earliest instance of a Canadian or Newfoundland government setting a minimum price for any fishery product. In future years, Newfoundland governments would many times intervene in the marketplace.

Lobster fishery booms and declines

Although never anywhere near the size of the Maritimes lobster fishery, still the one developing in Newfoundland was important. Lobster canning first took place in 1858. In 1879 came the first live lobster exports, although it would be decades before quantities became significant.

In 1890, Newfoundland set a size limit for lobsters. The fishermen largely disregarded it. The government also forbade the use of berried (egg-bearing) lobsters and set closed seasons, which went through later adjustments.

Lobster was dirt cheap. Newfoundland fishermen in the early fishery received 70–80 cents per 100 pounds. Besides spears, hooks, and gaffs, early fishermen at first used the hand trap: a circular hoop with a net that the fisherman guided by hand. Some fishermen on the west coast of Newfoundland in the 1880's were too poor to buy nets. Newfoundland forbade hand traps between 1905 and 1910, by which time regular traps had come into use.

By 1898, Newfoundland had 1,020 lobster "factories," most of course very small, even one-man opera-

tions. The colony set up many lobster hatcheries, which must also have been tiny operations. In 1890, a regulation granted authority to require fishermen to bring berried lobsters to the hatcheries, so the government could incubate and release them. This attempt at re-planting the harvest continued for years, fading away in the early 1920's.

By 1900, the colony had 1,400 lobster factories, and by 1913, an amazing 2,762, but they put up an average of only six cases each per year. Lobster size too, was decreasing. The Newfoundland department now had only three lobster hatcheries, and officials were questioning their value. As years went by, lobster would become so scarce that the authorities shut down the fishery for three years in the 1920's.[21]

Seal fishery slows down

The seal fishery after the 1860's began to drop in production, value, and employment. Oil production fell in particular, as petroleum products replaced marine oils. Exports of seal products in the 1890's came to only one-third of their value in the 1850's; the seals' share of Newfoundland exports fell from 24 per cent to 9 per cent.

Operators in the 1860's had begun using steam engines to power the large "wooden wall" sealing vessels. By the 1880's, about 20 steamers dominated the fishery; these could be more than 200 tons, with crews of nearly 200 men. They had displaced scores of smaller vessels. Total manpower had fallen from more than

The famous sealing steamer *Terra Nova* also served in Arctic and Antarctic exploration. (Library and Archives Canada, C-2013779)

10,000 in the 1850's to 5,000. By 1900, exports of sealskins averaged only around 250,000, compared with more than 400,000 in the 1830's, 1840's, and 1850's. As money got scarcer, low pay prompted a sealers' strike in 1902.

In 1898, the government of Newfoundland began offering operators four dollars per ton for each vessel in the seal fishery. Intended to encourage sailing craft, the bounty made little difference. The trend to bigger vessels and less production continued. Steel steamers operated from 1906; the largest, the *Stephano*, was 2,143 tons.

Some fishing and trading schooners still chased seals in season, especially from the south coast in the Gulf, but they produced less than ten per cent of the harvest. Employment would keep dropping, to fewer than 2,000 people in the 1920's, out of a population that was now well over 200,000. Although the exploits and hardships of sealers had by then firmly embedded themselves in Newfoundlanders' culture and self-image, the seal fishery had lost its prime status in Newfoundland's economy.

As early as 1866, the government had begun legislating to deal with problems in the seal fishery. The major regulation came in 1873, intended to let white-coats grow as large as possible before being killed. Steamers could not leave port before March 10; sailing vessels, before March 5. Another act in 1879 forbade killing "cat" (immature) seals. Additional legislation followed on other aspects. But the gradual decline in the seal fishery seemed to be as much, or more, a matter of economics than of conservation and abundance.[22]

Newfoundland whaling exceeds Canadian

Newfoundland whaling enjoyed a brief boom, outpacing development in Canada, where whaling was a small, off-and-on business. On the St. Lawrence estuary, a subsistence fishery for white whales, using weirs, dated back to New France; and Gaspé-based schooners had long taken whales of various types, up until the 1890's.

Newfoundlanders did little whaling until the 1890's. Adolph Nielsen, the Newfoundland superintendent of fisheries, came from Norway, where whales were now depleted, thanks to the new harpoon guns. Nielsen in the late 1890's got some

of his countrymen interested in whaling at Newfoundland.

Although Norwegians on Norwegian-built boats did most of the early Newfoundland whaling, British–American capital also moved in, and British subjects manned the factories. Newfoundland whaling became a sizeable but brief effort. Markets were strong enough: even though kerosene and petroleum products had reduced the demand for marine animal oil, buyers still wanted baleen for corsets, umbrellas, whips, and other products.[23]

Tonnessen 's history of whaling recounts the story of an 83-foot blue whale that when harpooned off Newfoundland towed the 545-ton whaling vessel at two miles per hour, in and out among icebergs, from 7 p.m. to midnight, against the vessel's engine running half-open in reverse. It took the whalers ten and a half more hours to kill the whale. Another whale towed the vessel *Puma* in reverse, nearly pulling her stern under, against the steamer's engine running full ahead. This whale took 27 hours to kill.[24]

(Whale-killing sometimes excited the participants. Commander William Wakeham, an officer with the Canadian department, wrote of the blood frenzy of American whaling and walrus-hunting crews in Hudson Bay. In another story from the early years of the century, men at St. Mary's Bay in Newfoundland were carrying an old lady's coffin along the shore to the cemetery for her funeral when small whales struck in. The men attacked them with axes and whatever they could find, until finally someone noticed the coffin floating away on the tide.)

Humpback whale at Snooks Arm, Nfld., 1910. One fluke has been removed from the tail. (Photo by R.E. Holloway. Library and Archives Canada, C-76175)

From 1898 to 1911, about 20 whaling stations started up in Newfoundland. In 1904, the whalers took 1,275 whales. But by 1914, scarcity had closed most factories. A factory at Rose-au-Rue at Placentia Bay, operated by the Newfoundland Steam Whaling Company, would operate until 1946. Another at Hawke's Harbour would last to 1951, and revive briefly in 1956–1959. In the 1960's, a small revival saw two factories operate in Newfoundland.[25]

The decline took place despite seemingly strict Newfoundland laws. The Whaling Industry Act of April 22, 1902, issued permits for a limited time only, and also controlled the number of boats. But Tonnessen et al. point out that the law set no limits on the fishing season or the number of whales; and despite Inspector Nielsen's warnings in 1903, the government issued more permits. It seems a story of limited entry applied too loosely.

Fishery shows little dynamism

In the early years of the 20th century, the efforts of Newfoundland's new department of Marine and Fisheries failed to change the general gray picture of the fishery. The cod-trap had spread, and the bank fishery had grown. But Newfoundland turned more slowly than Americans and Nova Scotians to the banks, was too far from market to develop much of a fresh-fish trade, showed little interest in trawlers, had less luck in developing new species, and was in some ways more on the fringes of things, despite her abundance of cod. Even Canadian and American fishing captains were losing interest in Newfoundland fish.

If J.J. Cowie, who reported in 1909–1910 on Canada's Atlantic fisheries, had visited Newfoundland, he would have found it like the weaker sections of the Maritimes and Quebec. The "non-progression" of which he complained was even more evident. In the 1880–1910 period, Newfoundland had no new developments with the strength of the Maritimes' sardine and lobster industries. And Newfoundland was slow to develop alternative industries.

CHAPTER 10.
On the Pacific, 1867–1914

Salmon industry grows fast

British Columbia's great salmon rivers supported only a small commercial industry, for salted and barrelled salmon, until the tin can came along. In the 1860's, canneries had started up in California and the American northwest, and an experimental venture took place in British Columbia.[1] Then, in 1870, four partners—Messrs. Hennessy, Loggie, Ewen, and Wise—built British Columbia's first salmon cannery at Annieville, just below New Westminster on the Fraser River.

By 1880, the Fraser River in southern B.C. had eight plants, the Skeena and Nass in the north had three, and Alert Bay had one. Fishermen with rowboats and small sailboats used gillnets on the rivers. Many if not most were Indians in cannery-owned boats.[2] The plants used the laborious techniques common to lobster and sardine canneries. British Columbians in the 1870's also began fishing herring, halibut, sturgeon, whales, seals, and other species, including dogfish (for a liver-oil plant on the Queen Charlottes); but salmon dominated by far.[3]

The transcontinental railway operating from 1886 boosted the B.C. canning trade and created a new fresh-fish trade. Cold-storage plants went up on the Fraser from 1887. Immigrants from China and Japan, together with the native Indians, provided much of the labour to build and run canneries. Steveston in particular, at the mouth of the Fraser south of Vancouver, long retained a strong Japanese influence. The fishing fleet itself soon included immigrants from many backgrounds, including Scandinavians, Greeks, English, Italians, French, and increasing numbers of Japanese, as well as Native people.[4]

The number of salmon factories nearly quintupled from 12 in 1880 to 59 in 1899. Attention switched

Scenes from the salmon fishery early in the 20th century. (Library and Archives Canada, PA-32198)

from red spring (chinook) salmon to sockeye. On Vancouver's Burrard Inlet, one entrepreneur in 1883 started up a mobile cannery and oil factory, known as Spratt's Ark, a venture that lasted only two years. Canneries spread to every part of the coast, bringing settlement to such isolated areas as Rivers Inlet and Smith's Inlet. Abundance was great and wastage high. Pink, chum, and other less valuable species got discarded.[5]

The number of canneries grew much faster than the production of salmon. The number of cases packed per year rose to more than 400,000 by the late 1880's, and occasionally topped a million in the 1890's and early 1900's. Although higher packs were common in later years, this was largely because, after about 1903, the industry supplemented its catches of sockeye, coho, and chinook by using more of the lower value pink and chum salmon. The first shipment of cured chum salmon went to Japan in 1897.

Commercial freezing of salmon began on the Columbia River in 1892, and moved up the coast. By then, gang knives for cutting salmon and mechanical fillers for salmon cans had come into use. Electricity aided the industry. Machines appeared that could seal cans without soldering. Then the Smith Butchering Machine came on the market in 1905; it could cut off the salmon's heads and tails and gut them at a high and constant rate. Its nickname, the "Iron Chink," bears witness to the prevalence of Chinese workers in the factories, often hired on a group basis by Chinese contractors, and of course to attitudes of the time.

Fishing boats on the Fraser became bigger, as competition for fish pushed them more towards the river's mouth. The Fraser skiff gave way to round-bottomed boats. As on the east coast and the lakes, towards the end of the century steam-powered tugs towed boats to the factories. The gas engine changed all that, as small boats got their own power. By the time of the First World War, the bulk of the Fraser fleet had engines. From about 1913, the gas engine markedly increased trolling, which was cheaper than gillnetting.[6]

Fisheries service sets up in British Columbia

James Cooper served as the federal government's Fisheries Agent from 1871 to 1876. On May 8, 1876, the government of Canada extended the Fisheries Act to British Columbia. For the time being, the government exempted the Indian fishery from regulation.

The British Columbia pioneer and writer Alexander Caulfield Anderson became Fishery Overseer from 1877 until 1882. The title of the top job then changed to Inspector, a post filled by George Pittendreigh until 1886. Thomas Mowat became Inspector from 1886 until 1891, followed by John McNab, 1891–1900, and C.B. Sword, in 1900–1911. The fisheries service divided the coast into districts, and collected fees by district. District 1 covered the Fraser area; District 2, the coast north of Johnstone Strait; and District 3, Johnstone Strait and Vancouver Island. The district supervisors

had by 1910 earned a good deal of respect. Some made the most of it, lording it over the Indian paddlers who took them on annual upriver trips to check habitat and spawning grounds.

Fisheries branch encourages development

Thomas Mowat, inspector for 1886–1891, is one of those whose character shines from the dusty pages of the department's annual reports. His coastal and off-shore fishing expeditions, "prospecting for fish" as he called it, constitute the department's first exploratory fishing. He encouraged ventures by others, and in particular helped get a black cod fishery going at the Queen Charlottes. Mowat also tried transplanting lobster in 1888,[7] at the aptly named Cape Disappointment. (Eight decades later, the Department of Fisheries would try an equally unsuccessful transplant at Useless Inlet.) The main federal development effort was, of course, the hatcheries. By 1910, they numbered eight in British Columbia.

After the 1898 Imperial Privy Council judgement on fisheries, British Columbia took a bigger role in fishery management. John Pease Babcock, appointed B.C.'s Commissioner of Fisheries, took a strong interest in questions of biology and management. Babcock among his other efforts worked to count and assess salmon "escapement" to spawning beds in various streams, a practice that federal fisheries later expanded to cover hundreds of streams. This in future years led gradually to the management of harvests so as to allow sufficient escapement. In time, and bit by bit, officials would begin trying to forecast catches based on previous escapements. All of this would take place with little help from organized science, except the empirical science of the fishery officers' own observations. (As the jurisdictional picture gradually sorted itself out, the province would administer sport fishing in non-tidal waters, except for anadromous—river-spawning, sea-dwelling—species. The salmon fishery remained under federal control.)

The fishery officers meet the Indians

Decimated by disease and dislocated by the white man, the Native peoples of B.C. clung to their fisheries. They sold some salmon for commercial use. The department's salmon regulations at first largely exempted the Indian fishery.[8]

The Indians in turn, at least on one occasion, exempted the white man. In 1888, Inspector McNab reported that the chief on the Nass said he owned all the fish. The white fishermen would have to get licences from him. He would keep half the money collected. But, the chief said, he would let it pass for this year. Earlier, Thomas Mowat had run into difficulties with Indians on the Skeena. At Masset on the Queen Charlotte Islands, the use of Indian constables helped to ease tensions.[9]

141

Large numbers of Native people fished for the new canning companies. But gradually, the department put controls on the Native fishery. In 1880, Fishery Overseer Anderson found that canners were getting around closed seasons by purchasing fish from Indians. The department began restricting such sales on the lower Fraser.[10] In 1889, new licensing regulations then tightening up the commercial salmon fishery said that the Indians could continue to fish without a licence, but only for food. The General Fishery Regulations stated the following:

> 1. Fishing by means of nets or other apparatus without leases or licenses from the Minister ... is prohibited in all waters of the province of British Columbia. Provided always, that Indians shall at all times have liberty to fish for the purpose of providing food for themselves, but not for sale, barter or traffic, by any means other than with drift nets or spearing.

Notwithstanding the Indian liberty to fish for themselves, the department in 1894 began requiring Indians to get permission before fishing. Further regulations in 1900 strengthened this requirement. Departmental permits now could fix the area and time of fishing activities, as well as the gear to be used.[11]

From the department's point of view, the Native people could still fish; separating the food and commercial fisheries made it easier to regulate the latter. From the Native point of view, they were losing an ancient birthright. The restriction on Native fishing created an uneasy situation on the coast that would long continue.

As the department extended its control over the Native fishery, a 1904 incident, one among many, exemplified the conflicts. Northern canners on the Skeena had complained about Indian fishing. Fishery Inspector H. Helgesen set out with a fishery guardian into the Babine River, a major tributary of the Skeena. Seven miles up the Babine, "we found two huge barricades, in full swing fishing." He described a complex structure, with posts driven into the river bed, braces supporting them, stringers run across the top and bottom, "panels beautifully made, of slats woven together with bark set in front of all ... this made a magnificent fence which not a single fish could get through." Slides let the fish pass through to traps on the other side, or continue up the river, where there were more barricades.

Helgesen told the chief that the Indians could use no barricades, could only obstruct one-third of the river with nets, must observe the closed season, could not sell fish as they had done in the past, and could take only enough for themselves and their families. "The chief advanced many points and some of them were well taken, he said that they had an indisputable right, for all time in the past, that if it was taken away the old people would starve ... and he wanted to know, to what extent the Government would support them, he

thought it unfair to forbid them selling fish when the Cannerymen sold all theirs, ... that the Canners destroyed more fish than they."

Helgesen continued his journey, destroying several traps along the Babine. He found huge smokehouses with immense arrays of dried salmon. Salmon was "an article of commerce," both sold and bartered, and even served as a sort of legal tender. In his report, Helgesen recommended permanent guardians for the area.

Tensions continued on the Skeena, and eventually the matter reached Minister Brodeur in Ottawa. In a 1906 agreement, the department promised to provide every head of family with nets sufficient to take fish for personal use and even for trade. The government also promised land and schools. But the barricades disappeared.[12]

Salmon bring strong regulation

On the U.S. side, fishwheels were doing great damage to the river fisheries. These near-perfect fishing devices, mentioned earlier, looked a bit like Ferris wheels. The down-rushing water pushed the bottom baskets, which continually rose to scoop up fish and dump them automatically. Some fishwheels were as high as a house.

In British Columbia, the authorities got rid of fishwheels, pound-nets, and traps on the Fraser, and set a weekly closed time. In 1878, an Order-in-Council prohibited driftnets except in tidal waters. Drifting could obstruct no more than one-third of the tidal waters of a river. No one could fish between 8:00 a.m. Saturday and midnight Sunday.[13]

New rules in 1888 and 1889 set the mesh size for salmon driftnets, limited net length to 150 fathoms, readjusted the weekly closed time to 24 hours, and restricted the use of seines in British Columbia (although drag-seining continued in specified areas).[14]

Although banned in British Columbia, some fishwheels persisted in the Yukon; photograph is ca. 1948. (Library and Archives Canada, C-14102)

THE FISHERY LAWS OF THE DOMINION.

TABLE of Close Seasons in force on 31st December, 1889.

Kinds of Fish.	Ontario.	Quebec.	Nova Scotia.	New Brunswick.	P. E. Island.	Manitoba and N. W. Ter.
Salmon (net fishing)		Aug. 1 to May 1.	Aug. 15 to March 1.	Aug. 15 to March 1.		
Salmon (angling)		Aug. 15 to Feb. 1.	Aug. 15 to Feb. 1.	Aug. 15 to Feb. 1		
Speckled Trout (*Salvelinus Fontinalis*).	Sept 15 to May 1.	Oct. 1 to Jan. 1.	Oct. 1 to April 1.	Oct. 1 to April 1.	Oct. 1 to Dec. 1.	Oct. 1 to Jan. 1.
Large Grey Trout, Lunge, Winninish and Land-locked Salmon.		Oct. 15 to Dec. 1.	Oct. 1 to April 1.	Oct. 1 to April 1.		
Pickerel (Doré)	April 15 to May 15.	April 15 to May 15.				April 15 to May 15.
Bass and Maskinongé	April 15 to June 15.	April 15 to June 15.				
Whitefish and Salmon Trout	Nov. 1 to Nov. 30.					
Whitefish		Nov. 10 to Dec. 1.				Oct. 5 to Nov. 10.
Sea Bass				March 1 to Oct. 1.		
Smelts		April 1 to July 1.	April 1 to July 1.	April 1 to July 1.	April 1 to July 1.	
		Bag net fishing prohibited, except under license.				
Lobsters		July 15 to Dec. 31.	July 1 to Dec. 31.	July 1 to Dec 31.	July 15 to Dec. 31.	
			On Atlantic coast, from Cape Canso to boundary line, U.S., July 15 to Dec. 31, in remaining waters of Nova Scotia and New Brunswick.			
Sturgeon				Aug. 31 to May 1.		May 1 to June 15.
Oysters		June 1 to Sept. 15.	June 1 to Sept. 15.	June 1 to Sept. 15.	June 1 to Sept. 15.	

NOTE.—The following Regulations are applicable to the Province of British Columbia :—

1. Net fishing allowed only under license.
2. Salmon nets to have meshes of at least 5¾ inches extension measure.
3. Drift nets confined to tidal waters. No nets to bar more than one-third of any river. Fishing to be discontinued from 6 p.m. Saturday to 6 a.m. Monday.
4. The Minister of Marine and Fisheries to determine number of boats, seines or nets to be used on each stream.
5. The close season for trout is fixed from the 15th October to 15th March.

SYNOPSIS OF FISHERY LAWS.

Net fishing of any kind is prohibited in public waters, except under leases or license.

The seizure of nets is regulated so as to prevent the killing of young fish. Nets cannot be set or seines used so as to bar channels or bays.

A general weekly close-time is provided in addition to special close seasons.

The use of explosives or poisonous substances, for catching or killing fish, is illegal.

Mill dams must be provided with efficient fish-passes. Models or drawings will be furnished by the Department on application.

The above enactments and close seasons are supplemented in special cases, under authority of the Fisheries Act, by a total prohibition of fishing for stated periods.

The footnotes to the 1890 table of close seasons reflect the new limitations in British Columbia, including: "The Minister of Marine and Fisheries to determine numbers of boats, seines, or nets to be used on each stream."

Department limits salmon licences

On the Atlantic, the department had made little use of licences in salt-water fisheries. In British Columbia, however, according to Geoff Meggs's history of the salmon fishery, the department issued fishing licences from 1877, mainly to canners who then owned most of the fleet.

The licences authorized a person to fish, either as an independent or assigned by a cannery. After an 1881 suggestion by Fisheries Overseer A.C. Anderson, the government began licensing canneries themselves in 1882. First the fisheries service gave out licences with a free hand. But over-competition contributed to some cannery closures during the 1880's. Meanwhile, concern for the salmon was increasing.

In 1887, fishery guardian Chas. F. Green reported as many as 250 boats fishing in Canoe Pass on the lower Fraser. He suggested issuing only a limited number of fishing licences and allowing no cannery more than 40 boats, contract or otherwise. Another guardian expressed similar opinions. The annual report for 1888 referred to the "lesson of the Sacramento and the Columbia," with their depletion. The fisheries service wanted to avoid American mistakes. As well, there was concern among the public and canners themselves.

In November 1888, Minister Charles Tupper got authority to fix the number of boat licences in the Fraser. In 1889 and 1890, the government clamped down. Tupper limited the number of licences on the Fraser to 500. Of these, 350 went to canneries, the number varying according to their capacity, and 150 to freezing plants and independent fishermen. New canneries would receive fishing licences based on their capacity. Accompanying rules restricted gillnets to only one-third of the river's width.[15]

Licence limitation gets relaxed

The sale of licences sprang up almost immediately. The growing numbers of independent fishermen pressed for relaxation of the rules. The fisheries service in 1890 partly lifted the restrictions, to the alarm of some department officials.

The government set up a Commission of Inquiry into B.C. fisheries under Samuel Wilmot, Superintendent-General of Fish Culture. Wilmot spent only two days in the Fraser region. His report called for limits on both boats and canneries. His criticisms of improvident fishing practices and gross wastage in canneries brought an angry reaction, leading to another royal commission in 1892—with Wilmot again in charge.

Wilmot and two other commissioners had to keep in mind a whole set of questions:

- Both canners and independent fishermen complained about being unable to get boat licences. Meanwhile, what about people holding a licence and failing to use it?

- Canners feared that restrictions on the number of plants could create a monopoly on canning.

- Fishermen wanted open entry, fearing that otherwise the canners would get all the licences and monopolize fishing.

- Since canners tended to use Japanese fishermen, what about the rights of white fishermen? And there were 3,000–3,500 Indians on the Fraser who had received only 40 licences. (Wilmot: "They are preferable to Chinamen.")[16]

Wilmot favoured controls, but failed to make a big impression on the B.C. industry. (For one thing, he had the mistaken impression that Pacific salmon, like Atlantic ones, could live after spawning.) One of the two B.C. commissioners, the speaker of the provincial legislature, strongly opposed licence limitation and the prohibition on seines at river mouths.

Following the royal commission, the department passed new licensing regulations on March 3, 1894. There would be no limit on canneries. Controls remained on the number of licences held by each enterprise; but the overall limit on the number of boats on the Fraser vanished. Any bona-fide fisherman who was a British subject could get 1 licence, each shipper no more than 7, curers no more than 7, and canners no more than 20 "tied" or "attached" licences. There were to be no transfers of licences; the holder had to return any licence to the department.

Even if overall control was gone, still there were remnants of the principle of using licences to control fishing effort, by restricting the number of boats per canner. But this may have encouraged the building of new canneries. And enforcement of the new rules may have been slack.[17] The general effect of relaxed controls was to allow a great salmon rush on the Fraser.

Canneries went from 27 in 1892 to 54 in 1897 and 73 by 1901.[18] Where Tupper had tried to limit licences to 500, by 1893 there were 1,174 boats working on the river, of which canneries owned 909. Open entry to the fishery was dissipating cannery control of the fleet. By 1900, the cannery fleet fell to 450, out of the now huge fleet of 3,683 working the Fraser.

Japanese immigrants now manned much of the fleet; they held 1,804 licences. Licences for Native fishermen, mainly employed on cannery boats, fell from 850 in 1896 to 423 in 1900.[19] A 1905–1907 royal commission noted that Japanese had largely displaced white people and Indians in the Fraser River fishery.[20]

New rules restrict fishing

As the department loosened licensing rules in the 1890's, it tightened some other restrictions, especially after E.E. Prince in 1895 chaired one of many royal commissions on B.C. fisheries. New regulations affected mesh size, net length and depth, and disposal of offal. The department also used fishing boundary

Fishing on the Fraser, early in the 20th century. (Library and Archives Canada, PA-31855)

markers, to keep boats from penetrating too far into rivers.

The department was already restricting trapnets, big structures stretching from shore. It took special permission to get the first salmon trapnet at Boundary Bay in 1894, where Canadians had to compete with American traps taking Fraser fish in the Strait of Juan de Fuca. As noted earlier, because of the international situation the department from 1904 allowed some additional traps.[21]

By then canners were getting anxious about their supplies of fish, especially given the competition from Americans with looser regulations. The international rivalry prompted the Canadian government to allow more use of salmon drag-seines and purse-seines from 1904, for white people and Indians (though Native people faced restrictions in purse-seining).[22] All required licences and fee payments. Many drag-seine privileges along the coast were leased to canners, who employed Native bands to fish. The department in 1906 partly restricted the purse-seine, to certain waters only.[23] Its use remained common, however. Rules as of 1909 set maximums of 500 fathoms for purse-seines and 300 fathoms for drag-seines; both required three-inch meshes. Purse-seines made it easier to catch coho, chum, and pink salmon.[24]

Limited licensing begins for northern canneries

The number of canneries reached 73 in 1901, a new high. In 1902, the newly formed B.C. Packers Association, with financial backing from eastern interests, bought 42 of the canneries, closing some, and also took over two cold-storage plants. Thirty of the canneries were on the Fraser.[25] It would be the first of several major consolidations in the B.C. canning industry.

Salmon canners had fought some early restrictions. Minister Charles Tupper in the 1890's had found them insatiable in their "savage ... attacks upon my department and myself."[26] But they began to recognize that there were only so many fish. In the Fraser system, the loss of much of the great Adams River run to dams built by loggers, and of Quesnel salmon to dams built by gold miners, made matters worse.

The Fraser River Canners' Association tried in 1900 to limit each canner to 20 boats, on a voluntary basis. In 1904, canners did voluntarily restrict the number of their boats in District 2, northern waters. Then the government weighed in. In 1908, in response to B.C. perturbations, new federal regulations stipulated that salmon canneries must have a licence. The department announced it would issue no new salmon cannery licences for northern B.C.[27]

Limited entry was back, at least for northern canneries. At the same time, the canners agreed to stabilize their northern fishing. They set an overall limit of 850 boats on the Skeena and 750 in Rivers Inlet, and made themselves boat allotments, taking into account their capacity and previous production. But the voluntary agreement began to fall apart in 1909.

Meanwhile, the provincial government was taking a strong interest in fisheries (especially in the years before a 1915 court ruling reaffirmed federal authori-

145

ty). The energetic John Pease Babcock took charge of allotments for 1910.

Babcock also agreed to chair a two-man Dominion–provincial commission on "Boat Rating." The commission recommended, and the Dominion government enforced, a new scheme that specified the number of vessels for each cannery in the Skeena and Rivers Inlet areas, divided the Nass boats equally among the four established canneries, and outlined the number of vessels for other northern canneries. This was the most thorough limiting of effort thus far.[28]

Department loosens controls in north

Although the northern canneries had a limited number of boats, they could in theory buy extra fish from independent boats. In practice, however, cannery gill-net fleets, largely run by Native people, dominated the north, and the boat-rating system effectively limited the fleet.

In a further intervention, a striking example of technology control, the department in 1911 forbade the use of motorboats north of Cape Caution (that is, north of the Strait of Georgia and Johnstone Strait). The authors of this policy put it forward not for conservation reasons, but more for operational stability. Without motorboats unbalancing the fishery, Indian fishermen on cannery boats could compete more or less equally.[29] When the blanket prohibition on all motorboats proved ineffective, the department in 1912 banned motor gillnetters only, not trollers or seiners (this prohibition would last until 1924).

The northern restrictions soon got modified. In 1912, Dominion Superintendent of Fisheries W.A. Found and Chief Inspector F.H. Cunningham, the department's top man in B.C., together with provincial official D.N. McIntyre, visited the north coast. To develop the area, they recommended loosening the limit on cannery licences. The department soon allowed some additional plants, while still controlling the overall number. The officials also recommended issuing more licences to "*bona fide* white fishermen," who would be independent of the canneries, and increasing such licences over time. Some "unattached" licences free of the canneries were duly issued.[30]

B.C. pioneers licence limitation

Although licence control went through various modifications and false starts, still the Pacific fishery was pioneering in limited entry for plants and boats. And another form of limited entry had come into play even before the licence freeze on the Fraser.

The department gave out local fishing leases for certain waters, commonly at a creek. Usually, cannery operators holding the leases employed local Native people to fish them in their traditional areas, using drag-seines to encircle the fish and herd them to shore. They could seine up to 200 yards into the estuary.

Although these licences were small local monopo-lies, protests were few. The department obliged holders of exclusive licences to pay rent, obey regulations, and sometimes do more. When S.A. Spencer in 1902 received a nine-year exclusive lease for the tidal waters of the Nimpkish River and vicinity, the department obliged him to build and operate a salmon hatchery. Other licences stipulated, for example, that a cannery use local people in plant work.

To some degree, the thinking behind these leases was the same as Whitcher's original thinking: a lease-holder would, in his own interest, promote conserva-tion. To quote Cicely Lyons's history of the Pacific salmon industry: "This method gave a measure of sat-isfaction. As a system, however, its great defect lay in the fact that the door was left wide open to political preferment and, as might be expected, some abuses did result."[31]

Drag-seine leases increased after 1904. The royal commission of 1905–1907 noted that leases were undesirable; but they remained in place.

As is already apparent, licensing considerations in B.C. could go well beyond conservation. In 1911, for example, the fisheries branch justified granting a new cannery licence on the Queen Charlotte Islands on the grounds that the cannery would employ only Canadian or European fishermen, and thus would help settle-ment. In 1912, the branch rejected another applica-tion by the British Columbia Packers Association, because the company was too big and a monopoly could result.

Regulation reached also into marketing. Apparently to protect the processing industry from foreign com-petitors buying up their raw material, a 1904 regula-tion prohibited exporting fresh salmon caught in trap-nets. A 1907 regulation extended this policy, requiring that salmon must go through Canadian processing before export.[32] Future decades would see export con-trols sometimes relaxed, sometimes tightened.

Commercial herring fishery starts up

Native people in British Columbia had for centuries eaten herring as well as dried herring spawn, collected on cedar boughs at the shallow spawning grounds. A commercial herring fishery began by 1877, with the export of cured herring to South America.[33] In the 1880's, Italian immigrants were among the pioneer seiners, catching sardines. Shore-based drag-seines dominated the fishery. There were at first prodigious amounts of herring in the bays, but fewer as time went on. An early superintendent of fisheries, Thomas Mowat, reported that the increase in shipping had caused the herring to leave Burrard Inlet. Where they had once seemed inexhaustible, now seines could catch only a few. In 1905, the department began licensing gillnet boats.[34]

Herring remained a rather small, off-and-on fishery until the 1900's. It grew as the growing halibut fishery demanded bait, and as markets developed for dry-salt-ed herring in the Orient. In 1910, the department

Brailing herring near Prince Rupert, 1913. (Library and Archives Canada, PA-30017)

banned the export of raw herring for processing or making into meal, which perhaps helped develop the home industry.

Herring drag-seines required licences from 1908. Regulations set limits on mesh sizes and on the length of gillnets and drag-seines. The purse-seine spread in the herring fishery from about 1910; the department began licensing herring purse-seines in 1913. This method came to dominate a large fishery for herring.[35]

New Englanders pioneer halibut fishery

In the late 1880's, three small New England sailing vessels began dory fishing for halibut off the state of Washington. They shipped the fish in ice by train across the United States, mainly to Boston. In 1888, Sol Jacobs of Gloucester began fishing off B.C., within the Canadian three-mile limit. In 1889, Thomas Mowat wrote that there was still almost no Canadian halibut fishery, and would be none as long as the U.S. fishermen came into Canadian waters.

The C.P.R., from 1892 on, ran refrigeration cars to and from British Columbia. This encouraged a halibut fishery by Canadians and by Americans who landed their fish in B.C. Early in the new century, the halibut fishery pushed outwards from the thinned-out inshore banks.

While the herring and salmon fisheries generally employed the same boats and fishermen, the halibut fleet was a different entity. First, sailing schooners and large steamers prosecuted the fishery; later, gasoline-powered schooners with five to seven dories. From 1913, British fishermen introduced longlining directly from the vessel. Longlining was heavy work; it was said you could spot halibut fishermen by their big right arms, developed in hauling the big fish over the gunwale.

As more boats entered the fishery, often manned by Scandinavian or New England fishermen, department officials noted depletion of the banks. Some feared that halibut would become a thing of the past. But management remained minimal in this international fishery, until the 1920's and 1930's.[36]

Pacific whaling gets serious

On the Pacific as in the Maritimes and Quebec, whaling took place in intermittent fashion in the later 1800's. Newfoundland at the turn of the century undertook a bigger effort. As Newfoundland stocks dwindled, after 1909 a handful of whaling factories started up in British Columbia, some with Newfoundland whalers, vessels, and capital. Another plant operated in Quebec. Canada passed regulations basically copying those of Newfoundland.

All boats and all manufacturers needed licences. The minister had to approve the factory site, assure the satisfactory conduct of the business, get the plans of the machinery, and so on. Non-users would forfeit their licence in two years. Licence fees were $800 for the first year, $1,000 for the second year, and $1,200 for each following year; or the government could instead take two per cent of gross earnings, letting the firms pay according to production, as the lobster canneries did.[37]

Although the small B.C. boom in whaling would fade by the First World War, at least one whaling company would operate in the province every year from 1904 until 1942. The Gibson family, B.C. Packers, and Nelson Bros. revived whaling at Coal Harbour in 1947, and this operation continued in one form or another most years until 1967, when the Western Canada Whaling Company finally shut down.

A British Columbia halibut steamer, shown in the 1910-11 Annual Report. "These vessels fish off the Northern coast of British Columbia and run to Vancouver with their catches. Dories … are used for setting and hauling the lines on the fishing grounds as in the Atlantic cod fishery."

Among other species, the possible marketing of dogfish, a nuisance fish, drew attention from the royal commission on B.C. fisheries in 1905–1907, but little happened. The same report recommended a bounty for killing seals, seen as a destructive predator. Bounties would later come into force.

B.C. fishermen begin to organize

Atlantic fishermen were scattered along thousands of miles. But B.C. fishermen were grouped up at the river mouths, and saw each other at the canneries. B.C.'s population already had an urban character. Recent immigrants from Europe were used to organizations in their home countries. These circumstances helped fishermen's organizations to spring up more quickly than in the east.

In 1893, fishermen formed the short-lived Fraser River Fishermen's Protective and Benevolent Association. This organization turned down Japanese fishermen as members and tried to exclude them from the industry. Meanwhile, the Native people manning cannery boats lacked any organization. The Fraser union recruited some of them, and in 1903 launched the first of many strikes in the B.C. fishery, winning some concessions.[38]

In 1899, New Westminster fishermen formed a union. Vancouver fishermen followed suit a few months later; this union tried to attract Japanese members. In 1899, white fishermen in New Westminster formed the Fraser River Fishermen's Union, which spread along the Fraser delta. From that year on, Fraser fishermen were rarely without a vocal union or association. Also in 1899, the Fishermen's Benevolent Society started representing Japanese fishermen.

In *Tides of Change*, his history of west coast fishermen's organizations, A.V. Hill notes big strikes at the beginning of the 20th century. Federal militia called out on one occasion were dubbed the "Sockeye Fusiliers." Strikes in 1900 and 1901 saw the normally diverse groups of fishermen—whites, Indians, Japanese—at times pull together, but ultimately, at the time and in following years, both whites and Indians came to oppose Japanese in the fleet.[39]

Most actions were not big strikes but local boycotts by small groups of fishermen against some particular buyer. Hill notes that (as on the Atlantic coast) buyers often controlled fishermen through financing them. Along with a loan to buy a boat or nets would go an unwritten obligation to sell your fish where you got your loan. Fishermen beholden to the buyer were in no position to bargain strongly over prices. Still, on the Pacific coast, a lot more bargaining took place than on the Atlantic.

The Pacific Halibut Fishermen's Union started up in 1901. This group had many members of Scandinavian origin, and Scandinavians are generally more organization-minded than most Canadian fishermen. In 1912, the union became part of the new Deep Sea Fishermen's Union. Although the New England Fishing Company used strike-breakers in the halibut fishery in 1904 and again in 1909, the new group got recognition and made some price gains by 1912. With headquarters in Prince Rupert, the Deep Sea Fishermen's Union would last for many decades.

On the company side, after a forerunner organization died away, Pacific salmon producers in 1897 formed the Combination of Cannery Packers. In 1899, this group changed its name to the British Columbia Salmon Packers Association. In 1900, the Fraser River packers formed their own separate Fraser River Canners Association. In 1902, the two groups merged under the name of the Fraser River Canners Association. In 1909, the name changed to the British Columbia Canners Association. This association evolved through various names to become, later in the century, the Fisheries Association and then the Fisheries Council of British Columbia, representing mainly the more sizeable firms.[40]

B.C. industry more dynamic than Atlantic

The B.C. canneries showed more innovative spirit than those on the Atlantic. For example, they voted in 1901 to tax themselves an extra $7,500 for fisheries promotion. They also favoured using a government stamp on their products.

By the end of the period, the young B.C. fisheries had taken on some long-lasting features. Large operators, including B.C. Packers and the New England Fishing Company, were on the scene. The major fisheries—salmon, halibut, herring—were all under way. And the new industry had taken on an aggressive, articulate, organized character, as likely to lead government in management as to follow it.

CHAPTER 11.
The nature of early regulation

T he years 1867–1914 were the foundation period for Canadian fisheries management. The government brought in powerful legislation on fisheries, environmental protection, and protection against foreign vessels. The department strengthened Canadian sovereignty, set up a country-wide administration, and took clear control of fisheries management coast to coast. It set rules for all major fisheries, launched a massive hatchery effort, and sponsored other forms of development, including the fresh-fish transport subsidies. On occasion, it applied not just conservationist but social and economic reasoning to the fishery. Officials conceived and tried out almost every fishery theory that occurred in later years.

Understanding of the fisheries grew through the new biological stations, the statistical and other reports by fishery officials, and the many royal commissions setting the early rules for Canada's fisheries. Whitcher, Prince, and their colleagues discussed fishery matters at least as intelligently as most experts and inquiries in later years.

The fisheries service grew to about 1,200 persons by 1910, and it worked hard. Regulation touched every major fishery, though it was light for groundfish.

Regulations reach widely

In the 1880's, a synopsis of Dominion fishery laws and regulations had filled only one page. By 1911, a summary of federal and provincial laws took about 30 pages, and the trend to thorough regulation was well set. Licences got frequent mention. Other rules might set closed seasons or areas, specify or prohibit certain gear types, set mesh and net sizes and spacing, set minimum sizes as for salmon, charge fees according to length of net, and so on. In British Columbia, cannery licences were conditional on sanitary standards.

The minister could set aside waters for fish culture, and require a fishway for any dam. The department had authority to protect fisheries waters from deleterious substances, such as chemicals and sawdust. The department could also restrict exports of fish (and foreigners had to acquire a permit for angling in Canada). Policies set under the Fisheries Act could impose further restrictions. Department activities were starting to go beyond conservation to touch quality, development, marketing, and on occasion the status of fishermen, as in Prince's recommendations to protect independent fishermen in Manitoba.

The early officials had thus created a department of wide activities, which, however, mainly reacted to problems, rather than shaping conditions in the fishery.[1]

Regulation moves from inland out

At the end of the 1867–1914 period, the great bulk of fishery rules and regulations still dealt with inland fisheries. Of the coastal provinces, British Columbia had the most complex regulatory structure; but most rules stemmed from the salmon fishery, which pursued a creature of both fresh and salt water. With river and estuarial species, people could see their vulnerability.

Although Atlantic sea fisheries saw less intervention, still the expanding lobster fishery had created a flurry of regulations. Pelagic species such as herring and mackerel, with their migrations and their concentrated abundance often close to shore, also had a way of bringing fisheries questions to a head. Pelagic species prompted the first major ban of a gear type, the purse-seine; the Bait Act in Newfoundland; the 1890 Inspection Act, aimed at pickled fish; the Fisheries Intelligence Bureau; and the bait-freezer development.

Management differs by region

By the early 20th century, Dominion fishery management was taking somewhat different paths in different areas, reflecting regional characteristics.

The technology of early times had helped to fragment the Atlantic fishery. Processing salt cod mainly depended on local fish, and flakes took up a lot of room on the shore. As one cove became crowded, fishermen would move to another. The nature of the industry spread a thin layer of settlement all along the coast, which worked against consolidation and streamlining. The independence and community feeling of coastal villages shaded into fragmentation and frequent backwardness. Compared with British Columbia, there was less education, organization, and discipline, and fewer industries to provide alternative employment. People made various protestations about what should be done, often without result. Prince pleaded in vain for strict control of licences.

During the 1880–1910 period, the fisheries of the Great Lakes passed under provincial control. Management began to allow the fishing down of species after species, to be replaced by lesser ones. But the industry was relatively small, and alternative employment was plentiful. Producers were close to a strong fresh-fish market. The general strength of the economy buffered the fishery problems.

On the Prairies, the fishery was still young. With Winnipeg was booming and other new cities burgeoning, the often-remote lake fisheries soon developed problems. As on the Atlantic, fragmentation and lack of alternative employment compounded the difficulties. The 1880–1910 period started a pattern of potentially far-reaching regulations such as quotas, applied too loosely to do much good. Complaints about quality, American control of marketing, and other problems remained rife until after the Second World War.

By 1910, the B.C. industry, only about three decades old, already seemed more consolidated, better regulated, and more able to do things together than the Atlantic industry. And fishery managers on the Pacific, both federal and provincial, were most likely to take hold and do something thoroughly. What explains the difference?

Apart from anything else, B.C.'s general economy already seemed stronger. The greater regional circulation of money and know-how aided the fishery. The vigorous economy could more easily pull low earners out of the fishery towards other opportunities.

Depletion was still fairly new on the Pacific. People still could see or remember the ocean's original productivity—for example, when herring ran onto the beach at Nanaimo, to be left knee-deep for two miles.[2] Declines in stocks, and losses such as the salmon runs at Quesnel, struck the imagination and the conscience. The new fishery had a critical mass that reflected itself in management. The salmon fleets massed up on the doorstep of the cities. Catches came fast and furious, followed by declines, followed by far-reaching regulations including licence limitation. While Atlantic and Prairie managers were forever lecturing, cajoling, and bewailing the industry, on the Pacific the industry as often as not led the managers. The Pacific fishery would do better than the Atlantic, for most of the 20th century, at providing decent livelihoods and good-quality products across the board.

Information, organization aid B.C. management

Part of B.C.'s advantage over the Atlantic came from the better flow of information. More fishermen lived near growing cities: Victoria, Vancouver, Nanaimo, and Prince Rupert. Organization came easier. There was more willingness to take hold of the fishery and make it work.

This question of information and organization, and of connectedness rather than fragmentation, is of great importance to fishery management. Compared with the Atlantic fishery, the British Columbia fishery was more closely knit, for reasons including settlement patterns, fewer major species, and the migration patterns of salmon. There was more organization and more streamlining.

Fisheries administrators sometimes noted the issues of education and organization on the Atlantic coast, and made scattered attempts at addressing them, but rarely with lasting effect. The royal commissions of the late 19th and early 20th centuries made no provision for systematically informing or consulting the fishing industry, and this was to be a lasting defect. Still, the many commissions were themselves a rough and ready forum of information and consultation, more effective than in some later periods.

When Prince and his colleagues were writing all the rules and regulations, how well did the officers enforce and the fishermen obey them? Then as today, it is hard to know for sure. Enforcement was always a problem. Minister Charles Tupper noted in 1894 that "every time you punish a man you excite an enormous amount of sympathy in the district for that man," and that some people organized sentiment against what they saw as a "tyrannical" department.[3]

But a certain Canadian respect for authority goes a long way back in the fisheries. It shows up in the recollections of old people and in the records of old controversies. If regulations had been something to laugh off, industry people would have ignored instead of, very often, disputing them. The U.S. Commissioner of Fish and Fisheries in the 1870's noted that Canadian fishermen went by the regulations more than Americans (who had fewer regulations to start with).[4] That being said, there were always plenty of fractious fishermen, and fishery officers would have thousands of confrontations, sometimes with armed and angry men, in the generations ahead.

Licence limitation remains rare

Whitcher's push for licences and leases foreshadowed late 20th-century thinking. It already seemed clear that a common-property fishery could lead to uncontrolled competition, conservation problems, and lost profits. Whitcher touted licences and leases not only as a means of control but also as a way to utilize natural economic instincts. If leaseholders could profit directly and exclusively from the returns, they would no longer see conservation as a remote ideal pushed by bothersome government enforcers. Rather, they would have a direct monetary incentive to preserve and enhance nature's bounty.

But after the Dominion government lost control of fishing leases in non-tidal waters, it seldom tried to limit the number of fishing licences, except in the B.C. salmon fishery. The department continued to make use of licensing, though without limitation, in many other fisheries, particularly inland or estuarial fisheries using traps or bag-nets. There, one could see the dangers; and the connection to the shore put space at a premium and made order more necessary.

Around the beginning of the 20th century, licences on the Atlantic applied for angling, bag-nets, box-nets, trapnets, shad, smelt, sturgeon, bass-nets, herring weirs, clams, oysters, and salmon gear of various kinds. British Columbia had many kinds of salmon licence; others applied for angling, abalone, clams, crabs, and so on through a long list. Although B.C. had many licensing rules, there was no limited entry except for salmon, and in the Maritimes, little licence limitation at all. Whereas Whitcher had talked of economic benefits, now the main object of licensing was fishery protection. Licences brought some order to the fishery, kept out the most frivolous entrants, and gave the government the useful threat of licence cancellation.

A century later, it would become an article of faith among theorists and then among fishery managers that open-access fisheries led to over-subscription, over-dependence, and finally, overfishing and depletion. Advocates pushed licence limitation, overall quotas, and finally, individual quotas. Otherwise, they said, incomes would drop and abundant fisheries become depleted. Governments needed to make a great equation: they must balance the levels of fishing employment, resource yields and abundance, and monetary returns for the optimum benefit to society.

If the advantages of limited licensing already appeared clear to Whitcher and Prince a century earlier, and if early officials restricted licences in certain fisheries, then why did officials and ministers hold back from full control of licences for all major fisheries? No doubt for a number of reasons, including the fisherman's natural resistance to regulation. But a prime factor may have been that in the biggest fishery of all, that for Atlantic groundfish, what later economists saw as an inexorable pattern of depletion didn't fully hold true.

While river and estuary fisheries suffered from depletion, groundfish in the late 19th century still appeared almost limitless. (Even in the mid-20th century, it still appeared that the groundfish fishery had great potential for expansion.) That fishery's main problems were elsewhere, in quality problems, lack of co-ordination in production and marketing, and other trade factors.

Many people went into the fishery only on a narrow scale, with small open boats, not because they expected great returns, but because they had no hopes elsewhere. The shipping, trading, and lumbering industries were fading, and there was little other employment. As the 20th-century fisheries economist W.C. MacKenzie put it, "they were not poor because they were fishermen, but fishermen because they were poor." To have applied strict licence control, in the absence of a strong conservation reason, would have required almost dictatorial action. In the early 20th century, Prince did call for limited entry in the declining lobster fishery, but others held that it would survive without such measures.

Neither did turn-of-the-century managers often try to control the absolute amount of fishing gear (although they sometimes set licence fees according to the length of net, or restricted the length). Instead they often attempted to control effort indirectly, through closed seasons (which might also protect spawning fish) and other measures. And they made good use of the second basic tool of management: controlling size and age at first capture, through rules on net meshes and legal size of fish. Even when there is no limit on the number of fishermen and the amount of gear they employ, size limits can still afford fish the chance to grow towards full size and value and to reproduce.

Managers do little evaluation

A more pertinent criticism of management, in the 1867–1914 period and in following years, might be the lack of close evaluation of management measures.

In 1974, Gordon DeWolf of the federal fisheries department reviewed the history of lobster management. He said much the same thing that Prince's commission had said in 1898. It was difficult to quantify any benefits or losses from the previous regulations. DeWolf added that there were serious grounds to doubt the value of most regulations for conservation, and that they had led to economic inefficiencies.

DeWolf's remarks about the difficulty of quantifying the benefits of management could apply to most fisheries, then and now. The many variables would in any case make it difficult to evaluate fisheries management; but what was worse, hardly anyone tried, especially in the early years, to do so in a thorough way. The fishery managers did their rough-and-ready experiments without benefit of science, while Biological Board scientists, although documenting the natural history of fish, rarely took part in experimental management.

What about DeWolf's second point—that in the case of lobster, many management measures might have been unhelpful? Does this mean that for other fisheries as well, the myriad regulations enforced with such care and cost over the decades accomplished nothing?

It is more likely that, in total, they had some good effect, especially if compared with a totally uncontrolled fishery. The most basic principle in fishery management is to give the fish a chance to grow up and reproduce. The multitude of Canadian regulations worked in that direction, and helped keep the industry more stable.

All told, the fishery regulators in the 1867–1914 period left an impressive record. They believed in what they were doing, and instilled a conservationist attitude that lasted.

APPENDIX 11.1:
A statistical snapshot for 1909–1910

The fisheries service

From 1880 to the 1909–1910 fiscal year, employment in the fisheries Outside Service about doubled to 1,200; the number included 680 seasonal guardians and 255 men on the patrol fleet. This fleet now included 13 vessels and 255 men. Fishery officers also had small patrol boats. (In British Columbia, for example, a report during the First World War noted that each of the three districts had an inspector, and that the province had in total 19 overseers, 25 patrolmen and guardians, and 19 patrol boats. Hatcheries by 1909–1910 had nearly quadrupled to 37, including establishments for salmon, lobster, whitefish, and pickerel; they employed more than 100 people. The amount spent on hatcheries in 1909–1910, at $180,000, exceeded the total salaries of fishery officers.

Value of production

For the industry, statistics of the day were none too reliable, but at least they showed trends. The product value of Canada's fisheries almost doubled over the three decades, from about $15 million to $30 million. British Columbia led the way. On the Atlantic, the fleet increased much faster than the value of the fishery. As J.J. Cowie wrote in the 1909–1910 annual report, in his article "The Non-progression of the Atlantic Fisheries of Canada," "We find that the fisheries of British Columbia and inland western waters have been giving us the increasing totals, and further that the aggregate value of the fisheries of the four eastern provinces has almost stood still for the last twenty-five years."

In 1880, British Columbia had produced a value of about $700,000; by 1909–1910, it was more than $10 million, providing about one-third of Canada's total fishery value. The prairie and Yukon fishery had gone from zero recorded to $1.4 million. Meanwhile, the Maritimes and Quebec had grown little, from about $13 million to $16 million; this in spite of intervening growth in the lobster fishery.

Among products, the big change was that salmon now led by far in value, at more than $8 million. Cod came to $3.9 million. Lobster was close behind, with fresh lobster now taking one-third of the total trade.

Although cod were still mostly salted and dried, the fresh-fish trade was making some gains. The fresh haddock trade, at more than $300,000, was more than double the fresh cod trade. Haddock all told were worth $830,000, and hake $367,000. Herring, salted, fresh, and smoked, were worth more than $2.7 million, and sardines more than half a million. Beluga skins were worth $436,000; whale products more than $300,000; and eels more than $100,000. The industry was growing in variety: the 1909–1910 report showed 56 kinds of product, including halibut, swordfish, dulse, sea otters, squid, mackerel, and other saltwater and freshwater species.

But some older fisheries were stagnant or declining. Dried cod production was less than in the 1880's. Oysters, too, were well down. Lobsters were a well-known problem.

The fleet

From 1880 to 1909, the total number of small boats in the Dominion rose from 25,300 to 41,200. Although the number of boats increased by 63 per cent, the number of fishermen on them rose only 10 per cent, from 51,900 to 60,700. It seems that with gasoline engines doing more of the work, more fishermen were buying their own boats and operating with fewer crew, or as individuals. By 1910–1911, some 5,000 Canadian fishermen used gas engines; by 1915, about 15,000.

Among larger craft, the number of vessels rose from 1,200 to 1,750, but according to departmental figures, the number of vessel fishermen dropped from 8,800 to 7,900, presumably because of labour-saving engines. The total number of boats and vessels in 1909 came to 42,900; and of fishermen, to 68,700.

On shore, the number employed in lobster canneries went from 13,000 in 1895 to 18,700 in 1899, dropping back to 11,400 in 1908 as landings fell. (The cannery figures left out shore employment by people curing cod and other species in the traditional fisheries.) In 1910, the department listed Atlantic shore employment at 12,800; Pacific, at 8,700. Total employment on Canada's two coasts (not including Newfoundland) added up to 24,000. By comparison, Canada's agricultural workforce at the time was more than 900,000.

Some highlights by province

British Columbia now produced the highest value, $10.3 million in 1909–1910. Canned salmon alone came to $6.5 million; pickled and dry-salted salmon, to $1 million. Halibut were worth over a million dollars. Also important were fresh herring ($512,000), whale products ($315,000), coarse and mixed fish

($140,000), fur sealskins ($124,000), and oysters ($31,000), as well as fish oil and fertilizer.

The B.C. fishery now had 173 vessels, including 32 sealing vessels employing 886 persons and 5,635 boats employing 9,925 persons. The B.C. fleet was the most valuable, at $6.8 million. Canneries and fish-houses employed 8,689, far ahead even of New Brunswick, where sardine and lobster canneries abounded.

On the Prairies, Manitoba's fishery was yielding products worth $1 million, led by pickerel, whitefish, coarse and mixed fish, and pike. The province had 10 vessels, mostly tugs, employing 74 men and 288 boats employing 565 men. Canneries and fish-houses employed 200. In Saskatchewan, production value came to $174,000, with whitefish far in the lead. The province had 565 boats employing 565 men, according to the statistics, and no employment listed in canneries and fish-houses. Alberta and the Yukon Territory had production of $196,000, again with whitefish in the lead. Alberta had 362 boats employing 732 persons; the Yukon, 68 boats employing 136.

Ontario's product value in 1909–1910 came to $2.2 million, led by trout ($516,000), whitefish, fresh herring, pickerel, and pike. Sturgeon was still worth $33,000; and sturgeon caviar, $8,700. Ontario had 145 vessels, mostly tugs, employing 708 men and 1,623 boats employing 2,893 men. The annual report gives no figure for canneries and fish-houses; it may be that virtually all Ontario's fish were already going to the fresh market.

Quebec's fishery in 1909–1910 had fallen in value since 1880, from $2.6 million to $1.8 million. Dried cod led at $803,000, followed by canned lobster ($282,000), bait fish, fresh salmon, salted mackerel, salted herring, fertilizer, oil, and other products. Quebec now had 42 vessels employing 104 men and 6,133 boats employing 10,691 men. Canneries and fish-houses employed 1,259.

In the Maritimes, Nova Scotia's fishery for the 1909–1910 fiscal year yielded $8.1 million, up less than $2 million from 1880. Dried cod was still the leader, worth the same as in 1880: $2.5 million. Canned lobsters were worth $1.1 million; live lobsters, $771,000. Herring were now in third place (salted worth $564,000, fresh worth $166,000, smoked and kippered worth $43,000), followed by mackerel (salted worth $455,000, fresh worth $319,000). Other important products included haddock (fresh and smoked taken together now exceeded the value of dried haddock), pollock, hake, halibut, salmon, and a host of minor fisheries such as smelt, shad, flounders, and bait, oil, and meal. Swordfish was becoming important.

Nova Scotia now had 16,102 boats—a major increase. But the boat fleet employed 18,583 men—an actual drop in employment. The province now had 785 vessels, up from 731 in 1880, and included 50–100 larger dory schooners fishing the banks. But the vessels were employing only 4,575 men, a marked decrease. Production per man was up. Canneries and fish-houses employed 3,515.

New Brunswick's fishery value by 1909–1910 had climbed to $4.7 million. Herring in all its forms—salted, fresh, smoked and kippered; canned, fresh, and salted sardines—came to more than $1.3 million. Canned lobsters had slipped to $624,000; live lobsters were worth $146,000. Smelts were now worth more than cod. Other notable species included salmon (especially fresh), pollock, clams, oysters, and sturgeon.

As in Nova Scotia, the number of New Brunswick vessels and boats had increased remarkably (to 512 vessels, well over double, and to 8,414 boats, about double). Employment in both sectors increased, to 1,459 men on vessels and 13,366 on boats. The number of vessels increased faster than the number of men working on them. The fishery was producing no more per man, and less per craft. Nova Scotia had pulled ahead in production per man, less by increasing the output than by decreasing the number of men. New Brunswick's canneries and fish-houses employed 5,602, more than Nova Scotia's.

Prince Edward Island's fishery in 1909–1910 yielded production worth $1.2 million, well below that of 1880. Canned lobsters were worth $677,000, down somewhat; live lobsters were worth only $13,000. Oysters were up markedly to $94,600. Cod, hake, smelts, herring, mackerel, "clams, quahaugs, scallops, &c." and bait fish were also important to the island's small fishery.

Vessels now numbered 83, almost three times the earlier number of decked craft deserving to be called vessels; but they carried fewer people, 125. Boats now numbered 1,989, also a major increase, employing 3,278 men, a decrease. Canneries and fish-houses employed 2,429.

Newfoundland

Newfoundland's greatest fisheries, for cod and seals, both reached their peak in the later 19th century, then dropped to lower levels.

The lobster fishery was also dropping back. It had canned 632,000 pounds in 1875; almost 6 million in 1880; almost 16 million in 1890; 8.3 million in 1900; 7.2 million in 1905; and 5.7 million in 1910, in which year the fishery had 2,081 licences and 4,487 men. By 1915, production had dwindled to 1.4 million pounds. By 1924, it dropped to 759,000. A three-year closure of the fishery followed, bringing some recovery.

The bank fishery, concentrated on the south coast, by 1910 appeared stronger than in the 1870's. It now employed 101 boats and 1,567 crew. Vessel tonnage totalled 6,630 tons. Production of 144,500 quintals of dry fish averaged out to 92 quintals per man.

The Labrador fishery in 1910 employed 1,126 schooners, 12,050 persons, and an estimated 3,000 cod-traps. The Labrador coast itself had 750 liviers with 10 schooners and 150 traps. The steam sealing fleet had 19 vessels and 3,364 men and took 333,000 seals. The annual report gave no figure for small boats.

The labour force of men engaged in catching and curing fish numbered 38,578 in 1857; 45,854 in 1874; 36,694 in 1891; 41,231 in 1901; and 43,795 in 1911, apparently varying with the catch. Women (working at curing) were enumerated at 18,081 in 1891; 21,443 in 1901; and 23,245 in 1911. In 1857, the male fishing labour force made up 90.4 per cent of persons occupied; by 1911, it had dropped to 53.1 per cent.

In 1910, Newfoundland exported 1.5 million quintals of dried cod, worth $7.3 million. Sealskins were worth $460,000; seal oil, $460,000. Canned lobsters, worth less than $3,000 in 1876, now came to $338,000. Other leading exports by value were cod oil ($353,000), pickled herring ($157,000), whale oil ($147,340), bulk herring ($93,000), pickled salmon ($57,000), and frozen herring ($52,000). There were lesser amounts of other products, such as trout, sounds (air bladders) and tongues, squid, and dried capelin.

PART 3: 1914–1945
SHORT BOOMS, LONG DEPRESSION
CHAPTER 12.
National and international events, 1914–1945

During the 1914–1945 period, the Atlantic fishery was mainly a problem area, opened and closed by bursts of wartime prosperity. British Columbia, although having problems of its own, showed greater dynamism.

The First World War brought great prosperity to the Canadian fishery. The army needed food, as did world markets that lost supplies from Europe. The production value of the nation's fisheries almost doubled from 1915 to 1917; in the latter year it came to $52 million. The wartime Canada Food Board set prices and bought large quantities of fish, providing stability and good markets.

At the turn of the century, the fleet had still consisted mainly of thousands of row-boats and open sailboats, hundreds of schooners, and a few steam-powered vessels. By 1917, some 15,000 vessels in the Dominion used gas engines, triple the number in 1910–1911.[1] Easthope engines were in production in B.C.; Acadia and Stewart Imperial were well-known brands on the Atlantic. In the 1920's, some larger vessels turned to diesel engines. Engines extended the range of even the poorest boat, opened more grounds to stronger fishing pressure, and created new businesses providing fuel and repairs.

Even during wartime prosperity, the fishing industry showed many weaknesses. In 1915, D.J. Byrne of the Canadian Fisheries Association gave a long list of typical problems. Bait was frequently scarce, a situation only partly alleviated by the government-subsidized bait-freezers. Dogfish were a nuisance, despite the establishment of reduction plants to fish down the species. Transportation was a problem, with high costs of handling and delivering to the centres of consumption (although the department's subsidy on express shipments was a help). Fish consumption in

Fishing schooner in the Strait of Canso, N.S. Schooners dwindled during the Depression. (Library and Archives Canada, C-338, undated)

Canada was low; the public needed educating that fish was an excellent, nutritious substitute for high-priced meats and poultry. Retailers treated the low-volume fish trade as a necessary evil, giving fish poor care but a high mark-up. Department stores needed better awareness that they could easily preserve and display fish. The market needed enlarging in general; the Canadian Fisheries Association hoped for the government to send a trade mission on fisheries to England.

To Byrne's list of shortcomings, E.E. Prince of the Marine and Fisheries department and the Biological Board added his own. The fishermen lacked organization. The industry generally needed better understanding and government–industry co-operation. Food quality suffered from poor handling of fish by fishermen ("they seem to delight in knocking the fish about and jumping on them") and all along the water-to-consumer chain. Public support of the department's officers was weak, Prince said; "if a man shipped a load of illegal lobsters, everyone ... seemed to sympathize with the poor fisherman." The industry lacked respect for some regulations, an attitude that was in some cases justified.[2]

Inshore boats on the Atlantic. (undated photo)

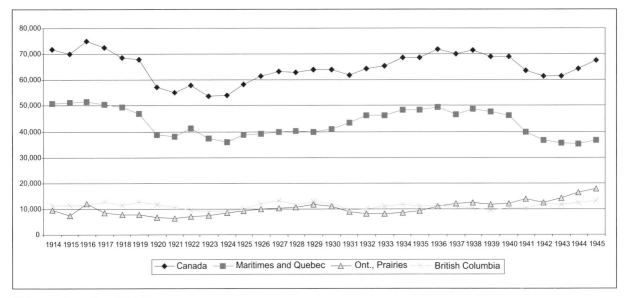

Fishermen by region, 1914-45.

Yet, wartime prosperity would give the illusion that the industry was now better able to take care of itself. The department abandoned most attempts at development, and stopped trying to limit the numbers of fishermen. It also gave up the attempt, embodied in Prince, to apply scientific analysis to management. W.A. Found became the ruling personage in the fisheries service. Administering in a no-nonsense, nose-to-the-grindstone way, Found imparted efficiency but little vision.

The Biological Board remained at arm's length from the department. This separation of thinkers from actors dimmed the chances of thorough analysis and creativity. The fisheries service had many people of fine intelligence, including Found himself; but, especially on the Atlantic, the emphasis was now on day-to-day administration and following the book. The Biological Board, with its brains trust of bright young people from the universities, rather than working on the resource and industry as a system, mainly followed the individual interests of the various researchers.

During the 1920's, the industry faced some external problems, including tariff difficulties with the United States. But, on the Atlantic in particular, internal troubles were as bad or worse. The whole fishery system was weak, as pointed out in a thorough study by Cockfield, Brown & Company, Ltd.

The Great Depression of the 1930's was writ large for fishermen, generally disorganized, and distant from the centres of money and power. The British Columbia industry suffered many cannery closures, job losses, and other difficulties. Native workers in particular suffered from lost cannery employment[3]. Yet B.C. as usual did somewhat better than the Atlantic. Canneries fostered the compulsory inspection that raised their product's reputation. Salmon had become scarcer after the 1913 Hell's Gate slide in the Fraser River; and halibut, too, caused concern. But major domestic and international efforts took place to improve both fisheries. The B.C. salmon industry carried out another major consolidation, while on occasion prodding the government towards better conservation.

Elsewhere, fishermen faced hardship and desperation. Few Atlantic fishermen made comfortable money. They could survive; most had their own houses, could grow food, and could catch fish to eat. But life was a constant struggle. In Nova Scotia, normally the most conservative of fishing areas, strikes and riots flared near the end of the decade.

On the Atlantic, little impetus for change came from the fragmented private sector or the department. Only when a price collapse caused an early depression, even before the Great Depression, did major actions emerge: a royal commission led to a ban on trawlers and to the department's helping to sponsor fishermen's co-operatives. These brought organization and hope to fishermen in parts of the Maritimes and Quebec. Worst off was Newfoundland, which not only suffered great hardship in the Depression, but lost its self-rule as the economy collapsed.

In the latter part of the decade, the Canadian and Newfoundland governments became more active. The Salt Fish Board in Canada and the Fisheries Board in Newfoundland were two important interventions. Then the Second World War changed the picture. Partly out of economic need, east coast fishermen flocked to join up. For those who stayed on fishing, wartime demand brought new prosperity. Again a wartime food board took over food distribution; the market worries of the Great Depression vanished.

Departmental set-up changes

Fisheries combines with Navy

From 1911 to 1917, Sir John D. Hazen continued as Minister of Marine and Fisheries, and *ex officio* as Minister of the Naval Service. The naval aspect took dominance as the First World War broke out; annual reports of the Fisheries Branch appeared under the rubric of the Department of Naval Service. As the war drew to its close, Charles C. Ballantyne of Ontario took over from 1917 to 1920, still as Minister of Naval Services along with Fisheries and Marine.

From 1911 to 1914, the minister had two top officials: Charles E. Kingsmill as Director of the Navy and Alexander Johnston as deputy minister for Marine and Fisheries. In 1914, G.J. Desbarats, a former deputy of Marine and Fisheries, became deputy of the Naval Service, now taking in fisheries. (Below the deputies, of course, W.A. Found was the real power in fisheries.) In 1920, the fisheries and naval agencies separated; the fisheries branch annual report for that year noted that "experience showed that there is really nothing in common in the duties of the Department of Naval Service and the Fisheries Branch."

C.C. Ballantyne continued as Minister of the restored Department of Marine and Fisheries until 1921. In that election year, Mackenzie King's new Liberal government appointed Ernest Lapointe of Quebec, a major figure in the King administration. P.J.A. Cardin, also of Quebec, took over as Minister in 1923. The brief Conservative government under Arthur Meighen in 1926 appointed no full-time Minister for the department. After an election, Cardin that same year resumed the post until 1930.

Found strengthens organization

E.E. Prince had become Commissioner of Fisheries in 1893; he retained the title until 1925. But by 1914, most administrative power had gravitated to William Ambrose Found of Prince Edward Island. A school-teacher in earlier life, Found had by 1898 become a secretary to the head of the fisheries branch, and by 1911, Superintendent of Fisheries. In this role he carried out a thorough administrative reform.

Fisheries Protection Service patrol vessel *Thiepval* in Hecate Strait, B.C. (Library and Archives Canada, PA-40990)

Fisheries patrol boat in Lockeport harbour, N.S. (Library and Archives Canada, PA-48057, undated)

The widespread fisheries service set up by Whitcher had its weaknesses. To quote the 1909–1910 royal commission on Manitoba and Keewatin fisheries: "[W]e do not favour the present system of a numerous staff of poorly paid fishery overseers, and a still more inadequately paid staff of fishery guardians. The whole territory should be under the supervision of six or eight active and properly paid fishery overseers." And the fishery officers should have something better than their one large and slow patrol boat, which poachers could easily see coming.

Complaints were frequent about the quality of fishery officials in the "Outside Service." Only chief inspectors, inspectors, and overseers had permanent employment. The branch hired guardians as required to assist overseers. Most jobs were part-time. An overseer might be a farmer first, a fishery officer second or third. Many if not most were excellent, dedicated officials; but politicians influenced their appointment, especially before the 1908 creation of the Civil Service Commission, and this was another cause of complaint.

The system was extensive on paper, but the annual report for 1917 stated that "the organization in the eastern provinces is, in most portions thereof, inefficient. The number of officers is unduly large, but they are paid mere pittances, so that it is unreasonable to expect that they can devote to their fishery duties the time necessary for their proper performance. It is essential that a complete reorganization of this portion of the service should be effected without delay."

Found began re-organizing in southwest Nova Scotia, and from there extended the same system elsewhere.[4] The organization changed from its provincial basis to three main divisions: Eastern, Prairie, and Western (B.C.).[5] Each division had a chief inspector. Under him, inspectors controlled wide areas, and below the inspectors, overseers controlled districts. Jobs became full-time but fewer. For example, the number of overseers in the Eastern Division went from 92 to 56.

Complaints about officialdom and enforcement seemed to become fewer after Found's reorganization. The overseers had a good degree of local authority. The size of areas remained daunting. British Columbia as a whole in 1918 had a chief inspector, three inspectors, 20 overseers (what we would now call fishery officers),

30 guardians, and four clerks.[6] British Columbia's District 2, the northern part of the province, in the 1920's had a supervisor, an assistant, and 10 people along the coast plus 2 working in the interior at river spawning grounds: total, 14, for thousands of miles of shores and rivers.[7]

This basic fisheries set-up, with its regional divisions, would last for decades. In 1919, a publicity and transportation division started up, producing magazine publicity, offering school prizes for essays about fish, and so on.[8] The increased attention to promotion sprang in large part from the work of the new Canadian Fisheries Association, in which Frederick William Wallace played a prominent part. In 1914, Wallace founded *Canadian Fisherman* magazine, for many years the bible of the industry, particularly among processors. Wallace wrote several notable books, including *Wooden Ships and Iron Men*, about the fishing and marine world.

Quebec takes over fishery management

Quebec, which had already taken over inland fisheries and the Dominion government's hatcheries, in 1922 took over management of the fixed-gear fisheries in salt water.

Quebec and British Columbia had been trying to increase their fisheries authority. Revenues were sometimes a factor. In one instance, the B.C. government had tried to impose licensing fees and a tax of $1 per 1,000 fish taken.[9] In 1913 and again in 1920, the Judicial Committee of the British Privy Council had rolled back provincial claims relating to tidal waters

and navigable streams.[10] The latter decision made clear that there was a public right of fishing in all waters, whether navigable or non-tidal; this right not being proprietary, "the Dominion Parliament has in effect exclusive jurisdiction to deal with it."

But the provinces still had jurisdiction over property and civil rights, and conflicts could still arise in situations where property covered by water was also the habitat of fish, as along a shore where fixed gear was fastened to the bottom. Much of Quebec's fishery then depended on fish-traps or nets fastened to the shore.[11] The situation became complicated. Fishermen frequently had to get licences from both federal and Quebec governments.

To ease the situation for fishermen, and to get Quebec's good will for salmon conservation measures, the federal government in 1922 agreed that Quebec would manage local fisheries. Under the administrative arrangement, the province also took over inspection.[12] For the Magdalen Islands, however, the federal authorities kept both management and inspection until 1943.

Thus Quebec after 1922 (and until 1984) controlled licensing, enforcement, rule-setting (though the federal side still had to pass formal regulations that the province recommended), development, and such activities. The Quebec fishery in the early 1920's generally accounted for about 15 per cent of Canada's Atlantic fishery value. The provincial authorities often proved active, for example in provision of cold storages. And when Gaspé dried cod faced high tariffs, the provincial government worked hard to get trade concessions.[13]

Cleaning fish, South Beach, Percé, Quebec, 1916. (Library and Archives Canada, PA-11352)

Federal government loses processing jurisdiction

The Dominion, one of the northern B.C. canneries. (Library and Archives Canada, PA-47826)

Through various federal–provincial and court battles, the Dominion government had kept active control of processing. That, too, would change.

In 1927, the Somerville Cannery Company sought to build a canning plant at Seal Cove, Prince Rupert, B.C. Although the principal figure, Francis Millerd, got a provincial licence, the Dominion government refused him a cannery licence, saying he could only salt salmon.

Successive court decisions ruled that the fisheries branch had no power to prevent the Somerville operation. The Supreme Court ruled on May 28, 1928, that the sections of the Fisheries Act of 1914 requiring that operators obtain an annual licence to operate a salmon cannery or curing establishment, or indeed any cannery, were ultra vires of the Parliament of Canada. "A fishing cannery is not, according to any of the definitions, or in practice, embraced in a fishery, sea coast or inland." Instead, canneries came under the provincial powers over property and civil rights.

In England, the Judicial Committee of the Privy Council, true to its tradition of enlarging provincial rights, upheld the decision.[14] So vanished the federal power over plant licences.

Francis Millerd wrote to the *Vancouver Sun* that "the functions of any government should only cover the conservation and the policing of fisheries. They will have to leave to the individual the right to do with these fish, after they are legally caught ... whatever he thinks is best, and generally speaking, I believe this will be found to be the best thing for the country. ... Ambition probably makes life worth living more than anything else, and if you destroy that you just about destroy everything."[15]

It was an eloquent statement for free enterprise. Only ten years earlier, however, other cannery owners in B.C. had wanted a tight rein kept on cannery licences, to protect their own interests. Over-expansion of fish plants, like over-expansion of fleets, was to remain a vexed question in Canadian fisheries.

The 1928 decision clouded the picture on inspection. To apply regulations on fish quality, the federal department now had to use its constitutional authority over interprovincial and international trade.[16] Shut out of processing jurisdiction, it now faced handicaps in influencing quality and marketing and in controlling over-expansion.

Department of Fisheries starts up

In 1920, Found became Director of Fisheries and assistant deputy minister, never having worked in the field outside Ottawa. The Canadian Fisheries Association and F.W. Wallace's *Canadian Fisherman* magazine kept lobbying for a separate department.[17] A 1927 royal commission on Maritime fisheries added its support; and finally, in 1928, Mackenzie King promised a separate department. Found in that year became deputy minister for fisheries. Finally, in 1930, Parliament created the Department of Fisheries.

The new department got a parade of ministers, some only part-time. Dr. Cyrus MacMillan, a P.E.I. native, former head of the English department at McGill University, and a member of the 1927–1928 royal commission on Atlantic fisheries, in June 1930 became the first Minister of Fisheries for the separate department, but he lasted only two months, until the election. The new R.B. Bennett government in August appointed Edgar N. Rhodes, a former premier of Nova Scotia. In 1932, Rhodes went to the Department of Finance.

E.N. Rhodes, an early Minister of the separate Department of Fisheries. (Library and Archives Canada, C-45314)

Meanwhile, with the new department, D.H. Sutherland had become assistant deputy minister to Found, and also served as head of the Eastern Fisheries Division. The Central Fisheries Division disappeared at the start of the 1930's, when the Prairie provinces took over their fishery management. A.J. Whitmore, who had started working with the fisheries service in 1919, from 1929 took charge of the Western Division.

In the early 1930's, fisheries was usually a part-time job for ministers also holding other portfolios. Alfred Duranleau of Quebec, Bennett's Minister of Marine, served as acting Minister of Fisheries from 1932 until November 1934. Grote Stirling of British Columbia, Minister of National Defence, then filled in until August 1935. William G. Ernst, representative for Queen's–Lunenburg in Nova Scotia, became a full-time Minister of Fisheries for two months, until the Bennett government fell in October. Some stability appeared with the new Mackenzie King government. Joseph E. Michaud, from inland New Brunswick, became Minister and stayed in the job until 1942, followed by Ernest Bertrand of Quebec in 1945.

Found controls all

For most of the 1914–1945 period, as ministers came and went, a single man dominated internal fishery administration: W.A. Found. In 1920, when Found, as Director of Fisheries, travelled west with Minister C.C. Ballantyne, B.C. observers noted that

W.A. Found

Found seemed to be the boss.[18] That impression continued.

To quote one long-serving fishery officer in British Columbia: "[I]n the days before the Second World War, we didn't know who the Minister was. We'd have a Minister of Fisheries for one or two years, just so the Prime Minister could get him used to being in Cabinet."[19] As another contemporary put it, "the department was W.A. Found."

Alfred Needler, later a deputy minister of fisheries, was in the early 1930's a young scientist working on oyster culture at Prince Edward Island. Needler recounted an instance of Found's wide reach. When the department encouraged leasing of oyster grounds, a particular fisherman "with a mad look in his bright blue eyes" objected to public fishing grounds becoming private. "He wrote to Found, and got an answer that gave him no satisfaction. He wrote to the Minister, and got another answer signed by Found. He wrote to the Governor General, and got an answer signed W.A. Found. Thinking I guess in terms of some colonial hangover, he wrote to the King, and got another answer signed W.A. Found."

Although Found was a kind man in personal dealings, his thrifty habits attracted some notice. Employees speculated on the ultimate life of the raccoon coat he wore year after year. Once, when fishery officials confiscated a fisherman's horse used during illegal fishing and the case was slow to come to court, the department found itself paying for feed for the horse. Letters flew as Found tried to get the fisherman to take the horse back for the winter, and the fisherman refused, recognizing free board when he saw it. Bureaucratic penny-pinching ran through the whole system. When a fishery officer in northern B.C. put in a 50-cent claim for his own lunch and another 50 cents for his horse's lunch, his supervisor disputed the expenses, and the whole matter got referred to Ottawa.[20]

Regional set-up continues

Besides the three regional divisions, Ottawa headquarters in the 1930's included Cowie's Promotion and Inspection Division (Inspection took over the publicity function in 1928), the Fish Culture Division, and an Engineering Division, which in 1935 became part of the Administrative Division. Occasional adjustments took place. From 1937, a Publicity and Statistical Division included a director of publicity, a publicity agent, and two demonstrators and lecturers. There was also a licence section, a records section, an accounts branch, an "establishment" section with clerks, typists, and stenographers, a correspondence section, and a librarian.

All told, the "Inside Service" in the late 1930's and early 1940's had 50–60 people. Once appointed, officials tended to hold the same position and salary for years and years. Found earned the relatively princely salary of $9,000 a year; Sutherland, as assistant deputy minister, in the mid-$4,000 range; and Cowie, as Director of Fisheries Promotion and Inspection, about $4,000. Joe Whitmore of the Western Division got $3,600 a year.

In the Maritimes and British Columbia, non-hatchery permanent positions in the Outside Service in the latter 1930's and early 1940's numbered somewhere around 150. As always since Whitcher's time, however, the employment of seasonal staff multiplied the number several times over.

Outside of headquarters, Halifax was head office for the Eastern Fisheries Division. The Division Supervisor earned between $3,500 and $4,020 a year. Nova Scotia had three districts, each with a district supervisor earning in the $2,100–$2,800 range, 8–14 fisheries inspectors, and a stenographer earning roughly $1,100–$1,400. New Brunswick, with three districts, and Prince Edward Island, with two, had similar set-ups (although one P.E.I. district had only a single inspector).

The title Inspector had formerly denoted a supervisory position; now it meant more or less fishery officer. Officers had no uniform, but wore a marine officer's cap with a badge.[21] The local inspectors usually would have charge of a few seasonal guardians or patrolmen, sometimes hiring not just their personal services but also their boat. The department also had its own patrol craft, small and large.

In British Columbia, the well-known Major J.A. Motherwell was Division Supervisor. The Vancouver head office had 11 positions, mostly clerical and stenographic. There were three B.C. districts, the number of inspectors therein ranging 9–13, mostly posted to sub-districts. The Canned Salmon Inspection Lab, the department's only inspection laboratory at the time (set up in 1935), had three chemists and a stenographer. The Prairie provinces had had some 24 personnel, until the provincial takeover.[22]

Hatcheries also accounted for scores of employees, but many hatcheries were closing in the 1930's, as their usefulness came into question.

Finn replaces Found

Found continued as deputy minister until he retired in December 1938 (he died in Ottawa in March 1940). J.J. Cowie took over as acting deputy minister until July 1940. Although Cowie's term was brief, some observers considered him an excellent deputy, because of his interest in practical results.

Donovan B. Finn took over from Cowie. Holder of a Ph.D., Finn had worked for the Biological Board, directing the technical station at Prince Rupert and then at Halifax. Finn was highly intelligent, typically thinking of the longer term. During the war he served on various national and international bodies; he represented Canada at the 1945 Quebec Conference which drew up the charter for the Food and Agriculture Organization of the United Nations (F.A.O.). In 1946, Finn left the department to become the Director of the F.A.O.'s Fisheries Division, where he served until 1964.[23]

Research board, department work at arm's length

Meanwhile, after the Biological Board got its own small budget in 1912, the board and fisheries service had drifted further apart. Prince remained a connecting figure, but his importance was fading. He retired as Chairman of the Biological Board in 1921, and as Commissioner of Fisheries in 1925.

In 1919, attempting to recover some scientific capability, W.A. Found had tried to set up a scientific division within the fisheries service. The department said that the board, using only volunteers from Canadian universities, was doing too little. Besides basic research, the department wanted more technical work and technical education. The Board fended off this challenge. The department had only one scientific position, employing Andrew Halkett as Associate Zoologist, but this job disappeared in 1932.[24]

Board sets up technological stations

The 1922 royal commission on B.C. fisheries said the board (with a budget of $40,000) was wasted money. It had no permanent scientists, only university volunteers, and it needed reorganization.[25] In following years, the board paid more attention to work related to the fishing industry.

After Prince retired in 1921, A.P. Knight of Queen's University became Chairman until 1926. Following discussions between Knight, the board, and the department, starting in 1923 industry representatives—generally fish processors—became members of the board. Prince's protégé J.J. Cowie also became a member, providing better liaison with the department. More of the board's work came from department suggestions.

In 1923, Parliament approved the setting up of technically oriented Fisheries Experimental Stations at Halifax and Prince Rupert; the board was also asked to do educational work among fishermen and processors. A.G. Huntsman, director at St. Andrews, also took charge of the Halifax technical station, which opened in 1924. The Prince Rupert station opened in 1925,

The St. Andrews Biological Station in June, 1920.

Specimens in museum at St. Andrews Biological Station, 1920.

under W.B. Clemens, soon followed by D.B. Finn, the future deputy minister.

In 1929, the Biological Board started a substation at Ellerslie, Prince Edward Island, dealing mainly with oysters. In 1936, another technological station, the Gaspé Fisheries Experimental Station, started up at Grande-Rivière, Quebec. The technological stations would work mainly on processing and quality, for decades to come.[26]

Thus, the trade needs of the industry got some attention. Back at the St. Andrews and Nanaimo biological stations, however, the directors, although they had to listen to the board, ran the show pretty much as they wanted. If not two solitudes, neither were the board and the department part of the same household; they were more like second cousins. Professors and students would arrive from their universities each summer, stay in dormitories at St. Andrews and Nanaimo, and depart at the end of the season. In later years, Huntsman himself acknowledged the board's weakness on industry matters.[27]

Science becomes full-time

All the work remained seasonal until 1925; Huntsman, director at St. Andrews, never spent a winter on the coast.[28] At the beginning of the 1930's, "volunteer" workers —professors and their students, often from the University of Toronto—still did most of the research.

From 1931 on, as the Great Depression bit into budgets, the board offered less help to volunteer workers, to the regret of Huntsman and others. Especially after A.T. Cameron of the University of Manitoba became Chairman in 1934, the emphasis changed to full-time paid staff, even though there was still no full-time Chairman. The residences for volunteer workers closed down.[29]

The Board's journal went through several name changes. From 1901 to 1925, it was *Contributions to*

Canadian Biology; in 1926, it became *Contributions to Canadian Biology and Fisheries.* In 1934, it changed to *Journal of the Biological Board of Canada*; and in 1937, to *Journal of the Fisheries Research Board of Canada*, after the Biological Board under a new act became the Fisheries Research Board (F.R.B.) of Canada.

The F.R.B. was undertaking extensive biological work, tracing the life history of fish and their migrations. And the technological stations were doing useful work on fish processing and products. In British Columbia, the Nanaimo station in the 1930's would bring in the fishery inspectors each year to fill them in on scientific progress.

However, little analysis got applied to the fishing industry as a system, with all its strengths, weaknesses, and conflicting interests. Neither did the department itself have a worthy economics branch or other such analytical service. The biological stations sometimes did work directly related to the industry, either because the department asked or a scientist took an interest. And J.J. Cowie kept up a good connection with the F.R.B. But thorough co-operation between science and management, and thorough departmental analysis of the workings of the industry, would lag until the Second World War and its aftermath.

Board builds biological knowledge

On the Atlantic, Archibald Gowanlock Huntsman of the University of Toronto was the leading figure in research, the man who knew more about more things than anyone else. Huntsman had worked at the F.R.B.'s biological station on Georgian Bay. He became curator at St. Andrews in 1912.

Although Huntsman wielded an influence, the university professors and students largely followed their own inclinations in research. The connection to immediate industry needs could be loose indeed. For example, the board in 1915 launched a major scientific expedition in the Gulf of St. Lawrence that was intended to be largely on herring, but the scientist in charge, the Norwegian Johan Hjort, changed its character more to an oceanographic expedition. And although the St. Andrews station was located in the middle of an intensive herring-fishing area, the board remained lackadaisical about the species because, as Huntsman later noted, a Board member from the industry had lost money on herring.[30]

Still, the board was accumulating fundamental knowledge on scores of species, for example with early tagging projects on lobster. Tagging also helped establish migrations of mackerel, shad, cod, bass, herring, and other species. Scientists in the 1920's learned that salmon from Cape Breton swam as far as the Strait of Belle Isle (later work would establish far longer migrations). The Board developed a checklist of species for Atlantic Canada and Newfoundland. Huntsman worked on the natural breeding of lobster, and recommended restocking parts of the coast with lobster and instituting higher size limits, though little came of this.[31]

International research begins

On the Pacific, sockeye tagging by Canadian and American scientists in 1918 constituted the first international research. On the Atlantic, European countries at the turn of the century had set up the International Council for the Exploration of the Seas. In 1921, following an informal conference of fishery experts in Ottawa, the United States, Canada, and Newfoundland set up an International Committee on Deep Sea Fisheries Investigations. In 1923, this became the North American Council on Fisheries Investigations (N.A.C.F.I.), which continued until 1938. France, too, became a member, and N.A.C.F.I. brought about various studies in fisheries and oceanography. In one instance, Biological Board and American scientists in the 1930's studied possible effects of the proposed international Passamaquoddy tidal-power project. Scientists warned of potential damage to fisheries inside the dams and unpredictable consequences outside.[32]

Other notable 1920's and 1930's efforts on the Atlantic included Alfred Needler's research on haddock and his promotion of oyster culture in Prince Edward Island. R.A. MacKenzie and others researched biology and operations of the deep-sea fishery. The department itself in 1931 started collecting trip reports for otter trawlers; and in 1939, for dory schooners and longline vessels.[33] On the technological side, Huntsman did work on freezing of fillets, and in 1937, the Halifax technological station designed an effective mechanical smokehouse for fish.[34]

On the Pacific, highlights included tracking the migrations of salmon, herring, halibut and other species, and "reading" scales and otoliths (earbones) to determine size and ages of salmon and halibut. Alfred Tester in the 1930's did important work on herring; among other things, Tester allied with fishermen who kept pilot-house logbooks for the board. R. Earle Foerster and colleagues did fundamental work on salmon, including a re-evaluation of salmon hatcheries of the day, which led to closures.

Overseas, the Russian Baranov developed the first theoretical models to calculate the waxing and waning of fish populations. This and other overseas work began to provide a framework for population analysis.[35] In Canada, little direct research went to the question of abundance. Scientists were conscious of the question; for example, J.L. Hart, later a Chairman of the F.R.B., noted in 1933 that it was possible to deplete fisheries, as had happened with sockeye, halibut, shad, Atlantic salmon, and herring. But generally, few people worried about overfishing and population dynamics. On the Pacific, there was concern about sockeye, and Foerster did pioneering work on hatcheries and populations; but Atlantic scientists often knew little of such matters. As Alfred Needler later recalled: "For someone like me

The Biological Board's *Zoarces*.

in the 1920's, a graduate student wandering around looking at haddock, the Pacific coast might as well have been China."

Oceanographic research grows

The government and Board pursued oceanographic work as well. A northern expedition in 1914 looked at the fisheries of Hudson and James bays. The Hjort expedition of 1915, with Canadian, American, and Norwegian scientists working in the Gulf of St. Lawrence and off Nova Scotia, did fundamental work on Atlantic oceanography. The government's Canadian Arctic Expeditions of 1913–1918 under Vilhjalmur Stefansson increased knowledge of the Arctic. In the 1920's, the board's 90-foot schooner-hulled *Zoarces* did oceanographic work in various Atlantic areas. H.B. Hachey, who joined the board in 1928, took charge of its Hudson Bay Fisheries Expedition in 1930; with that effort, Canada began to catch up with the Danes and Americans in Canadian Arctic oceanography. The expedition produced the first extensive account of water circulation and physical oceanography in the Arctic.

All told, as described in Ken Johnstone's *The Aquatic Explorers*, the board between the wars filled out much of the basic picture for Canada's fisheries and oceans and laid the foundation for a worldwide reputation.

Making treaties, losing jurisdiction

U.S. tariffs fall and rise

Dealings with the United States continued to preoccupy officials. The 1914–1945 period saw major treaties on Pacific halibut and salmon. But the most constant theme was tariff relationships. American restrictions stayed troublesome until the First World War, then loosened, then tightened again.

Since 1897, typical tariffs had run from one-quarter to one cent a pound on fresh and frozen fish, a bit higher for salted fish, and far higher – up to 30 per cent – for fish packed in oil. The Laurier government's abortive Reciprocity Treaty of 1911 would have admitted most fish free of duty. Gloucester fishing interests opposed it, fearing that the Canadian industry—with cheaper vessels, continuing bounties from the Halifax Award, and close proximity to fishing banks—would absorb a large part of the New England industry. Congress approved the treaty anyway, but Laurier's election defeat doomed it.

Even so, the fish trade came close to free trade under the U.S. Tariff Act of 1913. High duties remained for fish (except shellfish) in oil (25 per cent), canned fish not in oil (15 per cent), and skinned or boned fish (3/4 cents per pound). But practically everything else—shellfish, fresh and frozen fish, and prepared or preserved fish, including the major products of salt cod and salt mackerel—now went in duty-free.[36] The new rates coinciding with high wartime demand were a great help to the industry.

Having eased its tariffs, the American government in 1914 wanted Canada to loosen up its port privileges. The modus vivendi—the easy purchase of port privileges—only applied to the Atlantic; in 1904, Canada had limited it to sailing vessels, on the grounds that the original agreement had only these in mind. The U.S. wanted port privileges for powered vessels.

Canada had demands of its own. Canada since 1897 had allowed American halibut fishermen to land fish in British Columbia, where buyers would ship them in bond to the United States.[37] But the U.S. refused to let Canadian vessels land fish there straight from the fishing grounds, or to clear for the fishing grounds from a U.S. port.

The two countries convened a joint commission known as the American–Canadian Fisheries Conference, which lasted from 1917 to 1919 and held hearings in 1918. The prime result was the "Hazen–Redfield Pact," temporarily replacing the modus vivendi. Under this arrangement, fishing vessels of each country could enter the other's ports direct from the fishing grounds, dispose of their catch, take on supplies, and clear for the fishing grounds. This set-up lasted until 1922; and during the period, Canadian vessels landed about half a million dollars' worth of fish in Boston, Gloucester, and Portland.[38] With mostly duty-free fish and reciprocal access to ports, the two countries had practically recreated the earlier fisheries reciprocity, except, of course, that the Americans no longer held fishing privileges inside three miles.

The U.S. Tariff Act of 1920, aiming to protect domestic industries, changed all that, imposing painful new duties, higher than ever, on almost everything. Prices to fishermen in Canada dropped. After the U.S. action, Canada in 1924 ended the old modus vivendi whereby U.S. fishermen paid minor licence fees in exchange for port privileges. This move had little impact; the port privileges were now less important, as U.S. fishermen now fished mainly the deep-sea banks, returning directly to New England.

In the United States, intensive fishing drove down the Georges Bank haddock population to a new equilibrium. But the scarcity brought no lowering of tariffs; the New England fishing industry still wanted duties. The famous "Smoot–Hawley" tariff of 1930, often blamed for worsening the Great Depression, brought higher tariffs on some products.

An Imperial Economic Conference in 1932 created some small advantage for Canadian fish in England. But in North America, tariffs held back the Canadian trade. Even so, Canada in 1933 restored some privileges for American fishermen, permitting the purchase in Canadian ports of bait, ice, seines, lines, and other supplies, but not transhipment or hiring of crews. U.S. fishing no longer appeared a great threat, and the change brought some business to Canadian ports.[39] The *Halifax Herald* in December 1934 reported rejoicing in Yarmouth about the renewal of American port privileges for 1935.

Halibut Commission pioneers international management

Besides trade issues, the 1918 American–Canadian Fisheries Conference discussed conservation. Both countries lamented the drastic decline of sturgeon, recommending a ten-year shutdown. In the lobster fishery, the Canadians complained that American "well-smacks," rigged up to carry live lobsters, were fishing at southwest Nova Scotia just outside the three-mile limit during Canadian closed seasons, and using Canadian ports at night. The American industry itself ceased this practice, without government action.[40] Canadians also raised their concerns about the Pacific halibut fishery. Although the conference produced no agreement on the subject, it highlighted the issue.

Halibut grounds extended from California to Alaska, with the fish most abundant off British Columbia and especially in the Gulf of Alaska. Americans took the bulk of the catch. As time passed, vessels went further afield; a winter fishery developed at the Bering Sea. On both sides of the border, Scandinavian fishermen became prominent. Later, many Nova Scotians and Newfoundlanders joined the Canadian fishery.

Canadian participation, small at first, built up, especially after the completion of a rail link to Prince Rupert in 1914. Landings doubled by 1915. By then, despite a closed season, halibut were getting scarcer. In 1916 and 1917, W.F. Thompson highlighted the depletion in reports for the B.C. Commissioner of Fisheries. The 1918 American–Canadian Fisheries Conference drafted a treaty that dealt with conservation, tariffs, and reciprocal fishing privileges; the U.S. Senate blocked it.[41]

Steel vessels owned and operated by Canadian Fish and Cold Storage, 1920. (Library and Archives Canada, PA-95084)

Early halibut steamers gave way to dory schooners, which themselves got outlawed in the 1930's. (Library and Archives Canada, PA-40998)

The two countries held further discussions, separating the halibut question from the complex questions of port use and tariffs. W.A. Found took a key role in negotiations. In March 1923, Minister Ernest Lapointe and the American representative signed the Convention for the Preservation of the Halibut Fishery of the Northern Pacific Ocean. At Mackenzie King's insistence, Canada signed in its own right; it was the first treaty that Canada or any Commonwealth nation signed on its own, separately from Great Britain.

The treaty provided a closed season of November 16 to February 15, and established a four-member International Fisheries Commission. The commission was to carry out its own scientific work; W.F. Thompson became Director of Investigations. Studies showed that the Alaska and B.C. stocks were largely though not entirely separate. It became apparent by 1929 that the closed season had done little good; abundance was still dropping, and the fishery off B.C. was removing 40 per cent of the stock yearly.

The two governments in 1930 approved a renewed Convention, ratified in 1931, which created a stronger International Pacific Halibut Commission (I.P.H.C.). To support the commissioners, an equal number from both sides, the I.P.H.C. set up a permanent staff at Seattle, including scientific and technical experts. It came to enjoy a reputation as fair and professional. After the commission made its recommendations, the two countries normally put the requisite regulations or policies in place and carried out enforcement.

The I.P.H.C. also set up an industry advisory body—the Conference Board—and paid close attention to its views. In particular, Scandinavian fishermen with their cross-border relationships and their character influenced the whole development and operation of this venture in international co-operation. The commission would put its proposed regulations for the year before an open meeting of fishermen, thereby building up trust. With strong associations, halibut vessel owners had a big influence. The commission required licences and logbooks from fishermen.

Besides enacting closed seasons, the I.P.H.C. could establish regulatory areas. It set up three; British Columbia fell into Area 2, running from near Cape

Flattery in the state of Washington to Cape Spencer in southeast Alaska. The I.P.H.C. could also regulate the type of gear used; it insisted on longlines, and in the mid and late 1930's it outlawed dory schooners, which were fading anyway. The restriction to longlining also blocked out trawlers, then controversial on the Atlantic coast. The commission could also set the minimum size of fish to be caught, establish regulatory areas, and designate closed areas as nursery grounds.

As part of its control of the level of fishing, the I.P.H.C. could and did set quotas—the first quotas in any Canadian sea fisheries, and the first international quotas anywhere. Along with gear restrictions, area quotas became a prime means of management. The I.P.H.C. set seasonal openings timed to fit the quota. Landings, which had appeared on the decline, stabilized. Participation in the fishery rose during the 1930's from about 300 to 400 American boats, and from just over 100 to around 170 Canadian ones.[42]

Fisheries analysts later disputed whether the commission used exactly the right methods for the best biological and economic returns. In particular, some criticized the lack of limits on the number of boats. But it is generally agreed that the I.P.H.C. improved both catches and catch rates, pioneered new methods of sea-fisheries regulation, and wrote an early success story in international fisheries management.

International salmon regulation makes slow progress

Co-operative regulation of halibut progressed at lightning speed, compared with that of salmon. The Fraser River was the biggest salmon producer; here Canada paid for management, and tried to restrain her own fishermen. But Americans, subject to fewer restraints, could intercept migrating Fraser salmon on their way back through the Strait of Juan de Fuca. Indeed, the Americans tended after 1900 to take most of the Fraser catch, sometimes twice as many as the Canadians. Canadian concerns grew after the disastrous slides of 1913 and 1914 in Hell's Gate canyon

Hell's Gate Canyon on the Fraser River suffered a disastrous slide. Photo is from 1912. (Library and Archives Canada, C-6921)

blocked a large portion of salmon from their upriver spawning grounds. The 1918 American–Canadian Fisheries Conference said that failure to mend the situation would be "criminal."

The conference drafted an agreement to protect sockeye; the U.S. Senate blocked it, following opposition by the American canning industry, based in Puget Sound, Washington. The conference did succeed in prompting more research on salmon. In 1921, new negotiations with the state of Washington took place, with talk of a five-year closure to let runs recover and a permanent ban on purse-seines; but this too fell through.

All through the 1920's, neither fishing restraints nor river restoration work brought real recovery for the Fraser. In 1928, new negotiations produced another proposed treaty, under which Canadian and American fishermen would get roughly equal shares of the catch. Canada ratified the treaty in 1930. Again, the fishing interests of Puget Sound, fearing restrictions and losses, blocked the treaty in the U.S. Congress. Tempers in British Columbia reached such a point that the provincial Commissioner of Fisheries suggested abandoning the Fraser, and taking all the eggs to hatcheries on the Skeena and at Rivers Inlet.

But new support for a treaty came from the growing sport-fishery in the United States. In 1934, the sport fishermen successfully lobbied for a ban on fish-traps in Washington state; this sharply reduced the American catch to less than the Canadian. The ban on fish-traps ended a long feud between trap operators and purse-seiners; the latter now became more amenable to international regulation.

The province of British Columbia and the state of Washington arranged an informal conference on the treaty; and in July 1937, the United States finally ratified it—more than two decades after Hell's Gate and more than four decades since the first talk of international regulation, in 1892. Even then, the treaty provided for eight years of scientific research before the commission created by the treaty had power to make regulations.

In structure, the new International Pacific Salmon Fisheries Commission (I.P.S.F.C.) was like the Halibut Commission, with an international board, a director, and its own scientific staff. I.P.S.F.C. research went forward from 1938, dovetailing with other efforts by the two governments on migrations, prevention of overfishing, clearing of migratory obstacles, and ways to protect and assist propagation.[43]

Licensing becomes looser

While jurisdictional and international issues wound their way over decades to a conclusion, the Canadian fisheries service was changing some of its approaches to management.

Licensing became less of a preoccupation, most notably in British Columbia. In that province, as we will see, pressure from First World War veterans and other independent fishermen brought an end to licence limitation. But in other areas as well, the fisheries service seemed to lose interest in licences as a major tool of management.

At the outset of the period, many inland and shore fisheries required licences. The 1917 annual report noted that "no one is permitted to engage in most of the fisheries that will admit of only limited prosecution, unless he first procures from the department a fishery licence. ... During the present year a total number of 26,565 licences were issued." The total number of fishermen and boats was of course much greater. Still, the department was quick to use licences where it saw a need, and was inventive about related conditions.

For example, in the New Brunswick salmon fishery and the Gulf of St. Lawrence trapnet fisheries for cod and herring, licensees paid fees proportional to the amount of gear used. In Manitoba and Keewatin, various licences allowed different lengths of net. Salmon canneries, like lobster and whale factories, paid licence fees proportional to the amount of production, and in at least some instances, the department had to approve the plant and the conduct of business. Non-conservation rules included the ban on motorboats in northern British Columbia. That province also had instances of exclusive privileges (in effect, local monopolies) in the drag-seine fishery. Sometimes, area restrictions applied: for example, only residents could fish on the Fraser between New Westminster and Mission.[44]

After the First World War, many earlier licences continued, and in 1919, the department started issuing licences to Atlantic lobster fishermen. But rarely would licences limit entry to the fishery. The great Atlantic sea fisheries for groundfish and herring remained unlicensed. The fisheries service generally ceased to propound earlier theories of how leases or limits on licensing could benefit both conservation and profits. Broadly speaking, licences became a *pro forma* matter.

Transport subsidy ends

While loosening up on licences, the department under the spell of wartime prosperity also dropped an early development effort: the transport subsidy that had begun in 1907.

The one-third subsidy applied to express shipments of fish in less than carload lots. But by 1917, several full carloads were coming weekly from each coast without benefit of subsidy. On the Atlantic, the three-day fast-freight "Sea Food Special" took fish in refrigerator cars to Montreal and Toronto. The subsidy had been so successful, said the 1917 annual report, it had almost succeeded in "placing the fish business in a position where it can take care of itself."

The annual report looked forward to ever-expanding markets after the war. Only two things were necessary: to provide adequate rail transportation at reasonable rates, and to sell the public on frozen fish. Fish properly frozen as soon as landed, shipped frozen in refrigerator cars, and sold frozen was the "next best thing to

these fish right at the seaside" and much superior to fish packed on ice, even when these travelled in refrigerator cars. As it turned out, however, decades would pass before frozen fish became popular.

In 1918, the department dropped the main subsidy for Atlantic fish and for Pacific salmon and halibut, but continued temporarily a higher, two-thirds subsidy for shipments of Pacific flounders and cods, whether express or freight. This helped to develop the fishery for these then underutilized species.[45]

Although ending the subsidy, the department began paying some attention to promoting and marketing fish. The Canadian Fisheries Association, spearheaded by F.W. Wallace of *Canadian Fisherman* magazine, was lobbying the government to promote fish in Canada and send fisheries trade missions abroad.[46] Some advertising took place 1915–1917 through Cockfield, Brown & Company. And the department for several years mounted a fisheries exhibit at the Canadian National Exhibition in Toronto, including "a first-class fish restaurant" serving "a good fish dinner for 35 cents."[47]

Along with the transport subsidies, the department in this decade abandoned its bait-freezers. "The story of the bait freezers is a story of failure," the 1927 royal commission on Atlantic fisheries would later say. The uncertain supply and demand for bait had weakened the associations that were supposed to run the freezers. The normal Atlantic obstacles of isolation, fragmentation, and lack of education and organization no doubt played their part. Besides, private freezers were lessening the need for the bait-freezer associations.

Product-quality work increases

The department in the interwar years would make recurrent, though often weak, efforts to improve product quality, handling, and marketing.

Earlier laws had relied mainly on voluntary inspections, the common theory being that exporters would submit products for inspection, since a government stamp of approval would help them in marketing their products. This approach had no great impact. As well, the department in some instances had made factory licences conditional on sanitary operation. But there was no systematic inspection of products.

In 1914, the Fish Inspection Act provided stronger authority for inspection. The legislation set compulsory standards for pickled fish containers (pickled fish including such products as herring, mackerel, wet-salted groundfish, or anything else packed in brine), although inspection of products themselves remained voluntary. The following year, the department took more direct control over inspection of canning plants. In 1917, the government strengthened the inspection legislation.[48] Complaints by the troops about the quality of hake prompted some work by the Biological Board on quality.[49]

In 1920, a new Fish Inspection Act went beyond the pickled-fish containers, giving the department authority to inspect all pickled fish directly. Minimum standards also came into force for canned lobster.[50] In following years, new rules accumulated. The Fisheries Branch began inspecting dry-salt herring exported to China. In 1923, the Branch set inspection rules for smoked herring sold to the West Indies. The inspectors in 1924 got the authority to visit cooperages (barrelmakers); and in 1927, the department, in co-operation with the fish trade, set legal standards—not compulsory rules, but definitions—for dry and salt cod, haddock, hake, cusk, and pollock. In 1928, the Inspection Act was extended to cover all curing. J.J. Cowie helped train those working with saltfish.

At first, civilian appointees had done inspection. In 1920, the department put inspection under a fisheries service chief, still using hired workers. In 1929, regular staff took over the job. Over at the Biological Board, among other efforts by the technological scientists, G.B. Reed worked on the problem of lobster discolouring in the cans; the answer was speedier handling, to get the cans into the retorts sooner. In 1927, the Biological Board surveyed Canadian lobster canneries, and found most of them unsuitable for their work. In 1930, the board and the department began yearly surveys. By 1937, 95 per cent of canneries operated suitably, although some problems remained.

Meanwhile, the division of Publicity, Transport, and Marketing, which had started work in 1919, tried various promotions: pressing the Grand Trunk Railroad to improve its fast freight service for fish, talking up a National Fish Day, producing pamphlets, promoting school essay competitions on fish, and trying to educate fishermen away from using pitchforks to handle fish. In 1928, the department combined its promotion and inspection sections. Experts in fish cookery would visit larger centres to give lectures and demonstrations. The department sometimes gave fishermen instruction on handling and processing fish.

Coupled with such work were the efforts of the "Fisheries Intelligence" staff. More boats now had radios; intelligence work included passing on weather, bait, and ice reports by radio, from Halifax, Louisbourg, and Saint John. The department also telegraphed information on bait supplies to some ports.[51]

Old problems defy new laws

The department was working harder on fish quality. Still, few laws made product standards compulsory. Even those few stood more as guidelines than as strict rules. And the question that fishermen had posed to Moses Perley eight decades earlier remained unanswered: Why should they take extra care to protect quality if they got no extra money?

A 1932 report by the consulting firm Cockfield, Brown & Company noted widespread quality problems—bones, poor flavour, bad odour, lack of freshness—and outlined the unsatisfactory system of inspection, which the industry itself wanted improved:

The purchase of fish is not accompanied with such confidence as in the case of the purchase of meat. This is due to the fact that fish have hitherto reached the consumer's table in an unattractive condition, resulting in a large proportion of the consumer's market being apathetic towards the use of fish. ...

Whether the inspection of fish comes within the federal or the provincial legislative sphere or both, must, for the present, remain an open matter. ...

It is evident that existing legislation is not altogether successful in practice, and some attempt should be made to strengthen the entire system. ...

There appears ... to be a good deal of overlapping. For example, pickled and cured fish are provided for in the Fish Inspection Act, oysters in the Fisheries Act, canned fish in the Meat and Canned Foods Act and fish generally in the Food and Drugs Act. It would appear that the Food and Drugs Act provides a statutory basis for the inspection of fresh and frozen fish. One drawback, however, is the insufficiency of the practical inspection under this Act, for the reason that only twenty-five inspectors have been appointed. ...

While the provisions for adequate inspection may exist, the degree to which they are actually enforced is problematical.[52]

British Columbia spearheads quality work

Meanwhile, the B.C. canned-salmon industry, in hopes of getting tariff changes in Britain, was asking the government to do more on inspection. In 1932, an amendment to the Meat and Canned Foods Act provided for compulsory inspection of canned goods across the country. And in 1936, the department set up a canned-salmon inspection laboratory in B.C.—the first such establishment for thorough and organized inspection. Several more would follow in later years, especially in the 1950's and 1960's. [53]

Biological Board promotes frozen fillets

Freezing techniques were slow at first, and often produced a poor product. The 1920's saw major improvements in quick-freezing, with Clarence Birdseye of New England a leading figure. In Canada, Huntsman and his colleagues in the late 1920's worked on quick-freezing of fillets by air, brine, and other methods. They developed "jacketed" cold storage: double-walled freezers enclosing a layer of air. Huntsman and his co-workers arranged to market the "Ice fillets" through Toronto stores.[54]

Huntsman later said that they were 30 years ahead of their time. He blamed a Toronto dealer for killing the new method for fear of hurting sales of "fresh" fish.[55] It was years before the industry and railway companies applied jacketed cold storage and before frozen fish became common in Canada. After the frozen-fish experiment petered out, Huntsman paid less attention to technological work, and spent his great energies on biology.

Inspection, marketing make slow progress

The Second World War brought some changes. After the British embargoed Canadian canned lobsters, deputy minister D.B. Finn forced some reforms in production quality. On the prairies, where exporters had problems with the American industry's manipulation of U.S. Customs regulations, the provincial and federal governments in 1944 began joint inspection of whitefish. The Fish Inspection Laboratory in British Columbia, already dealing with salmon, began inspecting canned herring for export against compulsory standards. In 1945, the federal department set up a fish inspection laboratory in the Maritimes, a decade later than in British Columbia.[56]

Also during the war, after many attempts by many people, the Halifax technical lab of the F.R.B. developed what was considered the first low-cost artificial dryer for saltfish.[57] These dryers amounted to little more than a heating element and fan to blow hot air over the fish. Mechanical drying took place mostly in the cooler months, since in summer humidity from the outside air causes problems. Dryers were unable to produce the high-quality, light-salted product attainable through outside drying on flakes. Even so, mechanical dryers were a big help, and came to dominate Atlantic production.

Despite the occasional bright spot, the inspection system at the end of the war remained small-scale, based mainly on voluntary compliance. Complaints about fish quality remained frequent. On the Atlantic in particular, the whole chain that took fish from water to table appeared weak and fault-filled. It may have been easy for observers to overstate the defects. The Atlantic industry was supporting many thousands of people, and must have been doing something right. But in terms of product quality, rarely did it give an impression of real vigour and progress in the interwar years.

Report lambastes "disjointed" industry

The problems were widespread. The 1932 report by Cockfield, Brown & Company, quoted above on product quality, described a general malaise. The department commissioned the company to analyze the industry and its markets, towards increasing sales in Canada and especially in foreign markets. The multi-volume report called for thorough reforms:

> The disjointed condition of the Canadian fishing industry ... is a heritage, in part, of the geographical dispersion of resources, and in part of the complete lack of intercommunication in the days when each of its branches was first established. It is not a new problem, but has become steadily more acute over a period of many years. ... Return for effort and investment in every phase tends towards the vanishing point, and new capital cannot be directed into it for necessary improvements and additions to equipment, plant and organization.

The report said that the fragmented and uncoordinated industry needed basic improvements in quality and marketing, and control of over-expansion. Conservation policy should restrict seasons, places, and licences, but otherwise let people fish as they wanted. The industry needed more use of trawlers, refrigeration, and quick freezing. It needed better inspection. There should be systematic collecting of market information. Promotion was almost entirely lacking; practically the only form of advertising was to list fish prices. Bit by bit, there should be reforms, integration, and consolidation, until the industry worked as a system.[58]

The Cockfield, Brown report resembled many subsequent studies, both in the wide scope of the problems it identified and in the narrow results it produced. No wholesale changes followed to make the industry's entire system work smoothly. But limited improvements did take place.

Marketing work by government increased somewhat. The department and B.C. salmon canners undertook a joint campaign. In 1935, the department increased advertising for fish products, in Canada and overseas; Parliament set aside $200,000 for promotion. In 1937, the department surveyed foreign markets for dried and pickled fish. That same year, Minister J.E. Michaud announced that in conjunction with the last two months of an intensified advertising campaign, the department and the Canadian Fisheries Association would co-operate on an educational campaign to improve retail handling of fish.[59]

By then the country was in the full misery of the Great Depression, when nothing seemed to help that much. The federal government as part of its relief work sent considerable Atlantic salted, dried cod to the Prairie provinces, the idea being to help the industry and feed the hungry. But often this was little help, since many people had no idea how to cook saltfish. Some used the split fish as shingles for leaky houses.[60]

Alone among provincial governments, Quebec in 1932 started up cold-storage sites for the fishing industry. This helped to move the industry towards frozen fish. The Quebec cold storages, though generally small-scale, evolved into a large network. By the 1960's, Quebec would be operating 53 cold storages, 108 collection depots, eight artificial ice plants, six salt depots, and one drying plant.[61]

"Underutilized species" gain some attention

While trying with mixed results to improve old products, the department and industry made some attempts to produce new ones. Early in the 1914–1945 period, the department abandoned its dogfish-reduction plants. But J.J. Cowie resumed his promotion of driftnetting and producing "Scotch cure" herring, although the project never fulfilled his hopes. In following years, the industry began to use the innards of cod and hake to produce isinglass, as well as filters for beer and wine. On the Maine side of the Bay of Fundy, Americans began to use herring scales, often supplied by Canadian fishermen, to produce "pearl essence" for imitation jewellery. Marine oils from species ranging from herring to whales remained a common product, despite the inroads of the petroleum industry. Fish-meal, fertilizer, and cod-liver oil also served industrial purposes. Various other by-products came into use.[62]

Hatcheries slowly sink

At the start of the 1914–1945 period, the greatest effort at fisheries development still consisted not of new fishing, processing, or marketing approaches, but of attempts to increase abundance through hatcheries. At one point the hatcheries probably had as many regional staff as fisheries enforcement, if one discounts the temporary guardians hired yearly. The 1917 annual report counted 50 hatcheries in the Dominion, mostly for salmon but with considerable work on other species, including 14 hatcheries producing lobster. There were also 11 subsidiary hatcheries, six salmon retaining ponds, and one lobster pond. Artificial rearing ponds used gravel to reproduce natural spawning beds. Hatchery workers also used incubation boxes. The annual report stated that "evidence of the most satisfactory results from the department's fish cultural operations is apparent on all sides."

Stripping fish at a hatchery.

But there was no real evidence. Where fish catches did rise, the cause was uncertain. A.P. Knight of the Biological Board investigated lobster hatcheries and found little of value. The department in 1917 decided to stop operating them.

The joint fishery investigation by the Canadian and U.S. governments concluded in 1918 that sockeye hatcheries had as yet produced no good results. (Meanwhile, the department had closed three B.C. salmon hatcheries in 1914.) But in the department's annual report for 1919, Chief Inspector Cunningham, in charge of B.C. fisheries since 1912, shot back at the critics, speculating on the results if people spent as much energy in conservation as in criticizing hatcheries. Salmon hatcheries continued.

So did work on transplants. Japanese oysters were brought to British Columbia in 1912. Like hatcheries, transplants could inspire self-hypnosis, as in this 1911 statement about newly-attempted lobster transplants to B.C.: "[T]here seems little doubt that this valuable crustacean will thrive in its new surroundings."[63] As it turned out, those fish transplants that succeeded were likely as not to be harmful.

The department's hatchery work was a grand experiment without controls or analysis, and with little research. Only slowly did the board's scientists interest themselves in hatcheries and enhancement. And when Huntsman wanted to experiment with "artificial freshets" to help salmon get upstream, he found little interest on the department side.[64]

Doubts continued to surface here and there. The 1922 royal commission on British Columbia fisheries said there was no evidence that hatcheries helped the salmon fishery. The amount of fry released had gone up; the catch had gone down. The branch should evaluate the hatcheries, and it should do more at clearing streams. The same commission, which had considered the Biological Board's $40,000 budget to be wasted, noted far higher costs for hatcheries: $130,000–$140,000 in B.C. alone.

Still, the hatcheries persisted across Canada, though with a closing here and there. In 1926, the Dominion government still operated about 36 hatcheries, Quebec about 13, and Ontario about 18.[65]

By then, at the department's request, the Biological Board had begun studying the production of Pacific salmon hatcheries in comparison with natural propagation. R.E. Foerster, working at Cultus Lake, B.C., was the key figure. The study concluded that hatcheries offered little benefit in relation to their cost. (Work at Cultus Lake also provided the final proof of salmon homing to particular streams.[66]) At the beginning of the 1930's, British Columbia still had 14 hatcheries, but they were on the way out. In 1934 and 1935, workers ceased collecting salmon eggs for culturing. The remaining stations stopped operating in 1937. That same year, the federal government turned over responsibility for sport-fish culture to the province of British Columbia.

Besides its hatchery work, the fisheries service had long maintained fishways, cleared streams, and opposed or mitigated dams. This work was deemed more cost-effective and was supposed to increase after the hatchery closures in B.C.; but it was late getting going, then faded away during the Second World War.[67]

In the department's other attempts to aid nature in British Columbia, oysters transplanted from the Maritimes by 1936 appeared exhausted, as did native B.C. oysters. But the transplanted Japanese oysters had taken better hold and were to support a continuing small industry.[68]

The Maritimes in the early 1930's still had about 20 hatcheries. They escaped closure. Dr. Alfred Needler recalled that "Foerster's work at Cultus Lake [in B.C.] indicated that although there was some increase of abundance, it wouldn't pay. Found didn't like wasting money, and the hatcheries closed down. But those results didn't apply to Atlantic salmon. There was no big fuss about hatcheries at the time of the Pacific shutdown. There may have been some Atlantic hatcheries closed, but there was never a comparable investigation."

The Maritime hatcheries over time switched their emphasis from increasing commercial abundance to restocking sport-fisheries. The public supported such efforts. Needler recalled an instance when hatchery officials themselves proposed closures at Prince Edward Island, saying there was enough natural escapement and spawning; public sentiment stopped that idea. Still, he said, "hatcheries were by and large a waste of effort. There was a lot of poor management, shamefully poor. Most of the bred stock was released where salmon had easy access anyway. If they'd used the hatchery stock in other areas, they could have done some good."[69]

From 1930, when the Prairie provinces took over fisheries management under delegation, the federal government began handing over its hatcheries to them. With the B.C. hatcheries shut down, the federal government ended the 1914–1945 period with only its Maritime hatcheries running. The whole hatchery episode, if largely useless at the time, did provide information and also an orientation that helped the resurgence of hatcheries after the Second World War.

By the end of the 1930's, most national development efforts—the hatcheries, transport subsidies, promotional efforts—were weak or had vanished. And although direct subsidies for boatbuilding or fishing had been fairly common before Confederation, the Dominion government had never undertaken them since.

War brings developmental subsidies

The Second World War brought back boatbuilding subsidies, not so much because the industry needed assistance—demand and prices were high—but because the temper of the times changed. Suddenly, development became the predominant goal. The fisheries economist W.C. MacKenzie later recalled that starting with the war, "the task was to raise food production as fast as possible. This led to a reversal of the policy of restricting technology. The fishermen were earning as much as they'd be able to earn in most other jobs available to them, but the costs ate up their profits, and there was no chance for them to finance new vessels. This perhaps was the rationale at the time for subsidy and loan programs."[70]

In Nova Scotia, the main deep-sea fishing province in the Maritimes, the fleet of decked vessels over 20 tons, mainly schooners, had dwindled from 71 in 1939 to only 25 in 1943, largely because the war pushed many into freighting. Getting vessels built or repaired became difficult. And 1930's regulations to be described later had reduced the trawler fleet to only three.

The first subsidies came in 1942, when the government provided $165 per gross ton to build British Columbia packer-seiners of 72–78 feet long. Construction began on 11 such vessels. A similar subsidy began a few months later for building draggers on the Atlantic, although this one had less effect. The government also offered a two-thirds subsidy, up to $12,000, for converting schooners to draggers.

In 1944, the government began trying to encourage groups of fishermen to move up to bigger boats. A new Order-in-Council under the War Measures Act directed that the $165 per gross ton subsidy apply to groups of not less than four fishermen building draggers or longline vessels of 55 feet or more.

Boat-building subsidies would gain strength after the war, and become a prominent feature of the fishery until the 1980's.

CHAPTER 13.
The Maritimes and Quebec, 1914-1945

On the Atlantic, at the outset of the 1914–1945 period, there were now fewer accounts of great abundances of fish. But there was also less outcry about depletion, perhaps because fishermen along the coast and on the nearby banks had gotten used to fishing at a lower level.

Cod and other groundfish fisheries remained the most prevalent, with fishermen and their families salting their own catch in coves all along the long coastline. Lobster, too, were scattered coast-wide, never schooling on banks like cod or herring. The widespread nature of the fisheries tended against industry consolidation. For the most part, small boats and individual or family enterprises dominated the picture.

Schooners in Halifax harbour.

For the more than 40,000 local, smaller-boat fishermen, the gas engine and new wartime and post-war demand had brought more mobility and independence. They usually had their own houses and their own one- or two-man boats. The hundreds of schooners were a minority, and banking schooners a smaller minority. The latter could be sizeable, like the 161-foot *Bluenose*, later depicted on the Canadian dime. The larger schooners might be owned by a family enterprise or wider partnership. As well, then and later, processing–trading companies sometimes helped fishermen finance boats of anything beyond minimum size. This went together with an understanding that the fishermen under normal circumstances would sell his catch to that company.

Fishing remained a dangerous occupation; for example, the year 1918 saw 28 drownings on Canada's Atlantic coast and 19 on the Pacific. Newfoundland often had dreadful losses, the worst coming in 1914, when 78 men from the sealing vessel *Newfoundland* and 173 from the *Southern Cross* died.[1]

Wartime boom aids groundfish fleet

At first, the high demand generated by the First World War brought prosperity. The small town of Shelburne in southwest Nova Scotia had eight dory-building shops. Lunenburg became "wonderfully prosperous," with a shortage of fishermen.[2] J.J. Cowie told a conference that "on the Bay of Fundy there is a community of 2,000 or 3,000 people. They have splendid homes, and own twenty or thirty motor cars. ... Our sea fishermen, at least, cannot be classed as ignorant or uneducated."[3]

Table 13-1.
Vessels and boats by province (sea fisheries) in the Maritimes and Quebec, selected years, 1924–1944.

	P.E.I.			N.S.			N.B.			Que.		
	1924	1934	1944	1924	1934	1944	1924	1934	1944	1924	1934	1944
Vessels	7	4	8	378	303	520	314	340	335	7	50	123
Boats	1,902	2,420	1,854	10,232	10,502	8,645	7,649	7,389	6,220	3,939	6,104	5,424
Carriers	7	15	18	184	126	65	61	139	66	17	9	17

Table 13-2.
Total vessels and boats (sea fisheries) in the Maritimes and Quebec, selected years, 1924–1944.

	1924	1934	1944
Vessels	713	697	986
Boats	23,722	26,415	22,143
Carriers	269	289	166
Total	**24,704**	**27,401**	**23,295**

In the groundfish trade, saltfish, mainly dried salt cod, still dominated. Haddock, too, went largely for the saltfish trade, being dried for southern Europe. Pollock and hake also supported a big salting fishery. But the fresh trade was growing, with several trawlers now helping in supply.

When inshore fishermen protested against the trawlers, government figures questioned whether the protestors were concerned about conservation or simply with rivalry. But giving the benefit of the doubt to inshore interests, the department limited trawlers to 12 miles offshore,[4] a regulation which has endured. (Newfoundland had no such rule.)

After seeing steam trawlers at work, New England schooner fishermen with gas engines were by 1919 also using "drags." Enterprises at Lunenburg and elsewhere on the Canadian side began buying a few draggers, despite the opposition of hook-and-line fishermen. Meanwhile, larger gas boats from Nova Scotia began fishing the nearer Atlantic banks, usually with longlines, by 1918.

Nova Scotia continued to send larger schooners, usually with engines as well as sails, to fish the banks with longlines.[5] Many smaller schooners worked closer to shore, with longlines or handlines. In the Gulf, perhaps a hundred or more schooners operated out of the Acadian Peninsula, though this fleet was less pros-

perous; many schooners had no motors even in the 1920's. Here, a handful of companies—the Loggie, Robin Jones, Young, and Robichaud interests—controlled most of them.[6]

The Gulf, despite the wartime boost, still suffered from emigration and slowness of development. Indeed, Canadian fishermen from many parts of the coast, including Acadians from southwest Nova Scotia, sought work in the United States. The port of Gloucester in 1918 was said to have more Canadian (897) and Newfoundland (237) fishermen than it had Americans (754).

The Canadian fresh-fish trade grew further in the 1920's. Lunenburg schooners that salted fish in the summer would in the winter fish for the fresh trade. Vessels from places such as Lockeport would go as far as the northern Gulf of St. Lawrence or even Labrador for fresh halibut, which kept well on ice. Improvements in transportation aided the shipping of fresh fish to the interior. In 1926, the federal Department of Agriculture subsidized a large cold-storage depot at Halifax; the Harbour Commission bought it and leased space to fish dealers. So far, the groundfish fishery appeared to be making at least some headway.

Groundfish industry lags behind competitors

Still, the groundfish industry looked weak compared with its European and New England competition. New Englanders were quicker to use steam trawlers, starting in 1905. Schooner draggers, introduced in 1919, numbered 198 by 1929. Diesel engines spread particularly after 1928, increasing the power of draggers. Although New England still had old long-lining saltbankers like Nova Scotia's *Bluenose*, the new trawlers and draggers able to chase down the fish now dominated. By 1931, they caught 58 per cent of the fish landed. The New Englanders concentrated on nearer banks, particularly Georges. By 1929, only five per cent of the New England catch came from Canadian and Newfoundland grounds.

On shore, New Englanders introduced filleting lines in 1921, and began marketing packaged fillets. Before this, fishmongers or customers themselves had to do the filleting. Fillets created more market, which created more fresh-fish plants, which paid particular attention to haddock. Though a hard fish to split for salting, it was a fairly uniform size, and thus well suited for filleting. Interest also resurged in halibut, which keeps better on ice than most species.

After about 1923, improved techniques stimulated the frozen-fish trade. The amount of freezing increased. The General Sea Foods corporation organized itself to use the rapid-freezing, high-pressure Birdseye process. With dragging, filleting, and freezing, New England was laying the foundation of the northwest Atlantic frozen-fish industry.

While lagging in the fresh and frozen trade, Canada and Newfoundland were also falling behind the Scandinavians in the saltfish trade. Iceland, with a longer season than Newfoundland, was quicker to move to motor vessels and trawlers. That country, with a standardized cure for saltfish, improved its processing and marketing methods. Icelandic fish got higher prices than fish from Newfoundland and Canada. Norway, too, was a strong competitor, the government giving the fishery a good deal of support. While in Newfoundland, and frequently in the Maritimes, fishermen both caught and cured, in Norway merchants cured the fish.[7]

Groundfish troubles weaken entire industry

In Canada, saltfish quality had further declined during the First World War, when high demand made it easy to sell. Some saltfish producers had a low regard for the product they sold to the West Indies, still referring to it as "food for slaves." Tales were told of waste bits being swept up from the floor and mingled with the product.

As some other countries began to consolidate their marketing, Canadians sold saltfish independently, cutting prices. The year 1921 saw a sharp decline in prices, which helped discourage the Lunenburg bank fleet. In following years, not only the Inited States but

An early beam trawler, the *Promotion*, landing fresh fish at Liverpool, N.S.

Cuba, Jamaica, and the Dominican Republic raised tariffs, doing further damage. Meanwhile, as European trawlers took over European markets, Newfoundland, almost totally dependent on saltfish, had to sell more into West Indian and South American markets held by Nova Scotians. Around 1926 a supply war commenced that saw cod prices drop 50 per cent.

In the Maritimes, fish companies tried to readjust to changing conditions. But the Atlantic industry had no capacity to pull itself together for any major change, and no major initiative came from the government. While fish companies struggled to keep their heads above water, fishermen themselves had little room to manoeuvre. In Lunenburg County, the offshore fleet fell from more than 140 vessels in the previous decade to less than 100 by 1925. Funds for fishermen's relief and Workmen's Compensation came into play. In Canso on Dominion Day, 1927, fishermen flew the flag at half-mast, in mourning for their industry. Quebec had similar problems in the late 1920's. All around the Atlantic, the number of private wharves continued to shrink as schooner fortunes faded.

When the salt cod price dropped, it dragged down the price of "scalefish" (including haddock, hake, pollock, and cusk). Other salt products—pickled herring, mackerel, alewives—fell in price. Operators turned to fresh and frozen fish and to lobster, but the same low prices and over-competition weakened those fisheries. The saltfish problem ricocheted through the whole industry, even before the Great Depression.[8]

MacLean commission cuts back trawlers

The slump in both salt and fresh fish prices and the distress of 1927 prompted another royal commission— the so-called MacLean Commission, after its chairman—on the fisheries of the Maritime provinces and the Magdalen Islands. Operating in 1927 and 1928, the commission attracted complaints about anything and everything. But the chief issue was small-boat fishermen's complaints about the growing trawler fishery, for conservation reasons and because, they said, trawlers damaged the market. Trawlers were a new element; the price collapse was a new development; inshore fishermen linked the two.

Compared with the hundreds of longlining schooners and thousands of inshore longliners and handliners, the number of trawlers was small: 11 in the peak year of 1926.[9] But they had become a chief source of supply to some major plants, especially in the winter, when small boats were often tied up.

The commission's majority concluded that trawlers might indeed damage fish stocks, by harming habitat or, more likely, by destroying immature fish. But the bigger problems were economic. In that realm, the commissioners gave little attention to complaints that the lower quality of trawler-caught fish was damaging markets and consumption. Rather, "the heart of the problem" was that trawler firms, through their more reliable supply, had gained undue control of the market; when fish were abundant, the traditional producers competing with trawlers got lower prices or no sales at all. Declaring that "it is the function of industry in any country to produce men as well as goods, to make livelihoods as well as profits," the commission called for the prohibition of steam trawlers.[10] The result would be severe restrictions on the trawler fleet, lasting to the Second World War.

Royal commission finds many failings

Like most fishery inquiries, the 1927–1928 royal commission pronounced itself on a long list of subjects. Some would produce action; others would get lost in the shuffle. The commission called for a separate fisheries department (this came about in 1930); international regulation of trawlers; international study of cod, haddock, trawlers, and the possible use of fish sanctuaries; restored size limits for lobster (this occurred in 1934); studies on oysters and scallops; more departmental work on statistics and intelligence; a wider role for fishery officers; better enforcement and inspection (the department responded with new six-week training courses for some Atlantic officials); better quality products and use of grading standards; and more use of rapid freezing, rather than old techniques that produced "spongy, tasteless fish."

Like Prince and Whitcher decades earlier, and many reports in decades to come, the commission found fault with the level of information, education, and common understanding. There should be more educational work for the industry. There should be travelling instructors, instead of bulletins that went unread. The department in 1928 made some response, giving a grant to Dalhousie University for fisheries educational work. Otherwise, such efforts got little attention. The chief medium of information, independent of government, was the *Canadian Fisherman* magazine circulating to many in the industry.

The commission also called for support for fishermen's co-operatives. That recommendation would bring major changes, to be discussed later. But the main item at the time was the trawlers.

Trawler cutback creates long controversy

Following the commission's recommendation, the department required that all trawlers be licensed by the Minister and be registered in Canada. (Smaller draggers also required licences.) The new order of business effectively killed any chance for additional trawlers. Meanwhile, regulations made it more difficult to operate the existing fleet, which had to operate at least 12 miles offshore. On October 30, 1929, an Order-in-Council imposed special taxes on trawlers. The government also taxed landings to Canadian plants by foreign trawlers. In 1931, further regulations imposed a trawler licence fee of $500.

The National Fish Company in 1931 fought the trawler rules to the Supreme Court, and lost.

Meanwhile, the Maritime Fish Company abandoned its Canso plant; a New York-based firm, the Atlantic Coast Fisheries Company of New York, took it over. Trawler operations closed down at Hawkesbury. By 1933, the federal government was licensing only four trawlers. Further restrictions followed. By the end of the 1930's, only three trawlers were operating, for National Fish.

The trawler tax applied nationally. In British Columbia, trawlers, which had increased in the Strait of Georgia, now declined as in the Maritimes; they would become re-established only in the 1940's.[11]

It was all a blow to the offshore, big-vessel, year-round fishery that was beginning to build a stronger fresh- and frozen-fish industry. Meanwhile, the trawler restrictions brought no apparent improvement to the inshore fishery. Instead, conditions got steadily worse, as the fishery depression merged with the Great Depression.

The trawler restrictions ushered in years of controversy. The proponents of trawlers argued that the government measures only aided overseas competitors, worsened the technological lag of the Atlantic industry, and delayed the setting up of a modern frozen-fish industry. Meanwhile, the Europeans were operating thousands of trawlers and draggers (similar technology on smaller boats), with many of them fishing the Grand Banks. Another 300 operated from U.S. ports. Arguing from the opposite direction, supporters of the inshore fishery said that given their head, trawlers would displace the inshore fishery and destroy whole communities.

As with many fishery measures, it was difficult to evaluate precisely what effect the trawler restrictions had on prices, jobs, and communities. Even today, nobody has developed a precise measuring stick for the social and economic effects of different fishing and processing technologies.[12]

Fresh-fish trade grows slowly

Even without the trawlers, the fresh-fish and even the frozen-fish trade grew somewhat. Vessels from Halifax, Lockeport, and various other ports took part. By the early 1940's the Lunenburg fleet fished about

Baiting trawl (longlines) aboard a schooner.

Setting trawl from a dory.

half for fresh fish, especially in the winter, and half for saltfish. For saltfish, the schooners would make a winter trip to Newfoundland for frozen bait, then make a spring fishing trip and two summer trips to the banks. Some went farther afield, picking up frozen bait at Greenland and fishing the banks near there.

Despite American tariffs, Canadian railway rates made it possible to export fresh fish at competitive rates to the U.S. Midwest. Railways and trucks still relied on boxes full of ice and salt to lower the air temperature. This early method allowed quality loss in both fresh and frozen fish. At the Prince Rupert technical station, engineer Otto Young followed up Huntsman's work on jacketed cold storage and carried out other studies. But consistent application of good refrigerated transport was still years away.[13]

Science looks at sea fishery

Meanwhile, Biological Board scientists were beginning to look at sea-fishery stocks and the workings of the industry. Alfred W.H. Needler, a graduate student under Huntsman, started at the St. Andrews station in 1923 and 1924, working on fishery statistics. Data as then collected by the department revealed cycles in abundance, but showed nothing about where the fish were caught offshore.

Needler began working on fish migrations, especially of haddock. "I used to go out on dory vessels, which would have no power but a donkey engine. Also I used to go out on longline boats, which then had no gurdies." He sampled commercial catches all around the coast, for age, size, sex, and stomach contents of fish. "I nearly ruined my eyes looking at scales under the microscope, because your eye had to do some of the focussing work."

Needler established that haddock south and west of the Fundian Channel were distinct from Scotian Shelf haddock, which in turn were distinct from St. Pierre Bank haddock. The Passamaquoddy haddock were related to Georges Bank haddock, the Digby haddock to Brown's Bank haddock. "Information on cod and other species was then in the same state as information on haddock: limited. There was a lot of work sim-ilar to mine going on, sketching out the broad general background."[14]

Salt fish trade keeps falling behind

Haddock was the catch of choice. Needler recalled decades later that "there was a big increase in the trawler fishery from Boston in the early 1920's. The haddock catch on Georges and around there had been about 50 million pounds; it went up to 275 million pounds. It couldn't be held there. It dropped to a hundred million pounds. People thought it was overfishing; really it was just a new equilibrium. That drop in catch eventually led to a big increase in population studies in New England, and more attention to international regulation."

The New England fleet responded by using faster boats to fish more distant banks off Canada. In 1935, Americans started trawling for redfish off Canada; only in 1947 would Canadians get into this fishery. While Newfoundlanders still had little or no halibut fishery, Americans fished halibut 150 miles east of Labrador.[15]

The Americans charged more duty on dried fish than on wet-salted imports which they could process themselves, as boneless salt cod or in fish cakes. This reduced Maritime employment in drying fish. With Newfoundland selling more saltfish to the West Indies and Latin America, Maritime trade declined. Overall, although the Canadian "green-salted" trade kept up volume, the dry-salted trade fell. From 1929 to 1939, the total amount of groundfish put to salt in eastern Canada dropped from about 101,000 to 57,000 tonnes.[16]

Larger groundfish companies take lead

Particularly in Nova Scotia and the Bay of Fundy, most fishing and trading companies were small or medium-sized. The salt fish decline and the Great Depression weeded out many of them. Meanwhile, changes in the trade in some respects favoured larger

Fishing boats and freezing plant, North Sydney, N.S., ca. 1926. (Library and Archives Canada, PA-41860)

outfits. It took capital and know-how to organize trawlers and filleting plants. The fresh- and frozen-fish companies needed freezers and cold storage, not only for end products but also for the gluts of fish landed by their own vessels or independent boats.

Although shore fishermen took groundfish all around the Maritimes and Quebec, they were likely to fish other species, such as lobster and herring, as well. The biggest specialized operators for groundfish were Nova Scotia's schooners and few remaining trawlers. One study estimated that offshore fishermen on average took about 36 tonnes each per year, compared with less than 6 tonnes by inshore fishermen.[17] Schooners operated from ports including Shelburne, Lockeport, Halifax, Canso, and North Sydney, with Lunenburg the most important.

In Lunenburg, the Halifax *Chronicle* at the end of 1937 reported a fleet of 41 schooners (up from 25 reported fishing a year earlier). Of those, at least half were associated with three shipping and fish-dealing concerns: Adams and Knickle, Zwicker and Company, and W.C. Smith and Company,[18] with the latter particularly important in the fresh-fish trade.

W.C. Smith Company had started up at Lunenburg as a fishing company in 1899, outfitting 6 vessels the first year, 14 the next. It soon had 20 fishing in the warmer months, and in fall and winter trading locally or chartering to Europe and the West Indies. Typical cargoes included salt, saltfish, produce, and frozen herring. After the First World War, the company extended into processing saltfish; and in 1926, ahead of most competitors, it built a cold-storage and fresh-fish plant in Lunenburg. In 1928, the company built its first wooden diesel-powered trawler. In 1929, it opened a canning factory, and in 1930, a fish-meal plant. The Smith vessels got radio-telephones in the 1930's, again ahead of most competitors. In 1938, a holding company, Smith Fisheries Ltd., integrated the company's two main divisions, which now included holdings in seven ports, and wholesale houses in Montreal, Toronto, and New York.[19]

Meanwhile, the fresh-fish trade got another large organization: the Halifax-based Maritime National Fish Company, stemming from the merger of National Fish and Maritime Fish. Another merger in the 1940's would marry this trawler-owning company and the Smith interests, creating National Sea Products, which became a huge enterprise in the post-war period.

Depression hits fishermen hard

For most fishermen and most companies, conditions got worse and worse from the mid-1920's on. Prices were low, and marketing, to judge by most reports, was mainly a matter of cut-throat competition to sell products of inconsistent quality.

Articles in Halifax newspapers outlined the decline. The Lunenburg fleet had shrunk from its 1912 high of 136 sail to 25 vessels in 1936, landing fish with a market value of $350,000, one-tenth the 1919 value. Cod prices had fallen from $14 a quintal to about $4.50. Meanwhile, rubber boots had risen from $3 to $5.50 a pair; oilskins, from $2.50 to $5.00 a set. A vessel that cost $7,300 to build in 1912 now cost some $20,000; more with an engine aboard.

Fishing crews on schooners shared one-half the landed value; from the other half came the wages of the captain, cook, engineer, headers, and throaters (who dressed the fish aboard), and the returns to shareholders, if any. Vessels normally carried 20 dorymen or shoresmen, one cook, one engineer, and the skipper. The cook and engineer often received a dollar a day in addition to their share. One doryman acted as bosun. "Wages of fishermen were in a great many cases no more than $7 or $8 a week. They went over the side in dories holding torches in their hands at one and two o'clock in the morning to fish all day on the Atlantic Ocean in wintry cold. Usually they did not reach their bunks again until 11 o'clock at night."[20]

As some Nova Scotians abandoned the schooner fishery because of the low pay, Newfoundlanders often took their place, beginning a tradition of work in the Nova Scotia offshore fleet. In both Newfoundland and the

Illustration by Ritchie in Halifax *Mail*, January 7, 1937. (Courtesy of National Sea Products)

A fisherman's family near Gaspé, Quebec, ca. 1930's.
(Library and Archives Canada, PA-125344)

Fishing boats at Newport, Quebec, 1940. (Library and
Archives Canada, PA-57021)

Maritimes, many fishermen were still seeking seasonal work in the United States. In the early 1930's, however, new American restrictions on transient workers made it harder to cross the border, thus worsening conditions for many families.[21] Meanwhile, as fishermen tried to work to the south, some boats moved to the northward. New Englanders turning to draggers sometimes sold old saltbankers to Nova Scotians, and Nova Scotians sold old vessels to Newfoundlanders.[22]

Facing hard times, fishermen's families made all the use they could of fish and lobster for food. Those without good soil for a garden had a worse time. On Campobello, this writer's home island, people confessed years later to such small tricks as opening tin cans from the bottom, then leaving the empty can on the pantry shelf, so visiting neighbours would think they had a stock of food. Another trick was to put a square piece of turnip in a dish on a high shelf, to look like butter.

Yet fishermen in the Bay of Fundy and southwest Nova Scotia were probably a bit better off than those in eastern Nova Scotia, the Gulf, and Quebec. The southern areas were closer to American markets, and live-lobster sales contributed to cash income. (So had rum-running for fishermen in various Maritime locations in the 1920's, but that ended in 1933 when the United States lifted Prohibition.)

At Canso in the late 1930's, inshore fishermen were earning about 90 cents a day.[23] A Halifax newspaper described the arduous work done for minimal pay:

In that roaring, black December night, amid tumult and darkness, in a boat too small to permit of it having a stove or any other means of providing warmth, these men, after setting their trawl by torch light, would have to wait for several hours, perhaps, for the bait to be "taken", when the "hauling in" would begin.

There is no chance to sit down during all this time. The fisherman stands on a little wooden platform which does not look to be more than 20 inches square. He cannot walk up and down. He must balance himself there amid the heaving seas, swinging his arms and stamping his numbed feet to keep from perishing. Then, after an eternity of waiting, comes the arduous task of "hauling in", complicated by stiff fingers.

Conditions were no better in the Gulf, where some merchants on the Acadian Peninsula still compelled their fishermen to deal with the company store only. Generally, Atlantic Canada fishermen in the late 1930's earned perhaps $450 or less per year before the cost of the boat and operations, leaving little to live on. Another study put their typical net income at $300. Relief in Nova Scotia paid $1 a week for adults and 25 cents a week for each child.[24]

Co-operative movement aids fishermen

Particularly in the Gulf of St. Lawrence and eastern Nova Scotia, fishermen got help from the co-operative movement. The leading figures were Moses Coady, a priest and teacher at St. Francis Xavier University (St. F.X.), and his cousin Jimmy Tompkins, a priest who had lost his place at St. F.X. for his radical views. (Both were natives of Margaree, Cape Breton.) Tompkins helped establish some early co-operatives. He also helped foment the commotion that resulted in the 1927–1928 royal commission, to which Coady spoke passionately about education and organization of fishermen. The commission in turn recommended that St. F.X. carry out extension work, and that the department support fishermen's co-operatives.

The university was already looking in that direction, and in 1928 set up its Extension department with

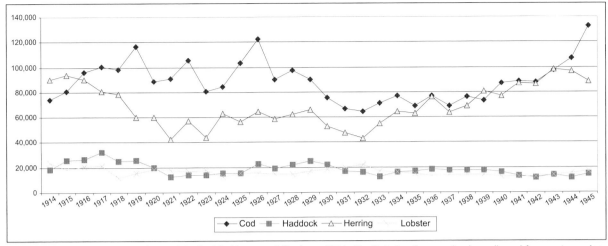

Landings of selected species (tonnes) in the Maritimes and Quebec, 1914-45. Both landings and values dipped for most species in the early 1930's, the low point of the Depression.

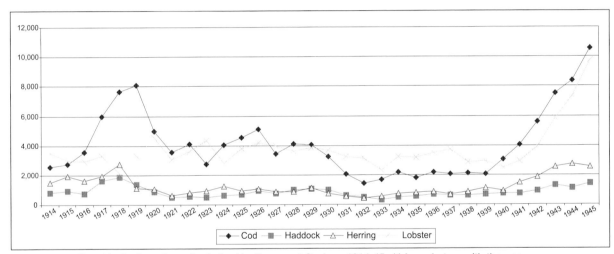

Landed value ($000's) of selected species in the Maritimes and Quebec, 1914-45. Values shot up with the war.

Coady as director. Deputy minister W.A. Found noted that organizations were useful to the government, but expressed his doubt whether fishermen could handle co-operatives.[25] Nevertheless, the department in 1929 hired Moses Coady as promoter of fishermen's organizations.

Co-operatives were common in Europe and parts of Canada, especially among farmers. Some were already taking shape in the Maritimes and Quebec. The organizations took various forms; they might operate stores, market farm or other goods, or provide credit. Common principles included one member, one vote; surpluses or profits distributed according to level of participation; and a stress on education.

Coady, Tompkins, and A.B. MacDonald and other colleagues at the new Extension department led what became known as the Antigonish Movement. Supported in part by Department of Fisheries funding,

they went to community after community, preaching self-help so that people could become "masters of their own destiny." The frequent pattern was a public meeting, establishment of a study club, and then a co-operative, whether a consumer co-op, a credit union, a housing co-op, or, increasingly, a fishermen's co-op.

In many places, the ground was ready. Coady wrote later that

[f]or decades ... the farmers and fishermen of Eastern Canada had endured great depression. There were a few notable periods in which they had fairly good times, but in the main the old order meant poverty and misery for thousands of them. ...

The lot of the industrial worker was not much better. ... In one fifty-year period Eastern

Canada lost four hundred and fifty thousand of its people to other parts of America. This acted like a pernicious anemia on hundreds of communities. For the people of Eastern Canada it was ... a whole series of depressions, of dark days when men lived in insecurity and fear, unable to concern themselves with anything but the grim struggle to keep body and soul together.[26]

Early fishermen's co-ops included that of Little Dover, near Canso, Nova Scotia, which Tompkins had helped organize in the early 1920's; the Tignish, P.E.I. co-op, begun in 1924; and the 1928 Gloucester Fishermen's Association, in New Brunswick. The movement accelerated after 1928: nearly 150 fishermen's organizations took shape, whether for producing or marketing.[27] They were especially popular in Acadian areas of the Maritimes. By 1942, there were 67 lobster-canning co-ops alone.

Some co-operatives survived, some failed. Gradually some coalescing took place. In 1930, the United Maritime Fishermen (U.M.F.) held a founding convention in Halifax, with Coady presiding and A.L. Barry, the department's Superintendent of Fisheries at Moncton, acting as Secretary. Representatives attended from around the Maritimes, the Lunenburg deep-sea fleet being a notable exception. The U.M.F. was to be a central educational body, amassing and distributing information on market conditions, fishing techniques, and co-operative organization. In 1932, the central organization began publishing the *United Maritime Fishermen*; this gave way to publications emanating from St. F.X..

In 1934, at the request of member co-ops, the U.M.F. went directly into the marketplace, buying and selling fish and purchasing supplies. Both the central body and the co-ops in general kept up strong momentum in the 1930's. But the war took away many workers, and post-war prosperity removed some need for the self-help movement. Co-ops lost some of their impetus. With radios, movies, cars, and then television providing new distractions, the local study groups that led to co-operatives became less workable. Other forces slowing growth were the individualism of fishermen, and the reluctance of some co-ops to follow good business principles, for example, by hiring good managers and firing bad ones.

(The U.M.F. would gradually consolidate and rationalize itself over decades. By its 50th anniversary, in 1980, it would have eight plants and around 20 sales offices, supply depots, fish markets, and so on run directly by the "Central." It also marketed fish for 27 member co-ops. The U.M.F. operated three large vessels of its own, and represented about 2,500 independent fishermen plus seasonal helpers. But a new financial crisis afflicting the Atlantic industry in the 1980's would bring its demise.)[28]

Price Spreads Commission calls for reform

As the co-op movement gathered strength, the federal Price Spreads Commission of 1934–1935 added a voice for reform. With R.B. Bennett's government under attack for inaction against the Great Depression, the maverick cabinet member H.H. Stevens headed this inquiry into business practices, including low returns to primary producers.

The Price Spreads Commission paid considerable attention to the Atlantic fisheries, and heard heated opinions. Suggestions from smaller operators included the abolition of trawlers. There should be a full-time minister of Fisheries. A federal fund should advertise fish as a food, and a board should have power to fix minimum prices. The government should equalize freight rates from all points in Nova Scotia to the markets of Montreal and Toronto. A Dominion and provincial fund should advance fishermen credit for purchase of gear and boats. And the government should use the Dominion Marketing Act to further co-operative measures within the industry.

From the large-processor side, H.G. Connor of Maritime National Fish Company opposed use of the Dominion Marketing Act or any other market control of fish. The government should instead subsidize fast freighters to take fresh fish to Britain. As for price-cutting and cut-throat competition by fish dealers, it was not the larger companies that did that, but the smaller ones. The Dominion fisheries department broadcast prices to schooner captains daily, and there was no company combine to set prices to fishermen. Mr. Connor added his opinion that "the day of the small, inshore fisherman was over."[29]

The commission's report called in strong terms for a better organized, better coordinated industry, with a strong government hand backing the changes. Government-supervised agreements between industrial units should "modify cut-throat competition and generally regulate the industry." Fishery changes should include uniform freight rates from the Atlantic, adequate and compulsory inspection of fish products, gradual discontinuing of all trawler operations, and further encouragement of the co-operative movement.

Moreover, a fisheries control board, with local advisory committees set up for fresh fish, cured fish, lobster, and so on, should run marketing. It should make every effort to raise prices to fishermen; see that grades of fish were accurate; establish a higher standard of quality; set up inspection for grading; eliminate consignment shipments; eliminate cut-throat competition; channel surplus fish production to bring the best returns; and survey possibilities for developing more export markets. As well, the Dominion government should extend credit to fishermen for purchase and repair of equipment, gear, and so on.[30]

It was a call for sweeping government intervention. But no major federal effort ensued. Any related changes came in bits and pieces, sometimes from the provincial side.

Nova Scotia sets up fishermen's loan board

Relief payments to fishing communities varied from place to place in the 1930's, welfare being primarily a provincial responsibility. The federal Department of Fisheries in 1935–1939 provided extra funding to fishermen, typically totalling $300,000–$500,000, administered through the provinces, except for British Columbia, which declined help.

Fisherman from Minudie, N.S., in the Bay of Fundy's Cumberland Basin, harvesting shad at low tide in this undated photo.

In 1935, the province of Nova Scotia prevailed on the federal government, rather than distributing a small amount of money to all fishermen, to give 25 per cent to genuine relief cases, lend 15 per cent to fisheries co-operatives, and use the rest for loans to fishermen for boats and gear. In 1936–1937, the province paid subsidies on dried cod and scalefish. That same year, a federal–provincial arrangement gave loans worth $157,000 to hook-and-line fishermen in Nova Scotia. In 1938, the federal government paid two-thirds of such loans to N.S. fishermen, which amounted to $450,000.

This was the start of the Nova Scotia Fishermen's Loan Board, which the province took over and has operated under various names since then. Other Atlantic provinces later set up their own fishermen's loan boards.[31]

Nova Scotia sees strikes and lockouts

Co-ops had the most effect in the Gulf of St. Lawrence, encouraging fishermen to buy and sell fish on their own. Elsewhere, in southern Nova Scotia, traditionally conservative and individualistic fishermen took the prices set by buyers. But those prices were minimal, and resentment grew.

In 1937, Halifax newspapers reported on fishermen earning roughly three cents an hour, and less than $50 a month. Why, asked the Halifax Herald, were fishermen's huge stores of pickled codfish going unsold? "What is back of this sinister opposition of the shore fishermen's efforts to market their own wares? ... Is it, as rumour declares, a matter of interlocking directorates, of mutual shareholding, of definite collusion in certain quarters to keep prices down...? ... Is there a devil fish somewhere in the industry, with slimy tentacles reaching into every cove and harbour in Nova Scotia, sucking the very life blood out of men, women and children?"[32]

Organization in Nova Scotia was minimal, despite some attempts. The Fishermen's Union of Nova Scotia, lasting from 1905 to about 1930, sometimes petitioned the government for benefits; a key figure was Moses Nickerson, a fish buyer and member of the Legislative Assembly from Clark's Harbour on Cape Sable Island. The U.M.F. attracted some Nova Scotia followers in the 1930's, but in Lockeport, for example, fishermen got discouraged and dropped out. Meanwhile, the Nova Scotia government in the late 1920's created a Fishermen's Federation. Nine "stations" organized along the coast, and brought some gains in price. Each station was an autonomous local co-operative, with no provision for collective bargaining. The federation comprised diverse interests: schooner captains, their crews, and inshore fishermen.

In 1937, Communist Party of Canada members helped organize workers at three Halifax fish plants into the Fishhandlers and Fish Cutters Union, an affil-

Fish-handlers at Halifax in 1939, unloading a schooner.

iate of the American Federation of Labor. That fall, the union tried to negotiate with National Fish; but the province's Trade Union Act of 1937 left certification procedures vague, and National Fish resisted negotiations. There were clashes and head-thumping on the waterfront.[33]

Fishermen then got into the act. The renowned Captain Angus Walters of the *Bluenose* called in December for a united front under the Fishermen's Federation, from Cape Sable to Cape Breton, to raise the price of haddock a quarter of a cent. When the Fish Buyers Association refused to negotiate, 800 fishermen tied up their boats, leaving nine companies in Lunenburg and Halifax with no fish. As new stations of the Fishermen's Federation formed, Captain Walters and the Fishermen's Federation allied with the Fishhandlers Union.

In retaliation, National Fish laid off 60 union men in Halifax. On January 7, 1938, the union hit back with a total strike. But within the allied Fishermen's Federation, strains were beginning to show between the schooner captains and the inshore fishermen. On January 17, National Fish offered a better price. The Lunenburg schooner fleet went back fishing; the Federation soon disappeared from view. The company also hired back the 60 laid-off Halifax workers and offered a vote to determine whether all employees wanted a union. Newly hired workers swayed the vote in the company's favour. The Halifax strike subsided.

Down the shore, in Lockeport, however, interest in organizing persisted. In 1939, the Canadian Seamen's Union (C.S.U.) of the Great Lakes shipping trade began organizing there. The Lockeport fishermen and plant workers in August 1939 formed the Canadian Fishermen's Union, affiliated with the C.S.U. But the town's two plants, Swim Brothers and the Lockeport Company Limited, refused to negotiate and locked out the workers. The "Lockeport lockout" dragged on into December. The union started up a fishing and processing co-op of its own. Tensions rose, amid stories of "Reds" running the union. The R.C.M.P. made itself visible.

Then the companies announced that they would be re-opening, but hiring no unionists with outside affiliations. The union called in everyone it could, some 700 people, to man the picket line. The people blocked the railroad tracks and routed some 50 Mounties with rocks.

Eventually, some 200 Mounties came to the small town to keep order. The whole affair had petered out by January 1940, when the Nova Scotia government stated that fishermen were not company employees, and thus could not organize under Nova Scotia's Trade Union Act. A fish-handlers union survived, but the companies refused to recognize the C.S.U.[34]

By the time the lockout ended, the Second World War had started. Ideas of organization mostly faded away for the time being.

Government sets up salt fish marketing board

Salted fish exports dropped by a drastic two-thirds in the 1930's, for various causes—competition, weak markets, the Great Depression in general. Frozen fish got a better price than salted fish, which went mainly to poorer countries. Atlantic Canada competed in the fresh- and frozen-fish trade less by modernization than by buying and selling at low prices.

Meanwhile, in the salted fish industry, the desperate situation drove governments to stronger intervention.

Curing cod at Yarmouth, N.S., in 1927.

The diesel-powered dragger *Cape North*, part of the switch from longline schooners to trawling for the fresh-frozen fish industry.

In 1937, Nova Scotia began supporting salt fish prices with "deficiency payments" – that is, subsidies to counteract losses. In 1938, payments were $1 per quintal of dried cod and an 86 $2/3$ cents subsidy per quintal of scalefish—other salted groundfish such as haddock and pollock. The federal government also provided aid in 1939. This appeared to relieve pressure in the salted fish market, and by so doing, to ease the strain in the fresh-fish, lobster, and mackerel markets as well.

In 1939, political pressures helped create the federal Salt Fish Board, to provide further such aid. Donovan B. Finn of the F.R.B. became Chairman of the Salt Fish Board; a year later he became deputy minister of Fisheries. Another member of the Salt Fish Board was Stewart Bates of Dalhousie University, later a great name in federal fisheries administration. Bates recruited a Dalhousie graduate and Cape Breton native, the young economist William C. (Bill) MacKenzie.

The Board had the power to subsidize exports up to 25 per cent of product value, but it never needed to. Bill MacKenzie recalled in 1987, "I joined the Board in 1940, after the war had started. I made a grid chart on the office wall and began to keep track of prices. In a few months, the prices went off the chart and right up to the ceiling."[35]

The Second World War ended competition from Norway and Iceland, and increased demand for fish in general. New employment drew away some of the people who had been trying to scrape a living from the shore fishery. Prosperity took hold, and the Salt Fish Board vanished into the general wartime system of controls.[36]

War speeds switch to frozen fish

As the war raised demand, prices soared, especially for fresh-frozen fish. In 1939, 34 per cent of Canada's Atlantic groundfish had gone into fresh or frozen, 54 per cent into salt, with such items as smoked fish and canned chicken haddie accounting for the rest. By 1943, frozen groundfish passed the 50 per cent mark. A big part of the change came in Quebec, with the province providing an extensive cold-storage network. All around the Atlantic coast, processing companies began switching part or all of their production to frozen. Salted fish remained important, however, especially dried salt cod from such areas as the Gaspé and southwest Nova Scotia.

The Nova Scotia schooner fleet made good catches during the war, there being so little other fishing effort. It was common enough for a schooner to land as much as 200,000 pounds of haddock in a week or ten days.[37] But trawlers were the better fishing machine. With productivity in mind, the government gingerly relaxed its restrictions on trawling. Prompted by wartime demand and government boatbuilding subsidies, the Smith interests of Lunenburg in 1943 contracted for two 132-foot trawlers, the *Cape North* and *Cape LaHave*, completed in 1945. It was the start of a fleet-wide change. A decade later, few schooners remained in the Maritimes, partly because trawlers and draggers did better fishing, partly because schooner operators found it difficult to recruit crews.

Wartime food board takes over marketing

Deputy minister D.B. Finn served from 1943 to 1946 as Chairman of Canada's Food Requirements Committee. The Wartime Prices and Trade Board, set up under the War Measures Act, controlled prices and wages. There was rationing of groceries, gasoline, and other goods. The board recruited businessmen to manage the industries they knew best. Fish processors had to observe ceiling prices on fresh, frozen, smoked, and canned production; the board controlled wholesale and retail mark-ups.

The Combined Food Board of the Allies, with headquarters at Washington, could set prices and production quotas, and controlled the flow of supply to troops and to the civilian population of Allied countries. The idea was to share scarce supplies as equitably as possible. Raymond Gushue of the Newfoundland Fisheries Board became chairman of the fisheries section; D.B. Finn represented Canada, but usually sent Stewart Bates in his place.

The Combined Food Board particularly affected Pacific canned salmon and Atlantic salted fish. Pursuant to the international arrangements, the Department of Fisheries administered the exporting of salted fish at agreed prices to various markets. As war turned into post-war, Bill MacKenzie began representing Canada at the Washington meetings on fisheries exports, until in 1947 everything reverted to the free market.

War had other effects on the fishery. On both coasts, the government commandeered some larger fishing vessels for various purposes. Military operations added to worries on the water. In the First World War, Canadian convoys had run down and sunk some schooners. Now, in the Second World War, Wendell Williams of Lockeport and other fishermen were sword-fishing on Georges Bank when a German submarine began to machine-gun their vessel. Mr. Williams and his mates abandoned the vessel, and had to row more than a hundred miles back to Nova Scotia.

Pelagics: the purse-seine returns

Groundfish tend to dominate historical accounts of the Atlantic fishery, and rightly so: no other species group yielded so much in the way of fish, jobs, and money. But thousands of fishermen switched back and forth between species, and some dealt only with shellfish, or with pelagics such as herring. The latter fishery saw major changes in the 1914–1945 period.

Although fishermen around the coast took herring, the single most intensive fishery was that of the Bay of Fundy, particularly on the New Brunswick side. The First World War brought high demand. But peace brought a steep decline; the price of sardines (juvenile herring) sold to Canadian and American factories dropping from $70 to $5 per hogshead.[38] The Fundy fishery persevered, supplying herring for sardines, for the smoked-herring trade that fed southern markets, and for other products. The main technology was weirs; inside these large fish-traps, despite the general ban on purse-seines, fishermen were allowed to use small purse-seines to take up the herring. Several hundred weir licences ("privileges") existed; it appears that at any given time, something over 200 would operate.[39] From 1926, the sardine catch fluctuated roughly between 10,000 and 45,000 tonnes; the Maine landings generally ran a bit lower.[40]

The Great Depression took its toll on the sardine business, closing many factories on the U.S. side that bought Canadian herring. When H.H. Stevens's Price Spreads Commission came to southern New Brunswick, representations from Campobello Island fishermen, aided by the Connors Brothers firm of

Clockwise from top left: Herring weirs, Grand Manan Island, N.B. (Library and Archives Canada, PA-41653) Purse-seining inside a weir. Packing sardines at Connors Brothers plant, Blacks Harbour, N.B., ca. 1925-1930. (Library and Archives Canada, PA-41675) Putting herring on racks for curing, National Fish Co. plant, Halifax, N.S. (Library and Archives Canada, PA-41878)

Black's Harbour, N.B., resulted in a lifting of the purse-seine ban, as the fishermen sought a better way to earn money. Campobello and later other Passamaquoddy-area vessels began purse-seining for sardines in the winter, first along the New Brunswick shore and from 1945 in Nova Scotia. In the latter province, the department allowed summer seining from the end of the 1940's.

The seiners began using fathometers for seining: these echo sounders sent down a sound pulse, which showed, on a moving roll of paper, not only the bottom contour but schools of fish beneath the boat. Echo sounders for fish detection had undergone testing as early as the late 1920's. In the 1930's, Albert Tester, of the F.R.B., tried them out for herring seining, and B.C. Packers put them aboard some vessels. Wartime advances in sonar would further transform the herring fishery after 1945.

Swordfish, tuna fisheries pick up

Since early in the century, and following the American example, Nova Scotians had caught swordfish for the fresh-fish market in New England. A "striker" aboard the boat would harpoon the swordfish from a "pulpit" projecting forward from the bow. Swordfish migrated north from tropical waters, supporting a fishery from the springtime until December, and from Georges to the Grand Banks. Nova Scotians fished mainly off Cape Breton. Some fishermen took the swordfish from small local boats; others migrated to Cape Breton from other parts of the coast. Landings were modest at first. Some growth in the fishery came through refrigeration and the fresh-fish trade. In the 1930's and 1940's, up to 400 vessels might take part in the fishery.[41]

Canadian waters also attracted large bluefin tuna. An important sport-fishery emerged at Wedgeport, a community near the southwest tip of Nova Scotia. From 1937 to 1976, the International Tuna Cup Match took place from the village, sometimes attracting

Swordfish vessel, with lookout on mast and pulpit on bow for harpooning. (National Film Board)

celebrity participants from Canada and the United States. Another sport fishery built up in Newfoundland. Tuna there seemed plentiful: around 1930, local people in Newfoundland's Bay of Islands mentioned seeing 200 tuna in one day.[42] In both Nova Scotia and Newfoundland, the tuna sport-fishery would attract well-off sportsmen for many years.

Another big sea creature, the beluga whale, got hunted not for sport but to rid the fishery of a competing predator. In the St. Lawrence River, a prime beluga ground, the Quebec government in the 1930's paid beluga hunters a bounty—an action that would have been unthinkable later in the century, as beluga became scarce in the St. Lawrence and in two other main grounds, Cumberland Sound and Ungava Bay.[43]

Lobster catches keep dropping

Working in a lobster cannery. (National Film Board)

In the lobster fishery, the department before the war had removed size limits except in the Bay of Fundy, Prince had failed to get a limit on the number of participants, and the number of Maritimes lobster fishermen had grown to 25,000 by 1913,[44] all of which did nothing to stop the decline in catches. Catches dropped from the 32,000-tonne range at the turn of the century to less than 20,000 by 1919. The number of canneries dropped from 760 in 1900 to 469 in 1919.

By then there was a new flurry of concern. The fisheries branch in 1918 proposed canning of lobsters only every second year, or else closing the fishery for a number of years. Although this came to nothing, the department did shorten the lobster seasons. It also, in 1919, started requiring lobster licences, without restricting entry; anyone could go lobstering. As in other licensed fisheries, the idea was to strengthen enforcement (an offender's licence could be cancelled) and bring a degree of order to the fishery. The licence fee was a nominal 25 cents.[45]

The commotion and concern about Maritimes lobsters subsided to a lower level, as did the fishery. Policing of lobster regulations continued to occupy many of the fishery officers and guardians. Seizures of boats for illegal fishing took place from time to time. In 1919 and 1936, royal commissions looked at the persistent problems of illegal fishing and canning in New Brunswick, a practice prevalent on the Gulf shore.

Catches stayed below 20,000 tonnes throughout the 1920's. Landed value was $11.5 million in 1920, dropping to $10.5 million in 1929, by which year the number of lobster canneries had dropped to about 300. The industry was switching more to live lobster, especially as competition increased from Japanese canned crab. The changeover began in the Bay of Fundy and southwest Nova Scotia and spread gradually to eastern Nova Scotia. The average size of lobsters caught rose, since the live-lobster trade demanded bigger animals than the canneries.[46]

Government assists lobster distribution

In a developmental effort, rare for the interwar period, the fisheries branch in 1927 began a collection service for fresh fish and lobster in eastern Nova Scotia, running between Port Bickerton and Canso. By 1929, this service grew to include six stretches of the coast; 20 boats carried ice, bait, and shipping boxes. The program ended after a few years. In 1930, the department began subsidizing transport of live lobsters from eastern Nova Scotia to Boston. This project led to private operations.[47]

Little change in regulation took place in the 1920's. The 1927–1928 royal commission on the fishery noted that there were too many lobster canneries, but that good manufacturing practices could increase sales. Although the commission suggested restoring size limits, nothing happened right away. In 1934, however, the department re-instituted size limits as far east as Cape Breton.[2] That same year, the department rearranged the lobster districts into some 17 areas, and forbade fishermen from fishing more than one district in a year.[48]

Catches rose briefly at the beginning of the 1930's, then began an alarming new decline to less than 15,000 tonnes. Increased fishing pressure perhaps influenced the decline. The economist H. Scott Gordon later described the 1930's situation in Prince Edward Island, where lobster rather than groundfish was by far the most important fishery. As the Great Depression took hold and employment shrank, more people entered the fishery, a job anyone could get into. At first the catch rose; then, the increasing numbers of fishermen brought no increase in catch. Too many people were in the fishery. Their situation became desperate.

The returns from fishing were hardly enough to sustain life and in some cases, receipts were insufficient to cover costs. ... An increase in the numbers of fishermen could only reduce the average yield per man. ... The lobster catch fell off sharply after 1932. In sum then, the increase in fishermen aggravated still further a situation that was already bad due to falling catch and falling prices. ... By 1939, there were more than 50 per cent more fishermen in Prince Edward Island than there had been ten years before. Within a year or two after the start of the war, these "extra" fishermen had disappeared from the industry.[49]

On the eve of the Second World War, the lobster fishery was doing poorly, with landings and prices no higher than in 1918. Catches picked up during the war, which also saw significant changes in regulation. In 1940, size limits spread to the whole Maritimes, as the government introduced a 6-inch limit on total length in the "canner" areas. These were mainly in the Gulf, which was further from the Boston live-lobster market. In 1941, the government raised the pre-existing size limit in "market" areas such as the Bay of Fundy from 3-1/16 inches to 3-1/8 inches carapace length. In canner areas, the total-length limit rose to 7 inches. Meanwhile, new regulations required operators of lobster pounds to liberate "berried" (egg-bearing) lobsters.[50]

In the canning industry, problems remained with product quality, despite decades of complaints and sporadic work by the department and F.R.B. As noted earlier, the British government at one point embargoed imports of canned lobster from Canada. Deputy minister Finn got involved, and the embargo prompted new controls on Canadian product quality. The canned-product difficulties helped switch more people into the live-lobster trade. By 1942, fresh lobster accounted for 45 per cent of Maritime production.[51]

Scallop fishery grows at Digby

The scallop fishery through the First World War remained a small operation, generating few departmental regulations. Participants were scattered, in the Bay of Fundy and southern Nova Scotia.[52]

As gas motors and dragging became more common, scalloping began to grow. In 1920, fishermen at Digby, N.S. found scallops on nearby inshore grounds, and started dragging for them. By 1929, some of the Digby fishermen had moved up to bigger boats, fishing farther offshore.[53]

In 1928 and 1929, exploratory fishing by the department found some scallops in other parts of Nova Scotia and at Prince Edward Island. But Digby would remain the biggest inshore scalloping ground.

Needler leads oyster-farming development

Among shore fisheries, clam-digging was significant, and generated a few mild regulations. The oyster fishery, though less important overall, continued to attract far more departmental attention. The main grounds were in the Gulf of St. Lawrence, especially Prince Edward Island. Officials continued to believe that oyster culture on private leases could remedy the poor conservation and other problems of the wild fishery.

By 1912, arrangements were in place throughout the Maritimes for the provinces to administer leases, resolving an earlier deadlock. Public beds remained a Dominion responsibility.

A flurry of farming began in 1913. But seed oysters imported from Connecticut brought a disease that started affecting beds in Malpeque, P.E.I. around 1914. The disease caused about 98 per cent mortality in the Malpeque area. It then cleaned out different places in the Northumberland Strait and in New Brunswick. An influx of starfish also damaged oysters.

Lease areas and wild grounds both suffered devastation. Oyster production in the Maritimes declined to 12,800 barrels in 1920. As the Malpeque disease waned, production rose slowly to around 20,000 in 1925–1928. Other problems continued, including reckless fishing, pollution, disease, and the jurisdictional situation itself, with neither government having full control.

The provinces, despite having argued for jurisdiction over leases, did little in the way of development. In 1928, Prince Edward Island agreed to hand back oyster-leasing responsibility to the federal government, which agreed to an attempt at development. Deputy minister Found, a P.E.I. native, thought oyster farming had good potential, and asked the Biological Board to help.

Some oyster research had already taken place. Besides the work of Ernest Kemp, there had been some "rule-of-thumb" oyster farming in the 19th century on P.E.I. Then Dr. Joseph Stafford, working for the board in 1904–1913, had gathered scientific knowledge and made management recommendations, mostly ignored by industry and governments. The steamer Ostrea had done some work, such as cleaning off mussels from the public oyster beds. But oyster culture only gathered force when the board in 1929 sent the young scientist Alfred Needler to P.E.I. to work on oysters. Needler's efforts gave a prime example of how development can proceed, coupling science with the practical and political world.

"My only instruction from Huntsman was, 'if you want to know how to grow oysters, grow them'." Needler picked a spot on the Bideford River in Malpeque Bay; this became the Ellerslie substation of the board. He and a few co-workers, including fisheries engineer Harry Lynch in the early stages, experimented with collecting spat and growing oysters in tanks on shore. Although Needler found disease-resistant oysters at the head of Malpeque Bay, these turned out to still carry the disease and to infect clean beds. Rather than trying to fight the disease by high-volume transplants, the workers let it run its course.

When oysters fill the water with spawn, spat collects on objects in the water. Needler used a newly invented collector: egg-carton fillers coated with concrete. This collector made it easy to separate the spat and to grow individual, well-shaped oysters. But typically, as soon as the culturists put the oysters on the bottom, silt covered them or starfish ate them. Needler and his co-workers devised floating trays with a screen bottom; these served to rear the oysters through their first summer. By fall, when the oysters were put onto the bottom, most were of bedding size and able to resist starfish. Needler later recalled,

I wrote a report in 1931 summarizing what we'd found and recommending leasing of oyster grounds. The department acted on it, late in 1931. Then there was hell to pay. The fishermen said, why not just let us fish the recovering stocks? There were speeches from the steps of country stores. The atmosphere was bad. Some people said leasing was illegal.

There was a flurry of leasing, despite the resistance. And around 1935 the atmosphere changed. An Oyster Growers Association formed. The Association held a field day and demonstrated what could be done: spat collection, killing starfish with quicklime and mops, what happens when you don't separate the clusters of spat, and so on. We, the Board workers, organized the exhibits. The church provided the dinner: scallops and so on, 35 cents a plate and they made money, all in the shade of the birch boughs.

Several thousand people came. One speaker was Thane Campbell, the Premier. But there was still resistance. The Premier wasn't too sure which way the cat would jump, didn't know what to do. So he read "the Walrus and the Carpenter" [the Lewis Carroll poem featuring oysters].

Later we had a second field day. Resistance collapsed with the field days, and everyone wanted to do it. Some got into leasing on a big scale. One guy over four years put in more than $3,000, with hundreds of trays.

Needler and co-workers kept working on practical problems: protecting the wooden trays from shipworm, putting out bulletins to oyster farmers, experimenting on starfish control, and more. In 1931, New Brunswick followed P.E.I.'s example and transferred leasing authority to the federal government; in 1936, Nova Scotia did the same. The transfers were for all bivalves, and Needler and co-workers also worked on clam and quahaug farming. By 1938, there were 594 leases on Prince Edward Island, and others on the mainland. Total P.E.I. production increased noticeably. Needler again:

It was all a going concern until 1939. Then came the war, and no more surplus labour. It's a labour-intensive technique. Some of the best people went into the war. Some of the lease operators had been farmers, or fox farmers, and the fox farming industry was going down.

The fishing people weren't as good, weren't as responsible as farmers. The farmers were used to visualizing years ahead, seeing what soil development and so on would do. That's bred into farmers from the time they're kids. Fishermen get bred in just the opposite. And the working capacity of fishermen varies from place to place. It makes a difference. Some people are still farming, and oyster farming could be made to pay.

The oyster business taught me a lot. I was there mostly alone, and saw it change. I found out you can't just tell people. You have to show them in a way that corresponds with their own knowledge.

Oyster culture was the department's most successful fisheries development project thus far. By study and persistence, and by winning co-operation from the Islanders, Needler created hundreds of person–years of employment in the hungry thirties. Later, in the 1950's and 1960's, he would oversee many fishery developments, approaching them as a gradualist and a bridge-builder, as he had done on Prince Edward Island.[54]

War rejuvenates fishery

For the Maritimes and Quebec, the interwar period was mainly a dismal story. It is no disgrace that departmental efforts failed to invigorate the Depression-era fisheries; the national government was equally helpless with the economy at large. That being said, the fishing sector was especially weak. The fisheries economist W.C. MacKenzie noted later that "the status of fishermen had, I think, been more or less in a slow decline during the twentieth century. The mature fisheries were all crowded. In the Depression, large numbers dabbled in fishing just as large numbers dabbled in subsistence farming: because there was nothing else to do."[55]

A 1940's study noted that with only $13 million output in 1937, fishing had become the smallest of the major Maritime industries. Boats were small and undercapitalized. Processing was underdeveloped.

The war brought perhaps a 40 per cent reduction in the fishing labour force, as other industries and the naval and merchant services attracted people.[56] Some veterans returned to the fishery after the war, but many others flocked to growing industrial centres, away from the Atlantic.

For the considerable number who remained in the fishery, the war was a great turning point, ending the constricted regime of the Great Depression and creating a mentality that, more than ever before, stressed productivity and development.

CHAPTER 14.
Newfoundland, 1914–1945

At the outset of the 1914–1945 period, Newfoundland's fishery benefited from the First World War, which interrupted European supplies of fish, raised demand, and increased the flow of money generally. Shipbuilding increased, and gas engines kept spreading through the fleet. More than a thousand schooners, large and small, thousands of smaller boats, and more than 40,000 men took part in the fishery.

Right after the war the number of banking schooners dropped sharply, from more than 100 to fewer than 50. But the total cod catch by Newfoundlanders stayed high, well above 200,000 tonnes annually and twice above 300,000. The northern and Labrador fishery remained strong. In the "floater" fishery, hundreds of schooners went north yearly. Thousands of stationers travelled by the new coastal steamers, taking small boats with them, or using craft they had left at their station. Fishermen could now ship their saltfish back on the same steamers or on collector boats.

Bait-freezers were more common by the end of the war. As in Canada, the Newfoundland government first assisted fishermen in operating bait-freezers, then abandoned the effort. Although there were some attempts to freeze fish for market by the end of the First World War, no significant trade took hold. Meanwhile, in the salt-cod trade, some operators now used simple indoor dryers: fish hanging from the rafters, and a big steel drum turned into a stove.

Except during the wartime boom, no great air of vigour surrounded the fishery. The industry had already developed serious problems by the end of the 19th century, being unable to support most of its people well; and the problems would get worse.

Every impression suggests that Newfoundland fishermen were falling behind those of the southern Maritimes, with Gulf fishermen somewhere in between. Many Newfoundlanders emigrated to central Canada or especially New England ("the Boston States") and New York. Sometimes they did so seasonally, returning home by train and boat for the summer fishery.[1]

Fishermen, boats, and vessels

The table below shows both the number of Newfoundland fishermen in selected years and their high, though decreasing, proportion of the labour force and total population. Many women and children also helped cure fish. (Note: the figures for 1935 include cod fishermen only.)

A fisherman ca. 1930. ((Library and Archives Canada, PA-148591, attributed to Fred C. Sears)

Table 14-1.
Male persons engaged in catching and curing fish, selected years.

Year	Number	As % of persons occupied	As % of total population
1911	43,795	53.1	18.1
1921	40,511	50.4	15.4
1935	35,018	39.5	12.1
1945	31,634	31.0	9.8

As for the fleet, the Newfoundland census (the source of the information above) in 1921 counted 1,881 vessels and 24,406 boats. The vessels are broken down as follows: 771 steam or gasoline vessels; 831 sailing vessels from 20 to 60 tons; 278 sailing vessels from 60 tons upward.

The huge fleet of boats, as many as in the Maritimes and Quebec combined, were classified by the quintals they handled. Most, 23,794 of them, were listed as 4–30 quintals. Boats from 30 quintals upward numbered 612.

The 1937 Commission of Enquiry on Newfoundland fisheries, using census data, tabulated 1,317 vessels, a sizeable drop. This total comprised 612 fishing schooners, 224 auxiliary schooners, 116 Labrador floaters, 300 Western boats (a type of small schooner), and 65 bankers. Bankers had numbered about 100 at the turn of the century, dropped to around 50 at the end of the 1920's, and increased since then.

While the vessel fleet had decreased, boats had increased by several thousand to 28,978. This fleet now included 18,454 dories, 898 motor dories, and 9,626 motor boats.

Fisheries service regulates lightly

The early years of the century had resolved international controversies with the Americans and French. Now government involvement with fisheries dwindled. Regulations under the Department of Marine and Fisheries typically concerned local matters, such as conservation on salmon rivers. The cod fishery was left largely as a matter of nature, although local regulations often affected the placing of trap berths and the like. In a colony where fishing was by far the main occupation, the government hardly dared consider limiting the number of participants, although incomes were typically low. Neither was it pushing hard for changes to the business side of the industry, which was weak in quality and marketing, though powerful socially and politically.

Coaker inspires huge fishermen's union

The great push for change came from fishermen themselves, led by William Ford Coaker. Born in St. John's in 1871, Coaker held various business jobs before starting a farm near Herring Neck in Notre Dame Bay, on the northeast coast of Newfoundland. Coaker read widely and thought deeply.

In 1908, after a disastrous year in the fishery, Coaker organized local fishermen into the Fishermen's Protective Union (F.P.U.), with the motto, "To each his own." The union spread quickly. In 1910, the F.P.U. started a newspaper, the *Fishermen's Advocate*. In its pages, Coaker asked if the fisherman received his own when

> he boards a coastal or bay steamer, as a steerage passenger and has to sleep like a dog, eat like a pig, and be treated like a serf? Does he receive

William Ford Coaker

his own at the seal fishery where he has to live like a brute, work like a dog...? Do they receive their own when they pay taxes to keep up five splendid colleges at St. John's ... while thousands of fishermen's children are growing up illiterate? Do they receive their own when forced to supply funds to maintain a hospital at St. John's while fishermen, their wives and daughters are dying daily in the outports for want of hospitals?[2]

The Fishermen's Union Trading Company started up in 1911, with four cash stores for the members. This activity grew until the F.P.U. had 27 stores and seven temporary branches. They bought waterfront premises in St. John's, a clothing factory (which, however, closed in 1914), and a steamer.

The F.P.U. wanted to change some of the circumstances surrounding the industry. This included such matters as reducing the number of French trawlers, curtailing the use of gas engines, and controlling the cutting of trees near the shore. But the union also wanted to change industry operations themselves. They wanted better conservation, a reformed fisheries department, and a state fund to supply credit. They wanted regular price information, trade agents abroad, cold-storage facilities for bait and fish, and an end to the tal-qual (single-price for a whole catch) system.

The F.P.U.'s Bonavista Convention in 1912 called for social reforms such as old age pensions, minimum wages, a night-school system in the outports, schools for every significant settlement, free and compulsory education seven months of the year, elected school and municipal boards, and long-distance telephones for all settlements. In the fishery, it called for far-reaching changes. There should be standardizing and a new system of culling and inspection for fishery products. A permanent commission should oversee culling of fish and the fixing of the price of fish shipped direct from the Labrador Coast.[3] The struggle for such reforms would dominate fishery politics for a decade.

By 1914, the union had 20,000 members, mainly on the northeast coast. The Roman Catholic church hierarchy was suspicious of Coaker's movement, and curtailed its spread on the important Avalon Peninsula. But to his followers, Coaker was god-like. When he visited outports, the people would build archways for him on the beach where he landed, and shoot off the big sealing guns in celebration. The union had its own hymn, one verse of which ran as follows:

> We are coming, Mr. Coaker, men from Green Bay's rocky shore,
> Men who stand the snow white billows down on stormy Labrador;
> They are ready and awaiting, strong and solid, firm and bold,
> To be led by you like Moses, led the Israelites of old.
> They are ready for to sever from the merchants' servile throng,
> We are coming, Mr. Coaker, and we're forty thousand strong.[4]

Union forms political party

The FPU tried to force up the price of fish and marine oil, and in 1912 broke an alleged combine that was trying to keep prices down. Coaker argued from the beginning that to win real reforms, the union must play a political role. The F.P.U. put up candidates in the 1913 election and elected 11 members. Their bills on sealing and logging improved labour conditions. In 1917, the union party became part of a coalition government. After the 1919 election, Coaker entered Sir Richard Squires's cabinet as Minister of Marine and Fisheries.

Meanwhile, wartime prosperity had aided the union. With shipping scarce, the F.P.U. in 1916 started a shipbuilding company. The union began bulk-buying supplies for fishermen. It bought fish on the Labrador; and later, in the 1920's, the Fishermen's Union Trading Company would operate sealing vessels. Coaker founded Port Union on Trinity Bay as a new base.[5]

Coaker tackles quality, marketing

By that time, Coaker as Minister of Marine and Fisheries had entered a great struggle to reform fish quality and marketing. The industry suffered disloca-

tions at the war's end, as currencies changed and competition resumed. Old problems remained; complaints about quality and market had been common for decades. The merchants all packed and exported their own fish, with no inspection. Fish generally went to market on consignment. Three or four steamers with Newfoundland fish might lie in an overseas harbour at the same time, creating a glut while the single agent looked for market.

The industry itself had already made some attempts to improve the system. Fifteen firms in 1911 grouped together to set up an agent in Spain. Another group did the same in Italy, and later set a voluntary price agreement. The agent firm of George Hawes and Company set up cold storages at Lisbon and elsewhere. But problems persisted, stemming in part from the lack of grading and inspection of Labrador fish.[6] Coaker decided, as he had earlier promised, to reshape the industry into an efficient one that would serve the fishermen well.

Coaker worked out a plan for better marketing. The government would raise and lower selling prices to best suit the circumstances—higher for more profit, lower for more volume, as needed. A proclamation on November 20, 1919, required that exporters get licences. The exporters fought this, and a court judgement declared the order illegal. Coaker and the government used the War Measures Act to force it through.

In December, Coaker set conditions for the price of exports, the terms of sale, and consignment arrangements for the various markets. The government posted fishery trade commissioners abroad. Exporters had to agree to ship such and such a grade to such and such a place. Each cargo needed a licence.

Further aggravated by a tax on exports, exporters continued to resist the reforms. Exacerbating the situation were disturbances in Mediterranean markets and exchange rate fluctuations. Under Coaker's controls, Newfoundlanders lost some markets to Halifax merchants. Meanwhile, some exporters continued making sales outside the new system.

In January 1921, the whole venture collapsed, and the government withdrew its fishery trade commissioners. The collapse of Coaker's scheme brought an immediate price drop. The idea of consolidated export marketing faded until the 1930's.

The government in 1921 passed a host of lesser regulations, to set minimum prices for bait sold to foreigners, to restrict longlines, to regulate drawing for trap berths, to set rules for processing cod-liver oil, and so on. Meanwhile, the position of the saltfish trade worsened. Fishermen had to pay high tariffs on their equipment. Newfoundland had no good-sized mercantile fleet. Shipping costs worsened the cost of living.

The F.P.U. had less political power, especially after 1924. The union continued its previous and widespread local activities. But nothing emerged to solve the fishery's basic problems, which were also Newfoundland's basic problems. In 1925, Coaker prophetically suggested a nine-man Commission of Government to carry out basic reforms. And in 1926, he predicted,

> In my opinion the day is not far distant when the country will be forced to decide, probably with its back to the wall, whether it will be governed by a commission elected by the people, by nominees of the British Government governing as a Crown Colony, or as a poverty stricken Godforsaken Island administered as a province of Canada.

In 1927, Newfoundland by a British court judgement won, against Canada, clear control of the Labrador interior. It was a rare bit of good news. Heavily indebted by railways, the war, the loss of fishery revenues, and general under-development, Newfoundland was losing the capability to run her own affairs. By 1927, Sir John Crosbie, Minister of Finance, stated that the colony was on the verge of bankruptcy. Coaker regained a share of power in the election of 1928, and in 1930 put forward another bill embodying his views, but it failed to pass the Lower House. Retiring as union president, Coaker took to spending more time at a mansion he had acquired in Jamaica. He died in 1938.[7]

Could Coaker's reforms have saved self-government?

Two historians of Newfoundland, Ian McDonald and David Alexander, consider the failure of Coaker's 1919 fishery reforms a turning point in Newfoundland's decline and eventual loss of self-government. In this view, the failure of reforms let the quality of saltfish slip further. Without a strong and well-organized fishing industry, and with the shipbuilding, shipping, and general economy also declining, Newfoundland would in the 1930's lose control of her destiny.

Coaker's idea of controlled and consolidated marketing sometimes occurred in Canada as well, usually arousing fierce opposition. Proponents of free marketing typically argue that the industry can react more quickly to changing conditions than a government-controlled corporation. Even if the government or an industry board only sets guidelines, still it obliges the industry to cope not just with the changing conditions of markets, but also with the obscure and changing ideas of bureaucrats, unlikely to be innovative. Far-reaching controls such as Coaker's are bad in general, and would have been bad in Newfoundland had they lasted.

Against all that, proponents of Coaker's plan can argue that observers of the time almost universally found fault with the existing system of saltfish trading. This known and traditional commodity was somewhat of a special case. The failure of the reforms did precede Newfoundland's economic collapse. And in later decades, first through its own government and then through the Canadian Saltfish Corporation,

Newfoundland did get consolidated marketing of saltfish, with results generally considered good.

Even without the disputed marketing reforms, Coaker had accomplished an astounding amount. Out of nowhere, his union by the time it was ten years old was buying fish, building vessels, bulk-buying supplies, publishing its own newspaper, and taking direct part in the country's government. These things alone would have made Coaker famous in Newfoundland's fisheries history. Add to that his nearly successful effort to change the whole saltfish system, and his stature becomes gigantic.

Newfoundland loses self-rule

As in the Maritimes, the Great Depression struck early for the fishing industry in Newfoundland. European fleets with modern trawlers were cutting into the saltfish market. In the latter 1920's, Newfoundland exporters poured more fish into Caribbean and Latin American markets, driving down prices for themselves and the Maritimes. Many planters with fleets of 10, 20, or more schooners lost them.[8] Fish prices declined to such an extent that in some places, the fishery ceased altogether. As if the sea gods themselves were angry, in November 1929 an earthquake on the Grand Banks sent a tidal wave into the Burin Peninsula communities, killing 28 people, and destroying houses, wharves, and flakes, and some of the saltfish production. Local opinion blamed the earthquake and tidal wave for a shortage of fish in subsequent years.

Circumstances in the early 1930's drove the government to take more action. In 1933, it announced financial aid: if a fisherman was unable to get credit advances from the merchants, the government would outfit him and provide cheap salt, then market his fish and get its money back from the returns. But this scheme proved too expensive.[9] Bait-freezers and cold storages had shown up here and there in preceding years, and in 1934 the government set up cold-storage bait depots at eight points around the island.[10] The Newfoundland Bait Service would last for decades to come.

In the crucial matter of selling saltfish, the government in 1931 and 1932 tried a partial revival of Coaker's fishery reforms, but exporters refused to cooperate.[11] Another 1933 bill authorized a Salt Codfish Exportation Board. Exporters were to hold a strong place in this scheme through an association whose members were named in the act. But matters lagged; the government itself was collapsing.

Affairs were in chaos, with most government revenues going for interest payments and for relief. The fishery was backward, other industries were weak, foreign trade was shrinking, the railway debt was high, everything was difficult. As Coaker had feared, Newfoundland was to fall under the rule of others. The United Kingdom set up a royal commission on Newfoundland affairs; the 1933 report by Viscount Amulree recommended government by commission.

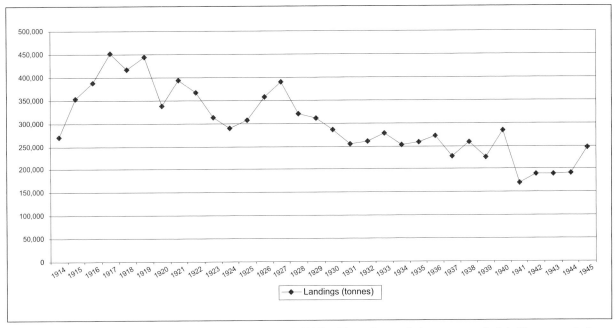

Newfoundland's cod landings kept dropping after the First World War. These figures, in tonnes, are calculated from exports for salt cod (conversion factor of 4.88), and thus omit catches for the fresh-frozen industry, which were growing by the Second World War.

Losing Dominion status, Newfoundland became a crown colony in 1934. Three centuries after the Western Charter of 1634, London was again setting the rules. Six commissioners, three British and three Newfoundlanders, ruled the country, with the Commissioner for Natural Resources controlling fisheries.

Thompson begins fisheries science

As Newfoundland was sliding towards collapse, the government in response to proposals from Memorial University College engaged a Scottish biologist, Harold Thompson. In 1930, Thompson surveyed the island's fishery, finding many instances of under-development.

Thompson pointed out faults in the drying of Newfoundland cod: sunburning, and so on. The trade needed more attention to processing, and needed specially trained fishery officers in the outports for educational work. In any case, Thompson said, the saltfish market, confined to poorer populations in the West Indies, southern Europe, and South America, would dwindle. The trade needed to increase its production of frozen and canned fish.

Some resources were almost unexploited. Newfoundlanders neither fished for flatfish nor ate it, to speak of. One outporter told Thompson that children suspected flounder of being poisonous. While U.S. vessels fished halibut miles east of Labrador, Newfoundlanders ignored that species.

In any case, Newfoundland had almost no fishing capacity for offshore flatfish. The one Newfoundland

trawler, the *Cape Agulhas*, fished mainly haddock for salting. U.S. tariffs held back the marketing of fresh haddock. Newfoundland needed trawlers, Thompson said.

In one advance, freezing had already boosted the salmon fishery. Before 1925, most Newfoundland salmon had been pickled in salt or exported fresh in ice, for example aboard foreign trawlers to England. Thompson reported that "sharp-freezing" (that is, quicker freezing for better quality) had now prompted the building of more salmon traps; the 1930 catch was up 50 per cent over that of 1929. Meanwhile, the salmon sport-fishery in 1929 had attracted more than 400 anglers from elsewhere. The colony charged them a "Rod Tax."

Despite earlier attempts to develop "Scotch cure" herring, Thompson noted, there had been little success in exporting to Europe, since Europe already had its own herring and its own packers to cure them. The herring industry in Newfoundland was less than prosperous. There was still some "klondyking" (over-the-side sales by Newfoundland fishermen to foreign vessels) at the Bay of Islands on the west coast. Newfoundland, Nova Scotia, and New England vessels fished there from November to February, and fished Fortune and Placentia Bays from January. But the speculation about huge herring resources should stop, said Thompson, until someone carried out a proper survey.[12]

After Thompson's report, the government in 1931 set up the Newfoundland Fishery Research Laboratory at Bay Bulls, and made Thompson director. The laboratory used the *Cape Agulhas* part-time. The fisheries authorities were said, however, to make little use of the research station. Thompson resigned in 1936; Dr. W.F. Hampton took over in 1937. That year the Bay Bulls laboratory burnt down. In 1937, the staff moved to St. John's, becoming part of a general government laboratory. Despite this turbulence, the Newfoundland researchers in the 1930's made a good beginning on scientific studies of the main commercial fisheries: cod, haddock, and Atlantic salmon.[13]

Templeman adds strength to science

Thompson gave his scientific survey a practical flavour: What could be caught? And what could be done with it? This concern with application continued in later science, especially by the renowned Wilfred Templeman. A fisherman's son from the northeast coast, Templeman studied at McGill and did one summer's work at St. Andrews. He returned to Newfoundland, eventually becoming head of the Biology Department at Memorial University of Newfoundland.

Templeman in the 1930's worked on various species. "We had a dogfish experiment in Placentia Bay. We caught a lot. It was to go into meal—but flies got into it, there were maggots for miles down the road. Maggots were the outstanding part of the experiment. We showed we could catch the dogfish, but we also showed there was no profit in them." Templeman's other work in the 1930's included lobster research, leading to a major publication in 1941 on the lobster fishery of Newfoundland, which also shed light on lobster in other Atlantic areas.

Templeman noted in a later interview that the war had had little effect on fisheries science in Newfoundland: "We had no research ships to work with anyway." That would change after 1944, when Templeman became director of the Newfoundland Government Laboratory, which did all science and chemical testing work for the colony. Templeman turned research as much as possible towards fisheries.[14] The lab began planning for a research vessel, the *Investigator II*, with which Templeman would later change the map of northwest Atlantic fisheries.

Hard times continue

In the mid-1930's, while Templeman was getting started, Newfoundland and her fishery were mired in Depression desperation. Production of salt cod had levelled off except for normal fluctuations. Back in 1891, 36,700 men had produced 1,244,800 quintals for export; in 1935, the figures were almost exactly the same: 36,500 men producing 1,233,200 quintals. This worked out to about 34 quintals (with prices ranging from $2 to $4 a quintal) or 3,800 pounds of product per man, enough for a small truckload, if one had a truck, which was rare, because money was hard to find.

In the 1930's, workers in Canada on average earned about $1,050; in Newfoundland they earned about $280. According to historian David Alexander, Newfoundland fishermen earned only half that, around $140, although their standard of living was about equal to that of other Newfoundlanders. After a half century of attempts at development, the colony was still far behind Canada: "Newfoundland had been running to stand still."[15] One person in five was unable to read or write. Government services were less than in Canada. Welfare relief paid only six cents a day, a figure burnt into the memory of Depression-era Newfoundlanders.

Handlines, longlines, and cod-traps still dominated the fishery, with gillnets scarce and trawlers almost non-existent. None did all that well for income. Don Jamieson, the Newfoundland broadcaster and politician, later recalled that "companies were often going to the wall, because of cut-throat competition. Schooners would go to Oporto and other places with no assurance of market. It wasn't uncommon for a cargo to rot while the exporter was seeking a purchaser."

Jamieson said that "most fishermen never got out of hock. The system was essentially feudal, and some business owners lived like feudal lords, skimming off what they could, living high on the hog."[16]

Cyril Pike, born in 1921, told the Memorial University publication *Decks Awash* about fishing salmon on the Labrador in the late 1930's and the 1940's:

> We'd leave around the first of June and our parents would take us from school about one week ahead of time. We never returned from anywhere between the 20th of October to the first week in November. So, you can see how much schooling we missed, and that's why a lot of people today, the older fishermen, have no education. ...
>
> If we struck any luck on the Labrador, by the last of June or the first week of July, we had salmon enough to ship it to the merchant. We'd come home in November. If we had money in the merchant's office, he wouldn't straighten us up until about Christmas, until he tried to drain every cent that he could out of us before we got our money. Now if this merchant had any work to give us in the fall of the year around his premises, or home, we never saw any cash. They had a little slip, like a receipt book, called the blue tail. You'd go to those merchants and you earn $20.00. They'd give you a little blue tail where you were obligated to come back to their store, you couldn't take it to another store. If you go into the store with that blue tail, and you spend $5.00, you pass in your $20.00 blue tail, and you'd get a $15.00 blue tail for change. This is why the merchant owned the fishermen. They never had a chance in the world to survive. ...[17]

Templeman gave a more nuanced view of the truck system:

My family was at Bonavista, a fishing family for hundreds of years. My father had four crew. In a bad year, the fisherman owning the boat was a bit of a merchant himself, in having to extend credit. He had to give the crew something. It was a hand-to-mouth existence, mostly all credit. I once asked my father what he'd cleared when all was sold and paid; he said "One dollar."

Except the merchant, the fishermen had no other source of money. No banks; the bank would have nothing to take away from a guy. The fishermen had to have credit; the merchant system was the only possible system. Some merchants did well, some went broke. When I grew up, it was customary to rail against the credit system. But what else was possible? It was a thing of its time.[18]

The decline in the saltfish trade weakened the bank and the Labrador schooner fleets, a slide that would continue. In the years 1935–1950, the floater fleet declined by 75–80 per cent.[19] As Templeman recalled decades later,

Our offshore line fishery for cod, on the banks, was dying in the 30's. A lot of schooners went to Labrador in the old days; but as the engines came, it led to smaller boats. There were always 'jack boats' in the 1930's, 40–50 feet long, on the south coast; these were schooners with two dories.

Before the 1940's, cod trap fishermen tended to be wealthier than others; a trap cost more, and needed four crewmen. Intermediate fishermen would either hand-line, with one baited hook, or maybe set one tub of trawl and then hand-line while it fished. They always tested the grounds by jigging before they anchored; jigging doesn't work as well when you're anchored. Sometimes they just jigged instead of hand-lining or setting trawl.

Trawlers could fish offshore with no opposition; but they were slow to build up, because there was no fresh-fish market. There was some salmon freezing, but not much groundfish.[20]

The fishery, although still the mainstay of the Newfoundland economy, was shrinking in its percentage share as other work increased, especially when the war brought construction of American naval and air bases. In 1935, Newfoundland's marine fisheries employed 45 per cent of the labour force; in 1945, only some 31 per cent.

Overseas, competing fleets were getting stronger. Norway had adopted Coaker-style reforms, including inspection, standardization, and later, price controls; these measures helped Norway push Newfoundland out of the Mediterranean market.[21] France had kept up a sizeable bank fishery in the northwest Atlantic, with schooners giving way to trawlers. In 1928, France had 47 trawlers; in 1934, only 37, but some were now as large as 2,000 tons.[22]

Frozen fish trade makes a start

Only a few bright spots lightened the gloom of Newfoundland's fishery. A rare case of strong markets occurred for cod-liver oil, an old product that gained importance from the 1920's. Fishermen or small processors would heat the livers and squeeze out the oil with a screw press, or else let the sun rot the livers, to bring out the oil. Arthur May, later a renowned scientist and fishery manager, recalled going as a boy in his uncle's boat in the 1940's to help collect cod-liver oil from stage heads for the Munn's company. "My uncle drank a lot of oil, right from the barrel. Many people then wore long underwear winter and summer, and his would be yellow, from sweating cod-liver oil." The cod-liver oil trade would fall victim to synthetic vitamins in the 1950's.

Some operators were trying to modernize. The Harvey interests operated the trawler *Cape Agulhas* from 1925 to 1930, and in 1935, the Crosbie interests acquired the steam trawler *Imperialist*. Some companies had earlier dabbled in freezing bait and fish, especially during the First World War. But headway in fresh and frozen fish was slow.[23]

Hazen Russell prefigured later developments. A native of Grand Manan, New Brunswick, he came to Newfoundland to open a branch of the Bank of Nova Scotia at Catalina on the northeast coast. In 1918, Coaker recruited him to manage commercial activities for the Fishermen's Protective Union. Russell got a small cold storage installed to freeze bait and salmon. In 1928, he moved on to Job Brothers Ltd. That company had bought a 5,000-ton refrigerator ship, the S.S. *Blue Peter*, to use as a floating cold storage. Russell had it re-rigged with a brine-freezing system and canning plant, to pack salmon on the northeast coast and at Labrador—an early factory ship. Russell converted other vessels to diesel and refrigeration, and in the 1940's formed Blue Peter Steamships Ltd. as a freighting operation.

Meanwhile, he had in 1939 formed Bonavista Cold Storage Company, a large fish-freezing plant. Now local fishermen could sell their fish either salted or fresh. As the war ended, the company was expanding, and would become a major operator in freezing and trawling.[24]

Even hard-working pioneers like Russell, who kept trying to innovate and produce high-quality products, found it difficult to build up momentum until wartime. Meanwhile, in the mid-1930's, the disastrous conditions prompted the government to new interventions, hoping against hope for a turnaround.

Engine-powered schooners used sails only as auxiliary. This post-1939 photo shows the dory schooner *Philip E. Lake*, registered in St. John's. (National Film Board)

Government takes firmer hold of industry

Under the Commission of Government, the Natural Resources department patrolled streams, ran the bait service, and the like. In the sea fishery, some conservation regulations affected the size of cod-trap mesh; others set the distance between cod-traps, and regulated the allotment of trap berths. But conservation seemed to pose no major challenge. Attention went rather towards the business side of the fishery.

Other countries suffering from the Great Depression were trying to put order into the fisheries marketplace. In 1932, Iceland put controls on the price of salt cod, as did Norway in 1933 and Sweden in 1934. Newfoundland, more than a decade after Coaker's attempt, began making its own stronger interventions.

In 1935, the government passed a new "Act for the better organization of the trade in salt codfish." This set up a Salt Codfish Board, which would issue licences to exporters, recommend standards to the Commission, and make recommendations on regulating shipments of cod. An association was attached to the board; this later evolved into the Newfoundland Fish Trades Association. Firmer control yet arrived in April 1936, when Commissioner P. Dunn and the

Department of Natural Resources created the Newfoundland Fisheries Board, which replaced the Salt Cod Board.

The Newfoundland Fisheries Board had wide regulatory powers to control the fishery and fish marketing. The board minimized competitive bidding by exporters and worked to develop markets and improve the quality of fish.

As for product quality, new inspectors and standards came on the scene. Although there had been earlier inspection laws, in 1920, 1932, and 1935, the Fisheries Board provided the first real order. By the early 1940's, the board employed 26 inspectors plus temporary helpers. Besides monitoring food products, the inspectors did lobster conservation work in season and other sea fishery work. Other officials monitored salmon rivers.[25]

Commission of enquiry supports strong government role

While the Newfoundland Fisheries Board was taking shape, a 1935–1937 commission of enquiry on the sea fisheries of Newfoundland exemplified the firmer attitude. Previous governments, it said, had paid too little attention to the all-important fishery. "The department of Marine and Fisheries," which had operated from 1898 on, "does not seem to have been able to make itself of any great importance to the Fisheries." It had compiled no statistics, and "there is even a doubt as to the laws and local regulations in effect for the control of fishing." The commission gave the following summary of the department's activities:

> From 1910 onwards, there has been evidence of some efforts to study the development of the fisheries, but these efforts were, in the main, spasmodic, and the possible benefits were lost through the influence of political expediency and the World War. Some encouragement has been given to the erection of cold storage plants; to experiments in driftnet fishing for Herring; to the establishment of a group of Fishery Inspectors for Cod Liver Oil, and later for pickled fish; to the granting of financial assistance to stimulate supplying, when the merchants showed a reluctance, or were unable, to supply; to regulation and control of prices of fish for export, and to the purchase, with Government monies, of fish, when it could not be sold to exporters at what was deemed a fair price. Most of the attempts were emergency measures, and, apart from the encouragement of Cold Storage Construction, and the establishment of Inspectors, there has been no constructive movement for the benefit of the Fisheries.

The commission recommended that "a broad, strong and liberal policy must be adopted without delay." The government should set up a separate fisheries department; state funds should aid fishermen to regain stages, boats, and gear; another fund should help them settle past debts; there should be emergency subsidies for the 1937–1938 season; and so on. Some recommendations went undone, including the separate department. Still, the commission underlined the Commission of Government's new assertiveness after 1934.

The chief government intervention was the board itself. Raymond Gushue chaired the board from 1936 to 1952, and as Don Jamieson put it, "became Mr. Fisheries for Newfoundland." The board could license exporters, set conditions for their operation, control the flow of product, and so on. Exporters had to get licences and pay fees, which went into funds used by the board. The board set up groups of exporters with exclusive rights for particular markets, including Spain, Portugal, and Puerto Rico, and posted trade representatives abroad.

Cod and herring got the main attention from the board. Other notable activities under Commission of Government included a bounty on dogfish, rebates on gasoline, and in 1939, through the Fishermen's Assistance Act, the setting of minimum cod prices.

Also in 1939, the government made an arrangement with the General Seafoods Company of the United States whereby, it was planned, Newfoundland would build a plant on the south coast for the American company to operate. The intent was to have the products admitted duty-free in the U.S., as the product of American fisheries. A similar arrangement already applied in the Magdalen Islands, where an American company employed about 200 local fishermen with about 100 vessels. Some 30 local shore workers salted the fish, which then went in American vessels to the U.S., as duty-free "American product." But when Newfoundland tried this approach in the frozen-fish industry, New England fishermen and plant workers protested. The U.S. Congress made it impossible to import Newfoundland-processed "American" frozen fish in the same duty-free way.[26]

The Commission of Government foreshadowed Joey Smallwood's "Resettlement" by promoting voluntary "land settlements" that were more or less equivalent to co-operatives. Some were based on fisheries, including a successful lobster co-operative on the west coast of Newfoundland.[27]

Twenty years after Coaker's attempted trade reforms, the Commission of Government had done just about everything Coaker had wanted. The Fisheries Board controlled the quality and grading of fish, the licences of exporters, the timing of shipments, and the prices to fishermen. But as in Canada, the Second World War made such efforts irrelevant. The building and operation of American bases employed Newfoundlanders by the thousands, easing the strain on the fishery. Relative prosperity ensued, even if conditions remained harder than in Canada. For the fishery, it was no longer necessary to set minimum prices to fishermen. With wartime demand, the problem was not to find market, but to allocate fish among the Allies, a task in which Gushue played a major role.

Lobster fishery recovers from closure

By the time the First World War began, back in 1914, Newfoundland had a sizeable lobster fishery, though much smaller than that of the Maritimes. Lobstering was strongest on the south and west coasts. As of 1913, there were still nearly 2,800 lobster factories getting annual licences. Most were very small operations, which would sell their products to larger firms. Quality was a frequent issue.

As initial abundances shrank, interest rose in conservation. Staggered seasons for different areas came into effect, the openings moving from south to north as the weather warmed. Newfoundland tried lobster hatcheries, but abandoned them after a Canadian study in 1917 showed little benefit.[28] In the years 1914–1920, Newfoundland's Marine and Fisheries department would buy berried (egg-bearing) lobsters that fishermen caught incidentally and return them to sea. But this prompted the fishermen to catch berried lobsters on purpose.

In the 1920's, rapidly dropping catches led to a drastic step: the authorities in 1925 instituted a closure that lasted until 1928. The effect on the resource was unclear, partly because the ban was poorly enforced. Only after the official re-opening did the Newfoundland industry begin exporting live lobsters in any significant quantity.[29] Such exports increased in the 1930's and 1940's, with the A. Northcott company of Lewisporte a leader. Meanwhile, the number of canneries dropped rapidly; only 73 of them still operated in 1942. Often the canneries put up undersized or berried lobsters. Templeman advocated their closure.[30]

Templeman did fundamental work on lobster biology in the early 1940's. Gushue of the Newfoundland Fisheries Board encouraged him, in the hope of higher yields. Templeman's experiments persuaded the Newfoundland government to keep the trap lath-spacing regulations devised by Superintendent Neilsen at the turn of the century. A federal task force on lobster in 1975 stated that Templeman's "pioneer work laid the foundation for subsequent lobster research in Canada."[31]

Sealing fleet fades

For the cod, herring, and lobster fisheries, the 1914–1945 years brought mainly bad news. For the seal fishery, the period was worse.

Pelt exports, which in the mid-19th century had run to 400,000 or more, had by the turn of the century declined to about 250,000; by the 1920's, they dropped to only about 145,000.

The steel sealing steamers were expensive to operate and, unlike the old fishing/sealing vessels, had only a short season. Finding it hard to make money, companies sold off some of the best steamers during the First World War, especially to Russia as Arctic ice freighters. "By the 1920's the fleet was reduced to eight or ten ships."[32] In the 1920's, some operators used aircraft to spot seal herds, even sometimes taking planes on ships to the ice, where the sealers would clear makeshift airstrips.[33]

People kept vying for berths on the shrinking fleet of sealing ships; the hunt was both a source of money during "the hungry month of March" and to some degree a cultural rite and affirmation of manhood.[34] Conditions could be horrendous. An example is this description of a young sealer's first trip aboard the *Terra Nova* in the early 1940's:

We had to continually build up our water supply ... by putting blocks of ice on board. We formed a chain and passed the broken ice from one to

the other. The ice was placed in a bin and steam was put on it to melt it. It tasted terrible. Usually it was mixed with salt and coal dust. We could not afford to use it to wash; as a matter of fact there was no place to wash. There were many lice on board. ... I remember we used to cut each other's hair and shave to try to get rid of the lice. ... The bunk was covered with coal dust, the floors were slippery with seal fat and our clothes were covered with blood. ...

[Back on shore] I went into the [liquor] controllers in the West End and asked for a bottle of rum. ... He said you can't buy rum until you are 21. ... I told him I was out to the ice. He asked me how much I had made and I said "$130.00 sir." He said, "if you were out to the ice you can buy a bottle of rum." ... When I got home in the morning I called my mother upstairs and gave her the money. She didn't speak. She looked at me and started to cry. I know that this was the most money that she had ever seen in her life.[35]

By 1939, only seven steamers remained; during the Second World War, the government commandeered most of them. By the end of the war, the old sealing-steamer fleet was gone. Unlike the case with the cod fishery, the war brought no particular prosperity to the sealing operators. As wartime spending changed the island, people were flocking to other jobs. The landsmen and small-vessel hunt continued on a small scale. But essentially, the seal fishery had faded to a ghost of its former self.[36]

War changes the landscape

The fishery from 1914 to 1945 had gone through a remarkable odyssey. Coaker-style reforms, which had seemed too visionary at the outset, by the later 1930's had begun to seem quite practical. The 1935–1937 royal commission considered it only sensible to devote government resources and attention to the most important industry. But it was still an industry that failed to support most of its people well. And by the time the government had decided on strong, thorough, comprehensive management, the Second World War and technical and economic developments inside and outside the fishery were changing the landscape. In following years, political union with Canada would erode some of the interventionist approach to fisheries management.

CHAPTER 15.
Freshwater, 1914–1945

Even after provincial authorites early in the century took over most fishery management in Quebec and Ontario, the federal government still ran the prairie fisheries. But in 1929, the Dominion agreed to transfer natural resources to the Prairie provinces. In 1930, these set up fisheries management authorities under one or another department of government, with powers delegated from the Dominion. This narrative, which deals mainly with federal management on the Atlantic and Pacific coasts, will only briefly summarize Prairie-province management up to and in some instances after the Second World War.

By the time of the First World War, royal commissions led by Prince and others had already set many regulations. The Prairie-province fisheries saw little change in management during the 1920's. The growth of settlements in both Canada and the northern United States widened fresh-fish markets. Railways spread their branches to new lakes, opening up new fisheries. Gas engines gradually replaced rowboats and sailboats after the First World War.

Fishermen face market problems

While the stronger economy of Ontario and the proximity to big-city markets helped Great Lakes fishermen, many in Manitoba faced poorer circumstances. As roads and railroads opened northern lakes, new fisheries would yield bumper catches, then drop back. Northern fishermen, often Indians, faced extra difficulty getting their catches to market. With a large proportion of the catch going to the northern United States, American ownership and American influence on markets were major, leading to frequent complaints of manipulation. As the Great Depression took hold, the industry appeared stronger in the south, weaker in the north, with general complaints of poor co-ordination.

In 1933, a royal commission on Manitoba fisheries called for a thorough overhaul. There should be a "clearing house" for fishery products, to bring more control and co-ordination to the market. There should be stricter control of fishermen's licences, and lower catch quotas on the lakes. Something should be done about the influence of the "U.S. racketeers." When American buyers had ample supply, they prevailed on the U.S. Customs service to turn back shipments from Canada on grounds of parasite infestation, whether or

Gillnet tug fishing out of Port Dover, Lake Erie, post-war. (Photo by Dr. W.E. Kennedy)

not there was a real problem. The province needed a Fishermen's Association, and more collective action.[1]

In those Depression days, little government action followed. The 1933 recommendation for better marketing prefigured a solution that was decades distant. As chronic complaints continued after the Second World War, the federal and provincial governments would eventually, in 1969, set up the Freshwater Fish Marketing Corporation (F.F.M.C.) to control production and marketing. The F.F.M.C. was to have a rocky start but then meet a good degree of success; it was still going strong as the 21st century began. Saskatchewan, with a smaller fishery, saw strong government intervention under C.C.F. governments in the post-war period.

Individual quotas, licence limitation emerge

Meanwhile, the prairies had experimented with limited licensing and individual quotas. Even before the First World War, the federal department had begun setting catch quotas for some lakes and species. Use of quotas continued and spread, under federal and then provincial management. Then, late in the Depression years, lake quotas began to metamorphose into individual boat quotas. In 1939, Manitoba restricted the number of gas boats operating on Lake Winnipeg to 65, and limited the catch per boat to 10,000 pounds, well below the previous year's average.

The idea was to improve conservation and, by cutting supply, to raise the price to fishermen. The government granted fishermen's licences by seniority and other factors—for example, the number of dependent children. The selection process caused great hard feelings among fishermen who lost licences.

Application of limited entry and individual quotas was less than tight, especially as wartime demand raised prices and eased conditions. By 1942, both the number of licences and the catch limit per boat had roughly doubled.[2]

Rising incomes spurred the demand for fresh fish, especially in the growing Chicago market. Production rose in Manitoba and Saskatchewan. After the Second World War, the Manitoba government returned to the charge on individual quotas, setting a ceiling on the number of licences on lakes Winnipeg and Winnipegosis, and a maximum catch per boat. According to T.A. Judson's comprehensive study of the prairie fisheries,

A maximum was established on both the number of licences (600) and the catch (five million pounds of all species except pike and mullets). These restrictions had no effect upon operations because the limit to the number of licences was set near the maximum ever issued on the lake and the catch limit had not even been reached in the record take of 6,372,900 pounds of all species in 1942–43. While there may have been some desire to prevent a further large deterioration in the catch per man, no serious management program was apparent.[3]

In 1946–1948, the catch of the more valuable species declined throughout the region. Though some discouraged fishermen dropped out, and others were still opening up new regions, overcrowding of lakes became more and more of a problem in post-war years. Licence limitation became a common practice in both Manitoba and Saskatchewan.

In the original thinking about licence limitation, one goal had been to increase the fishermen's independence and power vis-à-vis the buyers. But according to Judson, buyers anxious to ensure supply courted the licensed fishermen by offering them rental boats with expenses paid. The strong proportion of rental boats preserved the buyers' power over prices.

In the important fall fishery on Lake Winnipeg, Manitoba regulations aiming to protect jobs made the skiff (a flat-bottomed plank boat 18–20 feet in length, usually with an outboard motor) the basic unit of activity. This had its cost in quality of fish, since the skiffs had no capacity to carry ice, and often no protection from the sun.[4]

As it turned out, licensing remained common in Manitoba fisheries, but individual quotas faded away after the war, to return in the 1970's. In summary, the prairie lakes were early on the scene with overall quotas, individual quotas, licences, and attempts to manage marketing; but strong and fully enforced measures took decades to develop. Even so, the federal managers could have profited by paying more attention to the smaller inland fisheries, which were in a sense test tubes for management.

CHAPTER 16.
On the Pacific, 1914–1945

At the outset of the 1914–1945 period, the Pacific salmon industry encountered two great and opposing events: the increased demand brought by the First World War, and a major loss of supply.

The war hiked the market for pink and chum salmon, as well as sockeye, coho, and chinook. Pinks, which in 1913 had sold for three cents each, by 1917 fetched 32 cents each. In the fleet, gas engines made fishing easier; from about 1913 they particularly increased trolling, which was cheaper than gillnetting. The locally made Easthope and Vivian engines became popular.

Despite a consolidation by B.C. Packers in the previous decade, the number of salmon canneries rose to a new high. There were 56 in 1910, producing an average 13,600 cases each; 63 in 1915, averaging 18,000 cases; and 84 in 1919, averaging 18,800 cases. The peak year was 1917, with 84 canneries averaging 18,500 cases.[1]

The industry had begun on the Fraser River. Now, as canneries increased in more northern areas, the Fraser system suffered a loss that would haunt the industry for decades. Construction crews building the Canadian National Railway in 1913 and 1914 caused landslides into the narrow Hell's Gate gorge, about 130 miles from the Fraser's mouth. The canyon became a torrent of almost impassable waters.

The Hell's Gate disaster compounded previous losses. In the Fraser system's Upper Adams River and Quesnel watershed, a logger's dam had done long-lasting damage to formerly strong populations. The new Hell's Gate landslides were even worse. Despite attempts, led by department engineer John McHugh,[2] to remove the rocks and build flumes, the landslide continued to partly block upstream migration, more than was first realized. Heavy fishing pressure by the United States and Canada worsened the situation for migrating salmon. Fraser River runs plummeted. During the 1920's the average salmon pack from

Fraser fish was only about 16 per cent of its level before 1917.[3]

Since Canadians and Americans both fished Fraser stocks, management and remediation called for international co-operation. But effective measures would get going only in the 1930's and 1940's. Meanwhile, the Fraser remained well below potential.

Still, there were many other rivers, and wartime demand was driving up prices. The department was trying to control the number of canneries and fishermen. But to people on the outside wanting in, there seemed room to expand, especially in northern areas, where politicians and the public wanted to see more settlement. Landings were still climbing, although they would take a temporary plunge in the early 1920's.

The question of "who gets the fish" dominated the opening years of the 1914–1945 period. Back in the 1890's, the department had tried to limit the numbers of fishermen and canneries on the Fraser, for both conservation and economic reasons. Under pressure for

Towing salmon boats in after a day's fishing, Skeena River, 1921.

fish, the department had relaxed the Fraser controls, but in the early 1900's had placed restrictions on northern boats and canneries.

During and after the First World War, political pressure to develop the north and give veterans a chance to make a living would build. This would bring about the relaxing of controls on canneries and on white fishermen. The department would first retain, then relax, controls on Indian fishermen; but even with looser controls, Native people faced financial obstacles to getting their own boats. The department would also exert special controls on Japanese Canadian fishermen; these restrictions would last the longest of all.

Such is the summary of licensing in the 1910's and 1920's. The details constitute a complex and shifting story, most comprehensively laid out in *Salmon: The Decline of the B.C. Fishery*, by Geoff Meggs.

Limited entry for canneries fades

While Fraser canneries had increased up until the 1913–1914 Hell's Gate slides, the department since 1908 had strictly limited the number of canneries in more northern waters. It also allotted the number of boats each northern cannery could use, in proportion to conservation requirements and plant capacity. In some cases it specified gear: for example, at Namu and Smith Inlet each cannery could use 1 purse-seine, 8 drag-seines, and 25 gillnets. And the department continued to forbid motor gillnets in District 2, north of Cape Caution (a mainland point across from the north end of Vancouver Island).

Although the northern fishing licences were the most tightly controlled, there was also a degree of licence limitation elsewhere in B.C. The department appears to have controlled purse-seine and drag-seine licences closely, everywhere in the province. At times there were complaints that fishery officers had shown discrimination in issuing both fishing and cannery licences. Even in fisheries with open entry, the department often required licences, along with catch reports to the fishery officer.

In northern areas, the controls suited the department for conservation purposes, and the canners for stability and profits. But a railway was pushing towards Prince Rupert, and pressure was building to settle and develop the north. Accordingly, in 1912, the department had begun loosening the limits on northern canneries.

It also issued some fishing licences to "bona fide" white fishermen independently of the canneries, while keeping overall numbers under control. During the war years, canners found it harder to get labour for boats and plants, and recruited many Japanese. But in the independent fleet, the department favoured whites, at the expense of Japanese and Native people.

As demand for expansion grew, the department gave more ground. By 1917, it wanted to lift restrictions on the number of northern B.C. canneries. It also decided to untie the boats from the northern canneries, end-

The "mosquito fleet" at Prince Rupert, 1916. (Library and Archives Canada, PA-30248)

ing the "boat-rating" system; instead it would issue fishing licences to fishermen independently. And it intended to allow motor gillnets in the north.

The established canners fought back, saying that this policy would bankrupt them, lead to chaos, and put the fishermen in a worse position than ever. The government responded with another royal commission, led by W. Sanford Evans, an economist and statistician from Winnipeg.

Whitcher, Prince, and others had seen the benefits of licence limitation for both conservation and economic efficiency. Evans gave the concept new and clear expression:

> It is a clear public duty not merely to conserve the supply of salmon in its present proportion, but to increase it until each year it reaches the economic maximum and it appears to us equally clear that all the conditions surrounding the industry should as far as possible be stabilized and the excessive use of capital and labour obviated or prevented. ... The solution of this problem would not seem to be found in encouraging or permitting the employment of more capital or more labour than can efficiently perform the work. The public interest can be served in other ways. The privilege engaged by those who fish in tidal waters is not only fundamentally a public right but the public stand related to the industry as taxpayers and consumers. If the cost of production becomes too great all hope of advantage to the public as consumers will disappear.

Evans added that any excess profits in the cannery industry should go to the public. But his sentiments on controlling capital and labour got little attention.

Evans recommended restoring limited entry for northern canneries. But the federal and provincial governments both wanted to open new areas of B.C. So the department issued seven new cannery licences, while also imposing higher licence fees and a production tax on canneries. The principle of government

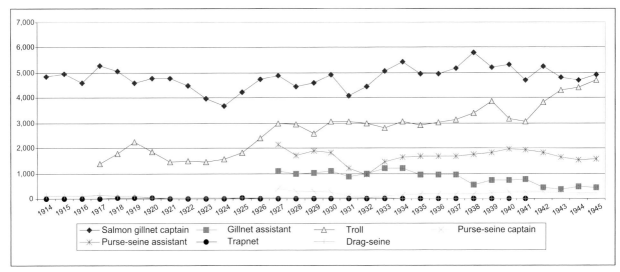

The chart shows salmon licences between the wars. The number of fishermen levelled off in the 1920's and 1930's.

control remained, but in practice, the limits were loosening.

Evans also recommended that the department continue limiting the number of fishing boats, but issue gillnet licences independently of canneries to British subjects who were "bona fide" fishermen. This change, which paralleled the department's original intention, came into effect; the boat-rating system of a set number of boats per cannery under "attached" licences vanished. The canners still owned most of the northern fleet, but the department would now licence a cannery boat or an independent boat as it saw fit.

Following another of Evans's recommendations, the department dropped its intention to allow motor gillnets in the north; instead, the ban would last until 1924. This suited the canners, who told a subsequent inquiry it would be difficult for them to finance fishermen (already a common practice) to buy motorboats rather than unpowered craft. Also, both white and Indian fishermen on the Skeena and Nass rivers feared that allowing motorboats would hike their expenses and let Japanese fishermen crowd them out.

It all added up to more canneries and more licences in the north, though the department was still trying to maintain controls. Chief Inspector Cunningham reported in 1918 that the pressure from returning soldiers for licences was becoming a serious matter. If one granted too many seine licences for one small area, and only a small run of salmon came in, fishing pressure could endanger the salmon.[4] Still, the department remained generous with licences for veterans.

Although the department was allowing additional canneries in the north, the total number of canneries operating in British Columbia would soon drop for business reasons. After jumping to 84 in 1917, the total fell to around 60 in the early 1920's, when catches were low, then resurged to 76 in 1926.

Limited entry for fishing vanishes

In 1919, the department allowed 46 new purse-seine and drag-seine licences to returned soldiers, out of 150 applications. Seine licences were supposed to go to white people only. The new seines brought extra fishing pressure on the west coast of Vancouver Island, so "hardly any chum salmon escaped." Veterans also increased the number of gillnet licences in northern B.C.[5]

Trafficking in licences, though illegal, was becoming a problem. The annual report for 1919 noted the possibility that as returning soldiers got licences but failed to make a go of it, Japanese fishermen might buy away the licences and get control of the fishery. There were complaints about discrimination in licensing, leading in 1919 to a royal commission on the subject.

Meanwhile, the Commissioner of Fisheries for B.C., William Sloan, was advocating that government itself take over and operate the salmon fisheries, thus protecting conservation, eliminating wasteful expenditure, and providing cheaper fish to the public. If the federal government was unwilling to do so, it should let the province take over. This suggestion of a crown-operated fishery went nowhere.[6]

Veterans continued to press for fishing privileges, wanting to end all remaining controls on their numbers. In one instance, returned soldiers who objected to a canning company's drag-seine privilege at Quashella Creek, Smith Inlet, destroyed several thousand dollars worth of company nets.[7]

With licensing subject to pressures and problems on every hand, the department gave up. In 1919, Found recommended that there be no restriction on cannery or fishing licences: "We must safeguard the situation by decreasing the fishing season where necessary and putting on sufficient Fishery Officers to prevent illegal fishing."[8] Abandoning efforts to regulate the number of boats on economic and conservation grounds, the department from 1920 would rely on such measures as regulating time, area, gear, and fish size.

The number of fishermen levelled off in the 1920's and 1930's to the 10,000–12,000 range. Even with freer entry, a natural balance was asserting itself; and even if licence limitation might have raised average incomes, still the B.C. fishermen already tended at least to do better than those on the Atlantic.

Drag-seines fade; motor gillnets arrive

As the department opened entry to the fishery, it began phasing out the old drag-seine leases. The department had issued such leases all along the coast, most going to a handful of canning companies, which left some leases unused. Often the department required the canneries to make arrangements with local Native families, if they claimed traditional fishing rights in a stream. The Native people would do the fishing for the cannery, in their traditional fishing spots. Some licences got issued directly to Native groups.

But the exclusive privileges went against the new trend of open entry. Independent, white fishermen fought them; some canners themselves found the old lease terms outmoded. Cunningham and Found wanted to eliminate them as far as possible.[9] Doing away with drag-seines would do away with a potential conservation problem at river mouths, and make more room in the fishery for the growing purse-seine fleet. So, while issuing more licences mainly to white fishermen, the department cut back on the leases. The number of drag-seine licences fell from nearly 140 in 1917 to fewer than 40 by 1921.

Most but not all of the remaining drag-seine privileges went to Native people for food fisheries. But new regulations moved them farther from the mouths of rivers and creeks, making it harder to fish well. A few drag-seine licences would last for decades. But the general cutting back on drag-seines opened up more fishing for independent fishermen. At the same time, it dislocated many Native people who had depended on the drag-seine fisheries, and who found it hard to get the new, independent licences. They did get permission to run cannery seine boats, because of pressure applied by the government's Indian agency.[10]

In the aftermath of the department's relaxation of fishing and cannery licences in the north, the B.C. industry in 1922 succeeded in getting another royal commission. Headed by William Duff, a member of Parliament from Lunenburg, Nova Scotia, the commission recommended letting matters lie for white fishermen. In northern B.C., where the main fleet still consisted of cannery rowboats and sailboats operated by Indians, the royal commission recommended allowing motor gillnetters, as the fisheries branch had earlier wanted to do. The branch did so from 1924. Sailing skiffs began fading away; more white fishermen with motorboats came into the fishery.

Department limits Japanese fishermen

Undated photo of troller shows poles that suspend hooks in the water.

The Duff commission also recommended reducing by 40 per cent the number of fishing licences, other than troll licences, issued to fishermen who were neither Indians nor Caucasians. This meant the Japanese (many or most of them now naturalized Japanese Canadians), who constituted a large portion of the fleet. Starting in 1923, the branch reduced their number as recommended. "Orientals" could neither operate nor work on purse-seiners. They could troll only within certain areas, not including the north and central coast.[11]

The new restrictions came on top of earlier limitations. The racial climate in British Columbia had long been uneasy, as shown by various riots and disturbances; the 1914 *Komagata Maru* incident, which saw a shipload of immigrants turned away; and the exclusion of Asians from voting, holding public office, or working as civil servants.

Native people drag-seining salmon on the Nimpkish River, B.C., ca. 1930. (Library and Archives Canada, PA-205827)

In the fishing industry, Chinese immigrants had helped to build and operate the canneries. Japanese fishermen had gained a major place on the water; they came to out-perform whites, for example, in fishing and dry-salting herring for the Orient. But racial consciousness kept cropping up. In the 1890's, Samuel Wilmot had said it was better to license Indians than Chinese. When Japanese fishermen tried to join the Fraser River Fishermen's Protective and Benevolent Association, whites turned them away. In 1911, after authorizing a new plant—despite general cannery controls—on the Queen Charlotte Islands, Ottawa said the plant was to employ only Canadian or European fishermen, so as to extend settlement of the area. Later, another new cannery was licensed to employ Indian fishermen only. From 1912, the department favoured licences for white fishermen in the north.

Many Japanese had by then entered the fishery, on their own boats or cannery boats. During the war, they had pioneered troll fishing on the west coast of Vancouver Island, where the department confined them to southern areas. As the department opened entry to northern fisheries after the war, it had issued new "independent" licences to whites only. Japanese, even if they had become British subjects, were excluded as a departmental policy, although Found recognized that the policy's legality was dubious.

Even with those restrictions, Japanese Canadians were already running about half the B.C. gillnet fleet and about a quarter of the troll fleet. British Columbia members of Parliament, white and Indian fishermen, and others demanded controls on the Japanese. The issue came to the federal cabinet's attention. The department in 1922 began reducing the Japanese Canadian licences. First came a one-third reduction in their troll fleet, to 334 boats. The department then moved towards a major reduction in other areas.

This effort was interrupted by the Duff commission of 1922, which found that the department was moving too slowly. Although Japanese Canadian fishermen made eloquent pleas and questioned "British justice," the commission recommended a 40 per cent cut in their gillnet licences. Among the remaining, "naturalized Orientals" should get the preference for licensing. Found recommended to Minister Ernest Lapointe an additional cut in the Japanese troll fleet.

In the fleet as a whole, the department in 1921 had licensed nearly 4,800 gillnet captains and about 1,450 trollers, for a total of about 6,200, not counting gillnet helpers. Japanese Canadians were the single biggest group, operating 2,600 boats, about 1,100 in northern waters, nearly 900 on the Fraser (where they outnumbered Indians and whites), and most of the rest on the west coast of Vancouver Island. (White fishermen then numbered about 2,200, including 1,600 gillnetters, and Native people about 1,200, of whom about 970 gillnetted for northern canneries.)

Successive licence cutbacks in different fisheries kept slicing into the Japanese fleet. Found reported in 1927 that the department had eliminated 1,374 of them. The forcing out of Japanese Canadians took place despite some opposition by Major Motherwell, the Pacific head of the fisheries branch, and by canners themselves.

Other restrictions applied to the Japanese. Their independent boats were disallowed from using gas engines in northern waters. In those waters (District 2), the Duff commission had recommended allotting "a reduced number of Oriental licences" to canneries, in proportion to their previous employment. There was no move to eliminate them, however; the canneries could get fish cheaper from the Japanese, who, according to Geoff Meggs, worked "in conditions of unparalleled subservience."

Eddie Moore, a fishery officer in the 1920's, said that "the Japanese weren't allowed seines in the north. The canneries had lists of Japanese; each cannery had a restricted number of Japanese they could employ on cannery boats." Those would include only gillnetters, trollers for certain areas, and "packers" and collectors. In another limitation, "Jap boundaries," as they were referred to, applied at the mouths of some rivers and inlets, to keep the Japanese farther offshore than other fishermen.[12]

The complexities of licensing all added up to painful cuts and severe restrictions for the Japanese Canadians, especially in the north. In 1926, the Japanese Canadians fought back with a legal case asserting their right to licences. Both Canada's Supreme Court and the British Privy Council upheld them. The Japanese Canadians also, by 1931, defeated the ban on their use of gas engines in the north, and the system limiting their numbers in cannery fleets. By then the political push to expel Asians had subsided. (White fishermen could still get angry, though; Ron MacLeod, the son of a fishery officer on Vancouver Island, recalled that when Japanese Canadian fishermen showed up in one new area after the lifting of the gas-boat ban, whites used rifles to drive them away.)

At the end of the 1930's, Japanese Canadian fishermen held about 15 per cent of commercial fishing licences, mainly for gillnetting and trolling: a significant number, but well below their earlier proportion. Japanese Canadians also worked in many canneries, especially at Steveston. White fishermen had emerged dominant; both Japanese Canadians and Native people were far weaker than at the beginning of the century.[13]

Fishermen's organizations build up

British Columbia drew fishermen from all over. A royal commission settling claims after the international sealing treaty recorded the testimony of fishermen from Yarmouth, Halifax, Shad Bay, Arichat, Digby, and other Maritime places; from Newfoundland; from Sweden (a Swede gave an account of making his living by beachcombing); and from Japan. One European, John Haan, told of how he came to North America, left on an adventure to go around the world in a 30-foot canoe, abandoned it, took up ranching, took up steamboating, lost all he had, became a cattle rustler, and wound up working on a fisheries patrol boat.[14]

British Columbia continued to breed fishermen's organizations. The Deep Sea Fishermen's Union represented halibut fishermen. Most organizations of course sprang up in the salmon fleet. The Fraser River Fishermen's Protective Association operated from 1914 to 1919, then became the B.C. Fishermen's Protective Association, which lasted till 1945 and its merger with the new United Fishermen and Allied Workers' Union.

Organizations often grouped up along racial lines. The Fishermen's Benevolent Society, representing Japanese fishermen, in 1926 merged into the Amalgamated Association of Fishermen of B.C., another Japanese organization with headquarters in Vancouver and four branches. Meanwhile, the B.C. Fishermen's Protective Association represented white fishermen in the south. Another group, the B.C. Fishermen's Association, started up in 1924, then merged with the Fishermen's Protective Association (F.P.A.) in 1928. The F.P.A. won a strike in 1928, raising the price per sockeye from 65 cents to 70 cents.

The Northern B.C. Salmon Fishermen's Association started up in Prince Rupert in 1920, but lasted only two years. Also headquartered in Prince Rupert, the Fish Packers' Union of B.C. operated from 1918 to 1935. The United Fishermen of B.C. started up at Sointula in 1917 and closed down in 1924. Sointula in 1929, however, gave birth to another group, the B.C. Fishermen's Co-operative Association, the first fishermen's co-op in B.C.

All this time, fishermen remained mostly unorganized in the Maritimes and Quebec (although Newfoundland saw Coaker's F.P.U. develop). Why were B.C. fishermen quicker to organize? The probably reasons include fishery and settlement patterns. British Columbia had more fishermen clustered together in sizeable groups, especially around the Fraser estuary. As well, the province had recent immigrants from countries where organizing was now widely practised. Another factor was the structure of the salmon-canning industry: companies were bigger and fewer than in the east, giving negotiators or strikers an easier target.

The canners themselves were always organized. In 1924, the British Columbia Canners Association changed to take in other processors, and became part of the Canadian Manufacturers Association. This arrangement continued until 1939.[15]

Fishery officers set up in remote areas

Major J.A. Motherwell in 1921 succeeded F.H. Cunningham as Chief Inspector for B.C. Motherwell would hold the position (later Chief Supervisor) until 1946, overseeing fishery officers spread along the coast. There were major concentrations of fishermen near Vancouver, Victoria, and other southern centres. By contrast, the central and northern coasts outside of Prince Rupert were sparsely settled. There were can-

Native families travelled north on coastal boats to the canneries for the fishing season. (Library and Archives Canada, PA-41175)

neries in most major inlets, but often no surrounding town to speak of. The fishery officer dealt with seasonal congregations of people and fish. During salmon season, cannery supervisors and workers would show up. Coastal vessels would pick up Native or other workers and take them to places such as Rivers Inlet. The workers would usually live in small buildings erected by the cannery. Cannery boats would operate in nearby waters, and independent seiners would congregate from elsewhere.

In District 2, north of Johnstone Strait, the department in the 1920's had ten fishery officers (one for each sub-district) on the coast, plus a supervisor and an assistant supervisor. Two people worked in the Upper Skeena, mainly patrolling river spawning grounds. By 1933, the province as a whole had 31 inspectors (fishery officers). Most officers would charter patrolmen (with their own boats) and hire guardians to help monitor the fishery. The fisheries branch in the 1920's began making some use of seaplanes, one of which unfortunately crashed in 1928. Don McLaren, a First World War ace, had a firm that hired out planes for patrol work. The department had sizeable offshore patrol vessels. These sometimes carried out sea-lion slaughters, to control salmon predation; crews would go to rookeries and shoot or club the pups.

As salmon migrated into river mouths, fishery officers had to set fishing boundaries, creating sanctuaries for the fish to school up before travelling upstream. The officers would keep track of the numbers of boats and of salmon, and often make on-the-spot judgements about whether or not the fleet could fish, and for what period of time. Dealing at close quarters, with salmon visible and vulnerable, the fishermen and the fishery officers built up a mutual respect—in modern jargon, a kind of love–hate relationship.[16]

This is not to say there were no problems. In 1918, accusations of corruption and incompetence triggered an inquiry; Chief Inspector Cunningham was exonerated, but the scandal forced him out anyway.[17] In 1926–1927, reports emerged of a canner-sponsored poaching ring at Smith Inlet, complete with financial and sexual payoffs to a fisheries guardian. Still, B.C. fishermen knew the officers, patrolmen, and guardians were on the side of the fish, which could otherwise be readily depleted.

The fishery officer's lament

From Whitcher's time on, fishery officers viewed with alarm the goings-on around them, including illegal fishing, habitat damage, and depletion. There was no let-up in the 1920's, 1930's, or 1940's. In 1944, Fishery Inspector J. Urseth, at Bella Bella, B.C., in his annual report put his complaints to paper. Excerpts follow:

> Present known supplies of salmon are being exploited to the utmost, yet in the face of this well-known fact, industry is undeterred in its continual investment in newer and more efficient gear of all kinds, all designed to land and process more and more salmon. Industry then demands extended fishing privileges on the ground that heavy capital investment may claim and is entitled to protection. The amount of gear in use is far beyond economic reason.

> If a good run of salmon appears at any point, it is immediately attacked by a large and highly mobile fishing fleet whose rapacity and disregard for future supply knows no bounds. This setup is further aggravated by the often bitter rivalry existing between fishermen belonging to different racial groups or organizations. The ultimate result is that any run of salmon, regardless of its extent or condition, can be exterminated in a few days or even hours. In the meanwhile the fishery officer responsible is compelled to make almost instant decisions as to opening or closing. The welfare of the run is contained within such narrow limits of time that there is little or no opportunity to study the situation. ... Under such conditions errors of judgement are inescapable relieved only by the choice to err on the side of conservation when there is bitter criticism charging that industry is prevented from taking a legitimate portion of the supply.

> It is fully realized that the situation just described is incapable of correction except by drastic means probably involving a system by which quantity and kind of gear is directly controllable in each area. The writer is of the opinion that present means of control are too slack. By virtue of a single inexpensive licence a fisherman, or seineboat, is allowed to roam at will in the coastal waters of the whole Province. Groups of men and boats can and do concentrate their operations on runs of salmon as they appear at various times and at various places with disastrous results to the fishery unless they are kept under constant surveillance.

> Fishermen's organizations and various unions are commendable in many respects but they show an almost total lack of interest in all questions concerning scientific investigations and problems related to future supply. Their vision does not extend beyond immediate gain.[18]

On the Atlantic, respect also existed, and officers had many similar duties, but there were differences. The Atlantic officer was likely based year-round in some small town or village, with another small settlement a few miles away, and so on all along the coast. Officers oversaw mainly a series of local fisheries. With a more diffused and varied fishery, less visibly vulnerable than Pacific salmon, it was less apparent that the fishery officer was helping to conserve fish. The officer did less monitoring and management, and more enforcement. There was less underlying sympathy for his work. Enforcement often became a game of hide-and-seek, with informants and stake-outs. Without overstating the differences—the Atlantic fishery officer was generally a companionable figure in the community—still the work on each coast had its own flavour.

Fleet becomes more mobile

Fishing techniques were diversifying after the First World War. As gas engines gradually took over, purse-seining and trolling became more common. The department divided the north coast and Vancouver Island into areas and licensed seiners for a single area only. In some areas, no purse-seining was allowed at all.[19]

After the department cut down the number of drag-seines, some drag-seiners gravitated into purse-seining. Around 1926 an influx of Yugoslavians, experienced net fishermen, added to the number. Power winches and rollers along the boat's side made hauling in easier. The department set maximum and minimum net sizes, to keep fishermen from using small nets for shallow areas at the mouths of rivers and creeks.

Purse-seines became a major element of the fleet; about 400 of them operated by 1927, although the number declined in the 1930's. In the fall, pilchard boats and larger trollers would switch to salmon purse-seining.[20] It became common to set boundaries for the purse-seiners a half mile offshore.

B.C. fleet uses smaller vessels

Although the fleet was getting more mobile and powerful, British Columbia never developed a large-vessel fleet to compare with Atlantic trawlers and large schooners, which were often over 100 feet long. There seem to have been several reasons for the difference.

Salmon were a lower volume fish than Atlantic groundfish or herring; they were more like lobster, in that small boats could fish them well. The winter halibut fishery required good offshore boats, and some were large at the outset. But halibut, too, was less of a volume fishery than Atlantic groundfish, and the halibut fleet settled out with smaller vessels than at the beginning. In that fishery, regulations in the interwar period outlawed both dories and drags, which worked against any move to larger boats.

In the herring fishery, B.C. fishermen used seiners in advance of the Atlantic coast, but no monster seiners developed. Many herring boats were primarily salmon boats, fishing herring when salmon were absent. Besides, the department limited the size of seines. And with much of B.C.'s coast sheltered by islands, the ocean though still dangerous was kinder to small craft. Whatever the exact factors, the B.C. fleet used mainly small and medium-sized vessels.

Some craft in this undated photo carry gillnet drums, developed in the 1930's.

A partial exception was the fleet of pilchard seiners that would emerge in the 1920's and 1930's, with vessels in the 70- and 80-foot range. These vessels also fished herring in winter. When the pilchard fishery faded after the Second World War, some of these vessels went into the salmon fishery, boosting its fishing power.[21]

For the most part, however, British Columbia never developed much of a large-vessel fleet. That being said, its medium-sized vessels were efficient, and B.C. had a higher proportion of vessels to boats than the Atlantic coast. The number of vessels dropped in the Great Depression but rose again during the Second World War, as shown in Table 16-1.

Table 16-1
Vessels and boats in British Columbia, selected years, 1924–1944.

British Columbia			
Year	1924	1934	1944
Vessels	382	286	448
Boats	5,238	8,305	7,671

Canners press for renewed limits on plants

After a slump in the early 1920's, the number of salmon canneries climbed back to 76 in 1926. In 1927, over-expansion and over-competition brought financial setbacks.

The B.C. industry began trying to think its way out of trouble. It seemed to some that too many canneries and boats were operating. They were duplicating their efforts in "packing" fish in carrier boats over long distances back to the plants. A high-powered delegation, including the premier of British Columbia, canner and fishermen representatives, and Major Motherwell, the department's regional supervisor, went to Ottawa to press for more order in the situation.

The delegation proposed cutting the fleet and dividing the coast into fishing areas, in such proportions as to support canning plants in each area. The fisheries branch should regulate the maximum amount of gear to be used in each area. Except for troll-caught salmon, or salmon destined for export raw, the fish caught within each area should be processed within that area. No more canneries should be licensed.

With a court case pending on jurisdiction—Francis Millerd's Somerville Cannery case—the fisheries branch put off any immediate decision. Meanwhile, Found travelled to B.C., and encountered less than general support for the proposals. The fisheries branch left the industry to work out its own solutions.[22]

When the courts in 1928 gave the provinces jurisdiction over processing, it ended any possibilities of cannery-licence control by the fisheries branch. That year, the principal canners themselves entered a five-year agreement to limit the amount of gear by area. They also made some associated transfers of boats and plants among themselves.

Area licensing begins for boats

Although cannery control was gone, some matters raised by the 1927 delegation remained within Dominion power. Delegation members had wanted the fisheries branch to divide the coast into fishing areas, and to limit the amount of gear within areas. If too many boats entered an area, the branch should extend the weekly or annual closed time.[23]

In March 1929, the fisheries branch divided the whole coast into 27 areas, and put a limit on the number of purse-seines in each area. As a new system took shape in following years, vessels on arrival had to transfer into an area. When the salmon run peaked in that area, the local fishery inspector's house could be full of skippers from other parts of the coast, waiting to get their licence endorsed for the area.[24]

The normal closed period for each area was 48 hours a week. If too many boats came into the area, the fisheries officer could extend the closure by 24 hours. On the east coast of Vancouver Island, the east coast of the Queen Charlottes, and Barkley Sound on the west coast of Vancouver Island, the new system cut back fishing by about 20 per cent, and the number of seiners dropped.[25]

The area system would last until the early 1950's. Although area licensing was imperfect and unwieldy, still it afforded a measure of control. And it showed once again that the B.C. industry was more likely than the fragmented Atlantic industry to lead government into wide-ranging action.

Canners consolidate, ask province for limits

Meanwhile, the B.C. Fishing and Packing Company (formerly, until 1921, the B.C. Packers Association) had over-strained itself by purchasing Wallace Fisheries Ltd. and opening two new canneries. Gosse Packing Co. Ltd. had also over-expanded with new plants. In 1928, the two companies did away with some duplication of effort by merging as British Columbia Packers Ltd. That same year, B.C. Packers took over the Millerd Packing Company Ltd., bringing in five more canneries and two salteries, for a total of 44 canneries plus various salteries and meal and oil plants. B.C. Packers immediately closed 8 of the 44 canneries, and in 1929 closed 4 more.[26]

The total number of canneries in B.C. dropped to 59 in 1930. Anxious to stabilize their positions, the canners that year presented arguments for licence control to Samuel Lyness Howe, Commissioner of Fisheries for the province of British Columbia. Their brief stated that although the Federal Government limits the number of fish to be taken each season, adjusting the quantity by closed periods, their present policy is nevertheless to issue an unlimited number of fishing licences to all qualified applicants. So long as that policy remains in force it will be difficult, if not impossible, to limit the number fishing in each area to the figure set for conservation, and consequently additional closed periods are imposed.

The canners are of the opinion that there are already too many plants in existence and the only way in which the present state of affairs can be remedied for the benefit of canners and fishermen alike is to limit the number of canneries as well as the amount of equipment to be used. This policy is not new but is the original method of control in force prior to 1912 and is very strongly recommended by each of four special Fisheries Commissions between 1905 and 1917.

Since the Federal Policy of 1912 of issuing an unlimited number of cannery licences and the policy of 1920 of an unlimited number of fishing licences, there has been a gradual increase of both, culminating in the disaster of 1927 when the industry lost in the neighbourhood of $2 M. due to intensive, unprofitable and reckless competition.[27]

The six o'clock gun: fishery officer with explosive, Skeena River, B.C. (Library and Archives Canada, Pak-41001)

Cannery at Butedale Lake, B.C. (Library and Archives Canada, PA-40989)

The province agreed to a five-year moratorium on additional licences.[28] Meanwhile, business factors themselves were affecting the canneries. The number operating dropped to 40-odd in the mid-Depression, less than 40 by the beginning of the Second World War, and only 30 by its end. Notable companies during these years included B.C. Packers, Nelson Brothers, Canadian Fishing Company (owned by New England Fishing Company), Anglo-British Columbia Packing Co., and J.H. Todd and Sons Ltd.

B.C. fishermen pioneer gillnet drum

During the 1930's the number of fishermen stayed in the 9,500–11,700 range. Gillnets and purse-seines accounted for most landings, with trollers a significant minority.

Motorboats now made up almost all of the B.C. fishing fleet. In the early 1930's, Sointula fisherman Laurie Jarvis pioneered the powered gillnet drum; mounted on the stern, the drum would unreel and then mechanically haul the net.[29] This became the prevalent method of gillnetting in B.C., and spread to other areas.

Fishery officers build up salmon database

Fishery officers in the 1930's continued monitoring fisheries all along the coast. Often, they were the main source of law and order in isolated communities, acting more or less as magistrates along with carrying out their fishery duties. Area controls, closed times, and judgemental use of boundaries and closures remained primary methods of management. At many fishing sites the fishery officers fired a six o'clock gun to signal the start of fishing after closed periods.

The fishery officers did considerable work patrolling and clearing salmon streams and monitoring runs and spawning areas. (The provincial Commissioner of Fisheries abandoned stream observations from 1932 on because of "financial necessity."[30]) The federal officials over time developed forms and procedures for observations of fisheries, streams, and escapements, building up a database. They noted the timing of runs, which of course the Indians already knew. Gradually, the officials worked out loose relationships between the amount of escapement and the size of future catches, a calculation made possible, though far from exact, by the relatively small number of eggs per salmon.

Ed Moore, the veteran fishery officer quoted earlier, had in the 1920's worked at logging and as a sawmill foreman, and at times on a chartered boat for the fisheries branch. In 1930, he became inspector for the southern Queen Charlotte Islands, then moved in 1935 to Butedale district on the central coast below Prince

Organizations keep forming

During the Great Depression, prices dropped for fish, as for most goods. The noted fisherman Jimmy Sewid later recollected that "we'd get five cents for a whole dog [chum] salmon, 15 cents for a sockeye. A great big white spring [chinook] was five cents, whatever size; a red one was 50 cents, whatever size. The prices were so low a lot of people gave up. From the first of June to the middle of November, $250 or $300 was big pay."[32]

For the 5,000 or so cannery shoreworkers in the 1930's, most often Chinese, Japanese, and female Indians, conditions could be hard indeed. Still, B.C. fishermen weathered the Great Depression somewhat better than those on the Atlantic. They remained far ahead in organization. Some previous organizations continued in the 1930's, including the Deep Sea Fishermen's Union, the B.C. Fishermen's Protective Association, and the Amalgamated Association of (Japanese) Fishermen.

Meanwhile, in 1931 the Native Brotherhood of B.C. started up, as did the long-lived Kyuquot Trollers Association. In 1932 came the United Fishermen's Federal Union; in 1935, the North Island Trollers' Co-operative Association; and in 1937, the B.C. Trollers' Association. The B.C. Cod Fishermen's Association, representing Japanese fishermen, started up in 1939 and lasted till 1944, when the federal government interned most of its members inland.

The Fishermen's and Cannery Workers Industrial Union organized strikes in the mid-1930's. This group and the trollers then became part of the new Pacific Coast Fishermen's Union. In 1940 and 1941, the latter group and the Salmon Seiners Union, which had started in 1937, merged into the United Fishermen's Federal Union, which had a marked communist influence. As other organizations joined, including the B.C. Fishermen's Protective Union, this grouping would develop into the powerful United Fishermen and Allied Workers Union.

Five fishermen's co-operatives were active by the end of the 1930's. The Prince Rupert Fishermen's Co-operative Association (P.R.F.C.A.), starting in 1931, was to grow into a renowned organization, dominating the north coast fishery in British Columbia. The P.R.F.C.A. absorbed some other co-operative associations. Often, such associations had no formal set of co-operative principles; they simply evolved as a means of collective selling, then might move into other activity.[33]

Rupert. As an inspector he got a net salary of $110 per month; after he put in three years this rose to the maximum, $115.

At Butedale, "I had over a hundred spawning streams. I couldn't cover them all. We had to rely on guardians. We had three patrol boats and three guardians on shore, maybe with rowboats. It was wild country so we had little worry about poaching in the streams, unless a gillnetter got up into a stream, but we usually kept good track of them."[31]

There was always a weekend closure in the salmon fishery, and additional closures if too many boats came to the area. To a degree, B.C. fisheries were becoming a stop-and-go operation.

Native fishermen lose ground, fight back

For Native people, misfortune followed misfortune. In the latter 19th century, the department had forbidden them to use traditional traps or to sell fish from their traditional fisheries; they could only fish for food, under permit. Many had joined the commercial fishery, only to be partly displaced from the Fraser by whites and Japanese. After the Hell's Gate slide, the department severely restricted food-fishing in the Fraser system, wreaking hardship on Native bands.

On the commercial side, Indians at the time of the First World War still fished many traditional areas with drag-seines under the cannery lease system; but that system faded after the war. Meanwhile, when the department authorized new licences and limited entry, it favoured whites.

The Native Fishermen's Association of B.C., formed in 1916, failed to win concessions. Meanwhile, the province of British Columbia urged Ottawa to purchase and eliminate Aboriginal fishing rights. Native fishermen were losing ground, although they still operated many gillnetters for canneries.[34] As well, Native shoreworkers would migrate to coastal canneries for the season. Some Native people hunted whales from their canoes for commercial companies.[35]

As the lease system for drag-seines fell away, the department licensed more purse-seiners. The Native people who had operated drag-seines were at first unable to get independent licences for purse-seines. Natives saw white seiners working in their traditional spots. When Found visited the coast in 1920, the Native Fishing Association made representations. As one chief said: "We come now to ask the Government to prevent fishing the creeks and rivers with purse-seines. ... We feel it is not right for us to have to buy a licence to fish at all, because the rivers belong to us."[36] This got little reaction. Another Native organization, the Allied Indian Tribes, lobbied for exclusive privileges in some areas, plus access to the commercial fishery on the coast. They, too, made little headway at first.

By the early 1920's, however, the department eased its policies, to some degree because of pressure from the Dominion agency responsible for Indians. Native people got permission to skipper company-owned purse-seiners, and then their own seiners. They also got licences for gillnets and trollers. But their share of independent licences remained low.[37]

As well, the department eased its food-fishing restrictions in the interior; some illegal commercial sales crept in as well. In the words of Geoff Meggs's salmon-fishery history: "A fishery that a century before had supported tens of thousands of native people, as well as the Hudson's Bay Company's export trade, was reduced to a closely watched food fishery which eked out a meagre harvest from the Fraser's shattered runs."

In 1931, Alfred Adams, a Haida commercial fisherman and Anglican lay minister, inspired the formation of the Native Brotherhood of British Columbia, which is today Canada's oldest Native organization. Although a general rather than fisheries body, the Brotherhood soon got involved, along with white fishermen, in bargaining and strikes during the 1930's.[38] As the canners faced hard times in the 1930's, they often found it more efficient to charter or to buy fish from independent boats. More Native people began getting their own boats, despite financial hobbles (for one, banks would not accept a reserve house as loan collateral). In the 1940's, they got many more boats, often taking over craft confiscated from Japanese Canadians.[39]

Jimmy Sewid

Jimmy Sewid, a Native fisherman from the Alert Bay area, became important in the Native Brotherhood, and a leading figure in the B.C. fishery. After fishing with his family and running a seiner for ABC Packing, he got his own boat in 1942.

Mr. Sewid recalled in 1979 that

Jimmy Sewid (National Film Board)

When I was a little boy, there were Indian boatbuilders all along the coast. Every village had two or three seines. They only used the drag-seines way up in the rivers, for food. The fish upriver would be spawning, and thin, so you could dry them faster and keep them longer.

We had big smokehouses. They'd slice the salmon thin. Before you eat it, you cut it into small chunks, soak it, boil it, then eat it. In the spring we got halibut, and did the same thing. Also there was eulachon, the chief food. For us, the eulachon oil is like olive oil to the Italians.

The Indians worked building all the salmon canneries for the companies. The companies would send a packer maybe to Bella Bella to get workers to build someplace in Rivers Inlet. They'd even get people from Vancouver.

There were canneries all over Rivers Inlet, side by side. And right up to Alaska. The Skeena and Nass were full too. They had houses for the workers, half the size of a small hotel room, for a whole Indian family. At the other end were the Japanese, with bigger houses that they built themselves.

My grandfather was one of the first Indian skippers for ABC [Anglo-British Columbia Packers], at Knight Inlet. The canneries had started to find out that Indians could make good skippers, because they knew the area. In the early days, Indians could only gillnet, drag-seine, and troll. Only Japanese and whites could seine. They used to seine off the river mouths. We in our area, my parents, got a line put on to restrict seiners in Bond Sound and Thomson Sound.

I think the restrictions on Indians all disappeared in the 1920's, when the canneries were fighting to get the Indian people. The first Indian seine licence came along when Moses Alfred, my wife's father, got a new boat. Some Japanese people encouraged him to get a seine licence, and they'd buy fish from him. He asked the Fishery Officer, who said no, it was against the rules. He went to the owner of the Nimpkish Hotel at Alert Bay; he had lent the hotel owner money. The hotel owner sent a telegram to Ottawa, he had some contact there, and Ottawa gave permission.

My grandfather was a net man. He used to gillnet in the Fraser; he'd get $2.50 for a 12-hour day. As I boy I fished with him around Alert Bay; we'd start fishing Monday at 6 a.m., and end 7 a.m. Saturday. We'd have a net on a table [turntable], and a small roller. The net would be two or three strips, cotton, with corks and a bit of leadline. The first nets I had cost $1,500. Now they're all nylon and cost $20,000.

The Indians used to fish more halibut and groundfish. I don't know why that changed. ...

My grandfather built the *Annandale*, and it was the first vessel I skippered for a cannery. The Italians were the first to start seining in our area. I began thinking that if the Yugoslavs and Italians could own their own seiners, why couldn't I? So in 1939 I went to the cannery manager and asked if I could buy a vessel from him. He said sure: $500.

I couldn't find any partners. I went to the bank, said I'd put up my house on the Indian reserve. As soon as I said 'Indian reserve,' he said 'no way.' Outside the bank I met a friend, the school principal; he said 'come back in, I'll co-sign.' That was the best thing that ever happened to me.

I always worked hard. I made more sets. I lost two or three nets my first year. I'd just take chances, and I found lots of new grounds. I don't watch the sounder for salmon, hardly. I just watch the kelp, and the tides—and I know where to set. I was always a high-liner among our people.[40]

War reinvigorates industry

In the late 1930's, the industry presented a picture somewhere between stability and shrinkage. Salmon landings and value were relatively level, with the number of canneries slowly dropping, to only 38 in 1940. The processing companies in British Columbia still owned many gillnet and seine vessels. The fleet still included a few sail skiffs. Homer Stevens, who would become a noted organizer of fishermen, recalled later

that in the early 1940's, "ownership was possibly half and half—but with company money into the private half. Trollers were small compared to the seiners, and historically more private. There were a lot of combination boats even then, able to troll and net-fish." But company ownership in B.C. was declining. "It was as if they decided to leave the troubles to the fishermen."[41]

On the Fraser River, landings were still only about a quarter of their level before the Hell's Gate slide. The International Pacific Salmon Fisheries Commission was finally getting into gear. A study starting in 1938

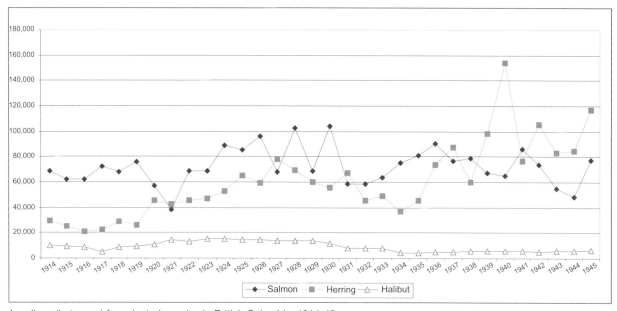

Landings (in tonnes) for selected species in British Columbia, 1914-45.

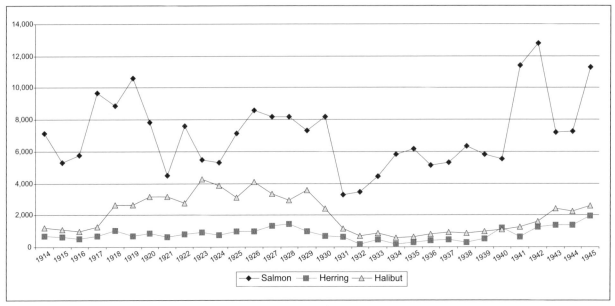

Values ($000's) for selected species, 1914-45. The Second World War brought a major boost.

showed that Hell's Gate, despite clearance work, was still a major obstruction to salmon escaping upriver to spawn. The two governments in 1944 began building the Hell's Gate fishways, completing them in 1946; and they built several other fishways in the Fraser system.

Then, the war boosted demand; and, with the ocean co-operating, a huge spike in landings and values occurred in 1940–1942.

During the war, both Dominion and provincial governments regulated for the maximum production of canned salmon. The federal government forbade the export of most fresh salmon, and discouraged freezing of chums. The province of B.C. refused to license salmon dry-salteries. From 1942 to 1946, the government allocated the B.C. fishery's entire production. About two-thirds of herring production and 80 per cent of the salmon pack went to the armed forces and Great Britain. In 1943, to take an example, 80 per cent of the salmon pack went to the British Ministry of Food and other such agencies, 16 per cent went for sale in Canada, and 4 per cent went to the Canadian Red Cross for prisoners of war.

Wartime controls had various repercussions in British Columbia. In 1940, the defence services took over about 90 boats, including some of the most efficient. Rationing of gasoline sometimes curtailed the range of fishing. In 1941, government got the canners to hold their price for canned salmon, even while fishermen won higher prices from the canners. The high-volume season, however, eased the strain on processors. In 1943, the government extended controls to the fishermen's price. The fishermen protested, and a fishermen–processor–government conference with D.B. Finn in Ottawa worked out a new arrangement that set minimum prices to fishermen, but allowed competitive bargaining to take prices higher. In 1944, with production low, government subsidized the canners.[42]

Government seizes Japanese fleet

The war, which helped most fishermen, brought bitter times for the Japanese Canadian fleet. Late in 1941, with Pearl Harbor freshly in mind and wartime suspicions high, pressure built up among B.C. politicians to do something about the many Japanese Canadian residents. The vocal groups in British Columbia included some white fishermen's organizations.[43]

To some degree, the federal government feared that white British Columbians would attack Japanese Canadians and their communities. The federal government began confiscating the property of some 22,000 people of Japanese descent in B.C., even though most were Canadian citizens. Some senior military and police officers opposed the move.

Ron MacLeod, whose father was a fishery officer, years later described the situation in his area, on the west coast of Vancouver Island:

> At Tofino and Clayoquot there were about 40 white and Japanese trollers, about 40 Indian netters, and 11 seiners. There was a quota on Japanese licences—34 licences. They could only fish in certain areas, and they could only troll. The Indians and whites did about the same in fishing. The Japanese did the best.

A roundup of Japanese vessels, December, 1941. (Library and Archives Canada, PA-134074)

After some big fish buyers pulled out, the Japanese formed their own co-op and had a buying station, open to all fishermen. They were good neighbours.

The Royal Canadian Navy moved into Tofino and other towns, rounding up Japanese and confiscating their boats. All told, Japanese fishermen in British Columbia owned 68 seiners, 120 trollers, 860 gillnetters, 148 packers, and 141 vessels fishing cod and the like. The government sold their boats and compensated them, but at a fraction of their value.[44]

Ron MacLeod again:

> The confiscation was sad. At Tofino, I'd gone to school with the Japanese kids. The Navy came in to take the boats. The Commander stayed with us, my father being the Fishery Officer. The Japanese were parading through the house, showing their documentation and so on. I remember one tall, grizzled fellow down at the end of the dock, tears streaming down his face. He'd been at Tofino 36 or 37 years.[45]

After the war, the Japanese boundaries and other restrictions disappeared. Many Japanese moved back into the fishery, some managing to buy boats, and many others getting modern boats built for them by processing companies, who knew their fishing ability.

Meanwhile, conditions were changing for Chinese plant workers. In earlier decades, cannery operators would contract with an English-speaking Chinese who supplied a gang of workers, and got paid by the case packed. The war, with its great demand for employ-

Unidentified Japanese-Canadian fisherman whose boat has just been confiscated, December, 1941. (Library and Archives Canada, PA-134097)

ment, saw many Chinese find other jobs. The "Chinese contract" system faded away.

Halibut fishery expands

The halibut fishery kept pushing into more distant waters. From 1913–1914, a large winter fishery grew up at Alaska's Aleutian Islands west of Kodiak, and in the Bering Sea. Fishermen would sail 2,000 miles from the south. Diesel-powered schooners became common, hauling longlines directly from the deck rather than using dories. After the railway reached Prince Rupert in 1914, the halibut fishery and "Rupert" grew

215

together. Many Seattle vessels shipped their halibut in bond through Prince Rupert to the United States. Canadian boats generally got one-quarter to one-half of the total halibut catch, which came mostly from Alaskan waters.[46]

Canadian catches rose to the 15,000-tonne range during the war, then slowly trailed off to less than 5,000 tonnes by 1934. As International Pacific Halibut regulations took hold, the catch slowly rose to nearly 7,000 tonnes by 1945. The size of fish also rose.

Fishing halibut

Norwegians, Nova Scotians, and Newfoundlanders dominated the halibut fleet between the wars. Hans Underdahl, a veteran halibut captain, Mrs. Underdahl, and crewman Olaf Bendikson in a 1979 interview recollected the early days:

Mr. Underdahl: Why do so many squareheads fish halibut? Because we're stupid. But fishing is independent. I started in the 1920's with Captain Salvesen on the *Flamingo*. She was 150 or 160 feet. She was a beam trawler, but switched to dory fishing for halibut. Also I fished on the *Sebasa* with Captain Johnson, and on the *Kelly*. In the dories we used four or five hundred fathoms each, in six skates, that would reach about a mile. It took us two or three hours to bait. We used net bags to handle the fish in the dories. The steamer and dory fleet ended in the 1920's because the independent guys got numerous and provided fish cheaper. I had the *Twin Covenant* built in 1927, and sold her in 1963. The companies still had a lot of boats then. The *Cape Beale* started the fishing in the Bering Sea in the 1930's. You could keep the fish on ice 20 days, to get them back.

Mr. Bendikson: Some people came each year to fish, from Norway and the Maritimes.

Mr. Underdahl: We made the discovery of using devilfish for bait. An old guy who lived on an island around Prince Rupert, he'd come up from the States in 1896, told us. Devilfish eat crab; you have to catch them at low water with a hook and poles. It was good exercise. We'd catch devilfish for two or three days, then get a quick trip.

Mrs. Underdahl: We kept it secret a few years—the crew was sworn to secrecy. Boats began to spy on Hans, to find out his fishing grounds.

Mr. Underdahl: There was one place with no fish smaller than 100 pounds. A place no bigger than the yard of this house.

Mr. Bendikson: And another place, you could look down and see hundreds of halibut. I said, "Don't tell them back home in Norway, or they'll all be over."

Mr. Underdahl: I belonged to the Deep Sea Fishermen's Union, and after it got going, to the Prince Rupert Fishermen's Cooperative Association. The Co-op let us find out what the actual market prices were. The Co-op started in salmon, on a small scale, in the early 1930's. Then in 1938 it started a halibut plant. In the war, the government put in a quota: you could buy fish only according to the amount that you bought before the war. Jack Dean [a Co-op founder] got around that regulation; after that, the Co-op handled most of the halibut. That was the turning point.

Mrs. Underdahl: It got so the Co-op had all the boats.

Mr. Bendikson: No, Vancouver was always rebels, out of the Co-op. But a nice big fleet was Co-op. Prince Rupert's full of millionaires from fishing.

Mr. Underdahl: Millionaires from fishing? I don't believe millionaires.

Mr. Bendikson: Well, a lot more made less.

Mr. Underdahl: The Halibut Commission has been a good thing—the backbone of some kind of rule. All over the world, they're ruining fisheries. You need a head, a boss, strict rule.[48]

In 1933, 384 boats and 1,903 fishermen took part in the fishery. But higher catches attracted more and more boats, including salmon trollers and gillnetters looking for off-season work; this resulted in gluts and shorter seasons. On both sides of the border, industry advice led to "lay-up" schemes, to reduce fishing time. In B.C., vessel owners and industry organizations managed the lay-ups. From the early 1930's to 1942, the voluntary program obliged vessels to lay up ten days between trips, and limited the catch of each boat according to the crew size. Fishermen thought the lay-ups and trip limits might help the price, and they appeared to do so. The halibut fishery had already pioneered international quotas; the trip limits appear to have been another first in Canadian sea fisheries.[47]

Quotas appear in herring fishery

By the First World War, herring were becoming the third mainstay of the B.C. fisheries, after salmon and halibut. Fishermen used gillnets, but herring were a high-volume fish, suitable for seining. The depletion of the Fraser after the Hell's Gate slide pulled more salmon fishermen to the west coast of Vancouver Island, where they also found herring. A fishery grew in Barkley Sound. The war boosted demand.

For herring, no one worried overmuch about depletion. One captain reported steaming for one and one-half hours through a single school. The fishery in the 1919–1928 period grew by virtue of dry-salt herring. Although that market dropped off, reduction plants had also appeared; they offered market especially after the decline of the pilchard fishery in the 1930's. Some herring also went for canning by the late 1930's.

Landings, which had reached more than 60,000 tonnes in the late 1920's, dropped during the Great Depression, then picked up in the late 1930's and with the war. The fleet expanded from 17 seiners in 1933–1934 to 44 in 1941–1942.[49]

Albert Tester of the F.R.B. worked on herring migrations, and got fishing vessels to carry logs and enter

information on herring. Some monitoring of spawn deposits also took place. Around 1937, Tester and co-workers of the board carried out exploratory herring seining; an increase followed in seining on the central coast.[50]

About the same time, Tester, the department, and the province co-operated in working out new regulations regarding the length of seines; mesh size; closed times and areas, especially to protect spawners; and quotas. Quotas commenced in 1936, for parts of Vancouver Island, and later covered various independent runs that Tester had identified in various parts of the coast. The department had earlier closed some areas to reduction fishing, for conservation reasons; quotas enabled their re-opening.

Except for the quotas in the lake fishery and those under the Halibut Commission, this was the first systematic use of quotas in Canada. Tester and the department had coupled science and management, although the link was loose. Herring conservation was no great issue. Quota management of herring would last into the 1950's. But by then the department, at industry request, was often extending quotas until they lost all meaning.

Pilchard fishery grows and fades

At the outset of the 1914–1945 period, an important fishery was developing for another herring-like species. The pilchard, or California sardine, supported a fishery both off the U.S. coast and, at the northern edge of its migrations, off British Columbia.

Canning of pilchards began in 1917. The B.C. catch increased through the 1920's as larger seine vessels became more common. In 1925, the industry began using pilchards for reduction: cooking, pressing, and drying the fish to produce fish-meal or fertilizer, as well as oil.[51] Soon B.C. reduction fisheries, sustained by pilchards, were moving ahead on an industrial scale, with several factories opening and cutting into the markets of smaller Atlantic fish-meal producers.

Pilchard catches climbed from, on average, 2,600 tonnes in 1916–1920 to 65,000 tonnes in 1926–1930. But pilchards off B.C. were at the margin of their range, far from their California spawning grounds, and there was little regulation in either Canada or the United States. Ron MacLeod, who grew up on the west coast of Vancouver Island, recalled that

> [i]n the 1930's there'd be a fleet of pilchard boats off Tofino in the summer—about 30 seiners and about the same number of packers [carrier vessels]. When there was a westerly wind they couldn't work, and they'd come into Tofino. We kids would go aboard, meet the Yugoslavs and so on. That fleet was separate from the salmon fleet. Some of the boats were big, as long as 72 feet; later some got to 85 feet. They'd fish pilchards in the summer, herring in the winter.

Purse seining at Quathiaski Cove, Campbell River, B.C., 1930. (Library and Archives Canada, PA-40979)

The famous lumberman Gordon Gibson, B.C.'s "Bull of the Woods," worked for years in the fishing industry. He recalled in his memoirs that pilchard fishing first took place in inside waters, moving offshore in 1928:

> The boldest skippers of our group ventured into open seas, beginning at the mouth of the inlet and then fishing progressively farther into the open ocean to get bigger catches of pilchards. ... In October 1931, just after the equinox, came a solid seething mass of pilchards about two or three miles wide ranging from one end of the coast to the other. There were literally millions and millions of tons of fish and every boat in the area loaded up.

But the pilchards would soon play out. Mr. Gibson in his book described the first big failure:

> There were a couple of hundred boats with 1,000 or more crew aboard as well as 400 men at thirty-five fish reduction plants scattered along the coast from Barkley to Quatsino Sound. The government sent out its fishing patrol boat, the *Givenchy*, with six skippers including myself, to locate the pilchards. We ranged as far south as the Columbia River, zigzagging back and forth, making about a 1,000-mile run searching for the fish. We didn't find them.[52]

J.L. Hart of the Biological Board had predicted the problems afflicting Gibson and other producers. Pilchards were in for a long decline. Wide fluctuations began hitting the fishery, with catch failures in 1933 and 1939. Reduction plants began to close. Average catches fell to 38,000 tonnes, and less than that in 1936–1940. After a recovery to an average 5,600 tonnes in 1941–1945, the fishery dwindled to an average 2,000 tonnes in 1946–1950. In following decades there was no pilchard fishery in B.C.

Scientists concluded that a combination of overfishing and environmental changes destroyed the fishery. Heavy exploitation of the pilchards may have let California anchovy take over the ecological niche.[53] However, after a long drought, the 1990's would see a modest revival of the pilchard fishery in British Columbia.

Herring fishery expands

As the pilchard fishery faded, herring picked up, first to feed the reduction industry, then to feed wartime demand. Landings reached more than 150,000 tonnes in 1940, then fell back to the 100,000 range.

Great Britain in particular needed inexpensive, nutritious canned food. In 1941, the provincial government refused to license dry-salteries, in favour of canned herring. In any case, even before Pearl Harbor, the Japanese market had dwindled because of curren-

cy difficulties. Post-war, the dry-salt industry never recovered, and pickling of herring also died out.

The canned-herring pack rose more than 60-fold, from 23,000 cases in 1938–1939 to 1.5 million cases in 1941–1942. Most of it went to the British Ministry of Food. But the quality of the canned herring was suspect. As a major processor of the times noted 35 years later: "They used to say more people in Europe died from the herring than were shot."[54] Canning fell off after the war. But the reduction industry would keep growing.

The war stimulated other fisheries, including for dogfish and shark, the innards of the latter being useful in the manufacture of airplane wings.[55]

Just as the war started in 1939, B.C. fishermen made their first commercial catches of albacore tuna. This fishery would become an intermittent one, with tuna migrating into B.C. waters every few years. Meanwhile, to satisfy demand, B.C. canners in 1948 began importing and packing Japanese tuna.[56]

1914–1945: Management in retreat

In the 1914–1945 period, apart from the extraordinary circumstances of the two world wars, the picture was mainly grim. Both coasts had fishery difficulties; neither showed much inspiration in management, which was mainly reactive.

After Prince and the turn-of-the-century royal commissions wrote the policies and regulations, the department had settled down to enforce them, with inconsistent success. It seemed to plod along from problem to problem, showing little creative force. The department never made a concerted attack on chronic problems such as, on the Atlantic, poor quality and general disjointedness.

The splitting of the Biological Board from the department perhaps stultified the application of science to management, weakening the "critical mass" for thoughtful regulation. Few if any people on the department side were paid to think and analyze. The freshwater fisheries time and again foreshadowed sea-fishery developments and management methods; these early warnings generally went ignored. Little study went to basic questions, such as fishery volume and value and the incomes of fishermen. Nobody measured the impact of regulations. Nobody measured the effect of hatcheries, until Foerster looked at their costs and benefits, half a century after they began. Nobody studied the exact effects of different technologies on jobs and industry efficiency. Initiatives tended to come from outside the department: from royal commissions on the Atlantic, or industry pressures on the Pacific.

On the Atlantic, the industry was falling behind the New England and European competitors. The chief new and vigorous elements were the trawler and the fresh-and-frozen fish industry. The main government intervention, following the 1927–1928 royal commission, was to cut short this development.

The same 1927–1928 royal commission backed co-operatives, which gave fishermen some hope. Still, during the Great Depression, many Atlantic fishermen were desperate, and the industry was weak. The department stuck mainly to policing inshore fisheries, with little attention to strengthening the overall situation. Could and should it have done more?

The department's posture was natural enough. Conservation was its main job, and there was no great conservation problem in the Atlantic sea fisheries. On other matters affecting the lives of fishermen, one would naturally look to the political side, not the department, for leadership and change. As for conservation itself, even if regulations sometimes seemed dubious in either their basis or their enforcement, still the fisheries branch must have been doing something right. There were fewer complaints about disappearing fisheries than in Whitcher's and Prince's time.

Although some studies called for thorough intervention to streamline the industry's producing and marketing system, no powerful politicians seemed to have similar views (except Coaker in Newfoundland). Industry interventions can be highly dangerous. Besides, the fishery was serving another purpose by providing jobs, even if marginal, to all comers.

One can thus explain away the department's interwar lack of initiative on the Atlantic. Still, compared with preceding and following decades, it seems a flat period in management.

If neither coast had a joyous time during the interwar years, still British Columbia did better. Fishermen and processors made more money, were better organized, and particularly in the case of processors, had more influence on management. The industry itself pushed initiatives such as area licensing and inspection laboratories. The Department of Fisheries was a national organization, with no major variation in its regional arms. But different regional circumstances produced a somewhat different approach in management.

One B.C. issue was too much for the department. The central conundrum of fisheries management, from today's point of view, is matching the number of participants and their fishing power to the size of the resource, for the best benefits to society. From the 1880's to the First World War, the federal fisheries branch in British Columbia had led the efforts to control participation. From the First World War to the Second, the government backed off. In acceding to demands to open the north coast to settlement, it managed to mistreat both Japanese Canadians and Native people and to abandon a prime tool of management—licence limitation—which would take decades to restore. Even though one can understand the department's actions, still the loss of licensing control was emblematic of an unimaginative period in federal fishery management.

PART 4: 1945–1968
THE AGE OF DEVELOPMENT
CHAPTER 17.
National and international events, 1945–1968

In the years 1945–1968, the fishery went through what was almost a delayed industrial revolution. It began with new visions of development, encouraged by government; it ended with fears emerging about conservation, as the department worried about bringing fleets under control.

The fishery changes formed part of a bigger transformation. In 1945, on the Bay of Fundy island where I grew up, few people had cars to travel the dirt roads. There was no electricity. We would hand-pump or carry water from the well, and use outdoor toilets. Chopping wood, oiling and cleaning lamps, and feeding the hens all took time daily. Families would salt fish, put up preserves, lay in potatoes and root vegetables for the winter, and often run credit at the grocery store until fish came in the spring. Telephones were fairly new and were party-line, with people frequently listening in. Some entertainment came from battery radios; most was self-generated, from weekly dances, card games, parties, and conversation at the stores and post office, or on the road and at the wharf. Medical care was distant, and no one dreamed of Medicare. The school in my village went to Grade Four. For Grades Five to Eight, you took a bus to another village; for high school, you either took a daily boat to the mainland or moved away. Few young people thought of university or other post-secondary education.

Most fishermen used small, open boats to handline groundfish or tend lobster traps and herring weirs. The occasional small boat still used sails. Only a few boats had an under-deck hold of any size, although many had an awning over the steering wheel and engine-box, and a cuddy forward. Fishermen often did other work such as cutting wood in the winter. There was no unemployment insurance for fishermen. Some people managed to find city work in winter and came home summers to fish. Other families moved away permanently.

By 1968, almost every family on our island had a vehicle for the paved roads, with cars now displacing the pickup trucks that had dominated in the 1950's. After 1966, Medicare made it easy to see a doctor. A bridge now connected us to the mainland. The houses had electricity, hot and cold running water, and bathrooms. Radio and television were displacing home-made entertainment. You could dial long distance on your private phone line. There was far more cash around, from fishing itself, family allowances, unemployment insurance, and old-age pensions. Most students were getting at least some high school, and a fair number were going on to post-secondary education. The three villages on the island had lost a bit of community closeness, and people often faced worrisome debts along with the higher incomes; but overall, we were better off.

By this time, few fishermen cut wood in winter. Fishermen could draw unemployment insurance from New Year's Day to mid-April. Although many small boats remained, now we also had powerful, sometimes far-ranging boats with diesel engines, radios, sonar, radar, hydraulic systems, haulers and power blocks, and nylon ropes and nets. Boats were more expensive, but you could probably get a construction subsidy from the federal government and a loan from the province, and perhaps also a loan from a processing plant.

Many fishermen were optimistic, although some worried that "nowadays the fish don't have a chance." Mostly, they watched for new developments and tried to get in on them, partly for the money, partly for the pride that came with a better boat and bigger catches.

The same pattern was appearing in hundreds of other communities. The biggest fishery, groundfish, led the changes in the industry. From salted fish for European, Caribbean, and Latin American markets, the emphasis switched to frozen blocks, fillets, and then fish sticks for North America. New draggers and trawlers, although costly, could catch with great efficiency, and provided year-round fish to plants. In the post-war "cold chain," trucks transported chilled or frozen fish to cold-storage warehouses and stores, whence customers took them to freezers in the home.

Pelagic fisheries, mainly herring, boomed in some areas. Shellfish expanded with new scallop grounds and, by the late 1960's, with new fisheries for crab and shrimp. While some old trades fell off, including smoked, salted, and kippered fish, new opportunities arose. Although the fishery is never without problems, the 1945–1968 period on the Atlantic was a time of relative optimism.

The Pacific also saw growth in groundfish and other species. But it took its main hopes from the salmon fishery, where remedial work on the Fraser River, higher prices, and the federal department's modernized management corps promised new efficiency. On both coasts, foreign fleets were lurking over the horizon; but they became an alarming threat only towards the end of the period.

Walter Scott, an alumnus of the federal fisheries department, around 1980 illustrated the chief types of Atlantic fishing craft, as they had developed in the post-war period. The drawings above show common types in the groundfish fishery, where open boats became fewer. Especially in the Maritimes, smaller boats now tended to have a wheelhouse, like the handlining or jigging craft shown above (top left). Many boats used longlines (top right). Gillnetters (middle left) became more popular post-war, particularly in Newfoundland, where cod-traps (middle right) also remained numerous. The most dynamic trend was the rising power of draggers and trawlers, culminating in large, company-owned otter trawlers (bottom).

Sinclair stands out among ministers

James Sinclair

During the 1945–1968 period, everybody believed in progress. The federal Liberal administrations in the 1940's and 1950's gave major assistance to education for veterans and others, while increasing pensions and providing hospital insurance. The Conservatives, in power for 1957–1963, developed national programs such as ARDA (the Agricultural and Rural Development Act, in 1961) and the Atlantic Development Board (1962) to help farmers and rural regions. Provincial governments sought growth through industrial developments, hydro-electric projects, or whatever they could lay their hand to. Productivity and development became the war cries. Officialdom saw the Atlantic fishery as a hard-working but backward, low-paying, often unstable industry, which needed to gear up and exploit the resources at its disposal.

In the immediate post-war period, federal fisheries ministers changed frequently. Under the Mackenzie King government, H.F. Bridges, a New Brunswick lawyer and teacher, headed the Department of Fisheries from late 1945 until his death in August 1947. Milton F. Gregg, a winner of the Victoria Cross in the First World War and a brigadier general, took over for five months before moving to Veterans Affairs. James A. MacKinnon, member of Parliament (M.P.) for Edmonton West, served as minister from January to June 1948, then went to Mines and Resources. Robert Mayhew, M.P. for Victoria, became minister under King and then Louis Saint Laurent until August 1952, when he became Canada's ambassador to Japan.

Many of these ministers were well esteemed, and they always reviewed important issues. But the first to leave a major mark was James Sinclair, who held office in 1952–1957. Sinclair himself remarked, "[B]efore me, they were mostly there for a year or two—young ones on their way up or old ones on their way down." Sinclair proved a dynamic minister, visiting the coasts, going out on fishing boats, tackling issues head-on. Development work speeded up; several major programs, including fishermen's loans, boat insurance, and unemployment insurance, came into force. Sinclair lingered in departmental memory as an impressive minister, able to deal with fishery issues and perform brilliantly in Parliament without breaking stride. He won respect from fishermen and processors alike. After leaving politics, Sinclair served as president of the Fisheries Association of British Columbia from 1958 to 1961.

When John Diefenbaker's Conservatives took over in 1957, J. Angus MacLean of Prince Edward Island, a war hero, became minister. In his career, which included later service as premier of P.E.I., MacLean was considered a man of great integrity. He was also, as federal fisheries minister, more than cautious. To quote a senior official of the day, "He was a fine gentleman who made one decision: that he would make no decisions."[1]

With the Liberal return in 1963, Hédard Robichaud of New Brunswick took over until 1967. Robichaud's family had been in the fishery; he himself had worked in fisheries for both the provincial and the federal government, and he proved a capable minister.

Stewart Bates stamps departmental mindset

Stewart Bates

In the immediate post-war period, ministers were changing frequently. With the partial exception of British Columbia, fishermen had no sustained and

223

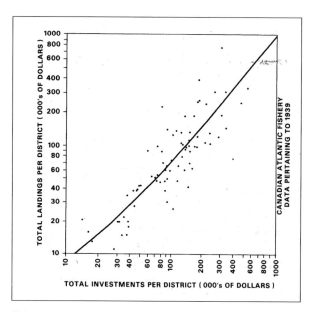

This graph, redrawn from Bates's report on the Atlantic fishery, shows the relation between landings and investment in boats and gear, for the Maritimes and Quebec. The implication was, more investment brings more fish.

concerted influence. Processors were better organized with a national council, but concerned themselves mainly with short-term issues. Nor was there any strong attention to fisheries in the academic world. It was left to department officials to do the main thinking and acting on fisheries issues.

Stewart Bates, an economist and university professor of Scottish origin, became deputy minister in January 1947 and served until December 1954. More than anyone else, Bates set a strong stamp on the post-war department, fostering a mindset of growth and expansion that lasted until the late 1960's.

The deputy minister impressed everyone he dealt with. Persons as diverse as Homer Stevens, a Pacific fishermen's leader and staunch Communist, and the leaders of the company-oriented Fisheries Council of Canada (F.C.C.) spoke admiringly of Bates. The departmental economist W.C. MacKenzie later said that "Bates was even more brilliant intellectually than Finn; and more brilliant than any of his successors. Some of them were extremely able men; but they didn't have the sparkle within that Bates had." Alfred Needler, himself deputy minister in 1963–1971, said that "Bates was the best of the lot, in both administration and ideas."[2]

Bates had already amassed considerable fisheries knowledge. While still a professor of commerce at Dalhousie University, he became consulting economist for the federal Salt Fish Board, set up in 1939. In 1942, he began a three-year stint as special assistant to federal fisheries deputy minister D.B. Finn, and he bore much of the responsibility for wartime administration of fish supplies. During the same period, he wrote *Report on the Canadian Atlantic Sea-Fishery* for

Nova Scotia's royal commission on provincial development and rehabilitation.

Bates's report described a generally poor and backward industry. For fishermen in 1939, after expenses "the net income remaining for living was inadequate, no matter how it is estimated." Bates wanted "a fuller use of the fishery resource" of the Maritimes, with operations better integrated from sea to marketplace. "This revolution can be achieved only by new methods that will increase its efficiency, and its productivity per man." He noted that "on this matter the industry itself is unlikely to voice any definite policy, both fishermen and firms being individualistic in outlook, and being inclined to regard any imposed action that requires a change in their traditional ways as 'interference' or 'regimentation'."

British Columbia and the United States were already using fishermen as "net-minders with machinery," Bates said. But eastern Canada, as in the trawler controversy, had tried to keep men at work rather than to modernize. This was trying to turn back history. The times demanded boldness for expansion. Government should subsidize prices after the war to keep them at war levels. It should provide aid for boats and plants, whether by loans, tax concessions, or subsidies. Frozen-fish plants should perhaps receive aid; these, with an assured supply of fish from draggers, would speed the transformation of the whole industry. Fish landings needed grading, inspection, and proper care, with the first grade going to fresh-frozen, the second to salt, and so on. The fish trade needed to organize itself for information, for advertising, for collaborating on research, for dealing with government. "Experimentation in organization and in selling fish is just as urgent as it is in the catching of fish." The industry needed to get closer to consumers, and needed more collaboration among firms in marketing.[3]

Once in charge, Bates re-oriented the department into promoting change, and linked it more closely with science and the Canadian industry. Bates helped unleash productive forces that would transform old fisheries and develop new ones.

Bates strengthens the organization

Since W.A. Found's time, departmental headquarters had revolved around the Eastern and Western fisheries divisions. In the early 1940's, offices took up a floor or two of one wing in the West Block of the Parliament Buildings; staff numbered about 60. A single man with two or three assistants handled personnel, finance, and the like. The work week was five and a half days.[4] Although some staff dealt with fish culture, licences, inspection, and publicity, the main work went to conservation and protection: that is, overseeing regional regulations and enforcement on size limits, seasons, and so forth, and dealing with those questions that floated up to headquarters.

In the regions, the regular department staff in the "Outside Service," exclusive of guardians and patrol boat crews, numbered 140 or so. Most were fishery officers. (The official title had been Fisheries Overseer until 1929, when it changed to Fisheries Inspector. But often they were referred to as fishery officers, and in the early 1950's Fishery Officer became the official title. At the same time, fishery officers got their first official uniforms.)[5]

One officer might use three or four seasonal patrolmen and guardians. Most protection work went towards shellfish and salmon on the Atlantic, and to salmon and herring on the Pacific. Besides enforcement, regional staff did a limited amount of product-inspection work. Fish-culture activities included clearing streams and maintaining fishways and, on the Atlantic, the remaining hatcheries. There were no scientists, economists, or technologists in the regions, except those under the Fisheries Research Board (F.R.B.).

Bates strengthened the department in every respect. He recruited Alfred Needler in 1948 to serve as assistant deputy minister (A.D.M.) and to help set up the new organization. When Needler in 1950 returned full time to the St. Andrews Biological Station, George Clark, recruited from the Pacific fishing industry, became A.D.M. Clark succeeded Bates as deputy minister in 1954, holding the post until 1963. Clark was active and practical, and pleased many industry people with his direct and forthright dealings. He also busied himself on the intergovernmental scene, both internationally and in Canada, where he fostered the Federal–Provincial Atlantic Fisheries Committee, created in 1958.[6] But it was his predecessor Bates who most marked the period.

In December 1949, following discussions led by Bates and minister Mayhew with representatives of the different fishing-industry sectors, the department announced ambitious plans to apply more science, biological engineering, economics, inspection, capital assistance, and international effort to its work. Headquarters would change from its Eastern–Western set-up to a "functional" basis. Its "Services" would now include Conservation and Development (incorporating a revitalized Fish Culture branch), Inspection and Consumer, and Markets and Economics. Other headquarters sections included Administration, Information and Educational, Legal, and a Newfoundland Service for the new province (this latter would later be absorbed in the general system).

Under Bates, the economics section would build up analytical power to a degree seldom equalled since, with figures like Ian MacArthur, W.C. MacKenzie, and H. Scott Gordon in the forefront. In the 1950's, a separate industrial development service would develop great strength, and the inspection service would greatly expand its mandate.

Work on marketing, economics, information, and even industrial development took place mainly from Ottawa. In the regions, besides conservation and inspection, fish culture remained important. Such work now increased in British Columbia. On the Atlantic there were still 13 hatcheries and additional rearing stations, retaining ponds, and egg-collection sites. Projects on salmon, oyster and clam culture, and other species were going forward on both coasts, often in co-operation with the F.R.B. The department was also working with the board on exploratory fishing.

Regional headquarters were supposed under the new set-up to get more power for "line" activities. Heads of branches in the regions were responsible to regional supervisors for operations (what they actually did) and to Ottawa for effectiveness (how well they did it) Contemporaries had divided views on whether the regions gained or lost power.

Part of Bates's new approach was an increase in "educational programs," which meant mainly publicity and demonstrations to increase fish consumption. The department set up an experimental test kitchen in the West Block of the Parliament Buildings, mounted exhibits and demonstrations, and distributed fish-cookery information across the country.[7]

As for more general information, reports, studies, and government–industry conferences became more common. The information branch under Tommy Turner in 1948 began putting out a small monthly magazine, *Trade News*, to many fishermen, processors, and other interested parties. (The magazine in 1966 got renamed *Fisheries of Canada*, the name it kept until its demise, in 1971.) The department also, from 1946 to 1969, published *The Canadian Fish Culturist*, on fish breeding and related work. Fisheries Research Board scientists and department officials used the publications to spread information, flag new developments, and talk about their work. Press releases became more common, and the department supported "Fishermen's Broadcasts" on the Canadian Broadcasting Corporation.[8]

Although these efforts had some impact, there was no sustained effort to inform, educate, or organize fishermen. Education was in any case a provincial field, and high schools were spreading fast after the war. When it came to spreading information on new fisheries or gear, personal contacts by officials and pilot projects could do the trick.

The increased activity came at a cost. From the early 1930's until 1945, departmental expenditures ranged from $1.6 to $3 million annually. By 1955, they came to $10 million; by 1968, $51 million.[9]

Fisheries Research Board broadens activity

The post-war age brought new efforts in science. The Fisheries Research Board was still its own entity, but it reported through the fisheries minister. The Board of Directors included some industry representatives, although university professors dominated. The F.R.B. had no full-time chairman until J.L. Kask took on that role in 1953, and "forged it into an integrated national organization."[10] The board under Kask acquired two large research trawlers: the *A.T. Cameron*, built in 1958, for the Atlantic; and the *G.B. Reed*, built in 1963, for the Pacific.

F.R.B. research in the Northwest Territories included studies at Great Slave Lake. Photo shows winter fishing there in 1952. (Photo by R. Wheaton)

The new research trawler *A.T. Cameron* cruising past Montreal. (Photo: National Film Board)

At the biological stations in St. Andrews and Nanaimo, some directors and scientists sought closer contact with the fishing industry. The technological stations at Halifax and Vancouver were already working with processors. (The Pacific station had moved from Prince Rupert to Vancouver in 1942.) The Grande Rivière, Quebec, station burnt down in 1943, but reopened and would keep operating until 1969. And in 1947, the board sponsored a headquarters for Eastern Arctic Investigations at McGill University; this later evolved into the Arctic Biological Station at Sainte-Anne-de-Bellevue, on Montreal Island.

Meanwhile, the F.R.B. in 1944 set up the Central Fisheries Research Station in Winnipeg. That same year, the F.R.B. started research in the Northwest Territories. The research station moved to London, Ontario, in 1957; it became the centre for sea-lamprey control in the Great Lakes, as well as other research. The station went back to Winnipeg as the Freshwater Institute in 1966.

Alfred Needler, the director at St. Andrews from 1941 to 1954, was one who encouraged on-the-ground work. Needler considered arguments about pure versus applied science a false distinction; if researchers pursued the interests of the fishing industry deeply and intelligently, science would benefit.

Needler later recalled that when he arrived at the St. Andrews Biological Station, "the budget was about $55,000; when I left in 1954, it was more than ten times as much." In the 1930's there had been only a few full-time scientists. But the war brought money. "Fish, for example, lobsters, helped buy guns." Interest continued after the war, under Bates.

Needler noted that "when the Board started, mainly the work came out of the scientists' own initiatives. In the 1930's, we were unsure if we were appreciated or wanted by the department. The 1940's were a turning point. By the 1950's, the F.R.B. couldn't keep up with the questions the department was asking."[11]

Board scientists and station directors still had a large amount of freedom. Some scientists pursued fishery questions of direct utility; others went where scientific curiosity led them. To work with industry, Kask had wanted a development branch within the F.R.B.; instead the department's Industrial Development Service (I.D.S.) came into being. The two groups often co-operated, particularly in testing new fisheries.

On the technological side, the Halifax lab after the war became an early leader in developing fish protein concentrate (F.P.C.). This was a dry, easily preserved powder intended to be incorporated into other foods as a source of protein, to fight malnutrition. A 1967 conference in Ottawa drew many international representatives to discuss processing and marketing. F.P.C. in various forms found some use in following decades, but because of problems of cost and acceptance, never lived up to early expectations.

The remarkable Alfred Needler

In the 1950's and 1960's, a great crop of scientists and managers came to the fore: economists like H. Scott Gordon and Bill MacKenzie, scientists like William E. Ricker and Wilfred Templeman, managers like Joe Whitmore and Cliff Levelton, and many more. Even in such company, Alfred Needler stood out.

Needler joined the St. Andrews station in 1924 as a student biologist. He did important work on haddock and other groundfish, helping to delineate stock boundaries. In the 1930's, Needler took charge of oyster investigations at Ellerslie, P.E.I., where he moved oyster culture forward. As chief zoologist at the St. Andrews Biological Station from 1939, and director from 1941, he turned research more towards work of practical benefit to the industry. In the years 1948–1950, Needler did double duty, serving as director for St. Andrews and as A.D.M. in Ottawa. In 1954, he became director of the Pacific Biological Station at Nanaimo and in 1963–1971 served as deputy minister of fisheries. Needler could communicate well with staff at all levels.

During the 1950's and 1960's, and for much of the 1970's, Needler represented Canada in international negotiations affecting both coasts. On the Atlantic, he

Alfred Needler

helped to strengthen international management and make Canada the leading force in the International Commission for the Northwest Atlantic Fisheries. That work helped prepare the ground for negotiation of the 200-mile limit. In Pacific salmon negotiations, Needler so frequently out-maneuvered the Americans that it became their burning desire to defeat him just once.

As deputy minister, Needler built bridges to provincial governments, giving impetus to such mechanisms as the Federal–Provincial Atlantic Fisheries Committee. He oversaw a huge development effort. Needler's tenure also saw the start of licence limitation in lobster and Pacific salmon; limited entry would become the cornerstone of modern management. After leaving his deputy minister post, Needler continued to lead various international negotiations.

He also served as executive director of the Huntsman Marine Laboratory in St. Andrews. When Needler retired from that position in 1976, fisheries minister Roméo LeBlanc noted that Needler's influence showed up internationally "like a watermark on fine paper." David Wilder, another renowned scientist, pointed out that Needler took an enormous hat size; "he needed that huge skull to keep all the brains from bursting out."[12] Needler helped to civilize the seas and to modernize Canada's fisheries, and usually made it look easy.

Science tackles stock assessment

As biological understanding of the various fish species continued to grow, the Fisheries Research Board also got involved in questions of population: How many fish were there? And how could you best count them?

Russian and European scientists had already explored population dynamics, developing basic models of how fish populations behave under commercial exploitation. Canadian work had supplemented their findings. By now scientists were thinking in terms of "stocks": distinct populations of a species, with their own locations and characteristics. A fundamental insight was that any exploitation would reduce a stock; but that didn't necessarily mean danger. A virgin stock

subjected to normal fishing would simply drop back to a lower but stable level. Like trees or hayfields, fish stocks needed cropping to yield product and make room for new growth. Maximum sustainable yield (M.S.Y.)—getting the most from a stock without inflicting damage—became the guiding idea.

Scientists developed mathematical models that generally fell into two groups. The first tried to get an idea of the general production of the whole biomass and its response to fishing effort. The second broke down the stock by individual year-classes, assessed the numbers and volume of each, and added up the results. The second approach came to dominate, but demanded large amounts of statistical data from catches and other sources. The number of fish caught from the different year-classes by commercial and research vessels, together with educated assumptions and many calculations, shed light on the number of fish remaining in the water.

The F.R.B. produced a great pioneer in population dynamics. In publications from 1954 to 1958, William E. Ricker outlined practical means to compute fish populations. "In fisheries biology, the formal study of stock and recruitment started with Ricker's [1954] paper."[13] Beverton and Holt in Britain and other researchers further elaborated theories and methods of estimating populations. But stock assessments still remained, in Ricker's words, as much an art as a science, yielding only loose approximations of reality. In future years, fishery managers would sometimes give scientific assessments more credit for accuracy than they deserved.

In the 1940's and 1950's, there was little concern about overfishing. Scientists shared the chronic optimism that has historically infested the fishery. A.G. Huntsman, the grand old man of the F.R.B., wrote in 1943 that "'depletion,' except as an uneconomic condition which may exist when fishing starts, is little more than a bogie to frighten the credulous."[14]

Frank McCracken, a well-respected F.R.B. scientist who made it his job to work closely with the industry, said that everyone in the 1940's and 1950's tended to dismiss old tales of historical abundance—"walking across streams on the backs of the fish"—and of their depletion through overfishing; there was little scientific data to support the anecdotes. Besides, in the postwar period it seemed possible to catch a lot more fish. In 1961, a resource analysis overseen by Ricker found important potential to increase catches on both coasts, including a doubling for some species of Pacific salmon.[15]

By the latter 1960's, however, more concern was emerging about overfishing, especially after Pacific herring stocks collapsed in 1967. Meanwhile, some additional attention was going to the ecological interplay among species and environment. In 1965, the F.R.B. opened the Marine Ecology Laboratory, housed at the Bedford Institute of Oceanography in Dartmouth, Nova Scotia, then operated by the Department of Mines and Resources.

Under F.R. Hayes, chairman of the Fisheries Research Board from 1964 to 1969, the board stressed fundamental research and co-operation with universities. Some long-standing work of more practical orientation got downplayed or even dropped. The department meanwhile was building up a semi-scientific arm of its own. The fish culture branch was active on both coasts, doing research with a view to immediate application on practical problems. In 1965, this metamorphosed into the resource development branch, working on salmon, trout, and shellfish. This branch by the early 1970's intended to broaden its activities further. But to some in government, the research situation seemed unco-ordinated. Pressure for change was mounting; the department would eventually take over the F.R.B.[16]

Economic understanding builds up

As biological knowledge grew, so did economic understanding of the industry. The department's economics section undertook more analyses of the industry and prepared more reports on markets. A new effort went towards producing reliable statistics, for both economic and biological purposes. The department had long relied on figures that fishery officers collected from plants and buyers. These data were weak, and showed nothing about the offshore location of catches.

The problem was an old one. In the late 19th century, the far-sighted official Andrew Gordon of the Fisheries Protection Service had suggested using a statistical grid for noting catch locations. Nothing developed at the time, perhaps because of logistical difficulties: the Maritimes had hundreds of ports with inshore fleets and more than 80 ports that sent vessels on long trips to grounds further offshore. In the 1920's the North American Committee on Fishery Investigations (N.A.C.F.I.) pressed the Canadian government for better data, but results were few.

From 1931, N.A.C.F.I. (now a "council" rather than a committee) set up a grid system for the northwest Atlantic, defining broad sub-areas to reflect the main ocean boundaries affecting fish populations and migrations. N.A.C.F.I. kept revising its grid in the 1930's as the Biological Board of Canada did more research. For example, Needler established that southwest New Brunswick haddock were related to those on Georges Bank, and Nova Scotia haddock, to Brown's Bank. This resulted in a sub-area boundary line drawn up the middle of the Bay of Fundy. In 1938–1940, R.A. MacKenzie collected data on catch origins from trawler and schooner captains. Such efforts clarified some of the offshore picture.[17]

Still, statistics remained weak. N.A.C.F.I. closed down in 1938. After the war, the new International Commission for the Northwest Atlantic Fisheries (I.C.N.A.F.) adopted the N.A.C.F.I. areas as the basis for its own sub-areas and grid system. Partly motivated by I.C.N.A.F., the F.R.B. and the department made new efforts to fill out the picture. At the St. Andrews station, W.R. Martin led the data work.

Frank McCracken, a young scientist at the time, recalled, "[W]e had port technicians collecting biological samples, and we used them to get statistics as well. We got 80 or 90 logbooks out aboard practically all sizes of vessels, although the rate of return was less for small vessels." Much depended on the individual fisherman or the local fishery officer. "Once we had offshore statistics, we could figure out stock distributions. We also did a lot of tagging, to find out where the stocks were and what percentage got taken. The whole effort represented an enormous increase in knowledge."[18]

In 1951–1952, the department changed its basic system. Instead of fishery officers getting numbers from plants and making monthly reports, the buyer would fill out a purchase slip when the fish came to the dock. The fishery officer, however, was often still involved, checking sales slips, keeping catch records, and verifying statistics for the economics branch.[19]

Defects persisted in the system. Purchase slips were more voluntary than mandatory. Buyers might fill out only part of the purchase slip. Fishermen might treat logs cavalierly, and give imprecise information on catch areas. Fishermen generally got no guarantee of confidentiality that would protect them from prosecution for a fishery offence recorded in their slips or logs; this affected the quality of information. Gaps existed in the herring fishery and in some smaller ones such as swordfish and shad. Still, the 1950's changes produced far more statistical information, and scientists began to rely on it, supplemented by other data, for stock assessments.

(The statistical system for catches would evolve further. A 1982 report described the intricate operation then prevailing in the southern Maritimes. The purchase slip recorded the captain, the boat, the I.C.N.A.F. division or lobster district, the buyer's name, and the quantities and values of different species. Vessels more than 14 metres in length or 25.5 gross tons in volume maintained a fishing log giving details on types of gear, fishing time, discards, and so on. This was an outgrowth of the F.R.B.'s voluntary system; the logs had become compulsory for larger vessels in the early 1970's.[20] Smaller vessels in some fisheries such as salmon, crab, and tuna also maintained logs. Various officials—fishery or inspection officers, statistical coordinators, or other personnel—collected the data, often filling in missing items. It took the equivalent of 18 people working full time to cover the area westward from Cape Breton along the Nova Scotia coast and on both sides of the Bay of Fundy, including 400 landing ports, 330 plants and buyers, and 6,700 craft.)[21]

Research by the economics service surprised officials by revealing, especially in the 1960's, a higher-than-expected amount of part-time and casual work in the fishery; this pattern seems to have increased with the availability of other work. As well, the department instituted costs-and-earnings studies on vessels, on a sample basis. These went forward from 1950 to the late 1960's.[22]

Gordon clarifies common-property mechanisms

On the analytical side, the economist H. Scott Gordon, a Bates recruit, issued a warning about overcrowded fisheries that remains famous in fisheries literature. In the late 1940's Gordon had moved from the Fisheries Prices Support Board to the academic world, but still did frequent work for the department.

Gordon's 1952 paper on the Prince Edward Island fishery, noted in Chapter 13, foreshadowed his ideas. From 1929 to 1939, the number of fishermen on Prince Edward Island had increased by more than 50 per cent, as desperate people indebted themselves for boats and gear. But the lobster catch, already at its limit, began to fall; ultimately, net incomes probably declined by at least 50 per cent. "By the fundamental nature of the case, an increase in the numbers of fishermen could only reduce the average yield per man."

In the 1952 paper, Gordon stopped short of recommending limits on entry to the P.E.I. fishery. He took a stronger tone in his famous 1954 paper, "The Economic Theory of a Common-Property Resource: The Fishery." Gordon pointed out that what counted was the number of fish per person. The idea of an inexhaustible sea had dominated until late in the 19th century, and still had advocates. But mankind, like any other predator, reduced the stock of fish. Fishing at an increased level would drive a fish population back to a new equilibrium, at a lower level of abundance.

Gordon wrote that "fisheries management" (a relatively new term) had focussed on attaining the largest sustainable catch (or M.S.Y., as it was coming to be termed), but neglected "the inputs of other factors of production which are used up in fishing and must be accounted for as costs. ... In fact, the very conception of a net economic yield has scarcely made any appearance at all." In reality, "the optimum economic fishing intensity is less than that which would produce the maximum sustained physical yield." The best profits occurred well below M.S.Y.

But a fundamental factor prevented attaining the best economic yield. "In the sea fisheries the natural resource is not private property. ... Each fisherman is more or less free to fish wherever he pleases." The result is a pattern of competition among fishermen which dissipates profits. Eventually, the cost of catching an extra fish, the "marginal cost" in economic terminology, is more than its value. "This is why fishermen are not wealthy, despite the fact that the fishery resources of the sea are the richest and most indestructible available to man."

Gordon was applying standard economic theory to the peculiarities of the fishing industry. In most occupations, extra effort would bring extra production. In the fishery, after a certain point, extra effort would only reduce catches; less effort would increase them. Yet one's only chance was to fish hard to compete, even if that competition ultimately meant disaster for many.

Although fishing required more skill and involved more hazards than many other occupations, fishermen earned less. And they had no easy access to other jobs. "Living often in isolated communities, with little knowledge of conditions or opportunities elsewhere; educationally and often romantically tied to the sea; and lacking the savings necessary to provide a 'stake', the fisherman is one of the least mobile of occupational groups." On top of that, fishermen are "incurably optimistic. As a consequence, they will work for less than the going wage."

Gordon cited the Pacific halibut fishery as an example. Despite favourable reports, the International Pacific Halibut Commission's influence on increased catches (through quotas and other means) was debatable. More important, there was no clear evidence it had made halibut fishing more prosperous. Fishing costs had risen.

> Since the method of control was to halt fishing when the limit had been reached, this created a great incentive on the part of each fisherman to get the fish before his competitors. During the last twenty years, fishermen have invested in more, larger, and faster boats in a competitive race for fish. In 1933, the fishing season was more than six months long. In 1952, it took just twenty-six days to catch the legal limit in the area from Willapa Harbor to Cape Spencer, and sixty days in the Alaska region. What has been happening is a rise in the average cost of fishing effort, allowing no gap between average production and average cost to appear, and hence no rent.

Gordon referred to the problems of common pastures in the medieval manor economy, and the elaborate rules that emerged to stint the number of cows and hours of pasturing. He concluded,

> There appears, then, to be some truth in the conservative dictum that everybody's property is nobody's property. Wealth that is free for all is valued by none because he who is foolhardy enough to wait for its proper time of use will only find that it has been taken by another. ... The fish in the sea are valueless to the fisherman, because there is no assurance that they will be there for him tomorrow if they are left behind today. ... Common-property natural resources are free goods for the individual and scarce goods for society. Under unregulated private exploitation, they can yield no rent; that can be accomplished only by methods which make them private property or public (government) property, in either case subject to a unified directing power."[23]

Although Gordon's basic idea had occurred earlier—that fish and profits per man were the main consideration, and that open entry to the fishery drove down both abundance and average incomes—his brilliant exposition gave new validity to an inexorable scenario. Gordon's ideas eventually became common currency in fishery management circles. Canadians led the way in spreading the new thinking and eventually in its practical application.[24]

But this acceptance took many years. Gordon's theory was only an academic paper, with no great army of supporters. Managers and industry were long reluctant to accept the full implications that attempts at development could soon turn futile, or that protecting private earnings meant applying a "unified directing power." The fishery had long survived without undue interference. (For that matter, so had common-property pastures, where users often worked out controls among themselves.)

Despite Gordon and others who shared his views, departmental reports in the 1950's and 1960's were still projecting ever-greater growth, without much worrying about possible bad effects. The department's approach on the Atlantic was sometimes unintegrated. The conservation service might create new restrictions even as the development service subsidized new technologies for growth.

By the end of the 1960's, the temper of the times was swinging more to Gordon's thinking, even among those who had never heard of him. Eventually, managers would limit the number of fishermen, control the size of their boats, and explore quasi-property rights. Even then, however, they shied away from full use of a "unified directing power."

Gordon told this author at the turn of the 21st century, "The basic point I have tried to stress is the one that was central to my ... paper: the exploitation of fisheries resources will be wasteful unless one can find a way to satisfy the requirement that factors of production be employed so as to equate the marginal cost of production with the price of the product. This is, of course, a standard theorem in economics and I pointed out that the ownership regimes in most fisheries prevent it from being realized."

> He added that

> Limiting entry will not suffice if those who are licensed to fish may buy larger and larger boats and more and more gear. For the same reason, catch limits will not succeed. Moreover, even if the total inputs are restricted, there will be no 'surplus' or 'rent' unless they are distributed so as to make marginal costs equal for all exploitation efforts for each species. In effect, this is what a farmer does in deciding how much plowing, fertilizer, etc. to put on each plot of land. He is able to do this because he has full ownership rights to his land (except in some cases, like open-range cattle grazing, which is like fishing in important respects).

> For some species, such as lobsters and oysters, it is possible to establish unified control regimes but I am not optimistic about most open-sea fisheries.[25]

One can overstate the seeming inexorability of Gordon's model. After all, the fishery had lasted for centuries; and if prosperity was uncommon, still, some people always did well. In Gordon's time, however, the growth of fishing power was about to worsen the consequences of uncontrolled competition.

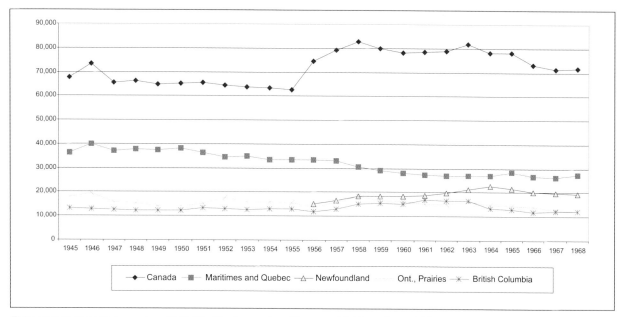

Fishermen in Canada, by region, 1945-68. Gordon spoke of increased fishing effort, which some people equate to increased numbers of fishermen or boats. The real relationship is more complex. Bearing in mind that Newfoundland fishermen appear on the chart only from 1956, the number of fishermen stayed fairly stable in the period. But fishing pressure rose apace, because of better boats and gear.

Class of '47 enters field

As understanding of the fishery grew, regulatory strength increased. In post-war administration in the Maritimes, separate directorates existed in each province, with a small co-ordinating staff at Halifax. Later, Halifax absorbed the separate regional directorships, although there was always some form of district superintendent. On the Pacific, regional supervisors continued to operate at Prince Rupert, Nanaimo, and New Westminster.

In 1946, the department made a special effort to recruit ex-servicemen, and in 1947 put them through extensive training courses, including classroom work.

(Before and after this special effort, on-the-job training was the rule; regular classroom training would reappear only in the late 1970's and early 1980's.) Operating on both coasts, the Class of '47 and a follow-up effort in 1948 produced a great and long-serving crop of fishery officers. Some, such as Cliff Levelton of British Columbia and Lorne Grant of New Brunswick, became leading figures in management.

Members of the Class of '47 later recruited new fishery officers in the same mould: often ex-servicemen, outdoor types who liked to be around the wharves and boats, up the streams, or in stakeouts watching for offenders. The fishery officer's life offered something out of the ordinary, even if it was hiding long hours in

The *Howay*, one of the large patrol vessels on the Pacific.

The medium-sized *Cumella* of the Atlantic patrol fleet usually worked the Bay of Fundy.

the dark to nab salmon poachers, or counting a bucket of rotting seal noses to give a fisherman his bounty payments. Postings could be isolated, to a half-frozen outport or to a B.C. cannery town that imported workers every summer and died every winter. The new officers had a fair amount of leeway, once out on the job. Often, an officer could make his own judgements.

By 1949–1950, the total "field staff" numbered 1,500, counting guardians and patrol vessel crews. The protection fleet was growing. The department had 19 craft in the Maritimes, mostly in the 34–65-foot range, but also including the 155-foot *Cygnus*. Newfoundland had eight vessels, including two in the Bait Service; the Northwest Territories had one; and British Columbia had 33, including two newly acquired large vessels: the 116-foot *Howay*, serving 1947–1981; and the 113-foot *Laurier*, serving 1946–1986. The 80-foot *Kitimat* was also at work. The department chartered about 100 smaller vessels on the two coasts.[26]

International fishery intensifies

While fishery officers monitored Canadian fishermen, an international fishery grew with few checks. Some European fleets had traditionally crossed the ocean from Britain, France, Spain, and Portugal. Now distant-water vessels became far more powerful. In 1954, the U.K.-based company Christian Salvesen launched the first factory freezer trawler, the *Fairtry*. No more salting fish: the 280-foot, 2,600-ton *Fairtry* dragged huge nets to bring in big catches over stern ramps, and fed the fish through automatic filleting machines into multi-plate freezers, with waste going into a fish-meal plant.[27] With the fish protected from spoilage, the vessels could stay on the grounds longer.

Other countries began developing factory freezers for the northwest Atlantic, notably the Soviets, Norwegians, Poles, and West and East Germans. The factory freezer trawlers could act as mother ships to feeder vessels, so a whole surrounding fleet could stay on the grounds longer. The Soviets and Japanese also sent distant-water vessels to the eastern Pacific.

In the fisheries department, officials including Needler, MacKenzie, William Sprules, E. Blythe Young, and Sam Ozere dealt with the new international realities, which became more complex as time went on. With a small whale fishery, Canada was part of the International Whaling Commission from its 1946 inception. In 1950, the Inter-American Tropical Tuna Commission began dealing with tuna fishing in the eastern Pacific. Canada, having few if any tuna-fishing interests in the tropical Pacific, waited until 1968 to join, then dropped out in 1984.

International North Pacific Fisheries Commission begins

From the 1930's on, Japanese fishing vessels had taken Pacific salmon on the high seas, prompting American and Canadian research on salmon migrations and interceptions. The post-war expansion of Japanese fishing alarmed west coasters. In 1952, Canada, the United States, and Japan formed the International North Pacific Fisheries Commission (I.N.P.F.C.). The parties drew an oceanic line at longitude 175°W, with Japan agreeing to fish salmon only

Factory freezer trawlers from overseas became a major presence off Canada's coasts. Here, a Soviet trawler enters St. John's harbour in 1973. (Photo by Kevin McVeigh, federal fisheries)

on the western side. Later investigations showed the line to be substantially correct, with little crossing by salmon. Sockeye and chum swam out towards the line; pinks, coho, and chinook stayed closer to home.[28] The I.N.P.F.C. sponsored research by member countries and followed the important "abstention principle": the parties would abstain from entering any existing fishery that participating countries considered fully utilized and under management.[29]

Canada, United States draw "blue line"

In the 1950's, an unexpected threat emerged from South American vessels fishing salmon off North America. The United States and Canada decided to restrict their fishermen from net-fishing in offshore waters; they figured that if they stopped their own people, they could also stop others, deeming their fishing under such circumstances to be an unfriendly act.

In 1957, the two countries drew what become known as the "surf line" or "blue line" at a certain distance offshore; seiners and gillnetters had to stay inside. From a Canadian perspective, this made fishery management easier by confining the fleet, thus eliminating, at least for the net fishery, the tit-for-tat game of offshore interceptions. But one element of the agreement caused problems later. Off Alaska, the negotiators drew a line that allowed Alaskan fishermen to intercept salmon migrating to British Columbia. and the state of Washington.[30]

Sinclair wins pink salmon conflict

Further south, the International Pacific Salmon Fisheries Commission now regulated the fishery for Fraser sockeye in a generally acceptable manner. But pink salmon were still as vulnerable to American interceptions as the sockeye had earlier been. The Americans were fishing them intensively in Puget Sound and the Strait of Juan de Fuca before they got to inner Canadian waters, outfishing Canadians by a two-to-one margin. Canadians grew resentful, and concerned about conservation.

In a remarkable move, fisheries minister Jimmy Sinclair encouraged Canadian fishermen to fish pinks offshore, before they got to the American grounds. Sinclair told fishermen he wanted them out there seven days a week if necessary. Soon the Canadian catch of pinks was approaching that of the Americans. The minister also eased the way, in 1956, for several B.C. fishermen to import large American purse-seine boats, idled by the collapsed pilchard (California sardine) fishery. This added strength to the Canadian pink fishery.

The United Fishermen and Allied Workers' Union (U.F.A.W.U.) doubted the wisdom of the fish war, which did some damage to the pink-salmon stocks, particularly American runs in Puget Sound.[31] But it worked. As a result, Sinclair said later, "the Americans came to us on their knees."[32] The result was a new protocol to the sockeye salmon treaty, in 1957, providing for bilat-

eral management of Fraser pinks and a 50–50 catch split, just as for sockeye.[33]

I.C.N.A.F. starts up

On the Atlantic, Canada in 1968 joined the International Commission for the Conservation of Atlantic Tunas (established by convention in 1966), and became an active member. The most prominent organization, however, was the International Commission for the Northwest Atlantic Fisheries. Already during the Second World War, officials in Great Britain had looked towards an international organization for the North Atlantic.[34] Work continued after the war, and on July 3, 1950, the International Convention for the Northwest Atlantic Fisheries (I.C.N.A.F. – the same initials were commonly used for the working body, the International Commission for the Northwest Atlantic Fisheries) came into force, with Canada a signatory. Other members included the United States, the United Kingdom, Norway, Iceland, Denmark, France, Spain, Italy, and Portugal. West Germany and the Soviet Union joined later in the 1950's, Poland and Romania in the 1960's, and Japan, Bulgaria, East Germany, and Cuba in the 1970's.

Until the latter 1960's, I.C.N.A.F. confined itself mainly to research and to mesh-size regulations, made in pursuit of the "maximum sustainable catch." But its members were getting used to working together, and as overfishing became clear, I.C.N.A.F. would pioneer the use of multinational catch quotas.

Pressure grows for offshore jurisdiction

Meanwhile, a second track of international diplomacy was opening up. Nations were beginning to cast an eye at sovereignty on the high seas. In 1945, the United States through the "Truman Proclamations" declared its exclusive right to mineral and hydrocarbon resources on its continental shelf, and stated its policy to develop internationally agreed conservation zones for fish stocks off its coasts. The latter declaration implied a custodial concept for itself as the coastal state.[35]

In the 1940's and 1950's, some Latin American countries laid claim to offshore zones. In 1958, a United Nations Conference on the Law of the Sea (U.N.C.L.O.S. I), at Geneva, dealt with such issues as freedom of navigation and control of pollution. Resulting agreements provided for the use of straight baselines in drawing the territorial sea, rather than following the sinuosities of the coastline. And the conference agreed that "[t]he coastal State exercises over the continental shelf sovereign rights for the purpose of exploring it and exploiting its natural resources." Seabed control was now unclear.

There was no agreement at U.N.C.L.O.S. I, or at U.N.C.L.O.S. II in 1960, on coastal states extending fisheries jurisdiction. But pressure was building in various countries. In 1964, Canada took a major step

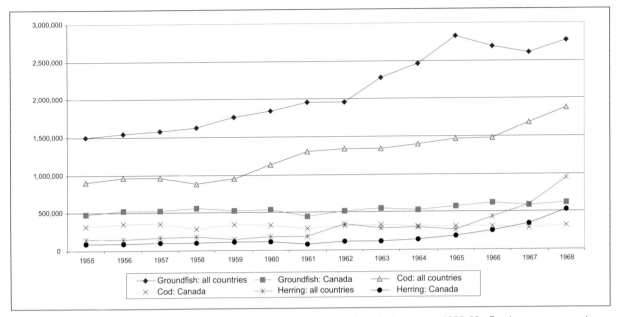

Foreign and Canadian catches of groundfish and herring in the northwest Atlantic, in tonnes, 1955-68. Foreigners were running ahead. Canadian fishing power was growing faster than its groundfish catches, which remained flatter than the foreign ones. But Canadian herring catches were rising fast.

offshore. The new Territorial Sea and Fishing Zones Act authorized a nine-mile fishing zone contiguous to the three-nautical-mile territorial sea. (Before, foreign trawlers had been able to fish up to the three-mile sea, even when Canadian regulations kept domestic trawlers out beyond 12 miles.) The legislation also provided for drawing the seaward limits from a straight baseline, headland to headland.

Both elements displeased the Americans (although they too would declare a 12-mile fishing zone in 1966) and some other countries. Canada delayed proclaiming exclusive fishing zones within the new limit, pending phase-out agreements with foreign fishing nations.[36] These would be slow in coming.

Department pushes development

Conservation was never absent from departmental thinking, especially for inshore and river fisheries. Safeguarding the resource was generally a rough-and-ready matter; if enough people complained, a closed season or area might go into force, without benefit of science.

But development could be as important as conservation, especially after the Second World War. Governmental thinking had swung towards increasing productivity. Several programs came into play.

Boat subsidies build up fleet

Canada needed a better fleet. During the war, the trawler ban ended on the Atlantic, and on both coasts the department offered $165 per ton boatbuilding sub-

sidies for larger vessels. War veterans with fishing experience could get other help.

The subsidies at first had limited impact on the Atlantic, but became more popular after 1947, when the federal government started delivering money

Launching a multi-purpose boat at Pictou, N.S., in 1960. Hundreds of new craft joined Atlantic and Pacific fleets during the period.

through the provincial loan boards. Nova Scotia had set up the first of these in 1936; the other Atlantic provinces and Quebec followed in the post-war years. The loan boards offered low rates and were generally lenient about payments, making allowances for poor years. Some fishermen let the loans run on and on.

From 1947, a fisherman getting a loan for draggers or longliners of 55–60 feet automatically got the subsidy. Groups of four or more fishermen, or fishermen's co-operatives, could also request the subsidy for larger vessels. The federal subsidies aimed to move fishermen up into a larger class of boat. But Newfoundland politicians, notably Gordon Bradley, Newfoundland's first federal cabinet minister after Confederation, lobbied to extend it to smaller vessels.[37] In the early 1950's, the size limit dropped to 45 feet.

Then and later, some provinces supplied smaller subsidies of their own on boats or equipment. In Newfoundland in the early 1950's, the federal and provincial subsidies for a 50-foot longliner ($165 federal per gross ton, and $70–$90 provincial, plus another subsidy if the vessel used a diesel engine) might add up to 35 per cent of the cost; and if the fisherman put up 25 per cent, the province would lend him the balance at low rates.[38]

Federal spending on subsidies increased, especially after 1956. The restriction to draggers and longliners vanished in 1961; now other vessels as well, if longer than 40 feet and less than 100 tons, could get a $250 per ton subsidy. In 1964, the subsidy began taking in craft down to 35 feet and paying a percentage of costs: at first 25 per cent for vessels 35–55 feet, and 30 per cent on larger vessels up to 100 gross tons. The rules changed from time to time. In 1965, a 40 per cent subsidy began supporting wooden vessels over 100 gross tons.

The 1940's and 1950's program had subsidized the building of more than 400 vessels. The broader programs that followed from 1961 to 1968 subsidized nearly 1,200 vessels, both wood and steel. The total $16.5 million spent from the 1940's to 1968 gave a major boost to the independent fleet, especially on the Atlantic. Bigger and better decked vessels dominated the wharves where 20 years earlier, open boats had been the main presence. The program also affected inland and Pacific fisheries, although in British Columbia it would come to an earlier end than elsewhere.

Fisheries department subsidies supported small and medium-sized vessels with mainly independent owners. Larger vessels, usually owned by processing companies, got help from another program. In the 1950's, companies usually got trawlers built in overseas yards. The government wanted to build up the fishery and help Canadian shipyards. From 1961, trawlers built in the five Atlantic provinces got a 50 per cent subsidy, which dropped to 35 per cent in 1968. This stimulated more Canadian construction. The Canadian Maritime Commission and then the Department of Industry, Trade and Commerce admin-

istered the shipyard subsidies. As of the late 1960's, non-wooden trawlers of at least 76 feet were still getting 35 per cent; vessels ineligible for that subsidy could still, if over 100 tons, get 23 per cent, the rate for general shipping.

The large-vessel program helped build 266 vessels across Canada in the years 1961–1969, providing a stupendous increase in fishing power. The Atlantic coast got most of them. Elsewhere, the program subsidized eight large inland vessels, including steel fishing tugs on the Great Lakes. Pacific steel vessels included seiners, gillnetter–trollers, and other combination boats. The large-vessel subsidies in the years 1961–1969 totalled upwards of $67 million; this was more than four times the money going to smaller vessels in the whole post-war period.[39]

Other assistance applied through the tax code. Starting in 1947, the rules let owners deduct from enterprise income up to one-third of the capital or conversion cost of their vessels. Another provision protected owners from government "recapture" of depreciation costs when the vessel was sold. The idea was to encourage the sale of older vessels and the building of new ones. Although the tax rules applied equally to all, they benefitted most the enterprises with a higher cash flow, that is, bigger boats.

Back in the 1940's, federal subsidies had aimed mainly to increase productivity by helping independent fishermen move up to medium–large vessels. Along the way, the program had got broadened to vessels both smaller and larger. Governments saw the subsidies as encouraging both the fishing and the vessel-construction industries. And in the 1960's, many people were anxious that Canada build up her offshore fishery before the foreigners took dominance right up to the shore.[40] The whole program had no relation to conservation; it was an effort at development, which turned dangerous.

Department provides insurance, loans

Owners of smaller vessels often found it difficult to get insurance coverage; private firms found them too risky or too small to bother with. In 1953, the department under Sinclair brought in the Fishing Vessel Insurance Program (F.V.I.P.). Operating until the 1990's, the F.V.I.P. provided thousands of boat-owners, particularly on the Atlantic, with protection at reasonable premiums (one per cent of value at the outset). Up until 1968, fishermen could also insure lobster traps under the F.V.I.P.

The Fisheries Improvement Loans Act in 1955 brought further aid. Fishermen could now get a federal guarantee for bank loans to improve their operations. In the late 1960's, coverage went up to $25,000. This program found special acceptance in British Columbia, where fishermen tended to be more cash-oriented, businesslike, and willing to invest if they could find a lender.

Industrial Development Service tests grounds and gear

From 1950, development work got extra money. And from 1954, a separate Industrial Development Service (I.D.S.) came into place, splitting off from the conservation and development branch. In the coming two decades the I.D.S. would undertake hundreds of development projects, trying out new gear and new fisheries. Sometimes new techniques came from other countries; at other times, I.D.S. officials came up with new ideas (in practice, usually a twist on an old idea). Finding fish with sonar, catching them with mid-water trawls, bringing them in with mechanized haulers—such developments often got their impetus from the I.D.S. The branch helped develop bottom and mid-water trawling on both coasts; smelt trawling on the Great Lakes; and on the Atlantic, purse-seining for herring, and new crab and shrimp fisheries.

The I.D.S. worked directly with hundreds of fishermen in scores of communities. Word of mouth and power of example rippled out to spread new methods. As well, by the mid-1960's, the department was frequently holding major national and international conferences on aspects of development: on fish protein concentrate, fishing vessel construction materials, automation and mechanization, the herring and crab fisheries, and other subjects.

Often, the Industrial Development Service worked with the Fisheries Research Board, particularly in exploring new grounds or under-utilized species. As well, the board carried out its own development-related work, a prime example being the pioneering work at the Vancouver technological station, in the late 1950's, in using refrigerated sea water aboard boats to preserve fish quality.[41] But the main developmental effort came from the I.D.S., especially in the 1960's.

For the year 1967 alone, a governmental study listed nearly a hundred projects of one kind or another; and that was for just one year out of many. The same document, *Trends in the Development of the Canadian Fisheries*, called attention to a certain lack of co-ordination, in both development efforts and industry growth. In fact, the I.D.S. tended to run off in all directions, tackling projects wherever it saw an opportunity. As with the boatbuilding subsidy program, there was little thought of conservation or the

The I.D.S. encouraged the growth of Atlantic purse-seining. Larger seiners began appearing in the 1960's and especially the 1970's. (Walter Scott drawing)

possible consequences of fishing species such as herring at lower levels of the food chain.

Scientific methods, such as having a standard "control" against which to compare new methods, played little part in development work. Officials like Wes Johnson and Jack Rycroft, coming from the fishing industry, tended to bull ahead and try the new technique, whatever it was. Often, this approach paid off; sometimes, it failed. Even the less successful efforts kept the pot boiling.

Canada's total sea-fisheries catch grew from about 900,000 tonnes in 1955 to 1.4 million tonnes in 1968—an increase of some 55 per cent. But the development workers were never satisfied. At the end of the period they still saw opportunities at every hand. The 1967 study forecast a further doubling of sea-fishery landings, on both coasts.

What did the Industrial Development Service do?

The 1967 study *Trends in the Development of the Canadian Fisheries* listed nearly a hundred current I.D.S. projects for 1966–1967, costing more than $3 million. The work involved 28 fishermen hired for the purpose, 3 technologists, 2 boatbuilding specialists, 11 foreign specialists, and 14 observers. The single biggest item was the community-stages program for Newfoundland, at about $900,000. This program had begun in 1959–1960, originally as a "winter works" program; the aim, as recommended by the earlier Walsh Commission on Newfoundland fisheries, was to give saltfish producers in isolated settlements better landing and processing facilities. Later, the program provided fresh-fish holding facilities as well.

Most I.D.S. work came through cost-shared programs with the provinces. In Newfoundland, work went forward on such items as combination fishing boats, Scottish seine demonstrations, the shrimp and squid fisheries, snap-on longline gear, synthetic cod-traps, cod-seining, food processing, various Labrador projects, and conversion of a longliner to a herring seiner. Among the notable efforts, four vessels using B.C. and Icelandic methods scouted the "virgin" herring stocks of southern Newfoundland; their catches prompted B.C. Packers to set up a reduction plant at Harbour Breton. The I.D.S. introduced new trawls for deep-sea fishing and B.C.-style "combination" boats able to pursue different fisheries.

Every province had its long list of I.D.S. projects. Nova Scotian efforts included scouting herring stocks, converting a scallop dragger to a herring seiner, demonstrating groundfish gear, experimenting in offshore trawling of certain species for reduction, doing crab and shrimp work, introducing hydraulic haulers for lobster traps, exploring for scallops, helping to re-establish whaling, and so on. In New Brunswick, among other projects, the I.D.S. provided new-type trawls for several vessels, converted a dragger to seine netting, experimented with small-boat purse-seining in the Northumberland Strait and with drum trawling, converted many lobster boats to multipurpose fishing, and surveyed for Irish moss.

In Prince Edward Island, the department experimented with plastic lobster traps and did experimental fishing for herring, scallops, crab, and shrimp. In Quebec, the many projects included longlining for swordfish and sharks, experimental fishing for shrimp and crab, surveying molluscs in inshore waters, surveying seaweed, demonstrating a portable echo-sounder for inshore boats, using gillnets for halibut, and, as elsewhere, encouraging purse-seining for herring.

Besides the cost-shared work with the provinces, the I.D.S. carried out many projects on its own. On the fishing side, this included 25 items such as trawler design and development, electrical winches, offshore lobster studies, underwater gear studies, a herring conference, and on and on. In B.C., the I.D.S. chartered a vessel to help the F.R.B. explore areas for the growing groundfish fishery. Seventeen more projects, in processing, packaging, storage, and transportation, went forward across the country.

The I.D.S. was helping development practically everywhere, and as Cliff Levelton, a top manager, later put it, "Nobody wanted to question Santa Claus."

Inspection service takes stronger hold

After a century of complaints about quality, and many reports calling for change, the department's inspection work in the 1940's was still small, mostly based on voluntary compliance. Inspection laws were still an unclear federal–provincial hodge-podge.

The 1940's saw some advances. In 1945, the department set up a fish-inspection laboratory in the Maritimes. In British Columbia, a lab already operated for canned salmon and, starting in the war, canned herring. The labs also inspected imported products. In 1948, the department got more regulations for pickled fish, more authority to enforce canned fish and shellfish inspection, and more staff. Meanwhile, the F.R.B.'s technological labs on the east and west coasts continued to work with the processing industry. The board also helped the department inculcate better practices in plants and boats. As refrigerated trucks after the war took the fish trade away from trains, stricter rules about transport temperatures came into force.[42]

In 1950, the inspection and consumer branch stepped up its previous promotion work. The new test kitchen developed recipes; the branch disseminated brochures and such with recipes and handling and storage instructions. Home economists out of Toronto, Montreal, Halifax, and other locations promoted fish to media and consumers.[43] (Later, in 1968, October was christened Fish 'n' Seafood month, but this effort had little impact.)

Inspection official checking Pacific salmon at plant, 1963. (Ted Grant photo)

In fish quality itself, the department's efforts were still less than thorough. Many other countries had introduced minimum standards and systematic inspection. In Canada, it became clear that the level of quality in fresh and frozen fish and the sanitary conditions in plants were sometimes poor. The industry itself suggested grading for some products.[44]

In 1953, the department and the F.R.B. surveyed more than 500 plants across the country, finding many failings. A federal–provincial conference in 1954 led to a national Fish Inspection Program starting, complete with a strengthened federal law and new administrative agreements. The federal side provided manpower and money, and could enforce provincial legislation. (Federal rules applied only for products crossing provincial boundaries.)

The department, in consultation with industry, developed compulsory minimum standards, and for traditional products such as pickled and salted fish, a grading system as well. Inspection became more thorough and systematic. The Inspection Service in the field became a distinct entity, separate from the fishery officer corps. Inspection officers not only monitored products for wholesomeness but also oversaw the conditions of raw materials, vessels, holding facilities, plants, and transportation. Officials, including the renowned Charlie Castell of the Fisheries Research Board's Halifax station, encouraged, nagged, and helped processors to improve product quality. More inspection laboratories came into place, examining imported as well as domestic fish products.

By 1964, most products faced compulsory inspection. Requirements differed. All canned fish and shellfish to be exported from a province required inspection, which was intensive for Pacific salmon and herring; otherwise, only spot sampling took place. For salted, dried, and pickled fish (for which demand was declining) both grading and inspection applied. Oysters got graded for the shape of the shell, but there was no inspection for shucked oysters. Scallop meat had standards for freshness and cleanliness, without grading. Some lobster products required inspection; others (including live and cooked in shell), none. Major frozen-fish plants had departmental inspectors assigned practically full time. Apart from the department's regulations, the Food and Drug Act of Canada applied against unfit fish.

By the end of the 1945–1968 period, hundreds of inspection officers were monitoring plants, pushing, prodding, and encouraging better practices. The program helped ensure that Canadian fish was always safe, even though producers rarely aimed for the higher niches of the market. The industry itself had few people with training in food technology.[45]

The department had hoped in the 1950's that the national Fish Inspection Program would lead to more self-regulation by the industry itself, which in a general way wanted better quality. The Fisheries Council of Canada had called for more federal inspection. But individual companies sometimes resisted specific measures. Only a few pursued top standards and competitive advantage through quality; fish was generally more of a commodity. A voluntary grading program for fresh and frozen fish, introduced in 1958, found few takers.

Problems remained with inconsistent quality, notably in the Atlantic groundfish industry, where producers aimed at the medium- and low-price trade. Several obstacles held back quality. Fishermen and processors generally had no strong interest in improving quality. Small and remote landing points, often distant from the processing plants, had poor facilities. Most fishing craft were too small to carry ice, and many that could didn't bother. Generally, landing better fish fetched no better price; and even in cases where incentive pricing was offered, it often proved difficult to produce better quality because of the industry's physical set-up.[46] But in general, quality improved during 1945–1968, with serious health problems from fish products almost unknown.

Prices Support Board aids marketing

Besides boosting quality, the department was now providing a form of support in the marketplace. In 1944, Parliament passed legislation for and in 1947 appointed members to the Fisheries Prices Support Board (F.P.S.B.), which could cushion the shock of market slumps by such means as subsidizing fish prices, or purchasing fish for later sale or for international food aid. The F.P.S.B. found frequent use, especially in inland fisheries and the Newfoundland saltfish industry. Groundfish markets seemed to weaken on a six- or seven-year cycle, in 1953–1954, 1960–1961, and 1967–1968.[47] Frequently, the F.P.S.B. administered purchases of mackerel or other species for world food aid.

Federal–provincial co-operation builds up

By the 1950's, the Atlantic provinces all had fisheries departments or at least divisions, and were paying more attention to the industry. The government of British Columbia took the opposite track. B.C. had had a fisheries commissioner since 1901 and a Department of Fisheries since 1947. But jurisdiction was federal, and W.A.C. Bennett, premier from 1952 to 1972, disliked fields where he lacked control. In 1957, the department shut down; fisheries work got subsumed under the Department of Recreation and Conservation. The federal side was more clearly than ever the main force in management.

On the Atlantic, the I.D.S. often worked up projects with the provincial governments. From 1958, federal and provincial deputy ministers met frequently through the Federal–Provincial Atlantic Fisheries Committee. Counterpart committees came into place for the freshwater and B.C. regions.

Fisheries Council dominates industry–government relations

Fishermen always had access to officials and M.P.'s. But only in British Columbia did they have an organized voice—in fact, several of them, sometimes contradicting one another. But they were far ahead of the fragmented, unorganized east coasters. The largest Pacific organization, the United Fishermen and Allied Workers' Union, grew powerful indeed.

On the processing side, the national Canadian Fisheries Association, formed during the First World War, had by the Second lost its earlier prominence. After the war, processors regained a strong voice. In 1945, the Fisheries Council of Canada organized itself as a federation of regional associations, with an Ottawa headquarters. (Frederick William Wallace, editor of *The Canadian Fisherman* and important in the original Canadian Fisheries Association, had long pushed

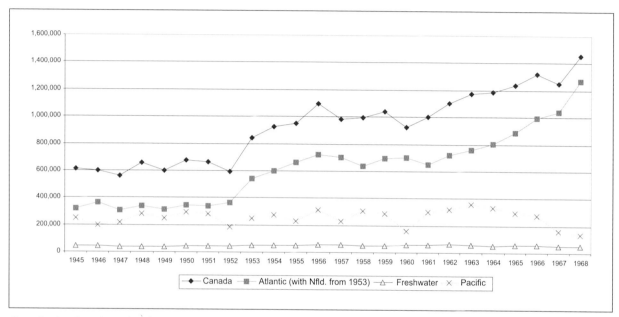

Canadian landings by region, in tonnes, 1945-68. Herring explained much of the rise in landings on the Atlantic, and of the late-1960's fall on the Pacific.

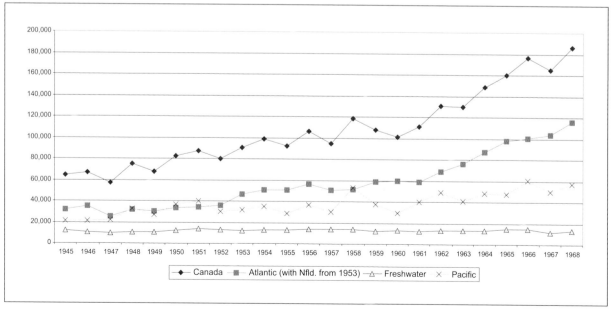

Canada's landed value by region, 1945-68 ($000's).

for the new arrangement.) Bates encouraged the new organization, telling the F.C.C. annual meeting in 1946 that as a group they could counteract the smallness of Canadian enterprises by providing some of the attributes of bigness: doing research, providing information, sponsoring advertising, working with government, and taking the longer view.

From time to time, Bates prodded the industry, telling the F.C.C. annual meeting in 1950, "The Federal Government has framed a programme largely based upon the request of industry. It is now our turn to ask—what is the programme of the fishing industry?" While all desired expansion, "[t]he sort of expansion awaited is a business expansion and that is the job of business men."

The F.C.C. made occasional statements in support of better inspection and grading of fish and, as foreign fishing became an issue, in support of more Canadian control. But their efforts at combined promotion in the 1950's, including a National Fish Week, petered out. The council, however, remained an effective advocate for business interests, which were already powerful. The fisheries economist W.C. MacKenzie later noted that "for a long time, the corporate sector had almost a veto over fishery policy. Not always—they lost out in the trawler controversy of the 1930's—but mainly, they were a very powerful voice in the councils of fishery administration. They weren't always well-organized, but compared to the fishermen ..."[48]

Unemployment Insurance includes fishermen

Back in 1940, the government had set up the Unemployment Insurance Program (U.I.), to help people who temporarily lost their job. There was no provision for self-employed workers, such as fishermen. Most fishermen operated as "co-adventurers." Typically, no one got a wage; instead, the crewmen and

---

Aid programs cover the waterfront

By the late 1960's, the government was assisting the industry every which way. There were the departmental and F.R.B. services in science, conservation, exploratory fishing, development, inspection, and marketing intelligence and promotion; and besides these, the price support, subsidy, insurance, and Fisheries Improvement Loans programs. In 1960, the federal government had initiated a cost-shared scheme in which provinces could propose to fund specific projects.[51] Self-employed fishermen could receive U.I. The government provided wharves.

A number of other programs provided financial aid. A 1969 booklet from the Fisheries Council of Canada included the following, which were of most interest for the better-organized companies with the time and talent to make use of them:

- *Regional development incentives*—A 1969 act set up this program under the Department of Regional Economic Expansion (D.R.E.E.), to assist manufacturers. The government would provide up to 20 per cent of the approved cost for expansions and up to 25 per cent for new facilities, as well as $5,000 for each job created. D.R.E.E. grants would become a staple of the fish-processing industry.
- *Cost-sharing agreements*—Under the Fisheries Development Act (F.D.A.) of 1966, the minister could enter into agreements with anyone for joint projects.
- *Fishery Products Storage Regulations*—This program under the F.D.A. could pay up to 30 per cent, or $150,000, for new facilities.
- *Program for the Advancement of Industrial Technology (P.A.I.T.)*—This Industry, Trade and Commerce (I.T.C.) program could pay 50 per cent of the cost of projects to develop products and processes.
- *General Adjustment Assistance Program (G.A.A.P.)*—This I.T.C. program could insure loans and provide other assistance arising out of "Kennedy Round" tariff changes.
- *Industrial Research and Development Incentives Act (I.R.D.I.A.)*—This I.T.C. program provided assistance for long-term research in science and engineering.
- *Small Business Loans*—Under a 1961 act, this program provided small businesses with loans up to $25,000.
- *Industrial Development Bank*—This forerunner of the Business Development Bank provided "last-resort" loans to enterprises unable to get credit elsewhere.[52]

Many such federal programs would continue for years to come. Atlantic provincial governments also provided aid. In 1982, for example, Newfoundland was offering fishermen's loans up to $50,000; guarantees for loans from chartered banks; boatbuilding subsidies of more than $1,000 per under-deck ton; subsidies on longlining gear; subsidies on inshore gear for Labrador fishermen; and replacement lobster traps at reasonable costs to replace those lost to ice or storms. Prince Edward Island was offering assistance for on-board fish-handling systems; conversion from gas to diesel engines; aquaculture; and retail and wholesale marketing of fresh fish.[53]

captain–owner shared proceeds from the catch. Most commonly, the captain would take a certain percentage "for the boat" and split the rest with the crew.

Atlantic fishermen in their small communities could see plant workers and others around them getting the new U.I., and political nature took its course. In Newfoundland, Premier Smallwood had formed an alliance with federal mandarin Jack Pickersgill, who ran for office in the new province and became Newfoundland's representative in the federal cabinet. Seeing the conditions in hard-pressed fishing villages, Pickersgill pushed for an extension of U.I. to fishermen. Fisheries minister Sinclair himself disliked the idea of U.I. for fishermen, and there was no pressure for it from fishermen in his home province of British Columbia. But he went along with it.[49]

Pickersgill, working with Sinclair and labour minister Milton Gregg, outmaneuvered opponents and outskirted the doubts of prime minister Louis Saint Laurent.[50] In 1956, new rules came into play, giving fishermen coverage for up to 12 weeks in the wintertime. Sinclair had wanted a "drop-off" mechanism to avoid subsidizing the better-off fishermen; nothing came of this.

In hundreds of fishing households, especially on the Atlantic, U.I. was cause for celebration. As with boat-building subsidies for small boats, Newfoundlanders had distorted a national program beyond the original intentions, in order to help the small-scale fishermen. Participation in the fishery picked up in Newfoundland and British Columbia, although other factors might have been at work as well; indeed, in the Maritimes and Quebec, the number of fishermen was still falling.

As decades passed, many observers, including even some fishermen, would see U.I. as more of a curse than a blessing, as it attracted more marginal participants into the fishery. Still, it is safe to say that the program helped keep enterprises and whole communities operating that might otherwise have faded from sight.

Calls emerge for centralized development

Bates had wanted to drag the fishery into modern times as a productive industry. The development programs begun in the 1940's and 1950's had accelerated the natural growth of the post-war industry. Meanwhile, H. Scott Gordon's 1954 paper suggested the futility of unchecked development. In a common property open to all, fishermen and processors would keep competing, unable to stop themselves, until the increased costs for an extra pound of fish would outweigh the increased value, especially as fishing pressure drove the resource down to lower levels.

The trend of thought so well expressed by Gordon persisted in some quarters of the department, notably in the work of W.C. MacKenzie, the leading departmental economist. But the development mentality was stronger. The need for progress was obvious, with the industry still below its potential, and foreign competition eroding Canada's position on her own coasts. The idea of restraint was a hard sell, and Gordon's related idea of a "unified directing power," harder still. So long as people were finding new resources, there seemed no need to confront the issue of fully exploited fisheries and their gradually vanishing profits.

Still, even in the 1950's, some officials bruited schemes of concentrating the Atlantic industry into a small number of major ports, and building up the "offshore" industry.[54] That sentiment grew in the 1960's, reflected in studies both external, by the Atlantic Development Board and Atlantic Provinces Economic Council, and internal. The 1967 interdepartmental planning document, *Trends in the Development of the Canadian Fisheries*, dealt with both coasts but, like most "national" studies, stressed the Atlantic. The document recognized problems of over-dependence and called for cuts in the number of fishermen. But the main answer to everything was more development, albeit on a more co-ordinated basis. In effect, the document called for Batesian development with a smattering of Gordonian controls.

The authors of the *Trends* document had canvassed provincial and industry officials. Their projections looked forward to further growth, even a doubling of catches on both coasts. But this alone was insufficient for a proper industry, the document said. There should be a radical cutback in the number of Atlantic inshore fishermen, by 10 or 15 thousand or more; half the 45,000 or so fishermen could take the same catch. Expansion had been too scattered. Growth should concentrate in a small number of major, centralized ports and should proceed through larger vessels, particularly offshore vessels. Trawlers were implicitly equated with efficiency.

The planning document laid out no precise road to what came to be termed "rationalization." It briefly mentioned licence limitation and Newfoundland resettlement, without proposing a concerted program. It skipped over the question of governance for the cantankerous industry. There was little suggestion of bringing fishermen and processors into the decision-making apparatus, or of building up information, education, and consensus.

Nor did the study address the question of ownership, even though a major shift from the inshore to the offshore would cut to the heart of well-being and pride for hundreds of communities and thousands of people. With no thought-out program for controlling the industry or building a consensus for that control, rationalization so far seemed like wishful thinking. The trend was still to develop now, worry later.

"Grow or die"

The 1967 document *Trends in the Development of the Canadian Fisheries* involved fisheries department and other officials, and reflected industry and provincial government projections. Department officials, while optimistic, were less convinced of ever-booming growth than the other parties. Following are some points in paraphrase:

Groundfish and herring fisheries were expanding, partly because of federal fisheries development work, provincial promotion, area development incentives by the federal department of industry, and work by the Atlantic Development Board in providing water supplies to fish plants. Strong markets and foreign investment had helped.

At the same time, however, foreign fleets were rapidly expanding. The Canadian catch in the I.C.N.A.F. area had dropped from 37 per cent in 1954 to 27–29 per cent in 1964. "The question then is, will Canada passively succumb to this aggressive expansion by others and let them take over this growth sector, or will we ... attempt to take advantage of our locational situation to compete effectively and thus maintain our foothold and even increase our share." There was considerable justification for rational expansion. "The situation is virtually one of grow or die."

(The authors were reflecting a widespread sentiment about competing with foreigners. Fisheries minister Angus MacLean had warned in 1960 that Canadians must improve their efficiency in order to compete.[55] His successor, Hédard Robichaud, noted in a 1965 speech that "to compete with them, indeed to outfish them, we must modernize both our inshore and our offshore fishing fleets."[56])

But development if undertaken required efficiency, the document stated. The inshore fishery would provide decreasing support for the industry. Offshore expansion would provide new jobs, ease the inshore adjustment, and more than offset the inshore drop in catches.

The offshore fleet was already growing. In the period 1966–1968, the Atlantic large-trawler fleet was expected to increase from 146 to 217–220, and by 1975 to 297. Development efforts should favour major, centralized ports in a rational way. Rather than small and scattered I.D.S. projects, there should be more effective concentration on larger concepts, along with additional funding.

The number of Atlantic inshore fishermen dependent on cod should drop, by unspecified means, some 10,000–15,000. Remaining inshore fishermen should either extend their season with better equipment or seek work elsewhere. In areas such as the lobster fishery, licence and gear limitation could help. Half the fishermen could take the annual crop, and with double the catch per person, they would have very adequate incomes. The problem would be dealing with the displaced. "If small-boat fishermen are to become masters or crew members of large efficient vessels—in those fisheries where such vessels would be the most efficient—they must be trained or gradually acquire the necessary skills and acceptance of a new way of life."

Plant work would increase sharply. Of the 520 fish processing plants in 300 communities on the Atlantic outside of Quebec, more than half were small enterprises and fairly rudimentary. The average wage was 95 cents an hour. From 1965 to 1975, plant employment should increase from 7,100 to 12,800.

All told, from 1965 to 1975, Atlantic groundfish landings would double, and pelagic landings would nearly triple. The herring fishery alone, as projected by the Federal–Provincial Atlantic Fisheries Committee, would multiply more than fivefold to reach a million tonnes by 1975. Shellfish would rise nearly one-third. Great progress was taking place in tuna, crab, shrimp, whales, squid, marine plants, and the new product of fish protein concentrate. Total Atlantic sea fisheries would more than double, from about 763,000 to nearly 1.6 million tonnes.

On the Pacific, most salmon runs could produce more than at present. Catches from the upper Fraser could double. Pacific groundfish landings would at least triple. Pacific herring catches, already nearly 250,000 tonnes, would also triple (this forecast came on the verge of a herring collapse and shutdown). Pelagic and estuarial landings all told would more than double. Shellfish would triple. Total Pacific landings would grow at least by half, and might more than triple.

Such were the projections of the study. Made public, it would have confirmed the worst suspicions of inshore fishermen. It projected enormous growth but mainly for the benefit of larger interests, while many small operators would vanish. The next two decades would work out differently.

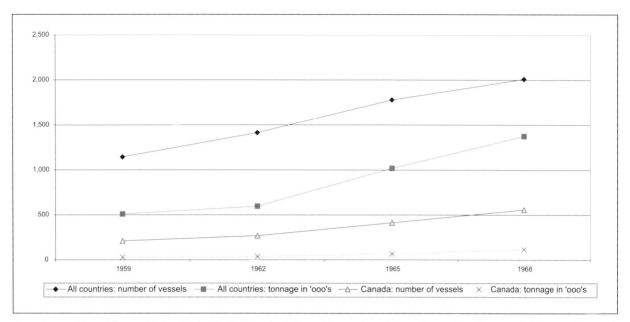

Canada raced to catch up with the foreign fleet. The chart shows the number of Canadian and foreign vessels 50 gross tons and over in the northwest Atlantic, selected years, 1959-68.

The rationalizers versus the sympathizers

The *Trends* document foreshadowed an issue that would later become more sharply articulated, especially on the Atlantic: the so-called "economic" versus the "social" fishery, or what might be termed the rationalizers versus the sympathizers.

An example of the rationalizer view appeared in a 1969 study by the Atlantic Development Board. The board said that "the inshore fishery throughout the Atlantic Provinces is an instrument of poverty. Far too many men, boats, and equipment are being applied to a basically limited resource."[57] Already then, but especially in the 1970's and 1980's, some advocates, often associated with the processing and vessel-owning companies, deprecated the social fishery as ne'er-do-well small boats subsidized by government programs, especially Unemployment Insurance. The rationalizers identified efficiency with fewer and larger boats, which could catch fish faster; and with fewer and larger plants at major ports, which would provide better paying jobs over a longer season.

Questions of true cost-effectiveness—for example, whether a small boat could produce better return on investment or more benefits for the region—often got left aside. Such analysis was not simple. Some studies, contrary to popular opinion, suggested that intermediate-sized vessels were the most cost-effective.[58] But in a race for the fish, the bigger boat could prevail whether or not it was more cost-effective in the normal sense. As for the best manner of fishing to benefit the regional and national economies, comparative studies were scarce, if any existed at all. It was, however, clear that fishing and processing had a strong "multiplier

effect": in Nova Scotia, a fishing job in the early 1970's created 0.7 jobs elsewhere; a processing job, 2.3 jobs elsewhere.[59]

Opposing the rationalizers, the sympathizers identified with small boats and small communities, talked about the fishing way of life, and sometimes suspected large corporations of manipulating the department and destroying the resource. Still, the loudest arguments came later. During the 1960's, although some small-boat fishermen and small communities were suspicious of the large-trawler companies, the accent was more on survival and growth than on any particular small-boat or big-boat vision for the industry.

The conflict of big boat versus small, and of mon-eyed mobility against static communities, was never quite as strong in British Columbia as on the east coast. In the mainstay fishery, that for Pacific salmon, the advantage of big boats over small was less marked than in Atlantic groundfish. Many operations took part in sheltered, "inside" waters. There was no common counterpart of the large, threatening Atlantic trawler. Still, especially in later years, powerful purse-seiners created somewhat of a split in the Pacific salmon fishery; owners would sometimes consolidate several small-boat licences to create bigger vessels.

Regarding the Atlantic, some department officials wanted centralization, and promotion of offshore companies. Personalities could come into play; certain officials might by nature be more comfortable dealing with company representatives, and thus gravitate to their way of thinking; to such officials, fishermen were rather an untidy and wild element. Other officials identified more with fishermen. Especially in the regions, some wanted earnestly to preserve communities and the smaller, independent fishermen. In the middle was a softer mass with no particular ideology, public servants trying to do what seemed reasonable at the time. Most ministers fit the same middle-of-the-road pattern. Subject to conflicting pressures, they tried to steer a reasonably safe course.

One clear fact was that the fishery, at least on the Atlantic, was providing only low incomes for most participants. In the 1950's and 1960's, only one direction seemed to promise good news for all parties, and that was development.

Recreational fishery gains strength

What about the sport-fishery from 1945 to 1968? The federal department of fisheries held a strange position with regard to the recreational fishery: partly hands-on, partly distant. It conducted a good share of freshwater research, with universities doing most of the remainder, and the provinces a small part. In the Maritimes and Newfoundland, federal officials made and enforced the rules, often taking advice from rod and gun clubs, as recreational fishing and hunting organizations used to be called. Indeed, until well into the post-war period, rivers and lakes probably took the most federal enforcement effort (with most sea-fishery effort going to lobster).

But even in the Atlantic provinces, the provincial governments issued freshwater sport-fishing licences, helped enforce the rules, and suggested regulations. From Quebec to Alberta, the provinces managed freshwater fisheries. Still, the federal side managed all fisheries in the Northwest Territories. As for British Columbia, the department policed the salmon fishery even in the far mountain regions of the interior; the province managed freshwater sport-fisheries and other inland fisheries.

The federal department gave its main attention to the commercial fishery in salt water. Only a few people, mostly in the economic section at Ottawa, thought much about the recreational fishery. But with more anglers and higher spending, that fishery was becoming an economic powerhouse. By 1975, anglers owned 1,644,000 boats. That same year, the recreational fishery generated about 26,000 jobs. About a million foreign anglers were visiting Canada yearly, and five times that many Canadians went fishing. The marketed value of the commercial fishery was $694 million; the expenditures of anglers, $1,022 million.[60]

When sport-fishing advocates began talking about their industry's greater importance, others sometimes questioned them. Economists trying to calculate the recreational fishery's value would add up the many expenditures of a fishing trip. Commercial advocates would argue that if the sport-fishery weren't there, people would spend their money on some other form of pleasure, from bars to bubble gum, and the economy wouldn't be shaken. The commercial fishery, on the other hand, produced food from the ocean, and provided vital support to huge regions through jobs that would not otherwise exist; thus, it was more important.

No matter the arguments, the sport-fishing numbers were highly impressive. In years to come, sport-fishing interests in some areas would begin challenging commercial.

CHAPTER 18.
The Maritimes and Quebec, 1945–1968

This chapter attempts to treat Atlantic-wide and then Maritime and Quebec issues, to the extent they can be disentangled. At the start of the 1945–1968 period, Newfoundland and Canada were separate countries. In both, the many fisheries each had its separate history of growth, stability, or decay. But the various fisheries were often interrelated; most boats fished more than one species. Groundfish in particular had an interregional and even international aspect.

The 179-foot patrol vessel *Chebucto*.

In the Maritimes, although inspection work increased especially after 1954, the big job was still enforcement. The Class of '47 and newer fishery officers had good authority in the field, without some of the detailed directives and paperwork that appeared later. Often, they employed temporary "guardians" to help them, mainly in lake and river work. Some officers were seasonal. Scattered along the whole Atlantic coast and in some inland areas, the fishery officers would know everyone in the community: who might cause problems, who could generally be trusted to obey the rules, who would co-operate and provide information.

The Maritimes developed a Special Force, nicknamed the Goon Squad. When local fishery officers found the situation getting troublesome, they would call for temporary assistance. (An officer's job had its share of confrontations and dangers; in the 1950's, poachers killed a fishery officer in Prince Edward Island)[1] Special Force members would come into the community; make get-acquainted visits with local authority figures, such as priests, ministers, and town officials; and then raid fleets or illegal lobster canneries. Stories circulated about such incidents as officers hauling the bedspreads off couples to find illegally canned lobster hidden below.

Fishery officer Stan Dudka of the Special Force, tall and good-humoured, gravitated to trouble spots. It was said that in one episode in the Gulf of St. Lawrence, surrounded by boats and with a gun leveled at him, Dudka picked up a shotgun and said, "Okay, you shoot first." The poachers backed down. In another instance, fishermen let Dudka know that they had chipped in to buy him a graveyard plot. Stories about Dudka were legion. He was the epitome of the friendly and fearless fishery officer, although sometimes confrontational. The department on occasion kept him clear of certain areas.

In the immediate post-war period, the department had 19 or so patrol craft on the Atlantic; more appeared in the 1950's and 1960's. At first, the 155-foot *Cygnus* was the only "offshore" vessel over 100 feet; a new, 153-foot *Cygnus* replaced her in 1959. The 179-foot *Cape Freels* began working out of St. John's in 1962, and the Arctica in 1964. In 1966, the *Chebucto*, sister ship to the *Cape Freels*, joined the *Cygnus* in Halifax.

Fleet grows faster than catches

Fishing power was growing far faster than catches. And in the offshore race for fish, foreigners were winning.

Within Canada, the rise in fishing power started in the 1940's, continued in the 1950's, and accelerated in the 1960's. The greatest Atlantic-wide fishery was that for cod and other groundfish. The strongest fishing machines for these species were draggers and trawlers, which towed large sack-like nets along the bottom. This sector grew the fastest.

Although the technology was the same, departmental terminology at some point began dividing these vessels by length, referring to those below 100 feet as draggers, and to those above as trawlers. (It appears that the department or its political masters encouraged the term "dragger" to avoid the controversy earlier associated with trawlers.)[2]   As the fleet developed through the period, most draggers stayed below 65 feet, though several were longer. Meanwhile, some trawlers reached the 150-foot range.

Right after the war, in 1947, only a couple of dozen trawlers operated on the Atlantic.[3]   By 1955, Nova Scotia had 19 trawlers and 82 draggers. Newfoundland in 1956 had 12 trawlers and 18 draggers, recording high landings. Only processing companies or associated enterprises could afford trawlers, which then cost $200,000–$300,000.   Although trawlers often incurred losses in fishing operations, the processing and marketing arms could absorb them.

Rapid growth continued. Nova Scotia by 1968 had 134 vessels longer than 100 feet; the other Maritime provinces and Quebec had 20; Newfoundland had 75. Outside of Newfoundland, some of these over 100-foot vessels might be scallop draggers or herring seiners; in Newfoundland, virtually all were trawlers.   Stern trawlers replaced side trawlers, particularly in the 1960's.

Atlantic-wide, large groundfish trawlers accounted for much of the fivefold rise in the total fleet's value, which climbed from $31 million in 1953 to $168 million in 1968. In the years 1959–1968, the Atlantic fleet of vessels over 50 gross tons—which would have been upwards of 50 feet—more than doubled, from 211 to 558, and their tonnage more than quadrupled, from 27,000 to 112,100.

Side trawlers (or "draggers") dominated at first, giving way to the stern ramp trawlers shown earlier. The broad doors hold the net open underwater. (Walter Scott drawing)

A post-war Halifax side dragger on the Grand Banks.

New "boats" equal old "vessels"

Not just trawlers and draggers but the whole fleet got more powerful after the war. Even small open boats might carry mechanical haulers and better lines and nets. Medium-sized vessels, from about 35 to 65 feet, might also carry winches, hydraulic systems, radar, sonar, Decca Navigators, and loran. Some might stay out fishing for a week or more, and travel widely over the fishing grounds.

Statistics for the period, although they changed in format and can be hard to decipher, show a huge hike in fishing power. In 1944, the Maritimes and Quebec had 986 "vessels," which meant, essentially, decked craft. By 1955, the number of vessels had grown to 1,519.

By the latter year, the department was classifying craft by tonnage, and defining vessels as anything over ten tons (that is, 1,000 feet of enclosed space, the same as in a box measuring ten feet per side). Although there is no precise length-to-tonnage ratio, craft over ten tons in those days were generally upwards of 35 feet. The great majority were less than 65 feet. Relatively small numbers would have fallen into the 65–100-foot and over 100-foot classes.

By 1968, statistical returns had dropped the term "vessels," but still classified craft by tonnage. The number over ten tons—vessel size—by now had nearly tripled over the 1944 level, to 2,567. These included 132 vessels in the 75–100-foot range and 154 larger vessels of more than 100 feet, mostly trawlers.

What about smaller craft? Back in 1944, the fleet of "boats," generally small open craft, had numbered about 23,200. By 1955, "boats" appeared in the statistics as craft less than ten tons, which generally meant less than 40 feet. By then, the number of boats had dropped to about 18,400; by 1968, it fell further, to 16,200. In particular, the smallest class of boat—unpowered sail and row-boats—was diminishing, from about 11,900 in 1944 to 8,900 in 1953 and 5,300 in 1968.

The number of motor boats was more stable: about 10,240 in 1944, 9,510 in 1955, and 10,863 in 1968. But the new boats had better gear, and were likely to have at least some enclosed space. Although

Smaller boats were still plentiful, as in this 1961 photo at Maces Bay on New Brunswick's Bay of Fundy shore, but the trend was to bigger craft. Most of those shown here fished lobster. The boat with a sluice-tank on top was a pumper for the herring-weir fishery; the pump stripped off the herring scales for separate commercial use.

"boats" by size, they could have as much fishing power as pre-war "vessels." Every class of fishing craft was getting stronger.

For years to come, departmental people and others would talk about the fishery in terms of the "inshore," which called to mind open boats making day trips; and the "offshore," as in company-owned trawlers. But particularly in the Maritimes and Quebec, the new medium-sized draggers, seiners, and scallopers represented something different. Although growing in number and packing high fishing power, they often got lumped with the inshore. Only in the 1980's did the term "midshore" become common. By then, these sometimes-ignored vessels had changed the balance of the fishery.

Table 18-1 shows changes in the Maritime and Quebec fleet from 1955 to 1968. (For 1955, the number of "vessels > 10 tons" includes trawlers and draggers. By 1968, statistics no longer broke out trawlers and draggers by gear, but showed craft by length and tonnage. Vessels over 100 feet were almost all trawlers.)

Table 18-1. Maritimes and Quebec fleet, 1955 and 1968.

	Year	N.S.	N.B.	P.E.I.	Que.
Trawlers	1955	19	0	0	0
	1968	n/a	n/a	n/a	n/a
Draggers	1955	82	77	13	19
	1968	n/a	n/a	n/a	n/a
Vessels >10 tons	1955	860	349	14	99
	1968	1,310	1,016	38	203
Vessels >100 ft.	1955	n/a	n/a	n/a	n/a
	1968	134	9	6	5
Motor boats	1955	4,529	2,057	1,403	1,524
	1968	5,326	1,947	1,671	1,919
Sail, row-boats	1955	3,481	3,569	724	1,131
	1968	2,982	1,307	567	445

Note: n/a, not available.

Larger vessels gain lion's share of landings

In the 1945–1968 period, "vertically integrated" processing companies with large processing plants and large trawlers became the main force in the groundfish industry, and for much of the industry in general. By 1961, the 57 or so large trawlers then operating in the Maritimes and Newfoundland were already taking about 40 per cent of the Atlantic groundfish catch.

In the years 1962–1966, the number of large vessels over 100 tons on the Atlantic more than doubled, from 111 to 273. (Vessels of that tonnage then ranged from 82 to 168 feet.)[4] Landings by trawlers over 100 tons quadrupled from 44,000 tonnes in 1960 to 185,000 tonnes in 1966. In the same period, landings by intermediate vessels, under 100 tons but over 25 tons (which meant, most often, over 50 feet long) doubled from 175,000 tonnes to 412,000 tonnes.

As of 1968, draggers and trawlers over 70 feet took 43.5 per cent of the groundfish catch in the Maritimes, and probably a similar proportion in Newfoundland. Back in the late 1920's, trawlers had taken much less of the catch; yet the inshore fleet had sufficient political sympathy to win the trawler ban. In the 1960's, with the trawlers taking far more, there was no move to ban them. Governments were bent on development. The trawlers were feeding more than 20 large plants, each employing hundreds of people, providing good year-round jobs.

As for landings of all species, back in 1957 the 32,000 Atlantic vessels less than 25 tons had taken about 487,000 tonnes of the 700,000-tonne harvest, leaving less than one-third for the 547 intermediate and large vessels. Over the following decade, those proportions roughly reversed. In 1966, the smaller vessels, now numbering 36,000, took only about 40 per cent of the now-higher total catch of 996,000 tonnes; the 829 intermediate and large vessels took 60 per cent.

A 1970 study by Carl Mitchell and H.C. Frick, the source of most of the statistics just noted, reflected the 1960's intention of building up the offshore fleet. "Subsidization of ... smaller craft did enable many inshore fishermen to get more efficient and comfortable boats, but it ran counter to the primary objective of the program, which was to aid in the expansion and modernization of the Atlantic fishing fleet by helping inshore fishermen to move into employment in larger, more productive offshore fishing vessels." The 1964 extension of subsidies down to 35-footers was a mistake, the authors said. "In the light of the existing excess of men, boats, and gear in the inshore fisheries, subsidization of small craft detracted from the drive to channel labour and investment capital into the offshore fisheries."

Boatbuilding subsidies were to reach a peak in the 1970's. The subsidy programs during their life produced various complaints, including that they helped boatyards more than fishermen and led to over-expansion. The charges had some validity. Especially in its earlier days, when it was channeled more to mid-size vessels, the boatbuilding subsidy answered prayers for many fishermen trying to move up from small or open boats. By the 1960's, most money was going to trawlers, thus shifting the balance of the fishing industry.[5] Growth in fishing power was taking place generally, but particularly at the upper end of the fleet.

Earnings rise

In boats of all sizes, the work, though still rough and demanding, was less back-breaking. Some lean fishermen developed bigger bellies. Trawlermen got home less often then they used to on other vessels, but they made better money. There were still many low-income fishermen, including a good proportion of the part-timers and transients. And independent operators now faced higher costs, and bigger debts to the banks or to the processing companies that often financed them. Still, full-time fishermen tended to be making headway, in earnings and in the pride that comes with better boats.

In the late 1950's, W.C. MacKenzie of the department's Economic Service wrote that incomes varied greatly, "from an average of $250 for a shoreman in the small-boat fisheries of Newfoundland or Quebec to an average of $5,000 for the skipper of an Atlantic Coast dragger or a Pacific Coast seiner." A 1962 study reported that the average fisherman in Nova Scotia made $2,260, up from $1,405 a decade earlier. (The average in a sense skewed the picture; many of the more serious fishermen would have been doing much better than the part-time and casual ones.)

By 1966, the average captain of a Nova Scotia trawler made over $14,000; of a Nova Scotia dragger, more than $7,000; of a P.E.I. dragger, about $4,200; of a New Brunswick Danish seiner, about $2,600; of a Quebec steel dragger, about $5,400. Non-captains generally made less. Deckhand shares on draggers, trawlers, and seiners 48 feet and up, in Newfoundland, Quebec, and the Maritimes, in 1967 were in the range $1,100–7,800.[6]

Although some fishermen were doing better than ever, most made less than people in other occupations. Federal taxation statistics for 1967 gave an average income of about $4,200 for Maritime and Quebec and $3,700 for Newfoundland fishermen, compared with about $6,700 as the national average for business proprietors in the retail trade, $5,500 for provincial government employees, and $5,300 for employees in general. Even those figures may have presented a rosier picture than the reality; at the time, the tax statistics tended to capture only the upper edge of fishermen. Many never filed taxes; others filed them but made more income from other occupations, and so never showed up as fishermen in the statistics.

Better jobs in bigger plants

Although catches rose during the period, fishing employment stayed relatively stable on an Atlantic-wide basis: declining in the Maritimes and Quebec, rising in Newfoundland, but totalling about 48,500 in 1954 and 45,700 in 1968, according to yearly counts by the federal fisheries department.

On shore, the trawler fleet by 1968 fed more than 20 large plants on the Atlantic, each employing hundreds of people, usually year-round. Although these were better jobs, it is unclear, at least to this writer, how much more shore employment there was in total.

Statistics Canada figures show no increase from 1953 to 1961. In the latter year, the department changed its method of calculating plant employment to a form of "full-time equivalency": adding up the number of plant employees reported for each month of the year, and dividing by 12. Calculated by the new method, their numbers rose from 10,500 workers in 1961 to 15,100 in 1968; the actual number of people employed at one time or another during the year was higher.

Until that mid-1960's increase, there seems to have been no truly radical jump in processing employment. Earlier, it seems that the modern groundfish industry meant less of a massive expansion than a reorganization from saltfish to frozen, by the same people and their children. Back in the 1930's, plant workers in the Maritimes and Quebec had already ranged around 8,000 or more, processing lobster, herring, groundfish, and so on, but not counting saltfish processing outside of plants, which was common in most Atlantic areas. In Newfoundland, back at the turn of the century, more than 20,000 women had worked curing fish on flakes and beaches, and many women still did so in the 1940's and 1950's, uncaptured by statistics.

One study in the later 1960's said that although the Atlantic provinces had 520 fish-processing plants, "any observer ... would probably classify at least half of them as improved fish houses or stores." The number of more solid "establishments" came to 256: 116 in Nova Scotia, 72 in New Brunswick, 51 in Newfoundland, and 17 in P.E.I. This study put plant employment at 11,000, with employee incomes ranging from an average $1,870 in Prince Edward Island to $2,650 in Nova Scotia.[7]

Catches increase less than hoped

Everyone had eyed groundfish with high hopes. It was groundfish that supported the most boats, plants, and people, groundfish that attracted the foreigners. And the Canadian catch did rise, the growth coming less in cod than in trawler fisheries on other stocks such as redfish and American plaice. But the increase was only 29 per cent from 1955 to 1968, from 486,500 tonnes to 628,500 tonnes. This was far behind the growth in fleet investment.

The average price per tonne of groundfish rose from $60 in 1955 to $93 in 1968, well above the 31 per cent inflation rate for that time span. Total value of groundfish climbed from $24.5 million to $49.4 million. Again, it was good but less than one might have hoped. Canada remained a commodity producer, plowing much of the catch into medium- or low-value products.

With pelagics in particular adding to the mix, Canada's total Atlantic catch increased more sharply than the groundfish catch, roughly doubling from 664,100 tonnes in 1955 to 1,267,539 tonnes in 1968. Total value more than doubled, from $49 million to $110.6 million. In volume, the greatest increases came in less valuable species. Herring more than quintupled

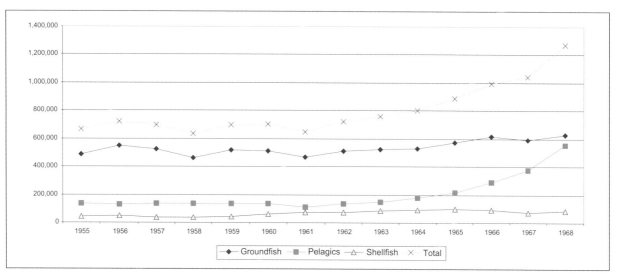

Landings in Atlantic Canada, 1955-68, by species group, in tonnes. Pelagics more than groundfish pushed the rise in landings. The pelagics boom would soon fade.

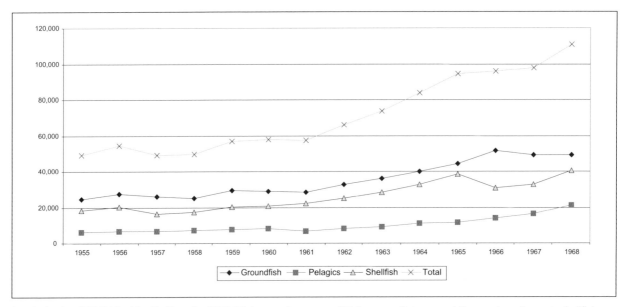

Landed value ($000's) in Atlantic Canada, 1955-68, by species group. While groundfish caused the most excitement, shellfish rivaled them in landed value. Groundfish created more processing jobs.

from 91,000 tonnes to 528,000 tonnes. But the herring bonanza would soon fade.

All told, the great development effort had positive but somewhat disappointing results. Although for some fishermen the 1950's and 1960's were bonanza years, overall investments were growing faster than profits. It was as if H.S. Gordon had written the history in advance. As a future fisheries minister, Roméo LeBlanc, would put it, "the technology has outrun the biology." But even in LeBlanc's time in the 1970's, the believers in development still foresaw great advances, if they could get rid of the foreign fleets.

Foreigners outrace Canadians

The post-war foreign fleets would fish an area until catch rates began to drop, then move elsewhere. Their numbers kept growing; by the late 1970's, some 1,500 distant-water vessels might cross the ocean yearly. In addition, American vessels fished redfish and other species, mainly on the Scotian Shelf and in the Gulf of St. Lawrence.

From 1955 to 1968, total catches of groundfish by I.C.N.A.F. members including Canada nearly doubled from 1,499,000 tonnes to 2,769,000 tonnes. But Canada's share, though rising in volume from 482,000 tonnes to 621,000 tonnes, fell from 32 per cent to 22 per cent of the total.

While making their own fishing explorations, foreign interests also followed the investigations of Newfoundland scientist Wilfred Templeman. For various newly explored stocks, foreigners beat Canadians into the fishery. Foreign catches flourished from smaller stocks such as Grand Banks haddock, which Burin Peninsula trawlers also fished heavily. This haddock

harvest rose to more than 60,000 tonnes, fell to low levels, and never recovered.

But the foreigners' greatest feeding ground was in the cod fishery. The biggest cod population was what came to be called the "northern cod" stock complex, found in I.C.N.A.F. divisions 2J and 3KL, off southern Labrador and eastern Newfoundland, and on the northern Grand Banks. While Canadians fished inshore, foreign trawlers hit the same stock offshore. Total catches of northern cod by I.C.N.A.F. members climbed from about 250,000 tonnes in the mid-1950's to more than 800,000 tonnes in 1968.[8]

For all northwest Atlantic cod put together, the catch enumerated by the International Commission for

Under I.C.N.A.F. procedures, Canadian patrol-vessel crews began checking foreign vessels. Here a crew departs the *Cape Freels* in 1975. (Kevin McVeigh photo)

the Northwest Atlantic Fisheries (I.C.N.A.F.) doubled from 902,000 tonnes in 1955 to 1,876,000 tonnes in 1968. But Canadian catches stayed almost flat, ranging between about 290,000 and 360,000 tonnes. By 1968, Canadians took less than 20 per cent of the cod.

I.C.N.A.F., which officially pursued M.S.Y., asked its member nations to set minimum-mesh-size regulations in their trawls, to let smaller fish escape. (Mesh-size regulations always sparked some skepticism among fishermen, who said that fish bunching up in the trawl, especially in the cod end, would block escapement.) Until the latter 1960's, however, nobody worried overmuch about conservation.

In 1967, an I.C.N.A.F. working group supported suggestions by Templeman and other scientists that catch quotas come into force.[9] Even then, opinions were somewhat mixed. A Canadian official wrote in 1968 that "in general, the European scientists are of the opinion that over-exploitation is now under way. Canadian scientists are of the opinion that conditions differ in different parts of the [I.C.N.A.F.] area." Although the Canadians thought the yield might yet be increased, over-exploitation was a clear threat.[10]

For most people in the Atlantic groundfish industry and the federal fisheries department, optimism still ruled, together with the urge to show some fishing muscle and claim a good share of the catch.

Department, industry push expansion

The preceding part of this chapter addressed some widespread aspects of the Atlantic fishery; the following will cover elements more specific to the Maritimes and Quebec.

At the outset of the 1945–1968 period, it was clear that the region lagged behind central Canada, and the fishery trailed within the region. Among young people, a large out-migration continued. Although high-school education became far more common after the war, many students destined for the fishing life dropped out around grade eight or nine. They could make money fishing; the necessary education came on the job.

Provincial governments in the 1950's and 1960's kept promoting industrial development projects, some successful, some almost laughable in retrospect. The federal government injected funds through mechanisms such as A.R.D.A. (the Agricultural Rehabilitation and Development Act of 1961), the Atlantic Development Board, and other precursors of regional expansion and equalization programs. The federal government encouraged fisheries projects almost at random. Provincial governments let processing plants spring up at will, and sometimes supported them with loans or other aid.

Groundfish: frozen trade takes over

Deck scene on a side trawler. They gradually took over from schooners; then stern-ramp trawlers took over from the sides. (National Film Board)

The Maritimes and Quebec depended primarily on groundfish and lobster, with smaller but important fisheries for herring, scallops, and other species. For groundfish, the frozen-fish trade expanded during the war to take more than half of production. Especially after the war, multiplate freezers became common. Refrigerated trucks offered better transport. More houses had refrigerators, and more supermarkets had frozen fish counters, which gradually displaced specialized fish stores. By the end of the 1940's, frozen groundfish was far ahead.

Various forms of salted fish, chiefly salted dried cod, remained important, especially in such areas as the Gaspé and southwest Nova Scotia. But salted fish went through painful market changes. In most areas, it was declining to a supplementary status, and in some places disappearing.

The dwindling fleet of diesel-powered dory schooners was centred in Nova Scotia, where by 1943 only a couple of dozen operated. As noted earlier, during the war they could still make good catches, landing as much as 200,000 pounds of haddock in a week or ten days.[11] (In later decades of less abundance, this would have been a decent catch for a modern trawler.)

After the war, with competition from trawlers and the general economy, schooner operators found it hard to get crews. The schooner fleet died away. Schooners had been within the reach of medium-sized companies or even individual families. After the war, large steel trawlers were generally out of reach. Although trawlers took more financing, the companies that could afford them caught more fish. Besides taking offshore cod, haddock, and pollock as schooners had done, trawlers opened up the fishery for flounders, barely touched by hook-and-line schooners. And starting in 1947, some 12 years after the Americans, Canadians began trawling for redfish in the Gulf of St. Lawrence. [12]

National Sea leads expansion

As time passed, large plants and a few large companies would dominate Maritime and Quebec groundfish production, outproducing the hundreds of small operators. Besides catching their own fish with large trawlers, the big operators often marketed fish for the smaller. Chief among the large companies was National Sea Products.

Back in the 1920's, the Smith interests of Lunenburg had owned only 20 of more than 100 Lunenburg salt-bankers. But even among Lunenburgers, the company had special vigour. In 1944, Bates's report on the sea fisheries gave impetus to the ambitions of Nova Scotia businessmen trying to put together a major fish company. In September 1945, the Smith interests merged with the other main producer of fresh and frozen groundfish, the Maritime National Fish Company of Halifax, to form National Sea Products.

National Sea Products produced not only fillets and blocks for export and secondary processing but also, and increasingly, retail packs for Canadian stores. The company soon bought a processing plant in North Sydney; expanded existing plants; got more heavily into fish-meal; bought more wholesale houses in Toronto and Montreal; built a new plant in Louisbourg, N.S., and acquired seven new trawlers. It would expand into Newfoundland and maintain its lead into the 1970's.

Other important companies included the Nickerson, Acadia, and O'Donnell-Usen interests. Foreign ownership increased as the trawler fishery built up. By the late 1960's, the region had several major year-round groundfish plants at such ports as Lunenburg, Canso, Louisbourg, Mulgrave, and North Sydney; and seasonal ones at such Gulf ports as Shippegan and the Magdalen Islands. The Maritimes and Quebec fleet by then included 154 vessels over 100 feet long, mostly groundfish trawlers. National Sea Products alone had about 40 trawlers.

All the large groundfish companies produced mainly frozen blocks and fillets. Blocks had become prominent with development of the "fish stick" or "fish finger" around 1953: fillets and smaller pieces of fish would be pressed into a flat block and frozen, after which machines would cut them into stick-like portions to be breaded, battered, and packaged. Typically, the secondary processing of blocks and fillets took place in the United States, the destination for the great majority of Canadian-caught groundfish. A few Canadian companies acquired plants in the United States. The others sold fillets and blocks to American secondary processors and distributors, mostly in Massachusetts. About

five large American companies repackaged the frozen fish in retail, brand-name packs for the supermarket trade. Other fish found their way into the restaurants and institutions of the food-service trade. Besides frozen fish, some fresh product went to the U.S., particularly from southwest Nova Scotia and both sides of the Bay of Fundy.

Fillets and blocks were predominantly a commodity-type product, with little brand-name identification (although National Sea Products established strong brand identity with retail packages in Canada). Marketing was less a matter of niches and value-added than of scoping out the overall supply-and-demand picture, and pursuing personal contacts with American buyers. Federal fisheries often helped Canadian producers with big-picture studies, forecasting the state of American demand and of supplies from overseas competitors, all this in relation to pork, beef, and other commodities.

More fishing power per boat

Among vessels less than 100 feet in length, and especially those less than 65 feet, owner–operators remained the rule. Boats less than 35 feet still made up the great majority of the fleet. Even at this smaller end of the fleet, as many fishermen moved up to more substantial boats (small "vessels" in the old terminology), they tended to get more financial backing from processors, as well as from federal subsidies and from newly active provincial fishermen's loan boards.

Over time, some medium-size boats took up dragging.

Among owner–operators, hook and line remained the dominant mode. Users included thousands of handliners in open boats, and thousands more setting "trawl"—that is, longlines with hundreds of hooks attached. With the advent of nylon, good nets had become cheaper, and gillnetting became popular on parts of the coast. (Gillnetting brought certain conservation problems—discarding of fish that spoiled in untended nets, and "ghost fishing" by lost nets—which received little attention until later.) Gradually, some of the independents took up dragging, the most powerful technology.

As already noted, radar spread widely in the 1950's, in independent as well as company boats. By the 1950's, practically all boats except the smallest used radios. The new communication meant that boats could help one another find fish, if they wanted to. (Warring with this impulse was the natural instinct to keep good fishing spots secret, which was now harder to do.) Decca positioning systems found use by some larger boats in the 1950's and 1960's. Later, loran sets meant that captains could retrace their course to the exact spot where they'd caught the most.

Governments push new technology

In the Maritimes and Quebec as elsewhere, the Fisheries Research Board, the department, and the provinces encouraged new fishing methods. The St. Andrews Biological Station helped develop an inshore flatfish fishery, experimenting in 1947–1949 with a 40-foot dragger, and gradually getting some fishermen and provincial governments, especially New Brunswick, interested. The station also worked on Danish seining, a trawl-like method of bottom-fishing, which the New Brunswick government followed up from 1957 on, encouraging the technique in the Gulf of St. Lawrence.

W.R. Martin and others at St. Andrews began demonstrating gurdies, already common on the Pacific coast, for hauling back longlines. Frank McCracken later recounted,

> The department of Fisheries built the *J.J. Cowie* as a longliner, B.C.-style. The *Cowie* demonstrated the haulers everywhere, with no result. Someone suggested just giving the haulers to fishermen. We put them aboard three boats, in return for the fishermen keeping records. The fishermen sold the idea. Within two years, no one was without a gurdy in Nova Scotia. That showed us we couldn't demonstrate gear on a government vessel. The mechanized haulers let the fishermen fish more gear and fish deeper. Eventually they led to lobster trap haulers.[13]

Gillnet haulers also appeared. Synthetic gillnets, available by the late 1940's, in the next two decades displaced longlines in some areas.

Meanwhile the Industrial Development Service helped spread the use of side and then stern trawlers and other new methods. The federal department made dragger captains, like lobstermen, buy a licence, but there was no restriction on the number of licences.

Future minister helps develop fleet

The New Brunswick government was quick to use federal subsidies and promote new technology. The key figure was Hédard J. Robichaud, later a federal Minister of Fisheries. Robichaud was born in 1911 in Shippegan, on the northeast coast of New Brunswick, where his father ran a saltfish business. "Shippegan, Caraquet, and Lamèque were probably the poorest part of the Maritimes," Robichaud recalled in the early 1980's. "The fleet was mostly schooners, without motors even as late as the 1920's. The smaller lobster boats had motors.

"The schooners were 50 to 65 feet, heavily built, and seaworthy. There were lots of them; in those three areas, perhaps over a hundred. Most of them were controlled by three or four companies: Loggie's, Robin Jones, Youngs, and us—we controlled six or eight."

Robichaud first worked with the federal Department of Fisheries. In 1946, when the provincial Department of Industry and Development set up a fisheries division, he became New Brunswick's Director of Fisheries. Robichaud immediately scouted around other provinces.

Hédard Robichaud

At Quebec, I found nothing. They spent more for fisheries without results than the province's entire landed value, on cold storages, boats, and so on. At Nova Scotia, I spent a day with the provincial director of fisheries. I was interested in their Loan Board, which started in the 1930's; though I wasn't totally sold on the fixed yearly payment they had. When I got back to New Brunswick, I prepared a report and recommended a Fishermen's Loan Board with a different system than in Nova Scotia. Fishermen would pay on the basis of their gross earnings. It took me nearly a year to sell the idea of a Loan Board to the government. The system I recommended is still in use. It works well, with minor losses to the Loan Board.

Meanwhile, I didn't believe what people had told me [in Nova Scotia and the Gulf] about having to have an 80 to 95-foot boat for dragging. I was authorized to find a naval architect to design a boat less than sixty feet, but for longlining only. Most of the fish dealers were against dragging in the Gulf.

I approached Walter McInnis [a well-known designer in Massachusetts] and had him design a 57–59 foot boat which would be suitable for longlining but easily converted to a small dragger. That fall, I influenced five fishermen from the Caraquet area to take this kind of boat, at about $25,000 each. When the boats were being built, I called the fishermen to a meeting at a hotel in Caraquet, in early March, and told them longlining would bring them too little money to pay for the boats.

A Caraquet wharf scene, 1948. (Library and Archives Canada, PA-115452)

Robichaud persuaded them to try dragging. "They started fishing in July and August, 1947, with longliners fishing beside them, and the draggers caught four times as much."

With the help of federal subsidies for medium-sized boats, small draggers began to spread. The Quebec government later borrowed McInnis's New Brunswick design. Robichaud himself designed a 48-foot dragger for the Bay of Chaleur fishery, and this too became popular. Robichaud again:

Before, the fleet was pretty much either schooners or open boats. The province had subsidized a few longliners which had a covered deck, and could take ice. Then the draggers started to displace longliners. First the co-ops

A Caraquet wharf scene after the buildup. Boats are decorated for a blessing of the fleet.

opposed them, but then the Lamèque Co-op accepted the idea of some draggers; they created more shore work. From about 1949–50, the dragger fleet was growing and you could see a change—fishermen with better houses, better cars.

Another aid to the industry came in 1947–1948, when Robichaud helped get private backers and a special federal grant for a cold-storage plant in Shippegan, available for use by various companies. "The cold storage did well; in six or eight years, it was buying fish of all kinds."[14]

Hédard Robichaud later became federal Minister of Fisheries for the years 1963–1968, presiding over a period of development and fast-rising catches.

Quebec runs active program

In the Maritimes, the federal government ran the fishing side, provincial governments the processing side, although the provinces also pushed fishery development where they could. In Quebec, although feder-al officials carried out fish inspection, the province administered most of the sea fishery. The province was active, helping fund vessel construction and sponsoring some development projects. It also operated a network of cold-storage plants, more than 50 of them by the late 1960's, for fish, bait, and ice. The provincial

Boats small and large at the Magdalen Islands.

Department of Industry and Commerce, then responsible for fisheries, also operated more than a hundred landing stations in small ports, to keep fish in proper condition while awaiting transport.

A staff of fish wardens, technicians, and technologists worked on fisheries. A fisheries training school operated at Grande Rivière, and the department also carried out consumer education and promotion. The province (helped by federal funds) encouraged the Quebec United Fishermen co-operative. A credit system helped fishermen buy boats and gear. Quebec also carried out some biological and hydrographic research.

Pelagic fishery sees huge increase

Herring were the main pelagic species, with the Gulf of St. Lawrence and Bay of Fundy the biggest producers. In Fundy after the war, more than 200 weirs of stakes, brush, and twine braved the tides of southwest New Brunswick, from the Maine border to above Saint John. Dozens more lined the shore of St. Mary's Bay in Nova Scotia. As well, fishermen sometimes "shut off" small coves with nets, to take up the fish at leisure. The weirs caught mostly small juveniles for sardine factories in New Brunswick and Maine.

Larger herring from the weirs got sold to smokehouses, where they were salted, then strung through the gills on herring sticks and hung in the rafters above slowly burning fires. The smoked herring went particularly to West Indies markets. Herring also went for lobster bait and pet food. As well, plants in Eastport, Maine, used herring scales to make "pearl essence," a decorative product that imparts shininess to materials. In both the Maritimes and Maine, relatively small amounts went for "reduction" to fertilizer or animal food.

As noted earlier, the department from the late 1930's allowed purse-seines. A small fleet from Grand Manan and Campobello fished, at first, mainly for sardines. Purse-seiners initially used fathometers: echo sounders, which beamed a sound pulse downward. On Campobello, Medford Matthews and his brothers learned of another electronic device: "searchlight" sonars, developed during the war, that pinged outward almost horizontally to detect submarines. Since herring schooled close to the surface, a searchlight sonar might find them better than a fathometer. With help from the St. Andrews Biological Station, the Matthews brothers around 1950 got a war-surplus sonar and began using it for purse-seining.

Seeing their success, other fishermen acquired sonars. The first boats functioned as informal schools; crew members would learn how it was done and get their own sonar-carrying seiners. The Matthews and Savage families helped spread the new method to Grand Manan and Nova Scotia.

Seiners found that a major body of large herring schooled from June to October off southwest Nova Scotia. They could catch high volumes, but needed market. New, high-volume reduction plants sprang up, particularly in Nova Scotia; they soon took the bulk of the Bay of Fundy catch. Sardines continued to provide the most employment, and gave fishermen a far better price per tonne; but fishing reduction herring, especially in the 1960's, brought bigger catches and more excitement. By 1968, more than 20 seiners operated from Campobello and Grand Manan, with a similar fleet in Nova Scotia.

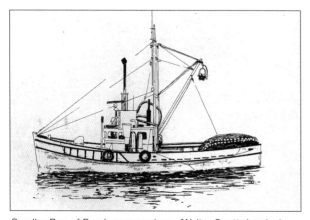

Smaller Bay of Fundy purse seiner. (Walter Scott drawing)

A pumper inside the weir and a carrier entering. (Walter Scott drawing)

Meanwhile, the use of searchlight sonar for seining spread from Campobello along the coast and around the world. In the United States, the Wesmar marine-electronics company gained impetus from the new development. On the Canadian side, Medford Matthews later worked with the C-Tech company in Ontario on omni-directional sonars. Instead of sending out a narrow beam that scanned back and forth, the "omnis" spread a pulse all around, like the ripples from a stone. Ontario manufacturers ultimately exported omni sonars to other countries including Japan, a fishing and electronics giant that was happy to purchase Canadian fishing electronics.

Bay of Fundy seiners also adopted the power block, invented in California in 1953 by Mario Puretic. This was a V-shaped pulley suspended from the end of the boom, resembling two truck tires fastened together. Hydraulics gave it great power. The fishermen after setting their seine and completing the circle used the power block to "dry in." They rove one end up through the block; the machine hauled in the netting, and the fishermen spread the descending folds on the deck below. Bringing in seines had been hard labour; the power block made it easy to "dry in" hundreds of tonnes.

At first in the Bay of Fundy, the herring carriers—single-masted schooner-like vessels without the sails—"tended out" on the seiners as they did the weirs, buying fish on behalf of the factories, and brailing them aboard with jignets, long-handled hoop-nets that pursed up at the bottom, then opened to release the fish. Then a third type of herring vessel appeared, the pumper, which pumped the herring from seine to carrier while removing the scales for pearl essence. Finally, in the late 1950's and the 1960's, the Matthews, Savages, and other fishermen began building "combination" vessels, some of them steel and as large as 100 feet long, that could seine, pump, and carry. In following decades the combination vessels became dominant, though most stayed more or less in the 65-foot range.

As purse-seiners multiplied, weirmen complained about seiners taking the fish before they could get inshore, or seining too close to the weirs. The department put in closing lines to keep seiners clear of weirs and out of certain areas. But the new seiners faced no major obstacles. In the Maritimes in the 1950's, regional director Loran Baker would tell the weirmen, "I've already got huge books stuffed full of regulations and you want me to make more of them?"[15] The department wanted productivity, and herring appeared plentiful, with seiners roaming into new grounds. The Matthews family eventually seined everywhere from Newfoundland and the Gaspé to south of Cape Hatteras.

Federal fisheries took note—this was *development*—and in the 1960's the Industrial Development Service helped spread seining techniques to the Gulf of St. Lawrence, sometimes contracting vessels and fishermen from British Columbia, where sein-

ing had never undergone a ban. Wes Johnson, a former fisherman from B.C., was particularly active in trying and promoting seining technology. In 1966, the department sponsored an Atlantic herring conference in Saint John, N.B. Provincial fisheries departments vied with one another in projecting great harvests. Their estimates added up to more than a million tonnes a year; a top Nova Scotian official said that that province's projections were conservative.

By that time many fishermen were beginning to worry that "the fish don't have a chance." Stan Savage, a high-line captain from Campobello, said that the conference was perhaps finding ways to "kill off a nice little fishery." Federal fisheries scientist Noel Tibbo expressed caution. But the dominant spirit was that of Alfred Needler, deputy minister for most of the 1960's and a giant figure among officials, but generally considered to have had a blind spot on herring. (On one occasion during the herring expansion, two senior officials went to talk to Needler about the need for caution and conservation. But as one recounted, "he got that hooded-eye look and practically threw us out of his office."[16]) Needler took the line, which he adopted from department scientists on the Pacific, that it would be hard to deplete the herring, and even if that happened, catches and therefore fishing would drop off before there was real biological damage. In the words of one purse-seine captain, "he told us you'll never make a dent in 'em."[17]

The 1966 conference encouraged further expansion. Reduction plants and seiners were now becoming numerous in the Gulf and Newfoundland; this rising fishery exploited two unusually strong year-classes, from the late 1950's, in a large stock migrating between the Gulf and southwestern Newfoundland.[18] Further impetus came when a number of large seiners migrated from British Columbia to the east coast, fleeing the herring depletion and closure on the Pacific. Maritime and Quebec catches of herring rose from the 90,000–100,000 tonne range in the late 1940's to about 380,000 tonnes in 1968.

By then, seiner captains in the Bay of Fundy and some in the Gulf were kings of the wharf, many captains buying big cars, building big houses, and taking thousands of tonnes a year. They could catch fish in volume probably as well as any captains in the world in vessels their size. But, just as in groundfish, huge foreign vessels were appearing in the late 1960's, seining or trawling for herring. With fishing pressure building inshore and off, and federal and provincial governments encouraging headlong expansion, the stage was set for the herring crises of the 1970's.

Commercial tuna fishery begins

The Matthews family did some tuna seining off the American coast. In nearby St. Andrews, the New Brunswick government, in one of the development ventures common in the 1960's, attracted and aided a tuna plant operated by the American company, Ocean Maid. (Ordinarily, American canned tuna faced a sub-

Rod and reel fishing for bluefin tuna. (Walter Scott drawing)

stantial tariff; producing tuna within Canada obviated the problem.) The company operated half a dozen large tuna seiners, which fished in tropical waters and shipped the fish to St. Andrews.

After the war, the sport-fishery for bluefin tuna was still going strong in southwest Nova Scotia, where the Wedgeport competition attracted an international crowd. In 1949, sport-fishermen took nearly 1,800 bluefin. For Wedgeport, the fishery largely faded away by the 1970's. Tuna by then were popping up in a hybrid, sport-cum-commercial fishery in Prince Edward Island. This would in the 1970's flower into a vigorous fishery with customers in the Far East. An intermittent sport-fishery for tuna also developed on the east cost of Newfoundland, where some sport-fishing had taken place decades earlier.

Longlines transform swordfish fishery

From the 1940's to the 1950's, swordfish landings by the Nova Scotia fleet roughly doubled, to about 3,000 tonnes. Values rose to more than $1 million. Like other fisheries, swordfish benefitted from subsidies, loans, and other development efforts. Larger vessels began going further offshore.

Canadian fishermen adopted night longlining from 1962, perhaps after observing swordfish catches by foreign vessels longlining for shark and tuna. Longlining soon displaced harpooning. By 1964, 100 boats were longlining. Landings more than doubled again, to reach more than 7,000 tonnes in 1964, after which they dropped back to around 5,000 tonnes, despite increasing effort.

While longlines brought in far more fish, they also took a far higher proportion of young ones. Back in 1909, the average size of swordfish landed had been estimated at 300 pounds; in 1959, the average was 264 pounds dressed; by 1970, the average was 88 pounds. As vessels moved further offshore, the Cape Breton fishery largely faded away. It is unclear how much of the decline related to fishing effort (swordfish migrate widely, and other nations fish them), climate fluctuations, or other factors.[19]

Shellfish: Trap limits, licences stabilize lobster fishery

Lobster fishing in the immediate post-war period remained a generally low-income proposition. But prosperity would improve market prices, as more consumers could afford a special treat.

The interwar period had seen the gradual spread of size limits throughout the Maritimes. As well, back in 1934, the department had prohibited individual fishermen from lobstering in more than one district in any one year. In 1945, it strengthened this limitation by forbidding the use of any boats or gear in more than one district. These regulations prevented boats massing up to the detriment of the resource, but also, in the doubled-edged nature of many regulations, worked against specialization and a potential means of reducing costs.[20a] Territory remained a sensitive issue with lobster fishermen. When fishermen spotted boats from another community setting traps in local waters, they might make their displeasure known, sometimes even cutting the traplines or destroying the traps of their rivals.

The already extensive set of lobster regulations would see major changes at the end of the 1945–1968 period.

Attempts to change lobster seasons fail

During the war, Needler recruited David Wilder, a graduate of Queen's and the University of Toronto, to work on lobster at the St. Andrews Biological Station, and gave him six technicians. Wilder's work stretched from 1942 to 1975. He became the F.R.B.'s leading expert on lobster biology. His studies showed that lobsters were mainly local (although later studies in the 1980's showed mature lobsters in the Gulf of Maine sometimes travelling more than 100 kilometres).[20b]

Departmental officials were asking the F.R.B. for advice on making the best use of lobster. Wilder took up the question, working at times with economists Jack Rutherford, Gordon DeWolf, and H.C. Frick. "To get answers," Wilder said in an interview years later, "you had to get into biology, economics, and sociology."

The lobster seasons had been set by groping and common sense, according to such factors as ice conditions, seasonal presence of lobsters, protection during egg laying and moulting, and quality of lobsters caught.[21] But Maine, with a year-round fishery, appeared equally productive. And Maine escaped the constant tension of fishery officers looking for off-season poachers. Wilder and colleagues concluded that fishermen could benefit from a change in lobster seasons. New timing could take account of such factors as the competition from the United States, which was strongest in July–September; the wear-and-tear of winter fishing; the yearly occurrences of soft-shell lobsters; and the best catch rates. But in a conservative fishery, change was to be a hard sell.

The summer-time canner fishery in the Gulf of St. Lawrence took too many lobsters, from a conservation point of view. The big summer fishery also glutted the market and reduced prices, provided a cloak for selling lobsters caught illegally in nearby closed areas, and made necessary the tarring of traps to repel shipworms. But the department feared that if it closed that season in favour of another, it would be unable to control the poachers.

So Wilder and colleagues looked to Nova Scotia. In southwest Nova Scotia, with a December–May season, the fishery was best in the first three weeks. "After that most of the fishermen sat around, cursed the weather, and so on. Then in March and April, they had a lovely fishery." The F.R.B. talked up the idea of going to a spring fishery only, which would reduce effort, yield bigger lobsters, and perhaps increase the whole area's production. "But the fishermen opposed it, partly since the fall fishery gave them some money just before Christmas." A different set of objections scuttled an attempt to adjust seasons in Queen's County.

The upshot: no change in seasons. "Above everything," Wilder said in retrospect, "the seasons are ingrained in people's minds. I'm not sure yet what would be the ideal seasons. The answer isn't in biology. The present seasons aren't too bad a balance, but they have no biological effect."[22]

Wilder and his colleagues returned to the old idea of designing traps that would avoid catching small lobsters. In 1949, the department set new regulations for minimum spacing between the two bottom laths on lobster traps. But again fishermen complained, especially at Grand Manan, N.B., and the department in 1955 revoked the regulations for the Maritimes and Quebec.

On the whole, Wilder and his colleagues made important contributions to lobster science. But they met great difficulty in making changes based on an economic rationale, without a majority of fishermen on side.

Trap limits come into force

The post-war economy was creating more non-fishing jobs, but some of the people with other jobs fished part time. "Moonlighters" raised the ire of full-time fishermen. Teachers, with their long summer vacations, and railway employees stirred particular wrath. Meanwhile, landings, which had stayed fairly stable in the 1950's, tended downward in the 1960's. Catches collapsed along the eastern shore of Nova Scotia. Overall, Canadian yields were dropping towards the near-record lows of the early 1970's, despite increased effort.[23]

More fishermen were using mechanical haulers, which let them fish more traps, as many as 700 in some instances, although the average may have been around 200.[24] In 1966, experimental regulations began to limit the number of traps per fisherman in Northumberland Strait, between New Brunswick and

Prince Edward Island By 1968, trap limits, ranging from 250 to 400 traps per boat depending on the district, spread to all Maritime areas. Quebec by then already had trap limits. For the time being, Newfoundland escaped the rule.[25]

"Limited entry" appears in lobster fishery

In 1967, departmental economist Jack Rutherford and colleagues published a major study of the lobster fishery in all its aspects. Noting the same pattern of futile over-expansion that H.S. Gordon had lamented, Rutherford called for consideration of property rights, and auctioning off the fishing privilege.[26] In effect, more than a century later, he was re-advancing the ideas of W.F. Whitcher and Richard Nettle. His suggestions passed with little notice at the time.

But a milder approach to providing security was taking shape. The department would soon limit the number of fishermen in major fisheries. "Limited entry" hit its stride from 1968 on, and belongs mainly in the next part of this book. But it began earlier in the lobster fishery.

The department by 1966 was getting worried about overcrowding in the lobster fleet, and lobstermen were complaining about moonlighters. In the period 1966–1969, minister Hédard Robichaud and his successor Jack Davis announced a series of restrictive measures, first affecting P.E.I. and the Northumberland Strait, then spreading to the southern

Workers separate meat from shells in a Barachois, N.B. lobster canning plant.

Gulf and all the Maritimes. By the end of all the announcements, every lobster boat had to be registered, each trap needed a tag, and the operator and his helpers had to pay licence fees. As of 1969, a personal lobster licence cost $2, and the boat registration cost $5.

As well, the new system established Class A licences for those who had been fishing more than a certain number of traps, and Class B for those (mainly seen as moonlighters) who had fished fewer. Class B licences would be non-transferable and would eventually die out.[27]

Most important, no one could get a licence unless he had held one the previous year. The fleet would get no bigger than its present size—about 10,000 vessels and 23,000 fishermen landing $25 million worth of lobsters in the Maritimes—and the department would look at ways to reduce it. The licence would be on the boat: "from now on a licensed lobster boat is the passport to the lobster fishery," Jack Davis said.[28] Licences would be transferable with the boat. In effect, fishermen would sell the licence with the boat.

It all meant that no longer could whoever wanted go fishing for lobster. Although the department had since the 1930's limited entry in Atlantic salmon driftnets and set up an eligibility list, the limiting of lobster licences marked the real beginning of modern-day licence control on the Atlantic coast.

Cliff Levelton, who had started as a fishery officer in British Columbia and was by the 1960's a senior manager in Ottawa, later said that limited entry in the lobster fishery was a departmental initiative, sparked by the anti-moonlighter sentiment. But officials also drew on British Columbian thinking about limited entry, ideas with which deputy minister Alfred Needler, economist W.C. MacKenzie, and some other officials were familiar. Still, said Levelton, "I'm not sure we understood all the implications."[29]

Those implications went much deeper than excluding moonlighters. Before, the government's main job had been to keep a supply of fish in the water; anyone could go fishing and try to catch all he wanted. In deciding to limit the number of fishermen, the department took on the additional obligation of deciding—in the absence of an auction, lottery, or other such mechanism—who would get the licence. For the time being, the answer was obvious: the existing fishermen, on whose behalf the department was acting. But what happened to the licence when the fisherman retired?

The fishery could support only a finite number of people. Limited entry suited economic theory. But with the government saying who could fish, many people saw themselves as deserving a licence. "My family always fished," "this community was built on the traditional fishery"—such would be the war cries of the future.

The questions of who should have a licence on what social, economic, or moral grounds; the frequent pressure to increase the number of licences; the difficulty of controlling or cancelling a transferable licence—all

would bedevil future fishery managers. So would the growth in licence values. A lobster fisherman who in the late 1960's paid $2 for his licence might by the turn of the century be selling the privilege for half a million dollars or more.

The department was edging into this new world without a clear idea of what it all meant. Nobody went back to research the complex licensing struggles in the old Province of Canada and in freshwater and B.C. fisheries after Confederation, which could have shed light on the new initiative. Still, the main goal was clear: to give existing fishermen more security. The department would generally cling to that rule of thumb in future.

Scallop fishery expands

The scallop fishery centred at Digby, Nova Scotia got stronger after the war. By the late 1970's, close to a hundred boats, some specialized, some multipurpose, would hold licences for the Digby and full Bay of Fundy fisheries. Others from New Brunswick held licences for that shore only. Production would run to several hundred tonnes of meat weight, equivalent to about eight times that in live weight.

By that time, a new fishery was outdoing the old Digby fleet. In 1945, Captain Johnny Beck, fishing out of Halifax for the Clouston company, tried dragging for scallops offshore around Sable Island, where groundfish draggers sometimes found them. He then switched to Georges Bank, where American draggers were already scalloping.[30] Other Canadian scallopers followed Beck to Georges, fishing especially the north-

Scallops coming aboard the dragger *Singer*, 1951. (Library and Archives Canada, PA-141288)

east peak, some 90 miles south-southwest of Nova Scotia.

The fishery did well. By 1963, some 50 large draggers, usually 90 feet and upwards, operated from such ports as Lunenburg, Riverport, Liverpool, Yarmouth, and Saulnierville. By the early 1970's, the offshore scallop fleet counted more than 70 vessels. Some belonged to independent operators; most, to a handful of corporations including National Sea Products.

Other scallop fisheries sprouted in the Northumberland Strait, elsewhere in the southern Gulf, and in western Newfoundland. Gulf landings rose sharply after 1965, reaching more than 1,100 tonnes by 1970, but then declined. Nova Scotia's offshore fishery remained the biggest producer, providing 50–90 per cent of landings, followed by the Bay of Fundy. Atlantic scallop landings overall saw a phenomenal increase, from 6,100 tonnes (round weight) worth $730,000 in 1955 to 56,800 tonnes worth $7.8 million in 1968—in little more than ten years, a tenfold increase.

Governments sponsor crab and shrimp experiments

Meanwhile, crab began to draw attention. In the early 1960's, some processors in the Shippegan area of New Brunswick tried processing crab taken as bycatch by groundfish draggers. In the mid-1960's, federal–provincial development efforts encouraged a crab fishery in the Maritimes and Quebec, using both drags and traps. In 1965, a Danish seine fishery began off Cheticamp, Cape Breton, expanding into New Brunswick and Prince Edward Island in 1966 and Quebec in 1967. By 1968, about 60 boats took part in the Gulf crab fishery, and in New Brunswick ten plants were processing them. The New Brunswick government was helping to develop specialized crab boats.[31]

Spider crab, the main species, had been deemed a nuisance; fishermen might occasionally cook a few, but throw the rest overboard. As markets developed, the name got changed to queen crab and then to snow crab. No one would have imagined at the time that by the end of the 20th century, the Atlantic coast crab fishery would be far bigger than the historic but decimated cod fishery.

Shrimp-fishing experiments also took place in the 1960's, but significant developments would come only in the 1970's.

Irish moss harvest grows

Before the Second World War, some people in Antigonish County, Nova Scotia, on the Gulf of St. Lawrence, collected the seaweed Irish moss (*Chondrus crispus*), and sold it for use in blanc-mange desserts. But Irish moss had many other uses, including as a stabilizer or clarifier in beer, wine, ice cream, coffee, and other food products; and as a substance in shampoos, ointments, or insect sprays. When the Second

Irish moss, raked from the bottom of the sea, is spread to dry on flakes (wooden racks) at Miminegash, P.E.I. The helpers are removing weeds and other impurities. (National Film Board)

World War cut off traditional supplies from Europe and Japan, Canadian production expanded.

Fishermen would rake moss from boats or gather it from beaches after storms, sometimes using a horse and cart in earlier years. They sold the moss to buyers, who dried it mechanically and shipped it out to extraction plants.

The Maritimes became the world's main supplier of Irish moss. By 1974, the harvest of more than 50,000 tonnes was worth nearly $6 million, although a drop followed. About 2,000 harvesters from several dozen communities took part, with Prince Edward Island producing about half the crop. Southwest Nova Scotia also produced Irish moss. Grand Manan, N.B. harvesters collected and dried another seaweed, dulse, as a food delicacy sold in the Maritimes. In the 1980's, the private company Acadian Seaplants, in Nova Scotia, began cultivating seaweed as well as harvesting it from the wild.

Salmon face new dangers

While new fisheries developed, a historic one was in decline. The famed Atlantic salmon had fed pioneering settlers in the New World, supported early commercial fisheries, and given impetus to the sport-fishing industry in Atlantic Canada, which grew with the economy. Anglers loved the salmon for its beauty, its fighting spirit, and its wooded upriver hideouts. Sport-fisheries for salmon developed in all the Atlantic provinces. Particularly in New Brunswick, wealthy pursuers of leisure activity from central Canada and the United

Specially-designed tank trucks take salmon collected at the Mactaquac dam to the hatchery brood-stock ponds, or to the upper waters of the Saint John River.

States came to backwoods lodges for idyllic river excursions. On the Restigouche and Miramichi, membership fees for the more exclusive lodges by the late 1960's were said to run as high as $20,000.

Meanwhile, a commercial fishery kept going, concentrated in Newfoundland and New Brunswick, and using trapnets and setnets in estuaries and driftnets further offshore. Starting in 1931, the department controlled the number of salmon driftnets on New Brunswick's Gulf shore. This had been an early example of Atlantic licence limitation.

In the Maritimes in the 1950's, besides the trapnets and setnets, 130 driftnet boats operated in the Gulf of St. Lawrence. Quebec had a smaller fishery. Gulf of St. Lawrence salmon migrated past Newfoundland and Labrador, where fishermen took them mainly with surface gillnets set out from shore, as well as with some inshore traps. A driftnet fishery operated off Port aux Basques, catching Gulf salmon on their migrations. In many places, Native people took part in the fishery, for their own uses or commercial sale.

The commercial fishery after the Second World War still took the majority of the catch. Sport-fishery advocates, especially in New Brunswick, argued that angling was a better use of the fish, bringing more money into the region. Commercial-fishery supporters

retorted that their trade provided food and jobs, and rounded out the season for fishermen fishing other species. Still, most observers came to agree that the salmon sport-fishery was worth more.

Salmon had to run a long gauntlet, from streams to the far ocean and back again. After the war, threats were thickening all along the route. Fishermen chasing salmon had better boats and gear. In the rivers, pollution and damage to habitat were increasing. Offshore, other fisheries were taking more salmon bycatch. And a new offshore fishery was targeting salmon directly.

Studies by the F.R.B. and others established that most Canadian Atlantic salmon migrated to west Greenland, as did Atlantic salmon from Europe. At west Greenland, a new commercial fishery got going in 1959. Greenlanders used setnets from shore, and an international fleet fished offshore with driftnets. Landings there shot up from 14 tonnes in 1959 to nearly 2,700 tonnes in 1971.

Canadian catches were already dropping before the war. In the Gulf, matters got worse in the 1950's. Landings fell from about 6,100 tonnes in 1930 to about 1,200 tonnes in 1955. In New Brunswick generally, the spraying of forests with DDT in the late 1950's and early 1960's probably hurt the stocks. On the Saint John River, draining into the Bay of Fundy, the provincial government in 1967 built the Mactaquac hydropower dam, just above Fredericton. The department in 1968 opened a hatchery and rearing station below the dam, and began trucking returning salmon around the dam. (Around the same time, the department closed two older hatcheries in New Brunswick and two in Nova Scotia.)

Quebec catches fell by about half in the 1950's and 1960's, to about 170–190 tonnes in the late 1960's. Only in Newfoundland and Labrador did catches increase, rising from 727 tonnes in 1956 to 1,814 tonnes in 1967. Generally, salmon were heading into trouble.[32]

Higher catches, more fishing pressure

From 1945 to 1968, the Maritime and Quebec fleet became far more dynamic. Decked vessels largely replaced open boats. Diesels began replacing gas engines in the larger craft. Nylon brought better nets for drags, gillnets, and seines. Mechanical and hydraulic haulers made nets, lines, and seines easier to handle. Before the war, fishermen had used dead reckoning through night, fog, and bad weather. Now they could tell exactly where they were, through Decca Navigators, loran, radio beacons, and radar. Through radio, they could help one another to find fish or market. With fathometers and sonars, they were beginning to see what was in the water.

Catches grew with fishing power, but not as fast, climbing from 374,200 tonnes worth $36.4 million in 1955 to 802,000 tonnes worth $86.9 million in 1968. Nova Scotia led among provinces, followed by New

Brunswick, Quebec, and Prince Edward Island Among species, pelagics led in volume in 1968, followed by groundfish and shellfish; but shellfish still led in value, with groundfish well behind. Despite their great volume, pelagics were worth less than half the shellfish harvest.

The resource still seemed abundant, with a few exceptions, such as Atlantic salmon. For many fishermen, far more cash was coming in with the higher catches, and flowing out with the bigger expenses. But fleet capacity was outrunning landings.

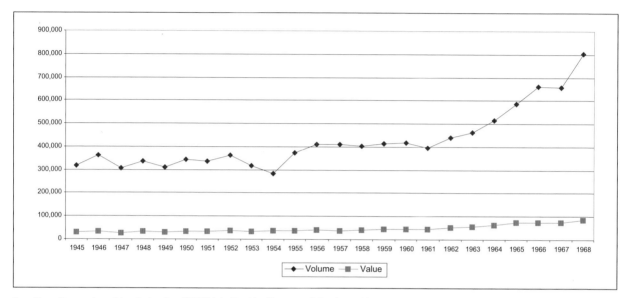

Landings (tonnes) and landed value ($000's) in the Maritimes and Quebec, 1945-68. Pelagics accounted for much of the rise in volume; their lower prices meant that value stayed flatter.

CHAPTER 19.
Newfoundland, 1945–1968

For Newfoundland in the late 1930's, the fishery still meant everything. Apart from St. John's and a handful of pulp-mill or mining towns, the colony consisted of some 1,300 outports living mainly from saltfish. The fishermen now had gas engines, and a few companies used mechanical dryers for cod. Otherwise the industry at the start of the Second World War operated much as it had a century before. Individuals or small concerns put up dried salted fish and sold it to an intermediate company or directly to a large St. John's company for export. Annual backing for operations came from those same large St. John's corporations—"the merchants," such as Job Brothers—and trickled down through the credit or "truck" system, via local merchants, to the fishermen at the end of the line.

Stowing fish on a side trawler on the Grand Banks, 1949. Draggers held great promise for Newfoundland. (Library and Archives Canada, PA-110810)

The whole system operated on little cash and little sophistication. Exporters were largely price takers rather than price setters, especially since the quality of products had declined in the 19th century. A vessel from Newfoundland could arrive at a foreign port to find the market flooded and the cargo worth little.

The Commission of Government in the 1930's had tried to come to grips with the fishery by various interventions, including the 1936 setting-up of the Newfoundland Fisheries Board, with wide powers over production and marketing. Then the war drove prices up, and showed the potential demand for fish. The building of American bases added to wartime prosperity.

Meanwhile, the Commission of Government wanted to encourage a more modern Newfoundland. In 1943, the government began lending money at low interest for companies building frozen-fish plants and buying offshore trawlers. Recipients in the next few years included Fishery Products Ltd. (led by the Monroe family), John Penny and Sons, and Job Brothers, sparked by Hazen Russell, who branched off to lead Bonavista Cold Storage.[1]

P.D.H. Dunn, one of the commissioners, in 1944 gave a radio address outlining a shining new fishery, which would be far stronger and more productive. Dunn said that the rising cost of living had long outpaced fish prices. In the period 1870–1938, the standard of living had declined, resulting in malnutrition and other ills. Now Newfoundland needed new fishing centres, which would use the frozen-fish trade as a base, and would dry only the surplus cod. The industry needed to centralize at about 15 main points. It needed fish-storage facilities, draggers, collector boats, roads, and cash payments for fishermen. It also needed co-ops, although these latter took a good deal of education and work among members. People should stop trying to square the circle by saying to the government, do this, do that, but stay out of trade. In fact, the government should license processing operations. It should influence marketing through the Newfoundland Fisheries Board. For faster growth, the government should finance private enterprise by buying shares. And the government would invest $4 million in a reorganization that would in all cost $10–$15 million.[2]

Although the Commission of Government kept helping companies to acquire trawlers and freezing plants, nothing like Dunn's outlined reorganization took place at the time. But Dunn's address prefigured future developments. The demand for frozen fish was by itself bringing great changes. And the idea of government intervention for development would persist.

Dunn was expressing a vision similar to that of Stewart Bates and many observers of the time. Progress was taking place all around, through mechanization and industrialization; it should also happen in the fishery—*especially* in the fishery, which had great resources and potential, if it could just zoom forward from the antiquated structure of small communities, small boats, and old-style operations.

The current situation seemed clearly undesirable. No one praised it; many condemned it. A federal fisheries official writing some years later described post-war Newfoundland in these words: "Complete dependence on the local merchant for credit and supplies, coupled with the physical limitation on his productivity of saltfish because of traditional and obsolete catching and curing techniques, condemned fishermen to a lifetime of unremitting and unrewarding labour. The lack of alternate employment further restricted and bound him to his little world of poverty and toil."[3]

Even if it still had many problems, Newfoundland emerged from the war with more money than it had enjoyed for generations. Great Britain was now a weakened force, exhausted by two world wars and the Great Depression. In both the United Kingdom and Newfoundland, people questioned the need for British administration, which had prevailed since 1934. A National Convention from 1946 to 1948 debated Newfoundland's future; two referenda resulted in a narrow vote for confederation with Canada in 1949.

Confederation brings five-year fishery transition

The epic and complex struggle over Confederation took place mainly on other battlegrounds than the fishery. Don Jamieson, then a broadcaster and anti-confederate, later a renowned Canadian cabinet minister, had this to say years later about the fisheries aspect:

> Sometimes we used scare tactics about the fisheries. But fisheries wasn't a big issue in the Confederation row. They sold Confederation basically—and it was only a one percentage point victory—on social security, the welfare net. The baby bonus would give $5–$6 a month, and the old age pension $40.

> Most fishermen voted for Confederation, despite all the warnings. On the south coast, there were a lot of connections with Nova Scotia and all the Maritimes. Some fishermen from the south coast served on Lunenburg bankers, they sold fish to Lunenburg vessels, they had a real linkage with Nova Scotia. On the south coast, there's a lot of Nova Scotia blood. But the Roman Catholic areas [such as the Avalon Peninsula] were practically 100 per cent anti-Confederation, largely based on religion. The Archbishop was vehement against it, because of fear of divorce, then nonexistent in Newfoundland, and of damage to the denominational school system.[4]

Before Confederation, the Newfoundland's Natural Resources department had carried out stream patrols, run the Bait Service, and the like. Under the same department, the Newfoundland Fisheries Board had dealt with the business side, including fish inspection. The inspection officers dealt with other sea fishery matters as well; the conservation officers dealt with fresh water.

After Confederation, the federal and Newfoundland governments provided a five-year transition period for enforcement, food inspection, and other aspects of fishery management to switch over to the federal system. Newfoundland employees became federal employees. The "inspectors" monitoring food production continued doing conservation work in the sea fisheries, until an organizational change years later.

Raymond Gushue, a chief figure in Newfoundland's fisheries administration, headed the federal department in the Newfoundland region; later, Harold Bradley took over. L.S. Bradbury of Newfoundland moved to Ottawa to lead the department's Industrial Development Service. The St. John's laboratory became federal, under the Fisheries Research Board, and researcher Wilfred Templeman became one of Canada's most renowned fishery scientists.[5]

Smallwood promotes development

An immediate effect of Confederation was more cash for individuals, through family allowances, old-age pensions, and unemployment insurance; and more cash for the province, through various federal–provincial arrangements. But what would be the overall direction? A leading force was Joseph R. Smallwood, the chief promoter of confederation, who became premier in 1949. Smallwood was a man of great knowledge, unquenchable energy, and grand dreams that sometimes shoved reality aside. He had once been a disciple of William Coaker, had written a book about the great union leader, and had tried his own hand at organizing fishermen, with far less success than Coaker.

Smallwood had been a socialist and retained some of those sympathies. At the same time, he believed in industrial growth and economic productivity. Despite Newfoundland's remote and rocky situation, there were some encouraging economic signs. The war had drawn many people out of the fishery; after the war, new forms of employment appeared. Smallwood gave great new impetus to education. Trying to diversify away from the fishery, he chased every kind of industrial development for Newfoundland. Roads linked scores of communities for the first time; other infrastructure came into place. But some development efforts were nearly ludicrous in retrospect.

Smallwood is reputed to have said in the 1950's that the cowboy would be more important to Newfoundland's future than the fisherman. It is often stated that he advised fishermen to "burn your boats." But he also made major efforts for fishery development.

The fishery's share of the provincial economy was dropping. In 1935, Newfoundland marine fisheries had employed 36,900 people, 45 per cent of the labour force. By 1961, the industry would employ only 18,800, 15 per cent of the labour force. But this was still major; Newfoundland depended more than any other province on the fishery.

Frozen food displaces saltfish

Before Confederation, Smallwood had watched with great interest the growth of the frozen-fish trade—first, with some doubts about its effect on fishermen; later, with full-bore enthusiasm.[6] Frozen-fish plants were taking more and more of the catch. Exports were going less to the kitchens of the Mediterranean and more to the refrigerators and restaurants of North America.

Before the war, Newfoundland processors had frozen salmon, but very little groundfish. Two side trawlers were operating by 1937, but were uneconomic. Aidan Maloney, a plant manager in the 1940's, provincial Minister of Fisheries in the 1960's, and later president of the Canadian Saltfish Corporation, described the changeover:

> In the late 1930's, frozen fish was just starting, by one or two firms; it was very modest. The fish

Workers at a Job Brothers plant wrapping fillets in cellophane for quick freezing, 1949. (Library and Archives Canada, PA-142652)

companies would have gone rapidly into frozen fish during the war, but there were problems of getting engines, getting captains, and so on.

I went to John Penny's in 1944, at Ramea on the southwest coast. There were no cod-traps on the southwest coast, and no gillnets. It was entirely a hook and line fishery, until 1946–47. Summer was the ideal time there for groundfish. There was also a lobster fishery, and a fresh salmon fishery, and a squid fishery for bait.

In 1944, we were doing mainly salt. Salting was a cottage, family industry. But by 1949, when Penny had changed to frozen, there were no more flakes [wooden racks for drying cod] along our shore. A lot of the people who'd been drying fish at the flakes were working in the plants. The women had become filleters.

Salt was still dominant for the most part in Newfoundland. But it became apparent that you couldn't be a successful producer for the U.S. if you were seasonal, or if you were doing cod only. To get flounder, sole, and so on, you needed offshore draggers, year-round. One by one, they began getting them. Penny's had one dragger in the 1940's. In the 1950's we got two more, wooden ones.[7]

The Monroe family, originators of Fishery Products Ltd., were early leaders in frozen fish, as were Hazen Russell and his family at Bonavista Cold Storage. Flatfish such as American plaice and yellowtail flounder became important, especially on the Grand Banks; trawls could take them in large numbers. By 1956,

Newfoundland had about 50 frozen-fish plants of one size or another, with a handful of larger companies operating draggers and trawlers. These vessels rose in number from 13 in 1949, already taking about 14 per cent of the catch, to around 30 by 1956. Some of the trawlers (which cost $200,000–$300,000) were making world-record catches.[8]

Meanwhile, the schooner fleet was dying. Historic old names in the saltfish trade—Job Brothers, Baine, Johnston & Co., Harvey—were fading from prominence. By the mid-1950's, the bank and Labrador schooner fleet had shrunk to a low level, although some schooners still went to Labrador in the 1960's. Meanwhile, catches by draggers and trawlers rose; by 1967, they would exceed those from the far larger fleet of small boats.[9]

Exports went mainly to the United States, especially after British purchases fell off after the war. In 1945, Canada supplied 90 per cent of frozen groundfish imports to the United States; Newfoundland, the rest. But Newfoundland's share was rising; by 1956, the province supplied 50 per cent; mainland Canada and Iceland, some 25 per cent each.[10]

In comparison with the foreign fleets, however, whose main fishing grounds were off Newfoundland, the province was well behind.

Walsh report calls for centralized development

The historian Miriam Wright has detailed Smallwood's approach to the fishery. In 1949, the premier set up a Department of Fisheries and Co-operatives under minister William Keough, who had a deputy for each mandate. Although Keough was concerned with ownership questions, no strong movement for fisheries co-operatives emerged; in 1957, the ministry became simply the Department of Fisheries.

Meanwhile, in 1951, the Smallwood government prompted fishermen to form the Newfoundland Federation of Fishermen. The organization never took independent flight and depended on government funding until it died in the 1970's. The larger processors had a representative organization from 1944 on.

Both federal and provincial governments were encouraging the rapid development now taking place, mainly in the form of new frozen-fish plants and large trawlers. The ice-free south coast, close to the Grand Banks, got more groundfish plants than ever before. Governments subsidized wharves, community stages, and water supplies.

Some of the underpinning for the governmental push came from two government reports: the Walsh report in 1953 (officially the Newfoundland Fisheries Development Committee Report, led by former Chief Justice Sir Albert Walsh) and the South Coast Commission of 1957. In the Walsh report, the first and bigger of the two, W.C. MacKenzie of federal fisheries, who had worked on the earlier Bates report in Nova Scotia, took a prime role.

The Walsh report called for industrialism in the fishery. The great problem was low individual productivity. The report looked with disfavour on the tendency of fishermen to earn their living from a mix of occupations: sealing, fishing, growing or hunting food, cutting firewood, and picking up other work if they could. Some fishermen should leave the industry. The report called for a major federal effort to build up a more specialized and full-time industry, both in the boats and in the plants. The women who so often helped cure saltfish would be better off if they could "devote their time to their household duties and ... live in an atmosphere of human dignity as wives and mothers." The report called for government loans for modern freezing plants, a direction in which Smallwood was already headed.[11]

Federal government resists Walsh report; Smallwood acts alone

The Walsh report among other recommendations called for large new plants, for both frozen and salted fish, at centralized sites. The federal government resisted, feeling that a special development program for Newfoundland alone was inappropriate, and anyway private industry should do the job. The federal fisheries department did help build a number of community stages for processing and holding fish, and sponsored some experimental processing projects for the saltfish trade. It also offered some short courses for fishermen on fishing methods, engine repair, and the like. But no large amounts of development money were forthcoming, except through the boatbuilding subsidy and other regular services.

Smallwood accelerated his provincial assistance, which in the years 1949–1967 totalled $53.5 million, including $33 million in loans and loan guarantees. The fishery needed to move ahead, "cease to be a slum industry and become as modern as the great paper mill in Corner Brook." The main beneficiaries of the loans were large, trawler-operating companies. Among the large operators were Bonavista Cold Storage, the newly prominent Lake Group Ltd., John Penny and Sons, and Fishery Products. The number of larger frozen-fish plants increased; these now included some on the northeast coast, despite its shorter season. From fewer than 2,000 after the war, plant workers increased to more than 7,000 by the mid-1960's.[12] (As noted earlier, the question of a genuine increase in shore employment is a bit complex. In some instances, the plant jobs replaced earlier salt-fish processing work on the flakes and beaches, which in some cases went uncaptured by statistics. What is clear is that the new plant jobs were better.)

Resettlement strengthens trawler ports

Closely entwined with fishery development was the resettlement effort, which in the 1950's and 1960's closed down hundreds of outports and moved their people to larger "growth centres," including among oth-

Moving a house in 1961 during the resettlement. (Library and Archives Canada, PA-154122)

ers such trawler ports as Trepassey, Marystown, Burgeo, and Harbour Breton. The thinking behind it was that schools, medical care, and other services were difficult and costly to provide in isolated outports. Economic growth would proceed faster with the critical mass of larger settlements, especially in the new trawler ports, which needed a pool of labour.

The first resettlement program began in 1954 (earlier, people had abandoned some communities on their own). The province provided subsidies of a few hundred dollars per family; funding eventually rose to $600 in communities where all the people voted to move. In the mid-1950's, that was more than a year's income for most fishermen; the South Coast Commission reported average earnings of less than $500. As for families, the earlier Walsh report put total income at about $1,300 for all members from all sources.[13]

In 1965, a major injection of money through the federal Department of Fisheries raised the ante. For the next several years, federal fisheries money funded 70 per cent of resettlement costs, running to more than $5 million by 1968. Now families could get bigger subsidies of well over $1,000, have their moving expenses paid, and get help towards a new residence. No longer did everyone in the community need to endorse the move; the qualifying percentage of votes was reduced in stages to 75 per cent of a community's population. Resettlement speeded up. Provincial responsibility for the program moved from the welfare department to the fisheries department.

All told, the resettlement efforts from 1954 to 1975 moved more than 250 communities, about 4,200 households, and nearly 21,000 people.[14] Many of the moves came from island settlements, particularly in Placentia Bay on the south coast and Notre Dame and Bonavista bays on the northeast coast. Of course, hundreds of settlements still endured on the long Newfoundland coastline.

The 1960's push in particular created problems and blunders, with some people complaining of being forced into resettlement, or of being unable to find jobs. The growth centres were too few, and were growing too slowly, to handle everyone well. The program caused even more controversy in retrospect. In the 1970's a romantic lament arose for lost communities and their values. But for most resettled people life was easier, with more conveniences. Decades later, at the turn of the century, Newfoundland opinions on resettlement remained mixed. Some abhorred it; others thought it was necessary and that most families were satisfied with their move, although it could have been handled better.

N.A.F.E.L. tries to revive saltfish trade

Meanwhile, the saltfish trade lingered on, still serving a major market. In Spain and Portugal particularly, saltfish remained an important food item, prepared according to hundreds of recipes and served up on special occasions. For Newfoundland, saltfish in

Loading salt codfish for the Portuguese market (vessel from Aveiro) at Fortune on the Burin Peninsula, 1949. (Library and Archives Canada, PA-110813)

1947–1948 still supplied 60 per cent of fishery exports. At the same time, the problems that Coaker had tried to address—poor quality, poor marketing—still hobbled the industry.

The Commission of Government had made some efforts to strengthen the trade. Then the wartime boom had solved market problems, but only temporarily. After the war, the exporters wanted to keep some of the momentum. Raymond Gushue of the Newfoundland Fisheries Board helped work out a new scheme. In 1947, 32 exporters banded together in a marketing agency, Newfoundland Associated Fish Exporters Limited (N.A.F.E.L.). The Commission of Government gave N.A.F.E.L. the exclusive right to export salt cod. N.A.F.E.L. could not, however, buy directly from fishermen or set the prices paid to them. The 30-odd exporting companies took care of that, buying saltfish from other companies or directly from the thousands of fishermen and families producing saltfish. The British trading house Hawes and Co., long associated with the Newfoundland trade, became N.A.F.E.L.'s foreign representatives.[15]

Single-desk marketing had arrived. N.A.F.E.L. worked reasonably well for several years. The organization gained control of nearly 25 per cent of world saltfish trade, and dominated the Caribbean market. But in the larger picture, saltfish markets were shrinking as other food sources came on stream after the war. In 1949, 1950, 1953, and several times in succeeding years, the Fisheries Prices Support Board provided saltfish subsidies.[16]

Meanwhile, Nova Scotian exporters opposed N.A.F.E.L., and the federal government gave it little sympathy. N.A.F.E.L. lost its export monopoly in the mid-1950's, first losing control of saltbulk exports in 1953, then being shut down in 1959. Smallwood in 1953 had pressed Ottawa for a National Fish Marketing Board, but nothing came of it.[17] A N.A.F.E.L. corporation remained in place as a marketing and promotion agency, but it was weaker than before; anyone who wanted could sell outside N.A.F.E.L.

The loss of single-desk marketing further weakened a declining trade. It was still of great importance; as late as 1960, 50 per cent of inshore fishermen and their families still produced saltfish exclusively, although they now tended to produce saltbulk for mechanical drying rather than shore cure.[18] But the individuals and companies producing saltfish frequently ran into problems. The industry remained a subject of concern.

Besides saltfish, another old trade declined in the post-war years. Production of cod-liver oil had been significant, especially from the 1920's. Like the dogfish-liver trade in British Columbia, the cod-liver trade fell victim to synthetic vitamins in the 1950's.

Abundance seems unlimited

The Newfoundland inshore cod fishery had always known resource cycles and "fishery failures." In the post-war period, bigger boats and better gear let more fishermen range further afield, including the offshore banks. As in other Atlantic regions, it now seemed to many that during previous scarcities, the fish had been there all the time, just hiding at a greater distance offshore. Through optimistic post-war eyes, the resource seemed huge, with new grounds and species coming to the fore, especially in Newfoundland.

Wilfred Templeman, working first for the Newfoundland government and then for the Fisheries Research Board of Canada, led resource explorations from the deck of the research vessel *Investigator II*. Templeman identified resources from Labrador to the Gulf of St. Lawrence. Besides mapping out cod stocks, he prompted new or enlarged fisheries on redfish and other species. Templeman won great renown not only as a fish-finder but as an all-round biologist. He was the chief figure in building up knowledge of Newfoundland's fishery resources.

Canada's greatest fish-finder

When Wilfred Templeman in 1944 took over as director of the Newfoundland government laboratory, he increased its research on fisheries. Templeman himself produced hundreds of scientific papers and an authoritative, still-used volume on the marine resources of Newfoundland. But much of his renown came from direct investigation of the groundfish resources.

Templeman in the early 1980's recalled his first years at the government lab:

At the time, research was pretty well dead. We began planning for a vessel. The shipyard at Clarenville built the *Investigator II* there, in 1946. She was a B.C.-type vessel, 82 feet long. Our minds didn't run very big. We thought we could use her for longlining and seining. We found out we had to trawl, that was the best way.

We studied natural history, everything about the fish. Besides that, we did some hydrography—temperature measurements and so on—although we never had a hydrographer. There was no real population work until the recent guys, Arthur May and others. We knew it had to be done sometime, and we were laying the groundwork.[19]

Templeman in the lab, looking over a Chimaera fish. (Photo courtesy of Ben Davis, D.F.O.)

Templeman's fishery explorations made him the greatest fish-finder Newfoundland and Canada have ever known. In 1946, the *Investigator II* found pink shrimp, previously thought rare, in deep water. Other explorations in following years by Templeman and colleagues (sometimes with the Industrial Development Service of federal fisheries) included

- Redfish, "All the major redfish areas now fished by American and Canadian trawlers in the Gulf of St. Lawrence, St. Pierre Bank, Southwest Grand Bank, Northeast Grand Bank, and the south coast of Newfoundland with the exception of the area near Ramea were first found and fished successfully by the *Investigator II*."[20]
- Cod in deep water off Newfoundland and Labrador.
- Greenland halibut and roundnose grenadier off Labrador and Baffin Island.
- Capelin on the southern Grand Banks and at Trinity Bay.
- Local stocks of witch, Greenland halibut, scallops, and other species.[21]

Both scientific knowledge and the fishery gained, although the Soviet Union and other foreign nations, Americans in the case of Gulf redfish, were sometimes the first to make use of the new knowledge. But Canadians soon followed.

After the Fisheries Research Board took over the Newfoundland lab, Templeman instituted an F.R.B. scholarship program and fostered scientists with a practical bent. Some, like Art May and Scott Parsons, would become well-known figures in both science and management.

The *Investigator II*, outside the entrance to St. John's harbour.

"Longliners" join the fleet

Longliner bringing fish into Bonavista.

While the trawler fleet bloomed, Newfoundland's fleet of medium-sized vessels was also growing, though less dramatically than in the Maritimes. Changes in gear were accelerating the growth in fishing power. The Industrial Development Service worked on fishing methods, as did Templeman and colleagues at the F.R.B.

Although fishermen in some areas used longlines, most Newfoundland fishing still took place by handlines or cod-traps. In the early 1950's, the "Bonavista experiment" introduced longlines and mechanical haulers in Bonavista Bay on the northeast coast. Vessels in the 50-foot range set trawl in the area, with reasonably good results. (On the south coast as well, where longlining was already popular, the F.R.B. helped spread the use of mechanical haulers.) In the case of Bonavista, the success of the experiment attracted foreign vessels to the area, negating much of the benefit. Catches and fish sizes dropped; many Newfoundland longliners left the area.[22]

As in the Maritimes, the spread of gillnets to some degree undercut the longlining work, especially after fishing pressure in the 1950's and 1960's lowered average sizes in many groundfish fisheries. Hook and line fishing became less effective as the proportion of large fish declined. The use of gillnets grew rapidly, increasing productivity while creating attendant problems of wastage and ghost fishing.[23]

Meanwhile, the province hired a naval architect to design some fishing vessels, including medium-sized decked vessels. The provincial Loan Board by 1971 had provided loans for 47 longliners, 10 draggers, and 30 combination trap skiff–longliners.[24] The new vessels were more likely to use gillnets than hooks. But in a lingering effect of the Bonavista experiment, the medium-sized vessels continued to be called longliners. Thus it happened that in Newfoundland, a "longliner"

typically fished with gillnets. The new gear increased catching power. Soon the many gillnets set off Newfoundland added up to thousands of miles.

Consultants, Templeman warn of over-expansion

Smallwood had appointed the Newfoundland Fisheries Development Authority (N.F.D.A.) to oversee changes following the Walsh report. In some instances, the N.F.D.A. built plants itself, then leased them to private operators.[26] But by the late 1950's, some companies, notably Fishery Products, were running into trouble, and asking for more loans to tide them over. The fish stick had failed to produce the hoped-for boom in consumption. Few of the frozen-fish plants were operating at full capacity. The Newfoundland government in 1957 hired a Boston-based consulting firm, Arthur D. Little Inc., to examine the situation.

The 1958 Little report said that the industry had expanded too quickly, creating more plants than needed, with the development program a factor. New England buyers were playing off Newfoundland producers against one another. The report suggested some form of co-ordinated marketing strategy, and perhaps a common "Newfoundland brand." The report brought little action.

In Miriam Wright's analysis, Smallwood's fishery development program erred on several counts. It poured so much money and energy into the frozen-fish plants, it had little left for other efforts. It overfavoured Fishery Products. It created a vicious circle of over-capacity: the more money you spend on capital development, the harder you must fish and process to pay for it. And it fostered "a cycle of dependency between capital and the state."[27]

Meanwhile, fishing pressure was increasing from all sides. The number of trawlers over 50 gross tons, Canadian and international, in the I.C.N.A.F. area rose from 540 in 1953 to 975 in 1962. Vessels were growing not only in number but also in fishing power. By 1962, Wilfred Templeman was warning that the catch per person and size of fish would decline. The inshore catch depended mainly on cod spawned offshore; now trawlers were catching them before they could migrate to the coast. Templeman would later note that from 1957 to 1964, inshore fishermen, vessels, and gear had increased by more than 50 per cent, while landings had stayed relatively stable. Other instances emerged of increased effort and falling catches.[28]

National conference produces few results

By the early 1960's, trawlermen were making better money, and thousands of people were getting year-round work in the plants. Yet progress was less than Smallwood had hoped. The incomes of Newfoundland fishermen still lagged behind those of mainland fishermen.[29]

Small boats, large trawlers dominate Newfoundland fleet

At the beginning of the Second World War, Newfoundland still had several hundred larger "vessels," although the numbers were dropping. Nova Scotia had fewer: as of 1939, only 71 vessels over 20 tons. By the mid-1950's, with the saltfish trade declining and new fishing technology advancing, the schooner fleet in both provinces had practically vanished. Nova Scotia was making up for it with an increase not only in trawlers and other large vessels, but also in medium-sized decked craft of about 35–65 feet. Some smaller craft were also at least partly decked, and looked substantial. Newfoundland saw a lesser increase in trawlers, and lagged far behind in medium-size vessels. Relative to Nova Scotia, Newfoundland had become more the land of little boats.

As of 1956, the statistics for Newfoundland listed 12 trawlers, 18 draggers, 8 Danish seiners, 4 purse-seiners, and 41 longliners, for a total of 83 vessels over ten tons. This was small compared with the Maritimes and Quebec, which together had around 1,300 such vessels.

Twelve years later, in 1968, Newfoundland's fleet of vessels over ten tons had increased to 497. This was now more than double the number of such vessels in Quebec and about 13 times the number in Prince Edward Island, but still less than half the number in either Nova Scotia or New Brunswick. In the overall number of vessels—anything over ten tons—Newfoundland was still lagging.

The gap was greatest in medium-sized vessels between ten tons (typically 35–40 feet at the time) and 100 feet. By contrast, in larger craft over 100 feet, Newfoundland had a powerful fleet of 75 vessels. This was nearly four times the number in New Brunswick, P.E.I., and Quebec, which combined had only 20 vessels over 100 feet. Newfoundland trailed only Nova Scotia, which had 134 vessels of that size.

As for small boats under ten tons, Newfoundland consistently had high numbers—about 29,000 in 1935, 11,800 in 1956, and 14,900 in 1968—figures comparable to those for the Maritimes and Quebec put together.

Why was Newfoundland's fleet of medium-sized vessels weaker than that of the Maritimes? One might consider several factors.

Newfoundland had fewer multipurpose craft, which might in some instances need more deck space. Coastal fishing grounds on the northeast coast of Newfoundland, where cod migrated in summer, had rough, untrawlable bottom; this held back growth of a dragger fleet. Cod-traps dominated and required only small boats. Draggers would have been better for winter, when the fish migrated to better grounds further offshore; except, ice then became an obstacle. Smaller boats using cod-traps and gillnets were perhaps more cost-effective for a short season. (Dragging did become popular on the northwest coast, which also suffered from ice but had better bottom.)[25]

As well, the Maritimes and Quebec had stronger economies (notwithstanding their problems), more financing ability, and better proximity to market. The greater variety of species allowed more vessels to switch around from fishery to fishery as the season progressed. Finally, in Newfoundland, far more federal and provincial money went to help large-trawler plants and vessels; perhaps their expansion took away opportunities to grow from the relatively small midshore fleet.

Table 19-1. Newfoundland boats and vessels, 1956 and 1968.

	Year	Nfld.
Trawlers	1956	12
	1968	n/a
Draggers	1956	18
	1968	n/a
Vessels >10 tons	1956	83
	1968	497
Vessels >100 ft.	1956	n/a
	1968	75
Motor boats	1956	7,083
	1968	10,451
Sail, row-boats	1956	4,683
	1968	4,439

Note: n/a, not available.

Premier Smallwood organized a fisheries convention in 1962, bringing together industry representatives and various officials and experts. He followed up with a Fisheries Commission to work out a new program for development. Smallwood then pressed the federal government to organize a national fisheries conference, held in Ottawa in 1964. He set great store by this initiative, where Newfoundland pushed for a national program for community-based fisheries that would include marketing boards, price support, better credit, and infrastructure.[30] He was largely disappointed.

The conference did, however, give impetus to discussions and studies that helped prepare the ground for two federal crown corporations, which would appear later: the Freshwater Fish Marketing Corporation, for the Prairie provinces and northwest Ontario; and the Canadian Saltfish Corporation, which provided single-desk exporting for saltfish from Newfoundland and the lower North Shore of Quebec. As well, more federal money appeared for cost-shared development programs and for resettlement.

Meanwhile, the frozen-fish industry kept expanding. In that market, Scandinavians dominated the high-quality niche. For Newfoundland as for the Maritimes, frozen groundfish was a commodity, where they competed with each other, with the Scandinavians, and with anyone else in the growing fleets. By 1968, Newfoundland was heading into a market crisis.

Smallwood saw limited opportunity in fisheries

In a 1979 interview, Newfoundland's former premier Joseph Smallwood recalled working with Coaker in the 1920's. The views that Smallwood then formed about better-organized marketing influenced his later approach as premier. "Competition has a disastrous effect on all primary producers," Smallwood said in the interview. He had supported N.A.F.E.L. (a government-sponsored saltfish marketing agency) as a means of orderly marketing. After Confederation, "I had to fight to retain N.A.F.E.L., and having won that, had to fight to get fishermen into N.A.F.E.L." (Despite Smallwood's efforts, as already seen, N.A.F.E.L. faded away.)

Shortly after he became premier, Smallwood set up the Department of Fisheries and Co-operatives. The co-operative element never worked as hoped. "The wartime had attracted people out of co-ops. But also, Newfoundland fishermen were highly individualistic. It was easy to get them to join a party or a crusade, but to put money and effort into a co-op was different; it never appealed to them."

The idea behind the Walsh commission on Newfoundland fisheries was, Smallwood said, to sell the Canadian government on the idea of capital aid to fishermen. But results were less than hoped. Later, in 1962, Smallwood convened a three-day conference of fishermen, processors, and experts from elsewhere. From this emerged the idea of pursuing "what the government had done 50 years earlier for farmers: crop insurance, capital aid, marketing assistance, maybe 15 headings under which government helped farmers. Why shouldn't they do it for fishermen?"

Smallwood then approached the federal government to sponsor a national conference, which took place in 1964; but "they watered everything down and couldn't get us out of Ottawa fast enough." The new program for fishermen never appeared. Still, as already noted, the conference planted the seeds for the Canadian Saltfish Corporation, which some years later essentially replaced N.A.F.E.L.

Smallwood denied ever having told fishermen to "burn their boats." Instead, he said, he had used that phrase in a speech, saying that if Confederation failed to do justice to fishermen, then they might as well burn their boats. He pointed to the context of the times. The industry's share of employment had dropped steadily as decades passed. By the post-war period, "the fishery was dying, at death's door. Fishermen advised their sons to get into *anything* else. I inherited, as premier, what all my predecessors had had: an industry of torment and troubles."

Smallwood wanted the industry to provide a decent livelihood for those remaining in it; meanwhile, other people would move into other lines of work, which government would help develop. He harboured no hopes that the fishery would create prosperity for the bulk of the province.

As for the resettlement, Smallwood said, it was the right thing to do and a great success. Before Confederation, it would have been useless and unnecessary to centralize. But after, it was urgent to bring children to the schoolteachers, who were no longer willing to go to every little settlement. "Education was the mother and father" to the centralization move, to good schools, to electric lights, and so on. Women wanting the best for their children made resettlement succeed. Before, Smallwood said, there had been thousands of illiterate people; "today they would be condemned to a life of servitude."[31]

Like most other Atlantic provinces, Newfoundland got a fisheries training establishment: in 1964 the Smallwood government set up the College of Fisheries, Marine Navigation, and Engineering. The idea was in part to produce trawlermen, but many graduates migrated instead to Great Lakes freighters.[32]

Purse-seining for herring in Bonne Bay on the west coast of Newfoundland.

Pelagics: herring catches boom

In Newfoundland, the legions of capelin were the stuff of legend. The sardine-sized fish were considered the main food of the cod. Every spring, capelin moving inshore to spawn would throw themselves on the beaches. Families might collect them off the beach, or catch them with castnets. But capelin, although plentiful, were a thinner fish than herring, and had never attained more than local popularity for food.

Herring dominated the commercial pelagic fishery. They were at the edge of their range in Newfoundland, most plentiful on the south and southwest coasts. The up-and-down herring fishery had yielded 75,000–80,000 tonnes a year in the years 1945–1950, when food demand was strong. In the 1950's, it reverted to mainly a bait fishery, yielding around 10,000 tonnes a year. Small boats did the harvesting.

As purse-seining re-established itself in the Maritimes, some Bay of Fundy seiners and then other vessels began fishing off Newfoundland. As well, the I.D.S. sponsored new purse-seining experiments. Local captains, such as the high-liner Kirk Anderson, acquired large seiners, and reduction plants set up operations, taking high quantities for low value. As in the Maritimes, some large purse-seiners from the exhausted B.C. fishery joined the fleet. Herring fishing became intensive on the south and southwest coasts.

Catches rose to 316,000 tonnes in 1968. This was an enormous increase, accounting for a major part of the rise in Newfoundland's overall catch. It turned out that the huge fishery was supported by two abnormally strong year-classes from the late 1950's. Only in the late 1960's did research under Templeman establish the unusual circumstances.[33] The expanding fishery was riding for a fall.

Lobster fishery holds steady

Although herring provided high volumes for southern areas, Newfoundland remained mainly the captive of cod, with some fishermen switching seasonally into other fisheries. Squid was intermittent; crab and shrimp were still unexploited.

In Newfoundland's relatively small lobster industry, live lobster exports dominated by far after the Second World War. Small canneries, once numbering several hundred, had fallen to several dozen during the war. Some had problems with quality as well as profit. They vanished after the war.[34]

As for the fishery, lobster landings in 1968, at about 1,800 tonnes, showed a slight decline from the early 1950's. Value however had climbed from $900,000 to $2.4 million, something less than ten per cent of the total landed value of Newfoundland's fisheries.

Seal fishery rebounds

Another fishery was now resurging, that for seals. The demand for seal oil had dwindled in the 19th century and for pelts for leather in the 20th century. But after the Second World War, the fashion trade stepped in. The pelts provided fur coats, boots, and fashion accessories.

Between 1949 and 1961 the Canadian take of seals averaged 310,000, reaching more than 400,000 in 1951. The most valuable pelts came from the "whitecoats"—harp seal pups just recently born. Although hooded seals also contributed, harps supported the main hunt. Pupping took place on the ice floes in the Gulf of St. Lawrence and on the "Front" off northeast Newfoundland.

The growing intensity of what came to be called the "seal hunt" (though Newfoundlanders still knew it as the seal fishery) brought conservation concerns. In 1961, the department set opening and closing dates for the Gulf and Front. In 1964, it began licensing vessels and also spotter aircraft, which had recently re-entered the hunt. In 1965, the department set a quota for the Gulf; quotas would later spread to the Front. The department also prohibited killing adult seals in breeding or nursery areas. A full complement of enforcement staff covered the hunt.

By the late 1960's, federal fisheries was issuing more than 6,000 licences annually, the great majority for landsmen from Quebec (mainly the Magdalen Islands) and Newfoundland. Landsmen would use small boats or else walk out on the ice. Hunters used rifles for adults and clubs or hakapiks, a Norwegian gaff, to crush the skulls of pups with a blow. Most landsmen took only three or four seals, in many cases for food.

On the Front off northeast Newfoundland, seven to ten large Canadian vessels took part, employing 300–400 men; these took most of the catch. The total value of pelts in the late 1960's ranged from less than $1 million to nearly $2 million. Landsmen mostly earned less than $100; hunters from large vessels several hundred or sometimes more than $1,000.

By 1968, the rules were keeping the Gulf take below 90,000, which the department considered the maximum sustainable yield for the area. A similar estimate of M.S.Y. applied for the Front, but there, the combined hunt by Canadians and Norwegians sometimes took three times the proper harvest. (Norwegians had taken part in the hunt since 1938; because it was an international fishery, Canada successfully pressed to bring the hunt into the I.C.N.A.F. framework.) Meanwhile, the harp seal population had dropped from an estimated three million in 1951 to somewhere around two million.

As years passed, tightening controls would let the herds increase. The department's main problem would be not conservation but protests, beginning in the 1960's. In 1964 a film, which the department considered misleading, suggested cruelty at the seal hunt in the Gulf; this helped spark protests in North America and Europe. Many more would follow.

The department in 1965 and 1966 set strict regulations regarding killing methods, began licensing individual sealers, and confined the hunt to daylight hours. The seal hunt compared well with slaughterhouses on land. But the slaughter on the ice had no concealing walls. Clubbing seal pups looked cruel. Environmental consciousness was rising in the 1960's, sometimes associated with cute animals.

The department in 1966 began taking representatives of humane societies and conservation groups to the hunt; these gave mostly favourable reports. Many authorities on animal welfare who looked at the seal hunt, then and later, considered that clubbing of the seal pups' thin skulls provided as painless a death as one could hope for. But protests continued. Brian Davies, originally an employee of the New Brunswick Society for the Prevention of Cruelty to Animals, began lobbying against the hunt and publicizing its ostensible cruelty. He soon founded the International Fund for Animal Welfare (I.F.A.W.), which became a leading voice of protest. The I.F.A.W. would gather substantial revenue in contributions from sympathizers, enabling advertising and other publicity. Other groups would get in on the campaign.

Department officials kept consulting animal welfare groups, kept a close eye on the hunt, and hoped that if they did the right things, the protests would go away. The next three decades would prove them wrong.[35]

1945–1968: slow headway for ordinary fishermen

How did the 1945–1968 period turn out for Newfoundland? The general Newfoundland economy had diversified somewhat, and under Smallwood, education reached everywhere. Still, the province remained more dependent than any other on the sea for its economy and identity.

Within the fleet, progress had been lopsided, more on the large-vessel side. Large trawlers and large plants had bloomed on the east and south coasts, nourished by free-handed provincial government loans. Trawlers had built up to take about half the groundfish catch. Trawlermen's incomes had increased, and there were a lot more year-round jobs in the trawler plants.

The great majority of fishermen still worked from open boats. Their incomes increased but remained behind those in the mainstream economy. Despite their numbers and importance, fishermen remained fragmented and unorganized. Still, most fishermen now had more material goods, and their communities still had their ways and their colour. Many outports also had at least a bit more diversity in their incomes, with more government and service jobs appearing.

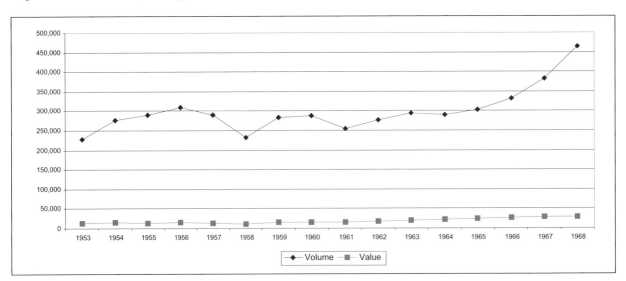

Newfoundland volume (tonnes) and landed value ($000's), 1953-68. As in the Maritimes, much of the increased volume came from herring.

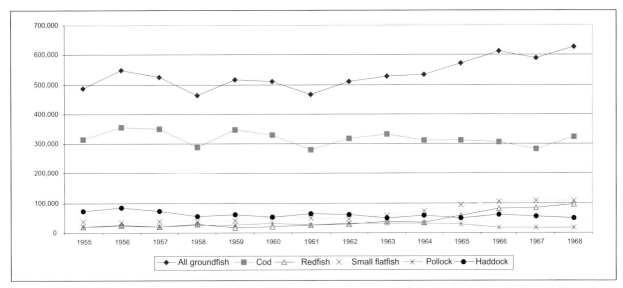

Canadian Atlantic groundfish landings, selected species (tonnes). Catches rose, but less than hoped, and cod stayed flat. Most pollock and haddock landings came in the Maritimes.

On the Atlantic: development without end

For the Atlantic coast as a whole, the fishery at the end of the 1945–1968 period had more muscle, but not a lot more brain. In sea-fishery conservation, there were few regulations to prevent an ever-expanding fleet. From time to time, warning voices like that of Templeman were raised. But many or most people in the industry, and many development-oriented officials, still thought there was room to expand. Many felt that, especially in the groundfish fishery, one must compete with the foreigners or else lose place and face.

Social considerations pointed the same way. Alternative employment was scarce; development could provide jobs. So long as there was a chance of more fish, it seemed there was no sense rationalizing people out of business. It was easier to go with the trend of the times, which throughout the economy was development and growth.

How ideas on "development" evolved

The economist William C. (Bill) MacKenzie worked in the Department of Fisheries from 1940 to the early 1980's. MacKenzie worked closely with Stewart Bates, H. Scott Gordon, and Alfred Needler. Involved in many policy developments, he was the key figure in writing the 1950's Walsh report on Newfoundland fisheries and the 1976 *Policy for Canada's Commercial Fisheries*. Many years later, MacKenzie outlined how thinking on development evolved in the 1945–1968 period:

> Back in the 1930's, nobody associated fisheries with development. The fishery was in a rear-guard action, a desperate struggle to maintain its position. The war led to a total change in mind-set. Everybody wanted maximum production. The pent-up energies of the Depression were freed by what amounted to wartime socialism.
>
> After the war, there was a general attitude in business and government that we had a lot of catching up to do. For the Atlantic fisheries, that meant moving from more primitive to technologically-advanced operations. The Pacific had already evolved that way.
>
> In the Atlantic region, first the government relaxed restrictions on draggers and trawlers. Then it got into active promotion of new techniques. The provinces also got into the act.
>
> Newfoundland seemed to have the most need for better techniques. When federal officials went there after 1949, what impressed us was the overwhelming dominance of cod. It appeared to many of us that a once-

great fishery had become degraded, and product quality generally was abysmal. The production system could no longer produce good quality. Part of the problem may have been a loss of morale. Small-boat fishermen appeared to be a down-trodden underclass. Average production per man was very low, for minimal prices.

In the Walsh report, we aimed for more production per person or enterprise and per port. The resource was plentiful. It was a question of poor equipment and poor quality. It seemed that development centred in larger ports could improve matters. We wanted more Lunenburgs and Marystowns. Over-dependence on the fisheries didn't seem to be the main problem, at first.

After release of the Walsh report, the provincial government introduced resettlement, bringing people more into central ports. But the whole scheme ran into an insurmountable obstacle: only a few of the growth centres, like Marystown or Harbour Grace, offered a real chance of developing an urban, industrial base. Otherwise, we were moving people into places where there was no employment. It became obvious that the problem wasn't so much the fisheries as the regional economy.

Meanwhile, Scott Gordon had written his paper on common property in fishery resources, and how it induces over-expansion and over-capacity. Tony Scott at U.B.C., Jim Crutchfield in Seattle, and a few others in the U.S.A., the U.K., and Scandinavia were thinking and writing along the same lines. As all these strands came together, one could see two powerful forces over-loading the fishery: the natural tendency to over-expansion, and the lack of alternative employment.

Moreover, our statistical research was beginning to show us that the percentage of full-time fishermen was surprisingly low. This became clear in the 1960's. A lot of people were part-time and casual, not really making much from the fishery, or contributing much to it.

Even with better boats and bigger ports, there wouldn't be enough revenue to go around. Besides better methods of production, the fisheries in Newfoundland and other Atlantic areas needed fewer, more professional people. But there was no real attempt to reduce numbers, or even control them, until the late 1960's and the 1970's.[36]

276

CHAPTER 20.
On the Pacific, 1945–1968

Post-war British Columbia was the golden land of Canada, beautiful, growing, and full of opportunity, attracting many newcomers. But the province was far more than a raw frontier. Along with an outdoor lifestyle, it had an urbanized, educated population. Conscious of the beauty around them and the salmon in the rivers, and seeing industrial development rise against a scenic background, British Columbians earlier than most North Americans developed a strong environmental consciousness.

From 1952 to 1972, the Social Credit party governed the province under Premier W.A.C. Bennett. Despite his media nickname of Wacky, Bennett was a shrewd operator: development-oriented but fiscally conservative and non-ideological. He presided over a period of great economic expansion, with hydro-electric developments, highways, and railways crossing the landscape.

As on the Atlantic, bigger, better boats were the trend. In the seine fleet, "table" seiners, with the net stowed on a turntable for easier running, dominated at first. Here, the *Western Ranger*, seining for sockeye in 1958, has completed a set. The power skiff tows the vessel from the side, to keep her from drifting into the floating seine while taking it in. (Library and Archives Canada, PA-146255)

If the province was a golden land, the fishery seemed prepared for a golden age. It had declined during the Great Depression as the number of canneries shrank from 59 in 1930 to only 38 in 1940. Then, as on the Atlantic, the industry got a boost from wartime demand and new technology. After the war, everyone looked forward to progress.

Salmon Commission starts restoring Fraser

In the mainstay salmon fisheries, after decades of Canadian–U.S. discord, the International Pacific Salmon Fisheries Commission (I.P.S.F.C.) seemed to promise increased returns. The Fraser system still suffered from the Hell's Gate slides. Returns in the early 1920's had been little more than one-tenth and in the early 1940's were still only one-fourth their level at the beginning of the century.[1]

A study starting in 1938 under the I.P.S.F.C. showed that Hell's Gate, despite clearance work, still posed an obstacle to salmon. The two governments in 1944 began building the Hell's Gate fishways, and completed them in 1946. They later built several other fishways in the Fraser system.

In 1945 and 1946, the I.P.S.F.C. began in-season management of Fraser sockeye in the "Convention

A fishway at Hell's Gate Canyon helped Fraser salmon runs.

area"—broadly speaking, the Strait of Juan de Fuca, parts of Puget Sound and the lower approaches to the Fraser, and the Fraser estuary. The I.P.S.F.C. commissioners included top fishery managers from both countries. Wearing their commission hats, they made recommendations to the two countries, and wearing their government hats, they carried them out, aiming to provide proper escapements and share the fishery equally between the two countries. The main method was openings and closings for different gear types. Special autumn closures helped rebuild the important Adams River run.

Fraser returns rose in the 1950's. Although many people credited the Hell's Gate fishways, the renowned biologist W.E. Ricker gave as much or more credit to management measures, starting with the belated U.S. closure of salmon trap fisheries in the 1930's. Hell's Gate, in Ricker's opinion, was only stopping five to ten per cent of sockeye, and was more important to pinks, which began to re-establish themselves in the middle reaches of the Fraser system.[2]

Despite the gains on the Fraser, total salmon returns in B.C. stayed relatively stable in the 1945–1968 period. The main growth would come in herring and halibut catches.

Salmon trade readjusts

Consolidation continued in the salmon fishery, dominated by B.C. Packers and Canadian Fishing Company. There were only 24 canneries by 1950, and none on the Nass after 1945, or at Rivers Inlet after 1952.

In the salmon trade, the wartime embargo on exports of raw salmon disappeared. This let American canners sometimes outbid Canadian companies for fish. Meanwhile, the Americans applied a 25 per cent duty on Canadian canned salmon. Following industry protests, the Canadian government in 1948 re-introduced some export restrictions. On the marketing side, the B.C. industry in 1949 launched a generic "no-

brand" promotion campaign, lauding salmon's economy, nutrition, and versatility. Although canning remained dominant by far, the trade in fresh and frozen salmon was increasing.[3]

Fisheries minister Sinclair in the mid-1950's intervened in the salmon trade. Austerity measures in the United Kingdom, a major market, had cut back imports and created a salmon surplus in British Columbia. Sinclair met with the Chancellor of the Exchequer, then got some Canadian companies to buy more British equipment. The U.K. in turn allowed more salmon imports. But such intervention was rare in the Pacific industry. Sinclair noted years later that "the East coasters were always saying 'give us more', and B.C. companies were saying, 'leave us alone'."[4]

Even before the war, trapnets and drag-seines had mostly faded from the picture. A handful of licences remained for each type after the war. Cliff Levelton, who was a fishery officer after the war, recalled two then operating in the Butedale area: "The licence holders, Luke Brown and the Robinsons, had letters from

On the cannery line at Steveston. (National Film Board)

Taking a break as a collector boat unloads.

278

Queen Victoria stating their privilege of drag seining. We used to ask them to shut down for a week to let the sockeye get up-river."[5] These survived into the early 1960's. The Sooke fish trap disappeared by the late 1950's. Another old tradition vanished when the department dropped the practice of firing a six o'clock gun to mark the end of closed periods.

As noted earlier, the "Chinese contract" system for cannery workers disappeared during the war. Chinese Canadians were finding other jobs, and minorities were getting fairer treatment. Similarly, restrictions on Japanese Canadians disappeared after the war. Many former fishermen came back, either buying boats or getting modern boats built for them by processing companies, who knew their fishing ability.

Fishing fleet builds power

The Pacific salmon fleet in 1949 included 12,200 fishermen, with about 5,400 gillnet, 4,900 troll, and 300 purse-seine licences. B.C. fishermen took full advantage of better gear, such as nylon lines and nets, and better engines. Sailing skiffs disappeared after the war. The size of gillnet boats increased in the late 1940's, typically from 28–30 feet to 36–38 feet. The department favoured gillnetters in some coastal areas by excluding competing gear types. Soon the gillnetters spread into areas the department had set aside for seiners. More gillnetters were putting on trolling poles, creating combination boats. By the late 1970's, nearly half the fleet would be combination boats.

The purse-seine fleet was powerful. A few hundred seiners in the 1950's typically took as much salmon as the more than 5,000 gillnetters, usually somewhere around 40 per cent or better for each fleet, with trollers usually taking less than 20 per cent. At first, most seiners used the "table-seining" technique: they would store the seine on a turntable on the after deck, to make setting easier as the boat turned. In the 1950's, seiners started using power blocks on the boom to haul in the seine.

Gillnet fishermen back in the 1930's had invented the gillnet drum, a big roller standing on the stern to spool up the nets. In the 1940's, some seiners found a way to use the drum roller to set and haul purse-seines (the Martinolich family led this development). Although it lacked the depth of table seining, drum seining was fast and labour saving, requiring only three or four men, and making twice the sets. By the 1970's, it would displace table seining to become the dominant seining technique in British Columbia.

In the 1950's, technological workers in the federal Department of Fisheries improved fish pumps, which replaced brailing in the herring fisheries; they also served to unload salmon "packers," which carried fish to the plants. And Stewart Roach of the F.R.B. helped spread the use of refrigerated sea water to chill fish in holds. His publications became standard references in the field.[6]

Vessel fleet doubles

As on the east coast, the fleet gained enormous strength in the post-war period.

Back in 1944, British Columbia had 448 vessels and 7,671 boats. By 1955, the vessel fleet had more than doubled, to 951, while the boat fleet dropped. (As on the Atlantic, "boats" less than ten tons were generally under 40 feet.)

By 1970, the vessel fleet had more than doubled again, while the small-boat fleet further declined.

Trollers at Ucluelet, 1962.

Table 20-1. B.C. boats and vessels, 1955 and 1970.

	Year	B.C.
Vessels >10 tons	1955	951
	1970	2,399
Vessels >100 ft.	1955	n/a
	1970	10
Motor boats	1955	6,299
	1970	4,552
Sail, row-boats	1955	785
	1970	24

Note: n/a, not available. Comparable statistics are unavailable for 1968. In the data for 1955, vessels over 100 feet are included in the total for vessels over ten tons.

Fishermen get stronger organizations

Although canners still owned a substantial minority of the fleet, fishermen may have had more true independence than on the Atlantic. B.C. fishermen traditionally made more money. They tended to stay in school longer than their Atlantic cousins. They often lived in cities or larger towns with a better flow of information. There was less fragmentation and more organization. Every fishery—trollers, gillnetters, seiners—seemed to have its association.

One multi-gear group would become a giant, frequently able to sway fishery management. In 1945, the United Fishermen's Federal Union, the cannery workers' union, and the old B.C. Fishermen's Protective Association (dating from 1919) merged to become the U.F.A.W.U. The new organization came to represent many plant workers, gillnetter captains, seine-boat crews, and others both in the company and in independent fleets. The U.F.A.W.U. gained the power to shut down the main elements of the industry through strikes. Its annual price negotiations with the larger processing companies became momentous affairs, affecting the whole industry. A 1968 strike in Prince Rupert, for example, became a storied episode in the community.

Homer Stevens and the U.F.A.W.U.

Homer Stevens was a major figure in the United Fishermen and Allied Workers Union from the late 1940's to the late 1970's. He was born in 1923, his ancestry a mixture of Indian, Yugoslavian, Greek, and Italian, with fishing families on all sides. He began fishing in 1936, first gillnetting in summers, then working on draggers. Four decades later, Stevens recalled the wartime situation among organizations:

Homer Stevens, ca. 1950. (Paramount Studios. Library and Archives Canada, PA-126358)

The seiners were the strongest, because they had less conflict. They could bargain with some success. But if they tied up, the gillnetters could still fish. There was no co-ordination. The organizations in general were weak.

I was in the B.C. Fishermen's Protective Association, and not too active. In the early 1940's, two things stirred me. One was a meeting at which Buck Suzuki spoke about the Japanese Canadian problems: the restrictions on fishing, on their families coming here, and so on. I thought it was horrible. I was interested in his words about people trying to understand each other, and using unity to solve problems.

The second thing was a meeting in the early 1940's in a parish hall. Scotty Neish and Gus Cogswell spoke about a drive to bring organizations together. They painted a picture of what it could be like.[7]

Negotiations to form a larger union at first failed to work out. But "in 1945 there was a merger convention, with unions plus shoreworkers. From then on, we were pretty well one."

The U.F.A.W.U. drew membership from both company-owned and independent boats, as well as plant workers. Japanese fishermen became members. Buck Suzuki, who had influenced Homer Stevens (and who as a Japanese Canadian soldier had spied for Canada in China during the war), became a member of the executive. Because fishermen were traditionally seen as co-adventurers, the union had no government certification as a bargaining agent, and had to depend on moral pressure to keep fishermen members in line. The plant workers provided a special lever. If during a fishermen's strike a company persuaded other fishermen to work for it, the UFAWU shoreworkers could bring the company's operations to a halt.

While organizing various strikes and tie-ups in the late 1940's, the union also gave attention to regulatory matters: conservation, environment, licensing, safety, marking of fishing boundaries, and so on. Homer Stevens said that "the Union was the outside writer of many regulations."

Often, the U.F.A.W.U. ran ahead of government. In 1948, the annual convention called for "a system of licence limitation for each branch of fishing, geared to be the conservation needs of the branch, and the provision of a decent livelihood for the consistent and regular commercial producers."[8] In following years, the union was a major influence towards licence limitation, which became the cornerstone of the fisheries management system.

Native people gain some ground

As the U.F.A.W.U. gained strength, the Native Brotherhood often allied with the union in negotiating prices. Indians had already been numerous in the fleet before the war, but faced obstacles in buying their own vessels; in particular, those who lived on reserves were unable to put up their houses or land as security for bank loans. During the war, the expulsion of the Japanese let many Native people buy boats at bargain prices. A fleet of Indian-owned seiners, trollers, and gillnetters emerged, although many Native people still worked as captains or crew on vessels owned by canneries or other fishermen.

After the war, Indian participation in the 1946–1962 period probably dropped from about 2,900 to 2,100 jobs (compared with, in the latter year, 16,400 Pacific fishermen in total). The decline came mainly in gillnets, which went from more than 1,600 to about 800, partly because of reduced fishing on the Skeena River system after a landslide. Native-held troll licences, however, rose from about 600 to 700; and seine licences, from 37 to 122. The number of Native skippers running company vessels was 119 in 1946 and 122 in 1962.[9]

All told, the B.C. industry outshone the Atlantic. Fishermen had more money and more power. Salmon processors had a healthy market, divided about equally between domestic and international trade. The surrounding society was stronger than on the Atlantic, with better job prospects and higher incomes. This lessened dependence and strain on the fishery. Rather than the employer of last resort, as was sometimes the case on the Atlantic, fishing was a trade that could hold up its head economically: outdoors and rugged, but self-reliant and generally prosperous enough.

Whitmore develops dynamic corps

Joe Whitmore

After Major Motherwell's 1946 retirement as regional director, A.J. (Joe) Whitmore directed federal fisheries operations in British Columbia from 1947 to 1960. Whitmore had entered the fisheries service in 1917 on the Pacific. In 1928, he moved to Ottawa, and in 1929 became head of the Western Fisheries Division at headquarters, where he served until moving back west. There he built an unmatched reputation.

Whitmore appears to have been the ideal fishery manager—knowledgeable, tough, and responsive. Officials in the region and in Ottawa, fishermen's representatives such as Homer Stevens, and processors all admired him. Jimmy Sinclair, the standout minister of the 1950's, saw Whitmore as his best official, even above Bates. "Whitmore should have been the deputy minister, but he looked the part less, with cigarette ashes dropping down his chin, and so on. Bates, being a Scotch economist, appealed to Mackenzie King."[10]

Whitmore liked to tackle the issues. At crowded meetings with the industry, he would remind fishermen and processors, "I'm the only man here who's paid to look after the fish." As one company president later said, "When Whitmore was here, fights were fun."[11] Jimmy Sewid, a famous high-liner among Native fishermen, recalled Whitmore showing up unexpectedly at Alert Bay and solving a regulatory problem on the spot. Whitmore commanded respect everywhere; speaking to conventions of the tough U.F.A.W.U., he could get standing ovations.[12] Alfred Needler said, two decades after Whitmore retired, "the Pacific salmon that we have today are the result of Whitmore's work." Rod Hourston, Whitmore's successor, considered him the architect of modern salmon management.[13]

Whitmore's contribution seems to have been less in policy than in active and thorough management, training, and example-setting. During the war, with too few fishery officers at work, excesses had taken place; Whitmore was determined to restore law and order. The post-war recruiting campaign brought in excellent Fishery Officers; in the 1950's B.C. had nearly 60 of them, along with seasonal guardians and patrolmen. The dynamic nature of the salmon fishery, with fast-moving runs now mingling in the ocean, now advancing into their hundreds of rivers and streams, led to a close monitoring and control system incorporating even remotely posted officers. Whitmore moulded the fishery officers and other officials into a corps that seemed to be everywhere, monitoring runs, observing salmon runs, watching fishing operations, and managing fisheries on the scene.

Fishery officers cover the waterfront

As of 1949, the department in British Columbia had 30-some patrol craft, mostly small or medium-sized with two to four men, but also including three larger vessels, the *Laurier, Howay,* and *Kitimat.* The more than 50 fishery officers were posted to 35 areas under three district supervisors. During salmon-fishing season, a fishery officer might hire four to seven guardians and patrolmen, the latter with their own boats to supplement the department's patrol craft.

Ron MacLeod, who served during the 1950's as fishery officer in Rivers Inlet, described the system under Whitmore. Before the season, the fishery officer would have forecasted the expected returns to the many spawning beds in his area (these might number 40–70 in a typical area).

The fishery officers and managers and in some cases F.R.B. researchers had for decades recorded observations of returning and spawning salmon for many of B.C.'s 1,600 spawning steams. From these the department had gradually devised rough forecasts based on the number of spawners. Since reproductive success was more predictable for salmon than for sea fish, this method had some chance of success, although it was far from precise. By the 1960's the department was issuing regular forecasts of expectations for the year. MacLeod said,

> I think the stream forecasts had their genesis under Motherwell, and got to their present state under Whitmore. The approach could vary for different streams. We'd do counts of escapement to the spawning beds, and sometimes downstream samples. Usually we'd go by escapement, and relate it to what happened in the past—what kind of run produced what amount of fish four years later. We'd also relate any extremes—rain, snow, drought—and then make a judgment. We'd say: "Based on the fishing patterns, I expect we could stand a two-day fishery, and I expect a catch of this order."

The local officer presented his views at an annual fishery officer conference. That session would co-ordinate seasonal regulations, taking pains to control intercepting fisheries that could over-exploit runs heading home to other areas.

During the active salmon or herring fishery, "[W]e'd watch the fishery, and adjust," MacLeod said. The fishery officer would monitor the fleet and catches, talk to the fishermen and packers, and sometimes reset fishing boundary markers to protect salmon in the estuaries. As time went on, more and more test fishing and acoustic soundings supplemented the fishery officer reports. "Sometimes we could confirm our forecasts with a counting fence [a V-shaped stream barricade used to trap, then release fish]. If the catch rate was up, we'd increase fishing; if it was down, we'd decrease it." The fishery officer would radio in his reports twice a day; these went to his district supervisor and the regional office in Vancouver, bringing together all the information. As well, field meetings brought together district supervisors and fishery officers from several areas, to compare notes and possibly change catch targets. If appropriate, officers on their own could make an unscheduled closure.

After the season would come the next coast-wide conference. As part of the exercise, the fishery officer would review the year's experience, and compare his forecast with what had actually happened. Whitmore knew the whole coast and its fish stocks. As MacLeod described it, "The greatest reward would be to receive Mr. Whitmore's rare accolade: 'Well done, Officer.' ... To make an annual forecast, to assess a fishery in progress, to regulate that fishery and then in the same year to measure the results on the spawning grounds

was what most Fishery Officers lived for. When he failed, the misery was terrible; when he succeeded the satisfaction was complete."

Despite the tight system, Whitmore gave his officers latitude, and the work had variety. Tasks could include inspecting and clearing streams; monitoring logging and other activities that could affect stream quality; surveying spawning grounds; and sampling eggs, fry, and alevins. At other times the fishery officer might be checking fishing gear, taking shellfish samples for toxicity tests, or monitoring the sales-slip system that started in 1951.

Predator control took attention; fishery officers might find themselves shooting birds (mergansers), in some instances, or paying a bounty on seals. At one point the department set up a gun at Seymour Narrows to shoot killer whales, then considered a menace; the gun, however, never got used. Shooting of seals and sea lions in B.C. ended in 1964. (Predator control applied on the Atlantic as well; the department paid a bounty on some seals, and for a time tried to keep down the numbers of cormorants and mergansers.)

Officers might speak at schools or local meetings, exchange information with fishermen and plant operators, and so on through a long list of activities. They also, of course, contended with poachers of various sorts, especially for salmon.[14]

Area licensing vanishes

The area-licensing system, whereby boats would check in and out of an area and too many boats would trigger a longer weekly closure, had worked adequately in the 1930's. The number of fishing craft was manageable; the Butedale area, for example, would get only 30 or 40 seiners in a whole season. By the late 1940's, it could have 250 seiners in a single day. Gillnetters could now number 400–500. The fishery officer had to deal with them all. Cliff Levelton, fishery officer at Butedale, described the system:

> If a fisherman had bought his licence in Area 12, he was licensed for there. If he came to Area 6, Butedale, he had to transfer in. I'd stamp the

Especially after area licensing ended, large numbers of boats could mass up in an area.

back of his licence. I had to sign 3,000 transfers in a season. It got to be such a pain that if a fisherman didn't bother to transfer in, we didn't bother to chase him.

If there were up to 25 seiners, they got four days a week fishing. Between 25 and 50, they got three days; over 50, two days. It was a convenient formula, but it was a lot of work, it was arbitrary, and it tied your hands. You applied it whether the run was big or small.

By the early 1950's, Whitmore and his officials had changed the cumbersome regime. "We went to a system where we'd assess the number of salmon and judge the fishing time we should allow," Levelton said. "For example, if in 1948 there'd been a good run of pinks, we'd expect a good run in 1950. Then if fewer showed up, we'd cut back to maybe two days a week. We considered all the factors, instead of an arbitrary formula. We could manage better—but as the fleet became more efficient and mobile, it created more pressure."[15]

Salmon science increases

Although the Fisheries Research Board took little part in developing the salmon-forecasting system, it initiated in late 1945 a large-scale study of the Skeena River salmon. This remained a major operation for some ten years, and included tagging and population studies. Then, in 1953, a landslide struck the Babine River, a major tributary of the Skeena. The landslide blocked much of the salmon escapement to Babine Lake, prompting a major effort by the fish culture branch and others to restore the system. The F.R.B.'s Nanaimo station, under Alfred Needler, got heavily involved. With tagging studies and a counting fence operating, the Skeena became a well-studied river, informative for salmon management in general.

The F.R.B. by now was tagging salmon from all along the coast, especially in the late 1950's. Some programs took place in co-operation with the United States and Japan. Scientists gradually filled out the picture of the far-flung migrations of the different salmon species. Bob Kabata of the Pacific Biological Station helped delineate fish populations by studying variations in the presence of parasites, a natural form of tag.

Besides the F.R.B. work, the departmental staff kept collecting and refining their information on salmon runs. By the 1970's, some had models of abundance, not just by week but by day.

Fleet gets more menacing

Neither the stream forecasts developed by fishery managers nor the later scientific models ever became precise. One official of long experience put their success rate at about two-thirds. The optimum degree of escapement for each stock to foster the best spawning and recruitment never became totally clear.[16] Meanwhile, the fleet was becoming more dangerous.

After area licensing disappeared, boats could go where they wanted, with hundreds sometimes massing up in a single area. If a fishery officer miscalculated and let fishing continue when it should be closed, the effect on a salmon run could be long-lasting. In the 1950's, the department often shortened the fishing week in various areas from the previously typical five days down to four, three, or even two. If it fully closed an area to protect salmon, the boats could congregate elsewhere.

Fish culture branch raises environmental consciousness

In the Fisheries Research Board, still separate from the department, the technological station located at the University of British Columbia was aiding the industry in fish handling and processing. Scientists at the Pacific Biological Station in Nanaimo were documenting the nature and behaviour of salmon, herring, and many other species. William E. Ricker was pioneering what became standard concepts of stock assessment.

But there was still somewhat of a gap between the F.R.B. and the Department of Fisheries. After the war, the department began recruiting more biologists and engineers for its fish culture branch. This group worked with new energy across the country. The emphasis changed from hatcheries, many of which still remained on the Atlantic, to such work as stream clearance and management, fertilization, and predation control. Like the management corps, the fish culture branch had standout officials, including engineer Charlie Clay, and colourful characters, like the biologist who, to cross a deep stream, would pick up a heavy rock under each arm and walk across under water.[17]

The fish culture branch had a scientific orientation different from that of the Fisheries Research Board. While the F.R.B. did bigger-picture work on fish behaviour and its causes and effects, the fish culture branch dealt with concrete problems requiring action. Work

The Pacific Biological Station at Nanaimo.

included building and maintaining fishways, and doing whatever else aided the fish.

The branch became especially important in British Columbia, where the salmon faced new threats from urban and industrial growth, hydro dams, and such affronts as loggers on bulldozers taking shortcuts down the beds of salmon streams. In 1947, despite local opposition, the John Hart hydro-electric dam went up on the Campbell River. In the early 1950's Alcan Corporation's Kemano power development, approved by the province despite the federal fisheries department's objections, diverted the Nechako River and damaged salmon runs.

Proposal after proposal came forward for B.C. hydro-power to serve economic development. Construction of the Columbia River and Peace River dams in 1960 failed to ease the pressure. W.A.C. Bennett liked such projects. Jimmy Sinclair, who after leaving politics became head of the B.C. Fisheries Association, recounted an early 1960's meeting with hydro proponents. "They showed me a map, and they had the whole province covered in dams."[18]

The growing threat to salmon galvanized conservationists such as the renowned Roderick Haig-Brown. Other events such as a D.D.T. spill in 1957 added impetus. Behind much of the public movement lay the scientific and technical backing of the fish culture branch. Its engineers and biologists gathered information; figured out consequences; spread information; and through public statements, court cases, and behind-the-scenes activity fostered the growth of environmental consciousness in B.C. Rod Hourston, who worked in the fish culture branch in the 1950's, said that "at first people thought the dam-fighters were crazy." But attitudes were changing. A major victory came when opposition prevented construction of the proposed Moran Dam on the Fraser system.

Salmon enhancement gathers steam

In 1953, engineer Les Edgeworth of the fish culture branch oversaw construction of an artificial spawning channel—the first in the world—at Jones Creek, near Hope in the Fraser Valley east of Vancouver. New construction was threatening salmon habitat in some areas of British Columbia. The idea of the spawning channels was to offset losses by creating habitat: scooping out channels for wild salmon to enter, and providing a suitable gravel environment and water flows for successful spawning and incubation. Jones Creek worked well, encouraging other efforts.

In 1959, construction began on the Big Qualicum hatchery and spawning channel, a major, modern project, feeding salmon into the Strait of Georgia. The next year saw the opening of the Robertson Creek spawning channel on Vancouver Island, another large project. Meanwhile, further improvements took place on Jones Creek. In the mid-1960's, large spawning-channel projects began in the Babine sub-system of the Skeena. Also in the 1960's, the I.P.S.F.C. carried out several

Part of the Jones Creek project.

spawning-channel projects in the Fraser system. Hatcheries and spawning channels were gaining favourable attention, setting the stage for further expansion.

Better management but fewer salmon

What were the results of post-war actions? The fleet had more fishing power, the department and the I.P.S.F.C. had a better management system, and some enhancement was taking place, but the catches were increasing less than hoped. Indeed, the total salmon catch, and notably sockeye, trended somewhat downward in the 1950's and into the 1960's. The value of salmon rose from $28.4 million in 1951 to $44.9 million in 1968, but that increase came from the market rather than management.

As on the Atlantic, a slight puzzlement may have crept into some apostles of development. But optimism still ruled. In 1961, W.E. Ricker and colleagues estimated there was room to multiply the sockeye yield by three or four times, and to more than double pink and chum catches, as runs recovered to historic levels through good management and enhancement.[19]

United Fishermen and Allied Workers' Union pushes for licence limitation

Meanwhile, some fishermen began to feel uneasy about post-war trends. Better boats were chasing the salmon, but with no major increase in catch to match the growth in investment. From 1948 on, the U.F.A.W.U. pushed for licence limitation, to help conservation and to protect incomes. "Conservation of fish resources is of no value unless it leads to conservation of fishermen," the union said. Resentment of moonlighters, people with no real stake in the industry, played a part. The U.F.A.W.U. wanted the licence to go "on the man," with entrants chosen from a waiting list, and no transferability of licences.[20]

Fisheries Minister Jimmy Sinclair promised action, and in 1958 commissioned a study by Dr. Sol Sinclair, a resource economist with the University of Manitoba. Sol Sinclair's 1960 report suggested higher licence fees and a five-year moratorium on licences. Then, government should fix an appropriate number of licences, and auction them off. Fishermen would be able to transfer licences among themselves.[21]

Various industry complaints emerged. Meanwhile, the government had changed. No action followed. Still, the idea of licence limitation was now percolating, at least in some quarters. Others still opposed government interference. To many, the industry seemed almost romantically free, an occupation where rugged individuals could test themselves against competition and the ocean itself.

Herring fishery crests and collapses

The herring fishery had boomed during the war. Canning declined after the war, and salting and pickling of herring died out, especially after Mao Tse-Tung closed the Chinese market in 1949. But the reduction fishery, having helped deplete pilchards, now turned to the richly abundant herring for fertilizer and fish-meal. Prices were low, typically $35 a tonne, but volume could compensate. One school of herring, for example, was reported to be six miles long and 30 fathoms deep.[22] In 1950, Captain Mel Stauffer of the *Maple Leaf C* took 1,260 tonnes in a single set.[23]

Landings had risen to an average 100,000 or so tonnes throughout the war. In the post-war years they rose again, yielding roughly 156,000 tonnes during the 1950's. Purse-seiners dominated the fishery. In the early 1960's, they began using high-powered mercury-vapour lamps to attract huge schools of herring at night. Catches increased again, to an average of more than 200,000 tonnes.

Since the 1930's, following work by Albert Tester of the F.R.B., the department had set herring quotas—perhaps not all that finely tuned, and sometimes extended at industry request (up to 250,000 tonnes in the mid-1950's), but at least a form of protection. Closures also applied during the spawning season. In the 1950's, a new generation of F.R.B. scientists took a different view, despite the cautioning voices of some department officials. All seemed abundant; were quotas really necessary?

As an experiment, the department removed the quota on the west coast of Vancouver Island; the stock still seemed to do as well as neighbouring ones. It was now thought that "the level of fishing was unlikely to reduce recruitment of herring stocks."[24] Fred Taylor, the lead scientist on Pacific herring, and Alfred Needler, director of the Pacific Biological Station from 1954 to 1963, supported the looser approach. It was thought that scarcities themselves would discourage fishermen well before there was any real danger to the stock. (Later, after herring collapses in various parts of the world, scientists would take a different view: that even when depleted, herring keep congregating in large schools, giving relatively good catches and the appearance of plenty.)

By 1965–1966, the catch was dropping fast. By 1967, the herring crisis was apparent to most. In hindsight, it seemed that heavy fishing and poor recruitment reduced the stock, while assessment techniques then in use failed to detect the decline in spawning escapements until it was too late. In 1968, the department closed the fishery. It would reopen only in the 1970's, with a much higher degree of caution. As a herring biologist put it years after the closure, "we're still fighting Needler's ghost."

Canadians, foreigners intensify groundfish fishery

The halibut fishery, under the International Pacific Halibut Commission, climbed slowly and steadily from the war until the early 1960's, then levelled off at around 20,000 tonnes. The I.P.H.C. controlled fishing methods, allowing no trawling; and set fishing times, areas, and quotas.

The war had stimulated demand for other groundfish. At the outset of the war, draggers were few, including 15–20 around Vancouver. The dragger fishery picked up during the war, especially as a market developed for dogfish, normally seen as a destructive pest but now recognized as a rich source of vitamin A (from their livers). Some seiners switched in winter to dragging for dogfish in the Strait of Georgia and other groundfish on the west coast of Vancouver Island. Draggers chased dogfish so much that, probably for the only time in history, worries rose about dogfish conservation. By 1950, however, synthetic vitamin A was destroying the dogfish fishery. Catches dropped from 9,000 tonnes or so in 1945 to practically nothing by 1965.

Taking herring with a table seiner in the 1940's. (Library and Archives Canada, PA-145356)

Halibut boat at Vancouver, 1961. (National Film Board)

Draggers turned to Pacific cod, rockfish, and other groundfish. The Canadian catch of all groundfish (including halibut and dogfish) ranged around 20,000 tonnes in the 1945–1955 period, rose to nearly 42,000 by 1966, then began to drop.

As on the Atlantic, foreign vessels had come onto the scene. Japanese vessels trawled in the eastern Bering Sea from 1954. In 1959, the Soviet Union joined them. By 1961, both countries were fishing in the Gulf of Alaska and off British Columbia and the northwest United States. Foreign fishing pressure kept growing during the 1960's. By 1968, Japan alone was catching more groundfish in the northeast Pacific (110,000 tonnes) than Canada (37,600 tonnes) and the U.S. (62,000 tonnes) put together.[25]

The foreigners, although not fishing directly for halibut, were competing with Canadians for other groundfish. Both foreign and Canadian by-catches affected halibut. As well, it became clear in hindsight that the I.P.H.C. had underestimated the efficiency of longlines and overestimated stock abundance. By the mid-1970's, halibut catches would fall to less than half their 1960's levels.[26]

Canada 1945–1968: More "development" than management

Fishery management in British Columbia differed in several respects from that in the Atlantic. The fleet was more modern, the fishermen more prosperous, the salmon more visible, the environment more vulnerable. Salmon management, based on ensuring up-river escapement, was fast-moving and dynamic, and influenced by a better-organized industry. Product quality presented fewer problems; price support was infrequent. The surrounding society was economically stronger, reducing dependence on the fishery.

But the two coasts shared many characteristics, including a great optimism that drove development. Both saw major fishery growth. Total sea-fishery landings climbed from 895,500 tonnes in 1955 to peak at 1,399,000 tonnes in 1968; landed value rose from $79 million to $173 million.

In the 1945–1968 period, understanding of the biological, economic, and social workings of the industry saw a great advance, though weaknesses remained. The policy and administrative system reached its strongest point to date. Some groundwork had been done for control of foreign fishing, although the main battles lay ahead. As for using the fish, market intelligence, public information, consumer promotion, and inspection of fish landings and products all came firmly into place during this period, together with a deeper knowledge of both how to catch fish and how to conserve them.

As for other aspects of management, questions of who should own the boats or plants were still mainly below the radar. As for the role of industry in making decisions: the Fisheries Council of Canada, a federation of provincial associations representing many large and medium-sized processing, vessel-owning, and fish-marketing companies, had great influence. Although B.C. fishermen had strong organizations, Atlantic fishermen had few; they were more likely to band together and make noise on particular issues. Government–industry seminars became more common in the 1960's, but they rarely involved debates, let alone votes. Rather, they tended to be like-minded people seeking the best means of development. There were no government–industry advisory committees to speak of, and little consistent attempt to involve fishermen in decisions. And no one gave much thought to the Aboriginal role in the fishery.

In the post-war period, the department and in many respects the F.R.B. saw development as a goal co-equal with conservation. Fish stocks were mostly thriving, with new fisheries coming on stream; it was the fishermen who seemed to need assistance, especially on the Atlantic. They got it; fishermen emerged from the period with better boats and better earnings, but at a future cost.

Bates had wanted to marry development and co-ordination, to make the industry work better as a system. In British Columbia, the elements meshed with some success, most of the time. But salmon catches never increased as hoped, and Pacific herring went into steep decline. The fleet was developing great power; conservation mechanisms sometimes failed to compensate. The Atlantic saw an even bigger build-up of fishing pressure by Canadian and foreign fleets. On that coast, over-investment, over-capacity, and lack of co-ordination helped to create, by 1968, an incipient crisis.

Some voices, notably that of the U.F.A.W.U. on the Pacific, were calling for controlled access to the fishery, both for conservation and for protection of livelihoods. In the department, economists tiptoed around the access question, making occasional suggestions like those of Rutherford for property rights or licence auctions in the lobster fishery. Meanwhile, scientists also expressed occasional concerns about this or that fishery. But the F.R.B. people were in one place, the department in another; there was no full-scale and continuing mechanism to bring conservation, science, and development together, let alone incorporating industry. The prevailing mood was less of co-ordination and planning than of adventure in development.

The year 1968 would prove to be a turning point. In the next decade, deepening problems of resource, market, and overcrowding, together with the presence of two decisive ministers, would transform fishery management into a comprehensive and complex system.

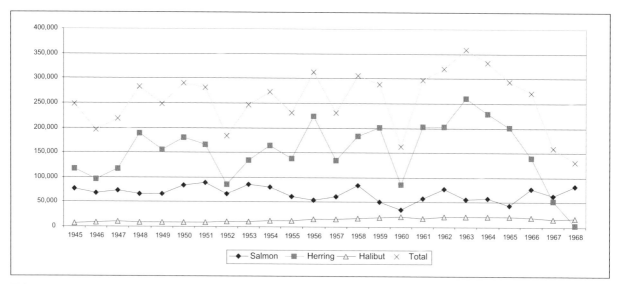

B.C. landings, 1945-68, total and selected species (tonnes).

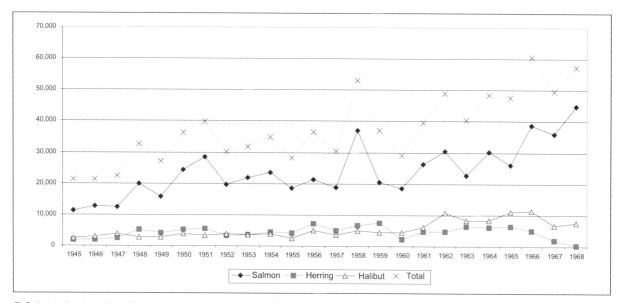

B.C. landed value, 1945-68, total and selected species ($000's).

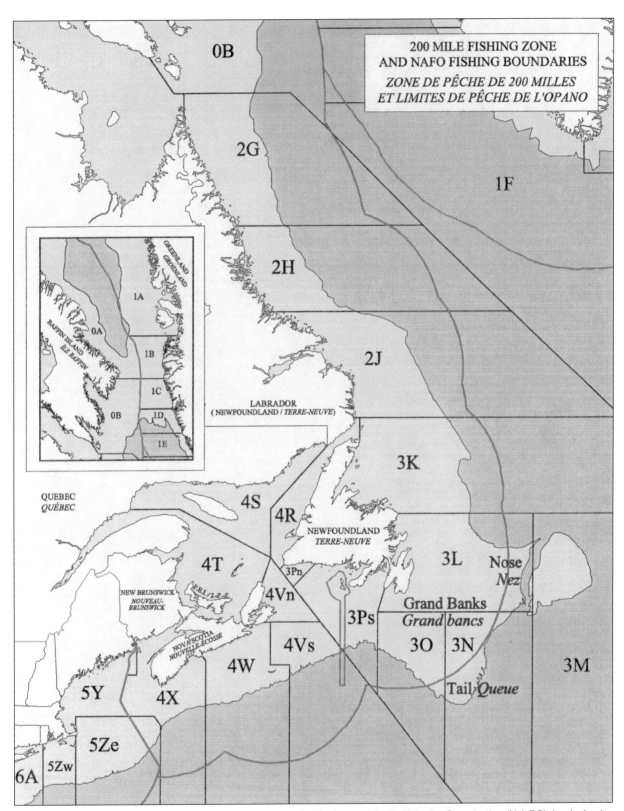

The fishing areas defined by I.C.N.A.F. would continue under the Northwest Atlantic Fisheries Organization (N.A.F.O), beginning in 1979.

PART 5: 1968–1984
COMPREHENSIVE MANAGEMENT BEGINS
CHAPTER 21.
National and international events, 1968–1984

The current system of fisheries management took shape in the period 1968–1984, chiefly under ministers Jack Davis and Roméo LeBlanc. Since Confederation, despite various experiments with leases and licence control, development and promotion, and such interventions as the trawler ban and the backing of co-operatives, the essentials of sea-fishery management had stayed the same. The fisheries generally remained open to all; the main goal was still conservation, although development had matched it at times; and the main means to conservation remained time, area, gear, and fish-size regulations.

From 1968 on, management grew far more comprehensive, leaving little untouched. As overfishing and overcrowding became more apparent, the government took the major step of controlling the number of licences and boats, though not of fishermen per se. It was thought that by stabilizing fleets, or reducing them in conjunction with other programs, "limited entry" could improve both conservation and average incomes. Strict licensing seemed to be the ultimate weapon against ever-growing fishing power. As well, the department brought in systematic use of catch quotas. These first applied for the total take in a fishery. Then came quota allocations by area and gear, and finally, by individual enterprises.

The new powers brought new burdens. Governmental control of licences and quotas implicitly meant control of fishing incomes and the fate of communities. Government wanted licences to curtail the "race for the fish." But equal importance would soon attach to the race for quotas and licences, a race run less on the fishing grounds than in meeting rooms. Interest groups multiplied their demands for special consideration.

Also during the 1968–1984 period, Canada helped form an international consensus on the Law of the Sea, and took control of fisheries within 200 nautical miles of the coasts. This brought hopes of doubling Canada's Atlantic groundfish harvest. On the Pacific, govern-

While boats continued getting bigger and stronger, electronics got more powerful. Wheelhouses could include radars, various radios, autopilots, loran receivers and plotters, sensitive sonars, and, as time went on, Global Positioning System receivers, electronic charts, and personal computers.

ment and industry aimed to double the salmon catch through the huge Salmonid Enhancement Program (S.E.P.). Meanwhile, crown corporations came into being to dominate the Atlantic salt-cod trade and the freshwater fisheries of the Prairies and northwestern Ontario. And fisheries personnel took on a stronger environmental role, as LeBlanc changed the Fisheries Act to give it new force for habitat protection.

As minister most of the time from 1974 to 1982, LeBlanc gave fishermen a bigger voice and more power, creating advisory committees and encouraging stronger organizations. Attempts to stabilize fishermen's incomes resulted in a sweetening of Unemployment Insurance benefits, attracting more fishermen into the fleet the department was trying to reduce.

In all, the federal department became more active than ever before, seeming especially on the Atlantic to bring new power and prosperity to fishermen. It took a giant step closer to Gordon's "unified directing power." But it was also immersing itself more deeply than ever not only in biology, but in the spheres of economics, politics, and social questions. Complexity sometimes brought perplexity; the department was biting off more than it could chew.

Government reflects activist age

As prime ministers come and go, government bureaucracies generally continue in their gray elephantine path. But Canada's federal apparatus seemed to respond to the spirit of the late 1960's and early 1970's, the days of rebellion, rock and roll, and new worlds waiting to be built. Prime Minister Pierre Elliott Trudeau pushed bilingualism, multiculturalism, and a Just Society, and had little compunction about intervening in business, through such measures as the Foreign Investment Review Agency and the National Energy Policy. Government was becoming more interventionist.

Yet government officials were also facing new constraints within the public-service system. The idea was growing that management was a skill one could transfer from department to department, without senior managers necessarily knowing much about a department's subject matter. Under Trudeau, organizational theories and flow charts became the rage. The "central agencies" of Finance, Treasury Board, and the Privy Council Office gained strength. The up-through-the-ranks career pattern for senior managers was shifting more to a pattern of move in, reorganize, and leave. This new approach was slow to take hold in fisheries, but as time passed it would become more apparent, especially in the 1980's and 1990's.

If the Trudeau-era government was ready to intervene, both the Atlantic and Pacific industries were ripe for change, for different reasons. The west coast fishery, despite its relative youth, was in some ways more mature. It had well-organized fishermen and a seemingly stable processing structure, dominated by a few large salmon companies. It enjoyed a reputation for good-quality products, and commanded a salmon market well balanced between domestic and export. The industry itself was pushing for progress. The main fishermen's organization, the United Fishermen and Allied Workers Union, wanted a new licensing regime. And all parties favoured using new and promising techniques to enhance the salmon resource.

On the Atlantic, there was less of that muscular and organized readiness for change. The industry was more fragmented. Groundfish and herring catches had increased; there were more mid-size boats; but incomes still lagged behind those of B.C. fishermen. Even after a quarter-century of development efforts, an air of backwardness hung over much of the industry. While the various fisheries pursued their own interests, the main push towards overall changes came from government.

The activist department under Roméo LeBlanc went beyond the quest for better conservation, more fish, and higher incomes. LeBlanc wanted to raise the status of fishermen. At the outset of the period, only processors had a strong organized voice and consistent influence. LeBlanc encouraged fishermen to organize, as some were already doing.

Large-trawler companies and other Atlantic processing interests were wary of government activism. Companies generally wanted non-interference, to the degree possible in a common-property resource. Most thought that federal fisheries should simply ensure security of supply and a competitive chance. But the larger companies themselves would wind up seeking federal intervention in the form of special-aid funding when they ran into near-bankruptcy twice in eight years.

Meanwhile, the department's new management system tried to control the number and then the length of boats. But fishing power kept growing, with better and costlier boats, gear, and electronics. "Limited entry" controlled the number of licences, but not the number of fishermen. A groundfish licence that once employed one person could come to employ several, especially as U.I. benefits improved. The number of fishermen and plant workers rose, especially as provincial governments encouraged more and bigger processing plants.

Fisheries department disappears and re-emerges

Fisheries becomes part of Environment Canada

The federal Department of Fisheries in 1968 had a potent combination of officials, experienced ones from the Class of '47 and its successors, and energetic younger ones. But they soon faced submersion in a bigger bureaucracy.

Jack Davis, a former Rhodes scholar with a Ph.D. in chemical engineering and a Master's from Oxford in economics, had worked in business and government

Jack Davis

before entering federal politics in 1962. He replaced Hédard Robichaud as Minister of Fisheries on July 6, 1968. Davis proved to be a clear and dynamic thinker and an energetic minister, although cool in dealing with people. Departmental lore had it that in six years as minister, he was known to laugh only twice.

The government soon gave Davis extra responsibility, on April 1, 1969 creating the Department of Fisheries and Forestry, in which fisheries became a separate service. In 1970, Fisheries and Forestry gathered in the Marine Sciences Directorate from the Department of Energy, Mines and Resources. Marine Sciences was responsible for oceanography and hydrography, the latter activity including navigational charts, tide books, and the Sailing Directions publications. Major centres included the Bedford Institute of Oceanography in Nova Scotia and the Marine Sciences directorate in Victoria, B.C.

Meanwhile, the public was becoming more conscious of environmental problems. Within government, Fisheries and Forestry officials helped lay the policy groundwork for a bigger agency. On June 11, 1971, the Department of Environment (D.O.E.) came into place, incorporating fisheries and forestry activities, as well as environmental research and management. The environmental side did little direct enforcement, relying more on the provinces for front-line work.

Alfred Needler retired from his deputy minister post in Fisheries and Forestry. Robert F. Shaw, an engineer whose reputation came mainly from his association with Expo 67, became deputy minister of the new Environment department. A reorganization in 1973 consolidated fisheries, oceanographic, and hydrographic activities under the Fisheries and Marine Service (F.M.S.) of Environment Canada.

Around the same time, the F.M.S. also took over hundreds of fishing wharves from the Department of Public Works; these went under the control of the Small Craft Harbours division. This was a major program worth $23 million in 1974. Wharves would become an important item for future ministers. Every fishing community worth its salt would lobby for new or improved wharves, while budget problems after the expansionist 1960's and early 1970's would make delivery difficult. Opposition M.P.s would sometimes accuse fisheries ministers of favouring the ridings of the party in power. Officials dealing with wharves gained a certain political wariness; some called the branch "Small Crafty Harbours."

The Fisheries and Marine Service in 1974–1975, now including fisheries research, accounted for 47 per cent of DOE spending; the money added up to $164 million, and person–years to 4,700.[1] (By comparison, back in the 1930's, the Department of Fisheries and the Fisheries Research Board had together spent less than $3 million annually; less than $6 million in the 1940's; and less than $20 million in the 1950's. Expenditures by 1968 had reached $50 million, a high figure compared with earlier years, but less than one-third the 1974–1975 level.) Activities were increasing, with wharves, oceanography, and hydrography; but at the top levels, attention paid to fisheries themselves was to a degree getting diluted.

F.R.B becomes part of department

The old Fisheries Research Board of Canada, which had been an independent agency, in 1973 became part of the Fisheries and Marine Service. The F.R.B. had mapped the migrations and natural history of fish stocks, fostered fishery development, and built up a worldwide reputation for Canadian fisheries science. But the F.R.B. and the department had never been that comfortable with one another. Particularly in the 1940's and 1950's, people of good will had worked together across the frontier, but institutional barriers always remained.

In the 1960's, F.R.B. chairman F.R. Hayes associated the agency more and more with universities, rather than departmental concerns. Misgivings about the board grew.[2] The governmental trend to consolidation prevailed; the department took over fisheries science and the F.R.B.'s research stations. F.R.B. personnel merged by 1975–1976 with Resource Development personnel; the new fisheries research arm on the Atlantic and Pacific reported through the regional directors-general (R.D.G.s) for fisheries management. The title "Fisheries Research Board" lingered on for a small rump agency turning out reports and offering advice, until it faded away in 1979. That same year a new

Bedford Institute of Oceanography

body, the Fisheries and Oceans Research Advisory Council (F.O.R.A.C.), took on a similar advisory role; F.O.R.A.C., too, would eventually vanish.

The merging of science and management marked a major change. Some observers still believe that the best science work took place under the independent F.R.B, and that a similar agency now could serve the country better. But most officials seem to think that management and science need a close and constant connection, best provided within a single organization.

At the time, some F.R.B. scientists looked forward to the amalgamation, only to voice mild complaints later that they were treated less as partners than as "tame scientists." Still, the amalgamation fostered more co-operation than before.

LeBlanc and Lucas fight clear of Environment

Ken Lucas, a fast-rising fisheries official from British Columbia, in 1973 became senior assistant deputy minister responsible for the Fisheries and Marine Service. The energetic Lucas had been one of the visualizers of the new Environment department. But for fisheries people, D.O.E.'s charms soon began to dim, as they encountered a surrounding bureaucracy with little knowledge of the industry. The fishery officials tended to fight for their own turf within the department, with a fair amount of success. Fisheries was a demanding field, and foreign fishing was raising its profile. At times the fisheries tail wagged the D.O.E. dog.

Jack Davis himself, the first Minister of Environment, spent the bulk of his time on fisheries matters, until his defeat in the election of July 8, 1974. Davis had angered fishermen in the Maritimes and British Columbia, though many accorded him a grudg-

ing respect. On the campaign trail, Prime Minister Trudeau let drop that Davis would likely be leaving fisheries in any new government. This was hardly a public endorsement of his minister. The B.C. electorate decided the matter—Davis lost his seat.

Lucas had appointed a strong new roster of regional directors; they, like Ottawa officials, chafed under the day-to-day obstacles of the D.O.E. bureaucracy. On August 8, they gained a champion when Roméo LeBlanc became Minister of State (Fisheries), under Environment Minister Jeanne Sauvé. A former school teacher and journalist, LeBlanc had become press secretary to prime ministers Pearson and Trudeau, then entered federal politics in 1972 as Member of Parliament for Westmorland–Kent in New Brunswick.

Eloquent, cultured, but down-to-earth, LeBlanc had grown up among small farmers and woodlot operators, and identified with the common man. When Trudeau asked him to be Minister of Fisheries, LeBlanc took a day or two to think about it, an unusual move for someone offered a cabinet position, then said that he would accept if he could behave as minister "for fishermen." Trudeau agreed, and for years gave LeBlanc practically everything he asked. Bringing to fruition initiatives begun under Davis and launching others of his own, LeBlanc would influence fisheries history more than any other minister since Peter Mitchell at Confederation.

Although technically a junior minister under Sauvé, LeBlanc took firm hold of fisheries, and soon began to look successful, especially with Atlantic fishermen. He gathered power accordingly. In September 1976, LeBlanc became Minister of Environment, and began calling it the Fisheries and Environment department.

Meanwhile, LeBlanc and Lucas fought within government to restore a separate department for fisheries.

As the idea of separation gained ground, infighting began over which future department would control what. Impatient with naysayers and obstacle-makers, LeBlanc went to Trudeau, who gave appropriate instructions to his Privy Council clerk.[3] On April 2, 1979, the Department of Fisheries and Oceans (D.F.O.) came into being.

Department of Fisheries and Oceans begins

Lucas reckoned that his battles for a separate department had burnt his bridges with the federal apparatus.[4] He moved on to the top fisheries posting with the Food and Agriculture Organization of the United Nations. Donald D. Tansley became deputy minister of the new department. Tansley had among other things fought in the Second World War, helped pioneer Medicare as a Saskatchewan government official, and served as deputy minister of finance in New Brunswick during premier Louis Robichaud's historic "Equal Opportunities" program. Like LeBlanc, he took a down-to-earth approach. Before reporting to work in Ottawa, Tansley took three months to tour the department's regions, talking to fishery officers, other officials, and industry members. As deputy minister, he ordered the upper level of department officials to make fishing trips with commercial fishermen. It was a shock to some; times were changing.

During the D.O.E. period, reorganizations had been frequent. For fishery officers and other field staff, the chain of command had gotten more complicated. Tansley created a more stable set-up for the new department. There would be four assistant deputy ministers, responsible for Atlantic fisheries; for Pacific and freshwater fisheries, including habitat; for fisheries economic development and marketing, including international work; and for ocean and aquatic affairs, dealing with oceanography and hydrography.

D.F.O. came into existence just before the May 1979 election; indeed, LeBlanc had to use connections and pressure to get the legislation through in time. When the Liberals gave way to Joe Clark's Conservative government, James A. McGrath, a well-respected M.P. for St. John's, Newfoundland, became minister. McGrath wanted to reduce the complexities stemming from departmental interventionism under LeBlanc. The Conservative regime would, however, last only about nine months before the Liberals and LeBlanc returned.

Gulf Region startles federal bureaucracy

The last notable change in departmental organization came on the Atlantic. In Quebec, the provincial government since the 1920's had controlled not only inland fisheries but also the fixed-gear salt-water fisheries. In the 1970's, the federal fisheries service had an administrative region in Quebec, working in French and controlling inspection, wharves, and some fishery matters.

Elsewhere in the Gulf of St. Lawrence, the French-speaking fishermen of New Brunswick and eastern Nova Scotia dealt at a distance with Maritime regional headquarters in Halifax, where most officials spoke only English. LeBlanc, an Acadian from near Moncton, identified with the sometimes frustrated Gulf fishermen; occasionally, he dropped references to Halifax as "the Kremlin."

After preparatory work by officials, LeBlanc in 1980 announced that he would create a new administrative region to take control of the entire Gulf, including Prince Edward Island and the bordering coasts of New Brunswick, Nova Scotia, Quebec, and Newfoundland. The main rationale was that the Gulf formed an ecosystem of its own, and needed unified management. But underlying that was the French fact.

Like many of LeBlanc's actions, the announcement caused consternation in the "central agencies" of Finance, Treasury Board, and the Privy Council Office. But Tansley and his officials worked the change through the system. A new headquarters took shape, first at Memramcook and then in a refurbished convent school at Moncton, under regional director-general Len Cowley and associate R.D.G. Jean-Eudes Haché. The new region officially came into being on April 1, 1982, as did the Scotia–Fundy Region, which took over the rest of the old Maritimes Region.

In the 1970's, the Gulf had at times been a wild place, with even a wharf-burning in one instance. Demonstrations and protests in eastern New Brunswick, many organized by the fledgling Maritime Fishermen's Union (M.F.U.), had become common. By the mid-1980's, they had almost disappeared, as Cowley, Haché, LeBlanc, and the presence of the new region improved relations.

After the 1979–1980 Conservative interval, LeBlanc remained as minister until October 1982, when he moved to the Department of Public Works. With more than seven years in power, he had been the longest-serving fisheries minister ever. Pierre De Bané, a Quebec M.P. and former Minister of Regional Economic Expansion, took over. De Bané proved to be a forceful

Besides fisheries science, oceanography, and hydrography, the new Department of Fisheries and Oceans kept responsibility for fishing wharves at hundreds of harbours, like this one at Gooseberry Cove in Newfoundland.

minister, launching initiatives on both coasts, with varied results.

For the Gulf of St. Lawrence, De Bané would strengthen federal influence. In Quebec, the province had long controlled most fishery management; the federal side, which LeBlanc had subsumed into the new Gulf Region, wielded only limited authority. That changed when, following a 1983 recommendation by a federal task force on the Atlantic fisheries, the federal government on April 1, 1984, took back management of Quebec's coastal fisheries.[5] De Bané also restored Quebec as a separate administrative region, which meant a weakening of the Gulf Region.

In June 1984, the new Liberal leader, John Turner, appointed New Brunswick M.P. Herb Breau Minister of Fisheries and Oceans. The energetic Breau served only briefly, as Turner launched an election campaign, losing in September 1984 to the Progressive Conservatives under Brian Mulroney. The Conservatives would leave departmental status untouched and make little change to the regional set-up.

The advance to extended jurisdiction

Cry arises for 200-mile limit

For most of the 1968–1984 period, two overriding goals dominated departmental thinking: first, to improve the fishery through a more comprehensive management system (as described later); second, to extend Canada's fisheries jurisdiction offshore to 200 nautical miles or more.

By 1968, foreign vessels in the northwest Atlantic took by far the most groundfish, and nearly half the herring. In the early 1970's, some 1,500 foreign fishing ships a year came into Atlantic waters off Canada. On the Pacific as well, Soviet and Japanese vessels caused alarm. Canadian fishermen on both coasts reported night-time fleets lit up like cities, just beyond the 12-mile limit.

Foreign vessels kept pounding away, as alarm arose in Canada. This Soviet vessel is landing silver hake. (D.F.O. photo by Michel Thérien)

Even so, departmental concern in the late 1960's was still rather muted. Gus Etchegary, a well-known industry figure with Fishery Products, the largest processor and trawler operator in Newfoundland, helped stir up public sentiment. In 1971, Etchegary and others launched the Save Our Fisheries Association; their presentations helped focus Ottawa's attention, as did the steady decline in catches after 1968.

In 1973, Icelandic patrol boats clashed at sea with British fishing vessels, cutting their trawl warps. Headlines about the "Cod War" further raised the pressure in Canada. Some South American nations had already declared 200-mile limits. Meanwhile, Canadian catches were fast dropping under foreign fishing pressure. The cry grew deafening, from fishermen's and processors' organizations, provincial governments, media, and public, for a Canadian 200-mile limit.

The long tradition of European fishing in the northwest Atlantic, the huge fleet launched since the war by the Soviet Union and its satellites, and the general practice of open-access fishing beyond 3–12 miles all complicated the situation. The Grand Banks alone covered some 36,000 square nautical miles, and Canada, it was declared in the 1970's when someone added up all the distances, had the longest coastline in the world, at 243,792 kilometres. Any jurisdictional move by Canada would have large international effects, and could provoke powerful opposition.

Some factors worked in Canada's favour. The International Commission for the Northwest Atlantic Fisheries (I.C.N.A.F.), with 16 members as of 1973, had at least gotten the different nations used to talking to one another. Alfred Needler had been a key figure. Although I.C.N.A.F. gained a reputation for toothless enforcement, still the organization helped to spread the idea of responsible high-seas management.

Canada asserts Arctic sovereignty

Meanwhile, the idea of offshore control was spreading. Back in 1964, Canada had passed the Territorial Sea and Fishing Zones Act and set up a nine-mile fisheries zone outside the three-mile territorial sea. The next move came from an unexpected direction. The United States in 1970 decided to send an oil tanker, the *Manhattan*, through the Northwest Passage. The U.S. considered the passage an international waterway; Canada considered it internal waters. The Trudeau government passed the Arctic Waters Pollution Prevention Act, asserting Canada's right to regulate navigation in an area extending 100 miles from its shores. Although the declared reason was to protect Arctic waters against pollution, the legislation implicitly made the point about sovereignty. As for the *Manhattan*, by agreement between the two countries, a Canadian icebreaker accompanied the American vessel through the passage. The issue passed away with no loss and probably a gain to Canadian sovereignty.

Boarding operations became common in the I.C.N.A.F. area.

"Closing lines" protect some fisheries

Also in 1970, the government unilaterally extended the territorial sea to 12 nautical miles. In March 1971, Canada established fisheries "closing lines" for the Gulf of St. Lawrence and the Bay of Fundy on the Atlantic, and for Dixon Entrance, Hecate Strait, and Queen Charlotte Sound (in effect, all the waters inside the Queen Charlotte Islands) on the Pacific. The necessary regulations came under authority of the 1964 Territorial Sea and Fishing Zones Act. On the Atlantic, the Fisheries department and External Affairs negotiated "phase-out agreements" with overseas nations (the United Kingdom, Norway, Denmark, France, Portugal, and Spain; Italy also withdrew) and with the United States; the Americans would thus abandon their important redfish fishery in the Gulf of St. Lawrence.

A special agreement with France followed in 1972. French fishing had long antecedents protected by treaty. Canada was eyeing extension of jurisdiction; an agreement with France might help pave the way. France accepted the 12-mile territorial sea, and in effect recognized that Canadian jurisdiction might extend further seaward in future. Canada in return promised France fishing rights, both in existing Canadian areas and in any future zone. The latter provision would cause problems later.[6]

I.C.N.A.F. pioneers multinational quotas

Meanwhile, Canada was pressing its fisheries-management case through I.C.N.A.F., the International Commission for the Northwest Atlantic Fisheries. Although I.C.N.A.F. had no direct relation with the Law of the Sea negotiations of the 1970's, still Canada was using both avenues to seek more control. Canada had been a founding member of I.C.N.A.F., which had its headquarters in Dartmouth, Nova Scotia The fish were adjacent to Canada, with hundreds of communities depending on them, and Canada kept claiming the high ground on conservation. All this gave Canada a certain moral authority, at least in Canadian eyes. At I.C.N.A.F. meetings Canada had a strong personal presence through Needler, Canada's chief commissioner to I.C.N.A.F. from the mid-1950's, and other officials. By the mid-1970's, dozens of industry members would form part of Canadian delegations.

From the mid-1960's, scientists through I.C.N.A.F. had warned that fishing effort would need restriction.[7] Now, the historic haddock fishery on Georges Bank was shrinking under high fishing pressure. Mesh-size regulations agreed to through I.C.N.A.F. had failed to stop the decline.

In 1970, I.C.N.A.F. applied overall catch quotas for western Scotian Shelf and Georges Bank haddock, the organization's first quotas. A yellowtail flounder quota followed in 1971. Then herring became the big issue. Scientists were piecing together evidence that the Bay of Fundy herring caught in summer and fall were part of the same migratory stock as those caught in winter and spring at Canso and Cape Breton. Meanwhile, foreign vessels were fishing the stock, which was weakening.

I.C.N.A.F. set a quota for herring. But for more orderly management, it seemed desirable to divide up the overall quota into national ones. This required a change in the I.C.N.A.F. Convention, which took place in 1971. In 1972, I.C.N.A.F. not only set an overall 150,000-tonne total allowable catch (T.A.C.) for herring from Georges Bank, the Gulf of Maine, and Nova Scotia stocks, but also divided up that overall quota among the different nations fishing herring. The various quotas added up to far more herring than the scientists wanted caught.

Still, it was a historic event. I.C.N.A.F. declared that "this marked the first time that national quota allocations had been agreed to in a multi-national fishery and demonstrated the ability of international bodies, such as ICNAF, to play an effective role in fisheries management." While making provision for multinational allocations, I.C.N.A.F. also replaced the goal of M.S.Y. with "optimum utilization," which could mean lower levels of fishing.

In 1973, I.C.N.A.F. began setting T.A.C.s for all major groundfish stocks, both inside and outside Canada's 12-mile zone and closing lines, and allocating quotas to member nations. Canada pressed for, and other members agreed to, a so-called 40–40–10–10 arrangement, with national quotas calculated from catches over the last ten years (40 per cent), catches in the last three years (40 per cent), coastal state preference (10 per cent), and another 10 per cent for contingencies. This seemed like progress.

As well, I.C.N.A.F. in the early 1970's made provision for inspections at sea. Canadian patrol vessels such as the *Chebucto* and *Cape Harrison* would launch fishery officers in small boats to go aboard foreign vessels to check gear, mesh size, logs, and the like. Other nations at first refused to allow fishery officers below decks to inspect catches, but eventually agreed to that as well. Davis and the department publicized every gain. Some Canadian officials felt that, as one put it, "We're writing the book on international fisheries management."[8] Even some industry members gained more faith in I.C.N.A.F.

Groundfish crash heightens pressure for jurisdiction

Meanwhile the push for a 200-mile limit gained impetus from fishery misfortunes. In l973, the Atlantic groundfish industry made the most money ever, as hungry markets more than compensated for declining catches. The next year, 1974, was its worst in history.

Although market fluctuations contributed, the worsening scarcity of fish loomed as the major factor. Groundfish catches by all countries in the I.C.N.A.F. area (essentially all the northwest Atlantic) fell from about 2,800,000 tonnes in 1968 to about 1,700,000 in 1974. Canada's take fell from 621,000 to 418,000 tonnes. For cod, the total take by all countries in the I.C.N.A.F. area fell from 1,876,000 tonnes in the peak year of 1968 to 791,000 tonnes in 1973, a 57 per cent drop.

Despite Canada's heavier fishing effort in the 1950's and 1960's, the Canadian cod catch had never really increased. Now it fell by more than half, from 323,000 tonnes in 1968 to 158,000 tonnes in 1974. A chief symbol of disaster was the inshore cod catch of only some 35,000 tonnes in northern and eastern Newfoundland, which normally yielded many times more. Foreign fishing got the blame, though Canadians, too, had been fishing all the cod they could. New minister Roméo LeBlanc set up an emergency-aid program.

Nations look towards Law of the Sea

Other nations were suffering from foreign fishing, though few had as much at stake as Canada. As more voices took up the cry for 200-mile limits, the United Nations began planning for the Third Law of the Sea Conference. U.N.C.L.O.S. III lasted from 1973 to 1982, with major meetings in Caracas and Geneva, followed by long sessions in New York. The conference dealt with a host of marine issues, including navigation, exclusive economic zones, seabed mining, and environmental protection.

There was no consensus on coastal states taking control of fishing or "exclusive economic" zones. Major powers such as the United States and the Soviet Union opposed such zones, fearing interference with commercial and military activities. Meanwhile, interest was growing in seabed minerals beyond the continental shelves, then seen as a potential source of vast riches. Advocates began to call for an international regime governing these mineral resources as "the common heritage of mankind." The Law of the Sea conference beginning in 1973 aimed to develop a consensus on all the issues, a "package deal."

Alan Beesley of the External Affairs department led overall Canadian efforts on the Law of the Sea; the federal fisheries department, however, dominated fisheries negotiations. Lucas built up the international branch under Leonard Legault, a bicultural Saskatchewan-born lawyer seconded to fisheries from External Affairs. Legault led a strong group including, among others, Mike Shepard, Bob Applebaum, Jim Beckett, Art May, Gary Vernon, Mike Hunter, and David Bollivar.

Canada tries species-oriented approach

In the early 1970's, Jack Davis and Canadian negotiators pushed for an ocean regime that treated different species in different ways—the "functional approach." Most fish spend their lives on the continental shelf of their origin. Canada pointed out that for a handful of countries, continental shelves reached out beyond 200 miles, as was the case with the Nose and Tail of the Grand Banks. The Canadians pushed for coastal-state control over the continental shelf and margin, which would take care of shelf-confined species. International arrangements should manage far-migrating species, such as bluefin tuna. For salmon, which could migrate hundreds and even thousands of miles, the state of origin should have a "special interest." (As part of the lobbying effort, Dave Denbigh of the department's information branch produced a boxed "salmon portfolio," an assembly of prints and a specially prepared book, which became a collector's item.)

The "functional approach" had biological and geographical logic, but found little international support. Two-hundred-mile limits were simpler, and there were already precedents in South America. The concept of coastal-state control out to the continental margin, which could have prevented foreign fleets' picking off stocks on the Nose and Tail of the Grand Banks, died an early death.

Canada presses I.C.N.A.F. for coastal-state priority

While pressing internationally for coastal-state jurisdiction, Canada also pursued coastal-state priority within I.C.N.A.F. A prominent bloc of member nations consisted of the Soviet Union and its sea-going satellite nations Poland, Romania, Bulgaria, and later, East Germany. Other members besides Canada included, as of 1972, Denmark, France, West Germany, Iceland, Italy, Norway, Portugal, Spain, the United Kingdom, the United States, and Japan. The various members had differing views on coastal-state jurisdiction under the Law of the Sea. But as they began to recognize that 200-mile limits were increasingly likely, they became somewhat more amenable to coastal-state claims within I.C.N.A.F. The organization moved slowly in Canada's direction.

By 1973 I.C.N.A.F. was, as already noted, dividing up major herring and groundfish stocks into national allocations. Canada began arguing that the coastal state should get all the fish it wanted from the Total Allowable Catch; other members could share the rest. At the same time, Canada pressed to cut fishing effort by as much as half to let the stocks recover.

Other nations resisted. Meanwhile, fisheries patrols and National Defence surveillance revealed that some were overrunning their quotas.

Bilateral agreements, port closures force the issue

Canada stepped up its overtures to allies and its threats to recalcitrants. For potential supporters, Legault and Mike Shepard held out the prospect of

Although Canada's new trawler fleet had great fishing power, small and medium-size boats dominated by far in numbers. Canadian negotiators stressed and bilateral agreements reflected the needs of coastal communities. Photo shows longlining off Souris, P.E.I., in 1977.

good fishing relations with Canada. Coastal-state control would mean better conservation, and access to surpluses for countries supporting Canadian objectives.

By 1975, it had become likely that the Law of the Sea conference would endorse 200-mile zones, but only as part of an overall treaty covering other issues. This would take several more years to develop. Canada and other countries found it necessary to consider unilateral action to establish 200-mile zones, in anticipation of their becoming part of the future Law of the Sea.[9]

Early in 1975, Legault and Norwegian fisheries representative Jens Evensen worked out a precedent-setting agreement. Norway and Canada agreed to respect each other's extensions of jurisdiction; Canada agreed to give Norway access, after extension of jurisdiction, to fish surplus to Canadian needs. Canada pursued further such arrangements. Such bilaterals could act as a self-fulfilling prophecy, influencing international opinion towards acceptance of 200-mile limits, and fostering reasonable dealings thereafter.

Meanwhile, problems continued with some distant-water fishing nations, including the biggest and the most powerful in world affairs, the Soviet Union. On July 23, 1975, LeBlanc made national headlines by closing Canadian ports to Soviet fishing vessels and warning Spain and Portugal that the same could happen to them. In fisheries diplomacy of the day, it was a thunderbolt.

Most Canadians cheered; some found fault. The Halifax *Chronicle-Herald*, which had loudly called for Canadian jurisdiction, began lamenting the lost business from Soviet vessels now shut out of Canadian ports. Naysayers said that the Soviets could get fuel, food, and all their needs equally well in St. Pierre and Miquelon. LeBlanc mused privately about, in that case, closing off St. Pierre's fuel supplies from Canada. On the port closure, there would be no backing down.

Intensive diplomacy took place, leading up to the I.C.N.A.F. meeting of September 1975, at the grand old Windsor Hotel in Montreal. Canada's delegation, led by Needler, was demanding a huge, unprecedented reduction in foreign fishing. At the same time, Legault and other officials were dealing with the Soviets on bilateral matters. Andrej Gromyko, the Soviet Foreign Minister, had come to Ottawa for discussions with Secretary of State for External Affairs Allan MacEachen. Starting in a tense atmosphere, the Montreal meeting emerged into triumph. The Soviets and their satellites supported Canada's demands in I.C.N.A.F., bringing a drastic T.A.C. reduction. Fishing effort for 1976 would be about 40 per cent below 1972–1973 levels.[10] In Ottawa the Canadian and Soviet governments announced their intention to conclude a bilateral fisheries arrangement, foreshadowing acceptance of Canada's future jurisdiction.

The Soviets and Canada made final their bilateral agreement in May 1976. Poland had concluded a bilateral a few days earlier, and Spain and Portugal did so in June and July. Those countries and Norway did 88 per cent of foreign fishing off Canada, and they were agreeing in advance to accept extended jurisdiction. Canada would provide them with surplus fish, but only according to the strength of the stocks, and after taking Canadian needs into account.

Canada extends jurisdiction

With the main fishing powers agreeing to extended jurisdiction, the path ahead was becoming clear. Meanwhile, the United States had a change of heart and announced in 1976 that it would enact a 200-mile limit as of March 1977. This further encouraged Canada. A team of officials, including Dick Roberts, Scott Parsons, and Dave Bollivar, developed plans for a new regulatory regime.

On November 2, 1976, LeBlanc and Secretary of State for External Affairs Don Jamieson announced the government's intention to act, under the authority of the Territorial Sea and Fishing Zones Act, to extend jurisdiction as of January 1, 1977. LeBlanc later declared his pride that "my department led the way."

Takeover goes smoothly

The 200-mile limit, later applied to Arctic waters as well as the Atlantic and Pacific, brought 3.7 million square kilometres of ocean—the largest such zone in the world, equal to 37 per cent of Canada's landmass—under Canadian fisheries control. A federal publication declared that Canada was "owner and manager" of the new zone.

Some commentators held that foreign vessels would continue to roam freely, and Canada would never be able to enforce the 200-mile limit. But the takeover went smoothly, bilateral agreements having paved the way. In 1976, federal fisheries and External Affairs notified foreign fishing nations that they would need licences to enter the new zone. They complied, requesting approvals through the Canadian authorities.

Still, there was tension at the outset. On New Year's Day, 1977, Canada had fisheries patrol vessels, Coast Guard vessels, and navy ships patrolling the zone. American author William Warner was travelling on a German trawler at the time, when they heard that the Coast Guard vessel *John Cabot* had apprehended a Norwegian longliner, was taking it into St. John's, and had threatened to call in the navy if necessary. "My God," said the trawler captain, "it's Chicago on the high seas."[11]

From 1978 on, foreign vessels fishing under Canadian licence within the 200-mile zone carried Canadian observers. Over time, several regional companies sprang up to supply observers, who were certified according to D.F.O.-set standards. The department advised on training programs. To stay on ships, deal with foreign captains and crews, and monitor all operations was no small challenge. The job attracted its own corps; the average observer was educated, intelligent, and adventurous. Some were filling in between other work; others tended to stick with observer duties.

Law of the Sea provides new framework

Always in the background as Canada moved towards extended jurisdiction was the Law of the Sea Conference in New York. The Canadian-backed concept of the coastal state managing the fish resources, taking all it needed, and sharing the surplus, gradually spread among nations, and was incorporated in the December 1982 Law of the Sea treaty. So was the coastal state's "primary interest in and responsibility for" anadromous stocks such as salmon, together with a ban on high-seas fishing of salmon. This provision, too, resulted in large part from Canada's efforts.

In addition, the treaty recognized the coastal state's right to manage seabed resources to the edge of the continental shelf, even where it extended beyond 200 miles: this included sedentary species such as snow crabs and scallops, as well as oil and gas resources. Among other provisions, the United Nations Convention on the Law of the Sea gave international blessing to 12-mile territorial seas and 200-mile economic zones, set rules for the international seabed, enshrined rights of innocent passage, and dealt with environmental protection. The Convention became international law as of November 1994, when 60 countries had ratified it; most of the world's nations followed.

Canada in the 1980's began abiding by most elements of the Convention, but would delay ratifying it until the new millennium. The chief problem for Canada was the lack of any firm control in the Law of the Sea regarding fishing of straddling stocks, those that overlapped the 200-mile limit. (Indeed, if Canada had ratified the Law of the Sea, it might have handicapped itself in a major fisheries dispute that emerged in the 1990's with the European Union.)

N.A.F.O. takes over from I.C.N.A.F.

The conservation cutbacks negotiated through I.C.N.A.F. and the new controls on licences reduced the Atlantic foreign fleet within the zone from about 1,500 to about 500 vessels, a number that would shrink as years passed. Foreign vessels, especially bilateral partners, retained substantial fisheries, particularly on the Nose and Tail of the Grand Banks and on Flemish Cap, which are outside the 200-mile limit. But the Canadian fleet, which before the 200-mile limit had taken well under half of northwest Atlantic groundfish, now took the clear majority, about 86 per cent of the desired species of groundfish and herring in 1978. Even when species such as capelin and silver hake were included, Canada's share of total finfish still came to 75 per cent, more than double the 34 per cent in 1974.[12] Canada's percentage would soon increase further.

D.F.O. official tracking foreign fishing vessels.

In the run-up to the 200-mile limit, the department and External Affairs had looked towards a "son of I.C.N.A.F.," to co-operate with Canada in managing the straddling stocks that overlapped the 200-mile limit and those still further offshore on Flemish Cap. The new organization came into being on January 1, 1979, under the name of the Northwest Atlantic Fisheries Organization (N.A.F.O.). Scientific and other committees came into place. N.A.F.O.'s Scientific Council could provide advice not only to I.C.N.A.F., but to coastal states as well. N.A.F.O. recommended Total Allowable Catches, quotas, and other regulations for several stocks.

Border disputes cause difficulties with U.S.

As Canada and the United States extended their 200-mile limits, their claims overlapped in four areas: the east coast on Georges Bank and around Machias Seal Island in the Bay of Fundy, the Beaufort Sea in the Arctic, the Alaska Panhandle (the old A–B line dispute), and waters running offshore from the British Columbia–Washington border. While the last three remained in abeyance, attempts to resolve Georges Bank led to a tentative agreement that later fell apart.

Georges Bank harboured rich grounds for scallops, herring, and groundfish. Immediately following extension of jurisdiction, the two countries made reciprocal fishing agreements, allowing pre-existing fisheries in one another's waters to continue. Then Canadian and American negotiators tried to skirt around the border question by negotiating a fisheries treaty. The theory was that if they could agree on management and sharing of stocks, the border question would become irrelevant.

Negotiators did come up with a draft treaty. But some American fishermen resisted, notably the Point Judith Fishermen's Cooperative of Rhode Island. Senator Claiborne Pell from the same state headed the Senate Foreign Relations Committee. Pell's influence helped to doom the treaty. The two countries in 1979 referred the boundary question to a panel of the International Court of Justice at The Hague, The Netherlands.

"Hague Line" bisects Georges

Although there had long been a significant Canadian fishery on Georges Bank, the Americans were used to thinking of the bank as their grounds. Many thought that under international arbitration they could get the whole of Georges and much of Brown's Bank, on the premise that they formed a natural prolongation of the American continental shelf.

Canada first called for a settlement based on equidistance, a common method of resolving borders. Then, responding to a larger U.S. claim, the Canadians widened their own, contending that an equitable equidistance settlement would follow the general run of the coast and ignore the protuberance of Cape Cod and the islands of Martha's Vineyard and Nantucket. The judges listened to the two sides presenting arguments about equidistance, continental shelves, historical dependence, and so on. They then, in October 1984, came up with their own solution based on a geographic formula of their own devising. The "Hague Line" gave Canada roughly the northern one-third of Georges Bank, the single richest part of the bank. The U.S. government pressed for a continuation of the reciprocal agreements, but Canada refused it. The new line came into effect, to the discontent of many American fishermen.

Canada arrests American tuna fishermen off B.C.

Meanwhile, in British Columbia, the American 200-mile limit of 1977 displaced 54 Canadian halibut vessels that had fished off Alaska. The Americans provided a two-year phase-out period. Canada's fisheries department in 1979–1980 provided nearly $7 million through a Halibut Relocation Program to help B.C.'s Alaska vessels shift back into Canadian waters, and to help the rest of the halibut fleet—some 350 smaller vessels—absorb the shock. The money went to compensate fishermen for lost income, and to buy or convert some vessels for the black cod fishery.

While the halibut fishery followed its peaceful international tradition, the Pacific tuna fishery shaped up differently. The United States had refused to recognize that 200-mile zones applied to highly migratory species such as tuna, or that coastal states had the right to exclude foreign vessels from fishing those species within the coastal zones. American fishermen continued to follow albacore tuna into waters off B.C., without Canadian licences and in violation of Canadian law. In August 1979, Canadian authorities apprehended 19 vessels and escorted them to B.C. ports. The U.S. retaliated with an embargo on Canadian tuna exports, affecting mainly the Ocean Maid plant in St. Andrews, New Brunswick.

The Conservative government was by now in power under Prime Minister Joe Clark. External Affairs, under Flora MacDonald, had grave doubts about the department's assertive stand. But D.F.O. minister James McGrath stood firm, and Clark supported him. Canada and the U.S. concluded a reciprocal agreement for licensing arrangements. American tuna boats thereafter respected the zone.

The 1968–1984 period ended with Canada strong and sovereign within the 200-mile limit. In future, the main problems with foreign fishing would occur outside the zone.

Department gears up for new regime

Science work increases

Throughout the departmental rejiggings of the 1968–1984 period, the accent was on growth, including in research. When the Environment department in 1973 took over the Fisheries Research Board, it gained the Newfoundland Biological Station at St. John's; an Irish moss research establishment at Ellerslie, Prince Edward Island; the technological station at Halifax, which did fish-processing research; the St. Andrews Biological Station; the Arctic Biological Station on Montreal Island; the technological station in Vancouver; and the Pacific Biological Station at Nanaimo. Many hatcheries were at work, including the giant Mactaquac hatchery for Atlantic salmon near Fredericton, New Brunswick. A Sea Lamprey Control Unit operated at Sault Ste. Marie. On the oceanographic and hydrographic side, the new department controlled the Bedford Institute of Oceanography at Dartmouth, Nova Scotia, a huge hydrographic complex in Ottawa, and the Marine Sciences Directorate at Victoria. It also shared the Canada Centre for Inland Waters at Burlington, Ontario.

Jack Davis, the first Minister of Environment, wanted new regional centres in Newfoundland and British Columbia. In 1974, D.O.E started work on a $10.4-million Newfoundland Environment Centre at the White Hills in St. John's. When Fisheries extricated itself from Environment, it kept the complex, which housed scientific and management staff and became known as the Northwest Atlantic Fisheries Centre.

In Vancouver, Davis acquired land near the Lion's Gate Bridge for a Pacific Environment Centre, which never came into being as such, but later provided space for the new West Vancouver Laboratory. More successfully, the department in 1974 began building the $11 million Institute of Ocean Sciences in Patricia Bay, near Sidney, B.C. The new institute took over oceanographic and hydrographic work on the Pacific, becoming a renowned oceanographic centre, and also gaining a reputation as an incubator for private-sector research enterprises. Davis also set in motion a $9-million expansion of the Bedford Institute of Oceanography.

Based at the Bedford Institute of Oceanography, the *Hudson* was the first vessel to circumnavigate the Americas.

The Northwest Atlantic Fisheries Centre, St. John's, Nfld.

The Freshwater Institute, Winnipeg, Manitoba.

The Institute of Ocean Sciences, Sidney, B.C.

Meanwhile, for the Fisheries Research Board's station in Winnipeg, an impressive new building, the Freshwater Institute, had appeared on the University of Manitoba campus in 1972. With the F.R.B. takeover in 1973, the institute came under the Fisheries and Marine Service, and served as regional headquarters as well as a research station. The Freshwater Institute would do significant work, including the Experimental Lakes project, which was to shed new light on aquatic environments worldwide.

In 1973, Davis launched a five-year, $50-million vessel construction program, for both enforcement and science. Work went ahead on several research vessels, large and small. At the time, the Fisheries and Marine Service fleet included, along with many smaller craft, 117 vessels of nine tons or more, of which 32 were longer than 70 feet. Fisheries patrol work employed 78 vessels, fisheries research 18, and marine sciences 21.[13]

A further build-up came with LeBlanc and the 200-mile limit. The department in 1976 and 1978 made major, successful requests for funding. Budgets for fisheries management increased, and research capability for sea fisheries roughly doubled. The department took on about 100 new research staff and boosted science spending to about $112 million.[14] Work went forward on two major research trawlers, the *Wilfred Templeman* for Newfoundland and the *Alfred Needler* for the Maritimes Region, both completed in 1982. Chartering of research vessels increased, including on the Atlantic the *Lady Hammond* and the *Gadus Atlantica*.

By the 1970's, Canadian researchers had filled out much of the picture for migrations and natural history of fish species and had contributed to international thinking on population dynamics of fish stocks. As various quotas came into play, scientists on both coasts increased their work on assessing and forecasting fish-stock abundances.

On the Pacific, techniques for estimating and forecasting salmon were already well established, though fishery managers and scientists kept improving them. The new emphasis was on enhancement. Work begun in the post-war period would culminate in the $150-million Salmonid Enhancement Program ("salmonid" because it also included trout) launched in 1977.

One science cutback came in the Halifax and Vancouver technological labs, which for decades had done research related mostly to fish processing and quality-keeping. By now, government and industry had solved the problems relating to safety and to general acceptability. Complaints persisted about the quality of Atlantic products lagging behind that of the competition; but this was no longer a question of basic goodness, but rather of degrees of quality for the higher paying markets. LeBlanc, whose sympathies ran far more to fishermen than to processors, converted the labs to work that was more directly related to fishing. "Processing and product development is first of all a private responsibility, secondly a provincial government responsibility, and only then a federal responsibility," he said in 1979. "We have done fish-product research because no one else did it." But now federal fisheries wanted to "ease out of a field where we don't really belong."[15]

Meanwhile, fish farming began to get new attention. In 1973, Ken Lucas chaired a national aquaculture conference in Winnipeg, and in 1974 appointed G.I. Pritchard as national co-ordinator for aquaculture. Scientists including Roland Brett in Nanaimo and Arnold Sutterlin in St. Andrews began to research the possibilities. Few imagined that salmon culture would one day dwarf Canada's wild salmon fishery.

The theory of optimal fishing

A 1978 D.F.O. leaflet, *Fisheries Science: How and Why It Works for Marine Fisheries*, outlined the thinking that had developed in the 1950's and 1960's on population dynamics and economic yields. The leaflet reflected the work of Gordon, Anthony Scott, and others, and drew directly on texts by fisheries department scientists R.G. Halliday, A.T. Pinhorn, W.T. Stobo, and W.G. Doubleday. Excerpts follow:

Virgin Stock
We'll start at a square-one situation. When a stock of fish has never been fished, scientists call it a virgin stock. The total weight of this stock is larger at this time than at any other time after it has been fished.

A virgin stock exists in balance with its environment. It's balanced because growth of individual fish in the stock and additions through reproduction **equals** the weight of fish which die from natural causes (natural mortality) such as predators, starvation or disease.

The net (total) growth rate of a virgin stock is zero because the weight of the fish which die from natural causes cancels out the weight of the fish being hatched and growing in the stock.

Changes Caused by Fishing
Fishing upsets the balance of a virgin stock. As soon as fishing begins, the death or mortality rate of the stock goes up, and because deaths by fishing are removing fish from the stock, the size of the stock goes down.

However, because there are less fish in the stock, there is more food for the fish that escape the fishermen. In other words, there is less competition for food. When there is less competition for food, individual fish grow faster and less fish die from natural causes.

The faster and more intensively a virgin stock is fished, the faster the remaining fish grow and replace themselves. So even though the stock size goes down because fish are being taken from it, the net growth rate of the stock as a whole goes up. The stock keeps trying to replace itself as though it was in the virgin state. It adds new weight to itself at a faster rate, trying to replace the weight the fishermen remove.

But this net growth rate does not continue to go up forever—it's related to the size of the stock. When the size of the stock goes down to a certain level, the net growth rate reaches its highest point and then drops off quickly. It will eventually diminish to zero, when the stock size approaches extinction.

This up and down relationship between stock size and the rate of net growth is shown on **Graph 1.**

Total Catch and Catch Rates
When a fisherman starts fishing a virgin stock, his total catch (total tons of fish caught) will be high and the catch rate (pounds of fish per hundred hooks fished, or pounds per otter trawl tow, for example) will be high at the beginning, too. At this time, the stock is being only lightly exploited, the net growth is increasing, and the size of the stock as yet hasn't gone down very much.

However, as more and more vessels fish the stock, the stock size is, of course, dropping down from fishing mortality, and the net growth rate of the stock is getting closer and closer to reaching its peak. Scientists call the level at which this peak or high level of total catch is reached the level of **Maximum Sustainable Yield or level of MSY.**

As more fishing brings the stock closer and closer to the level of MSY, the total catch will increase slowly, but a single boat's catch rate will really start to drop off. At the level of MSY, individual catch rates will have gone down 50% or more from what they were for the virgin stock.

Graph 2 shows how the total catch increases before the MSY level is reached and drops off after.

Graph 3 shows how both the size of the stock and the catch rates drop down quickly from the point when the stock is first fished.

To sum up: if a stock is fished at the level of MSY or close to it, the total catch will be maximized but individual catch rates will become lower and lower as the MSY level is approached.

If a fisherman's catch rate is low, he will soon be placed in a break-even financial situation at best, and will lose money at worst. He'll be making more trips out, for example, to catch basically the same amount of fish.

On the other hand, if the whole fishing effort is controlled (as it is under the limited entry program) so that fishermen take a certain amount of fish at some point before the level of MSY is reached, their costs will be lower and their returns higher.

In other words, if our purpose of fishing is to maximize the poundage of fish caught, **regardless of the cost of doing so**, then the MSY level is the most favourable approach. However, if **our purpose is to maximize earnings from our fishing effort** (as is the case in Canada), then we need to fish at a yield level where the costs of fishing are taken into account.

Let's look at these two approaches in terms of vessels, catch rates (or cost of fishing) and total catch, using an example stock. If the stock was fished at the level indicated by the arrow "a" on **Graph 4A** (which shows catch rates) and **Graph 4B** (which shows total catch), the results would be those shown on the table.

Worked out in percentage, it can be seen that from the level indicated by the arrow "a" to the MSY level, the number of vessels fishing has gone up 80% and the total catch has gone up 12%. However, the tons of fish caught per trip out, or catch rate, has gone down 37%. In other words, **the more the stock is fished, the faster it reaches the MSY level, and the closer one gets to the MSY level, the more it costs to fish.**

We can examine **Graph 4C** to more clearly see how the cost of fishing is higher at the MSY level than at a point before MSY is reached. Look at the distance between the dollar value of the catch and the cost line at the level indicated by the arrow "a" and the MSY level "b."

Marine fisheries management is now founded on the principle discussed above, commonly called a "best use" basis. The cut-off point for TAC's is now geared to the optimum sustainable yield level which corresponds to a point where net returns from the total fishery are highest—i.e. before the MSY level is reached.

In other words, **catch quotas are set on the basis of economic catch rates, with the biological facts setting the limits**. These economic catch rates will vary, of course, from species to species, and depend very much on costs and prices received.

— Excerpts from *Fisheries Science: How and Why It Works For Marine Fisheries*

In the now-standard thinking, scientists also categorized two types of overfishing. The first was "growth overfishing"—harvesting at such a rate that the average fish never reaches full size. This would result in loss of yield, the same as cutting trees when they're only half grown. The second was "recruitment overfishing"— fishing so hard you cut down the effective birth rate ("recruitment"), and threaten a particular stock's ability to sustain itself, just as continued harvesting did away with many pine forests in eastern Canada.

By the late 1970's, scientists and managers usually equated optimum sustainable yield (OSY) with, for Atlantic finfish, what was known as the $F_{0.1}$ level of fishing, which roughly worked out to catching one fish in five. But stock assessment faced several obstacles that received little attention at the time.

Scientists had no clear picture of the amount of fish thrown overboard as too small or otherwise unsuitable, or the degree of misreporting in order to evade quotas. For catch-versus-effort statistics, they often relied on data from large, mobile vessels of increasing efficiency, with no "control" statistics from passive fishing. Nor was a safety margin (except $F_{0.1}$) applied to the catch-quota calculations. In some instances, scientists and managers buoyed by good catches gave insufficient attention to the potential influence of natural cycles or habitat disruption. Even when scientists were conscious of weaknesses in their assessments, the message sometimes failed to get across to managers. In future years, such deficiencies would loom larger in the blinding glare of hindsight.

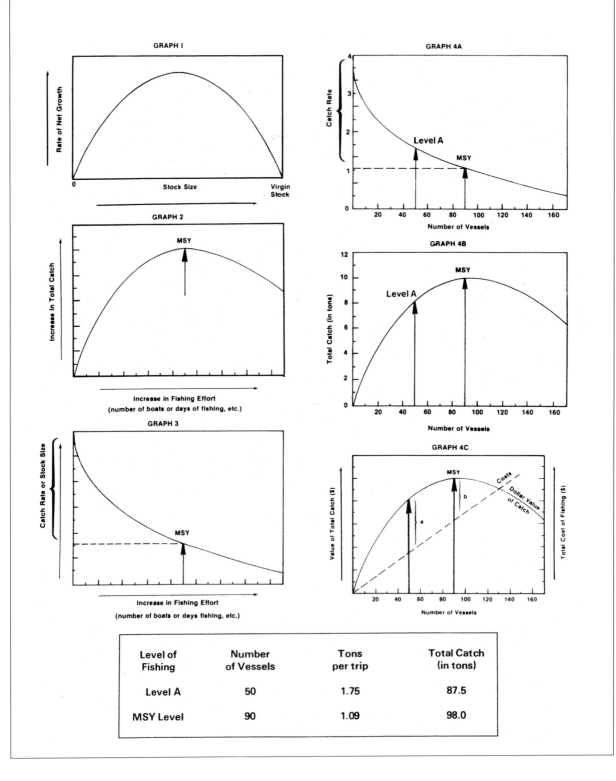

Level of Fishing	Number of Vessels	Tons per trip	Total Catch (in tons)
Level A	50	1.75	87.5
MSY Level	90	1.09	98.0

New ships, more fishery officers

Davis and LeBlanc added new ships to the offshore enforcement fleet: on the west coast, the *Tanu* in 1968 and the *James Sinclair*, to replace the *Howay*, in 1980; on the east coast, the *Pierre Fortin* in 1975, the *Cape Roger* and the *Louisbourg* in 1977, and the new *Cygnus* in 1982.

Spending surged for medium-sized patrol vessels as well. In 1974, Davis announced that the next five years would see 17 such craft replaced by new ones. By 1979, the department had 12 major patrol vessels and about 200 smaller craft on the coasts.[16]

A substantial increase in fishery officers took place in the late 1960's and early 1970's. For example, British Columbia in the 1950's had fewer than 60 full-fledged fishery officers, and in the early 1970's had about 75. Then, in the late 1970's, the department nationally replaced much of the guardian–patrolman system (although some still remained) with permanent positions. For B.C., the number of fishery officers on the payroll rose to about 125. (In addition, about 40 of the 150 or so officers and crew on B.C. patrol ships had designated powers as fishery officers.)

Nationally, the number of fishery officers rose to between 700 and 800. By comparison, back before the Second World War, the department had usually had fewer than 200 fishery officers (then called inspectors) in the field. But those officers would hire guardians and patrolmen (the latter term meant, in B.C., guardians with their own boats) to assist them.

The increased number of fishery officers meant an improvement, at least on paper. But, especially on the Atlantic, many of them were seasonal, as the old guardians had been. And the actual numbers of people patrolling, with reduced numbers of guardians and patrolmen, could in some instances be fewer than in the old days. Enforcement staff in British Columbia produced reviews contending that, compared with the 1950's, patrol work had actually declined.

Meanwhile, enforcement officers were beginning a long change from being a neighbourhood figure, with mainly on-the-job training supplemented by occasional courses, to a more police-like persona. In 1975, some fishery officers on the lower Fraser River in B.C. began to carry guns. In 1977, all B.C. officers started carrying arms.

Fishery officers in other regions began to ask for the same capability, at least for special areas and special circumstances. Officers sometimes said that the danger was less from legitimate fishermen or run-of-the-mill poachers than from those on the fringes, outlaw types who might get tanked up on drugs or alcohol. A slow transition was commencing that would, over the next two decades, see fishery officers everywhere carrying guns. They would gain in personal security but lose some sympathy among commercial fishermen, who often considered the guns an affront.[17]

Fishery officers get new training program

Along with the guns came more training. Since the ad hoc effort set up for the Class of '47, fishery officer training had mainly taken place on the job. But starting in 1977, those fishery officers using guns took firearms and related training at the Royal Canadian Mounted Police (R.C.M.P.) Academy in Regina. In 1981, deputy minister Don Tansley sponsored a fresh look at fishery officer training, led by Ron MacLeod and Bud Bagnell, an experienced official from Nova Scotia. This coalesced in a new Fishery Officer Career Progression Program, including formal training by the R.C.M.P. and by D.F.O. itself.

The program became well established over the next two decades. As of the late 1990's, the department was choosing recruits from among applicants with two years of post-secondary education in renewable resources, law enforcement, or the equivalent. A five-week program in Cornwall, Ontario, imparted national policies on fisheries, habitat, and departmental mandates. Classes also dealt with basics of the judicial system, cross-cultural sensitivities, and other such matters. Trainees then spent six weeks in the Regions, learning about fisheries and biology, and taking training in local regulations and enforcement scenarios. There followed seven weeks at the R.C.M.P. college in

Fishery officers check mesh-size and other regulations in the groundfish fishery.

The *Cape Roger*, built in 1977, patrolled off Newfoundland.

Regina, for training in firearms, self-defence, defensive driving, the Criminal Code, human relations, and investigative techniques. The final 18 months of training took place with senior fishery officers in the field. The R.C.M.P. program was deemed to be the most intense instruction, helping to jell officers together as a team. From 1977 to 2002, the R.C.M.P. trained about 800 fishery officer recruits.[18]

For all its work on enforcement, the department never sponsored a concerted publicity or consciousness-raising campaign to encourage conservation in commercial fisheries. It relied on regulations and surveillance. In some areas, however, the department set up telephone tip lines, with such names as Observe–Record–Report or Dial-A-Poacher, where members of the industry or public could report wrongdoing. Fishery officers frequently spoke to school classes about conservation. In B.C., the Salmonid Enhancement Program set up a major public-education effort on salmon biology, habitat, and conservation.

Fishery officers: Atlantic and Arctic work

As of 1983, D.F.O. had about 800 fishery officers all told, including eight women (their numbers would rise). Starting salaries were in the $20,000 range, top salaries in the $30,000 range. Earlier, this book highlighted fishery officer work on the west coast. The following excerpts from a 1983 article deal mainly with the Atlantic and Arctic:

Offshore patrol work

How do you tell a destroyer captain where to take his vessel? "Very politely," says a Fishery Officer who's done offshore work. The government used most of the Atlantic destroyer fleet on fishery patrol right after the 200-mile limit came into force, to get the idea across; and it still uses them for some far off-shore patrols on the Tail of the Grand Banks and on the Flemish Cap. ...

There's also an active boarding program on the West Coast, although foreign vessels there are fewer. Most boardings on both coasts take place from the department's own major vessels, such as the *Cape Roger* in Newfoundland (to be joined by a recent vessel named after the late Len Cowley, who was assistant deputy minister for Atlantic fisheries), the *Chebucto, Cygnus,* and *Louisbourg* in the Maritimes, and the *Tanu* and *James Sinclair* on the Pacific.

Boardings on the Atlantic coast of foreign vessels in the Canadian zone (licensed mainly for silver hake and other species unused by Canadians) numbered 315 in 1982; on the Pacific coast, 118. Fishery Officers found 68 violations.

On the Atlantic, foreign vessels also carry observers (hired by D.F.O. under a third-party contract) on most of their fishing days; and larger Canadian vessels also carry observers perhaps a quarter of the time. ... Observers can only report violations back for D.F.O. action later, as they lack the Fishery Officer's power to make arrests. Indeed, senior officials and the Minister lack that power, unless they've been sworn in as Fishery Officers. ...

Atlantic coastal work

On the coasts, the number of Fishery Officers in an area depends on the amount of fishing activity. Much of the officer's time goes to such matters as checking mesh and net size, watching closed areas and times, ensuring that fishermen use only the proper gear, collecting plant production statistics, and explaining regulations.

Poaching becomes a problem mainly in the lobster and salmon fisheries. Offshore, aircraft and fast patrol vessels make it relatively easy to spot the large fishing craft. Inshore, there are thousands of boats and coves, making it more difficult. Worse yet are the salmon rivers, where poachers may have a whole forest to hide in. This is probably the worst of all fisheries for hard feelings and violence against Fishery Officers. On the coast, the Fishery Officer most often is dealing with a legitimate fisherman who only broke a rule in a weak moment. On the salmon rivers, he can be dealing with any kind of a character, and the danger heightens.

Rocks, gunshots, slashed tires are a few of the hazards in the Maritimes and B.C. salmon fisheries. (Newfoundland has its share of poachers, but they are relatively gentle.) On one New Brunswick river, poachers stretched an automobile tire's inner tube between two trees, launching a boulder at a Fishery Officer that sent him to hospital for some time.

Despite increased co-operation from fishermen in recent years, the lobster fishery remains a hard one to police. Fishery Officers in the Gulf region have described how poachers set up signals (e.g. sheets on a clothesline to warn boats offshore), use their own radio codes, and keep searching out D.F.O. radio channels despite frequent changes in frequencies.

Almost every Fishery Officer who ever dealt with lobsters has a story about stakeouts and searches: stumbling across illicit lovers instead of poachers, searching houses for lobsters that might be in the baby carriage or in the bed, with "babies crying, the husband swearing at you, the wife looking daggers."

But the coastal officer's life has a much more pleasant side: passing the time of day with fishermen on the wharf; giving the signal on the first day of lobster season to let hundreds of boats loaded with traps steam out of harbour; feeding back fishermen's complaints to managers and actually getting something done about it.

Arctic work

Whatever the pleasures and pains of Fishery Officer postings in the south, few have to face anything as extreme as the Arctic's bitter cold and great distances. The Whitehorse district alone is said to be responsible for some 770,000 square kilometres. The handful of Fishery Officers scattered across the north often must move from place to place during the year following the native hunts for marine mammals. ...

The Fishery Officer carries information about new regulations and policies from the D.F.O. to the fishermen; and in turn transmits information about local conditions back to the area manager and regional headquarters. Besides enforcing regulations, the Fishery Officer often helps create them, in co-operation with the fishermen. Even if they blame the government for too many regulations, it is a fact that fishermen are constantly demanding them, usually against rival fleets. And the Fishery Officer will often advise departmental managers about what kind of measures might work and what might fail.[19]

Fisheries Act gets stronger

The Fisheries Act, which had gone through periodic updates since Confederation, still provided the framework for fisheries management and the main bulwark against pollution. In 1977, LeBlanc took major amendments through Parliament. The new legislation authorized much stronger penalties for habitat- and pollution-related offences. The slogan of the time was "the polluter must pay."

The amendments also provided new powers to protect fish habitat (essentially, the nurturing environment in streams or practically anywhere "on which fish depend directly or indirectly to carry out their life processes"). Environmental consciousness had grown in the 1960's and early 1970's; habitat issues were multiplying. In effect, the rest of the country was catching up with British Columbia, where the people wanted in particular to protect salmon-spawning areas and nursery streams. The powerful amendments to the Fisheries Act won LeBlanc a standing ovation when he spoke to the B.C. Wildlife Association; paraphrasing a remark of Prime Minister Trudeau, LeBlanc declared that "the state does have a place in the spawning beds of the nation."

The legislation opened the way for a major increase in habitat protection work. Work under LeBlanc, McGrath, De Bané, and subsequent ministers would in the mid-1980's culminate in a national habitat policy.[20] Habitat would become a growing preoccupation in later years.

LeBlanc clarifies policy goals

As science and management expanded, fishery managers got a clearer set of goals. Davis had given the first great push to comprehensive management. LeBlanc carried it through, and made it clear that the interests of fishermen would come first.

The department had always managed in the interests of fishermen, to a considerable degree, but the picture was sometimes cloudy. The Fisheries Council of Canada (F.C.C.), representing processors, was the only industry organization with a national mandate. Although processing was ostensibly a provincial responsibility, the companies knew where their fish came from; the F.C.C., a federation of regional associations, maintained its headquarters in Ottawa. Department officials were sometimes more comfortable with the better organized, better educated representatives of the F.C.C. than with the fragmented, hard-to-handle fishermen. To many people in the department, the words "fishing industry" meant large processing companies, particularly in Atlantic groundfish and Pacific salmon.

Roméo LeBlanc at the launching of the patrol vessel *Cygnus*, at Marystown, Newfoundland, 1982.

The new *Cygnus*, 207 feet long, was the third patrol vessel to bear that name.

Moreover, department officials in the post-war push for development often envisaged a "rationalized" Atlantic industry dominated by large offshore vessels, company owned. The department never wrote off the small-boat, independent fishermen; indeed, the fishery officers and those working closely with the industry had a good deal of fellow-feeling with the independents. But many of the upper-level thinkers in the department, those who wrote the plans, projects, and analyses, tended to look for progress among the larger enterprises, rather than the disorganized smaller ones.

LeBlanc administered a swift corrective. In his first major speech, to the Atlantic Provinces Economic Council, he put the independent fisherman first and foremost.

> The economists among you probably know a hundred different ways to analyze the fishing industry. Among all these sophisticated views, what often sounds unusual today is this: to take the view of the fisherman. ... To the tourist, the fisherman seems free. [... But there are cases] where the "freedom" is more like bondage. ...
>
> ... There is still room for expansion in the fisheries. [But] we need to think beyond catching the maximum number of cod per hour per man. When fish are counted, it's people that count. Any project or plan in the fisheries has really one basic criterion of judgement: does it improve life? ...

> If there is any hope for our fisheries, I believe we have to marry the two basic ideas I have tried to put forward: on the wide scale, the idea of opportunity for development; on the individual scale, the idea of a decent life for the fisherman. ...
>
> Nobody wants to see a lot of small fishing villages disappear in favour of larger ports. But are larger ports necessarily bad in themselves? Lunenburg is a large fishing port and Lunenburg is far from being an industrial slum. Nor are modernized fisheries bad in themselves. British Columbia's fishery is one of a relatively high technology, and the men on the B.C. boats still have the human qualities such as self-reliance that we associate with fishermen. ...
>
> [In B.C.] the fishermen and the processors know they are married, and they also know that in 1974 women's liberation has arrived. In the matter of prices and of what happens to the fish, the Atlantic coast fishermen also should have a bargaining voice. ...
>
> The law gives me a strong power regarding licensing of fishermen, and I intend to exercise it. Licences should be reserved to the man who earns the larger part of his income from the fisheries. ... Why should the fisherman not expect to be protected from the weekend sailors and the amateur lobstermen? ...
>
> ... In licensing and all these associated matters, I intend to listen closely to the bona fide fisherman, the man whose life is fishing. ... But I would remind the fishermen we can't consult every single one of them. In a word: organize. Be sure your voice is heard, and be sure that your spokesmen are properly mandated and accountable to you.

Make it possible for us to listen, and we will do our part. ... We can't hope to manage our fish resources without the help of the fishermen. On fish stocks, they have much knowledge; in any conservation policy that's to be effective, they are the caretakers; and finally, for us in the Fisheries Service as for all of you with the power to influence policies, the people whose lives we affect are and will be our judges.[21]

LeBlanc's call for organization made national headlines, giving encouragement to many groups, including the burgeoning Newfoundland fishermen's and plant workers' union. Processing companies began to wonder what was happening. LeBlanc drove the knife home in a Halifax speech to the Fisheries Council of Canada, in May 1975. The F.C.C. speech came in the wake of expensive emergency-aid programs for large Atlantic groundfish companies. It expressed not only an attitude but an extensive agenda, illustrative of LeBlanc's later actions:

Ladies and Gentlemen, I am glad to see you here in good health. When I visited this city last fall, some of you told me that, in effect, you might fail to survive the winter. The groundfish industry was sick.

Within six weeks of that meeting, the Federal Government applied the miracle drug: money. We announced a $20 million program to keep the industry—companies and fishermen—going through the winter. We'd given the fisheries a $15 million injection only three months before that. And two weeks ago, I announced another $50 million program. It has been enough to make one wonder: Are we dealing here with a hypochondriac, or an invalid, or an addict? ...

Undercutting each other on traditional products leaves our companies little margin to support anything new, or for that matter to bid up prices to fishermen. Instead, even ports with more than one buyer usually have only one price, whatever the quality of the catch. ...

One of the many conflicting myths about Canada's fisheries is that of the imminent death of the so-called inshore fishery—that is, the one which supports more than 34,000 boats in our sea fisheries, those fisheries which in turn help support over 2,000 communities in the Atlantic region, and form the only support of over one-fifth of them. ...

What the fishing communities want, I want—and so do you.

Goal number one: More offshore jurisdiction, for of all this industry's problems, the increased difficulty of catching fish is the worst. ...

Goal number two: A good life for the fisherman, with less economic fear. ... Why should insecurity always remind him that his boat is at one end of the highway to Toronto? ...

It is not particularly useful to go into the reasons why the industry has not grown in strength as it should. What is important is to strengthen it now. It is up to government to do this job. The government can. The government must. The government will. ...

We have [invested in the industry] through boat building subsidies, through DREE subsidizing new plants, through ports and harbours, through UIC, through compensation grants. We have done so through the biologists, market analysts, product inspectors, fishing gear technologists and all the rest whose salaries we pay to work for you; also, of course, through special programs such as all those announced within the last year. ... Your claim on the attention of the state implies the participation of the state in shaping the industry's future.

On [the fishermen's] behalf, I say again that public funds imply public responsibility. ... What I am advocating is the sharing of decisions by all those who are affected, and the full disclosure of facts on which judgments can be made. ...

... Get rid of the idea of fish being cheap raw material, and of the idea that you can make up a marketing loss by transferring it back to the primary level. I am speaking to the fishermen too; remember it is dollar bills that you have been ripping with the fish fork and tossing into the puddle of slime. Raw material today is precious. ...

In some cases we might want bigger boats, freezer-trawlers or freezer-carriers and the like, as in some foreign fleets. [But we might also] want to spread our fisheries out over a longer season for more boats, or to protect a local fishery from bigger mobile fleets. In this regard, zones and licences provide an instrument to balance proximity and economic efficiency, and also to reduce gear conflicts. ... Control of expansion should apply at the plant level also; ...separating fishing enterprises from processing enterprises might improve both prices to fishermen and quality of products: ... those products should include presently unused species, for in a world of protein shortage, a time of famine, we can no longer afford to throw away food.

About processing and marketing: I would like to see some ports with more diverse capabilities, more processing strength, able to use all that the fisherman lands. ...

Creating export groups through licensing, to limit the foreign marketers of groundfish from the present thirty-odd down to a handful, should hurt no one. It should help all producing companies, by stressing quality and stimulating sales. ... We are also considering, as you know, a stabilization plan whereby you and the fishermen would pay out in the good times, get back in the bad. ...

... The same two tendencies run through the whole fishing industry: the rugged individualist assertion that we should leave fisheries to God and the fisherman coexists with a cry for help. ...

To accomplish anything, we need to work together in a sensible, self-governing way. ... We need not rationalize the small family concern out of business, we need not separate human values from fisheries, nor on the other hand fisheries from progress. We can create a healthy mix, and with it strengthen the whole coastal economy.[22]

As the processors listened in shocked silence, LeBlanc added an aside about his intention to take firm control of the fishery: "There's an old expression in Newfoundland: 'leave her lay where Jesus flang her.' I don't intend to leave her lay."

The two speeches had spelled out almost all of LeBlanc's major projects: the 200-mile limit, comprehensive management including licences, quotas, and zones, control of expansion, improvement of quality and marketing, organization of fishermen and a stronger influence for them in management, and a strong role for government in industry operations. The only major missing item was the Pacific Salmonid Enhancement Program. LeBlanc and his officials were to achieve most of these goals, the major exceptions being export marketing consolidation. The F.C.C. speech also raised the idea of separating fishing and processing enterprises; in that, LeBlanc would achieve only partial success.

Policy document codifies thinking

It all meant a major change to department attitudes. Some thinkers in the department had long wanted firmer control of industry numbers and capacity; often, they identified progress with larger companies and "offshore" fishing. LeBlanc, too, wanted rational management and tight controls, but rather than corporate-centered development, it would be rationalization with a human face. Good management and the 200-mile limit would provide more fish; the benefits would start with independent fishermen. The department would build up the fishery "from the coast out." LeBlanc's approach took some getting used to, but by the late 1970's it was firmly if sometimes uncomfortably lodged in departmental heads.

The new approach got elaborated in a 1976 document, *Policy for Canada's Commercial Fisheries*. This remained at the end of the century as the only comprehensive national statement on fisheries policy in the department's history. The policy document, which owed much to W.C. Mackenzie, declared that "fishing has been regulated in the interest of the fish. In the future it is to be regulated in the interest of the people who depend on the fishery. Implicit in the new orientation is more direct intervention by government in controlling the use of fishery resources, from the water to the table, and also more direct participation by the people affected in the formulation and implementation of fishery policy."

LeBlanc himself noted in a later speech, "[W]e switched to managing not necessarily for the biggest catch of fish today but for the most benefit to people, to give them better incomes and better jobs and better lives, not just this year but next year and the years after that. It means replacing the preoccupation with volume by giving equal or greater attention to value and stability."[23]

Policy spells out goals, controls

The May 1976 *Policy for Canada's Commercial Fisheries* gave LeBlanc's approach an intellectual foundation. The associated summary had this to say in part:

The commercial fisheries of Canada often have been unstable, self-weakening, prone to crises, and providing low and insecure incomes for participants.

On the Atlantic coast, ... groundfishing fleets operating from some 2,000 locations deliver to over 1,000 landing points catches destined for over 300 processing plants, owned and operated by 200 private companies. The vast majority of these competing plants produce one product: frozen groundfish. They have plant capacity for about 340,000 metric tons (product weight) production yearly; they produce less than half that amount.

Despite the decline in resource availability, processing facilities proliferated in the late 1960's and early 1970's, often with the help of government incentive programs. ... "Gear conflicts" between local and mobile fleets increased. ... Deterioration in product quality contributed to market losses.

The same historical pattern of development has appeared in most of Canada's and indeed the world's fisheries. A brief period of prosperity while fish are plentiful attracts additional fishing craft and processing facilities. In the consequent over-expansion, fish stocks become smaller and profit margins shrink or disappear. ... The end result is severe social and economic distress ...

In [future] fishery management, the guiding principle would no longer be maximization of biological yield but the best use of society's resources (including labour and capital as well as fish stocks.) While private enterprise ... would continue to predominate ... fundamental decisions about resource management and about industry and trade development would be reached jointly by industry and government. ...
Necessary changes could include, for example, some consolidation of growth at ports where the services needed for a progressive, innovative industry could be provided by the local community. A healthy fishing industry can also sustain small ports, small companies, and the social and cultural values they represent. Economic progress can combine with the preservation of a way of life. ...

A more prosperous industry is likely to have fewer people in relation to output (not necessarily in absolute numbers) in primary production. Even with extended jurisdiction, the fishery resources off our coasts cannot provide an adequate living standard for an increased number of fishermen. Expansion should be based on the existing strength of manpower in the fleets. A prosperous, growing fishing industry can, however, produce more job opportunities in associated industries and services.

The document went on to list 25 strategies for management and development. Many were by then predictable, such as gaining control of the 200-mile limit, and applying entry control to all commercial fisheries. Others reached far into industry operations and people's lives. Some excerpts follow:

[In resource management]

• Provide for redevelopment of fish stocks whose natural habitat or environment is amenable to effective modification.

• [C]ontrol fishery-resource use on an ecological basis and in accordance with the best interests (economic and social) of Canadian society.

• [F]oster the development of successful aquacultural enterprise.

• Allocate access to fishery resources in the short-run on the basis of a satisfactory trade-off between economic efficiency and the dependency of the fleets involved.

[In fishing]

• Co-ordinate the deployment of mobile fishing fleets, over the fishing grounds and the operating season.

• Provide for the withdrawal of excessive catching capacity ..., and for the best possible mix of fleet units.

• Abolish the use of destructive and wasteful fishing gear and fish-handling practices.

[In processing]

• Facilitate price differentiation according to quality for fish landed.

• Provide for the allocation of landings (raw-fish supply) in accordance with the most profitable end use.

• Concentrate programs of technical and financial assistance for the processing sector on the up-grading, relocation and consolidation of existing facilities.

U.I. extension provides more benefits

Most of the management shifts under Davis and particularly LeBlanc are more easily considered in an Atlantic or Pacific context, and will be treated later. But some had national implications, including the changes to fishermen's Unemployment Insurance. The initial problem was Atlantic, but the program was cross-country.

The Atlantic fishery in the late 1960's and early 1970's still appeared a poor occupation with low and unstable incomes. Fishermen's U.I. added to those incomes. But the unemployment insurance authorities wanted to get clear of the program, which Atlantic politicians had imposed on them in the 1950's. To the U.I. authorities, the fishermen's program appeared to be an income-supplementation scheme, and it was inspiring other self-employed groups to demand similar treatment.

The department under Davis looked at national income-stabilization plans, consulting in particular with Richard Cashin of the new Newfoundland fishermen's union, but nothing germinated. LeBlanc inherited the issue. His officials considered a "net revenue stabilization" scheme for fishing enterprises. This would have counterbalanced the fluctuations of costs and prices, but it would have done nothing for a disastrous year in terms of catches. This idea faded away, as did suggestions of catch insurance. (According to

W.C. MacKenzie, such schemes tended to founder on the issue of cross-subsidization. Pacific fishermen did better overall, yet experienced violent year-to-year fluctuations in catch. They could have wound up being continually subsidized by Atlantic fishermen, who had steadier earnings, even though they were lower.)

It seemed likely that the management changes of the 1970's, together with the 200-mile zone, would lift incomes for the general run of fishermen. The main problem appeared on reflection to be a segment of fishermen, especially on the Atlantic, who were just plain poor and needed income support. New studies suggested that the most practical measure might simply be to extend the period of Unemployment Insurance benefits.

At the time, U.I. for fishermen lasted only 12 weeks. By contrast, regular U.I. could last up to 42 weeks, depending on the region. Plant workers on regular U.I. often got more benefits than the fishermen who supplied them, and who might work a longer season.

Some fishing people on the Atlantic got regular U.I. Those who fished only part time or casually sometimes got it from non-fishing work. As well, some 5,000 full-time fishermen got regular U.I. by working on year-round boats, or by working for wages under a contract of service rather than as a co-adventurer. But the large majority, more than 20,000, the most typical, genuinely self-employed Atlantic fishermen, used the special fishermen's program, with its shorter benefits.

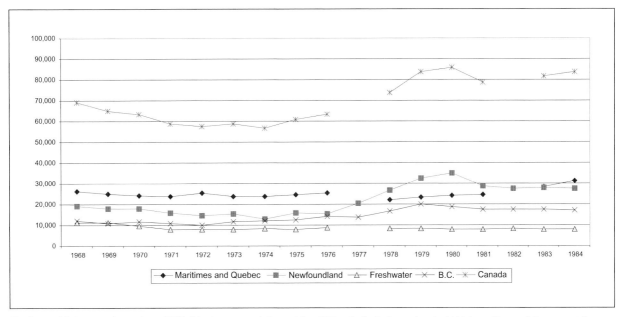

Registered fishermen by region, 1968-84. Increases followed the 200-mile limit, the extended U.I. benefits, and the generally brighter picture of the latter 1970's. Generally, less than half the registered fishermen got their biggest income from fishing.

LeBlanc won government approval and announced in August 1976 that fishing benefits would henceforth last longer in areas of high unemployment. That could mean, in most areas, up to 27 weeks of benefits.

The announcement made clear that the program was as much or more for income support as for insurance. The changes followed "extended consideration by government of means to give fishermen a more reliable income." Times had been difficult. "Adapting the unemployment insurance system, rather than building a new system, provides a workable way to compensate for some of the hardships at this time. The extension of Canada's fisheries jurisdiction to 200 miles, the already instituted limitation on the numbers of new entrants to the fishing industry in order to protect existing fleets, and other federal programs under the recently announced policy for Canada's commercial fisheries will solve many of the income problems of fishermen in future years."

The department thought that management changes would reduce the number of marginal fishermen. The U.I. changes would help support them in the meantime. But before many years had passed, common opinion held that the broadened program was doing the opposite. While raising incomes for fishermen, it was also attracting marginal participants to the fishery, and increasing the dependence on U.I. By 1981, the average self-employed fisherman on the Atlantic was getting about one-third as much from U.I. as he earned in total employment income ($2,500 compared with $8,100); by 1990, about two-thirds ($6,600 compared with $10,800). Meanwhile, an increasing number of fishermen were tapping into regular U.I. by fishing for wages. By 1990, more than 11,000 wage-earning har-

vesters would average $6,120 from U.I., compared with about $9,600 in total employment.[24]

Development programs take new tack

The 1976 policy stressed the futility of over-expansion, even with the 200-mile limit. LeBlanc gave the same warning over and over, often quoting himself from a 1975 speech in Lunenburg: "I fear that by gaining a zone, we will lose an excuse."

It had taken time to change the post-war development mind-set. Even after 1968, while Davis was giving new currency to an old phrase, "too many boats chasing too few fish," the industrial development branch was still working towards better boats to kill more fish faster. Wes Johnson, Jack Rycroft, and other experts worked to promote ferro-cement boats, pair-seining, midwater trawling, automated longlines, and other techniques. The department aided the rapid growth of the crab and shrimp fisheries on the Atlantic. And a large program in the early 1970's tried to promote the processing and marketing of capelin, with little result.

By the mid-1970's, however, the department was genuinely cutting back development work. The industrial development branch at headquarters closed down, most of its employees transferring to regional positions. The swaggering heyday of finding new stocks of fish and better ways to catch them was drawing to a close. Some development work continued, mainly in the regions, but with fewer major volume-oriented projects. Efforts shifted towards fuel-saving technology such as knotless netting for trawls, or more selective fishing methods, such as square mesh to let undersize fish escape more easily from trawls. The department did considerable work on automated longlining, which appeared promising but in the long run attracted few fishermen.

While development work slowed, the Fisheries Prices Support Board, Fishing Vessel Insurance Program, and Fisheries Improvement Loan programs all remained active during this period. The tax system's accelerated depreciation allowance still aided vessel buyers, as did boatbuilding subsidies.

The Fishing Vessel Assistance program in the 1960's had paid 25 per cent towards new vessels of 35–45 feet. In 1969–1970, subsidies came to $565,000 on 44 vessels. The program changed in 1970, the rules now allowing subsidies of 35 per cent for vessels of 45–75 feet. Typically, a few hundred vessels a year got support. (In 1971, for example, 286 vessels got a total of $4.2 million in subsidies.) In 1975, the department stopped subsidizing Pacific vessels, where the fleet was clearly overbuilt.

For vessels elsewhere, Jack Davis in 1973 changed the minimum size of qualifying boats from 45 to 35 feet. In 1976, new rules allowed subsidies for modifications and conversions, and the minimum size limit dropped to 30 feet. This got still further reduced in 1977 to 25 feet for sea fisheries.

LeBlanc and the department began to worry about the whole complex business of fishermen's subsidies and credit, which saw money flowing from federal subsidies, provincial subsidies and loans, and the Federal Business Development Bank. By 1979, officials were readying a cabinet proposal to create a new financial institution to consolidate and rationalize federal programs for loans. The new institution would also try to co-ordinate credit programs with the province. Then the Liberals lost power.

The new Conservative administration planned to get rid of the subsidy program and substitute a new kind of assistance package. But when the Liberals returned in 1980, the previous subsidy program continued. In 1979–1980, the 35 per cent subsidy still applied for Atlantic craft of 25–75 feet. In Ontario and the Prairie provinces, vessels 16 feet and up could qualify. About $11 million was allocated for the purpose, up from the $4 million level in the years 1975–1978. In 1981, however, D.F.O. reduced the subsidy from 35 per cent to 25 per cent. Provincial loan boards on the Atlantic also tightened their rules after 1980; LeBlanc had urged them to restrain expansion.

Meanwhile, a separate government program was still subsidizing shipyards, at 35 per cent, to build larger trawlers. In 1971, for example, the shipyard fund supported 13 trawlers for $5 million. From 1975, the Shipbuilding Industry Assistance Program (S.I.A.P.) lowered the large-vessel rate to 14 per cent, then raised it to 20 per cent until 1980, when it began to drop off. From 1980 to 1984, S.I.A.P. spent about $3 million a year on large fishing vessels. Shipyards also benefitted in the 1980's from grants by the Department of Regional Industrial Expansion.[25]

Thus, even though licence limitation in the 1970's stopped the numerical expansion of the fleet, and development work slacked off, still the subsidies helped build hundreds of bigger, stronger vessels.

Marketing studies, promotion increase

Among other national programs, marketing and promotion work increased under the marketing services branch , set up in 1970. Especially at the time of the 200-mile limit, it seemed that resource abundance was guaranteed; the problem might be selling the fish. On the Atlantic, the department for a time tried to promote export-marketing consolidation through a new agency or other arrangement, but never forced the issue.

The marketing services branch undertook a worldwide marketing study, analyzing prospects in many countries. Trade and sales missions got under way to various countries. This supplemented regular efforts,

Consumer promotional material, long a staple of the department, increased after the 200-mile limit.

which included frequent bulletins to industry on the major species groups, statistical reports, and studies on particular products and markets. At most Fisheries Council of Canada meetings, marketing officials would present their findings to attentive audiences.

To increase fish consumption in Canada, the department in 1978 began promoting November as "Fish and Seafood Month." Efforts increased in following years. National advertising took place. The promotion efforts seemed to help; supermarket sales rose. The initiatives coincided with the early stages of a shift in consumer attitudes. By the 1980's, a new consciousness of the health and diet benefits of seafood was seeping into the public mind. In the United States, per capita consumption ultimately rose from 12.5 pounds annually in 1972–1976 to 15.3 pounds in 1987–1991 (although this would later drop slightly). Prices were steadily rising.

Producers were gradually gaining sophistication. For international markets, the phone and fax were letting even smaller companies gather more intelligence of their own. Among larger concerns, the B.C. salmon processors already had well-developed markets, balanced between domestic and export. In the 1970's and 1980's, Atlantic groundfish companies increased their product-development work. The marketing services branch and the departmental test kitchen, with its promotional recipe booklets, were still valuable, but they were becoming less prominent in relation to the industry's own work.

Provinces push for more jurisdiction

For Atlantic provincial governments, the 200-mile limit brought new hope in their continuing search for jobs. Most provinces promoted development schemes and issued plant licences to virtually all comers. They also pushed to increase their powers in fishery management. British Columbia as well showed a new interest. Federal–provincial debates on the constitution were raging in the late 1970's and early 1980's; fisheries became part of the agenda.

LeBlanc did his part to fend off the provincial push for jurisdiction. "I've tried to examine the reasoning behind their demands. Is it because provinces control other natural resources? The other natural resources don't swim," he pointed out in a speech. "When one province will agree to reduce its fishing effort but four won't, when three provinces oppose foreign joint ventures and two clamour for them, when we start getting foreign nations playing off the provinces for special deals for this plant or that shipyard, then the day of a co-ordinated policy to help all Atlantic fishermen is over." Instead, he could foresee "five mushrooming bureaucracies, ... interprovincial conferences that settle nothing, while competing fleets try to catch the same fish five times."[26]

Industry organizations, including the Fisheries Council of Canada and the newly powerful Newfoundland fishermen's union supported federal jurisdiction. As a New Brunswick fisherman later wrote to a federal inquiry, "The feds are shortsighted, but the provinces are blind." Nova Scotia, at first supportive of a jurisdictional shift, changed its position. The provincial push for more control of fishery management and development subsided to a lower level. The most significant change went the other way, when the federal fisheries department in 1984 took back the management, ceded by the department in 1922, of fixed-gear sea fisheries in Quebec.[27]

Aquaculture was a different story. The constitutional discussions of the day helped cement the idea that provinces would hold the main power over licensing in that industry. Aquaculture would prove more important than most people expected.

Meanwhile, D.F.O. kept getting bigger. By 1980–1981, person–years came to 5,300 and total spending to $322 million, the latter about double the figure for the Fisheries and Marine Service eight years earlier. The department had facilities of one sort or another, from major research establishments to small local offices, at 1,200 locations across the country.

CHAPTER 22.
On the Atlantic, 1968–1984

This chapter deals with Atlantic-wide events, insofar as one can separate them from Maritimes–Quebec and Newfoundland issues, which appear in following chapters.

Groundfish dominate the picture

In the period 1968–1984, the Atlantic lobster and scallop fisheries were major, herring and other pelagics were important, and shrimp and crab were starting to take off. But the most fish, boats, jobs, hopes, and politics were tied up with groundfish. Although in every Maritime province more fishermen held licences for lobster than for groundfish, the latter supported far more plants. By the end of the period, groundfish would seem to be the great triumph of Atlantic fishery management.

But the period began with a groundfish market crisis, new only in its size. Ever since the war, the Fisheries Prices Support Board had been helping the groundfish industry, especially the salt-cod trade, with occasional support payments during market downturns. While Canadian catches of several groundfish species rose in the 1950's and 1960's, those of the prized codfish had stayed relatively flat. Landings in the 1960's never quite equalled the high point of 356,000 tonnes in 1955.

Still, Canadian cod landings remained high in 1968 at 323,000 tonnes, and foreign catches were also at their peak. The bulk of Canadian production went into cod blocks—frozen blocks that were later processed, chiefly in the United States, into fish sticks and the like. With the market treating blocks as a commodity like beef, pork, or chicken, prices fluctuated according to supply and demand. In 1968, market conditions produced a nightmarish situation for Canadian producers. A wave of fear spread through the industry.

Jack Davis was just taking power. Department officials, notably Lorne Grant and Bill MacKenzie, advocated a defence scheme: the F.P.S.B. would buy the whole pack, leave it in the companies' warehouses, and sell it as conditions improved. Such a purchase could restore confidence, prevent distress sales, and stabilize the situation. The officials modelled the concept on earlier programs by the Agriculture department. It would involve co-ordinating the move with other supplying countries (Norway, Iceland, Denmark) and keeping U.S. officials informed.[1]

In the 1970's and early 1980's, large amounts of money and effort went towards stabilizing the groundfish industry, especially the large-trawler sector, which underwent two crises.

Davis bought into the scheme, and the government bought the fish.[2] After a nail-biting period, markets did improve. The department, through the companies, sold off the fish with no loss to the taxpayer. It was a standout demonstration of market intervention, and a high point for the Fisheries Prices Support Board. The cod-block buy helped Davis's image in Newfoundland, where he gained more popularity than elsewhere on the Atlantic.

Canadian Saltfish Corporation starts single-desk selling

Davis pushed ahead with reforms to the marketing of salt cod. As noted earlier, Newfoundland's salt-cod trade had a history of chronic troubles. North Atlantic Fish Exporters Limited (N.A.F.E.L.) had in the late 1940's and the 1950's used a form of single-desk marketing, only to have the federal government remove its export monopoly in the 1950's. This further weakened the position of a declining industry.[3]

But single-desk selling could legally take place if legislation gave a crown corporation the power, as had been discussed at the 1964 National Fisheries Conference instigated by Smallwood. This proposal had slowly wound its way through studies and debates. A royal commission headed by former deputy minister D.B. Finn had found various problems with the idea. But considerable support remained in Newfoundland.[4] A salt-cod market crisis in 1967 made the time ripe for action when Davis arrived in 1968.

After federal–provincial discussions, the Canadian Saltfish Corporation (C.S.C.) came into being in 1970, with full authority over marketing of saltfish produced in Newfoundland and Quebec's Lower North Shore (that is, the most northeastern section of Quebec's Gulf of St. Lawrence coast).[5] The individuals and companies who produced saltfish now sold through the small C.S.C. organization, headed by president Aidan Maloney, a well-respected former fisheries minister in the provincial government.

Thousands of fishermen in Newfoundland and Labrador still produced salt cod, either drying it on flakes or selling the fish to processing companies with mechanical dryers. After the C.S.C. formed, a dozen or so previous exporters became agents of the corporation, buying fish on the C.S.C.'s behalf and drying it if required. Some fishermen at first kept drying cod themselves, partly because the labour involved qualified them for more "stamps" from the Unemployment Insurance system. But as years passed, outdoor drying of light-salted fish by fishermen practically disappeared in Newfoundland. (In Quebec, however, many fishermen continued to dry the "Gaspé cure.") Meanwhile, the frozen-fish market, which paid more to fishermen, kept growing to take the great majority of the Newfoundland catch.[6]

The C.S.C. worked as hoped, becoming stronger during the 1970's. No longer could buyers play off the different producing companies against one another.

Federal fisheries inspection officer checking salt cod. Saltfish remained an important, though shrinking, sector of the groundfish industry.

Prices and the trade became more stable. One clear loser was a Lunenburg, Nova Scotia, company run by the Zwicker family, which was said to be the oldest fish-exporting company in North America, but now lost its mainstay supply of Newfoundland salt cod. Some other interests complained about the government's being in the fish-selling business.

But the Newfoundland salt-cod producers selling through the C.S.C. appeared satisfied enough with the new corporation. The number of plants producing saltfish stabilized at less than two dozen. The C.S.C. seemed to do well enough from what was, in the long term, a declining market. Meanwhile, the Maritimes and Gaspé kept up their independent trade; as of 1983, southwest Nova Scotia had 123 saltfish plants.[7]

Davis, LeBlanc apply limited entry on Atlantic

From the start of his tenure as minister, Jack Davis plunged into the question of licence limitation on the Pacific, where the U.F.A.W.U. had for decades pushed the issue. The Atlantic coast had fewer such representations from fishermen, but had similar problems of overcrowding and over-capacity.

Licence limitation had already come into place in 1968 for lobster fishermen in the Maritimes. Meanwhile, large-vessel capacity on the Atlantic, mostly for groundfish, had multiplied several times over since the 1950's. Groundfish catches were, however, starting to drop. While foreign fleets rightly got most of the blame, Canadians, too, were rapidly increasing their fishing effort. The phrase "too many boats chasing too few fish" was spreading, although the problem was less in the number of craft, which had decreased, than in their fishing power, which was becoming huge.

In 1970, the department under Davis made a little-known attempt to cut back the industry. According to

Medium-size draggers were becoming more common around the coast, as the department tried to limit fishing-vessel numbers. (D.F.O. photo by Michel Thérien)

professor William Schrank of Memorial University of Newfoundland, the department told cabinet that the Atlantic fishery workforce should drop by 25,000. Regulation should keep down investment and employment, and look towards harvesting at minimum cost. Subsidies should disappear, and a new scheme should replace fishermen's Unemployment Insurance. As well, government should sponsor programs for relocation and re-employment of displaced fishermen. Davis apparently wanted rationalization through major surgery. But cabinet said no.[8] The department instead pursued a less sweeping way to control the fishery: licence limitation.

In 1971, limited-entry licensing began for the Maritimes salmon fishery (part of which was already controlled). The herring-seine and Bay of Fundy scallop fisheries followed in 1972, and the offshore scallop fishery in 1973. That year, Davis froze vessel-construction subsidies and then, following an industry–government seminar on licensing, announced in November a "new fishing fleet development policy." While lifting the current subsidy freeze, Davis said that future subsidies would apply only where there was room for expansion. All vessels would be registered and all operators licensed. Vessel operators needed to be Canadian or landed immigrants. The lobster, scallop, salmon, herring, and snow crab fisheries would all require licences in future.

By the time of the policy announcement, most of these fisheries had already come under at least partial limited entry. Groundfish had lagged somewhat. But Davis in 1974 introduced licences for fixed-gear groundfish fisheries in the Maritimes. Then LeBlanc in 1975 froze the number of large offshore trawlers. Limited entry followed in 1976 for 45- to 65-foot draggers.

Progress towards full-scale limited entry for groundfish was bumpy. For the Maritimes, the department in 1978 announced a six-month freeze on new entrants into the inshore groundfish fishery. But regional officials interpreted the freeze as applying only to draggers over 45 feet; they issued several hundred licences for smaller draggers. This became known as the "warm freeze," and had lingering bad effects in southwest Nova Scotia. Still, by the end of the 1968–1984 period, the groundfish fishery was as tightly controlled as the others.[9]

As in the Pacific salmon fishery, controls applied only on the fishing-enterprise licence; there was no limit on the number of personal fishing licences. The licence in most fisheries got attached to the boat rather than the person. A fisherman, with the department's blessing, could transfer the licence to someone else, normally when he sold the boat or gave it to a family member. In theory, the government always owned the licence, and there was no sale, just a transfer. But Davis himself knew that money would change hands, and felt that the sale would give a fisherman or his widow retirement money.

From boom to bust in one year

In 1973, the groundfish industry had its best year ever. Notwithstanding the declining catches (from 629,000 tonnes in 1968 to 540,000 tonnes in 1973), market conditions produced a record value of $81 million. Toasts were drunk and parties ran late at the Fisheries Council of Canada's annual meeting.

The next year, catches dropped again to 418,000 tonnes, about two-thirds the 1968 level, and below 1950's levels. Particularly poignant was the drop in inshore fishermen's cod catches in eastern and northeastern Newfoundland, to only 35,000 tonnes. In the late 1950's and early 1960's, their catches had run to more than 140,000 tonnes.

These thousands of small-boat fishermen depended mainly on what were becoming known as "northern cod." Over the years, F.R.B. scientists had analyzed the areas, behaviours, and biological characteristics of different "stocks," a rather loose term. A single stock could have many spawning areas and populations, not always known to scientists. They treated the "northern cod" in I.C.N.A.F. Divisions 2J3KL (off southern Labrador and eastern Newfoundland, and including the northern Grand Banks) as a distinct stock, the biggest groundfish stock in the northwest Atlantic.

The northern cod spawned in a series of inshore and offshore areas, ranging from Hamilton Inlet Bank in Labrador down through the Funk Island and Fogo Island banks to the northern Grand Banks. Many of these fish migrated inshore in the warmer months, often chasing capelin into the bays and coves. On inshore grounds of eastern and northeastern Newfoundland and southern Labrador, northern cod had supported first the historic hook-and-line fishery, then the trap fishery beginning in the latter 1800's, with gillnets also becoming popular after the Second World War.

Offshore, northern cod (as well as a different stock on the southern Grand Banks) had supported the bank fishery, carried out first by overseas fishing ships and then by Newfoundland and other North American vessels as well. This was historically the main offshore fishery adjacent to Newfoundland. After the war, distant-water factory freezer trawlers (F.F.T.s) began fishing the same stock further north, all the way up to Labrador, catching many cod before they came inshore.

The minimal inshore catch in 1974 made northern cod a symbol of Atlantic troubles. It seemed to both industry and government that this stock would be a harbinger of the future. If the depredations of foreign fishing could be reversed, and if Canada could get the benefits, northern cod could be the mainstay of a renewed and vigorous groundfish industry.

Resource problems bring emergency aid

The Atlantic groundfish industry had a familiar list of problems: fragmentation; no great market clout, in spite of high-volume production; less than a top reputation for quality; increasing over-capacity; and seasonal gluts, which hurt quality, markets, and prices. All those factors shared some of the blame for the chronic problems of low incomes and instability. But in 1974, the worst issue of all was the resource decline.

LeBlanc, the new minister, and his senior official Ken Lucas set up a task force led by Fernand Doucet, an official of Acadian origin and broad experience. Doucet had earlier worked with the Department of Fisheries and the Atlantic Development Board, and was currently chairman of the Freshwater Fish Marketing Corporation for prairie-province fisheries. LeBlanc made Doucet his special advisor. Doucet's team consulted around the coast, taking a thorough look at resource, markets, and the entire industry picture.

The government's main conclusion was to inject money while exerting firmer government control for a rational and prosperous industry. The Doucet group's various reports and a cabinet document melded with previous departmental thinking, such as that of W.C. MacKenzie, and with LeBlanc's own sentiments, to generate the 1976 *Policy for Canada's Commercial Fisheries*.

Money started flowing, with $50 million for emergency aid and other purposes. A Temporary Assistance Program (T.A.P.) came into place, and the tap stayed open: over the period 1974–1976, the government authorized some $200 million for special aid (about $140 million actually got spent), mostly but not all for groundfish.

Major funding went to a subsidy of $2\frac{1}{2}$ cents a pound on groundfish catches—an important amount, when prices overall averaged 9 cents per pound. In previous years, price-support subsidies had gone to processors, government relying on them to pass on the benefits to fishermen. LeBlanc insisted that the new payments go directly to fishermen. Processors got a separate subsidy of eight cents per pound of product. Funding also went to several other fisheries to support prices and special projects. Crab, lobster, and some freshwater fisheries got aid, as did Newfoundland fishermen affected by abnormal ice conditions.[10]

LeBlanc extends limited entry

Meanwhile, LeBlanc pushed ahead with limited entry. By 1979, he could say, "We have applied limited entry to almost every fishery, to protect conservation overall and the catch per boat." He added that "the romantics who see licences as an infringement on some fundamental freedom to fish will find few allies among the fishermen, no more than they would among doctors, lawyers, chartered accountants, or any profession that protects itself from moonlighters and economic loss by licensing."[11] He was right; fishermen generally favoured licences.

The main protests against limited entry would come from those part-time lobster fishermen whose licences LeBlanc was to downgrade and from university circles in Newfoundland, where some saw licences as interfering with the fishing way of life, or with their romantic conception of it. When John Crosbie as Newfoundland's fisheries minister offered a protest, LeBlanc did delay licence limitation for the inshore, small-boat cod fishery in northeast Newfoundland. This final loophole got closed off in the early 1980's.

Advisory committees, organizations get stronger

During the 1970's, fishermen's organizations grew in number and strength. Meanwhile, industry advisory committees, embryonic under Davis, expanded rapidly. The first permanent advisory committee—the Herring Management Committee—had come into place in 1973. Rudimentary committees had followed for several other fisheries. Under LeBlanc, advisory committees spread into every major fishery. Groundfish, the largest, had committees by region and sometimes by gear, as well as an Offshore Groundfish Advisory Committee for larger enterprises. (This later became the Atlantic Groundfish Advisory Committee, for all sectors of the groundfish fishery.) LeBlanc floated the idea "that as the fishermen get more organized, we institutionalize all-industry, all-province consultations."[12]

The committees spread for two main reasons. First was the increasing complexity of management regarding licences, quotas, and gear, which required more industry consultation. Second, and probably most important, was LeBlanc's push to give more power to fishermen. Fishermen's organizations took many of the seats on advisory committees. There were no elections, except for lobster advisory committees in some areas. Rather, the department asked obvious candidates, association heads or leading fishermen, to serve. Sometimes the industry itself put names forward, or representatives would just show up and demand a place. The committees gained strength over time; by the latter 1980's, more than a hundred of them operated on the Atlantic coast. In some fisheries, they gained almost a veto power.

Meanwhile, the Fisheries Council of Canada wielded far less influence under LeBlanc than previously. The organization remained strong, but would suffer a setback when, in 1984, the Fisheries Council of British Columbia split off as a separate entity.

LeBlanc hectors industry on quality, marketing

The department had long been prodding the industry to improve fish quality, often giving financial help. In 1973–1974, under Davis, several million dollars went to subsidize construction of ice-making, ice-storage, and fish-chilling facilities.

The 1976 policy called for price differentiation on fish landings according to their quality. Departmental thinking was that the Europeans and Japanese were out-competing Canadians, getting higher prices in their North American backyard. Canadians needed to do better. LeBlanc noted in a speech, "In the same way that people thinking of wines think of France, or for woollens think of Scotland and for cameras think of Japan, our export industry should make people the world over, when they think of fine seafood, think of Canada."

Nationally, the several hundred inspection officers kept pushing fish plants for improvements. Meanwhile, new regulations required improvements to fish holds. The prodding sometimes nettled processors. But improvements took place. Under a national Fish Chilling Assistance program, cost-shared with provinces, the department spent around $2 million in most years from 1974 to 1979. In another project, the department starting in 1977–1978 put new fish-handling systems into 200 Newfoundland outports and a number of Quebec and Maritime ports. Big gray container boxes, half as high as a man, became a common sight. Boxing and icing became the typical practice. Up till now, fish plants almost always stood at wharfside; in subsequent years, processors would often set them up far from the water, since it was easier to truck boxes to plants.

In 1980, D.F.O. launched an Atlantic Quality Improvement Program. At the moment, the department acknowledged, many sectors of the industry still handled fish poorly. Now, new regulations would ensure proper provisions for handling fish on vessels, in plants, and during transport. Fish when caught would be iced or otherwise chilled. Dockside grading was to begin—first voluntary, then compulsory. New standards would outlaw fish forks and set regulations for pumps and other handling and storage equipment. The department provided subsidies for quality improvement, began training industry graders, and ran pilot projects on grading. The intent was a rapid, major push forward.[13] In the event, however, industry resistance and the usual swirl of events and distractions would slow it down.

The department also pushed the idea of export-marketing consolidation, whether through single-desk marketing or other means. The 1976 policy called for both consolidation and "forward integration," with Canadian processors acquiring processing and distribution outlets abroad. LeBlanc noted that exporters in other major fishing countries worked together: "The Norwegian group Frionor represents some 300 independent processors." If Canadian exporters joined their efforts, "we could if necessary help them with legislation to group together for more efforts in common, such as establishing floor prices or bargaining for volume sales."[14] Privately, LeBlanc mused about some form of counterpart to the Canadian Wheat Board to market fish.

Partly to stave off LeBlanc, the Fisheries Council of Canada spun off a new organization, the Canadian Association of Fish Exporters, which did marketing intelligence and related work. Other than that, Atlantic companies guarded their independence, arguing the efficiencies of the free market. Besides, making their own deals was half the fun of the fishery. Conditions were improving; by 1979, Canada had become the world's number one fish exporter in terms of value. That same year, the brief Conservative government interrupted LeBlanc's hold on power. The department let slide its ideas of export-marketing consolidation.

"Fishing Plan" pushes trawlers to north

With the 200-mile limit in 1977, Canada began subdividing groundfish quotas by fleet sector, according to an elaborate "Fishing Plan."

Several factors produced the change. The coming 200-mile limit promised more fish, through rebuilding stocks to rebuild and displacing foreign vessels. Some large-trawler companies and most provincial governments were talking about huge expansions. LeBlanc worried about the trawler fishery growing to the detriment of the smaller-scale fishermen.

A triggering event took place in 1976. Biologist Scott Parsons had warned of impending reductions of the

Inspection labs as well as fisheries and oceanographic ones employed scientists and technicians.

Trawling in northern waters had its hazards. Photo shows National Sea's *Cape La Have* off Labrador.

redfish fishery in the Gulf of St. Lawrence. The department set a 30,000-tonne Total Allowable Catch (T.A.C.) for the stock, well below catches in previous years. But by the time it was set, unusually light ice conditions had let the large-trawler fleet catch most of the Gulf of St. Lawrence redfish quota that winter. This strained the groundfish situation for the rest of the year. LeBlanc got exercised, and looked for an improvement. It seemed that subdividing quotas could give each sector a guaranteed share, and space out the fishery over the longest possible season. Meanwhile, the department in 1976 was giving special assistance to help trawlers fish new grounds and species.

Out of this context emerged a detailed groundfish fishing plan for 1977, the first year of the 200-mile zone. Dozens of Total Allowable Catches already applied for different groundfish species in different zones. Within these overall quotas, the department hived off sub-quotas for the different sectors of the offshore trawler fleet, allocating them to vessels as defined by area, length, gear type, and even horsepower. (Smaller-vessel fleets at first got allowances: estimated catches subtracted from the T.A.C. Later, from 1981, quotas replaced most allowances.)

The first Fishing Plan pushed most large trawlers out of the Gulf of St. Lawrence, a historic change that irritated large processing companies. But the plan also subsidized them to fish offshore grounds off northeast Newfoundland and southern Labrador, where foreign trawlers had dominated the fishery.

Canadian trawlermen were unused to these offshore, often icy northern grounds, and unsure what to expect. But they learned fast. Over the next few years, large fisheries for northern cod built up off southern Labrador and around the Funk Island, Fogo Island, and other banks off northeast Newfoundland. As well, the Canadian fleet now dominated the cod, yellowtail flounder, and other fisheries on the Grand Banks.

LeBlanc stresses conservation, opposes over-expansion

After gaining the 200-mile zone, the department kept quotas low and hiked them only slowly, LeBlanc declaring he had no intention of "re-raping the fishery." Striving for Maximum Sustainable Yield had brought minimum stability, LeBlanc said. "The fish are forever there but forever fragile... . If we fool around with the foundation, the rest will topple."[15]

Departmental scientists in 1977 adopted the $F_{0.1}$ guideline for fishing mortality. This technical term meant a more cautious level of fishing, below M.S.Y. According to one definition, $F_{0.1}$ signified, for a stock coming under increased fishing pressure, the level of fishing mortality, or F, "at which the increase in yield by adding one more unit of fishing effort is 10 per cent of the increase in yield that would have been attained by adding the same unit of effort when the stock was lightly exploited." The concept had emerged earlier in connection with Georges Bank herring, and I.C.N.A.F. had used it for certain groundfish stocks in 1976. Scientists and managers never found a simpler term for $F_{0.1}$; the strange number would complicate many discussions with industry.

What $F_{0.1}$ meant in practical terms was that for most finfish stocks, the annual catch should be no more than about 20 per cent of the fishable biomass, or one fish in five. This would produce less than Maximum Sustainable Yield, but would reduce the risk of overfishing. It would provide 80–90 per cent as much catch with about one-third less fishing effort.[16] It would create denser concentrations of fish for easier fishing and better profits, thus bringing the fishery nearer to the inexact term "optimum sustainable yield."

As abundance increased in the latter 1970's, scientists became more confident in their stock-estimating

Side trawlers were still active, but giving way to stern trawlers and draggers. Photo is off Souris, P.E.I., in 1977.

and forecasting abilities. By the early 1980's, at least some thought that in some instances, they could get the numbers right within ten per cent.[17] Others were more cautious. Still, most department people believed that scientists had a clear picture of abundance.

Big companies, provinces push for expansion

The industry in general made no strenuous objection to the department's cautious, science-based approach to T.A.C.s. Nevertheless, some larger companies and most provincial governments wanted rapid expansion of fleets and plants to take full advantage of the new zone. Four large-trawler firms—the Nova Scotia-based National Sea Products and H.B. Nickerson and Sons and the Newfoundland-based Fishery Products and Lake Group—dominated the picture. The "big four" became three when Nickerson gained a controlling interest in National Sea. Those two merged companies printed up brochures and took display ads in newspapers, calling for freedom to operate and to grow, so as to take full advantage of the 200-mile limit. There was, they said, room for both the inshore and the offshore fleets to expand. The offshore fleet needed modern trawlers and freezer trawlers, the latter with access to traditional as well as under-utilized species.[18]

The fisheries ministers of Newfoundland and Nova Scotia, Walter Carter and Dan Reid, put forward a $900-million expansion plan, including new vessels and plants for $250 million and $110 million.[19] Carter sponsored a large study of Newfoundland prospects, and called for factory freezer trawlers. New Brunswick and Quebec also put forward development plans, each projected to cost several hundred million dollars. All of them wanted the federal government to pay for the expansions. LeBlanc resisted the expansion schemes, pointing out that the fleet already had enough capacity to take all the expected gains from the 200-mile limit.

LeBlanc resists large-vessel growth

LeBlanc's cautionary speeches, often made in their presence, left some provincial fisheries ministers with pained expressions. Nova Scotia's Dan Reid called LeBlanc "naïve" for postponing freezer trawlers. Newfoundland's Carter couldn't understand "why we have to go begging to Ottawa" about freezer trawlers; "How long do we have to wait?" Newfoundland should build at least ten trawlers a year, he said, and should have more constitutional control over fisheries. Carter also blasted the federal reluctance to approve joint ventures with foreign fishing firms.[20] The Halifax Chronicle-Herald launched a series of editorials, some on the front page, calling for LeBlanc to get busy, displace the remaining foreign vessels, and let private enterprise get to work.

LeBlanc began to reply in kind. In one speech opposing over-expansion, he mocked previous provin-cial development efforts that had failed spectacularly, such as the Bricklin car in New Brunswick and a liner-board mill in Newfoundland. This further inflamed provincial governments, who believed he was holding back development.

In 1978, LeBlanc told the Fisheries Council that the large-vessel fleet had the capacity to take half again as many fish, "if we increase the fish in the water and the catch rates. If we do it the other way around—increase the fleet first—we are like a man with an exhausted woodlot, who instead of planting more trees to get more growth, spends all his money on more chain saws to cut the shrubs. Massive fleet expansion at the moment would be a Titanic undertaking—and I use the word advisedly."[21] Gains would come instead from a changed management system of which the 200-mile zone was only a part—a challenge more than an opportunity.

The herring, redfish, and other fisheries gave examples of over-optimistic development, LeBlanc told another audience. "We see a history marked by boom and bust cycles, low incomes, and out-migration of fishermen and their children," caused by fragmentation and uncoordinated one-shot attempts at development. As stocks recovered, Atlantic Canada could probably increase its groundfish catch by 50 per cent or more, depending on the results from trans-boundary and beyond-the-zone stocks. "But a fleet the size of our existing one can probably take 50 per cent more groundfish. How do I know? Because ten years ago, we were taking almost that many fish with a smaller fleet." The need for a bigger fleet, for factory freezer trawlers, for foreign capital investment—all were myths. "The big money gains of the new zone will come less from volume of fishing than from value."

The fish were only beginning to rebuild—give them time, LeBlanc said. "In particular, why should we immediately, automatically, grant expansion licences to the three large companies who dominate the offshore fishery, and thereby close off that option for all smaller companies and for the many thousands of independent fishermen in the inshore and midshore fleets?" The F.F.T.s wanted by some parties could catch "probably triple what our best trawlers now can catch, and maybe 90 times what an inshore longliner can catch. ... I wish people would worry less about new factory ships, and more about the 29,000 existing inshore and midshore vessels, and the 250 or so offshore-sized vessels, that would have to compete with the new machines for the same fish, unless the freezer-trawlers were limited to non-traditional fisheries."

LeBlanc added that "in Atlantic fisheries generally, our policies support ownership of vessels by individual fishermen, or by the companies they form, rather than by processing companies. We have saved quotas and licences to give the inshore and midshore men ... the chance to move up to better boats and a better fishery. We want to build the 200-mile zone from the coast out."[22]

Joint ventures and "snake-oil salesmen"

In a subset of the fights over expansion, proposals surfaced for foreign fishing companies, in concert with Canadian interests, to bring over vessels or otherwise help develop the Canadian fishery. Since 1974, government policy had prevented foreign interests from holding more than 49 per cent of a company that held fishing licences. The new proposals often involved foreign charters to exploit the riches of the new zone. LeBlanc generally refused such overtures, publicly scoffing at "snake-oil salesmen," and declaring that he did not push for Canada's 200-mile zone just to see foreign interests take it over "by the back door."

But in limited cases under LeBlanc, the department allowed foreign-vessel participation to help develop fisheries new to Canadians. Developmental charters took place for such species as silver hake and squid, especially in 1979 for the latter. Japanese vessels would catch the squid and bring at least half of it to Canadian plants for simple processing and freezing. The foreign partner would then buy back the squid its own vessels had caught. In some cases, no fish was landed; the processor sold the quota over the phone, and a cheque got deposited to his account.[23] This arrangement of foreign fishing combined with light or even non-existent Canadian processing became known as "over-the-wharf sales." In other cases, the department let plants charter foreign vessels purely for their freezing capacity, as "plant extensions." Further charters helped supply "resource-short" plants with northern cod.

Most deals with the foreigners quickly faded, as the unusual squid boom of 1979 disappeared, and as the Canadian fleet did more of its own offshore fishing. An exception was the northern shrimp fishery off Labrador. The department allowed the use of foreign vessels, which took Canadians aboard to learn the fishery.[24]

While generally restricting processors from using foreign vessels, LeBlanc starting in 1976 allowed fishermen in the Bay of Fundy herring fishery, for the first time, to sell fish directly "over the side" at sea to foreign purchasers. This move got under the skin of processors, who had generally monopolized the market for Canadian fish. It added to their growing resentment of LeBlanc.

Fleet shares stabilize

The detailed allocations in the groundfish Fishing Plan brought out the industry's competitive instincts. At advisory committees, especially the Atlantic Groundfish Advisory Committee, representatives of fishermen, processors, and provinces argued over quotas. Inshore, midshore, offshore, fixed-gear, and mobile fleets pointed out the importance of their particular fishery and the disadvantages of others. But relatively little change took place.

LeBlanc had said in one speech that he had "a clear bias for the inshore fishery."[25] But rather than expanding inshore shares, he precluded the expansion of the offshore. Percentage shares of the allocations saw no great change.

In practice, the department generally made allocations on the basis of catch history. There was no going back to first principles to decide what fleet composition might produce the most economic and social benefits. Instead, the hope was to stabilize the fishery and increase landings for all. As Art May, a senior official, later put it, "[W]e froze everything and held on for dear life."[26]

This generally status quo approach had two major exceptions, the first involving "banked" licences. The trawler fleets had a number of licences previously granted, without vessels at the moment; with departmental approval, companies had kept the licences for when the fishery improved. LeBlanc removed these licences, which could have brought an important build-up in the trawler fleet. The second exception was the closing of the Gulf to most trawlers, while pushing them to the north. But even with this shift, the balance of groundfish allocations for "offshore" (longer than 100 feet) and smaller vessels stayed roughly 50:50, a split formalized in the 1982 Fishing Plan.[27] In actual harvests, though, groundfish catches by vessels under 100 feet rose from about 55 per cent in the late 1970's to about 60 per cent in 1984.[28]

Restrictions fail to control fishing power

It was clear to LeBlanc and leading officials that even with the 200-mile zone and new measures to increase abundance, the Atlantic fleet already had more than enough fishing power. Over-capacity and over-investment prevailed almost everywhere. The department wanted to rebuild fish stocks while keeping a lid on the fleet. Besides limiting the number of licences, it controlled the length of vessels. If a fisherman wanted to replace a vessel, he could increase its length only by a fixed, low percentage.

Besides vessel size, electronic gear added to fishing power. These smaller, multipurpose boats at Wedgeport, N.S., nearly all had radar antennas, simplifying navigation.

But vessels crept up in fishing power anyway. As limited entry and rising quotas and prices produced more money, enterprises looked for ways to spend it. Boatbuilding subsidies still applied, and an income-tax provision, the Accelerated Capital Cost Allowance, sometimes made it appear more sensible for a fisherman with high cash flow to build a new boat, rather than pay high taxes.

In 1981, the department tightened the rules. Replacements of craft of less than 35 feet could be no longer than 34 feet 11 inches. For replacements of vessels 35–65 feet long, a "five-foot barrier" applied. If someone had, say, a 41-foot boat, his new boat could be no more than 45 feet. Vessels 65 feet and over, previously allowed replacements up to 25 per cent larger, now could only be replaced on a foot-for-foot basis. And for all vessels over 35 feet, hold capacity could not increase. The rules raised vicious protest, but the department stuck to them, at least regarding length; fishermen often got by with increases in hold capacity.[29]

Still, the trawler companies, which in the 1970's owned upwards of 150 vessels, ordered many replacement vessels of greater power, often around 150 feet in length. In the much larger fleet of smaller vessels, owners stretched the rules as far as they could. Dragger licences issued in Nova Scotia's "warm freeze," noted earlier, were for boats less than 45 feet. To get more capacity, fishermen widened them. Among the "pregnant 44's" of southwest Nova Scotia, a 44-foot vessel could be 22 feet wide.

The department eventually, in the 1990's, would apply strict volume restrictions. But long before that, many wharves had filled up with vessels that were no more numerous than before but took up twice the space. A vessel the same length as one 20 years older might have twice the tonnage. Many new craft were high, wide, and bulky, some looking almost square, with bigger engines, and loaded with modern gear.

Loran navigators became common, telling fishermen exactly where they'd made the best catches. New colour sounders gave a far better picture of what was in the water. Some fishermen said that the new electronics multiplied their fishing power by four or five times. This writer once counted, aboard one herring seiner, 29 separate pieces of electronic equipment—radios, radars, positioning equipment, sounders—in the wheelhouse. Although the department was watching vessel size, the electronic increase in ability drew little attention.

Meanwhile, vessels often killed additional fish besides those they landed. Longlines and handlines were generally considered the more conservationist gear, though far from harmless. They tended to catch larger fish which stayed alive until taken aboard. Longlines were most prevalent in the Bay of Fundy and the Atlantic coast of Nova Scotia. Although some gillnetting took place in those same areas, gillnets were more popular in the Gulf and Newfoundland; they were fairly selective by size, but caused wastage in untended nets and ghost fishing by lost nets. Draggers usually frightened fishery managers the most. Especially strong in southwest Nova Scotia but common elsewhere, they could chase down fish in far-flung hideouts. Although mesh-size regulations applied, often small fish were unable to escape, because fish blocked the meshes and because, in some cases, fishermen put "liners" in the nets or otherwise tampered with them.

A departmental study in 1974 by Lennox Hinds and James Trimm found huge rates of discards, as fishermen threw back small fish or undesirable species. The amount discarded equalled about one-quarter of the groundfish landed. Of the discards, about one-third were desired species but undersized, and two-thirds were "under-utilized" non-commercial species.[30] But few people worried about discards at the time.

Boats keep getting bigger

From 1968 to 1974, the total number of Atlantic boats dropped from about 34,300 to 28,000. By 1984, it rose slightly to 30,300. These relatively stable numbers hid a major increase in fishing power.

The small-boat fleet fell sharply in number. In 1968, there were 31,200 smaller craft under ten gross tons, which generally meant less than 40 feet long. This class fell to 24,400 boats in 1974, and dropped slightly more to 23,300 in 1984.

The large vessels over 100 feet declined from 229 in 1968 to 204 in 1974, then increased moderately to 234 in 1984.

The big growth came in medium-sized vessels. The number over ten tons (almost all over 35 feet) and less than 100 feet nearly doubled from about 1,800 in 1968 to 3,300 in 1974, then more than doubled again to about 6,800 in 1984. A major part of the increase came in the 1976–1979 period.[31] Medium-sized vessels were the fastest-growing source of fishing power.

The terms "inshore" and "offshore" were by 1984 more misleading than ever. "Offshore" could mean, depending on the year or region, bigger than 25 gross tons, or longer than either 65 or 100 feet. The more than 230 vessels over 100 feet were truly offshore, able to go long distances and stay out for weeks. Another 138 vessels between 65 and 100 feet had broadly similar capabilities, though a good proportion fished within the Gulf of St.

Lawrence. But in fact, an offshore vessel in terms of length might fish close to shore, so long as it was at least 12 miles off. In real terms, "offshore" generally meant big and company-owned. At the other end of the scale, of the more than 20,000 craft less than 35 feet long, many were open boats and truly inshore, sticking close to home and making day trips only.

But what of the nearly 7,000 craft between about 35 and 100 feet? Many people still referred to them as inshore. Yet they were armed with the most modern equipment, and some could fish almost all grounds in almost all weather. They were in many ways the most dynamic part of the fleet, particularly strong in the Maritimes and Quebec. Gradually, they picked up the term "midshore" (especially those between 45–50 and 100 feet). But most of the time, the "inshore–offshore" terminology left them out, while calling up visions of giant vessels pitted against an old man in a dory. The misleading terminology sometimes affected public and departmental thinking.

"Separate fleet" proposal shocks industry

LeBlanc at one point threw a major shock into the processing companies that owned vessels. He worried about trawler operators increasing their share at the expense of the independents. At the same time, a theory was circulating among department officials about pricing practices within the large corporations. This went as follows: the processing arm of a company would pay artificially low prices to the trawler-operating harvesting arms; these low prices translated into low profits or even losses, and therefore a tax write-off for the harvesting arm. These same corporate low prices also pushed down the market prices paid, by the trawler companies and others, to the independent fishermen. Nobody produced solid evidence, but the theory had influence at the time.

In 1977, at a speech in Yarmouth, Nova Scotia, LeBlanc said that part of the reason Atlantic fishermen received lower prices than their counterparts elsewhere was that processors lessened fishermen's independence by financing their boats, and otherwise inhibited the free play of markets. He advanced a startling idea:

> I propose that in future we separate the fishing fleet from the processing companies in Atlantic Canada. ... Fishermen should own their own boats, and be able to sell fish where they want. ... Creating a truly independent fleet should improve the efficiency of vessel operations, improve the match of fishing and processing capacity, raise fish prices and fishermen's incomes, increase the fishermen's bargaining power, create a healthier balance of forces in the industry, and invigourate fleet development by the fishermen.[32]

The proposal caused consternation not only among the trawler companies, but also among many medium-sized and small corporations that owned boats. LeBlanc had hoped for support from the Newfoundland Fishermen, Food and Allied Workers Union, now a strong force under Richard Cashin. But the union represented trawler crews and plant workers as well as independent fishermen, and had at the moment a rea-sonable relationship with Fishery Products, the main employer. The union never came out strongly for fishermen taking over the trawlers. Indeed, there was no great wave of support from anyone. LeBlanc let the idea die. But the separate-fleet proposal, coming on top of other clashes, permanently soured LeBlanc's relations with the larger corporations.

Meanwhile, the minister enjoyed strong support from independent fishermen. In a report from Canada, the *Financial Times* of London noted, "[M]ention his name to fishermen in any small East Coast harbour and you will be answered in hushed, almost reverential tones."[33]

"Separate-fleet" and "owner–operator" rules protect independents

Although LeBlanc never took away the company boats, he prevented companies from taking over the independents. In 1979, what became known as the separate fleet policy forbade corporations from acquiring the licences of vessels less than 65 feet. Those that already held such licences (mainly small and medium-sized companies in Scotia–Fundy) could continue to hold them. Such "pre-1979" corporations could also, with D.F.O. approval, transfer licences to one another, or lease out their licences.

The owner-operator rule aimed to protect smaller, independent enterprises. Photo shows pair-trawlers at Prince Edward Island. (D.F.O. photo by Kevin McVeigh)

A second rule became known as the "owner–operator rule." For boats less than 65 feet, the licence-holder himself had to operate the craft specified in the licence. This provision seems to have come in over time. Such a rule applied early in Quebec, where the province still managed most fisheries in the 1970's. And in the Maritimes, a 1980 lobster-licensing policy referred to operators needing to be on board.[34]

In 1983, the owner–operator rule appeared in the "bona-fide" policy promulgated by the Maritime Fishermen's Union and approved by D.F.O. for the Gulf of St. Lawrence. A 1985 consultation document stated that everywhere except Scotia–Fundy, the department already required licence-holders to fish the licence personally.

In 1989, the owner–operator rule would become universal on the Atlantic for vessels less than 65 feet. Grandfather clauses applied for licence-holders, mainly in Scotia–Fundy, who had previously designated other operators. At first the fisherman was expected to own the vessel, but in most areas this policy was later relaxed, and the licence-holder could operate someone else's vessel. But the "owner–operator" term stuck.

As corporate influence crept into some fleets in the 1990's, independent fishermen increasingly viewed the owner–operator and separate-fleet rules as vital.

Plant capacity multiplies

The total number of Atlantic plants processing all species increased from 405 in 1973 to 700 in 1981, a major jump. Some were small; for example, a good number sprang up to serve the new roe-herring fishery in southwest Nova Scotia. Still, more plants meant more capacity and investment. LeBlanc questioned the lack of restraint by provincial governments on plant expansion. The provinces, wanting jobs, paid little attention. Meanwhile, many existing plants were modernizing and expanding. As of 1978, some 270 plants had freezing capability, and the number was growing.[35]

At the beginning of the 1970's several foreign-owned companies, including Acadia, Booth, and General Mills, had abandoned several large plants in Canada, including establishments at Harbour Grace, Canso, Petit de Grat, and the Magdalen Islands. Canadian takeovers restored some of them. After 1977, the large corporations expanded pell-mell, especially H.B. Nickerson and National Sea. Nickerson eventually took control of National Sea, aided by loans from banks and the Nova Scotia government. As of 1980, Nickerson had loans of more than $50 million from the province, and the provincial auditor was complaining about lack of information from the company.

Atlantic-wide, the year 1979 alone saw 17 new trawlers built for large corporations. Mostly owned by Nickerson, National, Fishery Products, and the Lake Group, trawler-fed plants at St. John's, Arnold's Cove, Fermeuse, Trepassey, Grand Bank, Fortune, Harbour Breton, Burgeo, Ramea, Louisbourg, Canso, Lunenburg, and other locations each employed hundreds of people year-round.

Besides the large corporations, many others were in the race for fish. Newfoundland, centre of the expected groundfish bonanza, saw the most expansion. The number of plants in the province went from 61 in 1973 to 225 in 1981. Many were "feeder" plants that did heading and gutting or even filleting, then sold off their fish to the larger corporations for further processing and marketing. In the period 1977–1979, the provincial government asked for the federal fisheries department's comments on 48 plant-licence applications; federal authorities recommended against 34 of them and refused any assistance; 22 plants nevertheless got built with provincial or bank support.

Despite LeBlanc's calls for restraint, the federal government itself, through the Department of Regional Economic Expansion and its successors, subsidized many plants. Indeed, D.R.E.E. during the 1970's put more than $130 million into fisheries development through federal–provincial programs and direct aid.[36]

The big companies were only part of the plant increase. That increase continued even after the trawler-company crisis of the early 1980's. From 1977 to 1987, the total number of plants in Atlantic Canada increased by 73 per cent, while landings increased only 26 per cent. The number in Newfoundland went from 147 to 250; in Quebec, from 83 to 111; in New Brunswick, from 80 to 190; in Prince Edward Island, from 40 to 65; and in Nova Scotia, from 169 to 337. The total for the coast in those years rose from 519 plants to 953, nearly a doubling in a decade. The actual increase in capacity is uncertain; some plants were for newly successful fisheries, such as crab, roe herring, and aquaculture in the Maritimes, and some herring "bloater" plants were now registered for the first time. Still, it is clear that investment was outstripping landings.[37]

The big companies become huge

By 1981, the large companies owned scores of plants all around the coast, and had established close marketing or other ties with many ostensibly independent plants, in such fisheries as lobster, herring, and scallops, as well as groundfish. Judging from trade newspapers, company annual reports, and federal fisheries documents of the day, holdings seem to have been roughly as follows (although never totally static):

National Sea Products had plants or divisions in Nova Scotia at Digby, Lockeport, Lunenburg, Halifax, Louisbourg, North Sydney, and Pictou; in New Brunswick at Deer Island, Shediac, and Shippegan; in Prince Edward Island at Charlottetown, Summerside, and Morell; in Quebec at Amherst, and Grande Entrée on the Magdalen Islands; and in Newfoundland at St. John's, Arnold's Cove, La Scie, Lark Harbour, Picadilly, and Burgeo. It also had operations in Maine and Florida. It controlled Scotia Trawling Equipment and other related companies. And it bought from, or packed or marketed for, more than two dozen other companies.

H.B. Nickerson and Sons Ltd. had plants or divisions in Nova Scotia at North Sydney, Dingwall, Canso, Aulds Cove, Lismore, Larry's River, Riverport, Port Mouton, Wood's Harbour, Wedgeport, and Yarmouth; in New Brunswick at Grand Manan and Deer Island; in P.E.I. at Georgetown; and in Newfoundland at Dildo, Charleston, Jacksons Arm, Lewisporte, Black Tickle, and Williams Harbour. Nickerson also operated a number of buying stations, controlled some fishing enterprises, and had operations in the United States, England, and other countries. It bought from or packed or marketed for 20 or so Canadian companies. As noted, Nickerson gained control of National Sea. Nickerson–National also had a stake in Versatile Air Services, Delta Transport Ltd., and a shipyard, Ferguson Industries Ltd.

Fishery Products Ltd. had plants or divisions in Newfoundland at Harbour Breton, Burin, St. Lawrence, Marystown, Trepassey, Holyrood, Catalina, Twillingate, St. Anthony, Port au Choix, and Bay Roberts. It had a plant in Massachusetts processing blocks from Canada, and other foreign interests. F.P. also bought from and marketed for other firms.

The Lake Group Ltd. had plants or divisions in Newfoundland at Englee, Bide Arm, Cook's Harbour, Gaultois, Fermeuse, Grand Bank, Bonavista, and Main Brook, and a subsidiary at Fortune. It had informal marketing relationships with other companies.

Besides the "big four," other substantial companies included Pêcheurs unis du Québec (Quebec United Fishermen. Q.U.F.), with five owned plants in Quebec plus subsidiary companies. The co-op mainly processed fish from other co-ops in Quebec. The United Maritime Fishermen co-op had five plants in New Brunswick, seven more in P.E.I. and Nova Scotia, and one in Newfoundland. The U.M.F. marketed for several other co-ops. Connors Brothers, based in Black's Harbour, N.B., was a strong and steady company, with a dozen or so plants for herring, groundfish, or shellfish in the Atlantic provinces. Newfoundland Quick Freeze, and Mersey Seafoods and Comeau's Seafoods in Nova Scotia, were also sizeable.

Number of fishermen grows

Limited entry had only limited the number of fishing licences, without regard to the number of crew. As fishing power and plant capacity increased, so did the number of registered fishermen on the Atlantic: from 36,400 in 1974 to about 59,000 in 1984, a two-thirds increase.

Rising catches helped foster the increase in fishermen, as did a short-lived squid boom in Newfoundland. As well, more generous fishermen's Unemployment Insurance benefits, starting in 1976, attracted some new entrants. Years earlier, U.I. recipients had been issued booklets of stamps to keep track of their entitlement to benefits. Now the term "fishing for stamps" became popular. Stories circulated about some cod-trap and other operations, which formerly used a handful of helpers, now employing dozens in the course of a season, to qualify friends and family for U.I. Similar antics took place in some processing plants.

The regular fishermen sometimes referred to the others as "free riders" or "moonlighters" rather than "real fishermen." But regular fishermen made sure to apply for their own U.I. benefits. The amount paid in by fishermen never came anywhere near the amount withdrawn from the program, which was well over $200 million by the latter 1980's. In some areas, particularly within P.E.I. and Newfoundland, fishermen by then were making more money from U.I. than from fishing. As stories spread about widespread use of U.I.,

some of the public sympathy for fishermen vanished.

In a typical year of the 1980's, out of some 60,000 registered and 45,000 active fishermen on the Atlantic, 20,000-plus would draw fishermen's U.I. Many others drew from the regular U.I. program. Increasingly, captains in some fisheries, particularly lobster, would hire helpers for the season at a fixed wage rather than a share of the catch, and thus open the way to "labour stamps."

Registered and "main-income" fishermen

A large number of part-time and casual fishermen took part in the Atlantic fishery. While D.F.O. counted "fishermen" its own way, the federal department of revenue counted them as those who got the single biggest part of their income from fishing. This number was always below D.F.O.'s number.

In 1984, the number of registered fishermen (those who got a personal licence from D.F.O.) came to 59,152; the number listed as fishermen by National Revenue (let us call them "main-income fishermen") came to 27,711, a little less than half the D.F.O. number. The National Revenue total corresponds fairly well to D.F.O.'s own count in 1984 of about 24,000 "full-time" and equivalent fishermen.

Conservatives take over

James McGrath

The Liberal defeat early in 1979 removed LeBlanc from office. While alienating many processors, he had pleased fishermen and reinvigorated federal fisheries staff. In an unusual move for cautious civil servants, department personnel of all ranks in Ottawa threw a big party for the departing minister. Richard Cashin, co-founder of the Newfoundland union, sent a telegram quoting a phrase once used about American president Franklin D. Roosevelt: "We love you for the enemies you made."

James McGrath, an experienced M.P. representing St. John's East and the fisheries critic for the opposition Conservatives, took office on June 4, 1979. McGrath had a reputation as being in tune with liberal and even Liberal sensibilities. He stuck mainly to LeBlanc's policies, which had generally protected the smaller, independent operators.

But McGrath looked towards a new management style. Under LeBlanc, the department had been intervening all over the place. More, and increasingly complicated, issues were floating upwards for ministerial decision. McGrath set in motion a thorough review of fisheries management, to culminate in a white paper that would, it was hoped, lead to a more readily manageable system. But the Conservative regime would be brief; no major reorientation would take place.

McGrath freezes factory freezers

Meanwhile, the usual swirl of events beset the new minister. Nickerson and National Sea Products were still pushing for the use of factory freezer trawlers, to take advantage of the huge and growing resources of the new zone. Foreign operators had used such vessels since the 1950's. Filleting, usually by machines, and freezing took place on the vessel. For Nickerson and National, the arguments in favour were many: foreign countries had long used them; they could allow quick freezing and better quality; and they were said to be essential for some under-utilized species, such as

The proposed step from trawlers like these to factory freezer trawlers caused great controversy.

hake, which spoiled quickly. The catch was that most proposals for the use of F.F.T.s also involved traditional groundfish species, the idea being that the groundfish money would offset start-up losses and lower profits on under-utilized species. Fishermen's organizations opposed F.F.T.s as a job-stealer.

Following a departmental review, a "factory freezer policy" allowed freezing but put a size limit of 200 feet on freezer vessels, while granting additional licences for four such vessels. Any filleting at sea would be for non-traditional species only. It all added up to freezers under restricted conditions, and no factory freezers for traditional species.

The new policy removed most of the economic appeal of freezer vessels. Only one company, Mersey Seafood of Nova Scotia, acquired a new vessel for non-traditional species. (As it turned out, the vessel did poorly in silver hake, and switched to freezing, but not filleting, northern cod.) But National Sea Products kept lobbying, and late in 1985, following a departmental review of the pros and cons, acting fisheries minister Erik Nielsen would grant a factory freezer licence for the 270-foot *Cape North*. The same decision opened the way for two licences to other enterprises.

But the F.F.T. licences were hedged with conditions: licence recipients had to retire equivalent capacity, keep their F.F.T.s out of the Gulf of St. Lawrence and Bay of Fundy, and get more than half their catch from previously under-utilized "enterprise allocations" (E.A.s)—that is, quotas allocated for specific large companies. No other companies asked for licences, and as the 200-mile limit's optimism faded, so did the push for factory freezer trawlers.[38]

Levelton report leads to "full-time, part-time" fishermen

After the Conservatives lost a vote of confidence in the House of Commons, the Liberals won the subsequent election, early in 1980. Political observers gave LeBlanc credit for having delivered, through his popularity, four ridings for the Liberals. Fishermen's groups lobbied the Prime Minister's Office to return LeBlanc to D.F.O. In March, 1980, he took up office again, for a term that would be more troublesome than the first.

As catches increased and profits soared in many fisheries during the latter 1970's, licensing had become more contentious. Fishermen generally favoured limited entry; a licence gave them more security, and they could sell it when they retired. But problems appeared among people who felt left out in one way or another. Some had no clear history of participation in the fishery, and thus failed to qualify for a licence. They could say they had "always" been a fisherman, like their father or grandfather; but in many instances, their participation had been long ago and transient, and the department was unyielding. Eventually the department set up appeal boards to hear complex cases.

In other cases, fishermen might hold some licences but feel entitled to others. In most cases, the licences were attached to vessels; so if, say, a herring fisherman

wanted to acquire a scallop licence, he might first have to buy a scalloping vessel and either operate it separately or hope the department would, after a period of time, let him combine the licences on a single vessel. Different practices sprang up for different fisheries; the licensing picture grew complex. Meanwhile, although one aim of licence limitation had been to get rid of "moonlighters," U.I. was attracting more of them in.

In 1978, the department assigned Cliff Levelton, a fishery manager of high reputation, to inquire into Atlantic coast licensing. Levelton's 1979 report, *Toward an Atlantic Coast Commercial Fisheries Licensing System*, made recommendations that helped clear up several nagging questions. One result was the categorizing of Atlantic fishermen into full time and part time.

Of the 60,000 or so persons who might register yearly with D.F.O. for a personal fishing licence, some floated in and out of the fishery as suited their purposes. Often, young fellows fished temporarily before moving on to something else. And some people just used the fishery for U.I.

Following Levelton's report, the department in 1981 began to distinguish between Atlantic fishermen, classifying them as full time or part time, depending on whether they fished the full season for their area. By 1983, after the first go-round and many hearings by licence-appeal committees, the numbers of full-timers and part-timers settled out at roughly 50–50. Without planning any major push against part-timers, the department was looking towards giving its main future consideration to full-timers. For example, the department, when cutting back the commercial salmon fishery in Newfoundland, allowed part-timers only half the salmon nets allowed to full-timers.

Levelton recommends against transferability

Levelton gave close attention to the question of licence transfers. During the late 1960's changeover to limited entry in British Columbia, to be described later, a leading fishermen's organization had contended that when a fisherman ended his career, the licence should go back to the state. Minister Jack Davis disagreed: the fisherman should be able to sell off his licence as a kind of retirement fund. That became the practice: the fisherman would recommend the recipient of a licence, and the department would re-issue it accordingly. The licence belonged to the government, and the department avoided the terminology of "buying" and "selling" licences. But everyone knew money changed hands.

Those already in the fishery when limited entry came into effect got the licence for nothing, and later could sell it, in some fisheries, for a hundred thousand dollars or more. ("Buy-backs" were another issue. The people of Canada through the department owned the licences, which the department issued for minimal fees. But when government wanted to reduce the number of vessels in a fleet, it often spent taxpayers' money to buy back what they already in theory owned.)

After listing the pros and cons of transferability, the Levelton report came down against the practice. It was questionable, the report said, that people should profit by selling a privilege granted them by the state. Free transferability made it difficult for new entrants to buy their way into the fishery; they faced higher fixed costs than previous entrants. As for the "retirement fund," fishermen in limited-entry fisheries already had a better earnings position than those in the many fisheries that were still open. Speculation in licences constituted an element of over-capitalization.

"It is therefore recommended that free or open transfers of licences not be permitted," Levelton wrote. "Limited transfers, to allow continuation of existing enterprises by other members of those enterprises, should be permitted through the licensing authority. Otherwise, ... licence holders wishing to leave the fishery should be required to relinquish their licences to the licensing authority who will determine the ultimate disposition." An improved administrative structure, with more direct involvement by fishermen, should establish objectives for each limited-entry fishery.[39]

The Maritime Fishermen's Union, a Maritime organization of growing strength, supported Levelton's views on transferability. Similar sentiments could be found on the west coast. The U.F.A.W.U. had earlier opposed transferability. Just before the Levelton report, a 1978 study for the department by Dr. Sol Sinclair, an important figure in the history of licence limitation, took the same stance. But the 1979 election interrupted any momentum for change. The Levelton recommendations against transferability went nowhere.

In theory at least, doing away with transferability could have solved a number of problems. Government and industry could have established a waiting-list system, with fishermen bidding for licences or qualifying by such criteria as their experience, or the cost-effectiveness or conservation value of their operation. To reduce fleets, the licensing authority could simply have delayed the re-issue of licences, or reduced their number. Instead, transferability remained, the fleet stayed over-large, and the government wound up in future years paying hundreds of millions of dollars to retire licences that were government property in the first place.

Transferability of course had its strong points. It seemed like business freedom, with people able to buy and sell as they wished. It could indeed provide a "retirement fund." Most fishermen seemed to favour the rough and ready system of transferability. Later in the century, however, some would come to question it, as licence prices continued to rise and the bigger bankrolls tended to acquire fishing privileges. But by then, transferability appeared to be a firm part of the fishery, difficult to challenge.

Individual quotas spread to trawlers

In the late 1970's, licences and quotas fused into a new mechanism: specific quotas for enterprises,

Icing fish aboard a trawler.

assigned under the licence and, eventually, transferable like licences.

Herring purse-seine fishermen in the Bay of Fundy in 1976 pioneered an individual quota (I.Q.) system, which appeared to have several advantages. Rather than racing desperately for a good share of the overall catch, each boat had its own, dependable percentage share of the quota. Before, a breakdown at peak season was a catastrophe, allowing other fishermen to take the whole pie. With I.Q.s, each fisherman had his fish to depend on. He could pace his fishing to suit markets. The switchover enabled a slower fishery, producing food rather than high-volume fish-meal.

I.Q.s began to spread, appearing in the northern shrimp fishery, the offshore lobster fishery, and the inshore Cape Breton crab fishery by 1981. There were still no I.Q.s for groundfish (although in southwest Nova Scotia, the haddock fleet was using monthly vessel trip limits, which had some resemblance to individual quotas).[40] A groundfish fishing frenzy would hasten their advent.

Initially aided by subsidies, the trawler fleet had found gold in the northern cod fishery. In early 1981, light ice allowed heavy fishing. Soon stories were circulating about fish being deck-loaded, crushed, and wasted. The large-trawler fleet took most of the northern cod quota by the end of February. The resulting commotion prompted LeBlanc and the department into setting company quotas, to foster a more orderly fishery.[41]

"Enterprise allocations" (E.A.s), worked out by D.F.O. and the companies and based mainly on catch history, came into place in 1982 on a trial basis only. That same year a special inquiry, the federal Task Force on Atlantic Fisheries, began looking at the fishery. So as not to pre-empt its conclusions, Minister Pierre De Bané decided not to sponsor E.A.s for 1983. The large-trawler companies continued E.A.s that year on a voluntary basis.

In 1984, following a recommendation by the Task Force on Atlantic Fisheries, the department re-instituted E.A.s. Other measures came into play, notably the establishment of percentage shares for inshore and offshore fleets.[42] Enterprise allocations initially came into place for a five-year period; in 1989 they became permanent.

A pilot project also commenced for individual quotas off western Newfoundland for groundfish vessels of less than 65 feet. Yielding to pressure from fishermen in that area, the department had granted 39 additional dragger licences. This increased competition and reduced incomes for existing draggers. The department and vessel owners, members of the Newfoundland fishermen's union, worked out an E.A. system based on average landings by vessel length-class.[43]

Meanwhile, the idea of individual quotas gained currency in economic circles. The Economic Council of Canada in 1981 published a report and a series of case studies by university professors, espousing "usufructuary rights," which translated into some form of individual rights, which could ideally be traded. In 1982, Professor Peter Pearse of the University of British Columbia in a D.F.O.-sponsored report recommended I.Q.s for several B.C. fisheries. The academic reports, and an Atlantic fisheries task force report appearing in 1983, added to the growing interest in I.Q.s.

Most academic and other theorists favoured the new mechanisms, often called quasi-property rights. Their rationale resembled Whitcher's arguments for salmon leases back in the 1860's and 1870's. Quasi-ownership could provide security and permanence. It would then be in the enterprise's interest to conserve fish and to invest money in conservation, rather than competing to grab all possible fish at the expense of the stock. Advocates added that transferable quotas could allow a painless rationalization, as a smaller number of operators bought up the fishing privileges that were unable to support the many now holding them. Although some people (including LeBlanc[44]) wondered about the long-term effects, relatively little attention went to possible drawbacks, such as the difficulty of enforcing a much larger number of specific quotas.

"Sector management" confines groundfish vessels

Meanwhile, the department was trying to control the newly powerful fleet. In 1982, after industry consulta-tions, D.F.O. began confining groundfish vessels of less than 65 feet to their home region: Newfoundland, the Gulf, or the Nova Scotia–Bay of Fundy area. Under the "sector management" scheme, some vessels got "historical overlap" privileges that let them cross boundaries where they had habitually fished. Most, however, now had to stay in their own home area. Licences became non-transferable between sectors.

From the department's point of view, "sector management" would curtail the increasing mobility of fleets, notably Nova Scotian draggers. It would also, at least in theory, give D.F.O.'s regional managers more freedom to make decisions affecting the fleet in their own area, without impinging on others.[45]

Large-trawler companies hit wall

By that time, the trawler companies in the early 1980's had entered a new crisis. Earlier, the years 1978 and 1979 had provided a rare golden age in fisheries management. All seemed well: research had increased, scientists confirmed that abundance was growing, and a new management system controlled the fleet. Atlantic coast catches had increased from about 850,000 tonnes in 1974 to about 1,270,000 in 1979; value had tripled from $170 million to $508 million. For large trawlers, catch volumes and catch rates had risen more than 50 per cent since 1976. Some of the "big four" corporations were looking for ways to expand.

Two years later, by the autumn of 1981, all were in desperate financial shape and facing bankruptcy. Plants began closing; by October, the big four (or three, since the Nickerson interests now owned National Sea) had laid off 4,500 plant workers and 1,000 fishermen.

The situation was both simple—costs too high, revenues too low—and complex. In the department's analysis, the disastrous combination of factors included changes in currency ratios that helped Scandinavian producers to the disadvantage of Canadians; high interest rates of 18–20 per cent that reduced the ability of wholesalers and distributors to hold inventories; a drop in the price of substitute protein sources, such as beef and poultry; a temporary lessening of per capita consumption of fish in Canada and the United States; higher costs for energy, raw material, and labour; and costly over-expansion leading to indebtedness.

Analytical fingers also pointed at the now-standard villain: the common-property nature of the fishery, which still encouraged over-investment, overcrowding, and over-competition in an optimism-fuelled race for more fish than were there. Licence limitation had never applied to plants, which were regulated by the provinces. The large processors had concentrated on throughput, and expanded accordingly. In the department's view, they showed too little concern for quality, with major deficiencies in harvesting, offloading, dockside handling, transportation, and processing. Marketing was fragmented and lacked innovation, making the industry vulnerable on price. Finally, said

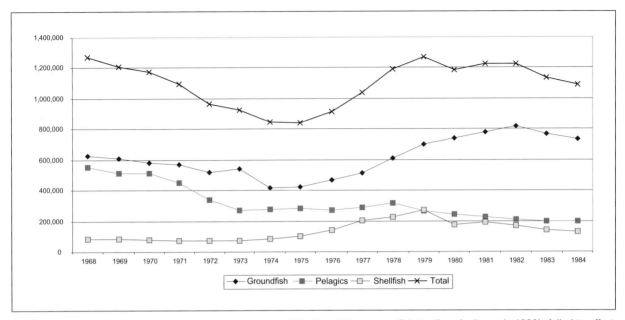

Landings by species group, in tonnes, for Atlantic Canada 1968-84. Higher groundfish landings in the early 1980's failed to offset cost and market forces for large-trawler companies.

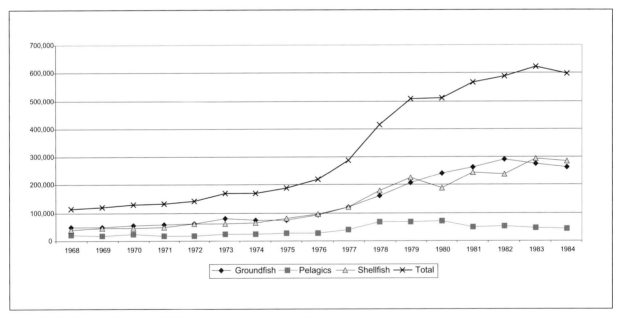

Landed value by species group ($000's), for Atlantic Canada 1968-84.

a departmental document, "in the industry itself, there is a history of poor decision making and backward management," reflected in unwise investment, poor planning, and too little attention to productivity and efficiency.[46]

The trawler-company crisis would create nearly three years of tension and turmoil.

LeBlanc loses dominance

LeBlanc wondered publicly why the trawler companies, well-positioned between rich fishing grounds and the world's biggest market, couldn't make money. His officials, led by deputy minister Don Tansley, considered the best solution—given the major problems of fragmentation, inefficiency, and over-investment—to be some form of consolidation. This would apply not to processing itself, but to harvesting by trawlers and to marketing. There were various ways for consolidation to take place, ranging from a voluntary approach to full government control. (At the same time, LeBlanc and officials considered buying up the major portion of the Lake Group, both to tide the company over with financing and to give government a window on the industry.)

Groundfish companies had got government help in the mid-1970's and on many occasions before. As one company official put it, the industry in moving forward on the road would every now and then strike a pothole; government's job was to help them past the potholes. But D.F.O. officials made clear that this time, any help would come with deep intervention. Alarm swept through the large corporations.

LeBlanc and his officials got ready to take their new scheme forward for approval. The powerful minister was used to getting what he wanted from Trudeau and the cabinet. But he had long ago alienated most of the large Atlantic corporations. He had also irritated the Atlantic provincial governments by resisting their expansion plans and generally paying little attention to them, preferring the advice of his officials and of fishermen. In addition, LeBlanc had peeved the federal government's own central agencies, Finance, Treasury Board, and Privy Coucil Office, by going around them to get what he wanted from Trudeau or other ministers. Finally, he now had the banks worried about what would happen to the large corporations they had backed with heavy loans.

The opposition accumulated force and penetrated the political system. The Fisheries Council of Canada at one point called on Trudeau to get rid of LeBlanc. In Nova Scotia, company officials made strong representations to other powerful ministers. In December 1981, LeBlanc and officials met with a cabinet committee, expecting approval of their interventionist direction. Instead they got rebuffed.

Major repercussions followed. The wounded LeBlanc fell back on the idea of a special inquiry. In January 1982, the Prime Minister's Office announced that Michael Kirby, a top official in the Federal–Provincial Relations Office, would lead a task force (mentioned above) on the Atlantic fishery. Although LeBlanc said that he had sponsored the inquiry, it was to report not to him but to a subcommittee of the cabinet committee on economic development. Tansley would retire later in the year; Kirby, it was planned, would then take over as deputy minister. The idea was that Kirby would not just offer advice that the department might leave on the shelf; instead, as deputy he would make sure that his findings brought action.

Kirby task force: new look at old themes

For the Task Force on Atlantic Fisheries, Michael Kirby recruited Art May, a senior D.F.O. official; Peter John Nicholson, a former vice-president of H.B.

Nickerson and Sons Ltd.; Father Desmond McGrath, co-founder of the Newfoundland union; Victor Rabinovitch, an official of the Canadian Labour Congress; Paul Sutherland, of D.F.O.'s Scotia–Fundy Region; and other officials from various agencies. Kirby reckoned that they had two jobs: a financial restructuring of the big four companies, which might take a few weeks, and reforms for the industry, which would take longer.

Instead, the reforms would take shape first; the task force would publish its report and recommendations in February 1983. The restructuring effort, foreseen as brief, would end only in 1984, after a series of battles pitting companies against government, communities against one another, and D.F.O. against the central agencies of government. This account will deal with the reforms first, then the restructuring.

As the task force began work, LeBlanc in 1982 gave what would be his last speech to the Fisheries Council of Canada. LeBlanc told the F.C.C. in effect that he had only wanted equality for fishermen, and that he still believed in that goal. "[But] when I look at some missed opportunities on the processing and marketing side, I wonder if government and processors tried hard enough to work together. Perhaps it is time for the pendulum to return to a balance."[47] But the time was past for any rapprochement. In October, LeBlanc became Minister of Public Works; Pierre De Bané took his place.

For years, LeBlanc and the department had urged improvements in quality and marketing, and cautioned against over-expansion. (Indeed, some efforts such as quality improvement slowed down while everyone was "waiting for Kirby.") The "Kirby report" would reprise those old themes, among others. But the thorough report would add a new depth of analysis, providing abundant facts, detailing the pros and cons of issues, and giving some reform ideas a stronger intellectual footing.

Clearer goals for the fishery

The 1976 policy had laid out a series of broad-scale goals, using terms such as "best use," "economic viability," and "optimal distribution of returns to ... labour, capital, and the natural resource." (A 1981 discussion paper, *Policy for Canada's Atlantic Fisheries in the 1980's,* had elaborated the policy for the east coast.) Now Kirby gave a new clarity to overall goals.

Viability would come first, followed by jobs, followed by Canadianization. To quote:

> (1) The Atlantic fishing industry should be economically viable on an ongoing basis, where to be viable implies an ability to survive downturns with only a normal business failure rate and without government assistance.
>
> (2) Employment in the Atlantic fishery should be maximized subject to the constraint that those employed receive a reasonable income as a result

of fishery-related activities, including fishery-related income transfer payments [this meant, for the most part, U.I. benefits].

(3) Fish within the 200-mile zone should be harvested and processed by Canadians in firms owned by Canadians wherever this is consistent with Objectives 1 and 2 and with Canada's international treaty obligations.

D.F.O. Minister Pierre De Bané announced that the government accepted the objectives, in that order of priority.[48] Officials would quote them for years to come. That being said, the succeeding Conservative government commencing in 1984 had a somewhat different orientation. By the 1990's, the Kirby objectives received little mention, and were left floating in policy limbo.

Quality push gets diffused

Although the Trudeau government accepted most of Kirby's recommendations, the delivery would vary.

On the chronic issue of fish quality, the department had long wanted a general improvement, together with some form of grading that would pay better money for better fish. The Kirby report recommended what amounted to a frontal assault on quality. The department should make it mandatory to bleed, gut, wash, and ice fish at sea, always allowing for "practical exceptions."

But all five Atlantic provinces declined the recommendation for matching legislation. Industry members complained, saying that in many cases fishermen already bled, gutted, washed, and iced their fish. In 1983, new regulations required all boats over 45 feet to carry ice. But the department backed off on mandatory bleeding and gutting.

Kirby also recommended compulsory grading, both at dockside and for final products. De Bané intended to push through with this, but after the Conservative government took power in 1984, the initiative faded away.

Before the Liberals left office, cabinet provided $11 million for ice-making and ice-storage equipment on the Atlantic. Another $9 million for similar purposes came from the Special Recovery Capital Projects Program, a broad-scale government program announced in 1983. The department also increased its "jawboning" work with the industry.[49] In effect, D.F.O. was resuming its previous approach, which featured encouragement and pressure more than compulsion. As it turned out, product quality did improve in the 1980's, partly from departmental efforts, partly from natural trends in the industry.

Marketing Council fails to fly

Fragmented marketing was another long-standing complaint about the Atlantic industry. The Kirby

Pierre De Bané

report gave lengthy attention to the issue. Kirby recommended an Atlantic Fisheries Marketing Commission to co-ordinate marketing strategies and carry out promotional work. Once again, the industry got money spent on its behalf. A five-year, $28 million campaign commenced to build up fish consumption in Canada and the United States. Advertisements and promotions featured the theme "Today's Dish Is Fish." The Program for Export Marketing Development provided another $25 million (mainly cost-recoverable) to match new marketing expenditures by groundfish and herring exporters.

But no marketing commission came into place. The department never accepted that recommendation as it stood. De Bané wanted to go further, with heavy government involvement in developing markets. The issue got ensnarled with the general question of big-company restructuring. The industry and some cabinet members balked at De Bané's assertive approach, and the 1984 cabinet shuffle and election, intervened.[50]

As with quality, changes were taking place anyway. Kirby's restructuring of trawler companies would eventually result in better marketing capacity by the major companies themselves. Meanwhile, the trade environment was changing. After the 200-mile limit, more international seafood shows took place, and trade information in general flowed more freely through phones, faxes, and eventually computers. Many companies became more alert. Some began to pool their sales and marketing efforts. And eventually, free-trade agreements with the United States beginning in 1989 would cut tariff barriers, making marketing easier.

Expansionist enthusiasm dies down

LeBlanc had often preached against over-expansion. Meanwhile, fishermen had used subsidies and skirted regulations to get stronger boats, and provincial governments had promoted plant expansion. The Kirby report once again pointed out the great problem of over-capacity.

Kirby recommended, and government agreed, that there should be no broad new assistance programs for fishermen or processors. D.F.O. rejected, for the time being, a recommendation to end vessel-construction subsidies. Still, Kirby's report took some of the air from the expansionist bubble, as did the groundfish crisis itself. The provincial push for more fishing jobs became more subdued.

To slow down the race for fish and provide more security, the Kirby report called for more use of individual quotas and similar arrangements. This endorsement strengthened the trend. Peter John Nicholson, the Kirby group's main proponent of this approach, later looked back on the support given to I.Q.s, individual transferable quotas (I.T.Q.s), and E.A.s as a chief outcome of the report.[51]

Foreigners get fewer fish

Kirby had said at the outset that he would take a reformist rather than radical approach. Small adjustments to the course, it was reasoned, could over time make a big difference to the destination. Besides the major though unimplemented proposals on quality and marketing, the report included many other recommendations, most of which had at least some impact.

Some recommendations affected foreign quotas, a controversial subject. The Law of the Sea consensus obliged coastal states to share fish that were surplus to their needs. Since the 200-mile limit, the department had allocated some fish, notably northern cod, to other fishing nations. Quotas were allocated to particular countries on the basis of the bilateral relationship, taking into account such factors as historical presence.

At first there had seemed to be room for such "surplus" allocations. Although total foreign allocations shrank in the years immediately following the 200-mile limit, they still ran to some 350,000 tonnes, including not only species of limited interest to Canadians but also some "traditional" groundfish. Abundance was increasing. The northern cod fishery in particular seemed to be a huge marine pot of gold. While the government encouraged more Canadian fishing in the north, it also allocated substantial quotas to foreigners.

Nations receiving such allocations were supposed to give something in return. Canadian policy first called for "commensurate benefits" and then for a "satisfacto-

ry trading relationship." This approach had some effect. Foreign fleets increased their use of Canadian ports for repair work and supplies. Portugal in particular increased its purchases of Canadian salt cod. Canada in 1981 entered a five-year "Long Term Agreement" (L.T.A.) with the European Community, trading quotas for tariff concessions on fish products.

But the Atlantic industry objected to the generosity, especially on northern cod. By 1981, Canada was telling other countries there was almost no northern cod surplus to hand out.[52] Meanwhile, despite the L.T.A., relations with Europe proved disappointing. The European Community was raising difficulties about Canada's seal hunt, and some countries, notably Spain, were being uncooperative on the northwest Atlantic fishing grounds.

In this context, the Kirby report recommended and the department adopted the practice of awarding allocations based not on promises but on past performance. Meanwhile, foreign allocations of northern cod dwindled. Totalling 93,000 tonnes back in 1977, they shrank by 1987 to 9,500 tonnes for the European Community only.[53] The department also laid out clearer rules for "over-the-side" and "over-the-wharf" sales to foreign interests. Both practices faded away in following years.

Many smaller, seasonal Canadian plants with no fleet of their own wanted a share of northern cod. The department after the 200-mile limit had allowed some chartering of foreign (and Canadian) vessels to supply "resource-short" plants with off-season fish. Following the Kirby report, new criteria defined the eligible plants and regularized the "Resource-Short Plant Program."

Kirby also recommended setting aside some northern cod quotas for possible use by new, Scandinavian-style longliners. A few of these vessels later came into the fishery, but fell far short of expectations. Finally, Kirby suggested percentage shares for the various users of northern cod; future groundfish allocations reflected his recommendations.

All told, the Kirby report resolved some nagging issues about foreign fishing, and reinforced the trend to tighten up on foreign allocations.

Miscellaneous recommendations, miscellaneous results

Among the report's 57 recommendations, one of the more innovative said that the Canadian Saltfish Corporation should become the agency of choice for strengthening the under-developed fishing industry in Labrador. But little happened on this front.

Some recommendations touched on independent fishermen. The Kirby report called on provinces to pass collective-bargaining legislation; Newfoundland had already done so, and New Brunswick was doing so. Following another recommendation, the federal fisheries department set up a new, higher-level review board to hear licensing appeals and grievances.

Kirby recommended a program to stabilize gross incomes. Nothing came of this, nor of a suggested "production bonus" system that would reward desirable fishing practices. A recommendation touching Unemployment Insurance did bring change. Before, benefits had been calculated on the last six weeks of fishing. Some fishermen would knock off fishing in the "shoulder season" to avoid reducing benefits. Now the system changed, to calculate benefits not by the last but by the best weeks.

Kirby also recommended a "sunset" provision to phase out fishermen's U.I. once the other income-stabilization measures were in place. But the new income programs never started up; U.I. remained a mainstay of the Atlantic fishery.

As noted earlier, the report said that the federal government should re-assert control over marine fisheries in Quebec, reversing a 1922 arrangement. Pierre De Bané, a Quebecer, was happy to do so. The 1984 re-assumption of jurisdiction made possible more consistent management in the Gulf of St. Lawrence.

The task force report also recommended that the department improve its consultative process and its dissemination of information among fishermen. Except for start-up funding for l'Alliance des pêcheurs commerciaux du Québec, nothing major followed. Many future reports would make similar recommendations on communication with fishermen, with similarly little follow-up. Kirby also recommended a small advisory group for the minister and senior officials; a new minister, John Fraser, would in 1985 create an Atlantic Regional Council, which operated for some years.

To carry through Kirby's recommendations, the government approved expenditures of $198 million over five years, not always spent as foreseen. D.F.O. was to oversee efforts on quality, marketing, fisheries management in Quebec, and various other activities. Other departments were to oversee U.I. changes and expenditures by the Program for Export Market Development.

What did the Kirby report change?

In the end, the reforms laid out in the Kirby report probably had less direct impact than the big-company restructuring. The major recommendations on quality and marketing never came into effect as planned. The rest of Kirby's reforms amounted more to housekeeping and small renovations than to house-building.

Yet the report had great value, in thinking through many Atlantic issues, in setting out the priorities for viability, jobs, and Canadianization, and in endorsing individual quotas. In 1983, D.F.O. minister Pierre De Bané commented that "in the end, the best value of the Kirby and Pearse [in British Columbia] reports might be educational."[54] He was probably right.

Observers have pointed to a major lapse of the task force report. Emerging from an economic crisis, it gave little attention to conservation. (Although Kirby recommended and the department enacted a coast-wide minimum size of 130 millimetres for trawl mesh, this was a quality rather than a conservation measure, to increase the size of landed fish.[55]) Kirby said in effect that there was no resource problem, that Canadian fishery management was the best in the world, that the northern cod catch alone could rise to 550,000 tonnes a year, and that groundfish catches all told would reach a million tonnes by 1987, double the levels of the mid-1970's.[56]

The figures came from D.F.O. researchers, some of whom later said they were pushed into making forecasts they would rather not make. But there they were, the best estimates of scientists. (For northern cod, they were based largely on average growth and recruitment rates seen in the 1960's, which were not to be the future norm.)[57]

Many people seized on the optimistic projections. Kirby's report gave credence to the idea of loads more fish to be caught and money to be made in the groundfish industry.

Central-agency officials take bigger role

Rather than becoming deputy minister as planned, Michael Kirby took a position with Canadian National Railway, although continuing to deal with the big-company restructuring until its conclusion. Arthur W. May, who had started as a research biologist, worked on international matters, and risen to become assistant deputy minister for Atlantic fisheries, took over as deputy minister in the fall of 1982. When John Turner succeeded Pierre Trudeau as Liberal leader in 1984, Kirby became a senator, as did LeBlanc and De Bané.

Meanwhile, a more cautious atmosphere was settling over the department, which had lost some of its confidence with Kirby's very arrival. Previously, as a contemporary report noted,

> [o]fficials and probably ministers of the so-called "central agencies"—economic development, treasury board, finance—considered Mr. LeBlanc, his deputy minister Don Tansley, and the entire fisheries department as mavericks. They were always doing things their own way, sometimes running roughshod over other departments. ... As one insider puts it: "what happened in December 1981 [when the Kirby task force was announced] was simply that nobody around town trusted Fisheries."[58]

With Kirby, more officials began coming from central agencies into top- and middle-level D.F.O. positions. The trend of importing managers would increase over time. The advantages were said to be that officials who moved around and gained experience would develop a broader picture and better management skills. The disadvantage for D.F.O. was the loss of subject-matter knowledge. Although up-through-the-ranks officials continued to hold most top jobs in the regions, they became rarer at D.F.O.'s top levels in Ottawa. (Still, experienced officials continued to hold many of those senior positions that dealt directly with fisheries management.) The maverick department of the 1970's became more like others, although it still kept a distinctive character.

The department's clout within government was declining. One indication came with a conflict over Gulf of St. Lawrence redfish allocations in the early 1980's, during De Bané's term as minister. After a low ebb in the late 1970's, the stock had increased sharply. Gulf- and non-Gulf-based interests were fighting about quota shares, while D.F.O. was calling for a lower-level fishery to spread the strong year-class over several seasons. As various politicians backed various gear sectors, De Bané at one point had to go to the Privy Council Office for resolution of an issue that Davis or LeBlanc would have settled by their own fiat.[59]

Restructuring brings years of turmoil

The focus of the multi-year restructuring effort was on the larger companies, whose virtual collapse had produced the crisis. The big four companies of the day—National, Nickerson, Fishery Products, and the Lake Group—accounted for 62.5 per cent of groundfish production, compared with only about 8 per cent for the next largest four companies. The big four also sold for many other firms, bringing their share of marketing

The trawler-company restructuring brought no significant reduction in plant or trawler-fleet capacity. The Kirby report looked forward to plentiful catches.

to 69 per cent. They had strong influence in the Fisheries Council of Canada, which represented some 200 enterprises across the country.

The large-trawler companies had made huge expenditures as early as 1978. In 1981, they had acknowledged they were in crisis. By the end of 1982, they reportedly owed the Bank of Nova Scotia $276 million, the Royal Bank of Canada about $75 million, and the Province of Nova Scotia about $50 million. Some large plants were closed, and everyone was "waiting for Kirby" to resolve the situation.

Complaints soon arose. By December 1982, Richard Cashin, of the Newfoundland Fishermen, Food and Allied Workers Union, was warning that there was "nothing in the Kirby approach for fishermen, no major change in marketing, and nothing much coming at all, except mergers and a bailout of the big companies, plus plant closures."[60] And in the words of one medium-sized processor and vessel owner, "The Kirby group is working on behalf of the Big Four plus Two [the U.M.F. and Q.U.F. co-ops], when the rest of us could survive. But the changes [they] cook up will affect all of us. Why don't we smaller ones speak up in the F.C.C.? Because the Big Four parcel out orders to some of us; they sell for some of us; and they're tied in with the federal and provincial governments that license us and control us totally."[61]

The Kirby task force came to conclusions broadly similar to those of LeBlanc's officials. Some of the large companies' troubles lay not in government rules or general industry conditions but in their own administration. They had, as Kirby later put it in a speech at Dalhousie University, over-expanded beyond their capacity to manage.

After Kirby presented his report in February 1983, most task force members went back to their previous jobs. Kirby and a core group including Peter John Nicholson continued with the restructuring. The government seems never to have seriously considered simply letting the big companies fail. For one thing, spokespersons said, a sell-off of big-company inventories would flood the market and damage medium-sized and small enterprises as well, along with the entire independent fleet.

The Kirby group wanted to consolidate operations, improve management, and bring viability to the large-trawler companies. It must have seemed simple: the federal government had the money to prevent bankruptcies, and he who paid the piper should call the tune.

But restructuring brought the threat that some large plants would stay closed and that company owners and officials used to making decisions would lose power. Communities and companies began to fight back. The restructuring effort that Michael Kirby hoped would take weeks or months instead lasted to mid-l984. It involved financial maneuverings, political infighting, and conflicts in the upper levels of government. The following sketches main events only. (Scott Parsons has written a fuller description.)[62]

As a first step towards new operations, National Sea and Nickerson (which owned National) merged their marketing operations in May 1982. Meanwhile, as the Kirby group began to look into big-company operations, industry disputes arose over the value of assets. The Kirby group set in motion a long study of the company assets and their value.

Quarrels abound over restructuring

Options for restructuring included organizing the companies on a provincial basis, or leaving Fishery Products and National Sea alone and merging the rest, or making one big company, or other combinations. Meanwhile, all three major fishermen's organizations, the Newfoundland union, the Maritime Fishermen's Union, and the Eastern Fishermen's Federation, supported LeBlanc's old idea of separating the trawler fleet from the processing companies. This proposal got little attention from the Kirby group.

Complex negotiations and bickering took place through 1983. The cabinet committee dealing with the restructuring decided on a provincially based merger, with the wrinkle that the Nickerson scallop interests in Nova Scotia would form part of the new Newfoundland company.

The Kirby group had put economic viability at the head of its list of objectives, and wanted to follow through on a market-driven, efficient industry. This meant in their eyes that, among other things, plants should close at Burin and at Grand Bank, the latter operation being only a short drive from another plant at Fortune. But the communities and provincial government protested, and eventually the Kirby group compromised. In the end, no major and symbolic closures took place in Newfoundland, despite the well-known problems of plant over-capacity.

Eventually, in 1984, the restructuring began settling out into two huge companies. A predominantly Newfoundland-based firm, it was reported, would take in Fishery Products, which had 6,100 employees, bought fish from 3,300 fishermen, and had 45 trawlers; the Lake Group, which had 3,000 employees, bought from 2,800 fishermen, and had 22 trawlers; and John Penney, which had 466 employees, bought from 336 fishermen, and had five trawlers. The Newfoundland company would also get several scallopers from the Nickerson fleet. The second company, National Sea Products, would take in existing assets of that firm (N.S.P. had 8,900 employees, bought fish from 6,900 fishermen, and had 45 trawlers, including 10 scallopers), and most assets of H.B. Nickerson and Sons Ltd., which had 5,500 employees, bought fish from 5,700 fishermen, and had 29 trawlers, including 15 scallopers.[63] But this plan still faced controversy.

Government acquires most of Newfoundland firm

In Newfoundland, it was clear that the federal government would own the main share of equity in the restructured Fishery Products International. The federal government put in $75 million and got 60 per cent ownership. This was purely an investment, nothing like the more active ownership stake that LeBlanc had visualized for the Lake Group. Additional funding support came from the Bank of Nova Scotia, converting $44.1 million of debt to equity, and the province converting $31.5 million, while taking 25 per cent ownership.[64]

When it became apparent that the federal group also planned majority government ownership in Nova Scotia, a storm of protest darkened the skies. Private investors stepped forward to create a "private-sector solution" that would prevent government ownership. The private parties put $20 million into National Sea Products. But government money would still keep the restructuring afloat.

The federal government provided $70 million ($80 million when interest charges over five years were added) to retire Nickerson debts to the Bank of Nova Scotia. The government also invested $10 million directly. Unlike the case in Newfoundland, the federal government got only 20 per cent equity in the restructured company. The Province of Nova Scotia converted $25 million of debt to equity, and on another $25 million postponed payments of principal for five years.[65] As business conditions remained poor in 1984, an additional $14.5 million of federal money went into price stabilization measures for frozen cod,[66] followed by $10 million of inventory financing for herring. In both cases, the idea was to buy inventory and sell it as markets improved.

Although the restructuring was finally taking shape, arguments continued. De Bané was still pushing for a major government role in marketing. Then Prime Minister John Turner, who had succeeded Trudeau, on June 30 appointed Herb Breau, a respected New Brunswick M.P., to succeed De Bané. Breau, too, believed in active government and wanted to strengthen the role of the Fisheries Prices Support Board in marketing. But the September 1984 election changed the government, and the Conservatives proved less interventionist.

Indeed, there now appeared less need for intervention. The financial aid had already put the big companies on a more solid footing. Business conditions were improving. And the new, restructured National Sea Products and Fishery Products International, with various management changes, did strengthen their marketing. A form of consolidation had taken place. Opinions differed on the hoped-for improvements in big-company operations. One industry official called the whole restructuring "a $200-million name change." But the two restructured companies did appear more viable. There had been no major reduction in plant capacity, but at least the push for expansion had subsided. Once more the industry seemed set for a period of stability.

Pêcheurs unis gets help; U.M.F. co-op fades

The Quebec fishery had its own restructuring saga. By spring of 1983, it was clear that the Q.U.F. (Pêcheurs unis du Québec), a federation of six co-operatives, was in serious trouble. The Q.U.F. employed 1,700 processing-plant workers, and accounted for nearly half the province's processing capacity. The company had defaulted on its debts and was unable to get credit.

Despite opposition from the provincial government, Pierre De Bané arranged for the federal government to buy up Q.U.F. assets, for $28.5 million, and form a new company, Pêcheries Cartier. The government upgraded the major plants, at Rivière-au-Renard and Newport. When De Bané left the fisheries portfolio in 1984, the company was transferred to the Canada Development Investment Corporation. The Parti Québécois government in Quebec consistently made life difficult for Pêcheries Cartier, whose assets were later sold off.

The other big Atlantic co-op, United Maritime Fishermen, also found itself in deep trouble in the mid-1980's. With several plants in the Maritimes, the U.M.F. was more of a shellfish than groundfish processor. It got no help comparable to that given to the trawler companies, and faded away by the later 1980's, although some of its component plants survived.

How much did the restructuring cost?

A September 1984 briefing note for the new Conservative government gave this federal tally for restructuring, counting both completed and pending disbursements: for Newfoundland, $135.2 million (for share purchases, a "social compact compensation" arrangement, and other restructuring costs); for Nova Scotia, $95.8 million (for share purchases and related costs, plus $2.3 million start-up funding for the idle Nickerson plant at Georgetown, P.E.I.); for Quebec restructuring, $28.5 million. The total federal commitment came to $259.5 million; allowance for contingencies brought it to $301.3 million. This was on top of the $198 million approved for Kirby's reforms. The same briefing note warned that Fishery Products International would need still more money, another $60–$100 million.

More money did go into F.P.I. in 1985; total funding to the company in the 1984–1987 period came to $167.6 million, although the government got back $104 million when the company went private in 1987.[67]

Salmon problems bring ban and buy-back

Atlantic salmon, like groundfish, were in many respects a coast-wide fishery. Salmon had long caused concern. In the 1970's, the problems got worse.

Fisheries work on salmon rivers included electrofishing; an electric current stuns fish for counting or other analysis. (D.F.O. photo by Kevin McVeigh)

In New Brunswick, several hundred commercial fishermen, mostly in the Miramichi and Saint John areas, fished for salmon. A strong sport-fishing industry also depended on salmon, and by the end of the 1970's it was generally agreed that the sport-fishery gave more benefit to the New Brunswick economy. (In 1982, for example, the entire Atlantic commercial catch of 1,555 tonnes generated about $7 million; the 243-tonne sport fishery, about $27 million.)[68]

Most commercial fishermen were in Newfoundland: as of 1968, about 9,000 of them took salmon when they could. The fishery was particularly important for local fishermen, including Native people on the coast of Labrador. Another important fishery took place on the south coast, where driftnet fishermen in the Port aux Basques area intercepted salmon on their way back to New Brunswick and Quebec.

Sport-fishing groups mounted a continuing lobby for salmon protection by the department. They also did conservation work of their own. In 1974, the International Atlantic Salmon Foundation with D.F.O. and other government help set up an Atlantic Salmon Centre at St. Andrews, N.B., to do research and related work.

The department did its best for salmon in both research and management, sometimes occasioning complaints that the species got overmuch attention at the expense of other fisheries. F.R.B. studies documented the effects of forest sprays and other chemicals. Hatcheries, including the world's largest at Mactaquac on the St. John River, tried to enhance the resource. But abundance never seemed to improve to any great degree; it was more a question of controlling damage from pollution, environmental changes, and predators, including anglers, poachers, and commercial fishermen.

The Greenland fishery, which intercepted migrating Atlantic salmon, grew rapidly, peaking at 2,700 tonnes in 1971. Canada and the United States through I.C.N.A.F. won an agreement that phased out the high-seas fishery in 1976. A quota came into place limiting the local west Greenland fishery to 1,190 tonnes, later reduced. Meanwhile, Canada lobbied internationally for the state of origin's special interest in salmon. The Law of the Sea recognized that the state whose rivers produced them should have "the primary interest in and responsibility for" salmon. Canada also helped set up the North Atlantic Salmon Conservation Organization (N.A.S.C.O.), beginning in 1983, with headquarters in Scotland. Agreements under N.A.S.C.O. would further curtail international interceptions.

Of the salmon's many enemies, commercial fishermen made the biggest and most obvious target. Decades earlier, in the 1930's, the department had frozen the number of driftnets. In 1971, it extended limited entry to the rest of the Maritime fisheries and to Newfoundland.

Then, in 1972, Minister Jack Davis announced a five-year ban on New Brunswick and Gaspé commercial salmon fishing and a permanent ban on the intercepting driftnet fishery at Port aux Basques, Newfoundland. This was the first major ban in postwar Atlantic fisheries. Roméo LeBlanc later extended the ban through 1980. The government paid compensation of some $1.5 million a year to about 700 commercial fishermen who chose to wait for a reopening, and it totally bought out 250 or so fishermen in New Brunswick. By the time the fishery reopened in 1981, only about 200 fishermen re-entered the fishery.

Despite the commercial ban, salmon still suffered from by-catches in the commercial fisheries of the Maritimes and Quebec (new restrictions eventually cut the by-catch). And while banning Maritime and Gaspé fishing, the department in Newfoundland in the mid-1970's allowed new salmon licences to people with a history in the fishery. The number of licensed salmon fishermen in that province rose from 5,050 in 1973 to 6,981 in 1975, with some of their fishing affecting Maritime and Quebec salmon. Still, other controls tightened up in Newfoundland; catches there fluctuated downward, from 2,000 tonnes or so in the early 1970's to less than 1,000 in 1984, worth $3.6 million.

Meanwhile, angling increased in New Brunswick, with the number of sport-fishing licences rising from about 12,000 in 1965 to 22,000 in 1982. Poaching also increased, as the commercial ban made contraband salmon more valuable. Native people with food-fishing permits sometimes took extra fish and sold them commercially. The many opposing factors undid much of the benefit from the ban; there was no major improvement.

Throughout the late 1970's and early 1980's, the department kept holding conferences and working out hopeful new plans for the salmon. A major attempt came in 1984, when the department under De Bané announced a five-year plan. It would cut interceptions, and catches in general, by shortening seasons; combat by-catches (it became illegal for any commercial fishermen except those licensed for salmon to keep a salmon); fight poaching by requiring commercial salmon to be tagged; pursue negotiations with Native

groups to lower their catch; and increase consultation through the Atlantic Salmon Advisory Board (which Minister James McGrath had set up in 1979) and zone-management committees. New rules required sport-fishermen to release large salmon and retain only grilse (smaller salmon that spend only one winter at sea, as distinct from larger, "multi-sea-winter" salmon).

D.F.O. also announced that it would issue no new commercial licences anywhere on the Atlantic coast, and would begin a licence buy-back program in Newfoundland and Labrador. The 1984 plan would prove to be the last major attempt to keep the commercial fishery alive.[69]

Seal wars get international attention

A perpetual sideshow on the Atlantic came from the seal hunt. For years, the Canadian government got more foreign mail about the hunt than any other issue. By the late 1970's, protest groups had turned the annual fishery into a frenzied "media circus," with hundreds of reporters converging on the outport of St. Anthony.

At the outset of the 1968–1984 period, the Atlantic hunt all told was taking well over 100,000 seals a year, yielding pelts worth over $1 million. Meat and flippers added some value. By the mid-1970's, pelt value had climbed to $2 million or more, total value of the hunt (primary and secondary operations) came to $5.5 million, and the seal hunt employed more than 4,200 people. Although returns were low compared with such fisheries as groundfish and lobster, the seal fishery gave important employment during "the hungry month of March" and into April.[70]

Among the protest tactics were attempted boycotts of Canadian fish products. The early 1980's postcard is addressed to a British supermarket chain.

Mother seal and pup, Northumberland Strait, March 1969. (Library and Archives Canada, PA-151864)

Total Allowable Catches (usually 150,000–200,000 harp seals and 15,000 hoods), licences, seasons, and restrictions on killing methods all controlled the hunt. Still, it was a bloody spectacle. Cute big-eyed harp-seal pups sometimes got clubbed a short distance from their mothers (which, however, typically showed no distress). Complaints arose that some sealers were failing to kill the animals properly, or were even skinning them alive. The latter charges in most cases arose from confusion; no one would want to try skinning a live animal when you could easily kill it, but even a dead animal's body could sometimes move from internal reactions.

As before, the department worked with animal welfare specialists to ensure good practices. Under Minister Jack Davis, the department set up a Committee on Seals and Sealing, including representatives of humane societies, to advise on the hunt. (Davis at one point apparently intended to ban the whitecoat hunt, but backed off.)[71]

Meanwhile, the protest groups were gathering steam. Besides Brian Davies and his International Fund for Animal Welfare, the new group Greenpeace campaigned against the hunt, as did Paul Watson, a former Greenpeacer who set up the Sea Shepherd Conservation Society. By the late 1970's, protest groups were coming to the ice every year, carrying out such media stunts such as spraying paint on pups to render the pelts worthless. Some tried to argue against the seal hunt on logical grounds: that it was driving the seals to extinction, or that it was cruel. Given the chance, the department could counteract such arguments. Other groups relied less on logic than on the emotional shock of clubs bashing the skulls of "baby seals." Supporters sent them large amounts; the protesting organizations were reported to collect far more money than the sealers themselves.

The small town of St. Anthony on Newfoundland's Great Northern Peninsula became a media centre, particularly in 1978 and 1979. Brigitte Bardot, the famous star of French cinema, and lesser celebrities showed up to give press conferences, along with reporters from Canada, the United States, Europe, and other parts of the world. Paul Watson went through notorious escapades. At one point he chained himself to a cable pulling seal pelts to a vessel. In another episode, he had to walk miles over dangerous ice to make land at Cape Breton, a feat that gained him a grudging admiration from some East Coasters. Watson was reported to have made one notable retreat, turning his protest vessel back from St. John's after a Newfoundland businessman threatened to use an aircraft to bombard the ship with chicken manure.

Mass mailings inundated the department, which gradually mounted its defences. The evidence showed that the seal herd, at 1.7 million the second largest in the world, was healthy and if anything increasing. And the Committee on Seals and Sealing and various animal-welfare specialists who visited the hunt testified that it was humane. The occasional prominent voice spoke out in support: Ronald Reagan, not yet president of the United States, used his radio broadcast in 1978 to support the hunt and point out the high revenues of protest organizations.

In Canada, the House of Commons unanimously supported the hunt. Jack Davis and Roméo LeBlanc both defended it, suffering bad press and insults for it, but making an impact. Many officials, notably scientist Mac Mercer and Charles Friend, LeBlanc's press secretary, spent large amounts of time explaining the situation to reporters. By the early 1980's, public opinion in Canada had gone through, if not a turnaround, at least a significant change. At times in earlier years, public opinion had clearly opposed the hunt. Now there was no groundswell against it; there was even some support.

But the whitecoat hunt would end, its demise stemming from overseas. The protesting groups, particularly the I.F.A.W., took their campaign to Europe. External Affairs tried to spread counterarguments through its embassies, but the brutal images of the hunt carried more weight. In 1983, the European Parliament, in one of its earliest notable actions, banned imports of seal-pup products. The main market was suddenly gone. The hunt dropped back to low levels; but it would eventually resurge.

Whale fishery ends

In the late 1960's, concern about whale fishing erupted in many parts of the world. In Canada, the Atlantic coast had never had a strong whale fishery, except for a brief period of intensive harvest in Newfoundland around the start of the 20th century. After that, operations were small and intermittent. From 1964 on, three small whaling operations started up, two in Newfoundland and one in Nova Scotia, mainly for fin whales. But scientific studies by the Arctic Biological Station and others showed only small and declining populations.

In 1972, the Stockholm Conference on the Human Environment called for a ten-year moratorium on commercial whaling. On December 22, 1972, Jack Davis announced that commercial whaling in Canada would end. The B.C. whale fishery had already shut down in 1967 because of poor economics. (On the Atlantic, where Davis had already banned much of the commercial salmon fishery, he picked up in some quarters the nickname of "Ban 'er Jack.")

The government provided compensation payments to plants, workers, and whalers. Native people in the Arctic were still allowed to take whales for their own consumption; beluga and narwhal catches amounted to several hundred a year. Canada stayed a member in the International Whaling Commission for several years, but withdrew in 1982.

CHAPTER 23.
The Maritimes and Quebec, 1968–1984

Fishery gains ground overall

Despite the turmoil of the two groundfish crises, a comparison of 1968 and 1984 statistics shows that the Maritimes and Quebec made major gains.

For the four provinces, groundfish landings went from 280,000 tonnes to 364,000 tonnes, nearly a one-third increase. Value shot up from $29 million to $146 million. Even with inflation ($100 in 1968 bought as much as $323 in 1984), this was a significant rise.

The heady days for herring were over; pelagic landings were 404,000 tonnes in 1968, only 139,000 in 1984. But value climbed from $17 million in 1968 to $28 million in 1984. Meanwhile, shellfish landings rose from 33,000 tonnes to 113,000 tonnes, more than triple, and value from $38 million to $258 million, more than sixfold. Although total landings for all species slipped from 717,000 tonnes to 640,000 tonnes, because of the decline in pelagics, overall value rocketed from $83 million to $436 million.

At the outset of the 1968–1984 period, people in the Maritimes and the Gaspé region still tended to feel that they lagged behind central and western Canada.

Provincial governments were striving to promote development, as were D.R.E.E., a federal department, and its successor the Department of Regional Industrial Expansion (D.R.I.E.). And within the lagging region, many people saw fishing as one of the poorer occupations.

But as time went on, there was less talk of poor fishermen. Boats in most ports got bigger and better. New fisheries emerged for crab and shrimp. The scallop fishery grew for most of the 1970's. Groundfish declined but resurged. And Unemployment Insurance poured increasing amounts into the fishery.

While the trawler fishery went through boom and bust, thousands of small enterprises kept working along. Many would gain value over the period, especially those associated with shellfish. Clockwise from top left, scenes at the Magdalen Islands, New Brunswick's Gulf shore (gaspereau fishing near Inkerman in 1977), Cape Breton, and Campobello Island, N.B.

Midshore fleet keeps expanding

Chapter 22 showed that Atlantic-wide, medium-sized vessels greatly increased in the 1968–1984 period. The biggest rise came in the Maritimes and Quebec, as reflected in Table 23-1.

The fleet of medium-sized craft between ten tons and 100 feet (mostly 35–65 feet) nearly doubled, from about 2,400 to 5,300, while improved electronics and other gear further strengthened their fishing ability.

Vessels over 100 feet fell slightly in number. These were generally groundfish trawlers, plus a few scallopers and purse-seiners. Even though their number dropped, this fleet was still increasing its fishing power through better-equipped replacement vessels and other improvements.

The number of small craft less than ten tons varied by province, but overall was declining.

This 65-foot crabber and Scottish seiner was one of many new boats built in the Caraquet, N.B., area in the early 1980's.

Table 23-1. Fishing craft by size category in the Maritimes and Quebec, selected years.

	Year	N.S.	N.B.	P.E.I.	Que.
Over 100 ft.	1968	134	9	6	5
	1976	111	14	1	7
	1984	118	6	1	7
10 tons to 100 ft.	1968	1,176	1,007	32	198
	1976	1,198	1,145	115	281
	1984	2,298	1,511	962	576
Boats <10 tons	1968	5,326	1,947	1,671	1,919
	1976	4,813	2,124	1,621	3,566
	1984	3,619	1,272	501	3,214
Row-boats	1968	2,982	1,307	567	445
	1976	2,086	1,077	463	157
	1984	n/a	n/a	n/a	n/a

Note: For 1968 and 1976, the table considers "boats" to be motor-boats, and lists row-boats separately. By 1984, departmental statistics no longer broke out row-boats; thus, the table for that year lumps together motor-boats and row-boats.

Fishermen's organizations make headway

The Maritime fishery included many species of fish, sizes of vessel, and types of gear. Conflicts between areas were rife, especially as boats got more mobile and began roaming further afield. At the outset of the 1968–1984 period, there were few mechanisms to bring fishermen together on a local level, let alone a wider basis.

Notable fishermen's organizations were scarce. Among the stronger were the Prince Edward Island Fishermen's Association and the United Maritime Fishermen's Co-operative, the latter a fishermen's organization in principle, although run mainly by a management corps.

Organizing efforts were taking place here and there. A fishermen's strike in the Canso area in 1970–1971 saw the B.C.-based United Fishermen and Allied

Workers Union trying to organize fishermen, especially trawlermen. Provincial labour laws hampered the effort; the law mainly considered fishermen as co-adventurers, not material for regular unions. To further complicate matters, the Canso effort involved an inter-union struggle. The Canadian Food and Allied Workers Union, deemed less radical than the U.F.A.W.U., won out; but the main company involved, Acadia Fisheries, closed down. The struggle, though murky in its outcome, made fishermen more conscious of organizations.[1]

LeBlanc's 1974 call for fishermen to organize stimulated more activity, for example in the Atlantic Fishermen's Association (A.F.A.), led by Dick Stewart in Yarmouth. This organization replaced the still-new Bay of Fundy Purse-seiners Association as the driving force in the Bay of Fundy herring fishery. The A.F.A. paved the way for two new organizations: the Atlantic Herring Fishermen's Marketing Co-operative (A.H.F.M.C.) and the Nova Scotia Fishermen's Association. The N.S.F.A. in turn metamorphosed into separate dragger and scalloper associations in southwest Nova Scotia.

Other formations and re-formations took place elsewhere. By the end of the period dozens of organizations, mainly small, existed in the Maritimes. Activity also increased in Quebec, where organizations eventually channelled themselves into inshore and midshore groupings.

Maritime Fishermen's Union becomes strong

Most prominent among the many new organizations was the Maritime Fishermen's Union. The M.F.U. began its official existence in New Brunswick in 1977, representing mainly multipurpose boats under 50 feet, fishing particularly lobster but also herring, mackerel, scallops, crab, and groundfish in the Gulf of St. Lawrence. Important figures in the organization's growth included Guy Cormier, Mike Belliveau, Reginald Comeau, Sandy Siegel, and the political and social activist Gilles Thériault. The M.F.U. took an assertive stance, holding demonstrations, provoking incidents, and gaining a radical reputation. LeBlanc soon recognized the organization's vigour and drive. D.F.O. decisions began to reflect M.F.U. influence. Relations with the department improved, especially after LeBlanc in 1981 created the Gulf Region.

In one early campaign, the organization set out to win a bigger share of the Gulf of St. Lawrence herring fishery, then dominated by large purse-seiners. In the early 1980's a buy-back of larger purse-seine vessels and accompanying quota adjustments transformed the Gulf herring sector into a mainly inshore fishery. Another M.F.U.-influenced change restricted groundfish fishing by vessels longer than 50 feet in the Northumberland Strait between Prince Edward Island and New Brunswick. Since LeBlanc had already pushed most of the large trawlers out of the Gulf, the smaller-boat, multipurpose fleet became more and

Three organization-builders of the period, shown in a later (2003) photo. From left, Father Desmond McGrath, co-founder of the Newfoundland fishermen's union, Gastien Godin of the Acadian Professional Fishermen's Association, and Gilles Thériault of the Maritime Fishermen's Union.

more dominant in the region. The exception was the growing crab fishery, which used midshore boats between 50 and 100 feet, mostly between 60 and 64 feet.[2]

M.F.U. wins bargaining legislation

The M.F.U. recruited a number of smart and influential fishermen as local leaders in both New Brunswick and Nova Scotia. The organization was less successful in P.E.I., although it made some inroads. The M.F.U. remained strongest on the Gulf shore of New Brunswick, where anglophone fishermen mixed with the majority Acadians.

In all three Maritime provinces, the M.F.U. lobbied for legislation to back unions. The most success came in New Brunswick. After a report by ex-D.F.O. official Lorne Grant recommended appropriate legislation, the province in 1982 enabled collective bargaining. Eventually the union won rights to bargain in the Gulf portion of the province. For some years, negotiations took place with fish-buying representatives. In the 1990's, however, the M.F.U. largely abandoned the complex government-backed process in favour of the regular rough and tumble of market bargaining, and gave its attention to other forms of representation with industry and government. Meanwhile, the M.F.U. had won a mandatory dues check-off system for their portion of the province.

At the end of the century, the M.F.U. claimed to represent more than 2,000 owner–operators, mostly in New Brunswick and Nova Scotia. Although this was only a fraction of Maritimes fishermen, still the M.F.U. was a strong force, especially in the Gulf of St. Lawrence.

M.F.U. pioneers "bona-fide" licensing system

For years fishermen had complained about "moon-lighters" and "free riders" cluttering up the industry, taking advantage of special programs, and by association pulling down the status of the more professional fishermen. In 1981, as already mentioned, the department began classifying Atlantic fishermen as full time or part time.[3] Around that time, the M.F.U. launched a campaign for the recognition of "bona-fide" fishermen in the southern Gulf. Fisherman Percy Hayne was said to have conceived the new approach, and fisherman Cameron MacKenzie helped spread it.

The scheme defined bona-fide fishermen as those holding key licences, including lobster, for their area, who made at least a certain set amount, and who got most of their income from fishing. Only bona-fide fishermen would have the privilege of accumulating licences. Since anyone could sell off a licence, but only bona-fide fishermen could acquire them, the system would eventually concentrate licences among a smaller number of bona-fide fishermen.

The new approach would require a change in the licensing rules. In many fisheries the licence still attached to the vessel. The bona-fide system would instead attach each licence to the person, and make licences transferable apart from the vessel. Thus, bona-fide fishermen could keep any licence without having to "use it or lose it"; this element favoured multi-species fishermen. The licence-holder had to be aboard during all fishing operations. Persons acquiring bona-fide status were supposed to have at least two years' fishing experience.

The department approved the new system in 1983 for vessels less than 50 feet.[4] Non-bona-fide fishermen in the Gulf were classed under what was called the "commercial" category. The bona-fide system proved popular among fishermen; its key features would later spread throughout the Atlantic, in what became known as the "core fisherman" system.

LeBlanc sponsors Eastern Fishermen's Federation

Though organization was increasing in the late 1970's, LeBlanc still had concerns about fragmentation among fishermen. In 1979, an unusual surge in the squid population was giving work to thousands of fishermen and shore workers in the Maritimes and particularly Newfoundland. But there were still huge masses of uncaught squid in the water. At the same time, co-operative arrangements with foreign interests were still common. Just before the Conservative government took power in 1979, LeBlanc took the unusual step, controversial among processors but popular among fishermen, of allocating quotas of squid to Japanese vessels in return for cash to be used for the benefit of fishermen's organizations. About $1 million went to the Newfoundland fishermen's union, and another million to a yet-to-be-created organization in

the Maritimes, the Eastern Fishermen's Federation (E.F.F.).

LeBlanc wanted to bring groups together under the E.F.F. Sixteen or so organizations joined together to take advantage of the new money. But the Maritime Fishermen's Union soon withdrew, and a number of others followed.

The E.F.F. soldiered on, under Don Landry, Allan Billard, and successive leaders. It has served as an umbrella group for, at times, two dozen or more Maritime associations. The E.F.F. undertook few major campaigns. Rather, it has served as a forum and representative body for smaller, more independent groups representing smaller boats. The E.F.F. remained active at the end of the millennium, still using the interest from the initial million-dollar squid allocation.

Community Service Officers start work

In the meantime, LeBlanc pursued other means of giving fishermen a voice and easing their dealings with government. The bureaucracy often baffled and sometimes enraged fishermen. They had to deal not only with the fisheries department in its manifold capacities—science, licensing, fishery management, Small Craft Harbours, development, marketing, inspection, and so on—but also with Canada Steamship Inspection, the Canadian Coast Guard, and others. It seemed to LeBlanc that local community service officers (C.S.O.s) in a non-policing role could cut red tape.

The department around 1977 engaged Harry Shorten to oversee the setting up of a network of community service officers. Shorten set up C.S.O.s at some 20 ports in the Maritimes. Some officials saw the C.S.O.s as a threat to fishery officers, who, they maintained, could serve the same purpose. But LeBlanc thought the department needed a friendlier face. The C.S.O.s proved useful. They never lacked for inquiries, usually on practical matters such as Small Craft Harbours.

After LeBlanc left, and as government cutbacks became the habit, the department decided it could do without the C.S.O.s; the positions disappeared by the mid-1980's.

Self-enforcement waxes and wanes

Under LeBlanc, the department undertook a new experiment in enforcement: using fishermen themselves to combat illegal fishing. The venture emerged in part from a debate over lobster size limits. D.F.O. and some in the industry thought that letting lobsters grow bigger would increase overall yields and values. Other industry members thought it was useless to talk about bigger carapace sizes if there was a danger that poaching might negate any gains.

A pilot project emerged: in 1978, D.F.O. and the P.E.I. Fishermen's Association hired several captains and vessels to help nab offenders. In 1979, the self-enforcement program spread to both sides of the

Northumberland Strait, with D.F.O. and the M.F.U. hiring 25 local fishermen as wardens for lobster District 8 in the Northumberland Strait. Results seemed good (although poachers attacked and vandalized some wardens' boats). The self-enforcement program faded away after a few seasons; again, budget pressures were a factor. As well, the union representing fishery officers was said to oppose the self-enforcement scheme.[5]

LeBlanc's various attempts to raise the status and power of fishermen saw some victories, some failures. Cumulatively, they had an impact. By the time he left office, fishermen held a new level of influence in fisheries management.

Herring fishery steers into crisis

Another venture put more power into the hands of fishermen, but over-reached itself and fell back to a halfway point.

The Bay of Fundy herring fishery exemplified the opportunities and problems of post-war development. A strong fleet of some 50 purse-seiners built up, about half on either side of the bay. Larger Fundy seiners roamed as far as the Gaspé and Newfoundland. The Gulf and Newfoundland had their own purse-seine fleets, fewer in number. Seiners mainly caught large herring for fish-meal and fertilizer, a high-volume but low-value market. By the 1960's, the high catches already worried some fishermen and scientists. When the Pacific herring fishery closed down in 1968, some 16 large seiners came around to Atlantic waters, putting more pressure on the fishery. In 1971, Jack Davis announced a freeze on licences; a limited-entry system came into effect in 1972–1973.

Meanwhile, foreign vessels were increasing their take. To cool off the international competition, I.C.N.A.F. in 1972 initiated the first multinational allocations. But within the national quota, Canadian vessels still raced to get the biggest share. Abundance suffered. From a peak of more than 500,000 tonnes in 1968, Canada's Atlantic herring catches fell to about 225,000 tonnes in 1973.

For the Bay of Fundy in particular, purse-seine catches dropped from 200,000 tonnes in 1968 to less than half that in the mid-1970's.[6] Catching ability had grown so much that seiners would take the quota in a

Soviet crewmen barrelling over-the-side herring in the Bay of Fundy, 1983. (Joint Trawlers photo)

few weeks, pouring huge volumes at low prices into the fish-meal plants. Meanwhile, problems emerged in the fish-meal market. The situation looked disastrous. The Bay of Fundy Purse-Seiners' Association, led by Stan Savage from the New Brunswick side, and southwest Nova Scotia seiners represented by Dick Stewart lobbied the department for action.

Large seiners like the *Lady Melissa*, based in the Bay of Fundy, entered the Maritimes fleet. (D.F.O. photo by Michel Thérien)

Herring-seine fishery undergoes revolution

In 1975, LeBlanc set up a team led by his special advisor, Fern Doucet, and including economist Carl Mitchell, scientist Derrick Iles, and other officials, along with Dick Stewart and fisherman Medford Matthews.

In the meantime, the Matthews family had initiated the first over-the-side sales in modern times, selling herring directly to foreign vessels on the fishing grounds. Herring were scarce in Europe, vessels there having overfished the North Sea herring; consequently, Canadian fish were gaining value. But the first over-the-side sale was an anomaly; no one was sure of the government stance, and there was no immediate follow-up.

LeBlanc travelled to Campobello to meet with Bay of Fundy seiners, and announced a subsidy of some three-quarter million dollars to tide the fleet over. In a new departure, the captains and department worked out the subsidy so that the main funding went to crew members. The Doucet group held a series of meetings with seiners, and eventually started a "club"—a marketing co-op—which at first attracted only a few, stand-fast fishermen, but finally gathered in the whole fleet. The bait was over-the-side sales: against the objections of some Canadian processors, the club got the privilege of selling substantial quantities directly to Polish vessels.

In 1976, a four-pronged program of major changes came together. LeBlanc later said he had butterflies in his stomach, but he followed the Doucet group's plan. First, the department banned the sale of herring for fish-meal, seen as a low-price market and a danger to conservation. The ban meant the end of several major fish-meal operations. But the job loss was relatively small, and the government paid compensation. The department also assisted fishermen in taking over several vessels belonging to a major fish-meal producer, SeaLife.

Second, the department authorized over-the-side sales to Polish vessels. The Poles were good sailors, bringing large vessels in amidst the tides and fog of the

Bay of Fundy. Later, Russian vessels as well bought herring for barrelling. The over-the-side sales more than made up for the lost fish-meal market. Meanwhile, most meal processors changed over to making food products from the large herring.

Third, in a pioneering move for Canadian sea fisheries, the department put the Fundy seine fleet on individual quotas, to give fishermen more security, cut down the race for fish, and let the fishermen pace their fishing to market. It was this slower-paced, longer-lasting fishery that let the processors switch over from meal to food.

Fourth, the department backed the fishermen to set up the Atlantic Herring Fishermen's Marketing Co-operative, which spread to include not only the Bay of Fundy fleet but also seiners in the Gulf and Newfoundland. Although weirs and gillnets provided some herring, the seiners' co-op now had a very strong position in the herring market, to the chagrin of some processors. In the Bay of Fundy, processors by 1977 had to do all their buying through the A.H.F.M.C. office. No longer were vessels associated with particular plants; the A.H.F.M.C. dispatched them to any and all as circumstances suited. Later, the A.H.F.M.C. experimented with fish auctions, getting processors to bid; this was another first for the Atlantic coast.

At the outset, Dick Stewart had made a toast: "Here's to a hundred dollars a tonne," an almost unthinkable figure at the time. By 1979, prices had reached more than $200. The fishery's value multiplied. As the European market declined to more normal levels, Fundy seiners began to tap into the Japanese roe herring market, although the Atlantic quality of roe fetched lower prices than the Pacific.

The Bay of Fundy venture worked for pretty well everybody, including the processors who switched over to food and benefitted from the hot European market. The project also helped to launch individual quotas as a serious management option.

But the traditional fragmentation of most Atlantic fisheries afflicted the co-op. By the early 1980's, the A.H.F.M.C.'s virtual control of supply vanished. Gulf and Newfoundland seiners returned to making their own arrangements with processors. In the Bay of Fundy itself, a group of seiners split off from the A.H.F.M.C. Central selling vanished; vessel captains returned to individual relationships with processors. Still, some permanent changes remained: a reasonably strong fishermen's organization, the switchover to food, and the I.Q. system.

In another first, the I.Q.s in 1983 became I.T.Q.s, which the fishermen could trade back and forth temporarily or permanently. This action was consistent with a recommendation by the Kirby task force on Atlantic fisheries. No licence-holder could acquire quota for more than four per cent of the Total Allowable Catch.[7] Over the next two decades, I.T.Q.s would reduce the active fleet to fewer than 30 vessels.

Meanwhile, the less numerous purse-seine fleet based in the Gulf of St. Lawrence, from both eastern New Brunswick and Newfoundland, ran into a grave resource problem in the early 1980's. Earlier, seiner quotas had run 40,000–50,000 tonnes. The Total Allowable Catch for 1981 came to only 15,000 tonnes, including only 3,000 for seiners. It was projected that even after recovery, the long-term yield would be only about 36,000 tonnes, well below earlier estimates.

After 1981, new regulations curtailed the roaming privileges of large seiners, confining Gulf and Scotia-Fundy vessels to their own areas. In the southern Gulf, gillnetters got some 80 per cent of the quota, regaining their old, pre-seiner dominance. Ministers LeBlanc and De Bané wanted to favour the smaller-boat fishery. (As already mentioned, the Maritime Fishermen's Union lobbied vigorously for the smaller boats.) The Kirby report in 1983 recommended a buy-back, aimed particularly at the nine or so larger New Brunswick-based seiners (Quebec had one seiner, and Newfoundland six). Kirby suggested that funding come from levies on the industry.[8] Instead, $6 million in buy-back money came from an Ottawa–New Brunswick economic agreement, which started reducing the fleet in 1984.[9]

Weir fishermen try catch insurance

Meanwhile, herring-weir fishermen in southwest New Brunswick, led by Ernest Wentworth, made representations about their own problems. Cheaper to build and operate than boats, the weirs—giant fish traps—fished passively along the shores, and gave the important sardine industry its main source of supply. But prices were weak in the mid-1970's, abundance seemed slack, and fish migrations within the bays and coves made catches unpredictable. Even weirs side by side could have entirely different results. A weir might have a bonanza one year and do poorly for five.

The Doucet group working with the seiners branched out to deal with the weirmen. The main results were some temporary financial assistance and

Brailing herring inside a weir. (D.F.O. photo by Bill McMullon)

what was reputed to be the first catch insurance program in the world. Departmental officials Ed Wong and Don Pepper worked out a rolling-average system to quantify landings. Fishermen were supposed to take money out of the fund in the bad years, put money in during the good years.

But the system was voluntary, and the owners of some of the best-earning weirs declined to join. The others soon depleted the fund, and the catch insurance program faded away. Still, the weirmen had got some temporary help, and they emerged with a stronger organization, which lasted well. Like the purse-seine fleet, the weirs would decline in numbers over the next two decades, but for a different reason: aquaculturists taking over many of their locations.

Over-the-side sales spread

Like individual and transferable quotas, over-the-side sales spread far from the Bay of Fundy. Fishermen in the Gulf of St. Lawrence asked permission to sell mackerel, and LeBlanc approved. Over-the-side sales also reached Newfoundland. Species involved at one time or another included herring, mackerel, gaspereau, turbot, and even cod during inshore gluts.

Processors objected, on the grounds that some way or other they might have used the fish to make shore jobs in Canada. They felt that their options were being precluded and their views ignored. They had traditionally done the selling; now the department and fishermen were short-cutting around them. To LeBlanc, the popularity of over-the-side sales with fishermen was political gold. He made it known that he "would sell to the devil" if it got a better price for fishermen.

In a few years, however, over-the-side sales slacked off. In the case of herring, Canadian processors were increasing their production, so there was less need for foreign buyers. For other species, over-the-side sales had a mixed and sometimes dubious record. In some cases, catches were less than local fishermen had thought they could supply. The department in 1984 promulgated a policy that set clearer ground rules for over-the-side sales, which by then were lapsing back into rarity.

Swordfish fishery declines

The swordfish fishery, significant in Nova Scotia since early in the century, had switched to longlining in the 1960's. Scores of boats took part. Landed values came to about $4 million in 1970.

By that time, reports had documented Minamata disease in Japan. In Minamata Bay, industrial plants had dumped mercury for many years. Mercury penetrated the fish, which local people ate daily. As the poison accumulated in their bodies, hundreds suffered disabilities such as deafness, partial blindness, and abnormal behaviour. Many died. As the Minamata problem became clear, other countries tested the level of natural mercury in fish. Tests showed that only a few species of large fish at the top of the food chain accumulated amounts that could be dangerous. One was swordfish.

The American government in 1970 put a 0.5 parts per million limit on mercury in swordfish for human consumption. Canadian fish sometimes exceeded that level. Exports to the United States practically vanished, at least the legal ones. American fishermen could and did still fish swordfish, through a legal loophole. In practice, many began buying swordfish from Canadian fishermen, who could no longer export them legally, but did so at sea, over the side, far from the eyes of Customs officers.

The Canadian government never banned swordfish fishing. The fishery continued at a low level. In 1979, the U.S. modified its regulations, raising the mercury limit to 1 part per million. The fishery built up again.

During the long migrations of swordfish, foreign vessels took most of the catch. Eventually, in the 1990's, the International Commission for the Conservation of Atlantic Tunas (I.C.C.A.T.) would begin setting swordfish quotas.

Tuna fishery becomes commercial

I.C.C.A.T. was already helping regulate Canada's bluefin tuna fishery. In the 1960's, with the Nova Scotia sport-fishery centred at Wedgeport dying away, a tuna sport-fishery surged at Prince Edward Island, especially around North Lake.

Meanwhile, improvements in air transport were making airplane shipments of fish more common in different parts of the world. Japanese consumers would pay extremely high prices for bluefin tuna. Soon Prince Edward Islanders were selling sport-caught tuna commercially. As soon as the sport fisherman, often from the United States, got his picture taken with the prize, the vessel captain would sell the tuna to an Island dealer, who would ship it by air to Japan. The recreational fishery was turning into a runaway commercial fishery. Jack Davis tried briefly to ban commercial fishing of bluefins. But, in time, the department began granting licences for commercial tuna fishing as well as sport.

P.E.I. lobster fishermen picked up several hundred licences, more than fishermen elsewhere. At first the department wanted fishermen to use rod and reel, but later permitted a restricted number of longlines. Bluefin tuna by the 1980's supported a significant Atlantic fishery; many multi-species fishermen would pursue tuna as part of their fishing year. Bluefin migrate long distances, from the Gulf of Mexico to Newfoundland, some even crossing the Atlantic. In 1969, Canada had joined with other countries in I.C.C.A.T. Bluefin tuna abundance fell off in the 1970's and early 1980's. I.C.C.A.T. in 1975 began restricting catch levels, and in 1983 started parcelling out national quotas to member countries, including Canada, the United States, and Japan.

Shellfish fisheries start to grow

Lobster fishery looks weak at outset

At the start of the 1968–1984 period, the Atlantic lobster fishery was large but in some respects languishing.

After the bonanza landings of the late 1800's, nearly a century earlier, lobster landings had never approached the same level. Limited entry and trap limits had come into place in the late 1960's, but departmental researchers remained concerned about conservation. The great majority of lobsters got caught before they ever had a chance to reproduce. Scientists warned of oncoming problems.

As of 1968, most people looked for growth in the big-boat, offshore fishery, not in the small-boat, inshore lobster fishery. Technology remained basic: the same simple traps that the fishery had used for many decades. The fishery had long seen a fair amount of turnover, with some fishermen trying lobster for a while, then giving it up. Observers looked towards a fishery that with luck might hold its own but would more likely decline. But the lobster fishery would surprise them.

In 1968, Atlantic lobster generated landings worth $24 million, well behind the $49 million worth of groundfish. Employment was high; a departmental study in 1974 found that 21,000 persons took part in the fishery, using about three million traps.

The same study found that incomes were low. As of 1973, about two-thirds of lobstering enterprises in the Maritimes grossed less than $5,000 from that species, and nearly one-third landed less than $1,000. While lobstering was not the only source of income for these fishermen, it was usually the main one. Maritimers often said that lobster was the backbone of the inshore fishery. Fishermen in Newfoundland and Quebec (except for the Magdalen Islands) fared poorer still. By contrast, salmon seiners in British Columbia grossed nearly $100,000, gillnetters nearly $15,000, and trollers more than $10,000.[10]

Licensing changes phase out part-timers

As already noted, back in the 1960's full-time fishermen had often complained about moonlighters with other jobs fishing lobsters in their off-hours or while on holiday. Licence limitation had then frozen the number of fishermen, created Class A licences for those who fished more traps; and relegated those with relatively few traps to Class B status. This put moonlighters in a secondary status. But they were still there; full-time lobstermen often raised examples such as schoolteachers or CNR employees who made a good living elsewhere. Moreover, licence limitation had had the perverse effect of increasing the number of traps in some areas, as Class A fishermen stepped up their fishing.

In 1973, the federal department set up a Lobster Task Force to examine and consult widely on the

Lobster trap limits per boat varied by district, with 300 a typical level. As of 1984, with nearly 7,000 lobster licences in the Maritimes and Quebec, probably close to 2 million traps went into the water.

Maritime fishery. Its 1974 report, besides making many "fine-tuning" recommendations, called for more research, an advisory committee system, and lower trap limits. Reduced fishing effort could improve returns, it said; effort should drop by 25 per cent or more. The task force recommended against a licence buy-back; instead, management committees and regulations should use attrition, through such means as restricting new entrants and cancelling fishermen's licences when they retired, rather than letting them sell the licence. Other licensing recommendations focussed on weeding out people with other full-time jobs, and promoting a "professional" fisherman group.

LeBlanc had the department mail out a survey to fishermen, asking their opinion about licensing changes. The department then moved ahead in 1976 with changes tightening up the fishery. People with a short fishing history and other full-time jobs got Class C licences, which were non-transferable and would lapse after two years. Those with regular employment elsewhere but more fishing history could get a Class B licence, good for as long as they wanted to fish; but the licence-holder could fish only 30 per cent of the maximum number of traps per boat allowed for his district, and the licence would die when he left the fishery. Only those dependent on the lobster fishery could get a Class A licence, which was transferable.[11]

The licensing changes aroused some opposition but more support, bolstering LeBlanc's reputation all around the coast. (Years later, however, some Native representatives would complain that Aboriginal fishermen who had been lobstering on a small scale lost out unduly, as the changes squeezed out the little operators. Even small barriers—the cost of a licence, the need of transportation to a fisheries office—could reduce participation.)

Buy-back further reduces fleet

The "A–B–C" changes had a major impact, especially when, in 1978, about 2,100 Class C licences vanished. By that time, a complementary program was kicking in. Prince Edward Island depended heavily on the lobster fishery. P.E.I.'s Minister of Fisheries, George Henderson, campaigned for a further fleet reduction through a buy-back of Class A licences.

LeBlanc in 1977 found money for a Lobster Vessel Certificate Retirement Program, intended to remove 400 licences from P.E.I. In 1978, the program got extended to New Brunswick and Nova Scotia, to remove another 1,100 licences. All told, the buy-back took out nearly 1,600 licences, at a cost of more than $5 million. By 1983, the number of Maritimes lobster licences all told was about 6,400, down from about 9,400 in the early 1970's.

The lobster buy-back may have been the most successful ever. The great advantage of a lobster fleet reduction was the controlled, measurable technology. In groundfish, herring, or Pacific salmon, advances in technology could increase fishing power even as the fleet shrank. But in lobster, fishermen could use only a certain number of traps per licence, and they had to be regular traps within certain dimensions. A fleet reduction meant a genuine reduction in fishing; fewer fishermen could share the same catch. (That being said, there were some changes in lobster gear. By the early 1980's, wire traps were beginning to replace wooden ones in some areas; they suffered less breakage, but otherwise worked the same way. Future years would eventually see better boats with better gear fishing harder.)

Following the licensing changes and buy-back, incomes improved. So did the lobster population. Scientists were reluctant to link cause and effect in lobster, and did not suggest that the lobster licence buy-back caused the increase. But for whatever reason, the 1980's would see a stupendous increase in lobster catches.[12]

Poaching subsides somewhat

Lobster poaching was an old tradition, often winked at within communities. House parties could involve feasts of short or out-of-season lobsters. But poaching seemed to begin a slow decline, even though it remained common. Observers say that when trap limits began in the late 1960's, lobster fishermen began to realize they could put out fewer traps, maybe dispense with one of their crewmen, and still make more money. Limited entry further increased their sense of proprietorship. It became clearer that the poacher was not so much outwitting the department as stealing from other fishermen.

The lobster fishery ended the period as the least "modern" of the major fisheries, still using small boats, fixed gear, and old technology (although sometimes fishing new grounds). In terms of the development mentality that ruled throughout the 1950's and 1960's, it was doing nothing right. But by the end of the century, it would turn out to be the most successful of the traditional fisheries

Offshore lobster fishery builds up

In the 1960's, an offshore lobster fishery built up off the east coast of the United States. Departmental research showed potential at Georges Bank off southwest Nova Scotia. Fishing began in 1971. The department initially limited entry to swordfishermen who had quit that fishery when American mercury regulations got in the way. By 1972, six swordfishing vessels had entered the new fishery. The department later granted two licences to other vessels. After 1976, the fleet stayed frozen at eight licences.

The size of vessels and traps, the number of traps allowed, the size of lobster caught offshore—all were larger than in the inshore fishery. Fishing took place on Georges Bank and along the continental shelf south and west of Brown's Bank. Regulations confined the offshore fishery to defined areas more than 50 nautical miles from shore, and from 1976 made use of quotas and size limits. Quotas applied first to the offshore fishery closer to Nova Scotia; after the 1984 Georges Bank boundary settlement, they were extended to include N.A.F.O. area 5Z (Georges Bank). From 1985, enterprise allocations" subdivided the overall quota among participants.

Despite increased understanding of the waters off southwest Nova Scotia, some aspects of lobster reproduction and inshore-offshore relationships continued to puzzle scientists. At times inshore fishermen feared that the offshore fishery would damage the breeding stock for inshore lobster. But as inshore lobster catches grew in the 1980's and 1990's, controversy about effects of the offshore fishery died down.[13]

Scallop fishery prospers

In the scallop fishery, the post-war period had seen the growth of an offshore fishery centred on Georges and Brown's Banks, which by the mid-1960's provided the great majority of Atlantic scallop landings. Major ports in the offshore fishery included Lunenburg, Riverport, Liverpool, Yarmouth, and Saulnierville.

Licence limitation starting in 1973 held the offshore scallop fleet at 77 vessels, mostly controlled by medium-sized or large companies. The Offshore Scallop Advisory Committee played a key role from the mid-1970's on, and gained a reputation as an excellent managerial group. Management problems in following years were relatively few. The main conservation tools besides limited entry were size limits in the form of meat counts (by 1976, no more than 40 meats per pound), and catch limits per trip and by four-month period.

In the latter 1970's, the American fleet was increasing its effort, with no regulation. Although 200-mile zones came into effect in 1977, the Georges Bank boundary dispute left no one in charge. Since increased American fishing left Canadians at a competitive disadvantage, D.F.O. relaxed its own regulations. Strong year-classes in the loosely governed offshore

Offshore scalloping on Georges Bank.

Inshore scalloping. (D.F.O. photo by Kevin McVeigh)

fishery raised total Canadian scallop catches to record levels of more than 100,000 tonnes (live weight) in 1977–1978. These subsided by 1984 to less than half that level. Meanwhile, Canada resumed strict management, especially after the 1984 Hague decision settled the boundary.

The U.S. scallop fishery had less successful management. American fishermen would sometimes sneak across the Canadian boundary line, because of the more abundant scallops on the Canadian side, and, in some instances, for the thrill of breaching the line.[14]

In the Bay of Fundy scallop fleet, independent operators continued to dominate. At the start of limited entry in 1972, this fleet had about 50 licences. In following years, representations and appeals raised the number to about 100, mostly on the Nova Scotia side. Meanwhile, more people began fishing scallops on the New Brunswick side. The department authorized a number of licences that allowed fishing out to seven miles from the shore; later these changed to "mid-Bay" licences, allowing fishing further offshore.

As for the Northumberland Strait in the Gulf of St. Lawrence, limited entry saw some 550 licences authorized. But these boats paid more attention to lobster, herring, and other species, and landed fewer scallops than the smaller, more specialized fleet fishing the Bay of Fundy and southwest Nova Scotia.

Shrimp fishery gains ground

Federal–provincial development work in the late 1960's and the 1970's helped launch a shrimp fishery in the Maritimes, mainly by aiding fishermen in testing new grounds and gear. Fishing began in the mid-1960's in the Sept-Îles area of Quebec and in the mid-1970's north of Anticosti Island and on the west coast of Newfoundland. A smaller fishery started on the Scotian shelf, mainly off southeastern Cape Breton. More fitful efforts took place to the westward and in the Bay of Fundy. The small-meshed shrimp trawls often took troublesome by-catches of groundfish, despite departmental attempts to develop selective gear.

In the Gulf, several dozen groundfish vessels were also fishing shrimp by the mid-1970's. Most were from Quebec, with New Brunswick in second place. Shrimp became the main fishery for some vessels, especially in Quebec. Limited entry prevailed by the latter 1970's, along with gear restrictions. Despite some mid-1970's difficulties, shrimp landings grew in the Maritimes and Quebec. Gulf landings from the Sept-Îles and Anticosti areas totalled about 4,500 tonnes by 1979.[15]

Crab fishery rises to prominence

The new crab fishery, first centred off the Gaspé and northern Cape Breton, by 1968 employed about 60 boats, 15 of them for the entire season. Landings came to 5,000 tonnes for nine processing plants (additional ones were being built). A 1969 government–industry seminar concluded that development efforts had been successful, and looked ahead with "reserved optimism." A tumultuous history would follow.

New Brunswick and Quebec boats, usually groundfish vessels between 50 and 100 feet, fishing with crab traps, provided most of the participation in what became known as the midshore crab fleet. These vessels sailed mainly from Gaspé ports and the Acadian Peninsula communities of Shippegan, Caraquet, and Lamèque. By 1979, they were landing nearly $11 million worth of crab. Under licence control, the midshore fleet would eventually level off at 130 vessels, about three-quarters of them from New Brunswick and the rest from Quebec. (At one point, Quebec unilaterally issued additional licences. This angered federal fisheries and contributed to the federal decision to reassume management of Quebec fisheries in 1984.) The inshore fishery developed mainly along Cape Breton's Gulf shore, where landings by 1978 reached nearly 2,000 tonnes, worth close to $7 million. Already the new Gulf crab fishery in total was rivalling the value of the historic Atlantic herring fishery.

The Atlantic Crab Association, comprising federal and provincial officers, processors, and marketers, helped guide the early fishery. This group served better for expansion than for regulation. In 1970 and 1975, the industry required emergency aid. Problems in the latter crisis included dropping catches and too much taking of "soft-shell" (close to moulting) crab; aid came to about $2 million.

The department set up a Gulf Snow Crab Advisory Committee. Regulations from 1975 limited the fleet, set a maximum number of traps per boat, and used seasonal closures to prevent landing of soft-shell crab. Of particular importance in the developing fishery were carapace size limits on the animals and mesh-size limits on the traps, designed to let females escape.

As the fishery grew, both Quebec and New Brunswick authorized new plants. The old boom and bust pattern seemed to be asserting itself. But a decline in Alaskan crab landings boosted the market, and catches kept rising. Regulations tightened up, against some resistance and malcompliance, and the department in 1984 instituted a catch quota. In the same year, the vessels switched to weekly trip limits, a forerunner of individual quotas.

A chief impetus for weekly trip limits was the demand for shore employment. The increase in plants, approved by the job-hungry provincial governments, had increased the appetite for fish and shortened the season. To enhance employment, New Brunswick enacted a law that obliged companies to put at least 60 per cent of their crab into canned or frozen products,

Fishing crab. (D.F.O. photo by Ted Grant)

rather than shipping it out with less processing. When plant managers tried to get around the rules, infuriated employees held them hostage. In this atmosphere, D.F.O. introduced weekly trip limits to avoid gluts and stretch the season. They would also improve chances for Unemployment Insurance benefits. U.I. was to be a frequent consideration in Gulf fisheries.

Meanwhile, the inshore fishery was growing in both Cape Breton and the North Shore of the Gulf of St. Lawrence. Rules limited the traps per boat to 30 and later 20, compared with more than 100 in the midshore fishery. Limited entry came into play in the latter 1970's. In 1979–81, the department instituted a Total Allowable Catch and individual boat quotas for Cape Breton areas. Boat quotas also spread to the Lower North Shore. The inshore fishery remained small, less than ten per cent of Gulf landings. Total catches for the Gulf peaked in 1982 at more than 30,000 tonnes.[16]

In the midshore fleet the Northeast Fishermen's Professional Association, later the Acadian Professional Fishermen's Association, grew up with the crab fishery and would come to wield a strong voice in management. The crab fishery would run into various storms, but more than in many other fisheries, fishermen and government would, while often bickering, work out a course together.

Biological station fosters salmon aquaculture

As noted earlier, the department was waging a long and difficult struggle to protect wild Atlantic salmon. Hatcheries formed part of the effort. In the 1970's, besides the big new Mactaquac hatchery, nine older, smaller salmon hatcheries still dotted the Maritimes and Quebec, left over from the hatchery mania of the late 19th and early 20th centuries. Although hopes of increasing abundance had faded, the salmon hatcheries still supplemented natural stocks, providing "put and take" fish for the sport-fishery, even though it was thought that "the results of hatchery operations, in the Maritimes at least do not justify the expenditures being made."[17]

An early salmon cage in Passamaquoddy Bay. (Stirling Lambert photo)

Meanwhile, research on salmon continued. Data on salmon physiology provided by Richard Saunders and others from the St. Andrews, N.B. Biological Station proved useful to Norwegian researchers, then doing work on salmon farming.

It was in Norway that salmon aquaculture took off. Arnold Sutterlin of the St. Andrews station picked up on the Norwegian experiments, and he and Gene Henderson pushed ahead with aquaculture work. In 1978–1979, a co-operative venture involving the federal and New Brunswick governments and Art Mackay, a businessman–scientist, succeeded in overwintering Atlantic salmon at Deer Island in Passamaquoddy Bay. Changes in the ocean helped them along; the lower water temperatures in the early 1970's would have killed salmon kept in net cages, but now they were surviving.[18]

Other entrepreneurs soon got into the act. By 1984, salmon farming was already a multi-million dollar enterprise. Several farms were operating in Passamaquoddy Bay, and the technique had spread across the border to Maine waters.

The Mactaquac salmon hatchery provided much of the eggs and juvenile fish for early operations while private hatcheries got going. Further research at St. Andrews helped support the new aquaculture industry, including Saunders's experiments with light ("photoperiod," the duration of daily exposure to light) to influence the growth of young. In nature, it took two years for a young salmon to smolt (that is, prepare itself for salt water); the St. Andrews station helped reduce the time to one year and even less.

Fish health units start up

Other forms of aquaculture were taking place here and there, including oyster culture in Prince Edward Island and New Brunswick, and "prairie pothole" fish farming in some western provinces. In 1973, the department held a national aquaculture conference in Winnipeg, and ten years later another in St. Andrews. But there was no major and continuing aquaculture program. Work went ahead, but mostly by individual scientists at biological stations following their own interests. A small aquaculture unit in Ottawa, headed by G. Ian Pritchard, kept an eye on developments.

As more businesses began raising and transporting live fish, a new concern arose: What if fish culture were to spread diseases? In January 1977, Fish Health Protection regulations came into effect, designed to prevent the spread of infectious diseases of salmonid fishes by controlling transport across provincial boundaries. In the Maritimes, a Fish Health Unit at Halifax now had the authority to certify establishments and approve or deny transport of cultured salmonids, eggs, or products into Canada or across provincial lines. Similar units began operating in other regions, to advise, approve, and otherwise work on fish health.

Highlights for the Maritimes and Quebec

In the period 1968–1984, the Maritime and Quebec fishery reached a new level. Traditionally associated with instability, small boats, and low incomes, the industry now looked more solid and more prosperous, even though problems remained. Landings in 1984, at 640,000 tonnes, were below the high 795,000-tonne level of 1968, but had recovered from the intervening crisis and were moving up. Value quadrupled from $103 million in 1968 to $436 million in 1984.

Incomes, too, were moving up. Back in 1968, most fishermen's incomes were clearly low (although some always managed to do well). Gross per capita landings averaged $3,900 for the 26,400 Maritime and Quebec fishermen registered by the department (not all of whom would have fished). Another indicator, the National Revenue statistics on tax-filers, showed about 7,600 persons in the Maritimes and Quebec deriving the biggest share of their income from fishing; their overall earnings averaged $2,400 that year and $3,700 in 1969, compared with $5,900 for similar "main-income" fishermen in British Columbia.

Incomes remained low in 1973. According to the *Policy for Canada's Commercial Fisheries*, average incomes of fishermen in the Atlantic region as a whole, including Newfoundland, came to $5,100 from all sources; of this, about 50 per cent came from fishing. In B.C., by contrast, the $12,800 average was more than twice as high, and fishermen got 70 per cent of their income from the occupation.

In the Maritime provinces, average household incomes that same year ranged about $7,000–$8,000. But in households where fishing was the first or second employer, the average fell to only $5,400 in P.E.I., $5,900 in New Brunswick, and $6,400 in Nova Scotia. In Quebec, the figure for fishing households was $6,200, around the same level as in the Maritimes, but the disparity with other occupations was greater; in that more metropolitan province, the income of the average household was $9,100, higher than in any of the Atlantic provinces.

By 1984, the average gross revenues for Maritime and Quebec fishermen, on a per capita basis before expenses, came to $13,800, more than triple 1968 levels. As for net incomes, the 16,600 "main-income" fishermen identified by National Revenue averaged $17,700 from all sources, about five times the 1968 level, outpacing the $14,300 average in B.C. Many trawlermen and some independents made well above the Maritime–Quebec average; many other fishermen of course made much less, in the highly varied industry. Still, a lot of fishermen were moving ahead.

Roméo LeBlanc dominated the period. LeBlanc elevated the status of independent fishermen, but without taking fish away from the larger, "offshore" companies. Instead, licences, individual quotas, and Enterprise Allocations stabilized catch shares. Groundfish catches increased for everyone; the herring fishery leaped ahead; lobster fishermen had fewer boats to compete with; and shellfish fisheries in general grew stronger. The period saw many controversies, but by 1984 most

things appeared to be working out. Abundance had increased, and it seemed that strong science and management could keep producing rich harvests. The fishery seemed at last to be reaching the long-sought goals of higher volume, higher value, and more stability.

Along with more money, fishermen had more power. The Maritime Fishermen's Union, the Eastern Fishermen's Federation, the Acadian Professional Fishermen's Association, the P.E.I. Fishermen's Association, the Grand Manan Fishermen's Association, the Bay of Fundy herring seiners' "club"— all had gained strength. Advisory committees had become a major influence on management.

The general situation looked as good as it ever does in the fishery. When I grew up on a Bay of Fundy island in the 1950's, grown-ups generally advised us to stay away from fishing and to move away where prospects were better. By 1984, more fishermen were happy to see their children choose the fishing life.

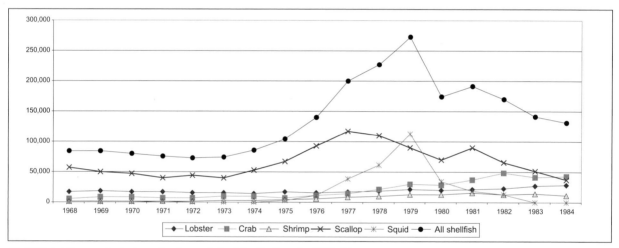

Shellfish landings, selected species (tonnes), for Atlantic Canada, 1968-84.

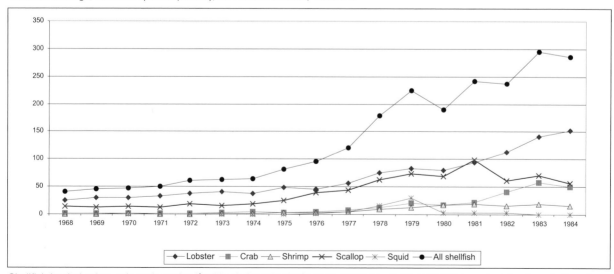

Shellfish landed value, selected species ($millions), for Atlantic Canada, 1968-84. Value kept rising even as landings flattened.

CHAPTER 24.
Newfoundland, 1968–1984

I n Newfoundland, the war and Confederation had brought more urban and business growth. But the fishery remained the fundamental industry, a constant source of despair and hope. Groundfish dominated by far, and the events tossing and turning that industry, already described, affected Newfoundland more than any other region.

After Smallwood's many development schemes, the Conservative government under Frank Moores from 1972 to 1979 created a calmer atmosphere in the province. In the fisheries, John Crosbie, serving in the provincial government as both finance and fisheries minister, exerted a steadying influence during the groundfish crisis of the mid-1970's, and had a good professional relationship with LeBlanc. When Crosbie moved to federal politics in 1976, Walter Carter became fisheries minister. At the time of the 200-mile limit, Carter allied with Dan Reid of Nova Scotia in pushing federal fisheries for a huge fleet expansion, which was refused.

Brian Peckford replaced Moores as premier from 1979 to 1988. As Moores had already begun to do, Peckford campaigned for more provincial control of the fishery, without making much of a dent on LeBlanc or his successors.[1] Jim Morgan, a long-serving fisheries minister under Peckford, proved colourful, congenial, and vocal. But provincial demands for expansion abated when the trawler-company crisis hit the industry. Although processors were a provincial responsibility, everyone looked to the federal government for the money and strategy to resolve the situation, as already described.

Development efforts find some success

While departmental development work in the 1968–1984 period was losing some of its vigour in other regions, it stayed strong in Newfoundland, on sea and shore. A "net-bag" system promoted by the department made unloading easier in some outports. Boxing, icing, and the use of containers increased. Many other projects went ahead, including significant improvements to dragging gear.

The provincial government also sponsored development projects—for example, helping to build up a fishery for lumpfish roe. In the wake of the Kirby report, the province got involved in trying out Scandinavian-style longliners in the cod fishery, although these never worked as hoped. Among fishermen, gillnets became ever more popular, attracting many away from hooks and cod-traps, and heightening problems of wastage from untended nets and ghost fishing by lost ones.

Capelin, which could be dipped from the surf during spawning runs, became the subject of a Canadian development effort and a substantial foreign fishery.

Newfoundland fishermen's union becomes strong

After the war, Smallwood had sponsored a Newfoundland Fishermen's Federation, which never took strong hold. The real successor to William Coaker's great union of the early 20th century emerged in the 1970's, founded by the unlikely combination of a lawyer–politician and a parish priest.

Richard Cashin

With groundfish prices sliding, Father Desmond McGrath in 1969 began working with fishermen in the Port au Choix area. McGrath called up Richard Cashin, a lawyer and former M.P., and invited him to the area. A descendant of the Newfoundland merchant and political aristocracy, Cashin decided to work with the fishermen. He proved to be an energetic leader and fiery orator. Cashin and McGrath took the new Northern Fishermen's Union into a 1971 merger with the Canadian Food and Allied Workers Union, which had organized a few trawler plants. The new entity became the Newfoundland Fishermen, Food and Allied Workers Union, affiliated with the Chicago-based United Food and Commercial Workers. Cashin became president; leading officials included Ray Greening, Kevin Condon, and Bill Short.

There followed a period of dynamic organization and dramatic strikes among inshore and trawler fishermen and trawler plant workers. In 1971, a drawn-out strike at the Lake Group's plant in Burgeo transfixed the province. Cashin in one instance leapt aboard a company vessel leaving the dock for the fishing grounds, and persuaded captain and crew to tie up. In another, the plant-owner's son cut a rope the strikers had strung across the harbour, almost causing a violent confrontation. Company president Spencer Lake said he would not "be dictated to by priests, lawyers or gangsters from Chicago." Lake also remarked, "In these isolated outports I contend there is no place for a union. ... You haven't got the local leadership to run them intelligently, with all due respect to the people—I'm very fond of them."[2]

But the union came out the winner. Lake left Burgeo; the new provincial government under Frank Moores took over the plant, signed an agreement with the union, and found a new operator.

Meanwhile, in the dying days of Smallwood's administration, the province in 1971 had given bargaining rights to inshore fishermen, adding strength to the union movement. A trawlermen's strike in 1973 turned into another victory. A provincial report by Dr. Leslie Harris deemed trawlermen to be employees rather than co-adventurers, and raised the idea of reasonable wages to a new level. Throughout the 1970's the union seemed to win every contest with plant owners. It also won better workmen's compensation coverage for fishermen, and set up a benefit system of its own for members.

The union took encouragement from federal minister Roméo LeBlanc's support for fishermen. Cashin established a good relationship with the minister. More often than not, the union got what it wanted; in one example, the department authorized the switchover of a number of longliners on the southwest coast to dragging, even though the latter was considered a less conservationist technology. LeBlanc also authorized over-the-side sales as in the Maritimes. LeBlanc told union members in a speech, "[Y]ou have more power than Coaker ever dreamed of." In turn, the union generally supported LeBlanc. When several Atlantic provinces pushed in the late 1970's for greater fisheries jurisdiction, the union, like other Atlantic fishermen's and processors' organizations, favoured the federal government.

Capelin fishery grows, herring fishery shrinks

Every year, capelin crowded Newfoundland bays and coves, casting themselves on the beaches as they spawned. Some were fished, mostly by traps, for food, fish-meal, or bait—perhaps more than 20,000 tonnes in the early 20th century, dwindling to less than 5,000 tonnes in the 1960's. Great abundances went unfished. Meanwhile, starting in 1971, Soviet vessels began to take major quantities on the Grand Banks.

Canada moved to get into the picture. Under federal minister Jack Davis, an effort went forward to develop the capelin fishery in the early 1970's. Experimental work took place on fishing, product

development, and promotion. But the products never caught on in any big way. Meanwhile, a strong constituency in Newfoundland opposed fishing capelin, on the grounds that it was the food of the cod.

A capelin fishery did develop slowly. The department set quotas to match the Japanese market; fishermen took only a small portion of the large resource. Canadian landings grew from the range of 4,000 tonnes in 1972 to the 20,000–40,000 range in the early 1980's.

The herring fishery, which had expanded hugely in the 1960's on the basis of a temporary super-abundance of Gulf herring, now dwindled. Landings dropped to the 40,000–60,000 tonne range in the early

1970's, and to less than 10,000 in 1983 and 1984. Meanwhile, large purse-seiners in the mid-1970's did more fishing on the east coast of Newfoundland. Smaller boats in the area took up the related technique of ring-netting. Rising prices for herring encouraged this movement. The department began restricting seiners to the south and west coasts.[3]

Crab and shrimp fisheries grow

Lobster remained a significant fishery in Newfoundland, though minor in volume compared with the Maritimes. A bigger shellfish fishery was coming on stream.

The crab fishery began in Newfoundland around 1968, later than in the Maritimes. The first commercial landings came as by-catch on the northeast coast. Fishermen then began targeting crab. The fishery spread in the early 1980's to southern Labrador and the south coast of Newfoundland. By 1984, landings amounted to some 9,600 tonnes, less than a third of the Maritime and Quebec catch. But for Newfoundlanders, crab were already giving three times the catch volume of lobster, although at $7 million, only half the value. By the end of the century, Newfoundland crab would not only out-value lobster many times over, but also outmatch crab in the Maritimes and Quebec combined.

Midshore fleet picks up speed

While the midshore fleet flourished in the Maritimes and Quebec, Newfoundland's fleet continued to be strongest at the upper and lower size ranges. But the mid-sized fleet gradually speeded its growth.

As of 1976, Newfoundland still had only about 570 vessels in the ten-ton to 100-foot range. By contrast, Nova Scotia and New Brunswick together had more than 2,300, and Prince Edward Island and Quebec another 400 or so. By 1984, however, Newfoundland had nearly tripled its mid-sized fleet to about 1,560. By then, of course, the Maritimes and Quebec mid-sized fleet had itself roughly doubled, to some 5,300. Newfoundland's mid-sized fleet was now growing faster than that of the Maritimes and Quebec, but still lagged behind.

As for large vessels over 100 feet long, as of 1984 the province had 90 such craft, far more than Quebec, P.E.I., and New Brunswick combined, and close to the 124 in Nova Scotia.

Newfoundlanders continued to operate a huge

Part of the growing midshore fleet: a 65-foot dragger built in the early 1980's and based at Port au Choix on the northwest coast.

fleet of smaller craft under ten tons. In 1984, these numbered about 14,700, far more than in the other four provinces put together (about 8,600). All told, Newfoundland in 1984 had some 16,000 fishing craft, more than half of the 31,000 in Canada's total Atlantic fleet.

Table 24-1. Newfoundland fishing craft by size category, selected years.

Nfld./Year	over 100 ft	10 tons to 100 ft.	Boats <10 tons	Row-boats
1968	75	422	10,451	4,349
1976	84	567	8,700	250
1984	90	1,561	14,674	n/a

Note: For 1968 and 1976, the table treats "boats" as motor-boats, and lists row-boats separately. By 1984, departmental statistics no longer broke out row-boats; thus, the table for 1984 lumps together motor-boats and row-boats.

Parallelling development in the Maritimes and Quebec, shrimping began on the west coast of Newfoundland in the late 1960's and early 1970's. The industrial development branch in 1967–1969 carried out exploratory shrimp fishing, ranging from St. Mary's Bay on the Avalon Peninsula along the south coast and up the west coast of the island. It found concentrations of shrimp near Port au Choix, and in 1970 contracted three vessels in the 40- to 50-foot range to trawl for shrimp. Other vessels moved into the fishery, 17 of them by the end of 1971, with the branch providing technical assistance. A plant started up in Port au Choix; it would be the only shrimp processor on the island for about 20 years.

Here as elsewhere in Atlantic Canada, the common species of shrimp was *Pandalus borealis*, "northern shrimp." But by the late 1970's, when people mentioned the northern shrimp fishery, they usually meant specifically in northern waters off Labrador and in the Davis Strait between Baffin Island and Greenland. Foreign vessels in the 1970's had begun fishing in the Davis Strait. The department in the mid-1970's sponsored exploratory fishing in what were about to become Canadian waters with the 200-mile limit.

The explorations found good amounts of shrimp off Labrador in the Hawke Channel. Meanwhile, foreign vessels had already fished further north in the Cartwright and Hopedale Channels. The department in 1978 contracted explorations still further north in the Davis Strait.[4]

The potential was clear. Department officials assessed the situation to figure out the best disposition of licences. The usual pattern had been first come, first served, but in this case minister LeBlanc and the department decided to spread the licences around the Atlantic provinces. By 1978, the situation settled out with five licences for Newfoundland interests (of these, two went to co-operatives in Labrador, of which one, the Labrador Fishermen's Union Shrimp Company Ltd., was affiliated with the Newfoundland Fishermen, Food and Allied Workers), two and a half licences for Nova Scotia, one and a half for New Brunswick, and two for Quebec. Thus every province got a share of the 8,100-tonne quota except for Prince Edward Island, which traditionally had no far-ranging fishery. In 1979, the department granted another licence to the Makivik Corporation, an Aboriginal organization in northern Quebec.

In Davis Strait, the shrimp were in part a transboundary stock. From 1977 to 1981, Canada pursued complex negotiations with the European Community (in respect of Denmark and Greenland). This produced a degree of shared management, which fell apart after 1981. The two sides then managed their own fisheries, seemingly with no adverse effects.[5] Meanwhile, the department allowed Canadian companies for the time being to charter foreign freezer trawlers; Canadian fishermen would accompany the foreign crews to learn the fishery, which they would later take over. Individual quotas applied from early on. By 1979, the Labrador

and Davis Strait fishery was yielding 5,500 tonnes, the Gulf of St. Lawrence 8,000 tonnes, and the Scotian shelf 800 tonnes, for an Atlantic total of 14,300 tonnes. In 1984, Newfoundland's local shrimp landings and her share of northern shrimp totalled 3,300 tonnes, worth $4 million.

Newfoundland moves slowly ahead

Comparing 1968 and 1984, and setting aside the major downturn in between, total Newfoundland landings rose slightly to 450,000 tonnes. But landed value increased fivefold, from $29 million to $164 million. The foreign trawlers that used to take more than three million tonnes a year in the northwest Atlantic were dwindling. Canada could now control the zone and, it seemed, keep building up the fish stocks.

The number of Newfoundland fishermen rose from 19,350 to 27,600. Despite more people splitting up the catch, the average return before expenses also rose, from $2,000 to $5,900.

What about net incomes? As noted earlier, the federal tax system yields data on those persons whose biggest single source of income is fishing. These "main-income" fishermen are always fewer than the number of registered fishermen, since many part-time and casual fishermen earn more elsewhere. As well, up into the 1960's, many Atlantic fishermen had never bothered to file income taxes. But as more cash circulated, the revenue department gradually caught up with them.

In Newfoundland, the tax system in 1973 showed 3,500 fishermen, averaging $5,800. By 1984, with the tax system logging far more returns, 11,100 "main-income" fishermen were averaging $10,100 from all sources. Although low compared with, for example, the national average for employees ($21,118), farmers ($15,855), government employees ($24,500), or fisher-

Squid drying. The brief squid boom of the late 1970's helped raise the numbers of registered Newfoundland fishermen.

men across Canada ($14,487), still this average was an improvement. And trawlermen in particular might make several times that level. Meanwhile, as earnings rose, the higher benefits from U.I. were softening the normal ups and downs of the fishery.

Atlantic seems set for prosperity

In 1984, the fishery in Newfoundland and all the Atlantic Provinces seemed on the upswing. Total catches in 1984 came to nearly 1.1 million tonnes, down from the nearly 1.3 million tonnes of 1968, but roughly 30 per cent higher than the mid-1970's low point. Meanwhile, landed value in the 1968–1984 period rose fivefold, from $116 million to $599 million. This well exceeded the inflation rate.

On the processing side, the restructured groundfish companies, Fishery Products International and National Sea Products, appeared solid. The newer Clearwater Fine Foods, which dealt mostly in shellfish, was becoming ever bigger. Medium-sized and small companies were enjoying a fair degree of growth and stability.

Management was more comprehensive than ever. Scientific research and knowledge had increased. The rising catches, especially for northern cod, seemed to validate both science and management. The uncertainties of scientific estimates, however, often went under-emphasized by scientists and overlooked by others.[6]

Knowledge of industry operations was also substantial. The department collected statistics on many aspects of the industry, constantly assessed markets, and through fishery officers, inspection officers, the Fisheries Prices Support Board, and the Canadian Saltfish Corporation, enjoyed a daily insight into operations.

The basic policing system appeared strong. Fishery officers now received classroom training. An area-manager system had moved more authority into a dozen or so local headquarters, at such places as Grand Falls, Tracadie, and Yarmouth; these in turn oversaw sub-offices and local stations. LeBlanc's amendments to the Fisheries Act had raised fines for pollution and such offences, and the department was working harder to protect habitat. In administration generally, the formation of the new department and the administrative set-up put in place by deputy minister Don Tansley had brought relative stability.

Questions of how to use the fish had grown more complex and argumentative. But improvements had taken place. It seemed that with limited entry holding the fleet steady, the 200-mile limit and comprehensive management could increase incomes all around. The Kirby report, with its objectives of viability, jobs, and Canadianization, had solidified the picture of an industry whose problems were those of growth rather than decline.

Most fish stocks seemed in reasonable shape. Optimum Sustainable Yield had become firm policy, expressed in finfish through the $F_{0.1}$ rule. Among shellfish, lobster still worried scientists, but licence limitation, trap limits, and size limits were providing control, and crab and shrimp looked in good shape.

With limited entry and quotas, access and allocation had become more stable, although highly contentious. Fishing success was beginning to depend not just on the abilities of the fisherman but also on government-given quotas, zones, and licences. Some of the competition in the fishery had shifted to bitter battles over allocation, although for the most part, once the industry and government had worked out percentage shares for different vessel classes, shares remained fairly stable.

The "rationalizers" in industry and government had begun to shift their thinking towards individual transferable quotas. I.Q.s and I.T.Q.s seemed to hold the promise of moving the industry towards better cost-effectiveness and more profits over time, with less government interference.

The handling and quality of fish had improved markedly. The pitchfork was at last disappearing from many fisheries; ice and containers were becoming the norm. D.F.O.'s marketing studies and promotional work, while difficult to quantify, helped create an attitude change that made fish and seafood more desirable.

Development work slackened in the period, and switched away from the volume-centred post-war approach. But the department was still helping to build new fisheries, such as crab and shrimp. The industry itself now had more of a critical mass to undertake new ventures.

LeBlanc in his first speech had spoken of obtaining "optimum social yield." After he left the portfolio, the fishermen-first ethos lost some of its primacy but remained a strong current in departmental thinking. The separate fleet rule still worked against corporate concentration. Fishermen now had higher incomes, and organizations and advisory committees had increased their power.

Before Davis and LeBlanc, the Atlantic fishery to most people looked to be a low-income occupation, the employer of last resort. If nothing else, you could always fish. After limited entry, one could often see fishermen making more money than people with year-round jobs on shore. In some cases people on the outside were clamouring to get into the fishery and couldn't.

As of 1984, the future seemed promising. Crab, shrimp, and other fisheries were growing, and groundfish remained the great hope. The opinion was widespread, not only in the department but in the Atlantic industry, that Canada had the best fisheries management in the world. But following years would rock the department.

Fishermen still lag in incomes

It was clear towards the end of the period that full-time Atlantic fishermen were doing better. Most owned their own houses, more could take vacations elsewhere, some did well indeed. Yet statistics showed them still lagging behind most Canadians in cash income.

As of 1981, four years after the 200-mile limit, with the Atlantic industry doing relatively well, the 35,850 self-employed fishermen averaged $5,800 from fishing, $8,100 from all earnings, and $11,900 from all sources, including $2,500 from U.I. The 4,600 wage-earning harvesters averaged $4,000 from fishing, $5,800 from all earnings, and $9,100 from all sources, again including about $2,500 U.I.

The 26,200 "main-income" Atlantic fishermen (those for whom fishing was the single biggest source of earnings) did better, as shown in the table below. As of 1981, they averaged $12,530 from all sources of income. But even they were far behind the national average for employees ($17,300), farmers ($15,500), or armed forces members ($20,300).[7]

Table 24-2. Earnings of "main-income" fishermen tax-filers, Atlantic and Pacific, 1968–84 ($000's).

Year	MARITIMES AND QUEBEC			NEWFOUNDLAND			BRITISH COLUMBIA		
	Total number fishermen	Average earnings fishing	Average income overall	Total number fishermen	Average earnings fishing	Average income overall	Total number fishermen	Average earnings fishing	Average income overall
1968	7,593	---	2.4	742	---	4.2	4,461	---	5.9
1969	5,476	---	3.7	981	---	4.3	3,186	---	5.9
1970	5,035	---	6.3	1,455	---	5.1	4,764	---	6.3
1971	4,724	---	5.0	1,279	---	5.9	3,775	---	6.6
1972	8,124	---	5.7	2,992	---	4.6	4,593	---	9.1
1973	8,408	---	6.6	3,546	---	5.8	5,400	---	14.5
1974	8,619	---	7.0	3,386	---	5.9	11,152	---	12.6
1975	8,601	---	8.3	2,708	---	5.4	3,588	---	9.3
1976	10,736	---	7.9	5,577	---	4.9	6,413	---	12.5
1977	11,526	---	10.0	6,609	---	6.3	7,082	---	13.8
1978	13,504	---	12.6	8,837	---	7.0	7,113	---	19.0
1979	15,127	9.3	13.1	12,458	4.3	7.5	7,914	15.1	21.0
1980	19,227	---	12.2	12,061	---	8.1	6,248	---	12.1
1981	16,095	---	14.6	10,120	---	8.8	6,632	---	15.1
1982	16,402	---	15.7	13,515	---	8.7	6,799	---	15.8
1983	16,264	---	17.6	11,666	---	9.8	7,351	---	14.2
1984	16,590	---	17.7	11,121	---	10.1	6,710	---	14.3

Note: This table represents tax-filers for whom fishing was the biggest source of earnings. The number of registered fishermen was higher, and included those who earned more elsewhere and some who never fished at all in a given year.

CHAPTER 25.
On the Pacific, 1968–1984

Two bold actions dominated Pacific fishery management in the 1968–1984 period: the imposition of limited entry, and the launching of the huge $150-million Salmonid Enhancement Program. Several factors coincided to bring action.

Strong voices in the commercial fishery were calling for limits on the fleet and a build-up of the resource. The recreational fishing industry, burgeoning with more and more boats and sport lodges, also wanted more fish. And increasingly, the public at large wanted salmon protection and enhancement. People had seen the threats to rivers and streams as humans altered temperatures, turbidity, flows, and chemical composition of the water. By the 1960's, public opposition was heading off major hydro projects. But the industry and public were also sensing promise from fishways, spawning channels, hatcheries, and other forms of enhancement.

The department itself was strong in the late 1960's, with a Whitmore-bred cadre of managers in place. The enforcement fleet included three ships, the *Howay* (till 1981), the *Laurier*, and the *Tanu*, and about 30 small craft. Rod Hourston had succeeded Whitmore as regional director in 1960, and he stayed in the post for 14 years. This meant that for almost all the time from Major Motherwell's arrival in 1919 through the Whitmore period to Hourston's departure in 1974, for all practical purposes only three regional directors ran the region. The stability seemed to work well; the accumulated knowledge of regional directors provided support and a safety check.

Glen Geen, brought in from university ranks, directed the region from early 1975 to 1977. Geen made various organizational changes, in particular trying to incorporate the Pacific Biological Station more closely into the management system. His departure came sooner than expected, after a contretemps with senior managers in Ottawa. Another official later noted, "They brought Geen in as a hatchet-man, but he turned the axe the wrong way." Wally Johnson, a distinguished scientist, then directed the region in 1977–1981. During those years, an exodus of senior management people—only a half-dozen or so, but important in the small world of fisheries—subtracted some ability from the region. Wayne Shinners, an experienced and capable manager from the Atlantic side, then served as regional director-general in 1981–1985.

Despite ups and downs, the Pacific Region organization retained good strength throughout the period. At the outset, under Jack Davis, it was at a peak. Davis himself was an action-oriented intellectual, and was part of a majority government under an activist prime minister. The time was ripe to do things.

Herring and salmon gillnetters continued to make up the bulk of the B.C. fleet.

"Davis Plan" tackles the salmon fleet

Since 1948, the United Fishermen and Allied Workers had pushed for licence limitation, both for conservation and to protect incomes. Resentment of moonlighters, with no real stake in the fishery, played a part. As noted earlier, Minister Jimmy Sinclair had in the 1950's promised limited entry, and a 1960's report by university professor Sol Sinclair had fleshed out the concept. The herring collapse in the late 1960's had shown the consequences of an overpowered fleet, while salmon enhancement work had raised hopes for resource growth. The idea took hold that catches could double, back to historic levels of some 300 million pounds (136,000 tonnes).

Part of the answer was to reduce fishing pressure—in phrases that were gaining currency, to solve the problem of "too many boats chasing too few fish" by "matching the fleet to the resource." In 1966, Minister Hédard Robichaud had announced that licence limitation might go forward for 1967, but the resulting questions and commotion in the industry had made him pull back.[1]

Davis forged ahead. In September 1968, he announced controlled entry to the salmon fleet as of 1969. The department would license vessels in two categories: Class A licences for those with a specified level of landings, "B" licences for those with less. Licences in both categories would be renewable annually, and for both, if the boat got sold, the licence could go with it. The department would allow no replacements for "B" vessels; when the vessel died, so would the licence.[2]

Under pressure from those who felt left out, Davis relaxed the program in November. Now, any vessel that had landed any species at all could get an "A" or "B" salmon licence, depending on landings. The announcement noted that "the Minister's statement opens the door to halibut, herring, groundfish and shellfish boat owners to fish for salmon." This would allow about 160 additional vessels to fish. Thus, some larger herring and halibut vessels with minimal previous involvement in salmon got "A" licences. A new "C" licence was to apply for non-salmon boats; owners could replace these boats without restriction.[3]

Limited entry soon followed for other fisheries: in 1974, roe herring ("H" licence); in 1976, trawl ("T") licences and the now-limited "C" licence (which allowed fishing a number of species not individually controlled); in 1977, abalone and shrimp ("E" and "S" licences); in 1979; halibut ("L"); in 1981, sablefish (black cod—"K" licence); and in 1983, spawn-on-kelp and geoduck ("J" and "G" licences). Also in 1983, a "Z" licence came into effect for certain specified fisheries.

Meanwhile, around the time salmon licensing began, Davis, in response to Native representations, created the A-I licence—a low-fee licence restricted to Indian fishermen (although some Native people just kept their previous "A" licence). And October 1968 saw the beginning of a new five-year Indian Fishermen's Assistance Plan, which would provide $4.6 million from the Department of Indian Affairs and Northern Development (D.I.A.N.D.). Loans and grants would go to B.C. Indians for building and buying vessels, gear, and shore facilities, and for training.[4]

License the vessel or the man?

At the outset of the limited-entry program, despite previous studies and discussions, Davis and his officials were still to some degree figuring things out as they went along. Should limited entry control the numbers of fishermen or of boats? Should it go by species or by gear type? When a fisherman ceased using a licence, should he be able to transfer it to someone else, or should it go back to government?

Sol Sinclair's 1960 study, mentioned earlier, had recommended limiting the number of licences, as well as hiking the fees to deter casual fishermen. The department should impose a five-year freeze, Sinclair said, and during that time remove any licences that went unused for two years. The department should also determine the proper number of licences for the fishery, and after five years auction them off. The licences should be transferable. New fishermen would enter by purchasing rights from old ones.[5]

This all seemed simple. But when Robichaud in 1966 had made his short-lived announcement foreshadowing licence controls, confusion had arisen about who would hold the licence. The owner and operator could be two different people.[6]

When Davis froze licences in 1968, the announcement said the licence would go on the vessel, and accompany the vessel if it changed hands. But the United Fishermen and Allied Workers opposed licensing "things" instead of "people." If the owner got all the proceeds from a licence sale, how was that fair to the

The Davis Plan aimed to reduce fishing power. Things turned out differently.

crewman who had spent 30 years in the fishery and was left with no equity? The union charged that the new measures would foster corporate control as processors bought up the fleet. The licence should instead go on the individual and be non-transferable.[7] The U.F.A.W.U. demanded that the whole issue go before the parliamentary Standing Committee on Fisheries and Forestry.[8]

Davis wrote a public letter to the union, making clear that the intention was to control vessels. By contrast, the union plan to license people would set up a select club, where seniority would rule and new entrants would be frozen out.

Davis noted that the department's plan would make vessels and licences more valuable; this would work to the advantage of older fishermen getting out of the business. And he disagreed with the union's concept that over half a fisherman's annual income should come from fishing. Salmon fishing was seasonal; if some fishermen could find more work elsewhere, fine. Davis added that he would be watchful about corporate concentration.

Davis believed firmly in his plan. At one point he told fishermen in Prince Rupert that "this commercial salmon fishing vessel licensing scheme is unique in the fisheries of the world. It is a world first. And you, I know, will make it a world beater."[9]

The licensing question did go before the Standing Committee on Fisheries and Forestry in 1969. Many fishermen's groups testified; they tended to favour controls, although some warned that the program would create a more powerful fleet. The U.F.A.W.U., making its arguments for licensing the man, said that waiting lists should apply; fishermen, government officials, and university people could work out guidelines for entry to the fishery. "We do not see why people should have to buy their jobs ... in the fishing industry." Nor should a licence-holder be able to dispose of that licence as he saw fit: "He has no right any more than any other Canadian, to a special right within a restricted resource which in the first place belongs to all the people."

Under questioning, union leader Homer Stevens acknowledged that the U.F.A.W.U. would like to unionize every boat, and if the union controlled who could fish, "it would not be a bad thing," although there was no closed shop at the moment. The U.F.A.W.U. never spelled out the mechanics of transferring vessels to new entrants.[10] Still, the union had put its finger on key points—transferability, and the status of fishermen—that would keep dogging the new licensing regime.

Davis stuck to his plan of putting the licence on the boat. There was no "owner–operator rule" like that which emerged on the east coast. The person holding the licensed boat was not obliged to fish it himself. Someone else could do that for him if desired, and might pay for the privilege of using the boat and licence.

When the fisherman sold the boat, the licence would normally go with it. Officials of the day recall that one of Davis's uppermost thoughts was that when a fisherman retired, or died, selling the licence would give him or his widow a form of pension.

A related point got little notice at the time. As licences took on a value, so would the government take on a moral obligation to compensate people in future fleet-reduction programs. As already noted, it would become common for government to "buy back" licences it had issued and still in theory owned.

Corporate takeover fails to emerge

As noted earlier, on the Atlantic from 1979 a "separate fleet" rule blocked corporations, including processors, from acquiring licences in the under-65' fleet. By that time, the west coast system had already taken shape, with no such controls. Anyone could buy a boat and licence, or several of them, and rent them out to whomever he chose. But at the time, there was no major corporate takeover such as Homer Stevens and the U.F.A.W.U. had feared. Why not?

The answer appears to be partly business factors, partly moral suasion. The department set itself a guideline that corporate ownership should rise no higher than its current level of about 12 per cent of the fleet. If company holdings seemed likely to exceed that figure, department officials would talk to company officials, and the companies would divest themselves of some older, smaller boats and their associated licences. This was easy to do in the 1970's, when purchasers were buying up old boats and licences to "pyramid" their fishing power on newer boats.[11]

Besides, although they competed fiercely for fish, the Pacific canners at the time never showed that strong a desire to run their own fleets. They could get fish from the largely independent fleet, without the headaches of maintaining thousands of boats. As on the Atlantic, processors backed many fishermen with financing for boats and gear; this gave them a certain power, without the responsibilities of ownership and maintenance. And where else could the fishermen sell their fish? They would have to come back to the handful of major companies. Whatever their exact reasoning, the salmon canners made no major effort to control the salmon fleet, and in some cases sold off vessels, although they maintained a powerful segment of ownership.

Reduced fleet gets stronger

The department began talking about the Davis Plan as a four-phase process: first, imposing the licence freeze; second, reducing the fleet's size; third, improving the safety and quality standards of vessels in the fleet; and fourth, moving towards an optimum fleet in terms of gear, area, and other arrangements.

Phase 2, reducing the fleet, seemed imperative, but the department's attempts would have unexpected results. As elsewhere, better boats, nylon lines and nets, power blocks and drums, and better electronics had strengthened fishing power. Already, from the

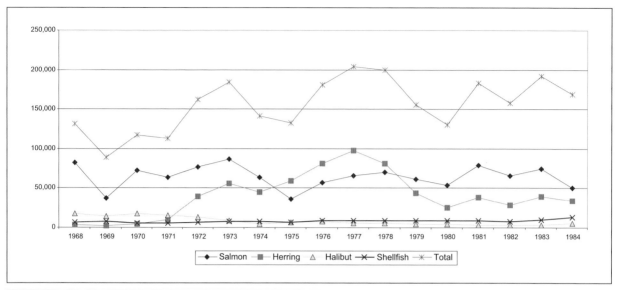

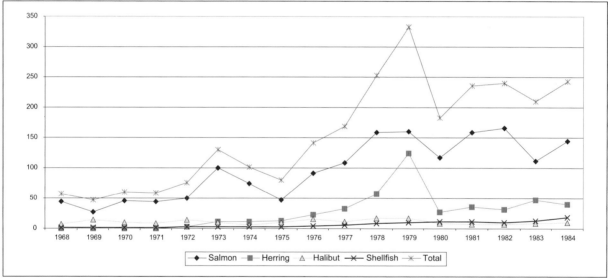

Top graph shows total landings and selected species (tonnes) in British Columbia, 1968-84. Bottom graph shows landed values ($millions). The latter-1970's jump in values boosted the fleet's power, with many new vessels.

early 1950's to the early 1970's, the value of the salmon fleet had roughly tripled to about $100 million, while catches rose only 89 per cent. During that time, new fishing techniques and freezing capabilities hiked the troll fleet's share of the catch from 20 per cent to 35 per cent, at the expense of gillnetters and especially seiners.[12]

Some fleet reduction took place at the outset of the Davis Plan, at least on paper. In November 1968, Davis pointed out that there were now only 7,000 of the category A and B salmon boats, a reduction of 1,200 since the licence scheme went into effect on September 6.[13]

In January 1970, Davis announced that "a sharp increase in category A commercial salmon licence fees

... will fund a buy-back program to reduce the salmon fleet." Licence fees rose to $100 for vessels less than ten tons, $200 for larger ones. "Company-owned boats, frozen at 800 last year, or about 12 per cent of the total commercial salmon fleet, will be reduced at a rate parallel to the reduction in the fleet." The same announcement put a ten-year limit on B licences, speeding up their rate of departure.[14]

The buy-back took out some 350 vessels, about 7 per cent of the A fleet, for about $1.5 million.[15] When, in 1973, excellent salmon catches and markets hiked the value of A licences, the buy-back came to a close. But in spite of the goal of matching the fleet to the resource, fishing power increased sharply.

The increase stemmed from the replacement rules. When limited entry began, the department applied a boat-for-boat rule, allowing fishermen to replace a vessel with a larger one. This soon got restricted; in 1970, a ton-for-ton rule came into effect, and in 1972 a foot-for-foot rule, for single-vessel replacements. But the department also allowed pyramiding on a ton-for-ton basis: if you had two 20-ton vessels and put them out of use, you could combine the licences on a new 40-tonner. The new vessel might have several times the fishing power of the two old ones. As business conditions improved, especially in 1973, many fishermen moved up to fancy new craft. Part of the motivation came from income-tax rules that made it advantageous to invest in a new boat, rather than seeing your money taxed away.

The overall number of vessels reporting salmon landings kept dropping, to about 4,700 by 1980. But the drop came in the gillnet and troll sectors, the least powerful part of the fleet. The number of seiners fishing only salmon rose from 286 to 316. That was only part of the story. The number of seiners fishing both salmon and herring multiplied from 83 in 1969 to more than 200 in 1980. And even the drop among trollers and gillnetters was deceptive, since the number of combination vessels carrying both types of gear increased.[16]

In the total B.C. fleet for all species, the number of boats dropped from about 7,700 in 1968 to about 6,700 in 1985. But the increase in fishing power far outweighed the reduction in numbers. Meanwhile, limited entry, by allowing the transfer of licences, had made fishing power difficult to remove. Instead of balancing the fleet with the resource, licence limitation had helped create a bigger imbalance.

The number of fishermen in the salmon fleet specifically dropped from about 9,600 before the Davis Plan to about 8,600 in 1979. This was a by-product of vessel reduction and in some cases of better, labour-saving gear. But there was nothing in limited entry to control the number of fishermen per se; it only controlled vessel licences. The total number of fishermen for all species dropped from 12,100 in 1968 to 11,900 in 1974, but then resurged. By 1985, the number of personal fishing licences had risen to more than 18,000. The B.C. fleet by then had twice as many people and several times more capital value and fishing power as when limited entry began.[17]

Fleet gets far stronger

In British Columbia during the 1968–1984 period, the number of craft was relatively stable, but fishing power was expanding. The fleet was strong to start with: in 1968, in the mid-size range of ten tons to 100 feet, B.C. had about 2,400 vessels, as many as the Maritimes and Quebec combined, and nearly six times as many as Newfoundland.

Growth took place particularly after the bonanza year of 1973, with powerful new vessels joining the fleet. Table 25-1 shows changes between 1970 and 1976; the increase in fishing power was, however, greater than the numbers suggest.

Table 25-1. B.C. fishing craft by size category, 1970 and 1976.

B.C./Year	over 100 ft	10 tons to 100 ft.	Boats<10 tons	Row-boats
1970	10	2,389	4,552	24
1976	26	2,807	4,642	19

Note: For 1968 and 1976, the table treats "boats" as motor-boats, and lists row-boats separately. By 1984, departmental statistics no longer broke out row-boats; thus, the table for 1984 lumps together motor-boats and row-boats.

Growth in vessel strength continued, especially after the roe-herring bonanza of 1979. Between 1976 and 1984, the B.C. statistical system abandoned tonnage, making exact comparisons with the above table difficult. In terms of vessel length, however, the increase was clear. Vessels in the 35- to 75-foot range went from somewhere above 2,500 to 3,479; those in the 75- to 100-foot range from 71 to 119; and those above 100 feet from 26 to 38.

Vessel standards rise

Phase 3 of the Davis Plan aimed to improve the quality of landings. This meant giving a new push to an old effort. Back in 1962, the department had launched a voluntary vessel-inspection program in the Prince Rupert area. Then, following a request by the Prince Rupert Co-op, deputy minister Needler had in 1967 set up a government–industry working group, which recommended higher standards for safety and quality and a quick test for vessel operators on the handling and storing of fish.

Safety was also an issue. Davis told the department in 1969 that vessels should meet both seaworthiness and fish-handling standards or lose their licences. In 1971, the department notified all vessel owners, not just salmon vessels, of the new standards. These mainly dealt with protecting fish from weather and contamination, and remained silent on refrigeration and time of holding. But even those un-exacting standards caused difficulties for some boats. After several extensions, more than a hundred craft lost their licences.[18]

Boats did improve, especially as many new ones entered the fleet. But safety remained a challenge; British Columbia would see several disastrous sinkings in the 1970's.

Department drops push for best fishing arrangements

Phase 4 of the plan, essentially designing a better fishery, had potentially deep implications. The department had never given any fishery a head-to-toe examination, then put in place the best methods and areas of fishing for optimum benefit. Instead, it would start with the existing situation, reacting to pressures and doing whatever seemed best at the time. The engineer Davis wanted a more rational approach. He stated that he would undertake a series of steps to "establish a firmer economic base for all segments of the industry." He wanted to modify area and gear regulations for the best performance.

Awareness was increasing that fishery management meant more than conservation; one had to consider size and condition of salmon and the market demand. Changes in troll gear were leading to increased catching of undersized salmon; these were thrown back dead or dying. Gillnets often lost catch through "drop-out" of fish. Seiners intercepted and killed immature salmon. And the end use made a difference: troll-caught fish sold fresh or frozen gave better prices to fishermen, but seine-caught fish fed the canneries. The same run of fish could yield different values depending on where and when it was caught. Salmon caught closer to river mouths lost colour, texture, and monetary value. And an optimum sized fleet could produce better returns.

But the phase 4 ideas spooked some members of the industry. There was already considerable tumult with the new licensing scheme and its appeal committees, and with the new vessel standards. Salmon enhancement was also emerging as a priority. By late October in 1971, it became clear a federal election was approaching. Davis delayed any concrete moves on phase 4. Committees geared up to consider the matter, but reached no consensus. Nothing major happened.[19]

Herring helps to fuel growth

Whatever its drawbacks, the Davis Plan gave the appearance of success. Soon salmon fishermen were making more money than ever. Coincidentally, in 1973, catches and markets boomed. Jack Davis noted in early 1974 that since 1968 the average fisherman's earnings had risen from $4,660 to $11,896, and the average income per boat had almost tripled, from $6,800 to $19,200.[20] Salmon values kept edging up. The average price of salmon in 1978 worked out to nearly $2,900 per tonne, five times the 1968 level. And the new herring fishery brought still more money into the fishery.

As noted earlier, overfishing had exhausted the herring fishery, which closed in 1968. The department reopened it in 1971 and 1972, using catch quotas, stricter than ever, for the different fishing areas. Landings rose to about 100,000 tonnes in 1977; after that year, the department took a more cautious approach, keeping quotas well under 50,000 tonnes.

Fish-meal remained disallowed. The food market remained small. But roe herring supported a strong new fishery. In late February and March, regular ranks of Pacific herring came inshore to spawn on kelp. They were easy to catch. In the plants, workers popped out the roe in neat little membranous units, like the segments of an orange. Carefully preserved, these provided a delicacy in Japan, associated especially with New Year celebrations. The fishery made a good complement to the salmon season; the same vessel could fish herring in winter and spring and salmon in summer and fall. Best of all, roe fetched a high price.

As the roe fishery began to take off, the department tried to deter overcrowding by charging higher licence fees: $200 for a gillnetter and $2,000 for a seiner. Managers turned again to licence limitation, beginning for herring in 1974. But instead of announcing limited entry as a fait accompli for existing vessels, the department announced weeks in advance that it would be coming into place. This was apparently a Davis decision, against departmental advice. Seiners and gillnetters rushed to get licences before the deadline. As well, Native people got an extended period to apply for licences.[21] The department had aimed for about 600 licences. By the time entry got fully closed off in 1977, there were about 250 seine and more than 1,300 gillnet licences.[22] Herring seiners had more than doubled from the 80–120 level seen before limited entry.[23]

The new money from salmon and herring produced a rebirth of the fleet. A frenzy of pyramiding, overhauls, and boatbuilding set in. Fishermen buying salmon seiners wanted them bigger, so they could also fish herring better. The seine fleet all told, for salmon and herring both, by 1980 had more than 500 vessels, a major increase. Boatyards churned out new aluminium and stainless steel vessels, with the latest electronics and gear. On seiners, the drum seine became ubiquitous, cutting crew size from six or more to as few as three. By the end of the 1970's, British Columbia had probably the newest, finest fleet of its size in the world, a menacing fisheries navy.

"Owner-operator" rule fades away

In a departure from the salmon-licensing system, the department when limiting roe-herring licences attached them to the person rather than the boat. The U.F.A.W.U., still lambasting the licence-the-boat system in salmon, prompted the differing rules for roe.[24] The department in 1974 said that the licence-holder should operate the vessel, as would become the case in most Atlantic fisheries. But this rule applied only to new licence-holders starting from 1974, not to previous participants. The attendant confusion made enforcement difficult. In 1979, the department dropped the restriction.

The initial licensing scheme for roe herring also ruled out transferability. Operators soon found ways around it. Since some licences were issued to companies, one could transfer them by selling the company. Also, after one fisherman died when his seiner sank, the minister allowed his widow to sell the licence—"what else could you do?" recalled a senior official—and this opened the door to other such cases. Then, after the department dropped the holder-operator rule in 1979, more loopholes came into play. Licences were changing hands, in spite of obstacles. (In 1990, following consultations with industry, the department would succumb and allow licence transfers.)

Meanwhile, leases became a notable feature of the fishery. Early restrictions on leasing got eroded. Licence-holders could then lease out the fishing privilege to another vessel owner or anyone else, sometimes for years at a time and for large sums. Licence-leasing caused complaints about "arm-chair fishermen" and "slipper skippers." The leasers were collecting money for a privilege belonging to the state and its citizen.

Roe frenzy leads to fleet "pooling"

In 1965, herring had fetched an average $30 per tonne; by 1975, about $230. Then, in 1979, came a legendary frenzy of roe buying. The average price rocketed up to nearly $3,000 per tonne, and sometimes as high as $5,000. Cash buyers on the boats and docks carried millions of dollars in their briefcases. Single sets by a seiner could fetch a million dollars. One opening of an hour and a half saw seiners take a catch reportedly worth $35 million. Total value that year reached $124 million. In one incident, when fishery

Herring-roe fishing became a high-pressure, high-payoff fishery.

officers confiscated an illegal catch and the associated money, it took them two days to count and validate the more than $250,000 cash.[25] A boat-load of Vancouver prostitutes sailed to the fishing grounds, hoping for a share of the wealth.[26]

Although prices dropped steeply the following year, roe remained a strong element of the B.C. fishery. Fleets ganged up at spawning areas. A hundred or more boats might wait for the fishery officer's go-ahead, then all jockey at once for a set. For the fishery officers, the situation was nerve-wracking. Allowing too little fishing might let millions of dollars go to waste for no particular purpose. Allowing too much might damage the stock for decades to come. It became common for the officers to set openings as short as 15 minutes, creating wild scenes on the water.

In 1981, the department moved to reduce the frenzy of the fishery. New rules divided the coast into three areas, each with its quota, and obliged the licence-holder to choose one. This limited the number of boats that could mass up at a single area. To make the most of the new system, fishermen began combining their efforts. Captains with individual licences for different areas would pool together, so as not to miss the best area, and split the earnings.

In a further development, not fully foreseen by the department, some licence-holders tied up their vessels, letting the others do all the fishing. The licence-holders and the remaining crew would then divide the earnings. Both pooling and leasing reduced the number of boats fishing. A captain with the money could lease as many licences as there were areas, and might tithe the crew to share the expense.

By the mid-1980's, of the 252 licensed herring seiners, about 180 would fish. By the early 1990's, the roe-herring fleet would drop by 35 or 40 per cent. For many crew members, this meant lost jobs. For the department it meant somewhat easier management. Still, it remained a fishery of high catching power and high intensity. The margin of error for department officials was small; a mistake could mean either lost earnings or a conservation disaster.

Managers seek stability

Herring landings never returned to levels of the 1960's, when they had reached a quarter-million tonnes. After building up to nearly 100,000 tonnes in 1977, they dropped back. In the 1980's, they would remain mainly in the 30,000- 40,000-tonne range, with values between $50 and $100 million. The stability of landings derived in large part from scientific and managerial caution after the 60's collapse. Research increased particularly after 1977, as did co-operation between researchers and managers. Herring forecasts improved, based on spawn deposition, acoustic soundings, test fishing, and computer modelling. The managers aimed to take no more than 20 per cent of the spawning stock. Conscious that higher landings might only lead to lower prices in the roe market in Japan, they set quotas below the scientific recommendations, according to market.[27]

Spawn-on-kelp fishery develops

The roe fishery developed an offshoot. Commercial interest rose in herring spawn dried on kelp. Prompted by a Japanese buyer, the department started feasibility studies in 1972. In 1974, a departmental biologist worked with the Skidegate Indian Band on the Queen Charlotte Islands to impound live, pre-spawning herring in enclosures where, when ripe, they would spawn on kelp. The dried spawn, sold on little pieces of seaweed, got even better prices than the roe segments.

In 1975, the department took applications and granted 13 spawn-on-kelp permits (temporary licences). Operations were located on the Queen Charlotte Islands, the north coast, the west coast of Vancouver Island, and Johnstone and Georgia straits. The department gave preference to persons or Native bands in remote coastal communities, and to those with previous experience in catching and impounding herring. The objective was to encourage a cottage-type industry that would favour remote communities. The department wanted about half of the permits to go to Native people.

Up until 1977, all Indians who applied for permits received them. The Native preference was part of the general desire of the time, as reflected in the Indian Fishermen's Assistance Program and other ventures, to build up the Native fishery.

The department allowed only slow expansion. Individuals were not allowed to hold both roe licences and herring spawn-on-kelp permits. In 1977, the British Columbia Fishery Regulations were amended so that a licence would now be required. Generally the spawn-on-kelp ventures prospered. Some Native communities used the profits to fund other economic development.[28]

Salmonid Enhancement Program promises more

After the war, salmon landings had cycled up and down in the regular fashion, with no great growth. But at the outset of the 1968–1984 period, there hung in the air the idea—blessed by Ricker among others—that rivers and ocean could support at least double the current yield.[29] Enhancement techniques were pointing in that direction.

Major fishways had come into place on several rivers, including the famous Hell's Gate fishway on the Fraser. Spawning channels, a pioneering new development of the 1950's, could multiply the size of spawning areas for natural breeding. After the first venture at Jones Creek, major projects had followed at Robertson Creek, Big Qualicum River, River Creek, Seton Creek, and the huge Babine Project, which also involved the Fulton River and Pinkut Creek. Fishways, too, were improving returns. A 1974 report noted that Fraser River fishways, with an investment of $2.3 million, had yielded $60 million in benefits over the last three decades, and that several spawning channels had paid back their capital costs in one or two years' operation.

Meanwhile, hatcheries were re-emerging after the 1930's closure. A new hatchery went up on the Nanika River in 1960. Much more important was the 1968 hatchery and spawning channel complex at Big Qualicum, which served as a testing ground for new methods. The improved techniques promised more fry and better survival.[30]

Jack Davis pressed for more hatcheries in the Strait of Georgia area, part of which was in his riding. The minister himself selected hatchery sides, and told biologists and engineers to get on with it.[31] Davis said early in 1974 that benefit–cost ratios "range from 3 to1 to as high as 20 to 1."[32] Construction of the Capilano hatchery began in 1971, to enhance a coho stock damaged by a dam. The Capilano hatchery would yield good

Large seiners increased in number after the Davis Plan, raising fishing power instead of lowering it. But the Salmonid Enhancement Program promised more fish to ease the pressure.

returns. Hatcheries at Robertson Creek and Puntledge started up in 1972, and at Quinsam River in 1974.

By that time, the department was thinking big. The idea of a huge program soon coalesced. A major seminar took place in early 1974, bringing together sport, commercial, Aboriginal, and other interest groups. Davis made it known that his planners were looking at a ten-year, $200-million program to double the salmon catch. Ken Lucas, the top department official from 1973 until 1978, made enhancement a high priority. Lucas appointed Fern Doucet to work with officials on the initial picture, and subsequently named Ron MacLeod, a former fishery officer who had become a top official, to lead efforts in B.C.

After Davis lost his seat in 1974, his successor Roméo LeBlanc sounded out industry and political support, and backed what became known as the Salmonid Enhancement Program (S.E.P.). LeBlanc, with his personal charm, helped energize B.C. officials, especially when he pledged to sell his office furniture if need be to help fund S.E.P.

Studies suggested highly positive scenarios; these increased the provincial government's interest. LeBlanc and the provincial minister of environment, Jack Radford, established a good relationship. The fishery had often appeared an impediment to development, interfering with hydro dams. Now it looked like an avenue of growth, for both commercial and sport-fisheries.

But there was more to S.E.P. than money. Ron MacLeod and others, working together with the provincial government and interest groups, imparted a strong emphasis on public education and participation—such things as stream clean-ups, local enhancement projects run by community groups, Indian economic development projects, and educational modules for the school curriculum.

Full S.E.P. effort gets underway

In 1975, Ottawa and British Columbia agreed on a co-operative process. A two-year preparatory effort got under way. Then, in 1977, LeBlanc announced phase 1 of the main program, intended to spend $150 million over five years. The goal was to double salmon and anadromous trout production to historic levels, increasing production by some 68,000 tonnes.[33] The means would include stream clearance, habitat restoration, fishways, spawning channels, hatcheries, incubation systems, and lake enrichment.

Construction went forward on more hatcheries and spawning channels. By 1982, the S.E.P. had 15 major and 14 minor facilities, along with about 100 small projects, and was fertilizing 12 lakes (lake enrichment proved to be a low-cost, high-return technique).[34] In addition, about two dozen smaller-scale community economic development projects started up over time. In many of them, Aboriginal people worked at stream clearance, running local hatcheries, or otherwise improving conditions for the salmon.

On top of that came public participation. By the time S.E.P. got into full swing, about 10,000 volunteers in any given year would be monitoring salmon runs, removing obstructions, and the like. Today, advocates of S.E.P. universally point to public participation and education—more than a quarter of a million students have participated in the "Salmonids in the Classroom" program—as a highlight of the program.

By 1984, problems and complaints were cropping up. Although spawning channels and stream maintenance work were probably more cost-effective, hatcheries had gotten heavy emphasis in the initial program. These seemed particularly useful for coho and chinook, two of the weaker species, which were important to sport fishermen. As well, hatcheries were something big and visible for the taxpayers' money. Some officials would have preferred, especially in hindsight, to emphasize the more natural forms of enhancement.

High inflation in the later 1970's and beginning of the 1980's ate into the program. S.E.P. lost more momentum when the $150 million was stretched to cover seven years rather than five. The industry was supposed eventually to help pay back the cost of salmon enhancement through a cost-recovery mechanism. But cost recovery never came to pass, as cabinet members resisted an implementing proposal.[35]

Concerns persisted that enhanced stocks would support more intensive fisheries, which would in turn damage weaker, unenhanced stocks. Some observers and officials began to mutter that it might be better to "enhance by management," on the premise that controlling fishing and preserving habitat could increase production just as well as hatcheries.

In 1982, a federal Commission on Pacific Fisheries Policy gave S.E.P. a scorecard to date. Phase 1 had aimed to increase annual production by 50 million pounds; projects able to produce 31.2 million pounds were already operating. Despite earlier hopes of higher benefit–cost ratios, S.E.P. had set a more modest target of 1.5 to 1. Actual returns were more like $1.30 for every $1 spent; "benefits to the target area will be about 40 per cent of the original target, at $78.3 million. The estimated continuing employment that will be provided to Indians is 32 person-years, half the target." And staffing the enhancement projects took 623 person–years, more than one-third higher than the number planned. Commission leader Peter Pearse wrote that "community development projects almost break even in terms of benefits and cost, and economically are expected to be as good as, if not better than, minor projects as a means of producing fish."[36]

Despite some shortfalls and questions, S.E.P. appeared to most people a strong and progressive program. Pearse recommended its continuation, though with caution and careful evaluation. Among industry and public, support was solid. S.E.P. was contributing to salmon runs that would in the later 1980's, helped by ocean conditions, rise to record levels.

LeBlanc heads off Alcan project

In the early 1950's, Alcan Corporation had launched a major hydro-electric project to provide power for smelting aluminum. The department had opposed it. When the province granted approval, department officials did what they could to mitigate the Kemano project's effects on the Nechako River, a northern arm of the Fraser system.

In 1978, B.C. Hydro began purchasing power from Alcan, causing the company to divert more water from the Nechako to its powerhouse. Then, in 1979, Alcan announced its intention to proceed with a "Kemano Completion" project. To power two more smelters, it would divert still more water from salmon and trout rivers in the Nechako system.

Department officials deemed the new project dangerous to salmon. The province was unsympathetic. Meanwhile, D.F.O. wanted more water flowing in the rivers, and looked to court action. In the federal Department of Justice, lawyers hesitated, wary of treading in gray zones that might be part of provincial turf. In 1980, Minister LeBlanc got exercised at the quibbling. In a departure from his usual habits, LeBlanc used strong language over the telephone to Justice officials, ordering them to pursue an injunction requiring release of more water to protect sockeye salmon.

By now the matter was a hot public issue. When Kate Glover of the communications branch passed word of LeBlanc's move to the Vancouver Sun, she heard the editor yell, as in the movies, "Stop the presses!"[37] LeBlanc's injunction worked, and the department kept renewing it every year. But the Kemano plan would resurface as a major issue later in the 1980's.

Problems arise in paradise

In the mid-1970's, limited entry, the salmon price hike, the promise of S.E.P., the dizzying rise of the roe fishery, the 200-mile limit, and the shiny new fleet all created as close to a glamorous atmosphere as the fishery ever achieves. Later in the decade, the main body of the industry ran into some of the same problems afflicting the Atlantic coast, along with additional ones peculiar to British Columbia.

The general economic climate changed, affecting markets and business operations. Fuel and other costs rose. Interest rates vaulted to the range of 20 per cent, unprecedented in recent history. Debt charges pried great strips of money from the over-built B.C. fleet. In some cases, fish prices dropped in the face of competition. Repossessions and bankruptcies loomed for many fishermen. Meanwhile, the powerful fleet represented a bigger-than-ever threat to the fish resources. Merely controlling the fleet, let alone the fine-tuning of fisheries, posed a growing challenge.

Department officials began to wonder where they'd gone wrong. The promise of licence limitation—a fleet matched to the resource, better conservation, better

average incomes—seemed to be evaporating. In the late 1970's the department commissioned another report on Pacific licensing by Dr. Sol Sinclair, who had authored the major 1960's study on limited entry. Sinclair now recommended against transferability: "when a fisherman wished to retire his licence would revert to the licensing authority and be cancelled." This would create a better balance between resource and capacity; although, he said, it would now be difficult to freeze transfers.[38]

Sinclair suggested licence fees based on vessel size and capacity, a royalty on fish catches, the extension of limited entry to fisheries not yet covered, a fleet buy-back, low-cost financing for young people trying to get established in the fishery, and a reconsideration of area licensing. The 1979 election pushed the Sinclair report off to the side, but some of its ideas would recur.

Meanwhile, the ideas of individual quotas and quasi-property rights were coming into vogue. They had appeared useful in the real-life case of Bay of Fundy herring, and were enjoying a strong theoretical life in academe, where professors in the burgeoning field of fishery-management studies saw them as an antidote to the "tragedy of the commons."

(An American biology professor, Garrett Hardin, had made that phrase popular in an often-quoted 1968 paper in *Science* magazine. Hardin used the example of a shared pasture open to all; if each herdsman keeps adding animals, a perfectly sensible action from an individual point of view, eventually the pasture and herds collapse through over-exploitation. "Ruin is the destination toward which all men rush ... Freedom in a commons brings ruin to all." Hardin was reinforcing what H.S. Gordon had said in the 1950's about the fisheries. Many economists began using the "tragedy of the commons" phrase. It was often overlooked that many forms of common-property usage had lasted for centuries, since people had developed rules of usage. Hardin himself later wrote that "to judge from the critical literature, the weightiest mistake in my synthesizing paper was the omission of the modifying adjective 'unmanaged' [in regard to commons].")[39]

While academics contemplated the big picture, fishermen were feeling desperate. B.C. politicians pressed for a public inquiry, and LeBlanc agreed.

Pearse Commission sets out to reduce fleet

In January 1981, Peter Pearse, a resource economist from the University of British Columbia, was appointed to lead the Commission on Pacific Fisheries Policy, noted above. One idea was uppermost in everyone's mind: find an acceptable way to reduce the fleet. Pearse undertook many consultations, and late in 1981 issued a thick interim report. Among many other recommendations (some similar to those in the recent Sinclair study), Pearse called for a huge fleet reduction, of dimensions to be determined; observers had suggested 40 or 50 per cent or even more. A major buy-back should take place. A special tax on salmon and roe-herring landings should help pay for the buy-back and for salmon enhancement. For some fisheries, although not salmon, individual quotas should come into effect; fishermen should be able to buy and sell those quotas. All forms of subsidy—for example, the rapid depreciation allowance in the tax system—should end. And the salmon fishery should, as possible and appropriate, shift towards inshore waters.

LeBlanc cautions Pearse

Alarm spread in the industry, as fishermen imagined one of every two boats vanishing. And who would determine the shape of the future fleet?

LeBlanc responded to the interim report in a B.C. speech early in 1982. Spotting Pearse in the audience, LeBlanc quipped that "you don't often see a man come to his own funeral."[40] LeBlanc expressed skepticism about putting a "blind-folded" buy-back mechanism in place. "Should we [instead] use our judgement about the right balance of gillnetters, trollers, seiners, large vessels and small, and the jobs and money they produce? ... Can we at least set rough goals acceptable to most fishermen?" Clearing out older, smaller boats, LeBlanc said, might give a disguised subsidy to new and large boats that helped cause the problem of too much fishing power in the first place. "One can easily sympathize with the idea of a more streamlined, efficient, cost-effective industry. ... One might however do away with a good many coastal communities in the process. Is that economic efficiency?"

As for individual boat quotas in the Bay of Fundy and some Manitoba fisheries, LeBlanc said, they had made regulation no simpler; instead they had increased problems of enforcement. And regarding vessel-licence buy-backs, maybe government should license people instead of boats, rather than buying back privileges that came "from the public's marine treasury in the first place." Instead of freeing up the buying and selling of licences, as Pearse suggested, perhaps government–industry committees or some other mechanism should decide who could fish. The industry should form a consensus. Meanwhile, "fishermen should lay to rest any fears of a wholesale rationalization against their will."

In February, LeBlanc announced a modest buy-back as a step towards readjustment. This removed 26 salmon boats for $2.5 million.[41] Meanwhile, Pearse paid little attention to LeBlanc's cautionary remarks. His full report, *Turning the Tide*, appeared in September 1982 and called for wholesale changes, with a commercial, you-get-what-you-pay-for orientation.

For the salmon and roe-herring fleets, the department should cut fishing power by half over ten years, reducing salmon gillnetters, seiners, and trollers by equal amounts. In future, limited-entry and quota licences should hold good for ten years only. The government should determine how much capacity was desirable in each fleet category, and make one-tenth of that capacity available for allocation each year. To get

new licences, enterprises should make competitive bids. Only existing licence-holders should have the right to bid during the transitional period.

As well, D.F.O. should divide the coast into three salmon zones, and licences should apply to only one. There should be no owner–operator rules. And a landings tax of 5–10 per cent should apply, to give the public a share of the resource value. Without such a royalty, windfall gains would foster additional fleet capacity.[42]

Fleet recommendations stall

Although Pearse had many more recommendations, the chief ones involved the time-limited licences and the 50 per cent fleet reduction. As LeBlanc had warned, this was a hard sell. Many fishermen opposed the Pearse scheme, with the United Fishermen and Allied Workers Union in the forefront.

When LeBlanc left the department in October 1982, the new minister, Pierre De Bané, inherited a tricky situation. Before he left, LeBlanc had appointed a Fleet Rationalization Committee, led by industry member Don Cruickshank, which recommended a major buy-back with a softer approach than Pearse's. Despite considerable industry support, Cruickshank's report got lost in the shuffle. De Bané decided on bolder action.

The department developed a series of far-reaching measures. Senior officials said that the government wouldn't just bail out the Pacific fishery, but would "buy change." The biggest problem was the excessive build-up of fishing power. Fishermen had invested some $300 million in the 1970's. "The result has been an overfished resource and a commercial salmon fleet that is technically bankrupt," said a departmental briefing note. "Also affected are Natives, who have seen their participation in the commercial fishery significantly reduced, and the sports fishermen who are facing declining opportunities."

Under De Bané, the department planned to initiate individual, transferable fishing quotas for the salmon fishery; this went even beyond Pearse's recommendations for that fishery. Other plans frequently reflected Pearse. Another buy-back proposed for vessels and licences would spend at least $100 million, to reduce the fleet as much as 40 per cent. Salmon licence-holders would have to select one area and gear type for their operation. More fish would be caught in terminal fisheries. There would be no bail-outs for financial institutions; other measures would, however, prevent temporary industry collapse and fishermen's bankruptcies. The government would protect Native participation in the commercial fisheries, and would institute a stabilization program for Native people. Developmental programs would come into play for the sport-fisheries and for aquaculture, and economic adjustment programs for displaced fishermen and affected communities. Legislation would be required. Finally, the government would apply cost recovery; licence fees would go up.

While traditional fishery officer duties, such as placing fishing-boundary markers at river mouths (left), continued during the 1968-84 period, habitat-related work (above) rose, along with developmental pressures.

B.C. fishermen's organizations protested vehemently. Still, by early 1984, the stage seemed to be set for action. Then John Turner became Liberal leader and appointed De Bané to the Senate. Herb Breau of New Brunswick took over as minister just as Turner called an election, which the Liberals lost.

Brian Mulroney's Conservatives had no interest in changing the Pacific fishery. The new minister, Vancouver M.P. John Fraser, had in Opposition criticized De Bané and his staff for blaming fishermen and sowing confusion and anger on the west coast.[43] Besides, conditions were improving. Landings and values jumped sharply in 1985. The Pacific turmoil appeared to be over, for now.

But gross over-investment and over-capacity still prevailed. Within the department, many officials held on and waited for the next Pacific crisis.

Pearse recommendations bring some reforms

Despite the failure of its central recommendations on fleet reduction, Pearse's report got wide praise for its intelligence and insights. The licensing recommendations took up less than half of the document; the rest discussed such matters as salmon enhancement, herring and halibut fisheries, habitat, and Native and sport-fisheries. Pearse endorsed individual-quota schemes (though not for salmon); his recommendation

Some of the Vancouver-area fleet in 1984.

raised their profile. I.Q.s and I.T.Q.s would in time spread to many Pacific fisheries.

Like Kirby, Pearse pointed to shortcomings in the consultative process, quoting criticisms of unclear direction, poor procedures, and "the widespread perception that advice is not seriously sought or listened to." He proposed reorganizing the Minister's Advisory Council, created earlier by James McGrath, giving it more power and a legal status. Pearse also laid out guidelines for a system of fisheries advisory committees, conservation committees, and special regional management committees. And he suggested that the department launch a high-quality periodical to inform the public and clear up misunderstandings. But the information and consultation recommendations came at the end of the book and got little attention.

Some of Pearse's recommendations had useful results. They pervaded the atmosphere and influenced thinking on such matters as habitat. But they brought no striking, immediate major changes. A less radical approach to fleet reduction, better sold by both Pearse and De Bané, might have done more good. Instead, the time bomb of over-capacity was left ticking. Two other ideas of the day—a landings tax or catch "royalty," and cost-recovery for S.E.P.,—also faded away without result. These might have dampened the gold-rush mentality of fisheries, and given industry more of a voice in management to go along with their financial contribution.

Pearse and Kirby: two versions of federal policy

In prodding government policy-makers, Richard Cashin of the Newfoundland fishermen's union often posed the question, "What are the fisheries for?" On the Atlantic, the Kirby report had given a form of answer, listing the fishery goals as economic viability, maximum employment, and Canadianization. In short, Kirby wanted to spread fishery money around, so long as the industry remained viable.

Pearse in an interview said that the uppermost goal was to benefit the national income, the revenue minus the cost. The fishery was too expensive, and could yield better returns. His emphasis was different from Kirby's. The differing tone of the two reports suggested that after more than a century, the federal government still had no clearly defined national answer to "what are the fish for?", other than the "best use" generalities of the 1976 policy document.

The Kirby report failed to produce some of the major results the writers envisaged. Subsequent changes in quality and marketing, although important, came nowhere near the proposals put forward. Perhaps the major result of the Kirby report was the impetus it gave to individual quotas. But these were a side issue in the report; and the department had already used I.Q.s widely.

That being said, the Kirby report in the short run had more concrete results than Pearse's. D.F.O. issued a checklist in July 1984, counting off successes in almost every category. These included, as noted earlier, clearer rules on international dealings, resumption of federal fishery management in Quebec, changes to the fishermen's U.I. system, and so on. Writing a decade later, Scott Parsons said that "about 65 per cent of the Task Force recommendations were followed as proposed, another 11 per cent implemented partially or in a modified form, for a combined total of 75 per cent."

By contrast, Parsons wrote, Pearse's main recommendations on licensing "[sank] in a sea of opposition." And "the radical reforms Pearse proposed may have delayed progress towards individual quota systems in some fisheries such as halibut." Pearse did sharpen thinking and encourage progress on habitat management, recreational fisheries, and Native fisheries. But in the aftermath of his report, Pearse had few changes that he could clearly call his own.[44]

Pearse's unpopular proposals for fleet reduction caught some of his more workable ideas in the downdraft.[45] But besides that, Pearse had a different mode of operating from Kirby. Pearse worked on his own, writing the recipe but leaving the kitchen work to others. Kirby had a bigger team, including strong officials who could take back to the department an understanding of the reforms and a desire to put them in place. He also had connections in the central agencies of government, with both mandarins and ministers. His proposals were more practical, and he had better means to implement them.[46]

Halibut fishery nosedives

By the start of the 1968–1984 period, the halibut fishery was declining. From a 1963 peak of 16,900 tonnes and $8.2 million, landings dropped to one-fifth that level: only 3,400 tonnes in 1974, worth $5.4 million. Catches remained low in following years, amounting to only 4,400 tonnes in 1984, although value rose to $9.4 million.

The International Pacific Halibut Commission had underestimated the fishing power of the American and Canadian fleets. Trawlers from overseas added to the damage. Even when foreign vessels disappeared after the 200-mile limit, Canadian and American trawlers still took many halibut as incidental catches. Although they were barred from keeping halibut, half those discarded died anyway.[47]

Fewer than 100 vessels now found their mainstay in the halibut fishery; others mixed halibut with other fisheries, particularly salmon trolling. In 1979, the department brought in limited entry for the halibut fleet. To qualify, vessels needed a certain level of land-

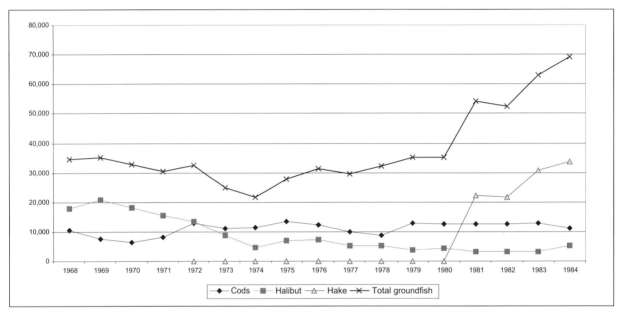

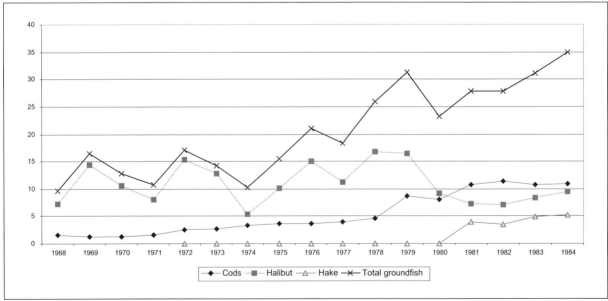

Groundfish landings (top, shown in tonnes) soared in the early 1980's, as over-the-side sales of hake came into play. Values (bottom, shown in $millions) rose accordingly.

374

ings. More than 300 vessels, with some 1,200 crew, got licences. But because of generosity during licence appeals, the number jumped to 422 vessels by 1981. Once again, limited entry created a bigger fleet than expected. The Pearse report in 1982 proposed a boat-quota and licence-auction system, together with catch royalties, but this idea went nowhere at the time.

As already noted, 200-mile limits had both helped and hurt the fleet. Some 34 larger vessels had fished the Alaska coast, producing most of B.C.'s halibut landings. The American extension of jurisdiction had forced them back to home waters. To absorb the impact, a "halibut relocation" plan shifted some larger vessels into the black cod fishery, and compensated others for giving up their halibut licences and gear. Meanwhile, the U.S. fleet was growing, adding to pressure on halibut stocks.[48] The Canadian halibut fishery would face a bumpy ride into the later 1980's.

B.C. waters held an assortment of other groundfish, such as Pacific cod, sole, ling cod, various rockfishes, and Pacific ocean perch. Catches of these stayed at modest levels. But a new fishery for Pacific hake started up. In 1978, the department began allowing over-the-side sales of hake, Atlantic-coast style. Canadian mid-water trawlers did the fishing. The Hake Consortium, an industry grouping, negotiated contracts with operators of foreign purchasing vessels, and oversaw the fishing. The department saw the co-operative arrangement as having several benefits: building up an offshore hake fishery, relieving the pressure on hard-pressed traditional groundfish stocks, and bringing some business to B.C. ports by foreign vessels. The foreigners, allowed to fish hake as well as buy it, purchased fuel, supplies, and ship repairs.

By 1980, 12 Canadian trawlers were delivering some 13,000 tonnes of hake worth about $2 million to purchasing vessels. Over-the-side customers in 1981 included Poland, the Soviet Union, and Greece. Such sales would become a fixture in the B.C. fishery.

Meanwhile, other groundfish had weakened in the 1970's. In 1980, D.F.O. instituted a groundfish management plan more like that on the Atlantic, allocating quotas for different species. The aim was to rebuild stocks and to develop fisheries for non-traditional species such as hake, pollock, and turbot. Groundfish landings all told, including hake, halibut, and other species, rose from 25,000 tonnes worth $17 million in 1973 to 69,000 tonnes worth $35 million in 1984. Much of the growth came in the hake fishery.

Shellfish multiply in value

Although far behind finfish, the B.C. shellfish fishery for crab, shrimp, clams, and such species was important. Harvests doubled from 6,600 tonnes in 1972 to 13,000 tonnes in 1984; value rose eightfold, from $2.2 million to $18 million. Two growing fisheries helped the increase.

From 1976 on, a dive fishery for geoduck and horse clams grew fast. Divers used high-pressure water

Fishery officer checking catch on a shrimp dragger, with patrol vessel *Atlin Post* in background. (D.F.O. photo by Michel Thérien)

hoses to loosen the sizeable creatures from the bottom. D.F.O. in 1979 applied limited entry and quotas.

As was typical in B.C. fisheries, an industry organization quickly formed to represent and help manage the fishery. Landings by the early 1980's ran to more than 3,000 tonnes, worth $2.3 million. Geoduck would come to be the west coast's most valuable shellfish.

Another dive fishery also expanded. Abalone, a low-level and intermittent fishery for years past, in the 1970's stimulated new interest. New markets were opening in Japan, salmon and herring fishermen were looking for ways to invest their money, and some people were looking for open fisheries where they could still get in without a licence. New operations started up, mainly in northern British Columbia, with scuba divers working off vessels that could freeze the abalone.

Worried about over-expansion, the department in 1977 instituted limited entry for the few dozen operators, and applied a size limit and closed season. But effort still increased, leading to a Total Allowable Catch in 1979. Part of the T.A.C. got subdivided into individual quotas. Yet, with all these management tools in place, the fishery would run into grave conservation problems in the 1980's.

What went wrong? One study laid out several factors. Licence-holders could lease out their licences, leaving fewer breaks in fishing pressure. A three-diver-per-boat restriction had been dropped. Poaching by unlicensed divers and under-reporting by I.Q.-holders became serious problems. Natural conditions may have helped to reduce the stock, about which knowledge was limited.[49] Finally, the department lacked resources for close monitoring and enforcement.[50]

Aquaculture begins to grow

Finfish aquaculture of a sort—the breeding of fish in hatcheries, and their enhancement in spawning channels—had long existed in British Columbia under government auspices. By the early 1980's, salmon-cage aquaculture by private companies was taking hold. As of 1984, nine or ten enterprises were said to be getting into the act, with Norwegian and U.K. interests looking to invest. By 1986, finfish operations were producing about 500 tonnes worth $3.3 million, with shellfish, mainly oyster-raising, yielding another 2,900 tonnes worth $2.5 million.

The early stages of aquaculture caused no major controversy. Salmon farmers started with Pacific salmon, but after Minister Tom Siddon in the 1980's allowed use of Atlantic salmon,[51] most would switch to that fast-growing species. B.C.'s temperate waters and many inlets would soon support a far larger number of cages than on the Atlantic.

Native fishermen strive to keep place

When the Davis Plan started in 1968, Native fishermen held a fairly strong position in the fleet, operating their own and several hundred cannery vessels. As some fishermen began to buy up and pyramid licences, Native fishermen were more often among the sellers than the buyers, probably because of straitened circumstances. In general, the First Peoples made less money (always with exceptions; they had their share of high-line fishermen). Among their financial obstacles, Native people who lived on reserve were unable to mortgage their houses as bank collateral to help finance boats or licences.

The department took several steps to protect the Native position in the fishery. The Indian Fishermen's Assistance Plan ran from 1969 to 1979, in partnership with the Department of Indian Affairs and Northern Development (D.I.A.N.D.). The program would spend more than $16 million in grants and loans.

In imposing licence limitation for salmon, the department in 1971 created an "A-I" licence at lower fees, which could not be purchased under the buy-back program, and could be transferred only to other Indians. Through a loophole, however, many Indians converted A-I licences to full A's, and then transferred them to non-Native people (until LeBlanc in 1978 forbade the practice).[52] In 1973, the department allowed conversion of "B" licences held by Indians, which would eventually have expired, to A-I licences. In 1977, LeBlanc authorized several additional roe-on-kelp licences for Native communities. And in 1979, D.I.A.N.D. began buying up some vessels to create a "tonnage bank" for Native persons seeking to enter the fishery.

Despite such efforts, Native participation in the salmon fishery declined. For 1969, the department had estimated 533 salmon vessels as being owned by Native people; the figure dropped to 410 by 1979. An accompanying set of estimates, for vessels either owned or operated by Native people, fell from 910 to 670.[53]

In 1980, D.I.A.N.D. launched an Indian Fishermen's Emergency Program, administered by Native boards. This spent $3 million in grants and loans. Another early 1980's effort placed a substantial fleet in Native hands. B.C. Packers was planning to sell off the rental fleet from its Port Edward cannery near Prince Rupert. In an initiative begun under LeBlanc and carried through under De Bané, D.F.O. and D.I.A.N.D. sponsored the Northern Native Fishing Corporation (N.N.F.C.), a corporation formed by three tribal councils, to take over in 1982 about 250 vessels and licences for $11.7 million. The N.N.F.C. kept the fleet roughly intact under Native control. The corporation held the licences, and sold or leased the boats to Native fishermen.

Meanwhile, the Pearse report of 1982 recommended an Indian Fishermen's Economic Development Program, as proposed by the Native Brotherhood of B.C. Supported by LeBlanc and De Bané, this came to pass in 1985 in the form of the Native Fishing Association, managed by Native people for Native people. The association had $11 million, used to relieve debt burdens, finance vessel and licence purchases and vessel upgrades, and support training. At the end of the century, the Native Fishing Association remained a going concern.

With the various programs at work, and after the shakeout of the early 1970's, the Native fleet now seemed to be holding its ground. As of 1985, Native persons owned or operated more than 900 licences, out of a total fleet of about 4,500. The Native fleet took about one-quarter of the catch; and the fishing industry accounted for 25–30 per cent of Native employment in B.C.[54]

Native food fishery creates frictions

In 1974, Minister Jack Davis stated clearly the pre-eminent position of the Native food fishery. Davis wrote to the Union of B.C. Indian Chiefs, "as I stated at our meeting of November 16, 1973, Indians have a right to fish for food but not for sale or barter. Furthermore, that in the exercise of this right the Indian food fishery has priority second only to conservation."

The department regulated Native fishing times and places, often where commercial fishermen were shut out. For generations, fishery officers had been filling out departmental permits specifying where and when the Native individual or family could fish.[55] By the late 1970's, it was also becoming fairly common that the department would make agreements with Native bands that enacted fishing by-laws of their own. Both kinds of permit irritated two groups: the Indians, because of the controls on fishing and selling, and commercial fishermen, because Native people could fish under special rules. Commercial fishermen often accused Native people of poaching, or selling food fish in the commercial fishery.

Tensions were recurrent. In one instance, D.F.O. mounted a "sting" operation against poaching and illegal sales. This lead to the issuance of 101 court summonses in January 1983, mostly to Indians. Native leaders accused D.F.O. of discrimination and terrorist tactics; they charged that the department had deployed

70 armed officers "ready to shoot and kill," while bringing along TV and press people for maximum publicity. Fishery officers took the unusual step of writing Minister De Bané, asking him to uphold due process of law and not drop charges. After complications in court, D.F.O. did eventually drop the charges, leaving a bitter taste all around.[56]

The Pearse report in 1982 suggested a number of reforms to give Native bands more power and responsibility in administering food fisheries. And D.F.O. deputy minister Don Tansley, in a speech, went so far as to suggest limited commercial fisheries for Native bands in various parts of the country, coupled with better reporting and compliance. Neither recommendation bore fruit at the time. Food-fishery tensions would continue to simmer, and Native land claims would become more prominent in the early 1980's.

Meanwhile, as noted earlier, the Community Development component of the Salmonid Enhancement Program enlisted Native people. Of the 14 projects active in 1981, nine involved Native communities. The spawn-on-kelp fishery also gave some bands a boost.

Comprehensive management promises prosperity

From 1968 to 1984, fishery regulators on the Pacific embarked on perhaps the most comprehensive management system in the Western world, with limited entry and salmon enhancement the main features. The ride was bumpy, but despite the turbulence, both the industry and D.F.O. management seemed in a strong position by the end of the period.

On the management side, for salmon and herring, probably no fishery authority in the world understood a major resource better. D.F.O. had determined ocean migrations through tagging, possessed almost a century's worth of stream records and catch statistics, and knew the day-to-day habits of every run of salmon at the river mouths and in the streams. The conservation and protection system was also strong. By the late 1970's, fishery managers were complaining of enforcement cutbacks; but offsetting that, officials expected the new licensing system to control effort, benefitting both conservation and incomes.

For the mainstay fishery of Pacific salmon, the department still aimed to double the resource to historic levels, helped by the huge Salmonid Enhancement Program. The harvest would increase for all. Fish-handling on vessels had moved ahead under the Davis Plan. The salmon-canning industry was producing good-quality products for hungry markets. For Native fishermen, special programs were providing help, S.E.P. incorporated Native elements, and in certain fisheries, the rules ensured that licences would stay in Native hands.

National picture looks promising

Both coasts had seen, in the 1968–1984 period, the most changes since the Confederation era. National measures included setting up the 200-mile zone; creating a stand-alone Department of Fisheries and Oceans, with wide responsibilities; strengthening science and enforcement; beefing up habitat-protection laws and activities; imposing licence limitation; and controlling the size and capacity of boats.

On the Atlantic, the department was now using catch quotas widely and subdividing many quotas for different sectors. Individual quotas, individual transferable quotas, and Enterprise Allocations were promising more security for fishing interests. The department had set up more than a hundred industry–government advisory committees, and fishermen's organi-

The offshore patrol vessel *Tanu* (left) joined the fleet in 1968, the *James Sinclair* (right) in 1980.

zations on the Atlantic had gained strength comparable to that of Pacific groups. The industry now had a more systematic influence on management. On the Pacific, enhancement promised a doubling of salmon catches. On the Atlantic, the Kirby report had forecast a doubling of Atlantic groundfish catches, from less than 500,000 tonnes in the mid-1970's to a million tonnes by 1987.[57] On both coasts, earnings, profits, and the lives of fishermen seemed bound to improve; and for problem cases, fishermen could make use of the generous U.I. system.

Altogether, the new management system instituted by Davis and LeBlanc had provided far more of the "unified directing power" that H. Scott Gordon had deemed necessary. Management seemed to be advancing towards the goals, reflected in the 1976 policy document and many speeches and statements, of higher volume and value, more stability, and a stronger voice for fishermen and others. But the next two decades would turn much of the Atlantic and Pacific industry upside down.

PART 6: 1984–2000
MAKING THE NEW SYSTEM WORK
CHAPTER 26.
National and international events, 1984–2000

At the outset of the 1984–2000 period, the Atlantic groundfish resource looked good. On the Pacific, the early-1980's crisis had passed; proposals to cut the fleet in half and introduce Individual Transferable Quotas to the salmon fishery had faded away. Consumer consciousness of diet and health was hiking the value of seafood. Canada was still the world's number one fish exporter. The horizon looked clear.

No one predicted the great changes that would come in the 1990's. The storied cod resource on the Atlantic would collapse, creating the biggest single job loss in Canadian history. People subsidized into the northern cod fishery in the late 1970's would now be compensated out. As other groundfish stocks weakened, the taxpayers would fund more than $4 billion in special aid. New fisheries that hardly existed in the 1960's—crab and shrimp—would become dominant.

On the Pacific, oceanic conditions in the 1990's would join with the fishery to deplete the salmon resource. Pacific groundfish and shellfish, formerly of lesser significance, would come to outvalue salmon. A new aquaculture industry, farming Atlantic salmon on the Pacific coast, would leave the wild-salmon fishery behind. As Pearse had wanted, the government would indeed cut the fleet in half, taking back licences and paying compensation.

Despite the turmoil, by the end of the 1984–2000 period, many of those fishermen who had stayed in the industry looked to be doing as well as ever, although some were on the edge. Licence-holders in many fisheries also had a bigger voice in management. Ever paradoxical, the fishing industry had turned itself inside out, yet kept surviving and evolving.

Government intervention lessens

Prime Minister Trudeau had run an activist government, with such interventions as the Foreign Investment Review Agency, the National Energy

High-technology electronics helped D.F.O.-chartered aircraft patrol the 200-mile limit. But navigational and fish-finding electronics also increased the fishing capacity of foreign and domestic vessels. (Todd MacMillan, Provincial Airlines)

Program, the patriation of the Constitution, and the Charter of Rights and Freedoms. The Progressive Conservatives, coming into power under Brian Mulroney in 1984, did less pushing and prodding of industry, and were less inclined to public-service activism. The Conservatives wanted to spend less, launching a campaign to reduce the federal budget deficit. Reflecting those changes, D.F.O. became less interventionist and abandoned some programs.

The new government set up a Task Force on Program Review under deputy prime minister Erik Nielsen, to look at the whole government system. The report on fisheries, while calling for resource conservation and protection, found government expenditures higher and the industry's efficiency lower than they should be. The task force recommended less intervention for such purposes as protecting jobs or communities. Market forces should be freer to work; other government departments than D.F.O. should deal with any problems of community adjustment. Peter Meyboom, a well-known administrator with a reputation for cost-cutting, led the review from the public service side; he would later, in January 1986, become deputy minister of Fisheries and Oceans.[1]

The public–private tension also emerged in the question of allocations. Especially on the Atlantic, a running struggle took place between advocates and opponents of Individual Quotas and Individual Transferable Quotas. The backers of "quasi-property rights," as they were often called, believed they would import private-sector virtues into the fishery, giving enterprises more security and promoting rational investment rather than destructive races for the fish. They would also foster conservation, since the fishery would be more orderly and the "owner" would take more care of the resource. Opponents, mostly in the small-boat sector, believed I.T.Q.s would privatize the resource to the benefit of larger operators who were already best-off. Still, as time went on, many fleets decided to give them a try. I.Q. and I.T.Q. fisheries would come to account for more than half the landed value.

Co-management, with the industry taking part in running the fishery, became another two-edged issue. Fishermen and vessel-owners in principle wanted more power, but sometimes charged that D.F.O., under the cloak of co-management, was offloading more work and expense on the industry. Despite occasional friction, however, co-management was clearly on the advance.

Otherwise, the period saw few major changes in management ideology. The major effort went towards making the system put in place by Davis and LeBlanc work as it should.

Fraser takes over amid high hopes

On September 17, 1984, John Fraser took over as minister. The popular M.P. for Vancouver South had served in the Joe Clark administration as Minister of Environment and Postmaster General, then become fisheries critic in the opposition. Fraser had opposed De Bané's plans to cut the B.C. salmon fleet and bring in I.T.Q.s. Now those ideas vanished, and Fraser was in no hurry to change anything else.

Fraser liked to deal directly with people; on both coasts he held many meetings with industry representatives. (On occasion he favoured industry requests by upping quotas beyond scientific recommendations, a fairly common practice in the 1980's.) In 1985, Fraser instituted an Atlantic Regional Council of 18 people, to offer advice on key issues. This body would soon lose influence, but it lasted to 1993.[2]

Pacific Salmon Treaty promises equity, abundance

The international scene was calmer than in previous years, with the 200-mile zone in place. But on the Pacific, Canada and the United States were disagreeing about salmon interceptions, with Canada feeling short-changed.

After they extended jurisdiction in 1977, Canada and the United States had concluded yearly reciprocal fishing agreements, which more or less preserved the status quo but also provoked new arguments. The two countries then set out to make a broader agreement that would resolve old quarrels. A 1985 agreement on a new Pacific Salmon Treaty brought general elation.

The previous treaty had covered only Fraser River sockeye and pink salmon stocks. The new one would cover salmon fisheries coast wide: Alaskan interceptions of salmon heading south for British Columbia, Washington, and Oregon; Canadian fisheries on coho, chinook, and other species bound for Washington and Oregon; and northern B.C. interceptions of salmon returning to Alaska. The two countries would manage interceptions on the basis of equity. Each would get benefits commensurate with the amount of salmon its rivers produced.

This meant that Canada could now enhance any and all rivers, even if they supported American fisheries, and still be confident of receiving the benefits. A new international body, the Pacific Salmon Commission (P.S.C.), was to replace the International Pacific Salmon Fisheries Commission. Advisory panels of the P.S.C. would put forward recommendations on all the interception fisheries; the countries would normally put those recommendations into effect.

Disputes creep into the picture

For a few years under the new treaty, the two countries managed with some difficulty to agree on yearly fishing arrangements. Then disagreements became more serious. What was equity? There were different ways to measure it. Alaska proved the main stumbling block; fishermen in that state intercepted many salmon as they headed south towards B.C. rivers. An old agreement was coming back to haunt the Canadians: the surf-line arrangement of the 1950's had allowed

Alaskans to net-fish further offshore than the Canadians thought appropriate. Now that offshore fishery was hard to control.

The Alaskans consistently argued for higher interceptions than the Canadians thought reasonable. The lower U.S. states also differed at times with Alaska. That state maintained that it deserved every fish it took: it was doing its part for conservation, and taking no more than its share. To Canadians the whole affair matched an old pattern. Agreements with Americans tended to change after the fact, because their federal negotiators lacked the legislative power to make recalcitrant states conform to the deal. The whole dispute would simmer on into the 1990's.

I.C.O.D. aids developing countries

Apart from the salmon treaty, the early years of the 1984–2000 period saw no major changes in international arrangements. Canada continued cutting back on foreign allocations within the 200-mile zone. D.F.O. officials pursued Canada's interests through the Northwest Atlantic Fisheries Organization; other such organizations including those for wide-ranging species like tuna; and many bilateral arrangements. Some problems lingered. Canada had difficulties with Americans fishing on Georges Bank and with overseas vessels fishing the Nose and Tail of the Grand Banks. But generally, the world of 200-mile limits was settling into place.

Meanwhile, the Conservatives in 1985 followed through on a Liberal initiative by putting the International Centre for Ocean Development (I.C.O.D.) into operation. In the lead-up to the 200-mile limit, developing countries had often allied with Canada; the new crown corporation was in part a way to pay them back. Gary Vernon, an assistant deputy minister in D.F.O., moved over to head the new agency.

Headquartered in Halifax, I.C.O.D. embodied a new approach to development, de-emphasizing physical projects. I.C.O.D. worked instead to impart management skills and knowledge that would help smaller countries, particularly island countries, deal with their new 200-mile zones. By 1991, I.C.O.D. had supported 270 projects involving commitments of nearly $46 million. For example, in the Caribbean I.C.O.D. organized a fisheries unit as part of the Organization of Eastern Caribbean States, to foster regional efforts in such matters as surveillance, statistics, and marketing. In the South Pacific, I.C.O.D. consultants worked on monitoring, control, and surveillance; set up a Pacific Islands Marine Resources Information System; and managed a Canadian International Development Agency (C.I.D.A.) project to cover all aspects of ocean development. In the south and west Indian Ocean, I.C.O.D. worked on such matters as tuna management, fish inspection, aquaculture, and coral reef research. In West Africa, support went to stock assessment, data gathering, and ocean research.

Other efforts went to non-fisheries matters, such as maritime boundary delimitation and offshore minerals.

Scholarships and training got close attention. In Canada, I.C.O.D. helped generate graduate-level courses in Marine Affairs at Dalhousie University and the University of Quebec at Rimouski, for foreign and domestic students.

To the distress of international-aid proponents, the organization fell victim to cutbacks in the federal budget of 1993. The Canadian International Development Agency took over the I.C.O.D. projects then under way.

Tuna controversy unseats Fraser

Minister John Fraser never got to oversee either the Pacific Treaty or I.C.O.D. In 1985, a headline-grabbing controversy brought the well-liked parliamentarian's term as minister to an end. Under a regional economic development scheme in the 1960's, the American company Ocean Maid Foods had built a tuna plant in St. Andrews, New Brunswick. Taken over by the StarKist corporation in the early 1980's, the plant was a major local employer. It operated six large Canadian-licensed seiners, which spent their time in tropical waters, shipping the fish back. Tuna also came from other sources.

In 1983, D.F.O.'s inspection officers found serious problems with StarKist practices. StarKist maintained it was following suitable methods; changing its ways to suit D.F.O. would entail serious expenses. Some New Brunswick politicians took the company's side. Departmental tests showed problems with quality, but no health risk. Fraser took no strong action.

By 1985, media were inquiring about the matter. Eventually, Eric Malling of the CBC program *The Fifth Estate* broke what became known as the "tainted-tuna" scandal. While Fraser insisted the tuna was healthy enough, the media and public almost gagged at the idea of impure food. Opposition parties raised a storm in the House of Commons. During a complicated series of events, Prime Minister Mulroney became embroiled in the issue. Mulroney had what appeared to be a public difference of opinion with Fraser, and on September 23, 1985, removed the minister, after only a year in the portfolio. (After a period on the backbenches, Fraser in 1986 became Speaker of the House of Commons, the first to be elected to that position by his peers.)

Erik Nielsen, a Yukon M.P. and the Minister of National Defence, took over as acting minister. British Columbia M.P. Tom Siddon then inherited the D.F.O. portfolio on November 20. An engineer and former university teacher, Siddon had during the 1979–1980 Joe Clark government served as parliamentary secretary to D.F.O. Minister James McGrath.

Around Christmas 1985, Peter Meyboom became deputy minister, succeeding Arthur May, who moved on to lead the Natural Sciences and Engineering Research Council, after which he became president of Memorial University of Newfoundland.

Meyboom carried through organizational changes, some of which had begun under May. The original D.F.O. set-up in 1978 had included assistant deputy ministers for Atlantic Fisheries, Pacific and Freshwater

Fisheries (including habitat work), Ocean and Aquatic Sciences (this meant oceanography, charts, tide books, and so on, as distinct from fisheries research), and Fisheries Economic Development and Marketing. As the organization coalesced under Meyboom, A.D.M.s remained for Atlantic and Pacific fisheries. Both fisheries and oceanographic research now came under an A.D.M. Science, Scott Parsons. There was no more A.D.M. for Fisheries Economic Development and Marketing; indeed, most marketing work disappeared. New A.D.M.s appeared for International (Victor Rabinovitch) and Policy (Louis Tousignant). An A.D.M. "Corporate" for administrative matters would appear later. In future rejiggings, the number of A.D.M.s would occasionally rise and fall.

Peeling back the programs

With government out to fight the deficit and the industry looking stronger, some programs launched during the post-war age of development now appeared less useful. Especially under Siddon, they began to vanish. Some changes reflected the views of the Nielsen task force.

Since the war, the department had subsidized construction of many new fishing vessels. In the 1970's, numbers could run to several hundred a year. Now, with fleet over-capacity widely recognized, the department in 1986 ended its boatbuilding subsidies. The Department of Industry, Trade and Commerce had also given a separate subsidy for vessels over 65 feet long, including many large trawlers. This program ended by 1985.

Since the mid-1950's, the Fisheries Improvement Loans program had been guaranteeing loans for fishermen who borrowed from banks and other creditors. In June 1987, the government stopped authorizing such loans, but amended the Small Business Loans Act to make fisheries loans available. Another post-war program, the Fisheries Prices Support Board, continued as an entity but lost most of its leeway for action. The F.P.S.B. made few interventions after the early 1980's, except for annual purchases of $2–$4 million worth of canned mackerel for the World Food Program. In 1995, the government repealed the board and its act.

The Fishing Vessel Insurance Plan, however, continued through the 1980's. In 1987, for example, it covered about 8,000 vessels, for more than $400 million.[3] And the special Unemployment Insurance program for fishermen proceeded unabated. Payments shot up from $20.4

The Maurice Lamontagne Institute, on the St. Lawrence River estuary, became an important research centre.

million in 1972–1973, before LeBlanc expanded the benefit period, to $270.1 million in 1988–1989.[4]

On the research side, in the early 1980's when money was still flowing, De Bané had initiated the Maurice Lamontagne Institute in Sainte-Flavie, near Mont-Joli, Quebec. This major establishment began operating in 1987. By the mid-1990's, it employed more than 250 personnel in fisheries science, oceanography, hydrography, and environmental sciences.

But elsewhere, budget pressures and reorganizations were affecting research. In the latter 1980's, the Marine Ecology Laboratory (M.E.L.) at the Bedford Institute of Oceanography closed down, causing a public stir. Although M.E.L. had its supporters, critics charged that it operated too much in isolation. Ecosystem work continued elsewhere and became more integrated with other science. The Arctic Biological Station near Montreal closed in 1992. Arctic research continued at the Maurice Lamontagne Institute and the Freshwater Institute in Winnipeg.

The D.F.O. science sector set up eight "centres of disciplinary excellence," to concentrate expertise in particular subjects at specific locations. At the same time, however, hiring of young scientists slowed down. Cutbacks were dulling the edge of D.F.O. science.

A research advisory council also closed down. After the Fisheries Research Board closed in the 1970's, the Fisheries and Oceans Research Advisory Council (F.O.R.A.C.) had begun reporting on subjects—aquaculture, for example—at the minister's behest. The principle seemed valid: that experts of long experience would offer sage advice. But F.O.R.A.C. never influenced events in a major and public way, although it sometimes helped to shape opinions within the department. F.O.R.A.C. closed down in 1995.

Quality, marketing become less interventionist

The department had launched an Atlantic Quality Improvement Program in 1980 under LeBlanc. The whole effort was supposed to result in compulsory grading of raw material and final product by 1985–1986. But delays had occurred. Officials had slowed down their activities while waiting for the Kirby task force to pronounce itself. Kirby had then come out in support of compulsory grading, along with mandatory bleeding, gutting, washing, and icing of groundfish at sea.

But De Bané had failed to push the changes through before the 1984 election. Meanwhile, fishermen and processors were resisting the new interferences. The department when looking at existing practices saw spoilage, waste, and foregone profits. The industry when looking at proposed changes saw more costs, with no guarantee of better prices. In the mid-1980's, inspection officials pressed for a decision on implementation.[5]

What occurred was a backing off. It appeared that departmental efforts, market demands, and industry changes were already producing improvements. The push to improve quality slowed down. There were fewer subsidy programs for ice, freezing, or other purposes, and no more talk of compulsory bleeding, gutting, or grading.

While continuing to preach quality, the department began to modify its inspection practices. The traditional practice had been for inspection officers to back up their persuasion with checks on both plants and products. Under the Quality Management Program begun in 1990, the emphasis changed towards approving the manufacturing processes within the plants. Inspection officers spent more time looking at the system, rather than the products.

The department had done consumer promotion work for decades, publishing cookbooks, advertising occasionally, and developing recipes at the Fisheries Food Centre. Promotion had taken a big jump under LeBlanc, and marketing studies had increased. But in 1986 under Siddon, the test kitchen and the whole marketing branch closed down. Some market analysis continued under other branches, but governmental fish promotion was belly-up, except for occasional special efforts in conjunction with trade shows. The intent was to privatize such activities; D.F.O. gave funding assistance to the newly established Canadian Seafood Advisory Council and to the Fisheries Council of British Columbia.[6]

The various cutbacks caused no great outcry. Governments had created many programs in the name of development. Now the industry seemed better able to stand on its own.

Tom Siddon

From "development" to "responsible fishing"

In the early 1970's, after the department shut down the industrial development branch at headquarters, regional work with some Ottawa co-ordination had continued on vessels, gear, and related technology. Over time the focus changed, with less attention to volume and more to value and costs. Projects often related to quality, selective fishing, and fuel-saving techniques. Somewhat of an upsurge of such work took place in the early 1980's, a chief force being Bruce Deacon in Ottawa. A 1983–1984 report noted that an old rule of thumb—increased quantity meant increased profits—no longer applied. "Advances in technology and improvements in fishing techniques have led to excessive harvesting capacity for the limited fish stocks available and the emphasis on quantity led to a reputation for inconsistent quality."[7]

In Newfoundland alone, as of 1983–1984, pilot projects were proceeding on fish containers for open boats and on containers and improved fish-pen systems for trawlers. A computer-equipped van was visiting inshore ports to help fishermen analyze fuel-saving factors such as propeller size. The department was equipping four boats with fuel-monitoring systems, and in conjunction with the College of Fisheries, Navigation, Marine Engineering and Electronics, was building two cost-efficient demonstration boats. Experiments were going forward on freezing excess trap cod in summer and processing it in winter. The department and industry members were evaluating inshore and middle-distance longlining systems; surveying for crab on the north and south coasts; dragging for groundfish on the Labrador coast; trying out shrimp pots in Trinity Bay; and experimenting with a mobile blast freezer, mechanical salting systems, and new kinds of saltfish drying trays. Other regions had their own long lists, except in British Columbia, where development work was now minimal.

The early 1980's push was development's last big fling. Most Ottawa work fell victim to cutbacks in 1986. As years passed, all the regional branches devoted to development closed down, except in Newfoundland. Elsewhere, related work continued under the aegis of other branches, but with less energy than before.

Still, some interest remained in selective and conservationist fishing methods, especially as groundfish conservation problems became evident. In Ottawa, from the early 1990's, a "responsible fisheries" unit sponsored regional work on selective gear. For net fishing, work went ahead on mesh sizes, square meshes, rigid grids to let undersized fish escape, and separator grids by which trawls could shunt aside unwanted

383

species. Other projects experimented with escape panels for trap fisheries, hook sizes, electronic devices to
avoid marine mammal entanglements, and so on.
More than 100 such projects went forward in the
1990's. The department helped sponsor a Fishing
Technology Network, centred at the Marine Institute of
Memorial University of Newfoundland and linking
experts across Canada.

D.F.O. also helped generate a Code of Conduct for
Responsible Fishing. Canada had worked with the
F.A.O. to develop an international code in 1995.
Follow-up consultations backed by D.F.O. in 1998 pro-
duced a Canadian code, the first such industry-devel-
oped effort in the world.

The department worked with training institutions to
impart responsible fishing techniques. Newfoundland,
Nova Scotia, New Brunswick, and Quebec all had fisheries colleges, and P.E.I. and B.C. institutions also did
some related training. In the latter 1990's, D.F.O. and
the New Brunswick authorities developed an industrial training program in responsible fishing, which
spread to some institutions outside the province.

Habitat policy prescribes "no net loss"

Another national approach came in the domain of
habitat. LeBlanc's 1977 amendments had strengthened the Fisheries Act's environmental power. As the
economy grew, habitat issues cropped up across the
country in relation to dams, chemicals, acid rain, land
use, foreshore development, offshore minerals, and so
on. The department produced discussion and policy
papers; consulted with builders, environmental
groups, and other parties; and in 1986, under Siddon,
published its *Policy for the Management of Fish Habitat.*

The overall objective was a net gain in fish habitat,
through conservation, restoration, and development.
The guiding principle was "no net loss." "Under this
principle," the document said, "the Department will
strive to balance unavoidable habitat losses with habitat replacement on a project-by-project basis so that
further reductions to Canada's fisheries resources due
to habitat loss or damage may be prevented."

This approach meant ever-increasing work on a long
list of issues. Over time, the department designated
many more officers to deal with habitat. Most of the
work came through "referrals." Other government
agencies would notify D.F.O. of new projects. Officials
then would consult with the construction company and
other parties, saying what was allowable and what
wasn't, advising on protective techniques, and suggesting mitigative measures and chances to create off-set-
ting gains. As well, fishery officers and habitat officials
sometimes scouted out violations or headed off threats
to habitat.

D.F.O. was now doing a better job on habitat, and
construction companies were improving their practices. But the threats and pressures were never-end-
ing.

Observers, aircraft, armed boarding enhance enforcement

After the Davis and LeBlanc build-up, the D.F.O.
fleet as of 1984 was substantial. Major fisheries
research vessels on the Atlantic included the *Wilfred
Templeman* (replacing the *A.T. Cameron*), the *Alfred
Needler*, the *E.E. Prince*, and the chartered *Lady
Hammond* and *Gadus Atlantica*; the latter was the first
fisheries research vessel able to work in ice. On the
Pacific, the *G.B. Reed* was soon joined by the *W.E.
Ricker*. The *Dawson*, the *Hudson*, the *Maxwell*, and the
Baffin pursued oceanographic and hydrographic
research from the Bedford Institute of Oceanography in
Nova Scotia; the *Parizeau*, from the Institute of Ocean
Sciences in British Columbia. Other sizeable hydro-
graphic vessels operated on the Great Lakes. Smaller
science vessels numbered nearly 20, divided between
the two coasts.

In the fisheries patrol fleet, major vessels on the
Atlantic included the *Cape Roger*, the *Chebucto*, the
Cygnus, the *Pierre Fortin*, the *Louisbourg*, and the new
Leonard J. Cowley, and on the Pacific the *James
Sinclair* and the *Tanu*, successors to the *Laurier* and
the *Howay*. About 70 intermediate patrol vessels, gen-
erally in the 40- to 70-foot range and built in the
1940's, 1950's, and 1960's, operated on the two coasts,
bearing such names as *Badger Bay*, *Cobequid Bay*,
Cumella, *Rustico Light*, *Sooke Post*, and *Thrasher Rock*.
Several hundred smaller craft made up the rest of the
fleet.

Canadian observers aboard foreign vessels kept an
eye on licensed fishing within the zone. But on the
Atlantic, foreign fishing still presented a problem out
beyond 200 miles, on the Nose and Tail of the Grand
Banks. Those vessels with authorization to fish under
N.A.F.O. might break the rules. Others might show up
under flags of convenience, with no obligations under
N.A.F.O.

In the mid-1980's, the department on the Atlantic
began leasing air surveillance craft and crews from private companies, to monitor fishing offshore and at the
Georges Bank line. In Newfoundland, the *Leonard J.
Cowley* carried a helicopter during a three-year pilot

The science vessel
Baffin, based at the
Bedford Institute of
Oceanography, did
multidisciplinary
work from 1957 to
1991, often in the
Arctic.

Built in 1984 and named after a highly
respected official, the 72-metre
Leonard J. Cowley patrolled offshore
waters out of St. John's, Nfld.

project. Although this effort lapsed, fixed-wing surveys continued, with excellent radar and computer equipment. (The department also operated an enforcement and search-and-rescue helicopter out of Yarmouth, N.S. from the late 1980's to the mid-1990's; in following years, a Canadian Coast Guard (C.C.G.) helicopter of lesser capability continued in the area.)

"Armed boarding" policy brings clash with United States

D.F.O. in 1987 launched an "armed boarding" program for offshore patrol vessels. The department mounted 50-calibre machine guns on patrol ships, and trained ships' crews and fishery officers to board recalcitrant vessels. A set of procedures applied: if a vessel resisted, the responsible A.D.M. could approve warning shots, and higher levels could approve stronger action. This brought about a tense situation with the Americans.

After the 1984 decision on the Georges Bank boundary, American captains sometimes sneaked across the Hague Line for better fishing. From 1984 to 1988, the Canadians took into custody about two dozen American vessels, bringing them into Nova Scotia ports and holding them until a trial took place. Most of the Americans on board were regular, good-humoured fishermen. But some were rough customers; one vessel carried an Uzi submachine gun.[8]

Donna Lynn incident brings gunfire

In October 1988, the Canadian patrol vessel *Cygnus* spotted the American vessel *Donna Lynn* on the Canadian side. The *Donna Lynn* refused to heave to. The patrol vessel set the new policy in motion. Regional officials got approval from Atlantic A.D.M. Wayne Shinners to fire warning shots well ahead of the *Donna Lynn*. When the patrol vessel fired the shots, the *Donna Lynn* still refused to heave to, and fled home to Massachusetts. American authorities later took the captain to court. But the shots had perturbed the Americans.

Meanwhile, D.F.O. enlisted the Canadian navy's help in patrolling the line. At times the navy sent submarines to the area. In December 1989, during a strike of civilian ships' crews, a navy surface vessel with a fishery officer on the bridge spotted the American vessel *Concordia* fishing on the Canadian side, and tried to bring her in. The naval vessel fired warning shots. Even this failed to overawe the rambunctious American fishermen. At one point, the *Concordia* caused consternation by allegedly trying to ram the thin-hulled frigate. The *Concordia*, too, got away.

By now the U.S. government was fully awake to the matter. The State Department got involved. In 1990, Canada and the United States signed a reciprocal enforcement agreement applying to both coasts, through which each country undertook to impose penalties for home-state vessels transgressing the other's fisheries regulations. As well, U.S. Coast Guard vessels began patrolling the Georges Bank line along with Canadian vessels. American poaching subsided. The system for the rest of the century appeared to work reasonably well.

Four ministers in less than four years

On February 22, 1990, Tom Siddon left D.F.O. to become Minister of Indian and Northern Affairs. New Brunswick M.P. Bernard Valcourt, a lawyer and former minister of Consumer and Corporate Affairs, took over for a year and two months. John Crosbie, the renowned Newfoundland politician, then became minister from April 1991 until June 1993. Ross Reid, another Newfoundland M.P. and a former assistant to Crosbie, served as minister for four and a half months (June 25–November 3, 1993), after which a new Liberal government took over. In the narrative of the period, many events overlapped ministers.

In 1990 Bruce Rawson, a well-known public servant, succeeded Peter Meyboom as deputy minister. Rawson brought in Art Silverman and Maryantonette Flumian to join the corps of A.D.M.s. The trio was hard-driving, controversial, and sometimes impatient with departmental practices and traditions.

Since D.F.O. became a department in 1979, separate A.D.M.s had overseen Atlantic and Pacific fisheries. In 1991, Rawson put fishery management on both coasts under a single A.D.M. First in the position was Jean-Eudes Haché, who after serving as Regional Director-General for the Gulf and Scotia–Fundy Regions had become A.D.M. Atlantic.

Sparrow decision endorses Native rights

Meanwhile, in 1990 when Bernard Valcourt was minister, a court decision re-stoked the smouldering issue of Aboriginal fishing rights. The department had long tried to separate the Native food and the general commercial fisheries, forbidding commercial sales from the food fishery. Particularly on the Pacific, fishery officers gave permits to individuals or families to fish in specified times and areas. By the latter 1970's and 1980's, D.F.O. sometimes made agreements on a band-by-band basis.

Many Native people felt they should be able to fish under their own authority. In some instances, they argued that treaties backed them up. Disputes emerged particularly over salmon, and particularly in British Columbia. Some incidents went to court. Until the 1984–2000 period, judgements generally supported the department and the powerful Fisheries Act.

From at least 1974, the department had considered Native food fisheries to have first priority after conservation.[9] It also agreed in some instances that bands could manage their food fisheries, within department regulations. But it considered food fishing a government-regulated privilege, while Native groups often considered food fishing a right. The department viewed its authority as essential; the alternative would be chaos.

In May 1990, the Supreme Court of Canada ruled on a Native fishing case in British Columbia. The Sparrow decision found that Native peoples had a right to take salmon for food, social, and ceremonial purposes. This strengthened the Native position. Their fishery wasn't just a matter of departmental discretion; it was a legal entitlement. D.F.O. responded in 1992, under minister John Crosbie, with an Aboriginal Fisheries Strategy (A.F.S.) aiming to foster greater Native participation in the fishery in an orderly manner, and to improve relations generally. Various regions set up "cross-cultural training" sessions to acquaint officials with the Native point of view and way of life. Some joint scientific and enhancement work took place.

Under an Aboriginal Guardian program, hundreds of Native people across the country took training and began helping to police the fishery. Sometimes results were good, sometimes poorer than hoped, especially since the program, like others, suffered from major cutbacks starting in 1994. The department also recruited Native persons to take training and become regular fishery officers. At the end of the century, about a dozen Native persons served as regular fishery officers, and about 250 as guardians.

As for fishing for food, social, and ceremonial purposes, D.F.O. took the stand that this would operate under regulation, and through communal licences granted to bands. The department made band-by-band agreements on the coasts, setting up trap limits, quotas or other fishing limitations, and in some instances providing help to improve Native fisheries. Inland provinces made similar arrangements for freshwater fisheries.

The department also encouraged Native participation in commercial fisheries. As part of the Aboriginal Fisheries Strategy, the Allocation Transfer Program starting in 1994–1995 helped Native people acquire about 200 licences in British Columbia, and more than 600 on the Atlantic. About 70 communal groups held licences for a range of species, including salmon, herring, crab, shrimp, and eels. The program to the turn of the century spent about $60 million, mostly in B.C., acquiring licences (the biggest expenditure), boats, and gear for Native benefit.

Commercial fishermen raised no objections to fishing on equal terms with Native people in the regular fishery, but often resented the food, social, and ceremonial fisheries taking place under different rules outside the regular seasons. In the Maritimes, commercial fishermen protested against increased Native effort in the lobster and other fisheries. In B.C., a specially formed industry lobby group throughout the 1990's criticized the conduct of the food fishery. On both coasts, non-Native fishermen charged that the enlarged food fishery served in large part as a disguised commercial fishery, with Native people selling their catch.

In British Columbia after the Sparrow case, the department worried about policing certain fisheries. A new initiative emerged. The department would grant some new commercial fishing privileges, while seeking Native co-operation to keep such fisheries under control. Experimental "pilot sales" programs went ahead, covering parts of the Skeena and Fraser systems and the Somass River on the west coast of Vancouver Island. Department officials worked with Native authorities to monitor the new commercial fisheries. The arrangement brought a storm of protest from some commercial interests, but continued at least until the early 21st century, when a court judgement called pilot sales into question.

For the most part, court judgements in the 1990's tended to back up Native rights. In some cases, the federal government made special fishery-management agreements with Indian or Inuit representative bodies. The two most notable were major agreements with the Nisga'a people in British Columbia, and with the administrators of what became Nunavut. The local authorities took on more power; D.F.O. adjusted accordingly.

Crosbie campaigns to control high-seas fishing

On the Atlantic, John Crosbie remains in memory as the minister who in 1992 announced the historic moratorium on commercial fishing for northern cod (the gigantic stock complex off southern Labrador and eastern Newfoundland). That subject is addressed

John Crosbie

later. But the codfish saga also involved international relations.

As noted earlier, right after the 200-mile zone came into place Canada often granted foreign allocations of "surplus" cod and other species. The Law of the Sea called on coastal states to allocate surplus fish, and in some cases, Canada's bilateral treaties obliged sharing. Such allocations were deemed to foster good relations, encourage foreign co-operation in conservation, and help open foreign markets.

But in the early 1980's, following industry pressure and a recommendation by the Kirby task force, Canada changed its approach. Quotas fell for such countries as Spain, a major fisher at the edge of the zone. Allocations now consisted mainly of less desirable species, such as redfish, turbot, silver hake, squid, grenadier, and argentine.

As of 1985, about 130 foreign vessels spent about 6,600 fishing days in Canadian Atlantic waters, taking about 280,000 tonnes, down from about 600,000 tonnes in 1977. Besides allocations of less-valuable fish, Canada still granted small quantities of "non-surplus" fish of more desirable species, particularly cod. Some of the latter allocations came from treaty obligations to France, some from the pursuit of particular conservation or trade objectives.[10]

Foreign allocations within the zone would keep shrinking, especially with the 1990's groundfish collapse. By the turn of the century, they would come to less than 2,000 tonnes.[11] Meanwhile, the bigger problem came from foreign fishing outside the zone.

After Spain and Portugal joined the European Community in 1986, their new influence made the E.C. fisheries authorities less friendly to Canada. The Northwest Atlantic Fisheries Organization regulated several fisheries for straddling and high-seas stocks on the Nose and Tail of the Grand Banks and the Flemish Cap. Besides Canada and the European Community, N.A.F.O. members included Bulgaria, Cuba, Denmark on behalf of the Faroe Islands and Greenland, the German Democratic Republic, Iceland, Japan, Norway, Poland, Romania, and the Soviet Union.

N.A.F.O., like its predecessor I.C.N.A.F. and most such organizations, had an objection procedure. Any member could choose not to be bound by a N.A.F.O. decision it didn't like. In 1986, the European Community began objecting to N.A.F.O. quotas and setting unilateral, higher quotas for its fleet. Then the E.C. vessels, mainly Spanish and Portuguese, often overran even their own unilateral quotas. In terms of the original, N.A.F.O.-set quotas, the E.C. vessels from 1984 to 1990 exceeded them fourfold. They were allocated 214,000 tonnes; they caught, by their own reports, 836,000 tonnes. Canadians said that the overrun was even greater: 911,000 tonnes.[12]

When John Crosbie became minister in 1991, Canadian officials were already engaged in a diplomatic and public relations campaign to curb the Europeans. Some D.F.O. officials were frustrated enough to talk of unleashing the Canadian fleet to carry out competitive overfishing. But most of the thinking went towards new international arrangements to curb overfishing.

Crosbie supported and took part in the international campaign. Among officials, Bob Applebaum of D.F.O.'s international directorate spearheaded the push for a new United Nations treaty on high seas and straddling stocks. The idea was that when a flag state took no action after a violation was cited, non-flag states would have authority to do so. At the 1992 United Nations Conference on Environment and Development, Canada won agreement for an international conference on high-seas fisheries. Those diplomatic efforts would bear fruit later in the decade.[13]

Canada–France dispute gets settled

One complex dispute got resolved in the early 1990's. Back in 1972, France had to withdraw its overseas trawlers from fishing in the Gulf of St. Lawrence. France also gave a partial nod to the idea of a further Canadian extension of jurisdiction.

Canada in return allowed French vessels to take 20,500 tonnes of Gulf cod until 1986. Under the agreement, even after 1986, up to ten trawlers from St. Pierre and Miquelon would be able to continue fishing in part of the Gulf. As well, vessels from metropolitan France would have allocations in the 200-mile zone. Neither right was quantified in the treaty.

When Canada extended jurisdiction in 1977, France made her own claim to an extensive zone around St. Pierre and Miquelon. This claim created a large disputed area off the south coast of Newfoundland. Then another disagreement emerged, about the workings of the 1972 treaty. In 1985, Canada licensed the St. Pierre-based factory freezer trawler *La Bretagne* to fish in the Gulf, but only to head and gut fish, not to fillet them in factory style. France protested, and an arbitration came down against Canada.

As 1986 and the Gulf phase-out of French vessels approached, the French stepped up their efforts around St. Pierre and Miquelon, in N.A.F.O. division 3PS. France was determined to maintain its newly increased fishery around "SPM," as the islands were known in Canadian official shorthand. The French increased their catches to well above the limits set in bilateral agreements with Canada, to a degree that threatened conservation and the livelihoods of nearby Newfoundland fishermen.[14] France also wanted more quotas in non-Gulf waters, in excess of its fishing levels before exclusion from the Gulf. Arguments over quotas became heated, with Lucien Bouchard, then Canada's ambassador to France, and officials in the Prime Minister's Office sometimes muddying the waters. (John Crosbie, though not yet D.F.O.'s minister, got involved as minister responsible for Newfoundland. Crosbie took to calling one P.M.O. official "Dr. Death.")

A proposed 1987 agreement, including the promise of northern cod quotas for France, produced outrage in Newfoundland, amplified nationwide by the outspoken premier Brian Peckford. The situation got still more complicated when St. Pierre and Miquelon authorities arrested a Canadian vessel fishing near their shores. Further negotiations produced, in 1989, a temporary agreement on fish quotas (including a small bit of northern cod for France). The two countries also agreed to send the boundary issue to international arbitration.

In 1992, the arbitration gave France a zone of 24 miles southwest of the islands and also tacked on an odd-looking corridor, only 10 miles wide, extending south to 200 miles. France got nowhere near its original large claim; nor did it get extensive fishery resources. France still had the benefit of the 1972 agreement, which allowed for continued fishing in Canadian waters. Crosbie, now Minister for D.F.O., took a hard line on French quotas, and the collapse of groundfish stocks quieted the issue.[15] An agreement in 1994 brought stability, with small French allocations in several stocks.

Crosbie tries to delegate quota-setting, sharpen penalties

On the domestic scene, Crosbie and the department had grown weary of quarrels about fish allocations. It seemed preferable to set up an arm's-length authority to deal with the matter, operating under ministerial policy. (The Kirby and Pearse reports had made recommendations in the same direction.) Crosbie had a discussion document prepared, and pushed the idea where he could.

Industry reactions were divided. An arm's-length board might have advantages, but it would also take away democratic recourse to the minister regarding management of a common-property resource. In most regions, fishermen and processors despite their complaints had established good working relations with departmental officials. Some feared that an allocation board could overrule the regions and bring in arbitrary, ill-considered decisions. Crosbie set draft legislation into motion; the 1993 election derailed it. But the idea of allocation boards stayed alive in departmental thinking.

Crosbie had little confidence in the virtuous instincts of fishermen regarding conservation, sometimes referring to them as "pirates." (Crosbie was always ready to take on all comers. As he once remarked to his assembled D.F.O. managers, "Every hand is against us, so shag 'em.") At one point he empowered regional directors-general to impose administrative sanctions, such as loss of fishing time or quotas, against rule-breakers. Legal complications later constrained this authority. Meanwhile, maximum fines increased in the early 1990's to as high as $500,000 for fishery offences and $1 million for habitat offences.

Fishery officers home in on enforcement

Enforcement and compliance had been a continuing problem ever since regulations were invented. Fishery officers remained the backbone of the enforcement system. About one-quarter of them were in British Columbia. Of the more than 400 on the Atlantic, most worked inshore, and about half were seasonal. In addition, D.F.O. regions sometimes used part-time

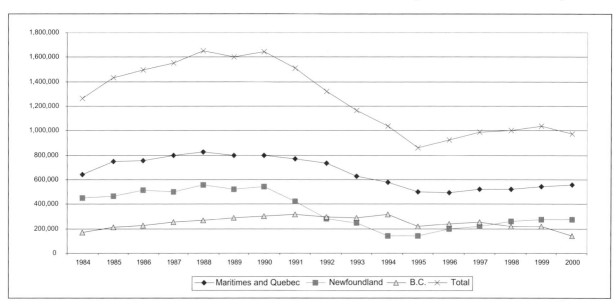

Sea-fishery landings in Canada, 1984-2000, by region (tonnes). Landings peaked in the latter 1980's, then fell into what was, for some fisheries, a disastrous decline.

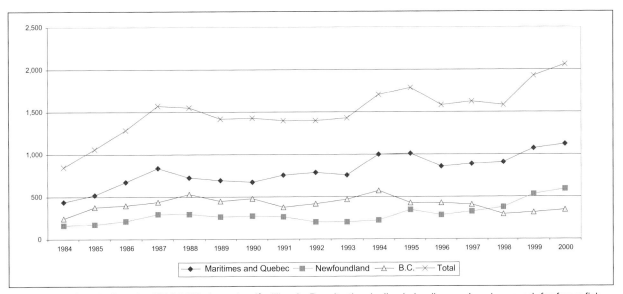

Sea-fishery values in Canada, 1984-2000, by region ($millions). Despite the decline in landings, values increased, for fewer fishermen and plant workers.

guardians, typically at peak season on salmon rivers. During the 1990's, the department more and more worked with Aboriginal guardians, as well as co-operating with provincial agencies and private conservation groups.

Although enforcement was his or her main job (female officers became more common in the 1980's and 1990's), the fishery officer had traditionally carried out many other duties, representing fishermen and the department to one another, conveying information, explaining policies, making recommendations, collecting statistics, and so forth as need be. Fishery officers often took a part in management, particularly in British Columbia, where they might sometimes open or close salmon or herring fisheries on their own judgement.

But being a generalist could cut into enforcement work. From 1993 on, partly for administrative reasons, the department channelled fishery officers more strictly into enforcement. The impact was strongest in British Columbia. There the number of fishery officers dropped, as some switched into new positions that specialized in habitat or resource management work and others migrated out of D.F.O.[16]

Some old-time officers complained about the changes. Still, statistics suggested an improvement. The national total of violations detected had already climbed from 2,400 in 1988 to 3,300 in 1993. In the next five years, it more than doubled to 7,200. By the year 2000, fishery officers were reporting some 9,900 offences. (Some of the increase came from increased use of ticketing in British Columbia for minor offences, particularly in the recreational fishery. An officer who might hesitate to haul someone into court for a minor offence felt less compunction about giving him a ticket.)

The gains seemed to outweigh the mild complaints of some fishery officers, who missed their more general duties and feared that their narrower role would impede their chances for promotion into management. In another 1990's change, the different regions moved away from the use of seasonal fishery officers, still common in many areas. Full-time employment became the general practice.

Beginning in the 1970's, fishery officers had carried guns in some areas or in special circumstances. Firearms policies clarified the situation in 1985 and 1991. Over time, some regions began issuing guns to all officers. Others, notably Newfoundland, Quebec, and the inland regions, at first opted out. But by 1997, fishery officers everywhere carried guns,[17] to the dislike of some fishermen.

Industry pays for "dockside monitoring" and observers

Meanwhile, the spreading use of individual quotas had complicated monitoring and enforcement. How to keep track of all the quotas?

In the traditional enforcement system, fishery officers had at times checked landings, but more for illegal sizes or species than for amount caught. At the plants, catches got recorded on a purchase slip. As Total Allowable Catches and then sub-quotas became popular in the 1970's, the statistics took on an enforcement use. Officials monitored them and would close a fishery when the overall quota was gone.

Over time, as the department began further slicing up the harvest into smaller portions including some individual quotas, misreporting became a problem. Fishermen might catch fish from one area and report them from another. Or, they might collude with buyers to report a lower catch or different species than actually landed. Misreporting could delay fishery closures, distort catch statistics, and contribute to false estimates of stocks by scientists. It could also enable the parties involved to evade taxes.

In 1989–1990, various groundfish vessels in Scotia–Fundy and the Gulf of St. Lawrence began converting to I.Q.s., worsening the monitoring problem. The department instituted stricter reporting requirements. These could include "mandatory hailing," with captains calling in their catches by radio, and dockside inspections. At the same time, the department shifted more responsibility to the industry. Under "dockside monitoring programs," monitors would go to the landing points to check the catch for themselves; fishermen would pay for the service. Private companies sprang up for the purpose.

As I.Q.s, I.T.Q.s, and E.A.s spread on both coasts, so did fisherman-funded dockside monitoring. By 1995, it had in most regions displaced most of the old purchase-slip system. The chief exceptions were such large, non-quota fisheries as Atlantic lobster and Pacific salmon.

The new system had its weaknesses. Dockside monitoring told nothing about catches taken in unauthorized areas or dumped at sea. As for reported landings, suspicions arose in some instances of "weighmasters" (as they were sometimes nicknamed, although checking was usually by eye only) colluding with fishermen to cheat the system. In the Bay of Fundy herring fleet, some owners built false holds to trick the monitors.

Reforms took place over time. The department introduced conflict-of-interest rules and required training of dockside monitors. At the turn of the millennium, some fishery officers remained suspicious of the new system's accuracy. But it appeared to satisfy most people.

As for monitoring at sea, the department in the 1980's had obliged certain Canadian vessels to carry observers. In 1989, a task force on Scotia–Fundy groundfish recommended 100 per cent observer coverage on groundfish trawlers, at industry expense. As the 1990's progressed, it became common for the department to require different fleet sectors to pay for observers, who would cover at least a percentage of their vessels. In some cases, notably the large-vessel fishery for offshore northern shrimp, all vessels carried observers.

The overall effectiveness of enforcement remains hard to quantify. Modern vessels have such powerful gear that even a small percentage of bad practices in dumping or discarding fish can have large effects. In groundfish and herring fisheries in southwest Nova Scotia and the Bay of Fundy, studies in the early 1990's estimated that less than two-thirds of groundfish landings and only half to three-quarters of herring landings got reported.[18]

Still, it may be that the increased concentration on enforcement, together with the dockside-monitoring program, has improved compliance. In some instances, the "conservation ethic" seems to have increased among industry members, even though the department never carried out a sustained campaign to promote conservation consciousness. Industry members increasingly helped to work out fishery-management plans, including conservation provisions; this may have changed some attitudes. The Atlantic groundfish collapse also raised conservation consciousness. That being said, there are still tales of widespread cheating.

Professionalization and certification take hold

As fishermen in the 1990's took on new responsibilities, in some regions they gained more professional recognition. Earlier, LeBlanc's 1976 policy had visualized professional status for fishermen through such mechanisms as training and certification. No concerted effort took place in his time, although related initiatives—the increase in the number and strength of fishermen's organizations on the Atlantic, the categorization of full-time and part-time fishermen, and the bona-fide licensing policy—moved in the same direction.

Fishermen had mixed opinions on professionalization, ranging from instinctive resistance in southwest Nova Scotia and the Bay of Fundy to divided opinions in British Columbia and strong support in Newfoundland from the Fishermen, Food and Allied Workers Union. In the late 1980's, as groundfish troubles became more and more apparent, special assistance programs started coming into play, sometimes to the benefit not only of "real fishermen" but also of "free riders." This further sharpened the Newfoundland union's desire to preserve the fishery for those who had invested their lives in it.

Father Desmond McGrath, co-founder of the union, pressed the department to take some form of action on professionalization and certification. This became a $5 million element of the Atlantic Fisheries Adjustment Program (A.F.A.P.), to be described later, which began in 1990. The A.F.A.P. money gave a boost to professionalization and certification, which eventually took effect in Newfoundland and Quebec, and is still gelling in the Maritimes and British Columbia.

Harvesters' organization groups fishermen nationwide

Although fishing came under federal jurisdiction, no fishermen's organization had an Ottawa office. By contrast, the Fisheries Council of Canada had maintained an office in the capital since the Second World War, even though fish processing came under provincial jurisdiction. The federal department controlled resource management and fish inspection, and processing companies wanted to press their points on those and other matters.

The F.C.C. went through some changes in the 1984–2000 period, as larger processing companies met difficulties. The Fisheries Council of British Columbia

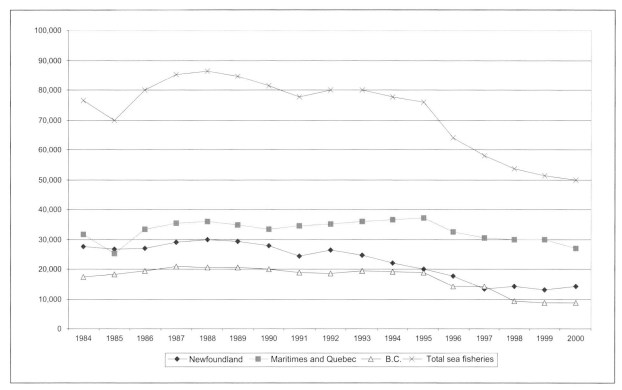

Registered fishermen by region, 1984-2000. Numbers rose in the 1980's, dropped in the 1990's.

pulled out of the F.C.C., then itself shut down in the 1990's. A new organization, the B.C. Seafood Alliance, took up the slack. B.C. processors still had links with the F.C.C. in Ottawa, but less representation than before. The F.C.C. continued, but with less prominence than in the earlier heyday of large corporations.

Meanwhile, fishermen got a new form of national representation, related more to human resources than to fish resources. As groundfish aid programs got under way, Human Resources Development Canada (H.R.D.C.) pulled together major organizations for advice on training and "adjustment." The initial groupings coalesced in 1995 into the Canadian Council of Professional Fish Harvesters (C.C.P.F.H.), operating with start-up funding from H.R.D.C.

At the end of the century, the C.C.P.F.H. included, in its own words, "representatives from the principal fish harvesters' organizations in the Atlantic Region, Quebec, British Columbia, First Nations fishers and Freshwater fish harvesters." The two biggest members were the Fishermen, Food and Allied Workers Union and British Columbia's United Fishermen and Allied Workers Union. The council aimed to lead the development of professionalization for fish harvesters, "to act as a National Industry Sector Council, to plan and implement training and adjustment programs for the fish harvesting industry in Canada," and to represent fishermen on national issues."[19] Along with its professionalization work, the C.C.P.F.H. issued policy statements relating to the role of fishermen, professionaliza-

tion and certification, the importance of the owner–operator rule, and other matters.

H.R.D.C. also in 1995 supported the setting up of a National Seafood Sector Council, one of many such economic "sector councils," to provide training and information for the processing industry.

Liberals return; Brian Tobin becomes minister

The Liberal party returned to power in November 1993 under Prime Minister Jean Chrétien. The government's approach to fisheries management underwent no great change. The department was still trying to make the system put in place by Davis and LeBlanc—licences, quotas, zones, advisory committees, and so on—work as it should. Costs were still an issue; under the Liberals, a government-wide Program Review would slash departmental budgets. Seemingly, there was little room for a new minister to manoeuvre.

But where there was an opportunity, Brian Tobin, the new Minister of Fisheries and Oceans, would make the most of it. First elected in 1980 at the age of 25, Tobin represented Humber–Port au Port–St. Barbe, on the west coast of Newfoundland. He had served for a time as parliamentary secretary to Roméo LeBlanc. The new minister was intelligent, eloquent, brash, and a master in handling the media.

A new deputy minister joined Tobin in 1994: William A. Rowat, a former senior official in D.F.O.'s Atlantic service who had since become associate

Brian Tobin

deputy minister of Transport. Pat Chamut, regional director-general on the Pacific since 1985, became the assistant deputy minister for fisheries management in Ottawa. The hard-working, broadly knowledgeable Chamut would be a key fisheries official in Ottawa for the rest of the century. Scott Parsons, another experienced official, returned from a four-year special project as A.D.M. Science. These officials and several long-serving R.D.G.s., including Eric Dunne in Newfoundland, Neil Bellefontaine in Halifax, Jim Jones in Moncton, were exceptions to the Ottawa trend to shift senior managers frequently. Although in the large department some senior positions changed rapidly enough, it became apparent that for fisheries management per se, knowledge and experience were valuable.

Tobin takes on foreigners

Tobin became a national celebrity through international assertiveness over foreign fishing on the Atlantic.

When John Crosbie had in 1992 announced the northern cod moratorium, many people had expected the closure to last only two years. But no rebuilding had occurred. Instead, more closures followed for other groundfish. The department was back into a conservation crisis, far worse than that which preceded the 200-mile limit. Meanwhile, foreign vessels were still overfishing just outside the zone to an alarming degree. Canadian inspectors boarding under N.A.F.O. authority often found Spanish and Portuguese vessels in non-compliance with N.A.F.O. rules. Violations

reported to the home authorities resulted in few penalties. There was, in effect, no deterrence for law-breakers on the high seas.[20]

The Liberals in opposition had pressed for a more aggressive policy on foreign fishing. Their Throne Speech of January 1994 promised "to take the action required to ensure that foreign overfishing of East Coast stocks comes to an end." But Tobin's first test came not on the Atlantic but on the Pacific, with Canada's American neighbours.

Since 1992, Canada and the United States had failed to conclude annual fishing arrangements under the Pacific Salmon Treaty. Instead, each governed its fleet as it saw fit. In Canadian eyes, the United States had abandoned the treaty's "equity" principle. The Americans, particularly the Alaskans, were intercepting what Canadians viewed as far more than their share of Canadian salmon. The Americans defended their practices. Matters came to a head in 1994. Tobin held conference calls with the many interest groups on the B.C. coast, finding broad support for an assertive Canadian stand.

Hundreds of American boats every year sailed from Washington and Oregon through Johnstone Strait, the "Inside Passage" between Vancouver Island and the mainland, to summer fishing grounds off Alaska. Tobin got the federal government to require American vessels to report in to Canadian authorities and pay a $1,500 fee before making the passage. As well, the department authorized the Canadian fleet to take a bigger than normal share of Canadian fish that were subject to American interception. That year, to Canada's good fortune, an unusually high percentage of Fraser River salmon returning from sea migrated home via Johnstone Strait, rather than down the west coast of Vancouver Island and through U.S. waters in the Strait of Juan de Fuca. This Canadian route aided the "Canada First" policy.

Tobin exercised his media skills ("spin, spin, spin," he chuckled to officials), getting wide publicity in both countries for his views. The department placed a long article in the *Seattle Post-Intelligencer*, laying out the Canadian case. Infuriated American officials responded with their own press conferences and fishing measures.

Caution moderates Canada First policy

Neither side backed down. The fish paid the price through higher than desirable harvests. Meanwhile, a sockeye-salmon shortfall on the Fraser raised conservation concerns and caused media headlines. As Canada–United States relations thawed somewhat, Canada dropped the licence fees and took a more cautious approach to harvesting.

Tobin said in a statement on July 4, 1995, that "the fact that the U.S. has chosen to ignore its conservation duty does not mean that Canada will do the same." The U.S. administration, he said, "is a hostage to the narrow regional interest of the state of Alaska. ... There is a fatal flaw in this negotiation process. It is not a weakness of the treaty—rather, it is a flaw in the U.S. system that allows a single interest to hijack the outcome. This is not acceptable."

As for conservation, "[R]ather than improving our ability to manage closer and closer to the edge, I intend to ensure that we move back from the edge. This means a more conservative approach, with a greater margin for safety." Tobin announced severe restrictions on chinook fisheries, especially on the west coast of Vancouver Island, and on coho fisheries. The department would reduce some fisheries by half, and close others completely. Tobin added that although programs elsewhere were shrinking, he had increased air and sea surveillance and the number of fishery officers.[21]

The Canada–United States frictions moderated somewhat, although the fishing relationship remained contentious.

Tobin arrests American scallopers

Meanwhile on the Atlantic, N.A.F.O. members and flag-of-convenience vessels were causing problems at the edge of the 200-mile zone. Following the cod collapse, N.A.F.O. had applied a moratorium on cod and several other species outside the zone, to match Canada's closures within. A Canadian-owned but Panamanian-registered vessel, the *Kristina Logos*, began fishing anyway. By a quirk, the vessel had formerly been registered in Canada, and that registration had never expired. This strengthened D.F.O.'s hand. In April 1994, Canadian patrols arrested the *Kristina Logos*.

On May 4, Tobin told the House of Commons, "I say to the pirates their day has come and we are going to stop that kind of predatory action." In May 1994, Parliament passed amendments to the Coastal Fisheries Protection Act, giving Canada authority to make conservation regulations to protect straddling stocks from commercial extinction. This meant new authority to arrest, on the high seas, vessels from specified countries that were violating N.A.F.O. rules. The amendments were aimed at flag-of-convenience vessels. Several such vessels now left the Nose and Tail of the Grand Banks.

Meanwhile that summer, two American vessels set out to fish Icelandic scallops outside Canada's 200-mile limit. International conventions in the 1950's and the U.N. Convention on Law of the Sea in 1983 had made clear that coastal states controlled the seabed of their continental shelves. But some Americans contested whether scallops were a "sedentary species" creature of the seabed. Yes, they rested there, but they also moved around in short spurts.

Tobin had no patience with seabed metaphysics. Knowing the American vessels were on the way, he made no major public protests before they arrived. When they started fishing, a patrol vessel took them in. Some controversy followed, but Tobin received fervent applause from most east coasters. D.F.O. later dropped its prosecution of the vessels, when American authorities agreed with the Canadian position about control of seabed species.[22]

Foreign fleets switch to turbot

After N.A.F.O. put cod and some other species under moratorium, some European fleets, particularly Spanish vessels, shifted their attention to turbot, also known as Greenland halibut. Long fished close to shore in Newfoundland, turbot also congregated in deep water offshore. By 1994, vessels from the European Union (known as the European Community until 1993) were taking more than 50,000 tonnes of turbot. That year, N.A.F.O.'s Scientific Council called for cutbacks. Canada slashed its own quotas from 25,000 tonnes to 6,500 tonnes and pressed N.A.F.O. members for similar action.

In January 1995, Tobin took part in N.A.F.O. meetings and helped win 60 per cent of the 27,000-tonne Total Allowable Catch for Canada. This was a victory for Canada and a setback for the E.U., which had been taking 80 per cent of the turbot.

E.U. overrides N.A.F.O. decision

Under pressure from Spain, the European Union decided to object and set its own quota: 69 per cent of N.A.F.O.'s T.A.C. on turbot. The Spanish went fishing for turbot; Canadian and E.U. diplomats went on the alert. Fishing interests and politicians pressed Tobin to stand firm. Tobin made direct representations to E.U. ambassadors, but failed to find a diplomatic solution.

The government's earlier regulations had targeted flag-of-convenience and stateless vessels fishing vulnerable straddling stocks. On March 3, 1995, the government passed new regulations extending the ban to Spanish and Portuguese vessels.[23] The regulations allowed Canadian officials to stop such fishing and to board and seize vessels on the high seas, if necessary, for conservation.

Tobin warned the European Union that such arrests would soon take place. E.U. vessels temporarily stopped fishing, but soon reappeared. D.F.O. deployed offshore patrol vessels from the Maritimes and Newfoundland to the area.

D.F.O. seizes Spanish trawler

On March 9, three D.F.O. vessels were playing cat-and-mouse with the Spanish trawler *Estai* in international waters outside the zone. The *Estai* resisted boarding attempts, cutting its nets to get away. The *Cape Roger* finally closed with the *Estai* and fired machine-gun bursts across her bow. The *Estai* captain gave up. A D.F.O. crew and an R.C.M.P. team boarded the vessel and brought her into St. John's.

The *Estai* coming into St. John's Harbour, under escort. (Canadian Press)

Brian Tobin speaking to media in New York, with the *Estai's* huge net and tiny fish. (Canadian Press)

A diplomatic storm erupted. The European Union's fishery commissioner, Emma Bonino, protested to every forum she could find. E.U. and Spanish officials made hostile statements. But in Canada support was massive. Fishermen in some other countries, who had often had their own sour experiences with foreign vessels, joined in. In the United Kingdom, whole fleets flew the Canadian flag.

Tobin held press conferences to show the evidence of *Estai* wrongdoing. Most of the catch was undersized. The vessel carried nets with smaller-than-regulation mesh, and net liners with even smaller mesh. D.F.O. officials found a secret compartment with 25 tonnes of Americn plaice, a species under moratorium.

On the fishing grounds, fishery patrols tracked down another Spanish vessel, the *Pescamar Uno*, and used an underwater device to cut loose her trawl nets. Spain sent armed patrol vessels to the area.

Prime minister Jean Chrétien and Foreign Affairs Minister André Ouellet by now had made representations to E.U. authorities. Meanwhile, in New York, a previous Canadian diplomatic effort was just now coming to fruition: the United Nations had begun its Conference on Straddling and Highly Migratory Fish Stocks. At the conference, the E.U.'s Emma Bonino denounced Tobin and Canada in fiery terms. Tobin defended Canada's actions on conservation grounds. "It's not the mark of a pirate to reach out in desperation to save the last fish stock. It's the mark of a patriot."

In a move that attracted hordes of media, Tobin had the *Estai*'s trawl net brought to New York harbour and hung from a crane for all to see. Displaying the tiny fish taken by the huge net, he told his audience, "We're down to the last, lonely, unloved, unattractive little turbot, clinging by its fingernails to the Grand Banks of Newfoundland, saying 'someone reach out and save me at this eleventh hour as I'm about to go down to extinction'."

Bonino and other E.U. officials were no match for Tobin, who picked up the nicknames of "Captain Canada" and "the Tobinator." International sympathies clearly favoured Canada. The *Estai* owners posted bond, Canada released the vessel, and negotiations resumed.

Prodded by Spain, the European Union at first remained recalcitrant. Spanish vessels were still trying to fish. While Foreign Affairs officials were cautious, Tobin and D.F.O. pressed their case. Canadian patrol ships hounded Spanish vessels, passing them at close quarters. Prime Minister Chrétien ordered naval vessels into the area and authorized firing directly on Spanish vessels that failed to heave to when ordered. Another high-seas confrontation with a Spanish vessel was less than an hour away when negotiations broke the impasse.[24]

In April, the European Union accepted a new agreement with Canada. All vessels in the N.A.F.O. area outside the zone would carry independent observers. Satellite surveillance, new fish-size limits, and new penalties would take effect. Canada returned the Estai's bond and stayed the charges. The European Union obtained a higher share of N.A.F.O.'s turbot T.A.C.

The *Estai* incident had several results. It soured relations with the European Union and particularly Spain for several years. But it also rallied Canadian public opinion and became a major symbol of strength and sovereignty. (As it happened, D.F.O.'s Ottawa operations centre during the *Estai* affair was in the Peter Mitchell boardroom, named after the Confederation-era minister who had enhanced Canadian sovereignty in clashes with the Americans.) Fishery conservation took on a higher profile, with Tobin making sure to guard the high moral ground. On the Nose and Tail of the Grand Banks, international enforcement improved. N.A.F.O. adopted 100 per cent observer coverage. Vessels began carrying monitoring devices trackable by satellite and calling in their locations every six hours. Violations dropped from some 130 a year to half that level.[25]

U.N. conference produces agreement for straddling stocks

The *Estai* incident helped energize the U.N. conference in New York. In August 1995, the session produced what Canada had wanted: a convention strengthening protection for highly migratory and straddling stocks. Commonly referred to as the United Nations Fish Agreement (U.N.F.A.), the convention stipulated a precautionary approach and an ecosystem concept. It obliged states fishing such stocks to take account of coastal state management practices and provided for compulsory dispute settlement. It gave states the right to monitor, board, and inspect vessels of another flag for compliance with internationally agreed rules. Finally, in a key point, it set out a step-by-step enforcement framework that enabled coastal states, as a last resort, to take action against offending vessels when the flag state failed to do so.

The U.N.F.A. would apply only to nations that ratified it, and only after at least 30 had done so. The 30th ratification took place in December 2001; the U.N.F.A. took its place as an element of international fisheries law. By then, five N.A.F.O. members—Canada, the United States, Norway, Iceland, and Russia—had signed on; the European Union would do so two years later.[26] But in doing so, the E.U. took a legal position that threatened to sap the authority of the coastal state to arrest foreign vessels, in the absence of flag-state action, outside the zone.

The U.N.F.A. changed little in the day-to-day operations of fishing and enforcement. Yet it marked a new departure. Flag states had agreed to give up a fraction of their sovereignty on the high seas. Coastal state authorities could under certain circumstances haul foreign vessels from outside 200 miles into port, unprotected by the flag state. It was a signal achievement for Canadian fisheries diplomacy. As with the 200-mile limit, D.F.O. led the way.

Meanwhile, some fishery frictions had re-emerged in N.A.F.O., where the Europeans were swaying votes. In Canada, especially Newfoundland, some politicians and members of the fishing industry and public were calling for Canada to assert "custodial management" of fisheries out to the continental margin.

Program Review slashes budgets

The Conservative administration under Brian Mulroney had fought the federal budget deficit; the Liberals under Jean Chrétien heightened the assault. A government-wide Program Review aimed to reduce departmental budgets by as much as 40 per cent.

In 1994 and 1995, Tobin and deputy minister Bill Rowat cut major chunks from D.F.O.'s budget. Tobin announced in February 1995 that "D.F.O. will fundamentally change the way fisheries are managed. We will focus our policies and programs on the department's core responsibilities of conservation and sustainability."[27] A departmental Strategic Action Plan listed the main intentions as getting rid of most freshwater responsibilities, moving towards shared management of the fisheries, and rearranging science work. There was to be more multi-disciplinary work and more focus on the key areas of fisheries science (including stock assessment), marine environment and habitat science, and hydrography. In practice, this meant budget cuts wherever tolerable. Money was also to be saved through changes in fleet management, and more fees and cost-recovery mechanisms.

The department set out to raise $50 million in licence fees from the industry. Part of the rationale was that it would be fairer for direct beneficiaries of the resource to bear more of the management costs, rather than forcing all taxpayers to pay. Like previous cost-recovery efforts, this ran into snags. For industry people, any increased charges always seemed like the last nail in their coffin. Disputes emerged, parliamentary hearings were held, but the department did manage to collect more than $40 million annually in licence fees by the latter 1990's.

As for saving money, several national programs created in the 1940's and 1950's—fishermen's loans, boat subsidies, the Prices Support Board—had already vanished, along with most development and promotion work. The government now zeroed in on one of the last such programs: the Fishing Vessel Insurance Plan, which got privatized in 1996.

The department also began working to get rid of recreational fishing harbours, reduce the number of commercial harbours, and increase harbour fees. Officials had already been encouraging local harbour authorities to take over management of wharves, collecting wharfage fees and overseeing maintenance; that trend now accelerated.

After years of budget pressure, D.F.O. had few big and easy targets for cost reduction. Rather than trimming fat, the knives sometimes had to slice into muscle and bone. Science took a one-third cut in personnel.[28] Both science and management had to abandon some previous work, with substantial consequences. This was the era of early retirements; to reduce personnel, the government provided extra payments—"cashouts"—for some retirees. While cutting internally, the department tightened up externally. Fewer industry groups came looking for special-project funding, as they recognized that "D.F.O.'s got no more money."

Coast Guard and D.F.O. merge

Meanwhile, despite cutbacks, the department grew bigger through a merger. In February 1995, Tobin announced the intention that D.F.O. and the Canadian Coast Guard (from the Department of Transport) would merge, "resulting in a combined fleet size of 168 vessels including 42 offshore vessels." The C.C.G. dealt mainly with marine safety, aids to navigation, and response to pollution. The merger, the announcement said, would cut costs and strengthen enforcement at a time of conservation threats on the high seas, and would help to consolidate ocean activities. "Savings will accrue by reducing overhead expenditures, consolidating hydrography (ocean mapping) and vessel management functions, and decommissioning excess vessels, bases, harbours and ports. The Coast Guard would retain its name and identity." And "[t]he strengthened

D.F.O. vessels changed colour after the merger with the Coast Guard. White oceanographic vessels, dark-hulled research trawlers, and gray patrol vessels all went to red hulls with a white slash. Above, the 38-metre patrol vessel *Louisbourg.*

department would focus on four main areas: safety, environmental protection, renewable resource management and facilitation of maritime industry and commerce."[29]

The union became official on April 1, 1995, making D.F.O. the fifth-largest department in government, with some 9,000 employees at the end of the millennium, about evenly divided between C.C.G. and the rest. In 2000–2001, D.F.O. had a total budget of about $1.6 billion. Although divisions of work were sometimes blurry, spending readily associable with fisheries included $149 million and 1,265 people in fisheries and oceans science, $409 million and 1,580 people in fisheries management, and $90 million and 107 people in provision of harbours. Habitat management and environmental science took another $132 million and 312 people.[30]

The department was harking back to the old days of Marine and Fisheries. Back then, the fisheries service and fishing industry had fought to separate the two mandates. The 1995 reunion prompted some skepticism from both the shipping and the fishing industries. Each felt that the additional duties would divert attention from their particular needs. But there was no strong resistance.

Within the merged organization, most activities remained separate. But D.F.O.'s science and patrol fleets got amalgamated under Coast Guard. This resulted over time in fewer sea days for D.F.O. vessels, because the C.C.G. union's agreement entailed more crew for vessels, raising daily costs. Both enforcement and research suffered further damage, on top of the Program Review cuts.[31]

Deputy minister Bill Rowat oversaw a reorganization that merged fisheries and C.C.G. activities under single regional directors-general. Most regions—Newfoundland, Laurentian (Quebec), Central and Arctic, Pacific—followed pre-existing boundaries. A new Maritimes Region incorporated the previous

Scotia–Fundy and Gulf regions. Within the new region, the Gulf retained somewhat of a separate identity. Moncton continued to control fisheries management for the Gulf area.

Fish inspection departs D.F.O.

As budgets shrank, D.F.O. lost its several hundred inspection officers. The department had traditionally felt that the fishing industry needed a dedicated inspection corps, familiar with its complex workings and special problems of quality preservation. Program Review swept that thinking aside. The government aimed to cut costs and gain efficiencies by amalgamating food inspection duties under a single agency. In 1997, the new Canadian Food Inspection Agency took over fish inspection. D.F.O. Inspection had encouraged good quality for its own sake and for market purposes. The new agency laid less stress on quality enhancement, and more on basic health and safety.

D.F.O.'s inspection corps, with officers constantly around the boats and plants, had provided insight into industry workings. And over the years, many Inspection officials had moved up to senior roles elsewhere in the department. Thus, in losing Inspection, D.F.O. also lost a window on the industry and a talent pool for management.

The number of fishery officers stayed around the 600 level. Tobin was able to state in February 1995 that D.F.O. had kept existing levels on the east coast—even though the groundfish fishery had collapsed—and in British Columbia had increased enforcement by 15 per cent.[32] Still, staff complaints about Program Review cuts were common.

D.F.O. tries to hand off freshwater habitat

As part of Program Review, the department tried to hand off most federal freshwater responsibilities. Tobin announced in February 1995 that "authority for freshwater fish habitat, protection and management will be transferred, following successful negotiations, to provincial governments."

But the provinces proved less than eager to take on habitat responsibilities in the way D.F.O. envisaged. Political complications ensued. Meanwhile, court cases in the mid-1990's highlighted federal governmental responsibilities in habitat. In the end, the federal government decided, rather than relinquishing freshwater habitat, to do a better job of it, and provided D.F.O. with additional funding.[33]

D.F.O. found itself facing still more habitat work. In 1995–1996, the department had worked on fewer than 300 environmental assessments across the country. By 1999–2000, the number nearly tripled.[34] D.F.O. took a more assertive role, notably in the Prairie provinces. In 2000 the department added 50 fishery officers in the Central and Arctic Region.[35]

Co-management increases with Joint Project Agreements

In the 1990's, the department talked more about "co-management." The first push for co-management had come from LeBlanc, who called in a 1975 speech for "the sharing of decisions by all those who are affected, and the full disclosure of facts on which judgements can be made." LeBlanc felt that fishermen should hold a high degree of power, using phrases like "a voice in their destiny" and "giving the fishery back to the fishermen." The 1976 *Policy for Canada's Commercial Fisheries* said that "fundamental decisions about resource management and about industrial and trade development would be reached jointly by industry and government."[36]

LeBlanc's era had seen growth in fishermen's organizations and advisory committees. The 1980's brought some further successes in co-management. But the push subsided; the department in the latter 1980's did less initiating of co-management. Then Program Review brought renewed attention, partly for cost-cutting reasons.

In February 1995, Tobin announced that the department would "develop partnerships and co-management arrangements with client groups in order to share decision-making responsibilities and benefits." D.F.O.'s Strategic Action Plan, featuring management phraseology of the 1990's, said that the department must "work more with people than for them, ... becoming less paternalistic and more collegial; less regulatory and more facilitative."

Key to all this would be "the generation of new trust between the department and its clients." Besides new co-operation in self-regulation and enforcement, there would be "new social awareness of the desirability of compliance." The science program would improve its communications and consultative mechanisms. There would be "an aggressive program of partnerships with the fishing industry and other clients for the collection of the information needed by the science program." There was an "urgent need to improve internal sharing of information and to restore public confidence in the department's management and research capability. ... Communications, both internal and external, would become an integral part of the everyday responsibility of all D.F.O. managers."

Rhetoric failed to transform reality. The department never followed through with any major communication or information-sharing campaign. But some changes did come in the 1990's, including more consultation on fishery-management plans.

Written-out plans of a sort went back a long way, for example in the B.C. salmon fishery in the 1950's, but the early ones tended to be rudimentary and to circulate little outside the department. In the latter 1970's, more formal public plans began appearing for Atlantic groundfish, and then for other fisheries. After advisory committee meetings and ad hoc consultations, most plans got approved at the regional level.

What changed in the mid-1990's was that the department began consulting more widely, turning out more elaborate Integrated Fishery Management Plans (I.F.M.P.s) for major fisheries. The I.F.M.P.s became more substantial, reasoned, and informative documents. As well, fishermen had to take more responsibility in developing "conservation harvesting plans" for specific fisheries.

A new mechanism, Joint Project Agreements (J.P.A.s, often referred to as "collaborative agreements" on the west coast), became common, most typically in smaller fisheries that were doing relatively well. J.P.A.s spelled out the roles of the department and the industry parties in running a fishery or other joint project. Typically, the enterprises involved would contribute money for research, enforcement, or other purposes. In some cases they would take part in research or monitoring. The enterprises in return got clear recognition and a bigger voice in management.

Some fishermen were suspicious, calling it less co-management than "co-payment." D.F.O. was increasing its demands that the industry pay for dockside monitoring and other work. But, even while complaining about "off-loading" of responsibilities, industry groups often took advantage of the chance for more direct influence and, at least by implication, more secure access to the fishery.

Meanwhile, in the advisory-committee system, industry representatives often complained that when their recommendations got referred up the departmental ladder, they had no way to keep track of who was calling for changes or influencing final decisions. Some industry interests suspected others of doing "end runs," going directly to senior officials and politicians.

Still, the advisory process in most fisheries was now a vital part of management. Industry was also, in the latter 1990's, participating more strongly in science, whether by funding or by direct efforts. It all added up to progress in co-management, though well short of perfection.

Oceans Act comes into force

On January 8, 1996, Brian Tobin left the fisheries and oceans portfolio, heading to Newfoundland to take over as Liberal leader and premier. Another Newfoundland M.P., former naval vice-admiral Fred Mifflin, became minister for the next year and a half. Coming after the spectacular Tobin, Mifflin was a quiet, low-key minister.

Mifflin's term saw the coming into force of the Oceans Act. Tobin and officials, notably assistant deputy minister Scott Parsons, had worked on the initiative. The act became law on January 31, 1997, 20 years and 30 days after the 200-mile limit, and coinciding with the United Nations' "Year of the Ocean." Canada's Oceans Act was the first such legislation in the world.

Part of it was marine housekeeping. Subsuming the 1964 Territorial Sea and Fishing Zones Act, the Oceans Act defined Canada's sea boundaries more clearly. In the territorial sea, extending 12 nautical miles from the land baseline, the government could operate just as if on land. In the Contiguous Zone (another 12 miles), Canada had authority relating to customs, sanitary, fiscal, and immigration laws. In the Exclusive Economic Zone, running out to 200 miles, "Canada may exercise its rights and responsibilities with respect to the exploration and exploitation of living and non-living resources of waters, subsoil and seabed. The E.E.Z. also provides Canada with the responsibility and jurisdiction to protect the marine environment, to regulate scientific research and to control offshore installations and structures." On the continental shelf beyond the zone, "Canada may exercise its rights and responsibilities with respect to the exploration and exploitation of mineral, non-living resources and living resources (sedentary species only—e.g. scallops)."[37]

The Oceans Act also put forward new provisions reflecting ideas developed in the 1980's and 1990's. It stipulated that management practices should follow three basic principles: sustainable development, integrated management, and the "precautionary approach, that is, erring on the side of caution." It gave the Minister of Fisheries and Oceans the chief responsibility for co-ordinating governmental activities affecting the oceans, and called for development of an Oceans Strategy. D.F.O. set up an Oceans section under Scott Parsons. The act also gave authority to set up Marine Protected Areas. M.P.A.s would become a frequent subject of worldwide environmental interest in the latter 1990's.

The Oceans Act said little about enforcing its principles, and as of 2001 had little quantifiable effect on fishery management. But it had helped to cement the precautionary approach and other principles into departmental and public consciousness. The Oceans Act had the potential to grow in influence, depending on how government and the public used it.

Anderson tackles Pacific conflict

On June 11, 1997, David Anderson, an experienced British Columbia M.P. and former minister of National Revenue and of Transport, took over at Fisheries and Oceans. Anderson had a reputation as an environmentalist. With major resource crises on both coasts, he declared that his three priorities would be "conservation, conservation, conservation." The minister made good on his words.

On the international front, Anderson brought a form of resolution to the long-standing conflicts with the United States over Pacific salmon. Since Tobin's clashes of 1994, rhetoric had subsided while difficulties continued.

In the mid-1990's, there came a salmon catch decline such as the coast had never seen. This would reshape the B.C. fisheries, as described later. Besides fishing pressure and habitat degradation, ocean conditions themselves assaulted the salmon. As part of the

David Anderson

oceanic changes, El Niño events warmed the water. Mackerel moved up the coast, their increased predation of juveniles weakening some salmon stocks. Chinook and coho, the sportsman's favourite species, had already suffered strains; the mid-1990's conditions put them in further peril. The department completely closed fisheries on some weakened runs, and started a major fleet-reduction program. Commercial fishermen's incomes plummeted, forcing many out of the business.

The Canada–United States conflict now seemed to some people like picking at sickly salmon runs to hasten their death. Anderson pushed for a solution to a chronic problem that was depleting salmon and fatiguing the public.

New agreement provides "abundance-based management"

Negotiations picked up speed through 1998. On June 3, 1999, Anderson announced an agreement. Ten-year fishing arrangements would cover northern boundary fisheries, transboundary rivers, northern boundary coho, southern coho, southern chum, and chinook salmon coast-wide. Twelve-year agreements would apply for Fraser River sockeye and pink. Anderson declared that the new set-up would reduce interceptions and bring more fish back to Canada. "Instead of fighting over a shrinking pie, we will now be working together to conserve stocks."

A new approach would govern the interception fisheries. Previously, the two countries had fished every run according to pre-set shares. When a particular run declined, the interception fleets would nevertheless pick away at it, since everybody wanted to get his full share. As the run approached home and its weakness became more apparent, the conservation burden would fall on the domestic fishery close to the river mouth—generally Canadian, since this country was the biggest producer south of Alaska. Even situations of high abundance could create problems. If fishermen reached the catch ceiling early, that could trigger an area closure, preventing fishing on other stocks.

Now, under "abundance-based management," the two countries would examine the health of the stocks at the outset. If a run was weak, the overall harvest would decline, and everybody along the fishing gauntlet would take a reduction. Different approaches would prevail for different levels of abundance. The new system required more scientific information and more pre-season contact between the two parties. But Canadian negotiators believed that it would provide more sensible management. A salient feature was an agreement to increase and share research.

The two countries also announced two Pacific Salmon Treaty Endowment Funds, for the north and for the south, totalling $209 million (U.S.$140 million). Canada and the United States were to administer each fund jointly, investing in habitat, stock enhancement, science, and salmon management initiatives. Canada and the U.S. agreed to do more on habitat protection and to co-operate better in general, with new procedures to resolve technical disputes. At the end of the century, no major disputes had cropped up.

Anderson drops Fisheries Act changes

Anderson calmed the waters on another front. As mentioned above, John Crosbie had tried to push through Fisheries Act changes that would set up arm's-length boards to parcel out catch shares. That bill died with the 1993 election. Another bill, hatched under Tobin, went forward in 1996 under Fred Mifflin, to allow "partnering." This approach, encouraged by department officials, would have enabled formal and binding arrangements, stronger than joint project agreements, whereby industry members could take formal responsibility for some aspects of management. As well, new tribunals would have gained authority to impose administrative sanctions, such as fines, quota reductions, or licence suspensions. The parliamentary Standing Committee on Fisheries and Oceans held hearings on the bill, some M.P.s taking a dim view of it. As time ran out, the bill fell victim to the 1997 election.

Coming at the same time as Program Review cutbacks, the "partnering" initiative had aroused some hostility among industry members. In 1998, David Anderson commissioned Donald Savoie, a well-known academic, to look at the issue. After many consultations, Savoie recommended in December that the department hold back on any such legislative changes. First it should get its house in order, improving existing arrangements for co-management and promoting transparency. D.F.O. should review and co-ordinate efforts to develop a "community-based management approach." Anderson shoved partnering onto the departmental back burner.

Dhaliwal meets Marshall

Herb Dhaliwal, a B.C. M.P. and former minister of National Revenue, replaced Anderson on August 3, 1999. Dhaliwal was a self-made millionaire. He had earlier served as parliamentary secretary to D.F.O. Minister Brian Tobin. Otherwise his only connection to the fishery, as he remarked, was that after his family emigrated from India, his mother worked for many years in a salmon-processing plant.

Only six weeks after his appointment, when Dhaliwal was still reading briefing books and meeting people, a Supreme Court of Canada decision dealing with the Maritimes and Quebec sent shockwaves through the fisheries-management system nationwide.

Native people since colonial days had never been front and centre in the Atlantic commercial fishery, but had taken some part in such fisheries as lobster and salmon. In the late 1960's and the 1970's, licensing restrictions and the salmon-fishing ban had squeezed some Native fishermen out. Although their lost commercial opportunities never became a major issue, they left an undercurrent of resentment among Native people. On the food-fishing side, the department had given some recognition to Native requirements, and in the 1970's issued special licences to several New Brunswick bands, which eased local tensions.[38]

After the 1990 Sparrow decision spelled out Native rights to fish for food, ceremonial, and social purposes, the department had adjusted with reasonable success, making new allowances for the food fishery, and helping some Native people gain additional access to the commercial fishery. Then, in September 1999, the "Marshall decision" seemed to throw open the entire commercial fishery.

Donald Marshall, a member of the Membertou band of the Mi'kmaq peoples in Nova Scotia, had earlier in his life been wrongly convicted and served prison time for murder before being exonerated. The story made him famous. In 1993, officials in D.F.O.'s Gulf Region charged Marshall with illegally fishing eels and selling them commercially. The case wound its way to the Supreme Court. The September 17, 1999, judgement declared that Marshall had a right to sell the eels. It carried the further implication that all the Mi'kmaq, Maliseet, and Passamaquoddy peoples—34 bands in the Maritimes and Quebec—had a right stemming from treaties in 1760 and 1761 to fish commercially not just for eels, but in general.

The decision stunned the fishing industry and caused controversy across Canada. Some commentators challenged it as bad law based on poor history, opening the way to a free-for-all on the water. They condemned the decision for providing no adjustment period and for failing to spell out the competing rights of Native people to fish and of the minister and department to manage. Some Native representatives in the Maritimes declared their right to fish as they pleased. Bands in other regions made pronouncements about similar rights, causing concerns among traditional commercial fishermen across the country.

Dhaliwal declared that the Aboriginal fishery would operate in a regulated way under departmental licence. But commercial fishermen remained upset. To them the situation seemed earth-shaking. D.F.O. officials suggested that only a few hundred Native people might want to fish, compared with more than 40,000 people already in the fishery. The commercial fishermen were unconvinced.

Already in past years, Indians had sold some "food fishery" lobsters commercially. Now, with the Marshall decision, non-Native fishermen could see no limits on Native expansion. Even a modest number of Native people, they feared, might somehow use their newly endorsed rights to take over such small but valuable fisheries as offshore lobster. And in local areas where Native people were relatively populous, they might mount a large fishery, displace non-Natives, and threaten conservation.

Miramichi Bay in New Brunswick had more Native people than most areas. When members of the Burnt Church band and other First Nations rushed to put out lobster traps, a form of marine riot took place. White fishermen destroyed hundreds of Native traps. Violent clashes took place on land and sea, complete with assaults and arson. Across the Maritimes, a "Fishermen's Alliance," representing many commercial groups, started up to oppose any loss of their own rights.

It was a confrontational, dangerous situation. But Native leaders themselves mostly took a moderate and responsible tone. Mike Belliveau of the Maritime Fishermen's Union defended his fishermen's interests but counselled moderation. Dhaliwal called on everyone for calmness while the department worked through the situation.

It was clear that the bands in question had the right to fish commercially. The court decision had mentioned Native people gaining a "moderate livelihood" from the fishery, but confusion surrounded the phrase. D.F.O. took the view that the fishery could provide opportunities for some to work toward a moderate livelihood, without providing a guaranteed annual income. The Marshall bands numbered some 25,000 people, counting every man, woman, and child, and probably less than a thousand would want to fish commercially. The department had both to make room for them and to regulate them.

Consulting with fishermen's groups and noting advice from the parliamentary Standing Committee on Fisheries and Oceans, Dhaliwal and his officials decided there would be no fleet expansion, nor would any-

Many Native people took well to fishing. Above, overhauling traps at Millbrook First Nation in Nova Scotia, 2005.

one be forced out of the existing fleet to make room. Existing fishermen who gave up their licences on a voluntary basis would receive compensation. The licences would go to bands; they would decide who would fish and how to share the benefits. The department got $160 million approved by cabinet for licence buy-outs and related purposes.

At the same time, the department set out to conclude fishing agreements with the 34 bands. James MacKenzie, a Carleton University professor of Cape Breton origin, became the Chief Federal Representative negotiating with the bands, helped by regional officials including Kathi Stewart, Frank Ring, and Gaetan Coté. Fishing agreements already existed for the food fishery; the new agreements would take in both food and commercial. Bands would receive appropriate fish allocations and, depending on their co-operation, additional help in rigging up for the commercial fishery. This could mean provision of boats, gear, shore facilities, and training. In commercial fishing, the bands would follow the same rules and seasons as other people. For the controversial food fishery, allocations and management would tighten up.

Among non-Native fishermen, fear and confusion stayed rampant. The situation eased slightly after the Supreme Court, on November 17, 1999, issued a related judgement clarifying the situation. Some observers called it an unprecedented backtracking for the court. The new decision spelled out clearly that the federal

government had the power to regulate the treaty-based fishery within reasonable limits.

Throughout the year 2000, federal fisheries negotiated interim fishing agreements with the affected First Nations, and worked to provide Aboriginal access to the fishery. Dhaliwal made clear that he wanted to create jobs. At the same time, officials launched a campaign to clarify the situation and calm nerves, with public bulletins, interviews, and sessions with newspaper editorial boards. By August, about 30 of the 34 First Nations had made agreements with the department. Some 200 commercial fishermen had retired their licences voluntarily. Since most of them held licences for several species, the actual total of licences came to more than 600. About 90 vessels had gone to the bands, some from the fishermen, others newly built. Meanwhile, some fishermen complained that the federal payments were driving up the price of licences, making it difficult for new entrants to buy their way into the fishery.

The department publicly posted the various agreements on its Internet Web-site. In a leading example, the Big Cove band near Richibucto, New Brunswick, would receive 30-odd lobster licences, 23 new-built vessels (at a cost of more than $4 million), 5 other lobster vessels, a crab vessel and quota, and the requisite traps and gear, including up to 7,000 lobster traps. The department also agreed to assist with training and to work with the band on research projects, resource enhancement, and feasibility studies for economic development. Total funding devoted to Big Cove came to more than $6 million, not counting the millions spent for the commercial licences that would go to the band.

The Department of Indian and Northern Affairs undertook to lead a long-term process to address the general situation of the bands in question. Meanwhile, D.F.O. worked to build up Native capability in the commercial fishery. Many existing fishermen gave informal training to new Native fishermen. D.F.O., Native representatives, and provincial fisheries schools worked out a more formal program. The money and licences took quick effect. Less than a year after the 2000 agreement, this writer remembers standing on the wharf at Richibucto, where Native fishermen had been rare, and watching an entire fleet of Native lobster boats arrive, with captains and crews maneuvering boats and handling gear as if they'd been at it forever.

None of it happened easily. Negotiations with bands were often problematic. Commercial fishermen and the Fishermen's Alliance both voiced and caused alarm. Gilles Thériault, a founder of the Maritime Fishermen's Union who had turned to consulting, became the Assistant Federal Representative dealing with fishermen. Thériault smoothed out a number of potential conflicts, about use of wharves and such matters.

Burnt Church clash sours atmosphere

By the late summer of 2000, most Marshall bands appeared reasonably satisfied with the progress. But the Burnt Church band on Miramichi Bay, scene of initial clashes, and the Indian Brook band in Nova Scotia disputed federal regulation and claimed their own right to manage fisheries. This brought renewed public clashes. Burnt Church band members declared their own lobster season during the regular closed time, and set traps, which fishery officers confiscated. Confrontations took place on the water. In August 2000, Native fishermen throwing rocks injured fishery officer Dominic Benoit, breaking bones in his face.

The next day, while some fishery officers hauled traps for confiscation, other D.F.O. craft patrolled a defence line to protect them. When a Native boat tried to breach the line, a D.F.O. boat bumped and overturned it.

Although no one got hurt, the dramatic television image went across Canada and around the world. Media from across North America and some from overseas showed up at Burnt Church, some bringing satellite vans to broadcast the conflicts. Activist groups and Aboriginal people showed up from other parts of the country to demonstrate solidarity with Burnt Church. Coastal fishing communities offered vocal support for conservation and the non-Native fishermen. Gunshots were fired on the water; some Native persons brandished firearms on land. A daring pre-dawn raid by fishery officers scooped up thousands of traps, and helped re-assert federal authority.

Most of the craft obtained under the Marshall agreements were in the 30-50 foot range, often fishing lobster. But some communities got larger, midshore vessels. The *Enmali*, shown in 2005, fished snow crab and shrimp for Gesgepegiag First Nation on the Gaspé Peninsula.

Eventually, the Burnt Church band declared their season over, without giving up their claims to management authority. A similar though lesser clash took place in southwest Nova Scotia, between D.F.O. officials and members of the Indian Brook band. The incidents created an ugly impression and caused concern across the country. But at both First Nations, the problems died down early in the new millennium.

For the most part, the adjustment to post-Marshall life was working. D.F.O. had made agreements with the vast majority of bands. Conservation rules and management authority were in place. No huge dislocation had taken place in the traditional commercial fleet.

There were still problems and questions. Different bands shared the fishery benefits in different ways. With stricter enforcement in the food fishery, an ordinary Native person who wasn't one of the band's designated fishermen might have less chance to go out and catch a fish than before.

Still, the whole affair seemed like progress. Many members of the affected bands had learned modern methods to fish lobster, crab, and other species. Aboriginal representatives said late in 2000 that they already had more than 500 people in fishing and related work. Compared with most other job-creation efforts for Aboriginal people, the initial stage of post-Marshall work had been the most notable success for years.[39]

Throughout the Marshall commotion, Herb Dhaliwal was patient and positive. Backing him up were deputy minister Wayne Wouters, who succeeded Bill Rowat in 1997 (he would stay until 2002, the longest-serving deputy since the 1960's, in an often-difficult portfolio), and associate deputy Jack Stagg, who took a strong role on the Marshall file. While keeping a firm line on enforcement, Dhaliwal pushed for economic opportunity for the Marshall bands, arguing strongly and successfully in cabinet, which approved more funding after the initial $160 million. Some observers felt that not only his character but his brown skin helped resolve the situation: it was hard for agitators or self-appointed commentators to accuse him of racism.

Aquaculture comes to the front

Dhaliwal made his presence felt otherwise, particularly in aquaculture. Since the early 1980's, fish farming had grown fast, especially for Atlantic salmon on both coasts. Mussel culture, which had started up in the 1970's in the Gulf of St. Lawrence, also became strong in the 1980's. Other cultivated species included oysters, trout, Arctic char, mussels, clams and scallops, with experimental work on Atlantic halibut, haddock, and cod, and on Pacific black cod (sablefish).

The federal department held less control than in the wild fishery. Aquaculture licensing and various other matters came under the provinces. Often, municipal governments got involved as well. Still, federal fisheries

Atlantic salmon cages became common on the west coast as well as the east. (British Columbia Salmon Farmers Association)

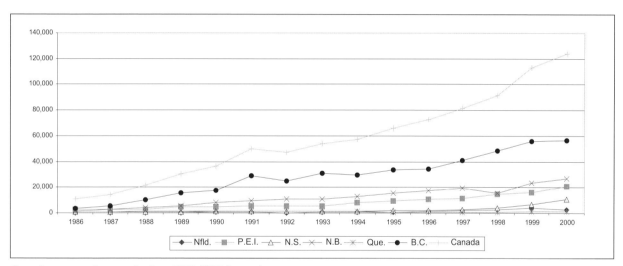

Canadian aquaculture production, 1986-2000 (tonnes).

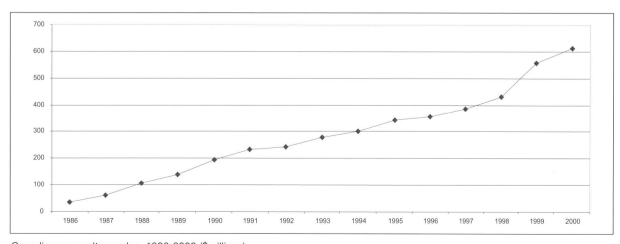

Canadian aquaculture value, 1986-2000 ($millions).

played its part. Research at the St. Andrews, Nanaimo, and other biological stations supported much of the growth in aquaculture. D.F.O.'s science side also monitored the environmental effects of cage culture. The department was spending about $3 million on aquaculture by 1984–1985. In the Maritimes, federal–provincial Economic and Regional Development (E.R.D.A.) agreements were providing additional money, including funds for cost-sharing with private developments.[40] Other federal departments also supported research.

In 1984, the government named D.F.O. as the lead federal agency for aquaculture. Two years later, the prime minister and premiers agreed to a statement of national goals and principles. Also in 1986, the industry set up the Canadian Aquaculture Producers' Council, later renamed the Canadian Aquaculture Industry Association. By that time there were already about 1,000 licensed commercial aquaculture operations in Canada, of one kind or another, including many inland efforts such as trout farming.[41]

In 1988, Robert Cook, director of the St. Andrews Biological Station, and other officials helped the Standing Committee on Fisheries and Oceans prepare a report, *Aquaculture in Canada*, recommending more activity by D.F.O. Various government–industry forums and consultations took place in the early 1990's.

In February 1995, Minister Brian Tobin announced a Federal Aquaculture Development Strategy (F.A.D.S.). The document noted that Canadian aquaculture produced $289 million in revenue, with supplies and services generating another $266 million. This was small by world standards; Canada ranked only 27th in aquaculture. Still, it represented an enormous increase from the starting-out levels of the 1970's. Canada's waters, market proximity, and research and management expertise gave it high potential. Tobin's announcement forecast that employment in aquaculture and related work could more than double the existing 5,200 jobs, and promised that F.A.D.S. would make aquaculture a federal priority.

The main push of the federal strategy was to improve co-ordination. F.A.D.S. laid out the main paths for research, training, and technology transfer and for working with other governments and industry through Aquaculture Implementation Committees. As well, a Liberal caucus task force recommended that aquaculture get more visibility.[42] Later, in 1998, Minister David Anderson appointed a Commissioner for Aquaculture Development to work with the provinces in following up the federal strategy.

When Herb Dhaliwal came into office, the aquaculture commissioner's office and the department's own aquaculture section grew in strength. Late in 1999, Dhaliwal launched a $600,000 program to encourage partnerships within the industry. In August 2000, he announced a Program for Sustainable Aquaculture, worth $75 million over five years. Funding went to scientific research and development, monitoring for health purposes, and improving the management and regulatory framework. In appreciation of Dhaliwal's support for aquaculture, the Canadian Aquaculture Industry Association named an award after him.

By 1999, saltwater and inland production of finfish (chiefly salmon, trout, steelhead, Arctic char, tilapia, perch, and walleye) and shellfish (chiefly mussels, oysters, and clams) came to $611 million, generating direct employment for some 8,000 workers. Shellfish operations tended to be owner-operated and small. Prince Edward Island, an important producer, in 1996 had 238 sites operated by 124 mussel growers. Prince Edward Island dominated mussel production, and mussels accounted for 52 per cent of all farmed shellfish. As for finfish, Atlantic salmon production was major in New Brunswick, with 40 companies operating about 90 sites in 1998, but a smaller number of companies dominating. In Ontario and other inland provinces, trout farms were highly popular.

British Columbia was the biggest producer. Fourteen companies operated 80–85 salmon sites, mostly for Atlantic salmon.[43] Farmed salmon in the 1990's became the single most important fish product in B.C., worth more than the historic wild-salmon fishery. Salmon dominated aquaculture production on both coasts; and larger corporations, many of them foreign-owned, dominated salmon farming. For the ordinary consumer looking for fresh fish in Canada's supermarkets, aquaculture by the turn of the millennium was often outdoing the historic traditional fishery, providing reliable fresh fish at a bargain price.

Will aquaculture dominate?

Ken Lucas, the department's former senior assistant deputy minister, remarked in the mid-1980's that in earlier years he had thought aquaculture could never compete with the wild fishery. Fish farmers had to lavish time and money on their little plots of water, while the boundless ocean fed and maintained fish for free. But over time, Lucas had concluded that all the wild fishery's management problems—many, at their base, problems of communication in the broadest sense—would forever hold it back. Aquaculture would come to dominate in many areas.[44] By the new millennium, it seemed that that might be starting to happen.

Still, even in British Columbia, the wild fisheries in the 1990's yielded more landed value than aquaculture. And some observers thought that eventually, aquaculture would have costly environmental impacts that would damage the new industry.

CHAPTER 27.
On the Atlantic, 1984–2000

A s earlier, this "Atlantic" chapter attempts to treat the main cross-regional fisheries. Later chapters deal with matters more specific to the Maritimes and Quebec and to Newfoundland.

Although lobster and other species were important, groundfish dominated at the start of the period. Left, a trawl net on the dock; right, small boats in Newfoundland. Between the two extremes of trawlers and small open boats, the medium-size boats kept building up their fishing power.

Confidence prevails at the outset

In 1984 on the Atlantic, researchers expected some trouble for lobster, but a continued bonanza of groundfish. The big-company restructuring had settled down. Science would set the harvests, with licence limitation, quotas, zones, vessel-size limits, and the F0.1 rule ensuring a safe level of fishing. With more fish to divide among a limited fleet, there should be more productivity per person.

But the system had loopholes, with regard to both people and fish. Limited entry controlled only the number of fishing licences, not the number of crew. With the fishery improving and Unemployment Insurance acting as a magnet, the number of registered fishermen shot up from 39,000 in 1973 to a peak of 66,000 in 1988. In the 1980's, boats got still bigger and better. Many fishermen switched from wooden boats to fibreglass, which needed less care. Colour sounders, global positioning system (G.P.S.) receivers, and other electronic devices kept boosting fishing power, a surge that sometimes got little departmental notice.

Meanwhile, stock assessments were less solid than most people in the industry thought. As for Pacific herring in the 1950's and 1960's, so now for groundfish: increased fishing power would bring good catches and contribute to overly optimistic assessments, preceding a steep decline. Meanwhile, lobster catches would multiply against all predictions.

I.Q., I.T.Q. disputes mark period

The industry had accepted licence control. As time passed, licences became a means not just to limit access but to control operations. A licence by the end of the 1980's might have several pages of "licence conditions" attached.

But controversy surrounded an outgrowth of licensing: "quasi-property rights." They were part of an ideological conflict that pervaded much of the period.

Individual quotas (I.Q.s) and their variations had spread rapidly. As noted earlier, Bay of Fundy herring purse-seiners used them from 1976, and switched to individual transferable quotas (I.T.Q.s) in 1983. For large-trawler operators, company quotas ("Enterprise Allocations," or E.A.s) had emerged in the early 1980's. In 1984, draggers in southwest Newfoundland moved to I.Q.s, later changed to I.T.Q.s. (Meanwhile, boat quotas were also spreading in the provincially managed Great Lakes fisheries.)[1]

All Atlantic groundfish vessels of 65–100 feet, many of them processor-owned, went to E.A.s in 1988. That

Despite the owner-operator and separate-fleet rules, control of some licences in southwest Nova Scotia began to pass into corporate hands.

still left a much more numerous fleet of craft less than 65 feet. In the Gulf of St. Lawrence, more than 700 vessels from the Maritimes, Quebec, and Newfoundland held dragger licences, although the great majority were multipurpose rather than groundfish specialists. From 1989 through 1992, many of the more specialized vessels in the Gulf fleet switched to I.Q.s and I.T.Q.s.[2]

In 1990–1991, the Scotia–Fundy dragger fleet began changing over to I.Q.s and I.T.Q.s. Meanwhile, the Scotia–Fundy offshore lobster fishery went to E.A.s in 1985. The offshore scallop fleet did so in 1986. More shellfish and other fleets would switch to I.Q.s and I.T.Q.s in the 1990's.

Departmental officials tended to look on I.Q.s and I.T.Q.s with cautious favour, treating them on a case-by-case basis and bringing them in where it was clear that the majority wanted them. Typically, government–industry discussions developed quota-sharing formulas based on catch history and vessel size. In most if not all cases where transferable quotas prevailed, regulations prohibited single enterprises from acquiring more than a specified, low percentage of the Total Allowable Catch. As well, restrictions generally applied on transfers of licences or quota between areas or provinces.

The main arguments over I.T.Q.s

The contending views on quasi-property arrangements became well articulated. Academic economists generally favoured the new trend, a notable exception being the well-known fisheries economist Parzival Copes. To many, I.Q.s and especially I.T.Q.s seemed able to square the circle of free enterprise and common property, giving the fishery a more businesslike orientation.

In the fishing industry itself, backers at first tended to be medium-sized or larger corporations. As the Fisheries Council of Canada put it in 1994, there should be better security of "investor access to the resource," by "allowing the market to set the value of fishing rights" and "allowing that right to be traded freely among economic units." The F.C.C. and other proponents associated a whole set of benefits with I.Q.s, I.T.Q.s, and E.A.s:

- More corporate freedom of operation, more secure access to the resource, and less government interference.

- Fewer regulations: if quotas controlled the amount taken from the water, many other effort-limiting regulations could go by the board.

- A leaner fleet, as trading and accumulation of quotas removed excess operators and reduced over-capacity.

- A more efficient and entrepreneurial system, with more "fishing to market," less seasonality (no need to race), and better profits for a viable, self-sufficient industry.

- A sense of resource ownership and stewardship which would help conservation. Dumping and discarding under individual quotas were no worse than in competitive fisheries; indeed, I.Q.-holders could take time to plan their fishery more carefully and avoid undersized or undesirable fish.

- An end to the "social" fishery, whose attributes included frequent subsidies or make-work programs, high use of Unemployment Insurance ("fishing for stamps"), and lower efficiency.

The Fisheries Council advocated an end to the "owner–operator" and "separate-fleet" rules, which prevented companies from taking over independent boats. If independent fishermen could buy processing plants, why couldn't plants take over smaller-boat fishing licences? The F.C.C. argued that "any quota-holder, whether individual fisherman or fishing enterprise or processing company, [should have] the right to hold a vessel licence."[3]

The opponents of I.Q. and I.T.Q. fisheries, typically smaller-boat fishermen and coastal community representatives (the latter sometimes self-appointed), tended to hold a different set of opinions:

- Far from fostering conservation, I.Q.s, I.T.Q.s, and E.A.s damaged it, partly by providing an incentive to dump or discard fish. The captain with a set quota wanted to get the best value for the restricted volume and so discarded undersized fish or unwanted species, a practice known as high-grading. Or, if the vessel caught too much of a desired species, the captain might misreport the fish as coming from a different area. As well, it was charged, plant operators specified "shopping lists" of species and quantities wanted for the trip; captains who caught too much of a species would dump the extra ones to avoid using up quota. As time went on, many longliners and gillnetters blamed trawlers and draggers using individual quotas for stock depletion.

- Smaller and midshore boats could be as cost-effective as trawlers, even if more seasonal. After all, the large-trawler corporations had gone virtually bankrupt twice in eight years.

- Smaller-boat fleets could provide as much processing employment as the larger operators, and more fishing employment.

- Fishing communities were the economic, social, and cultural foundation of the Atlantic coast. It was unjust that as companies accumulated licences and quotas, they would be able to concentrate operations and close down whole communities by their own say-so.

- The owner–operator and separate-fleet rules remained necessary on both economic and social grounds, to avoid excessive concentration in the industry.

- Individual quotas (or "Enterprise Allocations") held by large corporations amounted to the privatization of a common-property resource for the benefit of a few.

I.Q.s, I.T.Q.s create new twists

Over time, subtler aspects of transferable quotas became more visible. First, they could transmute the danger of over-capacity, in the sense of fishing power, into the danger of over-capitalization. Even when department rules managed to mitigate the race for more or bigger boats, money could still chase after licences and quasi-property quotas, driving up their value. Second, through behind-the-scenes legal agreements, corporations small or large could get around the separate-fleet rule and gain control of licences and quotas.

In earlier years, the independent fisherman seeking a boat had four main sources of financing: the bank, the federal boatbuilding subsidies, the provincial loan board, and the processing companies. The processor who lent money to a fisherman was essentially gambling on a captain to do a good job fishing. By their gentlemen's agreement, the captain would bring fish to the home plant, at least in most circumstances. Fishermen often maintained a margin of freedom and might sell elsewhere, especially after they had paid off their boats.

In the 1980's, federal boatbuilding subsidies ended, provincial loan boards became less generous, and banks sometimes got nervous about lending money. But companies remained a source of financing, which was badly needed. Boats were more costly. Licences, too, were rising in dollar value, and so especially were quotas, which were a virtual guarantee of fish.

Financing might come not just from older, established processing firms, but from new and aggressive ones. In some instances, enterprises primarily based on fishing were gathering together multiple vessels and quotas. Some such fishing companies integrated forward into processing. This caused complaints among older corporations. Blocked by LeBlanc's separate fleet and owner–operator rules, they couldn't integrate backwards into fishing; at least, not openly.

Companies of various genres, most notably in southwest Nova Scotia, found ways to control fishing licences despite the rules. When a company financed a fisherman to buy a boat or quota, the D.F.O.-granted licence would stay in the fisherman's name. But he could agree in writing, usually through a civil contract, to use the licence and any associated quota in a way that would benefit the company backing him. The agreement might even stipulate that if and when he transferred the licence, the company would approve the recipient. Through such arrangements, often termed "under-the-table," the backer was no longer just taking a chance on the fisherman, but was getting at least partial control of the licence and the quota.[4]

Since licences and quotas meant certainty, the Holy Grail of fish companies, their value kept rising. By the end of the 1990's, they often cost more than the boat and gear. A fisherman selling off his licences and quotas could garner a retirement fund, just as Jack Davis had wanted. Prices in the more lucrative fisheries ran into hundreds of thousands of dollars. (As some observers pointed out, this also meant unrecovered "resource rent"; while the state maintained the resource, the returns, or "rent," were going elsewhere.)

Meanwhile, for new entrants into the fishery, acquiring licences and quotas could mean a punishing debt load. Even licence transfers from father to son or daughter could become costly, for tax reasons. This raised the odds that new entrants would obligate themselves to processing companies. In southwest Nova Scotia, dragger and herring-seiner fleets that had been independent or semi-independent appeared by the latter 1990's to be mostly under corporate control. Such control was said to be spreading into other fisheries, such as lobsters, and into other parts of the Atlantic coast. A public policy on the separate fleet was being partly eroded, with little public debate until the turn of the millennium.[5]

Still, independent fishermen themselves in many instances came to favour I.Q.s and I.T.Q.s. Some of their touted advantages, such as greater security and ability to fish to market, seemed to be working.

Some opponents charged that transferable quotas were causing social divisions. As licences and quotas accumulated in the hands of stronger fishermen and processors, smaller-scale captains or crew members sometimes complained of "fish lords" asserting their superior status as "owners" of the resource.

Yet, the whole licence–I.Q.–I.T.Q. web seemed in some ways to strengthen the position of smaller-scale fishermen and their communities. The Kirby report had argued against "community quotas," charging that they would hinder natural efficiencies and market forces. But as quotas got subdivided over time, fishing arrangements in some instances began to resemble, if not quotas for specific communities, at least quotas by district.

Licences themselves still raise questions

Some I.Q.–I.T.Q. arguments linked back to the original question of licence transferability. A fisherman could in effect sell his licence to anyone, at least in his area. Although the Levelton report had argued against transferability, most Atlantic fishermen had accepted licence transfers with little argument. Transferable quotas caused more controversy. But they were in large part a natural outgrowth of transferable licences.

By the end of the millennium, with licence and quota prices skyrocketing, at least some fishermen were saying that when a licence-holder retired, the licence should go back to the state, as Levelton had recommended. But most continued to accept licence transfers, saving their complaints for quota accumulation and under-the-table arrangements.

By then, the federal government had spent hundreds of millions of dollars "buying back" licences and quotas that, in principle, it already owned. How else to reduce fleets? Cancelling inactive licences could stir up a political storm; cancelling active ones could cause a hurricane. Loss of a licence could wipe out a fisherman's whole career, unless he got compensated. The licence, ostensibly a privilege granted by the state, had become, in some circumstances, a moral and financial obligation on the state.

If getting rid of licences was a headache, so was the creation of new ones. An existing fishery might be making its licence-holders uncommonly wealthy from a common-property resource; yet, any suggestion of sharing the wealth through additional licences could cause controversy. Licence-holders would complain that yes, they might be doing well at the moment, but expenses were high and the resource was cyclical.

Nor was it simple to grant licences for new species or areas. The department might use lotteries, or call for proposals and rank them against criteria. Sometimes D.F.O. allowed only temporary licences that were supposed to vanish if the resource declined. This approach, too, had its problems. The "temporary" fishermen often lobbied to get the licence in their permanent grip; ministers sometimes found it hard to refuse.

In short, the minister and department, which ultimately handed out the licences, often found themselves handcuffed by transferability and other, customary aspects of licensing.

Groundfish: the foreign factor

The elements leading to the groundfish crisis of the 1990's included foreign fishing. N.A.F.O. controlled management for several straddling stocks, including yellowtail flounder and American plaice in divisions 3LNO (Grand Banks); cod, witch, and capelin in 3NO (southern Grand Banks); redfish in 3LN (northern and eastern Grand Banks); shrimp in 3L (northern Grand Banks); and squid and turbot (Greenland halibut) over wide areas. (N.A.F.O. also managed discrete stocks of

Difficulties with foreign vessels would continue off and on throughout the period. Early in the 21st century, Canadian authorities had difficulties with the Portuguese vessel *Joana Princesa*.

cod, redfish, shrimp, and American plaice on the Flemish Cap, outside the 200-mile zone.) But Canada controlled the bulk of the groundfish fisheries, including the giant one for northern cod.

Outside the 200-mile zone, as noted earlier, some N.A.F.O. members broke N.A.F.O. regulations, particularly Spain and Portugal. This problem worsened around 1986, when Spain and Portugal entered the European Community. The European Community as a body began to object to N.A.F.O. quotas. Fishing aggressively on the Nose and Tail of the Grand Banks, foreign fleets exerted damaging pressure on yellowtail flounder and other species.

As for northern cod, when in 1986 N.A.F.O. complied with Canadian wishes and recommended a moratorium outside the 200-mile limit, the European Community set its own quotas unilaterally. Although this was harmful, it was still the case that most northern cod stayed inside the zone. It was said that on average, fewer than five per cent migrated outside.[6]

That still could allow a considerable fishery, especially given year-to-year fluctuations in migrations. In the latter 1980's, foreign catches of northern cod ranged from 27,000 to 67,000 tonnes. But Canadian catches ran much higher, 179,000–207,000 tonnes. Thus it is hard to see how foreign fishing should get the main blame for the cod crisis that was to follow.

Industry appetite keeps growing

At the start of the 1984–2000 period, good catches and market conditions helped the "restructured" groundfish companies. They were sizeable enterprises. As of 1985, according to its annual report, National Sea Products had 59 large trawlers and scallop draggers, and 18 processing plants of various sorts in the Atlantic provinces and United States. It was the biggest seafood company in Canada and one of the biggest in the world. Fishery Products International was close behind.

The federal government in 1987 gave up its majority stake in Fishery Products International. The company went private, issuing shares on the stock market. The government received $104.4 million from the privatization, well below its $167.6 million in investments. Only in 1997 did the government sell off its minority stake in National Sea Products, for $5.8 million. Again, this was well below its $10 million equity investment, not to mention the large amount spent to retire debts associated with the Nova Scotia restructuring.

In the 1980's, provincial governments granted processing licences with a free hand, increasing the appetite for fish. From 1977 to 1988, the number of plants in Newfoundland went from something over 160 to nearly 250. In Nova Scotia, partly as a result of new, small herring plants, the number went from about 150 to nearly 350. New Brunswick plants went from less than 100 to nearly 200. Only Prince Edward Island and Quebec stayed reasonably stable. "The number of federally registered fish plants in the Atlantic region grew from about 560 in 1978 to over 1,000 in 1991; the workforce in fish plants increased by 50 per cent."[7]

Many D.F.O. officials had a vague idea that the number of plant workers was in the range of 20,000–25,000. Statistics Canada reported such numbers, using an averaging formula. But the actual number of individuals later turned out to be well above 60,000.

Abundance high, expectations higher

Groundfish catches had nearly doubled in LeBlanc's time, coming close to 800,000 tonnes. As already noted, the Kirby report in 1983 predicted that the increase would continue. Groundfish catches would rise by 50 per cent to reach more than a million tonnes, then level off. The biggest growth would come in northern cod, off southern Labrador and eastern Newfoundland. Even though fished at the conservative $F_{0.1}$ level, northern cod could well provide about 550,000 tonnes annually, a huge amount. But, acknowledging some uncertainty in estimates, the Kirby task force projected a Total Allowable Catch of only 400,000 tonnes by 1987.

In actuality, groundfish catches dropped slightly after 1983, and inshore catches of northern cod became chancy. Still, the northern cod fishery remained the wishing-well. The stock was yielding more than 200,000 tonnes, with large-trawler plants taking a big share. As well, seasonal plants were getting a share, with Canadian and foreign vessels catching 10,000 tonnes of northern cod and other species for "resource-short" operations. The Resource-Short Plant Program lasted until the groundfish quota cutbacks of the late 1980's and early 1990's.[8]

Structure of the groundfish fleet

A useful breakdown of the Atlantic groundfish fleet as of 1981 appeared in the Kirby report. Total catches that year came to 780,000 tonnes, worth $264 million.

About 150 trawlers more than 100 feet in length took over 40 per cent of the catch. Smaller vessels totalling some 17,800, more than a hundred times the number of trawlers, shared the remainder.

Below the trawler class, the 75 Atlantic vessels between 65 and 100 feet took only about three per cent of the catch. A bigger share, 36 per cent, went to the fleet between 35 and 65 feet. These craft had recently increased by about 25 per cent, and numbered around 5,300.

The fleet below 35 feet numbered nearly 8,300 craft. They made up about 60 per cent of the fleet by number, but took only 16 per cent of the groundfish catch. Within both this size class and the 35–65-foot class, about two-thirds of the boats took about 90 per cent of the catch.[9]

The picture outlined by Kirby stayed reasonably stable in following years, with trawlers and 35- to 65-foot boats taking the biggest bites of the catch, and smaller, more numerous craft taking much less.

Groundfish were important everywhere, but especially in Newfoundland and Nova Scotia. As of 1986, Atlantic groundfish landings came to 786,000 tonnes, roughly the same as in 1981, and value to $367 million, a nearly 40 per cent hike. Newfoundland took 50 per cent of Atlantic groundfish landings: 382,000 tonnes worth $137 million. Groundfish provided more than two-thirds of the province's landed value.

Atlantic fishermen in the 1980's were still building some side draggers, like the one on the left, based on the Gaspé Peninsula. But stern draggers like the one on the right, also Gaspé-based and built a few years later, were much more common. Large numbers of boats were multi-purpose, switching, for example, between trapping lobsters and longlining groundfish.

Newfoundland in the mid-1980's had about half the trawler fleet. Outside that fleet, small boats dominated, including thousands of undecked "day boats." The mid-size fleet, though growing, was far behind that of Nova Scotia. Draggers were few outside the west coast of the island, where several score draggers did well on groundfish and, increasingly, shrimp. Most craft used gillnets and cod-traps (although the latter were losing popularity to gillnets). The Newfoundland fleet all told had some 7,000 groundfish licences.

Nova Scotia in 1986 caught less groundfish, but it was worth more: 287,000 tonnes, at a value of $175 million. Groundfish provided only 43 per cent of the province's $407 million in landed value.

Nova Scotia had fewer but larger boats than Newfoundland: some 6,100, of which about 2,500 were longer than 35 feet. In the Scotia–Fundy Region in 1986, about 1,900 vessels held more than 3,000 groundfish licences, including some 2,600 for longlines, gillnets, and handlines. (Longlines were a popular gear, deemed more conservationist than gillnets. But when the department in the early 1980's tried to phase out gillnets in favour of longlines, resistance forced a halt.)

The growing force in Scotia–Fundy was the draggers. The region had about 450 dragger licences; only about half worked at it full time. But these boats now had great fishing power; a few of them could equal a large trawler. Department studies in the mid-1980's showed that the groundfish fleet fishing southwest Nova Scotia and the Bay of Fundy had four times the required fishing capacity. Draggers managed to make good money, but their strength caused a constant headache for fishery managers.

In Scotia–Fundy in 1986, some 560 vessels with groundfish licences reported no groundfish landings, and some 300 reported no landings at all. This was worrisome; the active fleet already had more capacity than required, and if inactive vessels should resume fishing groundfish, they would overburden the fishery. Accordingly, the department under Tom Siddon tried in 1987 to suspend the re-issuance of inactive licences, only to back off in the face of industry resistance.[10]

Quebec, New Brunswick, and Prince Edward Island took only 7, 3 and 2 per cent of Atlantic groundfish landings. Groundfish provided about one-third of Quebec's total landed value of $82 million. In P.E.I., it provided only 12 per cent of the $50 million landed value, and in New Brunswick, only 11 per cent of the $94 million value.

The Quebec fleet had a high proportion of small boats: 2,200 out of 2,800 were less than 35 feet. New Brunswick had bigger boats: about 1,300 out of 1,980 were longer than 35 feet. In P.E.I., where the great majority of the fleet was lobster boats, almost all the 1,460 craft fell between 35 and 50 feet. On the Gulf shore of Nova Scotia, more than half the 840 craft fell into the same range, with another 370 boats below 35 feet.

Thus, the Gulf of St. Lawrence was the home of small and medium-sized, multipurpose boats, with no overwhelming dependence on groundfish. But, for a great many of them, groundfish provided an important part of the year's fishery, as they switched from species to species. The Gulf and Quebec regions held some 5,200 fixed-gear licences and more than 700 dragger licences, although many went unused.[11]

Quota disputes pit independents against trawlers

Within Canada, vigorous disputes took place every year over groundfish allocations. Representatives of the less-than-65 foot fleet wanted less offshore fishing pressure, and a major transfer of quotas from the large-trawler fleet to the inshore and midshore. Allocation got more attention than conservation. Industry groups frequently pressed for larger T.A.C.s. But by the mid-1980's, it became clear that the projected increases in abundance were coming more slowly than hoped. The department began recommending quota cuts in some instances. Industry interests protested.

D.F.O. and industry representatives agreed on a "50 per cent rule." Beginning in 1986, if the scientific recommendation of Total Allowable Catch for a stock changed by more than 10 per cent, either upwards or downwards, then the department would move the quota only 50 per cent of the way to the new T.A.C. in that year. The object was to smooth out fluctuations in the fishery. In hindsight, the 50 per cent rule contributed to excessive fishing.

Fixed-gear fishermen in the Maritimes and Newfoundland sometimes complained that trawlers and draggers were depleting stocks. In their eyes, the mobile fleet not only had excessive quotas and fishing power, but a destructive manner of fishing. They charged that drags and trawls took more undersized fish and unwanted species. They recounted horror stories of wholesale discards, dumping, and misreporting by mobile vessels. Trawlermen tended to counter that they fished as cleanly as possible, and any fishing method could be destructive, depending on how it was used. (However, within the Newfoundland fishermen's union, trawler crews themselves raised the issue, though with no result.)[12]

The department had rules against wholesale dumping of fish. Until 1993, however, discarding unwanted species or undersized fish was legitimate in most groundfish fisheries. Nobody could precisely quantify the amounts of discarded fish or hidden catches. But D.F.O.'s observer corps, gradually increasing their coverage of domestic vessels, said the practices were widespread. A later report in 1994 noted that "Many captains are still hesitant to discard 'under the nose' of an observer; but their crews freely admit to massive dumping when observers are not aboard. ... Dumping, discarding, highgrading, and misreporting by species and area have been an ongoing problem since extension of jurisdiction in 1977. Furthermore, the introduction of E.A.s in 1982 generated more incentives to continue these activities as the fleet captains made specific requests for a species and size mix for each trip."[13]

Newfoundland fishermen accuse department of poor science

For northern cod, with the 200-mile limit, Canada had first cut the T.A.C. by almost half, to 160,000 tonnes. Abundance increased, though not to 1950's levels. The T.A.C. rose to 266,000 tonnes in most years of the period 1984–1988. The new offshore fishery was doing well, with catches rising from low levels in the 1970's to 110,000 in 1983. But inshore catches had never built back to the more than 140,000 tonnes typical of the late 1950's and early 1960's.

Then, in the mid-1980's, inshore catches of northern cod fell in coastal Newfoundland. Departmental scientists thought the decline probably had to do with transitory water temperatures in the "cold intermediate layer." Catch rates per unit effort helped researchers gauge abundance. For catch-rate statistics, they had to rely primarily on the offshore fleet, which fished all the time and yielded statistics in a systematic way. Catch rates in that fleet looked fine. Meanwhile, the statistical personnel in Newfoundland had abandoned monitoring the catch-per-effort of the far more numerous fixed-gear fleet,[14] the main source of complaints about dropping abundance.

Inshore suspicions coalesced in 1986 with formation of the Newfoundland Inshore Fisheries Association. N.I.F.A., a separate entity from the Newfoundland union, maintained that something was going badly wrong with cod stocks. The group engaged three scientists from Memorial University of Newfoundland to report on the issue. These scientists said that there was indeed a problem, and found fault with departmental methods. D.F.O. officials viewed the university report with some suspicion, since the researchers had been hired to prove a point.[15]

At the request of A.D.M.s Bill Rowat (fisheries management) and Scott Parsons (science), the Canadian Atlantic Fisheries Scientific Advisory Committee (C.A.F.S.A.C.), the department's peer-review system, took a close look at the northern cod.[16] C.A.F.S.A.C.'s 1986 report recommended a T.A.C. of 266,000 tonnes for 1987, which would approximate $F_{0.1}$. The report noted that the biomass had almost tripled from 1976 to 1980. Then growth had slowed. Still, by 1985, abundance had risen to four times the 1976 level (although this was little more than half the levels back in 1962).

As for the inshore difficulties, various factors could be at work, such as variations in migrations, temperatures, and the supply of capelin. C.A.F.S.A.C. recommended that offshore catches should be spread equally among divisions 2J, 3K, and 3L (vessels had been increasingly concentrating in 3L). C.A.F.S.A.C. concluded in retrospect that previous T.A.C. recommendations, thought at the time to be below $F_{0.1}$, had actually been above that level. This was the first sign that the stock might be rebuilding more slowly than thought.[17]

As well, the report said, discards in the offshore fleet were extensive, estimated at 24.4 per cent by number and 10.7 per cent by weight. Still, it suggested no real problem with abundance; the stock looked good.

The department under Tom Siddon decided to sponsor an impartial look at the issue. American scientist Lee Alverson led a panel of international experts known as the Task Group on the Newfoundland Inshore Fisheries. Meanwhile, D.F.O. followed C.A.F.S.A.C.'s advice and subdivided the northern cod T.A.C. among the 2J, 3K, and 3L divisions, to avoid overstraining any one area.

Alverson's report in 1987 agreed that the stock was increasing. The report essentially endorsed D.F.O. science, while differing on certain points and warning of potential overfishing in some areas. The task force found fishing mortalities somewhat higher than C.A.F.S.A.C. had estimated.

The department took due note while also defending its work. It distributed a booklet, The *Science of Cod*, outlining biological facts and stock assessment methods. The scientific consensus was that there was no major problem with northern cod.[18]

The shock of 1989

How did the science get done in the first place? For each stock, certain researchers had the lead responsibility. They would present their findings to the relevant subgroup of C.A.F.S.A.C., which had come into being in 1977. Although C.A.F.S.A.C. kept its meetings closed to observers, it distributed its research and advisory documents widely. Few industry members, however, had the expertise to critique them.

Early in the life of every year-class of fish, scientists estimated its numbers. They would base the estimate on previous experience, research-vessel trawl surveys and acoustic soundings, trawler catch rates, the logbooks required on larger vessels, and any other evidence they could gather, combining this with assumptions about cod biology and fishery behaviour.

Port technicians sampled catches to find the proportion of fish taken from particular year-classes. Applying these proportions to the overall catch gave the total take from each year-class. The prevalence of a year-class suggested its strength; scientists could estimate how many of its age group were still in the water. As time went on, the fleet would catch up more and more of that year-class. By the time it was almost gone, researchers could add up all the catches it had

Age	1978	1979	1980	1981	1982	1983	1984	1985	1986	1987	1988	1989	1990	1991
3	300363	152275	160671	359797	320664	349967	432460	338626	157808	129970	160619	182303	95684	37065
4	272786	244719	123630	129236	292600	260997	284190	353361	276655	128450	104303	128990	147723	71378
5	209089	207453	189174	90339	99320	211251	201366	219219	275894	212735	96826	72139	89647	84248
6	63704	135712	135923	128811	62028	64121	134410	136127	146352	185918	144911	61011	39925	40452
7	19658	33771	84689	84125	83021	37757	35281	75097	80757	78319	107825	75286	27139	12153
8	**7741**	9117	17713	53035	48956	44958	20015	17572	36144	42570	38362	46906	26614	7389
9	4072	**3553**	4334	10131	32701	24763	23506	9830	8008	16271	17205	14571	12616	5557
10	3176	1950	**1733**	2447	5422	15973	12191	11183	4577	3748	8057	5919	4061	2044
11	1054	1621	911	**948**	1299	2699	7340	6156	4485	1938	1850	2672	1552	733
12	635	467	820	536	**497**	757	1289	3715	2412	1883	975	815	654	290
13	267	321	217	464	263	**265**	396	647	1520	978	752	416	149	103
3+	882545	790959	719815	859869	946771	1013508	1152444	1171533	994612	802780	681685	591028	445944	261412

Scientists estimated the numbers in each year class. These figures show population numbers at age, in thousands, from the commercial cod fishery in N.A.F.O. divisions 2J3KL, for the years 1978-91. The bold numbers represent fish from the 1970 year class. C.A.F.S.A.C. figures reproduced from *A Glossary of Fisheries Science*, by Joseph Gough and Dr. Trevor Kenchington.

yielded, combine that with their estimates of natural mortality, and calculate with some certainty how many fish had been in the year-class in the first place.

In other words, the system was strongest for old and bygone year-classes (provided there was good catch data), and weakest for young ones coming on stream. On top of that, ocean conditions made the survival of young and their recruitment to the fishery highly variable among year-classes and impossible to predict.

Still, virtual population analysis, as it was called, appeared to be the best available method, and to be giving good results. Yes, there might be problems with discards and misreporting in the various fleets. Yes, recruitment of young fish might vary wildly. But stocks in the late 1970's and early 1980's appeared to be increasing right on schedule. To the department, scientific quota management and the $F_{0.1}$ guideline—roughly, taking one fish in five—appeared to be working. For northern cod, despite inshore complaints, scientists believed that the resource was still growing, only more slowly than hoped.

Meanwhile, although the catch rates from commercial trawlers had generally looked good, research-vessel surveys were sending mixed signals. As well, there were deficiencies in data, although these were clearer in retrospect. (In 1991, the year before the moratorium, C.A.F.S.A.C. noted that "2J3KL cod is one of the stocks for which the data were reasonably good.") There was no thorough set of numbers on inshore catch rates or on discards. There were no long-term research-survey data for 2J3KL as a whole. Indeed, the department had had no research vessel capable of operating in ice, to go north when the cod were offshore and the fishery occurring, until it chartered the *Gadus Atlantica* starting in 1978.[19]

Scientists in 1987 recognized a consistent trend to overestimate the population. Even then, however, northern cod looked to be in good shape.

C.A.F.S.A.C.'s report in 1987 said that the $F_{0.1}$ level for 1988 would be 288,000 tonnes. As it turned out, however, the T.A.C. stayed at 266,000 tonnes for 1988.

In that year, C.A.F.S.A.C. never prepared a regular assessment for 1989. In an advisory document, C.A.F.S.A.C. said it was having difficulty analyzing catch-rate data. It wanted to address stock-assessment recommendations made by the Alverson group and by C.A.F.S.A.C. itself. An assessment, it said, would be impossible until the end of January 1989.

Meanwhile, the scientists' view of the stock was changing. A research survey in 1987 had shown lower abundance than the 1986 survey. Then the 1988 survey confirmed the newly pessimistic picture.[20] In hindsight, the results for 1986 appeared unrealistically high.

Finally, in the January 1989 assessment, C.A.F.S.A.C. scientists concluded that they had made a grave overestimate. The high catch rates by increasingly efficient trawlers, whose captains were learning more about the stock, had helped to mask the reality. Fishing mortality rates were about double the earlier estimates. A 1986 drop in commercial catch rates, "which had previously been thought to be a 'blip' in the data, was real."[21]

Northern cod had grown until about 1986, but were now declining. C.A.F.S.A.C. said the Total Allowable Catch should for 1989 drop by more than half, from 266,000 tonnes to 125,000 tonnes, at the target level of $F_{0.1}$.

Struggle erupts over northern cod quotas

The idea of halving the quota of northern cod—the great support of trawler plants, and the vital crutch for "resource-short" plants—brought consternation. Because of other quota reductions, the major groundfish processors in 1988 and 1989 were already cutting back operations, even closing plants.[22] While minister Tom Siddon pondered the C.A.F.S.A.C. advice, John Crosbie, the powerful federal minister from Newfoundland, and other Atlantic politicians pressed him to soften the blow.

It is easy to imagine their thinking. The same scientists who this year recommended 125,000 tonnes had two years ago been recommending more than twice that. Were they any smarter now? Might not the truth be somewhere in between? People needed jobs. Meanwhile, as it happened, inshore catches in Newfoundland had resurged in 1988, lessening the inshore concerns.[23]

The department was now habituated to using the 50 per cent rule to cushion fluctuations. Siddon considered a lower T.A.C., but under pressure from Crosbie and others left it far above the scientific recommendation of 125,000 tonnes. It also stayed above the 50 per cent rule level, which would have been about 195,000 tonnes. The Canadian T.A.C. for 1989 got set at 235,000 tonnes.

Harris report: on the edge of a precipice

Siddon also commissioned Dr. Leslie Harris, a distinguished Newfoundlander and president of Memorial University of Newfoundland, to study the northern cod issue. Harris assembled a team of six and presented an interim report in 1989. While recognizing that C.A.F.S.A.C.'s 1989 warnings were in the right domain, the Harris panel recommended a higher T.A.C. for 1990, at 190,000 tonnes, than C.A.F.S.A.C. had wanted.[24]

Siddon, in January 1990, set the Canadian T.A.C. higher still, at 197,000 tonnes (a small French quota brought the total T.A.C. to 199,262 tonnes). Although higher than recommended levels, this was a significant drop from the previous year. Meanwhile, both Fishery Products International and National Sea Products were moving to close some major plants and retire some of their trawlers.

For the century prior to 1950, the final Harris report noted, the northern cod stock had yielded on average about 250,000 tonnes. Foreign vessels had driven the catch to three times that level, some 800,000 tonnes in 1968, followed by the 1970's collapse. Then had come the 200-mile limit and the department's efforts to rebuild the stock through the $F_{0.1}$ strategy (catching only one fish of five).

During the next seven years the euphoria that had been engendered by the declaration of the exclusive economic zone was reinforced by the steady growth of the stock, by continually improving catches, and by the belief that the F0.1 objective was, indeed, being met. In those circumstances, scientists, lulled by false data signals and, to some extent, overconfident of the validity of their predictions, failed to recognize the statistical inadequacies in their bulk biomass model and failed to properly acknowledge and

recognize the high risk involved with state-of-stock advice based on relatively short and unreliable data series. Furthermore, the Panel is concerned that weaknesses in scientific management and the peer review process permitted this to happen.

Harris said that in fact, as C.A.F.S.A.C. had already concluded, fishing mortality rates had been "at least double those projected in the F0.1 strategy." There were all kinds of scientific difficulties, Harris added. "Nevertheless, it is possible that if there had not been such a strong emotional and intellectual commitment to the notion that the $F_{0.1}$ strategy was working, the open and increasing skepticism of inshore fishermen might have been recognized as a warning flag demanding more careful attention to areas of recognized weakness in the assessment process."[25]

Often enough in fisheries, people got carried away by their ideas. Harris was suggesting it could even happen to objective scientists. D.F.O. needed, among many other recommended actions, to increase research and to move fishing mortality towards the $F_{0.1}$ level. Fishing had its risks, Harris said. "If we continue to insist upon walking the very edge of the precipice, the laws of chance ordain that we daily walk in greater risk of falling over." Harris noted that "the fishery ... will not be saved unless the spawning biomass is permitted to grow.[26]

The final Harris report in March 1990 called for an immediate reduction of the fishing mortality rate to a specified intermediate level, then, "at the earliest feasible date," to a mortality level ("0.2" in technical terms) that was right around $F_{0.1}$. The new minister, Bernard Valcourt, under pressure from Crosbie and others fearing further economic hardships, instead kept the 1990 level as earlier set by Siddon, at 197,000 tonnes.[27]

Scotia–Fundy task force calls for conservation measures

Meanwhile, in the Scotia–Fundy Region, scientists in the late 1980's were recommending quota cuts for several groundfish stocks and hoping for a recovery, but none was coming. Projecting next year's results on the basis of last year's data, they were playing catch-up, but the resource was falling behind. Over-capacity in the fishing fleet made the situation worse.[28]

The Scotia–Fundy dragger fleet, as already noted, had some 450 licensees, but only about 215 were active. Of these, about one-third were in the 45- to 65-foot range, the others smaller. By mid-1989, the dragger fleet had used up its quotas. The department closed the fishery. Industry representatives protested; Siddon stood firm.

The minister appointed regional director-general Jean-Eudes Haché to head a task force on groundfish, which held intensive hearings all around the regions. As usual when the department made such an effort, it calmed the waters.

413

Studies done for the task force showed that despite quota and other restrictions, fishing mortality in the 1980's for the prime species of cod, haddock, and pollock had been more than double $F_{0.1}$.[29] The Haché report recommended a mix of conservation measures relating to seasonal closures, size limits, hook sizes, and ghost fishing by gillnets. The department should make inactive licences—the region had some 1,200 of them—non-transferable, so they would die off as licence-holders retired.

As both the minister and the deputy changed in 1990, the Haché proposals lost momentum. A key recommendation was to increase the minimum mesh size of drags and trawls to 140 millimetres square mesh or 155 millimetres diamond mesh. But after the new mesh size came into effect, National Sea Products lobbied at high levels against the change, which got partly rolled back. Another recommendation said that vessels should carry observers more often, including 100 per cent coverage on trawlers. Trawler coverage to date had risen from 6 per cent in 1979 to about 20 per cent. But no major increase in the percentage of observers followed the Haché recommendations until 1993, when the number of trawlers dropped.[30]

As groundfish problems deepened, common opinion held that the recommendations had been good, the follow-through weaker than desirable.

Individual quotas come into play

The Scotia–Fundy task force recommended dividing the groundfish fleet into three groups: the smaller-boat, multi-fishery fleet; the specialist fixed-gear; and the draggers. The latter group, which presented the biggest problem, could then choose from various future options, including I.Q.s or I.T.Q.s. Instead, headquarters in Ottawa gave a conclusive push towards I.Q.s.

I.Q.s came into play for draggers in 1991, at first transferable during the year, and later transferable permanently. As was the usual practice, the department assigned quotas mainly on the basis of catch history, consulting and making adjustments here and there for other factors. No single operator could amass more than a small percentage of the quota. Through the rest of the 1990's, a combination of scarce resources and the transfers and combining of quotas reduced the active dragger fleet from more than 200 vessels to around 130.[31]

Vessel replacement rules get stricter

Since 1981, the department had relied on the "five-foot barrier" to restrain increases in vessel length. Other rules restricted increases in hold capacity. Yet, fishing ability was increasing greatly; a new vessel the

Small and medium-size boats built in the 1980's and 1990's were often bulkier than in the past, with more superstructure. Stabilizers became popular, as on this southwest Nova Scotia craft built around the turn of the millennium.

same length as an old one could have twice the fishing power. As efficiency grew, a few 65-footers could together catch as much as an offshore trawler.

The Haché report noted that the departmental rules that restricted increases in vessel size and fish-hold capacity were less than effective. Subsequent attempts to tighten the replacement rules caused controversy. But by 1993, after various disputes and confrontations, the department arrived at a new system, which applied coast-wide. Generally, if you were replacing a vessel of 35 to 65 feet, you multiplied its length, beam, and depth to arrive at a "cubic number." The replacement could have no larger a cubic number. Regions supplemented or modified the rule to fit circumstances. In the later 1990's, some exceptions emerged for I.T.Q. vessels. Still, in the overall picture, vessel replacement rules in the 1990's became much tighter.

Plants begin to close

Meanwhile, things kept getting worse for the groundfish industry. Catches all told fell from 820,000 tonnes in 1982 and 786,000 tonnes in 1986 to 685,000 tonnes in 1989. It was bad enough that quotas were dropping; in some instances, catches were dropping even faster. For the years 1985–1990 in N.A.F.O. divisions 4RS–3Pn (southwest Newfoundland and the northern Gulf), catches usually came nowhere near the Total Allowable Catch (although cheating may have produced higher than recorded catches). In 1990, of a 58,000 tonne T.A.C., fishermen took little more than half. The department finally reacted in 1991, lowering the T.A.C. to 35,000, a two-thirds cut from the 1985 level.

Several other important stocks of cod, redfish, pollock, and other species suffered major cuts in the years 1989–1991. The major groundfish corporations announced closures or projected closures at such places as Lockeport, Canso, Trepassey, St John's, Gaultois, and Grand Bank, and cutbacks elsewhere.[32]

Valcourt launches assistance program

In February 1990, the new minister, Bernard Valcourt, inherited the gathering groundfish crisis. In May, the government introduced A.F.A.P., funded at $584 million, which would run until 1995. A.F.A.P. covered Newfoundland and the Maritimes. An associated Quebec Federal Fisheries Development Program dealt with Quebec and brought the total to $637 million.[33] A.F.A.P. was the first major special-aid program since the 1984 restructuring. An alphabet soup of big-spending programs—N.C.A.R.P., A.G.A.P., T.A.G.S., and C.F.A.R.—would follow.

A.F.A.P.'s objectives were to rebuild cod and other groundfish stocks, adjust the fishing industry to "new realities," and diversify the fishing and related economy. Other departments were to deal with development outside the fishery.

D.F.O. itself oversaw spending of about $220 million of the A.F.A.P. money on about 600 projects. In large part, the A.F.A.P. effort assembled ideas that bubbled up from officials. Some funding helped improve regular operations, such as offshore surveillance and enforcement. Money also went towards new ventures, such as the spread of I.T.Q.s and the dockside monitoring program. Science also benefitted, notably in the five-year $43 million Northern Cod Science Program, and in a major effort to survey biomass for the expanding crab fishery.

A.F.A.P. also aided salmon aquaculture, and helped develop products and markets for under-utilized species, such as sea urchins. Work went forward on gear selectivity and conversions. In 1991, an Ice Compensation Program assisted fishermen affected by abnormal ice conditions. Sealing associations got help to keep their organizations and the industry alive after the European ban. And A.F.A.P. helped bankroll the licence-retirement program for salmon fishermen.

Overall, the A.F.A.P. money went to useful work, but did little to reshape the fishery in a major way. The department still hoped that groundfish would recover.

Professionalization gets a boost

As noted in the last chapter, A.F.A.P. also gave a boost to "professionalization and certification" of fishermen. Previously, federal officials had tended to be hesitant, wondering if they were getting into provincial-government jurisdiction. But now, Valcourt heeded representations from the Newfoundland union and approved a $5 million budget from A.F.A.P.

Meetings and seminars raised consciousness of professionalization around the Atlantic. The Newfoundland union stayed in the lead, holding more than 200 meetings around the province. Fishermen backed the idea, and professionalization moved forward in following years. After the Newfoundland government passed appropriate legislation in 1996, a Professional Fish Harvesters Certification Board came into place in January 1997. In the new system, fishermen got registered as apprentices (less than five years' full-time fishing), or as professional fish harvesters level 1 (five years) or 2 (seven years). Existing fishermen were "grandfathered" in; new entrants took training. Apprentices could move up to level 1 after a certain amount of fishing time and of land-based credit courses in such matters as safety, fishing methods, and fish handling (although not in conservation or fisheries management). More fishing time and more courses led to level 2 certification. Funds for the system came from registration fees paid by fishermen. The union exerted a strong influence.

Quebec also adopted a certification program, and provided some training in responsible fishing. As well, by the end of the century, professionalization appeared to be making headway in the Maritimes, especially in the Gulf of St. Lawrence, and in British Columbia.

Hard-hit communities get A.F.A.P. help

Under A.F.A.P., other federal agencies dealt with parts of the program that were less specific to fishery management. Owners of tied-up boats got income support. A Plant Workers Adjustment Program gave payments to older plant workers who left the industry. Retraining programs started up. Funding went to diversify the economy of hard-pressed communities, especially where major plants had closed. Newfoundland was the worst hit.

There were some small successes. The town of Grand Bank, for example, got a scallop plant, which offset some of the lost employment from its groundfish plant. But usually, communities where major groundfish plants had closed were unable to recover lost ground. They tended to be remote places where the workforce was not highly educated. There were few other industries around.

Back when they first took root, the more substantial Atlantic coastal communities had four bases of support: fishing, wooden boatbuilding, shipping, and trading. Now, for primary employment, they had little beyond fishing and were trying to balance on a one-legged stool. Transfer payments such as old-age pensions helped, as did jobs in schools, government agencies, and services. But nothing in sight could readily replace a major fish plant. For example, the town of Trepassey in the 1980's had more than 2,000 people; early in the new millennium it had fewer than 1,000.

Burgeo went from 2,500 to 1,700 or so. Many others, such as Marystown, Ramea, Burgeo, and Fermeuse, suffered declines.[34]

Canso rallies to survive

The town of Canso, on the Nova Scotia mainland close to Cape Breton, billed itself as the oldest fishing town in the Maritimes. The community was stuck out on the end of a peninsula at the northeastern extremity of the Nova Scotia mainland. In the sea-oriented world of previous centuries, Canso had enjoyed a central location for catching and drying fish and shipping them out in sailing vessels. In the continental economy of the late 20th century, Canso was remote and isolated. The roads to the town were no great shakes, nor was the surrounding countryside rich. No alternative industries built up.

With major quota cuts taking place, National Sea Products announced late in 1989 that the Canso plant would close, as would their St. John's plant, while operations at North Sydney would be reduced. Canso faced a life-threatening crisis. Mayor Ray White later said their first purchase was a fax machine to pump out press releases, urgent letters, and the like. The town held rallies, enlisted support, and intrigued the national media, becoming one of Canada's top news stories for the year.

Valcourt's personal staff took the lead in developing a survival scheme, working together with National Sea,

Far fewer trawlers operated from Canso and other ports, after the early 1990's.

416

the town of Canso, and other parties. In the deal that emerged, the government provided loan guarantees to help the new SeaFreez enterprise, formed by two established processing companies, buy the National Sea plant and take over eight associated trawler licences. As well, SeaFreez would get major quotas of under-utilized species, including silver hake. These were not just to fish as an experimental venture; an additional benefit allowed SeaFreez to trade off its silver hake and capelin quotas to the U.S.S.R. for more valuable supplies, such as groundfish caught in the Barents Sea or shrimp caught outside the 200-mile limit.

Among under-utilized species, silver hake stood high as a prospect for development. It was plentiful on the Scotian Shelf, and it was a groundfish, though hard to handle and quick to spoil. When SeaFreez got its silver hake quotas, rivals said that Valcourt had not only pre-empted their opportunity to develop under-utilized species, but also given the company an unfair business advantage in general. The vehement protestations from industry representatives chastened the surprised Valcourt. But the deal was done.

The advantages to SeaFreez diminished as silver hake quotas dropped in 1993 and following years. Meanwhile, Canadian experiments in fishing and processing silver hake went forward on a small scale. By the latter 1990's, a Canadian silver hake fishery was taking hold in Nova Scotia. Although it had earlier been thought that silver hake needed immediate on-board freezing to preserve quality, an ordinary dragger fishery was proving feasible. (As well, with traditional groundfish scarcer, the late 1980's and the 1990's saw more interest in fishing skate, dogfish, and shark.)

In Canso, the plant survived, operated by the Newfoundland-based Barry interests, a rising force in the processing industry. But it never employed as many people as before. The department worked out a deal with local interests that enabled a fisheries co-operative to operate. This helped the situation, but the spirited little town remained hard-pressed during the 1990's.

Northern cod moratorium begins

Two dates stand out in the post-war history of Newfoundland: March 31, 1949, when the colony joined Canada; and July 2, 1992, when John Crosbie put the historic northern cod fishery, the foundation of Newfoundland's history and economy, under moratorium.

As noted earlier, in 1989 C.A.F.S.A.C. had wanted a 125,000-tonne T.A.C.; Tom Siddon had set it nearly twice as high, at 235,000 tonnes. Then, in 1990, it had got set at 197,000 tonnes, higher than the Harris panel had wanted.

For 1991, C.A.F.S.A.C. advice said that stock abundance was still more than double mid-70's levels, and a strong 1986 year-class was waiting in the wings. It gave a range of options for the T.A.C., all considered "biologically acceptable," and ranging from $F_{0.1}$

(100,000 tonnes), to a higher level (150,000 tonnes) based on the 50 per cent rule," and to still higher levels of 170,000 or even 215,000 tonnes.

Meanwhile, Minister Bernard Valcourt in 1990 had assigned a task force under Eric Dunne, D.F.O.'s director-general for Newfoundland, to consult on northern cod. The Dunne group recommended using multi-year T.A.C.s to reach $F_{0.1}$. In December 1990, Valcourt announced a three-year management plan, with Canadian T.A.C.s of 190,000 tonnes for 1991, 185,000 tonnes for 1992, and 180,000 tonnes for 1993—all levels well above $F_{0.1}$. Large amounts of cod were still coming out of the water, with trawlers now getting only about one-third of the quota. Scientists still expected a rebuilding, but at that rate of fishing, a very slow one.

John Crosbie became minister in April 1991. The C.A.F.S.A.C. advice prepared that year for 1992 found some grounds for optimism. The stock might increase faster than forecast the previous year, because of the strength of the 1986 and 1987 year-classes. By 1993, the fishery might reach the desired $F_{0.1}$ level. But, C.A.F.S.A.C. acknowledged, even though data for 2J3KL cod were reasonably good, they were too imprecise to say with confidence that the stock was any bigger now than last year. So, the T.A.C. for 1992 should stay as laid out in the multi-year plan, at 185,000 tonnes.

The picture was changing rapidly. The 1991 fishery turned out poorly, both offshore and inshore, taking only 127,000 tonnes in total. In that very cold year, unusual numbers of cod migrated out beyond the 200-mile zone. Foreign vessels, mostly from Spain and Portugal, caught about 47,000 tonnes on the Nose of the Grand Bank, their second highest take since 1977.

In the winter of 1991–1992, research surveys showed a scarcity like nothing yet seen. Commercial trawler captains, who a few years back had argued that there were loads of fish to catch, now reported a near desert. Even fish that were caught were smaller than normal. Scientists found that the spawning stock biomass, of fish aged seven or older, had declined to only 130,000 tonnes, among the lowest levels ever seen.

Heeding C.A.F.S.A.C. and N.A.F.O. scientific advice, Crosbie cut the T.A.C. for 1992 to 120,000 tonnes. After even grimmer recommendations in June, the minister in July announced the moratorium for two years or longer.[35]

John Crosbie in his autobiography described the preceding string of events this way:

> By the late fall of 1988, it was apparent that the actual fishing mortality rate since 1977 had been at least twice the rate projected in the $F_{0.1}$ strategy. ... The scientists found that, over the previous ten years, the northern cod had not been reproducing nearly as rapidly as they thought. A quota designed to allow fishermen to take roughly 20 per cent of the stock each year was actually letting them take about 35 per cent.

417

The scientists thought they knew a lot more than they did. ... The problem with the fisheries scientists was they believed in themselves too much. ...

Siddon, as Fisheries minister, and I, after reviewing the situation with our officials, decided to reduce the TAC for northern cod to 235,000 tonnes for 1989 only. The scientists assured us that one year's fishing at this level would not endanger the stocks. ...

At the end of 1989, we took the total allowable catch for 1990 down another notch, to 197,000 tonnes, or 38,000 tonnes fewer than in 1989. We were on the right track, but we had to move slowly for social and economic reasons.

Any reduction in the allowable catch produced an angry reaction from people whose livelihood depended on the fishery. ... Although Siddon and I knew we were walking a very thin line between scientific advice and economic reality, we were trying to keep the TAC high enough to permit the continuation of part of the offshore fishery and save the jobs of people employed by at least one of the three threatened plants owned by Fishery Products International.

If we'd followed the advice of the Harris panel, the TAC for 1990 would have been 190,000 tonnes. However, I favoured a slightly higher number because I believed, if the quota was a bit larger, FPI might be able to keep its fish plant open at Trepassey in my constituency. ... [Note: In fact, Trepassey closed anyway, a bitter blow for Crosbie.]

At the end of 1990, Valcourt ... announced the 1991 quota for northern cod would be 187,960 tonnes. ... The picture grew bleaker month by month. [But then, D.F.O.'s A.D.M. Science, Brian Morrissey, said on the radio that] the stocks appeared to have increased since the Harris panel did its study and, in his view, the situation was not as alarming as Harris found it. Future events showed that Harris's forecast was far more accurate than the assessment of the scientific branch of DFO.

———

I was still struggling to keep the fishery open. In December 1991 [Crosbie was now fisheries minister], I announced that the total allowable catch for 1992 would be 120,000 tonnes, the lowest TAC ever. Two months later, however, ... the scientists were telling me there were fewer than half as many sexually active cod as I was told there were when I set the 120,000-tonne TAC for 1992.

It was obviously impossible for the department to manage the fishery properly when the advice it received from its scientists varied so alarmingly. I started to talk about the possibility of a moratorium on fishing for northern cod.[36]

The July 2 announcement of the moratorium set off protests and demands for aid. Crosbie announced some help at the time, and would soon go to cabinet for more. Fishermen blamed the department for mismanaging the fish. Crosbie lashed back during one wharfside confrontation, "I didn't take the fish from the Goddamn water."

Below the blame was a shocked wonderment among Newfoundlanders that the cod fishery could be shut down. Many expected it would soon rebound. But departmental scientists pointed out that there was little left in the water. Even if a strong new year-class appeared, it would take seven years to grow to a fishable age. Meanwhile, it was becoming clear that other groundfish were in trouble.

Crosbie creates F.R.C.C.

For groundfish, quota-setting had long been a battleground. When scientists recommended T.A.C.s, industry groups would lobby for the biggest share for their particular sector. They would also, in many cases, push for a hike in the overall T.A.C., to leave more room for fishing. Their arguments could sound reasonable—for example, that there was no sense putting boats or companies or communities out of business because of a temporary scarcity that would cure itself.

How much in those years did lobbying and "politics" result in the department and ministers setting T.A.C.s above the scientific recommendations? Higher-than-recommended T.A.C.s did occur, partly because of the 50 per cent rule, which applied from 1986: if new scientific advice recommended more than a 10 per cent rise or fall, the following year's T.A.C. would only go halfway towards the goal. In the case of northern cod in 1989, as already noted, the T.A.C. got set well above what the 50 per cent rule would have allowed. But in most cases, there was less divergence.

In the case of Scotia–Fundy cod stocks, C.A.F.S.A.C. recommendations and the subsequent T.A.C.s were reasonably close until 1986 and the 50 per cent rule. Discrepancies then became more notable. In 1987, for example, scientists recommended a 6,000-tonne quota for 4Vn (Sydney Bight) cod in the May–December period; the T.A.C. came out at 9,000. Other divergences were usually smaller, but they were there. To make matters worse, fishermen's catches tended to be well above the T.A.C. Setting quotas higher than recommended was only one factor amidst a panoply of weaknesses.[37] Still, it was a problem.

After the codfish collapse, a new body put more distance between ministers and T.A.C.-setting. Crosbie and the department in 1993 set up the Fisheries Resource Conservation Council (F.R.C.C.) to make recommendations on T.A.C.s and other conservation matters, including departmental research. The first chairman was Herb Clarke, a former vice-president of Fishery Products International. Industry members and university professors of biology or oceanography made up the bulk of the council. A few D.F.O. officials served on an ex-officio basis.

Meanwhile, deputy minister Bruce Rawson disbanded C.A.F.S.A.C. But departmental scientists would still do their research and discuss their findings in a similar way, through what became the Regional Advisory Process (R.A.P.).

The new F.R.C.C. began setting up public hearings where D.F.O. scientists would present their research results and assessments. Industry members and other interested parties would then offer comments and critiques, and the F.R.C.C. would recommend the T.A.C.s. By the time the F.R.C.C. was well under way, Ross Reid in June 1993 had taken over as minister. Reid gave the F.R.C.C. firm backing, telling it that he would follow its recommendations.

The F.R.C.C.'s first public sessions drew heavy turnouts, sometimes attracting five or six hundred angry, disputatious fishermen. In following years, attendance dwindled to the normal core of industry representatives and other parties. The council had some success in spreading information, opening a broader dialogue, and taking the politics out of T.A.C. discussions. Ministers followed its T.A.C. recommendations with only rare exceptions.

On other F.R.C.C. recommendations, such as reducing seal predation, action was sometimes slower. The council in the 1990's occupied itself mainly with groundfish, but would also, at Brian Tobin's request, carry out a major review of the lobster fishery.

F.R.C.C. calls for groundfish closures

In its early months in 1993, the F.R.C.C. brought no good news. It was now obvious that the groundfish problem was far wider than northern cod. Abundance had dropped all over the Atlantic coast, but particularly in waters north of Halifax, the ones most affected by temperature change. Indeed, eastern Nova Scotia (4VsW) cod had collapsed as badly as northern cod. The department had already by early 1993 closed an additional five groundfish stocks. Scientific assessments showed many more at low levels.

That summer, after considering scientific and industry views, the F.R.C.C. recommended shutting down commercial fishing for practically every major groundfish stock in Atlantic Canada. The main exception would be southwest Nova Scotia, which could continue at a reduced level. Atlantic groundfish catches, which had been 685,000 tonnes in 1989, dropped to only 153,000 tonnes in 1994. By 1995, the minister had closed the fishery for 25 groundfish stocks.

The "restructured" companies, National Sea Products and Fishery Products International, closed most of their remaining large plants. Those few that continued operating switched their emphasis to shellfish and to processing groundfish imported from over-seas. Other companies closed plants, or cast around desperately for other fish to process.

"A catastrophe of Biblical proportions"

The northern cod moratorium and the other groundfish closures meant that some 40,000 people lost work. It was said to be the largest such layoff in Canadian history. About half the job losses were among fishermen, half among plant workers. Of the plant workers, about half were women. Richard Cashin, leader of the Newfoundland fishermen's union, called it "a catastrophe of Biblical proportions."

Arguing for the 200-mile limit, Canada had declared that coastal-state management would bring better conservation. The department had doubled its science, imposed what seemed to be cautious quotas, and limited vessels in number and size. In the late 1970's and early 1980's, many D.F.O. officials and industry members had believed that Canada had the best fishery science and management to be found. Now the department had presided over the collapse of the world's most famous cod fishery, and other groundfish along with it. What went wrong?

The department never held a thorough and public investigation. The political emphasis switched to aid programs, and in the commotion, nobody pushed hard for an inquiry. Most people thought the groundfish would soon recover. But by the year 2000, Atlantic groundfish catches had only crept up to 149,000 tonnes. This was better than the 104,000 tonnes in 1995, but minimal compared with the more than 700,000-tonne harvests for most of the 1980's.

At the outset, some scientists thought a major ocean event, possibly related to unusually cold years of the early 1990's, had decimated the cod. Others suspected overfishing. And nobody could rule out the impact of the growing seal herds, or of foreign fishing on that fraction of the groundfish that swam on the Nose and Tail of the Grand Banks. (Some people believe that when ocean conditions in the late 1980's and early 1990's drove an unusual proportion of northern cod out beyond the 200-mile zone, foreign fishing helped precipitate the collapse. Still, other groundfish stocks nowhere near the edge of the zone also collapsed.)

In following years, most scientists swung around to the view that the main villain was too-heavy fishing. It appears that fishing pressure was already very heavy in the mid-1980's, with high quotas compounded by dumping, discarding, and misreporting by all sectors. The department never got a clear picture of the prevalence of such practices.

In the early 1980's, environmental conditions were more favourable. Then ocean conditions worsened, especially, it seems, north of Halifax. Indeed, the abundance and robustness of some finfish stocks declined even where there was no fishery. For the main stocks, continued heavy fishing brought on the collapse.

Even if overfishing was the main culprit, the ocean climate did serious damage. It appears that cold temperatures in the late 1980's and early 1990's pushed cod southward from their normal distribution and slowed growth rates, producing smaller fish at a given age. "Since fishing quotas were in terms of biomass, this meant that fishermen had to catch ... more fish per unit of quota. Also, because small fish are less valuable, fishermen resorted to dumping small cod overboard in favour of keeping larger cod." The ocean climate also reduced recruitment of young to the fishable stock.[38]

A decade after the moratorium, cod were still very scarce, and there were no immediate prospects of recovery. Some scientists warned that even with no fishing, recovery would take a long time,[39] and that seal herds were impeding the rebuilding by eating a significant proportion of the few young cod.

Environmental factors were a big question mark. Although waters in many east coast areas got colder in the late 1980's and early 1990's, the exact mechanisms linking temperature and other factors to the behaviour, food, growth, and predation of groundfish were only partly understood. Some speculated that a form of far-reaching environmental and ecological change had taken place, which would continue to discourage groundfish survival and stock recovery. But no one knew.

What went wrong in science?

How did some of the world's best fishery scientists overestimate the abundance of northern cod and most other stocks for most of the 1980's? Over-optimism appears to be part of the story. The Harris report suggested that scientists, influenced by good catches in the late 1970's and early 1980's, placed more confidence in their population models than they deserved.[40]

Still, scientists had frequently expressed caveats. For several years in the 1980's, the department issued a "Resource Prospects" publication projecting catches, but warning that uncertainties made the forecasts only a general guide to likely events. Such warnings often got overlooked. Scott Parsons, who served as an assistant deputy minister in both science and fisheries management, in 1993 wrote, "Scientists were aware that their estimates of population size and projected catch at particular levels of fishing mortality (e.g. $F_{0.1}$) have large error limits. Until recently, however, these shortcomings of the abundance estimation methodology were not communicated clearly to the fishing industry and fisheries managers."[41]

Population estimates were almost as much an art as a science, requiring assumptions based on statistical data. For northern cod and various other groundfish, scientists had little data on the catch versus effort of inshore, passive-gear fisheries, whose fishermen were complaining. Instead, they used catch rates from mobile vessels, which could hunt down fish wherever they concentrated. It may also be that researchers gave insufficient weight to the improvements in catching ability from new electronic and other gear, as captains shared information and tricks.

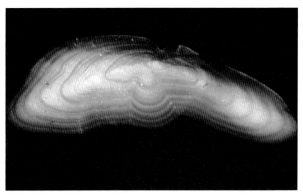

To determine the age and growth rate of fish, scientists often analyzed the growth rings in otoliths (earbones).

Existing data were sometimes weak. Especially in hindsight, many people questioned the accuracy of fishermen's logbooks and of departmental statistics on the size and location of catches. Even before the crisis, fishermen were skeptical. In 1989, the communications branch in the Scotia–Fundy Region surveyed some 200 captains on management issues. About three of every four captains thought it was impossible to estimate fish stocks accurately. Two of every three said D.F.O. estimates failed to reflect trends in their catches. As for sampling methods, they said, D.F.O. research-vessel surveys were the best indicator of abundance, with catch statistics trailing behind, and fishermen's logbooks the least reliable. The study noted that 31 of the captains said vessel logbooks were "a mockery."[42]

Observers on Canadian vessels were still rare in the 1980's. Discards were common both inshore and off. In most fisheries, the scientists had no firm estimates of the number of discards; they left them out of their calculations. Also, it became clear over time that adherence to T.A.C.s was poor. Meanwhile, scientists based next year's forecasts on last year's data, with the situation getting worse in the meantime.[43]

Oceanographic information was also scanty. The department had temperature sets and other data from some areas going back many years. But there was no comprehensive grid giving detailed information. As fish stocks shrank, a temperature change in a particular area could have had a major effect, with no one the wiser.

By 1987, scientists doing retrospective analysis of various population forecasts found major shortcomings.[44] But no safety margin came into play. The researchers thought they were improving their accuracy, and no doubt were, but not fast enough to counteract the advancing catastrophe.

Even at its best, population dynamics could only show part of the picture. To quote Scott Parsons, "in Canada, we have only begun to understand the complex interactions in the marine ecosystem."[45] Researchers operate without two normal facets of science: precise data, and experiments. Starting in the

1960's, departmental researchers at the Experimental Lakes Area in northwestern Ontario treated different lakes differently, and watched the results over long periods. The experimental lakes yielded fundamental information, helped bring about controls on phosphates in household detergents, and thus improved the health of lakes across North America and around the world. But in salt water, the department never set aside a stock of commercial fish for precise, controlled experiments on levels of fishing and types of gear. New regulations amounted to an informal experiment with many fluctuating variables, but without controls or precise measurements. Thus, they often yielded unclear results.

After the cod collapse, a story spread in some media that some departmental scientists had warned of the cod collapse even before 1989, but had been stifled by senior managers. No one appears to have offered solid evidence of this, and it sounds unlikely. Scientists have their own code of honour. It is a rare researcher who would, on the basis of management orders, conceal results. It is rarer still that one could muzzle a whole group of scientists. The entire C.A.F.S.A.C. group was endorsing the northern cod science, just as did the Alverson group.

Groundfish scientists, especially those in northern cod, made a collective misjudgement before 1989. They were working hard, using the best methods available. But data and compliance had their limitations. And Harris was probably correct: scientists let good catches lull them into overestimates. Even after 1989, when scientists called for a northern cod T.A.C. cut of more than 50 per cent, the C.A.F.S.A.C. advice was still nuanced. For 1991, C.A.F.S.A.C. said that catches as high as 200,000 or more tonnes could be "biologically acceptable," although well over $F_{0.1}$. The groundfish fishery was so old and seemingly resilient that few could conceive of a collapse. A good many fishermen, chiefly dragger and trawlermen, were saying that cod was abundant and inshore fears were groundless.

Some researchers believe that had scientists in the early 1980's known more about the discards and what was actually happening aboard the boats, they could have headed off the worst. At least one senior scientist later said that researchers had too little contact with industry.[46] Fishermen had no real place in the assessment process, and there was no reward in the scientific system for mixing with fishermen. Doing so might even cause problems for scientists. Recognition and promotion came largely from publishing. Many scientists to their credit did spend time on boats or mixing with industry people, but most, as was natural, gravitated to the lab.

The situation later changed at least somewhat. Before the groundfish closures, some D.F.O. scientists had worked with fishermen on "index fisheries." The scientists would make special arrangements with particular boats to get extra information. As the groundfish fisheries mostly closed, scientists still needed commercial information. Around the Atlantic cost, scores of "sentinel fisheries" came into being, the F.R.C.C.

encouraging this development. Scientists would contract fishermen, using various types of boat and gear, to fish particular fisheries according to strict scientific protocols.

When all is said and done, how much responsibility for the groundfish collapse attaches to science? Seen under the lamps and magnifying glasses of hindsight, they could have done some things better. But they were smart, competent people, many top-notch in their field, and working hard in the cause of good management. International scientists contributing to the 1987 Alverson report seem to have had broadly similar views. Canadian researchers emitted the 1989 call to cut northern cod quotas by more than half, which if followed might well have averted the collapse.

The scientists were only part of an overall management system with worse weaknesses than anyone suspected, compounded by poor communications between scientists, managers, and fishermen, and by the oceanic and political climate. As one industry executive later said, "There's blood on everybody's hands."[47]

What went wrong in management?

Many scientific difficulties emerged from management shortcomings and industry contrariness. To repeat, some scientists later believed that had they had better knowledge about discards and misreporting, they might have avoided many of the groundfish problems.[48] Scientists depended on managers to make sure that fishermen respected quotas and reported them accurately for a good statistical system. This was all the more necessary after I.Q.s and E.A.s broke big quotas into hundreds of smaller ones.

The department had at least four obvious ways to make sure the fleet caught no more than it should. Looked at closely, none was simple. The first was to "match the fleet to the resource," a 1960's goal still being cited in ministerial speeches of the late 1990's, by which time fishing power had multiplied several times over. In theory, if the fleet were matched to the resource, the department could save money in enforcement and even in science. Rules would be fewer; fishermen could fish more freely. But after bringing in limited-entry licensing, the department had never seriously tried to reshape the fleet. Transferable licences took on a hard-to-control life of their own.

A second way to control the catch was by regulating gear. Drags and trawls were generally thought to be destructive, even when mitigated by large mesh sizes. But the department never forced a major changeover to longlines. There were too many counter-arguments. Only drags and trawls could fish effectively for redfish and some flatfish species. Besides, even longlines could overfish. Gillnets were not much better; some considered them more destructive than trawls. Cod-traps, too, took undersized fish. As for drags and trawls, one could improve their selectivity. The department did encourage the use of square mesh, which let small fish escape more easily, kept its shape better, and prompted other improvements to trawls. But overall, despite some attempts at more selective fishing, there was no thorough push.

A third means to ensure compliance was enforcement against cheating on quotas, dumping and discarding, and other bad practices. The department did its best, but four or five hundred fishery officers on the Atlantic, with many other duties, couldn't monitor every one of the 29,000 Atlantic boats, inshore and off. Industry members and politicians sometimes called for more enforcement, but the bill was already hefty. Fishery officers and guardians made up about ten per cent of the D.F.O. workforce, and patrol craft were expensive. There was little prospect of channelling landings into central ports to make monitoring easier; that would have gone against the whole geography and culture of the coast.

A fourth method to control fishing was persuasion. One could try by information and education to raise fishermen's consciousness of conservation and their compliance with the rules. But here again, one could argue that common sense and respect for the law should by themselves ensure compliance. The department never undertook a thorough public campaign for Atlantic conservation, nor was there any demand for one.

Looking at the individual weaknesses in science and management, one can explain each of them away. Taken together, and coupled with industry behaviour, they destroyed the groundfish. Even today, nobody has definitively weighed the various factors. One effort to get at the roots of the groundfish crisis came in the Scotia–Fundy Region, where biologist Mike Sinclair led an inter-branch analysis. The officials found problems all along the line, with no single outstanding villain.[49]

Everybody's good intentions paved the road to codfish hell. D.F.O. minister Herb Dhaliwal had this to say in March 2000 about the causes of the groundfish collapse:

No one has nailed it down in detail, but we know the general picture. And I am not making excuses for my department when I say that environmental changes did some of the damage.

We did the rest—not just my department, but the whole fishing society.

As a department, we knew less than we thought. On top of that, fishermen often provided false or incomplete catch information, and dumped or misreported fish.

Too often everybody lobbied for higher quotas, and took whatever they could get. People fought for themselves; the fish lost; and we all paid the price.

The codfish collapse wasn't just an Act of God or an Act of Parliament. It was the actions of peo-

ple, in government, in industry, and in coastal communities, failing to work closely enough to protect the fish on which we all depend.[50]

For northern cod in particular, it seems that an initial overestimate was compounded by a political error and further exacerbated by oceanic conditions, all against a background of heavy fishing, with weaknesses in compliance and reporting. Still deeper in the background was the foreign onslaught of the 1960's that by itself took two million tonnes of groundfish a year. Ocean conditions then may have been unusually favourable.[51] The foreign catches had been far above what Canadians allowed themselves after the 200-mile limit. Canadians tended to believe the previous foreign catches were a promise of future Canadian abundance. Perhaps instead they imparted a lasting weakness, from which the stocks never got a true chance to recover.

Could it happen again?

If the northern cod and other collapsed groundfish stocks ever recover and the fishery resumes as before, what would then stop another horrendous collapse?

The department has changed its ways in some regards. As of the early 21st century, the F.R.C.C. process has at least reduced the previous lobbying for higher T.A.C.s. The F.R.C.C. usually recommends cautious T.A.C.s (although some departmental scientists thought the F.R.C.C. mistaken to allow small-scale resumption of cod-fishing in Newfoundland and the Gulf in the late 1990's). Ministers generally follow F.R.C.C. recommendations to the letter. The advisory committee system has become more elaborate, and industry members take more responsibility in co-management. Some fleet reduction has taken place.

But fishermen still have much more than the necessary fishing power. Some people question how much the dockside-monitoring program has improved reporting or compliance. Management has seen improvements but no major breakthroughs.

Scientists are trying to work more closely with fishermen in research and stock assessment. Industry members now take part in the Regional Advisory Process which replaced C.A.F.S.A.C.. Scientists take more pains to follow a strict precautionary approach. But since the groundfish collapse, there has been no major increase in research. Rather, it suffered from the mid-1990's cutbacks.

Nor has population dynamics yielded major new insights. There are still difficulties in calculating recruitment—that is, the survival of spawned fish and the entry of young into the parent, fishable stock. Although methods of catching and counting the very young have improved, there is no accurate way to forecast the unborn. Monitoring of ocean conditions remains less than desired. Some scientists say the best they can do in assessing even the best-understood stocks is to come within 25 or 30 per cent. Others say the best they can do is half or double; some outside the department put it at one-third to triple.

Could we, given the current state of knowledge, detect the early signs of another finfish catastrophe? Some scientists say yes, others say no. If it were detected, would management and ministers head it off? The jury is still out.

Meanwhile, various other countries have faced similarly grave problems in fishery management and conservation. New England and the European Union at the turn of the millennium were suffering drastic stock declines for some species.

A.F.A.P. to N.C.A.R.P. to A.G.A.P.

In the early 1980's, the Atlantic coast had about 20 major plants fed wholly or partly by trawlers; these plants were at La Scie, Catalina, St. John's, Fermeuse, Trepassey, Arnold's Cove, Burin, Marystown, Grand Bank, Fortune, Harbour Breton, Gaultois, Ramea, Burgeo, North Sydney, Louisbourg, Canso, Petit de Grat, Halifax, Lockeport, and Lunenburg. Each employed hundreds of workers. The quota cuts of 1990, the cod moratorium of 1992, and the groundfish closures of 1993 closed almost all of them. A few, such as those at Lunenburg and Arnold's Cove, stayed operating at a lower level. Many other medium-sized plants in such places as Souris, Port Bickerton, Liverpool, Port Mouton, and Clark's Harbour, and hundreds of smaller ones, felt the effects.

A.F.A.P. was still operating, but the moratorium took the crisis to a new stage. Crosbie on July 17, 1992, announced the Northern Cod Adjustment and Recovery Plan (N.C.A.R.P., pronounced "en-carp"). The projected cost of the 1992–1994 program was $920 million, of which D.F.O. was to spend $587 million.[52] As groundfish closures spread, the government broadened assistance with the Atlantic Groundfish Assistance Plan (A.G.A.P.), operating in 1993–1994 with $381 million. Both N.C.A.R.P. and A.G.A.P. provided income support for affected fishermen and plant workers, assistance to inshore vessel-owners to maintain their vessels and gear during the closure, training for work inside or "adjusting" outside the fishery, and financial incentives for licence retirement and early retirement of fishermen and plant workers.[53]

Taken together, A.F.A.P., N.C.A.R.P., and A.G.A.P. spent $510 million on income support, for which nearly 40,000 people qualified at the outset, although several thousand soon found other work. Most of the payments went to Newfoundland, where it was known as "the package" (some called it "the parcel"). Fishermen and plant workers initially got $225 a week, raised after vehement protests to $406 a week.

Another $281 million went towards adjustment, in the form of training, community economic development, and the like. The programs also devoted $17 million to vessel support, and $26 million to licence retirement. This all added up to roughly $834 million

in direct aid related to groundfish, not counting the A.F.A.P. money devoted to science, enforcement, and the like. Another $29 million from A.F.A.P. went for salmon-licence retirements, a separate project to be discussed later.[54]

Strength of fishing culture fades

The income support helped many towns get through the crisis. It even provided an unexpected boost for certain Newfoundland outports where fishery incomes had never been that great. The various training projects, for example in adult literacy and navigation, often provided valuable help. A few recipients even got to university through the programs. Community development efforts, including attempts to attract new food-processing or other ventures, also had some successes. But often they produced nothing substantial or permanent.

Meanwhile, the cod collapse brought a cultural change in Newfoundland. Traditionally in fishing communities, boats and plants had offered early employment to many young people, of whom a good number would stay on for a lifetime. The fishery was the source of songs and stories and family memories, the cultural rock of Newfoundland, the thing to tell strangers about when you travelled. All that was now thrown into question. More young people sought to complete high school and post-secondary education, and looked for jobs away. The population of many fishing towns dropped, as did the population of Newfoundland.

The struggle to downsize

A.F.A.P. had made little effort to reduce the fleet. N.C.A.R.P. and successor programs at least tried. It would be a long struggle. Reducing fishing power involved not only boats but licences, which were transferable. Fishermen could die, but the licence lived on.

With the switch to limited entry in the 1970's and 1980's, more than 50,000 licences of various sorts for

While government tried to cut down the fleet, fishermen kept hard at work. A D.F.O. surveillance aircraft took this shot of a longliner fishing at night off southwest Nova Scotia.

various species had come into play for the 29,000 Atlantic boats. Earlier licence buy-backs in the lobster and salmon fisheries had only nibbled at the edges of the total over-capacity. With no limited entry on people, and with U.I. supplementing incomes, the number of registered fishermen had nearly doubled in the years from 1973 to 1988, peaking at 66,000.

In 1992, with the groundfish crisis well under way, the department under Crosbie cancelled thousands of "inactive" groundfish licences, whose holders were doing no fishing. Atlantic groundfish licences dropped roughly from 17,000 in 1992 to 13,000 in 1993. Fishermen protested, often saying that they needed a set of licences to switch among as resources and markets varied. If one had to fish every species regularly to keep a licence, that would only worsen resource depletion. Under Tobin, the rules eased, and some licences got restored. The number of groundfish licences rose to about 14,000. Thus, many less-than-active licences still remained on the books.[55]

The fishery was over-stuffed both by hard-working boats with excessive fishing power, and by marginal participants who did less damage to the fish but over-burdened the U.I. system and other special programs. N.C.A.R.P. wound up tackling mainly the marginal participants, none too successfully. "Adjustment" aimed to help fishermen into other jobs, through retraining and other assistance. Although results were imprecise, this route probably removed only about 300 fishermen.[56]

The job-finding effort faced obvious obstacles. A common question was, "Retraining for what?" Small fishing towns offered little alternative employment. Yet people were reluctant to move away, having lived all their lives in close-knit communities, where they probably owned a house. Even with the adult-literacy and other programs under A.F.A.P. and N.C.A.R.P., finding work elsewhere would be difficult, especially since half the people concerned were age 40 or over, and three-quarters had never finished high school.[57]

Besides "adjustment," two other mechanisms removed fishermen: licence retirement and early retirement. A cap of $50,000 applied on licence retirements; thus, they appealed mainly to smaller operators. This part of the program spent about $25 million. Early-retirement stipends could go to older fishermen over 55. Together, licence retirement and early retirement removed 876 licences and about a thousand fishermen. Combining this with the 300 removed by "adjustment," N.C.A.R.P. took out some 1,300 fishermen. Others left on their own. The number of registered fishermen in Newfoundland dropped from about 26,500 in 1992 to 24,700 in 1993, a reduction of about 1,800.

The number of boats in Newfoundland dropped from about 15,000 in 1992 to about 13,100 in 1994. The decline came almost totally in small boats, less than 40 feet and especially less than 35. It all represented only a fractional drop in fishing power, possibly 5 per cent.

As for plant workers, N.C.A.R.P. planners underestimated their numbers. Instead of the expected 10,000, more than 15,000 took part in the program, most drawing income support. Only 30 per cent enrolled in programs leading to an exit from the industry. Of those, many never found other work. All told, through retraining or early retirement, N.C.A.R.P. removed about 1,760 people from plant work.[58]

Thus, in reducing numbers, N.C.A.R.P. had only minor influence. The program had additional problems. The Auditor General of Canada in 1993 complained about sloppiness, a shotgun approach, and a lack of focus and of clear legislative authority. Still, N.C.A.R.P. had buffered the immediate crisis, kept people going, and gotten at least some people out of the industry.

Cashin report calls for better programs, smaller fleet

In mid-1992, Crosbie commissioned Richard Cashin, leader of the Newfoundland union, to head a Task Force on Incomes and Adjustment in the Atlantic Fishery. The five-member group, drawn from industry and academe, began examining government programs affecting fishermen's incomes. A task force discussion paper noted that "there are even less fish than in 1974 [an earlier low point for groundfish], but more than 50 per cent more people trying to get an income from them."[59]

A prime consideration was the fishermen's Unemployment Insurance system. Critics questioned its costliness. In 1987, for example, fishermen nationally and fish-buyers as their "employers" paid in $17 million and received benefits of $223 million. Plant workers received U.I. as well; their benefits in 1988 totalled $226 million.

Most fisherman claimants used the main program of seasonal fishing benefits for self-employed fishermen. One could accumulate insured weeks in the May–November period and draw benefits in the November–May period, or vice versa. Fishermen needed to earn a certain minimum amount in at least ten weeks, six of them in fishing. Generally, benefits could run up to 27 weeks.

As of 1990, about 23,000 fishermen used seasonal fishing benefits of more than $300 weekly; individual benefits averaged about $7,900. Another 5,700 fishermen worked on boats for wages, gaining access to regular U.I. ("labour stamps"), from which they drew on average about $9,000.[60] Total benefits that year came to about $240 million, equal to more than one-quarter of Atlantic landed value.

Although successive governments had tried to tighten the rules here and there, politicians from fishing areas had resisted such efforts. Entry into the fishing U.I. system remained relatively easy. The Cashin report appearing in November 1993 noted that the fishery included thousands of individuals "who find their way into the official ranks of the fishery by doing just enough to meet their own objectives of topping up their other income and by qualifying for special Fishermen's U.I. benefits. These marginal participants have little long-term commitment to the industry, and contribute little to the total catch." The report added,

> For example, between 1981 and 1990, some 80,000 people in Atlantic Canada reported some self-employed fishing income in at least three of those years. However, only 14,000 of them fished in each of those 10 years, and only 36,000 fished in at least five years. ° By any standard, the core group of professional fishermen is a lot smaller than the registered total. Approximately half of those who register and two-thirds of those who actually fish at some point in the year would qualify as professional fishermen. These are the real fishermen who run almost all the boats and enterprises."[61]

The Cashin report proposed that the fishermen's Unemployment Insurance scheme channel its benefits more narrowly to the "real fishermen." Among other reforms, U.I. should switch from the awkward "insurable week" system and should gauge benefits by the whole season's results. Benefits should go only to fishermen whose earnings met a certain minimum level, which varied by region. Those changes did take place in 1997, coupled with other reforms that further tightened the rules for what was now called Employment Insurance (E.I.).

In related recommendations, the Cashin report proposed a new system of income stabilization, roughly comparable to farm programs. Like earlier such proposals, this had no result. Cashin also suggested an Integrated Registration and Reporting System, which would, among other effects, make it more difficult for fishermen to falsify landings and thereby evade quota, U.I., or income-tax rules. Part of the thinking was to require fishermen to give the same information on landings to all concerned agencies, including D.F.O., E.I., and tax authorities. Interdepartmental work began on this initiative, but it fell victim to Privacy Act concerns.[62]

During its work, as the dimensions of the groundfish crisis loomed larger, the Cashin group had given increasing attention to adjustment. The report called for income support to allay the crisis, more action towards professionalization of fishermen, and adjustment efforts that would almost amount to a Marshall plan. Cashin also called for a major reduction in fleet and plant capacity, by about 40–50 per cent.[63]

Appearing about the time of the late-1993 federal elections, the Cashin report added impetus to the Conservative John Crosbie's and the Liberal Brian Tobin's efforts towards fleet reduction.

T.A.G.S. provides another $1.9 billion

In 1994, the new minister, Brian Tobin, announced another $1.9 billion for The Atlantic Groundfish Strategy (T.A.G.S.), which replaced the previous programs. T.A.G.S. provided money for income support for affected fishermen and plant workers, labour adjustment through employment counselling and training, and long-term community economic development. In approving the $1.9 billion program, government set the goal of reducing fishing capacity by 50 per cent.[64] D.F.O. and Human Resources Development Canada (H.R.D.C.) led the program overall; the Atlantic Canada Opportunities Agency (A.C.O.A.) and its federal counterpart in Quebec led the economic development work.

Meanwhile, at least some groundfish jobs were returning. Starting in 1997, the F.R.C.C. recommended and ministers allowed some small-scale reopenings for northern cod and Gulf groundfish stocks, which gave work (and E.I. benefits) to hundreds of fishermen. Some people questioned whether the F.R.C.C. was being strict enough on conservation, or might be bending to socio-economic considerations.

"Core fishermen" come to front

The Cashin report's emphasis on "real fishermen" reflected an old issue. In the 1970's and early 1980's, some efforts had taken place to favour those who worked the hardest and depended the most on the fishery. Lobster licences had differentiated fishermen by extent of participation. LeBlanc's policies had looked towards a more professional cadre of fishermen. The department had differentiated "full-time" and "part-time" fishermen, and where appropriate favoured full-timers. In the Gulf, the department had adopted the Maritime Fishermen's Union "bona fide" licensing policy, favouring those with the greatest stake in the industry. But little had happened since.

Now, both D.F.O. and H.R.D.C. wanted to focus on the real fishermen. The department consulted with the Canadian Council of Professional Fish Harvesters to develop a set of "Special Eligibility Criteria," identifying those fishermen who could participate in the licence-retirement program, or, if they stayed in the industry, could receive licences transferred from retiring fishermen.

Officials also reviewed the licensing system, consulting widely. In 1996 the department put the "core fisher" policy into effect. This closely resembled the M.F.U.'s "bona fide" scheme. Only those fishermen who fished certain key species for their area and had a clear dependence on the fishery for their income, could qualify as core fishermen. And only core fishermen could receive licences through transfers. Non-core fishermen could only transfer them away. (For groundfish in particular, not even this was allowed; when the non-core licence-holder left the fishery, his groundfish licence evaporated.[65]) Over time, the number of fishermen would drop, as non-core fishermen transferred their licences to core fishermen.

To become a core fisherman, one had to take over an existing core fisherman's licence and core designation. In some regions, this could happen easily. Others,

425

notably Newfoundland, required a form of apprenticeship and certification.

During the early days of licence limitation, the department had sometimes applied "use-it-or-lose-it" participation rules. These had sometimes encouraged fishermen to specialize in one or two species, while giving up unused licences in other fisheries. The bona-fide and core fisherman policies of the 1980's and 1990's reversed that trend. Now, full-time fishermen could more readily collect a portfolio of licences, and switch from fishery to fishery as circumstances changed. Another change from the early days was that, through the core system and other changes, the licence was more clearly "on the man," rather than on the boat.

Although never the subject of headlines, the core policy was a landmark in fisheries history, promising a more stable and sensible industry.

The department's previous categorization of full-timers and part-timers gave way to the new criteria. Meanwhile, numbers were falling, especially for the "core" category. In 1990, the Atlantic fishery had included about 61,400 fishermen, with full-timers numbering slightly less than part-timers. By 1996, the total had dropped to around 50,000, and by 1999 to 42,700 registered fishermen. By the latter year, the 12,400 core fishermen were far fewer than the 30,300 non-core.

Although numbers were dropping, the 42,700 total in 1999 was still more than the 39,000 registered fishermen back in 1973, when "limited entry" was coming into force for licences but not for fishermen.

Co-management gets stronger

As Program Review cutbacks coincided with the groundfish crisis, D.F.O. had less money for its activities. It also had less confidence that it knew all the answers. Officials talked more and more about going back to the "core mandate" of conservation, sometimes forgetting that it had almost as strong a tradition of intervening in other matters. As the department tried to do less, it asked fishermen to do more.

As already noted, co-management got more attention. It was particularly strong in certain fisheries, such as crab in the Gulf of St. Lawrence and offshore scallops in Nova Scotia. In the mid-1990's, "Integrated Fishery Management Plans" (I.F.M.P.s) for many fisheries made the planning process more inclusive. Management plans often became multi-year. And the department and industry began using Joint Project Agreements (J.P.A.s), signed documents itemizing the responsibilities of D.F.O. and industry groups.

The snow crab fishery in Area 19, on the northwest side of Cape Breton, provided a notable example of co-management via J.P.A.s. A small-scale fishery had begun in the 1960's, supplementing the groundfish and lobster fisheries and expanding as crab increased in value. Individual quotas had come into play. By 1992, the number of licences had increased to 59 permanent and 15 temporary. As requests for access multiplied, fishermen at one point opposed a departmental plan and argued for their own approach. Discussions led to an I.F.M.P. for the period 1996–2001, incorporating a Joint Project Agreement.

The subsequent scheme allowed more than a hundred licences, but with reduced quotas for new ones. Along with quotas, a system of individual transferable

---

Key elements of the "core fisher" policy

The 1996 *Commercial Fisheries Licensing Policy for Eastern Canada* outlined the core fisher policy. In summary,

For the "inshore" sector, defined as vessels less than 65 feet long, there would be a "core" group of a maximum number of multi-licensed enterprises. To qualify, a licence holder was required to be the head of an enterprise; hold key licences (or, for some Scotia–Fundy fishers, a vessel-based licence); have an attachment to the fishery; be dependent on the fishery.

One could enter the core group only by replacing an existing enterprise. The new entrant had to be a "certified professional fisher." As only Newfoundland had certification in place, fishermen elsewhere needed to be full-timers or "bona fide."

Under the policy, most benefits would go to core members. "The Policy promotes the concept of multi-licensed enterprises while recognizing specialized fleets. Fishing enterprises are viewed as businesses with normal responsibilities such as selection of crew and reporting of landings."

Reaffirming existing practices, the policy stated that although licences strictly speaking were not transferable, the minister could prescribe conditions governing the issuance of a licence to a new licence holder, as a "replacement" for an existing licence being relinquished. Replacement licences could go to an eligible fisher recommended by the current licence holder.

Licences in the under-65-foot sector could only go to the head of an existing core enterprise or to an Aboriginal organization; or, all vessel-based licences held by the head of a core enterprise could go as a package to a qualified new entrant, who would then hold core status.

Vessels at Cheticamp on northwest Cape Breton, in crab-fishing Area 19. (D.F.O. photo by Michel Thérien)

traps came into play. Fishermen would have the main voice on several matters, including the exact level of fishing (within certain limits, and respecting science-based Total Allowable Catches) and the exact shares of the catch. When the value of the harvest rose above a certain threshold, temporary licences would provide access for others. Fishermen would pay D.F.O. to carry out research surveys, and would pay for monitoring costs. Little stayed undefined.

Many other fisheries worked out detailed co-management arrangements. Progress was particularly strong in more homogeneous fisheries, such as northern shrimp, crab, and offshore scallops and lobster. (A special case applied in several Aboriginal land-claim agreements, typically operating through joint management boards.)

Although some people in the Atlantic industry favoured a strong central authority and a no-nonsense D.F.O. approach, most wanted a good degree of co-management. There were also a few efforts to form broader-scale councils to look at wider issues. Although fisheries varied, in general the management system in the 1990's had arrived at a higher degree of industry responsibility than ever before.

Some fishermen charged, however, that co-management went together with individual quotas and privatization. Private interests, they said, were displacing D.F.O. and treating the common-property fisheries as their own.

Processing companies work through the crisis

Meanwhile, what of the processing companies after the groundfish collapse? The two large groundfish companies "restructured" in the 1980's, National Sea Products and Fishery Products International, closed most of their plants at the cost of many jobs, and sold most of their more-than-100 trawlers to foreign buyers.

This time there were no big-company "bailouts." Yet both N.S.P. and F.P.I. survived, mainly by processing and selling fish from overseas. Product came from such places as Russia, Alaska, China, Thailand, and

the Nordic countries, for further processing or marketing. Some smaller enterprises did the same.

Back in 1990, among the hundreds of Atlantic plants for all species, there were 61 employing 250 or more people, and 134 employing between 100 and 249.[66] After the groundfish collapse, such large workforces became much scarcer. Newfoundland, the most groundfish-dependent province, took the biggest hit. Before the 1992 cod moratorium, some 280 plants operated in the province. By the year 2000, only 125 operated; these now included 32 crab and 11 shrimp plants. The rest subsisted on the remnants of groundfish, along with capelin, lumpfish roe, herring, and whatever they could gather.

That being said, for Atlantic processing overall, the groundfish crisis although damaging proved at least somewhat less of a disaster than first appeared. Indeed, some smaller and medium-sized operations gained vigour with the shellfish boom.

Strange as it seems, Statistics Canada figures for plant employment ("Fish Products Industry") were nearly as high at the end of the century as before the groundfish collapse. For Newfoundland, the numbers were 9,400 in 1989 and 8,050 in 1999; for Nova Scotia, 4,800 and 4,600; for New Brunswick, 4,800 and 3,700. These figures were averages, derived by adding up employment through the year and dividing by 12; thus, they gave no precise information on the actual numbers of individuals working in fish plants at one point or another. Still, on an overall scale, they suggested decline more than disaster. While some groundfish-dependent communities might look more like ghost towns, other places might gain strength from shrimp and crab.

In Newfoundland, larger producers had long complained about the Canadian Saltfish Corporation, the federal crown corporation that sold all salt cod from Newfoundland and the Lower North Shore of Quebec. Other processors thought the C.S.C. was an interference with private enterprise, especially when, in the late 1980's, the C.S.C. had begun to sell some frozen fish and to subsidize vessels for its producers. Now the C.S.C. fell victim to the shortage of fish. It was losing money. The government closed it down in 1995.

Montreal Round Table clears air

In 1994, the Fisheries Council of Canada launched a public campaign for a revised approach to the fisheries. The F.C.C. wanted more secure access to the resource for processing and fleet-owning companies, notably through individual or individual transferable quotas. Minister Brian Tobin responded by calling all sectors of the Atlantic fishery to a "round-table" meeting in March 1995.

The Montreal Round Table had some success. Participants agreed on the objectives of a viable fishery, reduced harvesting and processing capacity, and reasonable incomes. Fishermen's representatives also agreed that individual and individual transferable quotas could be appropriate for fisheries where a clear majority (e.g., two-thirds) agreed and where other safeguards applied, including restrictions against the undue accumulation of quota.

The Round Table also backed professionalization and encouraged multi-species fisheries. The multipurpose boat was more common than not in the Atlantic fishery, but owner–operators sometimes felt threatened by larger, specialized vessels. The Round Table gave the multi-species fishermen a vote of confidence.

Such a gathering was by now an unusual event. Back in the 1970's and early 1980's, the department had frequently held policy conferences. These had now become scarce, reflecting either fatigue or some degree of acceptance of the state of the industry. Calls for massive transfers of access to either inshore, midshore, or offshore were dying out. Industry members were learning to live with the complex world of present-day management. But there could still be major flare-ups.

Nova Scotia squabble strengthens co-management

Even after the northern cod and other moratoriums, southwest Nova Scotia still had a significant groundfish fishery, supporting a hundred or more draggers, and some 3,000 licence-holders for longlines, gillnets, and handlines. In 1995 and 1996, the federal fisheries department was trying to subdivide quotas among sectors of the fleet, slicing fewer fish into more pieces. At the same time, changes in licensing policy were removing core status from many of the smaller-scale, handline fishermen. Among some of those affected, frustrations boiled over. Fishermen staged occupations of D.F.O. offices, and in one instance held a demonstration in Halifax, hundreds lining up behind one another to carry a long section of rope through downtown. Minister Fred Mifflin restored core status for handliners.

Regional officials convened a major workshop bringing together fishermen and community representatives, where people could vent their opinions and make suggestions. Out of the commotion came a higher degree of co-management. For the Scotia–Fundy fleet of fixed-gear vessels less than 45 feet long, nine groundfish "community management boards" each took over their own percentage of the quota, based largely on catch history. The boards dealt with such matters as fishing schedules, trip limits, and local I.Q.s and I.T.Q.s.[67]

Fleet-reduction money goes for income support

Meanwhile, for the Atlantic groundfish fleet generally, the great issue of over-capacity remained. The T.A.G.S. program had started up in 1994 with ambitions for a major reduction of fishing capacity. Harvesting Adjustment Boards came into place in the different Atlantic regions to run a Groundfish Licence Retirement Program (G.L.R.P.). Fishermen put in bids by a "reverse auction" process; the boards decided which offers constituted the most capacity for the money. As well, the Atlantic Fisheries Early Retirement Program (A.F.E.R.P.) gave stipends to fishermen aged 55–64 who wanted to retire with some dignity.

The large number of fishermen and plant workers eligible for benefits—T.A.G.S. dealt with some 40,000 people at the outset—surprised the planners. More people than expected wanted to stay in the industry. Many still thought the groundfish fishery would soon be reopening. T.A.G.S. wound up doing less retraining and licence retirement than hoped. Income support took far more money than expected, eating into the fleet-reduction budget. T.A.G.S. in its entirety ran out of money and closed in 1998 after only four years instead of the expected five.

By then, the G.L.R.P. had taken 478 licences out of the fishery, mostly in Newfoundland; fewer than a hundred came from the other provinces. Fishermen participating had to give up all licences and their Personal Fishing Registration, and leave the fishery forever. The cost came to some $60 million, or about $125,000 each. Of vessels involved, 76 per cent were under 35 feet, and 93 per cent under 45 feet. Some observers felt that, like N.C.A.R.P., T.A.G.S. busied itself overly with the more marginal operators.

The early retirement program, A.F.E.R.P., for its part retired 333 fishermen for $28.5 million, or about $85,000 each. Monthly benefits to retirees ranged from about $600 to $1,200, depending on previous earnings.

All told, T.A.G.S. removed about 800 fishermen from Atlantic fisheries, mostly in Newfoundland. This was significant, but far from the hoped for 50 per cent reduction in capacity. (Meanwhile, some training programs under T.A.G.S. gave some fishermen more efficiency in using new technology, thus working against any reduction in fishing power.)

As for plant workers, according to an evaluation by Human Resources Development Canada, officials in H.R.D.C., A.C.O.A., and, for Quebec, the Federal Office of Regional Development (F.O.R.D.) helped prepare some 10,600 people to "adjust out" of the fishing industry towards other jobs, whether or not they materialized.[68]

Anderson launches "final" fleet-reduction program

In June 1998, minister David Anderson announced approval for the biggest and, it was said, last fleet-reduction program for both coasts: the Canadian Fisheries Adjustment and Restructuring (C.F.A.R.) program. On the Atlantic, C.F.A.R. provided $730 million overall. Most of the money went for "adjustment" efforts. T.A.G.S. clients would get additional money: about $180 million in lump-sum payments, to make up for the earlier-than-scheduled termination of that program. Further measures to help them gain skills and work experience, become self-employed, or relocate would cost up to $135 million, and would be run through H.R.D.C. Another $100 million would go for community and regional economic development, through A.C.O.A. (Atlantic Canada Opportunities Agency) and Canada Economic Development (the new name for the federal development agency in Quebec). Up to $65 million, cost-shared with provinces, would go towards early retirement.

Under C.F.A.R., fleet reduction through the Atlantic Groundfish Licence Retirement Program (A.G.L.R.P.) got a budget of $250 million, to buy people permanently out of the fishery through reverse auctions. Again, the program went mainly for smaller operators. In Newfoundland, the Fishing Industry Renewal Board replaced the Harvesting Adjustment Board, with a similar mandate. In most regions, fishermen selling out had to yield all their licences to the government. In Nova Scotia, core fishermen could transfer non-groundfish licences to other core fishermen.

At the start of A.G.L.R.P., Atlantic fishermen (and a small number of licence-holding companies) held some 13,000 groundfish licences, down from 17,000 in the early 1990's. The reduction had come both through buy-backs and through cancellations of inactive licences. The new program aimed to remove another 3,000 fishermen.

A.G.L.R.P. concentrated on Newfoundland; the number of groundfish licences in the province dropped from about 6,700 in 1998 to 5,000 in the year 2000. Because retiring fishermen gave up all their licences as a package, the program took more than 6,200 other licences in Newfoundland. More than a third of lobster, capelin, and herring licences disappeared. Costs in the province came to $159 million.

In following years, the average value of Newfoundland landings rose. But it was debatable, a D.F.O. evaluation found, how much of the credit should go to licence-retirement efforts. The prime factor was the rising value of shellfish landings.

Elsewhere, the Scotia–Fundy Region took out 388 groundfish licences, many of them under-utilized, for $39 million. The Gulf Region took out 52 high-capacity fishermen, and Quebec retired a total of 129. In those regions, too, the effect on viability seemed unclear to industry members.[69]

All told, it appears that C.F.A.R. took out about 2,400 fishermen, T.A.G.S. about 800, and N.C.A.R.P. about 1,300, for a total of 4,500. But in every program, many of the retirees were smaller operators or even marginal. By themselves, the licence-retirement programs, for all the millions spent, probably had no major effect on the viability of the remaining fleet, and failed to achieve the long-held goal of matching the fleet to the resource.

The licence-retirement programs did, however, put money in the hands of recipients. They reduced over-dependence on the fishery. And the coincidental rise of shellfish catches and values saved the day for many remaining enterprises.

Fishermen, fleet drop by one-third

In the meantime, the fleet was shrinking for other reasons besides the licence-retirement programs. Between 1990 and 1999, the number of boats in the Atlantic fleet dropped from 29,200 to 20,400, and the number of registered fishermen from 61,000 to 43,000. Most of the boats were small. Many of the fishermen had never been all that active (only 40-odd thousand fished in a given year). Still, it was a major, one-third drop in numbers.

Several factors were at work besides the special programs. Many small-scale fishermen had fished groundfish. Some left as the resource shrank, without benefit of any special programs. The normal attrition was going on, as some people got too old to fish. In some cases, transferable quotas were being concentrated in fewer vessels. Meanwhile, new entrants faced new obstacles. Non-core fishermen could no longer transfer their licences to new entrants, but only to core fishermen. New boats and licences cost more to obtain, and D.F.O.'s licence fees were going up. Professionalization was imposing new requirements in some areas.

The atmosphere surrounding the fishery was changing. The old fishery was the employer of last resort, easy to enter, where you could always make at least a few dollars. The new fishery was bigger boats, and was more difficult and costlier to enter. Existing fishermen were more jealous of their status and the resource.

What did the fleet reduction change?

In the 1980's, the Atlantic fishery generally was over-subscribed, in several ways. The fleet was too powerful, threatening conservation. It was too costly, threatening profits. It had too many people, creating over-dependence.

In a typical year of the late 1980's, the landed value of Atlantic fisheries came to around a billion dollars. The Cashin task force offered a rough but instructive calculation. After expenses in those years, perhaps $600 million remained to share among 45,000 or 50,000 active fishermen. The average earnings from fishing thus worked out to some $12,600, a low income by Canadian standards for a highly skilled, hard-working occupation. Some fishermen of course did better, but many did worse. There was too little money to go around.[70]

That was still the case at the end of the century, even with the shrinking numbers of participants. By a similar rough calculation, the 42,700 Atlantic fishermen in 1999 averaged about $22,600. This was a significant gain, but still low compared with other occupations. (More detailed income statistics appear later.)

A good number of fishermen at the end of the 1990's were doing well, better than the averages would suggest. But for many, their good fortune depended on a shellfish boom of uncertain duration. The fleet still had

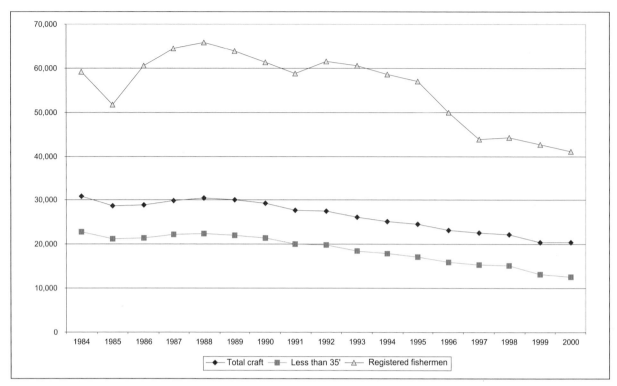

Fishermen and fleet both dropped in numbers. The lower line shows craft under 35 feet long, which still made up most of the fleet, although their proportion was shrinking.

more than enough power to devastate stocks, and was still over-capitalized.

That being said, there were fewer vessels sharing the catch. At the turn of the century, as far as one can forecast the unpredictable fishery, it appeared that the fleet might have a better chance at profits than at many times in the past, at least if shellfish stayed abundant. Meanwhile, the groundfish crisis and the aid programs of the 1990's went at least some distance to jolt east-coast communities out of their over-dependence on fisheries. That change of attitude, it appeared, might help both the fishery and the economy at large.

What did the special programs cost?

The budgets of A.F.A.P. and its counterpart Quebec program, N.C.A.R.P., and of A.G.A.P., T.A.G.S., and C.F.A.R. for the Atlantic added up to well over $4 billion dollars. This roughly equalled the entire landed value, combined, of Atlantic groundfish through all the years from 1984 to the late 1990's. The bonanza fish of the 200-mile limit and the Kirby report had swirled down the ocean's drain, taking billions of taxpayer dollars with it.

The bulk of the money went for income support, training, relocation, and community development. These costs related less to fishery management than to an attempt to reorient people and communities away from a fishery that in the 20th century had too seldom provided good and reliable incomes except to a minor-

ity. But whatever way one looks at it, the expenses were major, on top of the regular fishery expenditures. The costly programs caused many people to question government spending on fisheries in general.

What do the regular programs cost?

Setting aside the costs for special programs such as T.A.G.S., how much did D.F.O. spend nationally on fisheries management? The fiscal year 1995 provides one example. The groundfish crisis had hit, but shell-fish were moving up. Program Review cutbacks were starting to sink in, but had not yet reached their nadir. The regular expenses came to $82 million for fisheries research, and $160 million for fisheries operations. Another $55 million went for small craft harbours, for a total outlay of about $300 million.

That year the landed value came to $1.36 billion for the Atlantic and $1.78 billion for the country (not counting $342 million from aquaculture). Processing and handling added more value. Exports amounted to $3.1 billion, supplemented by Canadian sales. Value exceeded costs many times over, which is not that bad for managing a complex, common-property industry. Some of those costs would remain even without a commercial fishery. The country would still feel obliged to do fisheries and ocean science, to conserve recreational fisheries, and to protect fisheries sovereignty.

Outside of D.F.O. spending, however, one other major expenditure comes into the picture. In 1995,

Table 27-1. Volume and value of major species groups, 1989 and 1999.

	Volume (tonnes)		Value ($ millions)	
	1989	1999	1989	1999
Cod and other groundfish (e.g., redfish, flounder, pollock, haddock)	685,000	152,000	359	190.5
Herring and other pelagics (e.g., capelin, mackerel, tuna)	359,000	255,000	85	75
Shellfish (e.g., scallops, lobster, shrimp, crab, clams)	228,000	378,000	503	1,300

benefit payments through fishermen's Employment Insurance came to $227 million, about equal to the budget for fisheries operations and small craft harbours combined.[71] Fishermen's E.I. was the Canadian government's single biggest fishery expense.

Shellfish takes over

Across the country, Canadians conscious of the groundfish crisis tended to think the whole Atlantic fishery had gone to pieces. But by 1995, the fishery was setting new records for value. While groundfish, traditionally the biggest catch, went from the highest volume to the lowest, shellfish climbed from the lowest to the highest. And shellfish were worth far more per pound. Even in 1989, they had accounted for more than half of landed value. By 1999, they supplied four-fifths. Table 27-1 shows the great shift from finfish to shellfish.

Nobody had forecast the shellfish bonanza. Back in the late 1970's, when lobsters were seen as an important but low-growth fishery, researchers had warned of future problems. Instead, from 1972 to 1999, lobster catches almost tripled from 15,000 tonnes to 44,000 tonnes. Lobster value rose almost 15 times over, from $37 million in 1972 versus $537 million in 1999. Total shellfish catches multiplied fivefold, from 72,000 tonnes to 378,000 tonnes. Shellfish value climbed from $60 million to $1.3 billion, more than 20 times as much.

As shellfish flourished, prices rose, and fishermen found new grounds, the demand for access to the resource increased. In many cases, the department yielded. (The major increase would come in the mid-1990's, when the department would issue about 3,000 "temporary" crab licences in Newfoundland.) Processing companies built new plants and often helped fishermen fund new vessels. Several hundred million dollars went into the crab and shrimp expansion, raising fears of a new boom and bust.

What explains the shellfish boom?

Why did lobster, shrimp, and crab increase? The answers at the turn of the millennium were still unclear to scientists. The decline in groundfish might have cut down predation. Environmental effects also seemed to be at work, but it was difficult to pin them down. Snow crab and shrimp might have benefitted from cooler water temperatures in some areas, notably off northeast Newfoundland and Labrador in the late 1980's to the mid-1990's. But there was no readily apparent relation between recent temperature changes and lobster abundance.

Shellfish scientists never attempted forecasts in the same manner as for finfish. There was no comparable system of determining shellfish ages and estimating year-classes and biomass. Stock assessment did take place, using other methods, for snow crab, scallops, and, to some degree, shrimp. Quotas often applied for these fisheries, but not for inshore lobsters.

For crab, at the turn of the millennium, scientists appeared fairly confident that management controls were protecting stocks, despite concerns about discarding and wastage. But it was expected that changing ocean conditions would reduce abundance of both crab and shrimp, the key question being by how much.

As for lobster, the Fisheries Resource Conservation Council had warned of excessive fishing pressure. Most lobsters got caught before they could reproduce. D.F.O. had recently undertaken conservation measures to encourage more egg production, despite some industry resistance.

Commercial salmon fishery ends

In the period 1984–2000, the commercial salmon fishery, one of the oldest in North America, came to an end.

In the Maritimes, after a multi-year moratorium, the New Brunswick fishery had reopened in 1981 on a

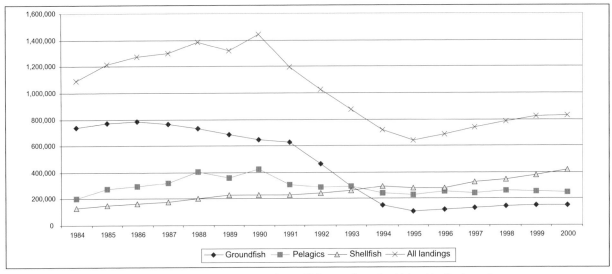

Atlantic landings (tonnes), 1984-2000, by species group. Groundfish went from the highest to the lowest.

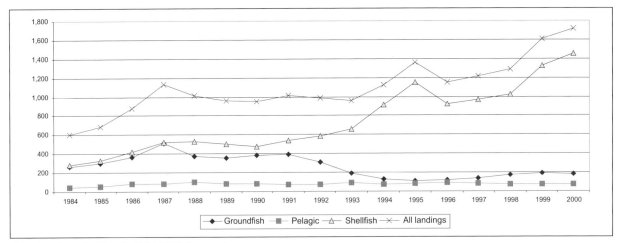

Atlantic values ($millions), 1984-2000, by species group.

tightly restricted basis. Then, in 1985, minister John Fraser closed all commercial salmon fisheries in the Maritimes. A buy-back followed for most remaining licences, at a cost of $2.2 million. No more commercial fishing took place in the Maritimes after 1985. The Maritimes and Quebec recreational fishery continued under close restrictions.

Atlantic salmon in the Maritimes and Quebec continued to decline despite all efforts. Sport fishermen lobbied, notably through the International Atlantic Salmon Foundation and its successor the Atlantic Salmon Federation, for further restraints. They called for anti-poaching patrols, more control of Native fishermen, and so on.

Conservation rules did get tighter. The department gave what some saw as disproportionate attention to Atlantic salmon. Scientists, fishery officers, and hatchery staff all worked hard. Guardian programs and toll-free lines to report poachers came into play. Canada and other countries through N.A.S.C.O. (the North Atlantic Salmon Conservation Organization) cut back on intercepting fisheries. But nothing seemed to help that much.

The sport-fishery remained important in New Brunswick. But in the Maritimes by and large, Atlantic salmon never came back to the flourishing levels of yore. In the inner Bay of Fundy, they kept declining. Early in the new millennium, the authoritative federal Committee on the Status of Endangered Wildlife in Canada declared inner Bay of Fundy salmon to be endangered. Ocean conditions got the main blame for this decline, with acid rain perhaps damaging the rivers as well.

Elsewhere on the Atlantic, no one pinpointed a single major villain. Atlantic salmon ran a long gauntlet, from the spawning grounds to Greenland and back,

Despite federal hatching and rearing facilities like this one at Mactaquac on the Saint John River, salmon abundance kept slipping in the Bay of Fundy.

past hooks and nets, pollution and predators, and water-threatening developments. With aquaculture growing, some people charged that escaped farm salmon could displace wild Atlantic salmon.

In Newfoundland, a 1984 buy-back took out more than 700 commercial fishermen. The commercial fishery continued on a smaller scale. In 1985, the department cancelled licences held by part-time fishermen. Another voluntary licence buy-back took place. Open seasons grew shorter. In 1989 a 1,300-tonne quota came into effect; this soon got reduced. In 1992, D.F.O. and the government of Newfoundland funded a $40-million licence-retirement program, which reduced the number of licences to something over 200. (Smaller programs followed to take all but a few licences from New Brunswick, for $1.5 million; and from Nova Scotia, for $1.3 million.)

The commercial fishery on the island of Newfoundland closed in 1992. In 1998, the Labrador fishery went under moratorium, and another licence retirement program came into play. Around the turn of the century, 81 fishermen in Newfoundland and 2 in Labrador were still holding onto their licences in hopes of a reopening.[72] But for all intents and purposes, the commercial fishery was gone.

Seal fishery declines and bounces back

In 1983, the European Parliament, influenced by sealing protest groups, banned imports of products made from whitecoats and bluebacks (harp and hood seal pups that have not begun to moult). With this major market gone, the hunt dropped back to low levels. From 1983 to 1995, the take of seals never reached 100,000, and fell as low as 20,000.

While hurting areas in Newfoundland and the Magdalens, the loss of market did even worse damage to Inuit communities. Many had relied on selling harp and ringed seals. The loss of sealing interfered with their economy and traditions and exacerbated social problems.[73]

The department in 1984 engaged Albert Malouf, a former judge, to head a Royal Commission on Seals and Sealing in Canada. In its 1986 report, the commission found that there was no convincing evidence of inhumaneness or conservation dangers. Nor was there a major ethical case against the hunt, since most people accepted the killing of animals. Even so, the report said, the hunt for seal pups should close down:

> Opinion polls, letter-writing campaigns and other measures of public feeling show that there is considerable opposition to the clubbing of seal pups. While this opposition may be largely an emotional response to the attractive picture of a white, dark-eyed "baby seal," or to the brutal image of one being clubbed and skinned on the ice, it is a very strong response, and it is unrealistic to consider any resumption of the whitecoat harvest. Whatever the facts about conservation or cruelty, a renewal of large-scale commercial hunting of seal pups would make sealing once again a matter of divisive public controversy. Consequently, the killing of the pups of harp seals (whitecoats) and hooded seals (bluebacks) for commercial purposes should not be permitted.

Malouf recommended that development work take place, to help market products from seals other than whitecoats and bluebacks. Assistance should also go towards Inuit seal products. The Inuit should also get relief payments of up to $4 million, for at least five years. As for the main body of sealers in Newfoundland, Quebec, and the Maritimes, a new fund in the order of $50 million should go towards development and retraining in sealing communities. A second fund of similar size should compensate sealers, plant workers, and plant owners. Among other recommendations was the typical one that the government do a better job explaining its policies and spreading information.[74]

Hundreds of seals on the ice.

433

A sealing vessel.

The federal government in 1987 followed through on the ban, prohibiting the commercial harvest of whitecoats and bluebacks. It also banned the hunt by vessels longer than 65 feet. But government held back on the big-money recommendations. It did, however, provide some funding to the Canadian Sealers Association, and it subsidized attempts to develop and market meat and other products.

The harp seal herd grew rapidly, to more than five million animals by 1996. Meanwhile, the industry at first seemed dormant, if not dying. But in the mid-1990's, markets resurged. For a few years, to the horror of protest groups, an oriental market developed for seal penises, used in making aphrodisiacs. But this faded. The main influence was the renewed interest in furs for fashion.

The hunt now took mainly beaters: harp seals just past the whitecoat stage. In the late 1990's, the take was running to some 260,000 seals, the highest level since the 1960's. Nine or ten thousand sealers were taking out commercial licences every year; others took out "personal use" licences. Besides fur, the industry was producing meat, leather, and health food supplements. The Newfoundland government estimated the total value of the industry at $25 million.

Sealing resurged among the Inuit as well, encouraged by the Nunavut Wildlife Management Board. Arctic sealers in the late 1990's were taking up to 20,000 ringed seals a year, and a few hundred harps.[75]

Some animal-welfare groups, particularly the International Fund for Animal Welfare, founded by Brian Davies, still protested the hunt. But the media frenzy of the 1970's never resurged.

What were the results of the three-decade campaign by protest groups? The initial campaigns in the 1960's may have speeded reforms that the department was already eyeing. But from that time on, expert observers consistently found the seal hunt to be well conducted. By winning the European ban, the protesters no doubt contributed to the seals' increased abundance. But again, was it necessary? There was never a conservation problem; the harp seals of the northwest Atlantic were among the most populous seal herds in the world.

As of the turn of the millennium, the tripling of the harp seal herd has probably damaged fish stocks, but to an uncertain degree. The F.R.C.C. believed that seals were a major factor impeding the recovery of cod stocks. They appeared to be major consumers, each of the five million animals eating 1–1.4 tonnes a year.[76]

Anti-seal-hunt groups downplayed the seals' impact on commercial species. Indeed, it was reckoned that commercial species such as cod made up only one to two per cent of the seals' diet. The harps spend much of their lives in Arctic waters that have no commercial fishery. Still, it was estimated in the late 1990's that harp seals could be eating as much as 140,000 tonnes of northern cod, 68,000 tonnes of northern Gulf cod, and more than 10,000 tonnes of southern Gulf cod. The F.R.C.C. noted that north of Halifax, harp, gray, and hood seals were killing more cod than any other known factor. The council called for a major cull, closely controlled and documented, to relieve the pressure on groundfish.

Apart from harp seals, in Nova Scotia in the 1970's and 1980's, the unhunted gray seals posed a serious problem for the then-energetic groundfish industry. The gray seals harboured a parasite—sometimes called codworm, sometimes sealworm—that was harmless but unappetizing. The problem was worst for codfish in eastern Nova Scotia and Cape Breton. Plant workers "candled" the cod, putting fillets over a light source to spot and extract the worms. To cut down the infestations, the department for many years subsidized hunters to kill a few thousand grays annually, but ended the cull in 1990.

Gray seals, too, affected fish abundance. In the late 1990's they were estimated to eat between 5,400 and 22,000 tonnes of eastern Scotian Shelf cod, from a total biomass thought to be as low as 32,000 tonnes. Cod in this area collapsed as badly as northern cod; seals appear to be a major obstacle to recovery.[77]

CHAPTER 28.
The Maritimes and Quebec, 1984–2000

I n the Maritimes and Quebec, while groundfish soared and crashed in this period, the shellfish industry grew steadily. Scallops and crab prospered, with good catches and good prices. Departmental research led to a new fishery for surf clams. For the old standby, lobster, which scientists had thought might lose ground in the 1980's, catches kept growing in a startling fashion.

Big-company picture changes

As always, the region had a vigorous and diverse mix of enterprises. Among the majors, National Sea Products survived the groundfish collapse by importing fish to process, emphasizing value-added, and diversifying. But it no longer dominated the industry as it had. The firm's plants dropped in number to only three at the turn of the millennium—at Lunenburg, Nova Scotia, Arnold's Cove, Newfoundland, and Portsmouth, New Hampshire. Its workforce dropped from 8,000 to about 1,600.[1] In 1999, National Sea Products changed its name to High Liner Foods.

Vessels in West Pubnico, Nova Scotia. Southwest Nova suffered less than other areas from the groundfish decline, and had strong fisheries for lobsters, scallops, herring, and other species. (Photo courtesy of Musée acadien et Centre de recherche, West Pubnico, N.S.)

The Clearwater corporation, built mainly on shellfish, kept growing. Early in the new millennium, Clearwater employed more than 2,200 people in Canada and other countries, operating 23 vessels and seven plants, at Clark's Harbour, Lockeport, Arichat, Glace Bay, and North Sydney in Nova Scotia, and at Grand Bank and St. Anthony in Newfoundland. Clearwater was a major force in scallops, shrimp, and inshore lobster. The company dominated the offshore lobster and the new surf clam and Jonah crab fisheries. In conjunction with D.F.O., the company like some others carried out considerable research, including bottom mapping of scallop grounds.[2] Clearwater also pioneered the use of dry-land lobster pounds, to reduce seasonality and provide year-round product.

D.F.O. Regions change shape

On the government side, D.F.O.'s regional set-up went through changes. Roméo LeBlanc, an Acadian, had in the early 1980's made Moncton the headquarters for the entire Gulf, including western Newfoundland. His successor, Pierre De Bané, a Quebecer, had withdrawn Quebec from the Gulf

Region. After John Crosbie, a Newfoundlander, became minister in 1993, he moved control of western Newfoundland back to St. John's. The changes whittled the Gulf Region down to Prince Edward Island and the Gulf shores of New Brunswick and Nova Scotia.

More was to come. As already noted, when in 1995 Brian Tobin merged D.F.O. and the Canadian Coast Guard, the Gulf and Scotia–Fundy areas got amalgamated under a new Maritimes Region, headquartered in Halifax and including the Coast Guard. The Gulf became a kind of sub-region, losing some of its functions and a lot of its status.

Meanwhile, Roméo LeBlanc in 1995 became Governor General of Canada. The fading away of the Gulf Region cut LeBlanc to the quick. Despite his non-political position, he could still express opinions to people he knew, and did so to Minister David Anderson. Others lobbied for the same cause. Under Anderson and Herb Dhaliwal, the Gulf sector regained full regional status. Control of the C.C.G., however, remained in Halifax, along with control of the old Scotia–Fundy fisheries area.

D.F.O.'s different administrative regions followed basic national policy, yet developed their own charac-

Fleet begins to shrink

After decades of expansion in fishing power, the fleet in the 1990's began to shrink, at least in terms of numbers. The trawler fleet dropped sharply after the groundfish collapse. The small-boat fleet also shrank. The mid-size fleet stayed strong.

Engines and gear kept improving. All but the smallest boats now had the advantage of G.P.S. receivers, which were replacing loran and enabled pinpoint navigation to fishing sites. Outside of the trawler fleet, it is hard to say if the drop in numbers represented much decline in fishing power.

Table 28-1. Maritime and Quebec fleet, selected years.

	Year	N.S.	N.B.	P.E.I.	Que.
Over 100 ft.	1984	124	10	1	9
	1991	96	9	4	13
	1999	54	3	0	3
35–100 ft.	1984	2,558	1,722	1,461	612
	1991	2,602	1,735	1,453	652
	1999	2,403	1,704	1,404	592
Under 35 ft.	1984	3,747	1,262	67	2,659
	1991	3,538	883	51	1,836
	1999	2,539	1,065	68	944

teristics, reflecting the surrounding industry and social context. In Newfoundland, where the Fishermen, Food and Allied Workers Union held major influence, officials tended to emphasize the common interest. In the Scotia–Fundy area, where fishermen's organizations were vigorous but fragmented, both industry and government held a more free-enterprise, survival-of-the-fittest point of view. The Gulf and Quebec fell in between. Those two regions had strong fishermen's organizations, notably the Alliance des pêcheurs professionnels du Québec (A.P.P.Q.), the Fédération régionale acadienne des pêcheurs professionnels (F.R.A.P.P.), the Maritime Fishermen's Union, and the P.E.I. Fishermen's Association, but none dominated the picture like the Newfoundland union.

In each region, fishermen's and processors' representatives built up good working relationships with D.F.O. officials. While quick to criticize the department in general, industry representatives often suggested that Ottawa give more say to the regions.

Lobster fishermen stick to traditions

In the lobster fishery, Maritime and Quebec catches grew from 26,000 tonnes in 1984 to a peak of 45,000 in 1991. Landings dropped in following years, significantly in some areas, but still amounted to 42,000 tonnes in 1999. Values shot up from $139 million in 1984 to $518 million in 1999. The cost of a lobster licence rose to several hundred thousand dollars.

As already noted, there was no clear explanation for the catch increase, nor was its strength apparent at the outset. Scientists continued to worry that the great majority of lobsters got caught before they had a chance to reproduce. In the 1980's, the Scotia–Fundy Region proposed a substantial increase in minimum size limits, to let more lobsters spawn. The Maritime Fishermen's Union led a campaign against any such increase, citing scientific uncertainties. No change took place in Scotia–Fundy.

In some Gulf lobster districts, however, minimal size increases took place. Other changes in lobster management saw use of escape vents to let small lobsters escape, and of biodegradable trap rings to prevent ghost fishing. Otherwise, there was no significant change in regulation in the 1980's or early 1990's.

Meanwhile, traditional wooden traps increasingly gave way to metal ("wire") traps, which reduced breakage. They also appeared to increase efficiency, as did improvements in vessels and electronic gear. Some concerns arose about another replication of the herring and groundfish patterns, in which increased fishing power masked a decline in stocks.

In 1995, a Fisheries Resource Conservation Council report, *A Conservation Framework for Atlantic Lobster*, concluded that young lobsters had too little chance to survive. In 1997, minister David Anderson asked fishermen through their Lobster Fishing Area advisory committees to develop conservation harvesting plans that would lead to doubling of egg production in the next two to three years. Although semi-voluntary, this was the strongest lobster-conservation measure since the buy-back of the 1970's.

Anderson's request, with the threat of mandatory measures in the background, produced some action. The F.R.C.C. report had laid out various conservation options, such as raising size limits, closing some areas to fishing, reducing trap limits, buying back licences, shortening the season, or V-notching the lobsters. (In the latter technique, fishermen cut a notch in the tail of an egg-bearing lobster. Even after she sheds her eggs and gets caught again, other fishermen can recognize her as a producer and throw her back.) Different areas applied different measures, including small increases in size limits in the southern Gulf. Results will take time to evaluate.

Offshore lobster controversy flares and dies

As the eight-boat offshore lobster fishery prospered in the 1980's, other parties watched with interest. The offshore fishery took place in a defined area off the southern tip of Nova Scotia, more than 50 nautical miles offshore. Minister Tom Siddon received representations to allow an experimental offshore fishery in an area running eastward along the shelf. Although regional officials counselled caution, it looked to Siddon like a clear opportunity for development. In 1987, he granted four experimental licences. Different enterprises geared up for the fishery.

Inshore fishermen assailed the move, saying that offshore lobster catches could well influence the inshore populations. After some backing and filling, the minister withdrew the permits. The government wound up paying compensation to the people who had prepared for the fishery, only to see their licences disappear. The offshore lobster fishery went on as before.

Poaching seems to decline

Nobody pretends to have reliable statistics on poaching. But over time, the impression spread that fishermen to some degree had improved their compliance with conservation regulations. Lobsters were valuable and were fished relatively close to shore, where fishermen could watch one another's behaviour. More and more they frowned at rule-breaking neighbours, or informed on them to the department. In some areas, it became an annual practice for D.F.O. officials and fishermen together to drag the waters for out-of-season traps, which they would confiscate or destroy. There is still poaching, but many observers believe it has diminished.

In the second half of the 20th century, the inshore lobster fishery ignored some trends in management. There was no stress on development or mobile fishing. The fishery remained a local, fixed-gear one, using regular traps. There were no quotas or virtual population analysis as for finfish. Managers judged the fishery by catch rates using standard gear. The only great changes came in the 1960's and 1970's, with fishermen's co-operation: trap limits, limited-entry licensing, and the phasing-out or buying-out of some fishermen. Otherwise, the lobster fishery stuck to the tried and true. At the end of the century it was, at least temporarily, the most successful of the old-line fisheries.

That being said, the fishery in some areas was beginning to take part in the technological race that damaged finfish stocks. In southwest Nova Scotia, bigger boats were staying out longer, making more hauls, and moving from place to place as they fished down local abundances.

Scallop fishery climbs in value

After a decline in the early 1980's, the scallop fishery in following years made gains. In Scotia–Fundy, a line drawn near Yarmouth in 1986 separated the inshore and offshore fleets. For the offshore fishery on Georges, Brown's, and other banks, Enterprise Allocations came into play that same year, negotiated among licence-holders mainly on a historical basis. The quotas were transferable. At the time, about 70 boats were fishing, for a small number of companies. The fleet owners reduced the number of vessels, especially after a resource downturn in the mid-1990's. By 1999, only 28 vessels for seven companies were fishing, and on average bringing in landings worth well over $1 million per boat. The fishery supported about 500 fishing and 150 processing jobs, full time. Companies in this fishery developed a market for scallop roe as well as meats.

A high degree of co-management prevailed in the offshore scallop fishery, which had a small number of like-minded operators. Quotas and size limits (meat-count regulations) controlled the catch. Vessel operators helped make management decisions, paid part of management and science costs, and in some instances conducted research in conjunction with D.F.O. Electronic monitoring devices helped the department to keep track of the fleet.

The Bay of Fundy fleet of about a hundred vessels had a choppier time of it. This fleet supplemented its landings with other species, depending on resource and market conditions. The Fundy scallop fleet first

Smaller boat rigged for scallops, and flying Nova Scotian and Canadian flags.

saw a sharp rise in landings, peaking in 1989 at more than 4,000 tonnes, followed by a major decline, and even by talk of closure. In 1997, the fishery went to quota control, supplementing seasonal, gear, and meat-count restrictions. The advent of Individual Transferable Quotas saw the fleet drop to 40-some vessels.

All told, scallop landings in the Maritimes and Quebec went from 34,000 tonnes (live weight) in 1984 to about 55,000 tonnes and climbing in the late 1990's. Most of the increase came in Nova Scotia; that province's share rose roughly from 80 per cent to 90 per cent of landings. Elsewhere, the Northumberland Strait supported a notable fishery. Value for the four provinces climbed from $55 million to $80 million. Scallops were a major fishery, although well behind the half-billion dollar lobster fishery.[3]

Crab fishery starts off strong

By the early 1980's, the snow crab fishery in the Gulf of St. Lawrence was in high gear. Some 130 midshore vessels were fishing crab: more than 80 from New Brunswick, another major fleet from Quebec, and a handful from Cape Breton. Each carried four or five crew. New Brunswick took the lead at first, landing 18,000 tonnes in 1984, with Quebec taking 13,000, Cape Breton smaller amounts, and Prince Edward Island very little. Total landings for 1984 came to 33,000 tonnes, worth $34 million.

A preventive Total Allowance Catch applied from 1984, and tighter quotas by the end of the decade. But landings nevertheless dropped; in New Brunswick, they fell about 80 per cent by 1990. Quebec took the lead in the fishery, but landings there dropped as well. In 1990, the total catch came to less than half the 1984 level. The dwindling catches raised concern. To share out the smaller harvest, individual quotas came into place in 1990 for midshore crab vessels, and in 1993 for inshore vessels.

Building back the stock

After 1990, catches built back to more than 25,000 tonnes, partly because of natural resource trends, partly because of improved science and management. The fishery commenced when the ice disappeared in April, and closed sometime in July. It had to close as soon as young, soft-shell crab made up more than 20 per cent of the catch. Other regulations, applying since the early years of the fishery, included minimum size limits and trap-mesh size limits, to let the females and young escape. A limit of 150 traps applied.

The 1990–1995 Atlantic Fisheries Adjustment Program supported a thorough survey of crab biomass, which gave scientists more data to go by. Fishermen's logbooks provided additional information, and annual trawl surveys helped researchers keep on top of trends.

Although the Acadian Professional Fishermen's Association in the 1990's went through a split and reconfiguration, co-management remained reasonably strong. The department held frequent public meetings with industry members on the crab resource. In the later 1990's, the industry was paying for dockside monitoring, port samplers, at-sea observers, and the annual research trawl survey.

In the 1980's, the catch had gone mainly into frozen crabmeat products for North America, with prices to fishermen generally running less than $1 per pound. In the 1990's, a strong market developed in Japan for frozen-in-shell crab sections. Prices reached several dollars per pound. Value in 1999 came to $122 million, three times the 1984 level. In addition, a fishery developed for other species of crab, formerly discarded, mainly rock and Jonah crab. Catches of such species in 1999 came to more than 7,000 tonnes, worth $5.6 million. In the late 1990's, more than 30 plants processed crab.

Inshore fleet presses for bigger share

By the mid-1990's, the midshore vessels in New Brunswick and Quebec were making good money. Some crab captains "integrated forward" to control crab processing plants. The high earnings of so-called "Cadillac fishermen" brought renewed calls from small-boat interests, particularly the Maritime Fishermen's Union, to share more of the resource with the 10,000 inshore fishermen in the area.

By that time, a substantial inshore fishery had built up. A P.E.I. fleet got permission to join the fishery in 1985, and by the 1990's counted 30 vessels. As well, more than a hundred vessels operated from areas 18 and 19 of northwest Cape Breton. In the mid-1990's, these vessels averaged some $67,000 each from the crab fishery.[4]

Five-year plan shares revenue

As noted earlier, Cape Breton's Area 19 snow-crab association and D.F.O. had in 1986 worked out a "threshold" rule that allowed temporary participation by other boats when the fishery's value exceeded a certain level. The midshore fleet in the Gulf developed a similar arrangement.

Plant workers in the 1990's complained that because of new conditions in the fishery—the groundfish moratorium, the lower crab T.A.C.s, the switch to frozen-in-shell crab sections, the sometimes shortened seasons because of changes in fishing patterns or the occurrence of soft-shell crab—it was difficult to get enough weeks of work to qualify for Employment Insurance. The plant workers proposed that the captains create special funds to pay for community work projects; these would provide the additional weeks for E.I. At the same time, other fishermen pressed for access to the crab fishery.

After negotiations, the Integrated Fishery Management Plan for 1997–2002 allowed for additional, temporary participants when vessel revenues exceeded a threshold. Vessels would also contribute to special funds. In New Brunswick and the Gulf shore of Nova Scotia, these were known as "Solidarity Funds." When the fishing season ended, the funds paid for work projects associated with the processing plants. The projects enabled plant workers not only to "make their E.I." but to get higher benefits through additional weeks. Crew members also got some help from the funds. This arrangement lasted through the five-year co-management plan, but then ran into difficulties as some fishermen questioned the program and a court case put it in doubt.

Meanwhile, inshore interests continued to press for a larger, permanent share of the lucrative fishery. The Maritime Fishermen's Union said in effect that although co-management sounded good, it sometimes meant sweetheart deals for the best-off fishermen or corporations. Those interests were willing to pay for science and enforcement because it helped perpetuate their privileges. The M.F.U. declared that since 25 per cent of the crab occurred in inshore areas, the inshore fleet should have at least that 25 per cent.

It amounted to a classic fishery puzzle. Allowing more boats into the fishery could share the wealth, but might also move the fishery back towards the old days of over-dependence, over-capacity, and overfishing; unless of course the department forced existing boats to reduce their catches. Dictating such a reduction could, however, derail business plans, reduce incomes, and diminish the security and freedom implicitly promised by the granting of the original licence. Alternatively, the department could use a licence-retirement program to buy back midshore licences and issue a larger number of licences to smaller boats, with no net increase in fishing power. But why should the taxpayers have to pay for licences that the department granted for free in the first place, and that made some of the licence-holders rich?

The department had never made hard and fast rules for such situations. Outcomes emerged from pressures and debate over time. [5]

Research fosters surf clam fishery

In the 1970's and 1980's, D.F.O. biologists surveyed fishing banks off Nova Scotia for offshore clams and quahaugs. Similar species supported a fishery in the United States; researchers hoped the Scotian Shelf might offer opportunities.

On Banquereau Bank off northeastern Nova Scotia, Terry Rowell and other researchers found commercial quantities of what came to be called the Arctic surf clam. In 1986, a few companies began fishing on an experimental basis. In 1987, a three-year trial fishery got under way, the companies operating under Enterprise Allocations.

At first, the companies found the going difficult. The expected U.S. market never developed. The companies instead found a market in Japan, selling the surf clams as "hokkigai." D.F.O. subsidized a successful marketing campaign. Catches rose to more than 20,000 tonnes, and landed values to more than $20 million. Annual product sales, highly dependent on Japan, grew to $30–$50 million a year.

At the turn of the millennium, Clearwater was the main participant. A few large, expensive vessels, fishing mainly Banquereau and the Grand Banks, used hydraulic dredges to stir up and gather in the clams. They were said to average the highest landed value of any vessels in Canadian fishing history. Crews of more than 30 people froze the clams at sea; further processing took place ashore. The industry employed some 600 plant workers and fishermen.

As in other offshore shellfish fisheries (which tended to have common interests and few owners), co-management was strong. The Offshore Clam Advisory Committee helped set Total Allowable Catches, bycatch controls, and other regulations. The industry shared research and other costs.

Nordmore grate boosts shrimp fishery

Shrimp shared in the shellfish growth. Landings in the Maritimes and Quebec rose six-fold from 8,500 tonnes in 1984 to 49,000 tonnes in 1999. Value rose ten-fold, from $12 million to $120 million. The figures included landings from the trawler fishery off Labrador, but the local fishery was also major. In 1999, Gulf and Scotian Shelf shrimp quotas totalled about 27,000 tonnes, accounting for more than half of total landings.

In the Gulf, most fishing took place in the St. Lawrence estuary. Total Allowable Catches had come into play in 1982. D.F.O. had sometimes restricted shrimp-trawler operations because of the by-catch of groundfish. A technical development solved this problem. Department officials encouraged use of the Nordmore grate, a device at the mouth of the trawl which deflected groundfish but let shrimp pass through. Common after 1990, Nordmore grates became mandatory after the cod moratorium, and speeded growth in the fishery.

Other management measures were typical. Individual quotas began for part of the Gulf fleet in 1991, became transferable in 1993, and spread to the rest of the fleet in 1996. Other requirements included minimum mesh sizes, logbooks, dockside monitoring, and observer coverage on a percentage of vessels. The industry paid dockside monitoring and observer costs. D.F.O. divided the fishery by area; vessels could fish only one area per trip. As of 1998, New Brunswick had 20 and Quebec had 51 shrimp trawlers fishing the Gulf, along with vessels from Newfoundland. As in the crab fishery, high abundance could trigger the issuance of temporary licences to other fishermen.

Off Cape Breton's Atlantic coast, the shrimp fishery stayed small into the 1980's, then ballooned as the resource grew and the Nordmore grate came into play. By the latter 1990's, quotas were totalling around 5,000 tonnes. Twenty-three Scotia–Fundy based vessels, all less than 65 feet, and six Gulf-based vessels, 65–100 feet, were fishing the area. I.T.Q.s came into effect for Scotia–Fundy vessels in 1994 and for Gulf-based vessels in 1996. In both Scotia–Fundy and the Gulf, independent operators dominated the picture.

How long would the shellfish boom last?

Would fishing pressure deplete shellfish as it had helped deplete groundfish? At the end of the century, no one was sure. Managing shellfish had important differences from managing finfish. For lobster and crab, fishermen used passive traps, rather than mobile gear that could chase down every last concentration. It was generally easier to control size limits. And shellfish were more sedentary and local, making it easier for fishermen to move from a hunting to a farming mentality. Co-management and the conservation ethic were increasingly strong, in at least some shellfish fisheries.

Shellfish in the Maritimes and Quebec had prompted major investments in vessels and plants. But the frenzy was less than that for groundfish at the time of the 200-mile limit, and the fishing power less murderous. It appeared that the real test of management would come when the fishery faced a downturn, either from fishing pressure or from changes in the ocean.

Herring fishery becomes more stable

While groundfish collapsed and shellfish boomed in the 1984–2000 period, the herring fishery saw reasonable growth. Landings rose from 120,000 tonnes in 1984 to 187,000 tonnes in 1999, and value from $18 million to $32 million, which was significant.

But herring had lost their previous excitement. For the Scotia–Fundy fleet, catches remained well below those of the 1960's heyday. After the Gulf of St. Lawrence herring crisis of the early 1980's, Bay of Fundy seiners lost access to the Gulf and Newfoundland, although some still fished off Cape Breton. Living on fewer fish, the fleet met downturns in markets. The fishery gradually changed from one of glamorous, high-volume fish-killers to a more modest and confined business. Some operators got tired of the fishery and sold off their quotas. As for licences, the number in Scotia–Fundy dropped from 49 in 1983, at the start of I.T.Q.s, to 40 by 1996, of which a smaller number were active. Average landed value in 1996 was about $332,000 per vessel.[6]

By that time, the fleet was generally considered to be under corporate control. The separate fleet rule still applied, individual fishermen still held the seining licence, and the licence-holder still had to operate the vessel. But the most recent licensing policy, that of 1996, made no stipulation that the operator had to own the vessel he was registering. Processing companies in the herring industry made side arrangements with most licence-holders that, in return for vessel financing or other financial considerations, gave the company a high degree of control of both the vessel and the licence.

Meanwhile, the main sardine producer, Connors Bros., kept consolidating and reducing its plants in New Brunswick and Maine, where it became the dominant producer, to a handful. The weirs which supplied

Vessels from the strong fishing port of Caraquet in northeast New Brunswick pursued crab, shrimp, groundfish, and other species. (D.F.O. photo by Michel Thérien)

sardines also fell in number, many giving way to aquaculture sites.

In the Gulf as well, the big excitement took place before 1984. Seiners had come on strong in the 1960's and 1970's, taking most of the catch. In the early 1980's, ministers LeBlanc and De Bané had shifted the situation, using licence-retirement programs, I.T.Q.s, and changes in allocation to give inshore fishermen most of the landings. At the turn of the millennium, 6 large seiners operated from New Brunswick, 5 from Newfoundland. As well, 32 small seiners held Gulf licences.

The department in 1984 began managing the Gulf spring and fall fisheries separately, and in 1987 divided the southern Gulf into seven management zones aligned with the major spawning grounds. As of 1996, herring gillnet licences in Quebec and the Gulf Maritimes totalled about 3,800. These licences were almost all on multipurpose boats. In a typical year, about half the licence-holders would go after herring.[7]

Swordfish and tuna continue at modest levels

After the mercury scare of the early 1970's and the near-disappearance of the swordfish fishery, the fleet kept inching back. By the mid-1990's, nearly 70 boats were longlining swordfish from the Scotia–Fundy Region, on the Scotian Shelf, Georges Bank, and the Grand Banks. Most swordfish boats also fished other species.

For tuna, Prince Edward Island had been the main base for the fishery in the 1970's, when limited entry began. In the 1980's, catches had surged off southwest Nova Scotia; Gulf boats would often migrate to grounds known as the Hellhole.

In 1999, the Atlantic region had about 800 tuna licences, the majority in the Gulf of St. Lawrence Region. There were also about a hundred in the Scotia–Fundy area, and smaller numbers in Quebec and Newfoundland. Boats from one region often chased tuna in another. As well, Japanese vessels fished tuna within the Canadian zone under I.C.C.A.T. (International Commission for the Conservation of Atlantic Tunas) rules. Fishermen usually sold the tuna by auction or on consignment. The tuna got packed in ice and air-freighted to Japan, where a single fish could fetch tens of thousands of dollars.

Typical Canadian management measures applied: limited entry, quotas (a complex system that sometimes caused inter-regional disputes), industry-funded observers and dockside monitoring, and so forth. The fishery also had an international dimension, since bluefin can migrate across whole oceans. International allocations took place through I.C.C.A.T., which assigned Canada a Total Allowable Catch in tonnes.

In 1999, Atlantic Canada's tuna landings, the majority of them in Nova Scotia, came to 900 tonnes, worth about $12 million. Catches included some bigeye, yellowfin, and albacore tuna, as well as bluefin.

Scotia–Fundy boosts co-research, communication with fishermen

Outside the world of regulations and development, the period saw another effort at working with fishermen. Back in the department's early days, both Whitcher and Prince had noted problems of communications with fishermen. Over the following decades, dozens of studies, for example the Kirby report, had made the same point. Their recommendations generally got tucked into the back of the report, behind the latest regulatory formula. But the Scotia–Fundy groundfish task force of 1989 paid more attention than usual to issues of information, education, consultation, and participation.

The task force recommended opening advisory-committee meetings to media, which happened. It also recommended that C.A.F.S.A.C. open its meetings to industry observers. Although C.A.F.S.A.C. made no change for the moment, such meetings opened up later, after the groundfish crisis worsened. In the meantime, Scotia–Fundy biologists under Mike Sinclair began holding more information meetings with fishermen and seeking more collaboration. The communications branch instigated a series of community meetings at which scientists and fishermen generated a new organization: the Fishermen Scientists Research Society (F.S.R.S.).

With D.F.O. and other government funding, the F.S.R.S. and department scientists set up a training program for fishermen. Captains and crews began collecting fishery data in accurate and precise forms that scientists could use. In 1994, the F.S.R.S. officially became a non-profit organization. It conducted research on such matters as fish migrations, habitat, diet, and lobster carapace size, and shared information through a newsletter, workshops, and other means. At the turn of the millennium, about 300 fishermen belonged to the F.S.R.S. It was an early venture in co-operative research, which later in the 1990's became a popular topic in Canada and some other fishing countries.

In another venture, the Scotia–Fundy Region funded an industry-run Communications Council, starting late in 1995. Fishermen's organizations in the region were fragmented; the council brought together the main groups and placed its own employee within department headquarters to distribute information. This "secretariat" helped to schedule advisory committee meetings and distribute related information. With help from the communications branch, the council built up a fax and computer network to spread general information. Fishermen's groups liked the scheme, but it got weakened by funding cuts at the turn of the millennium.

Aquaculture continues to grow

Salmon farming had begun in the late 1970's in New Brunswick's Passamaquoddy Bay, near the American border. Through the 1980's and 1990's, cages spread rapidly in the area's many bays and coves. By the end of the 1990's, about 90 sites were operating. Some operations became highly automated, with computers controlling the feed, and underwater cameras monitoring the fish. Salmon farmers sometimes set up small processing plants for local production. In Charlotte County, there were estimates that salmon farming supported one job in four. New Brunswick aquaculture, which also included trout and shellfish, was yielding about $200 million product value, versus $165 million landed value in the wild fishery. Atlantic salmon provided most of the province's aquaculture value.[8]

Federal–provincial agreements gave the provinces the primary power for licensing. Their regulations set minimum distances between salmon sites and other operations, and limited production capacity at a given site. D.F.O. played a part in site approvals, through its fishery role and also, after 1995, through the Navigable Waters Protection Act, administered by the Canadian Coast Guard. D.F.O. research, which had helped launch the industry, by the 1990's became more concentrated at the St. Andrews Biological Station. Researchers studied nutrition and other growth factors and experimented with new species, such as haddock, halibut, and lumpfish. They also monitored pollution and environmental effects, notably at the densely farmed L'Etang estuary, in Passamaquoddy Bay.

Industry members did research and development of their own. Although some commercial fishermen moved successfully into the field, many fish farmers came from other backgrounds (some had worked for D.F.O. science). They were a new breed, working outdoors like commercial fishermen, and combining business and biological skills. At St. Andrews, the biological station, the Huntsman Marine Laboratory (a research and educational institution partly supported by D.F.O.), and the New Brunswick Community College helped to gather and impart knowledge. The little town became widely known in world aquaculture circles.

Salmon farming soon ran into issues like those of the fishing and farming industries. A handful of larger corporations, some local like the long-established Connors Brothers and some based overseas, took over many of the early, small operations and came to dominate the industry. The remaining, smaller "independents" often sold through the larger firms.

Market problems cropped up. So did various diseases. In the late 1990's, Charlotte County farmers had to destroy many millions of dollars worth of infected salmon. Vaccinations of fish fended off some damage. Meanwhile, the provincial government and industry shifted to a system that isolated year-classes, to hinder the spread of disease from adults to young.

Opposition to the cages grew on grounds of pollution, both visual and environmental. Some critics feared that salmon escaping from cages would colonize rivers and displace the already-weak races of wild salmon. And if the water temperatures should ever revert to colder pre-1970's levels, the industry could

disappear. Even with those problems, however, salmon farming appeared to be a strong new industry, boosting the economy of New Brunswick.

Other aquaculture species also saw growth. Nova Scotia in 1999 estimated trout and salmon production at $28 million, with mussels, oysters, scallops, and other species bringing total production to $34 million. The industry gave work, whether full-time or part-time, to more than a thousand people.[9]

Mussel crisis shocks Canada

In the late 1970's and early 1980's, Prince Edward Island enterprises, with provincial government encouragement, had begun growing mussels on lines strung under water over natural mussel beds. The spat (larvae) spawned by the wild mussels would cling to the lines and grow. Some observers were delighted by the idea that a species long ignored could taste so good and become so popular. But part of the reason most coastal people had ignored mussels was probably the memory of people getting sick from them.

Bivalve (two-shell) species such as clams, mussels, and oysters feed by straining water and filtering out food. In the process, they can also absorb toxins. Paralytic shellfish poisoning (PSP) was a perennial problem for clams, particularly in the Bay of Fundy and parts of the Gulf of St. Lawrence; the department's inspection branch tested clams in the warmer months to detect toxins. Mussels had a far higher ability to accumulate toxins. In the Bay of Fundy, the department forbade mussel harvesting year-round.

In the Gulf of St. Lawrence, the new mussel industry progressed for years without major problems. But late in 1987, more than a hundred people in different provinces took sick after eating mussels. Some were disabled, and at least three died. The news made headlines across Canada. A full-blown crisis emerged.

University, D.F.O., and other government scientists spent days and nights in their labs. The National Research Council laboratory in Halifax made the definitive finding of domoic acid, produced by the *Nitzschia*

Mussel aquaculture buoys. Suspended mussels feed on nutrients in the water.

diatom, which had suddenly bloomed in parts of the Gulf. Research showed that besides the amnesic shellfish poisoning caused by domoic acid, another ailment called diarrhetic shellfish poisoning could occur. D.F.O. and the industry worked out new protocols for monitoring and testing mussel products and tagging the certified products. The department also pumped more money into inspection.

Although the mussel crisis dampened the market for farmed shellfish, the industry soon recovered. No major incidents occurred in following years. Prince Edward Island in 1999 produced $17 million worth of mussels and $5 million worth of oysters, along with small amounts of finfish. Shortly after the turn of the millennium, total landed value reached $30 million, and the industry employed more than 1,500 people.[10]

A mixed picture in the Maritimes and Quebec

What did the 1984–2000 period mean for fish, fishermen, and communities in the Maritimes and Quebec? One sees a varied but mainly positive picture, at least for core fishermen. The worst part was the nose-dive in groundfish, from 364,000 tonnes in 1984 to only 84,000 tonnes in 1999. But total landings suffered less of a drop, from 640,000 tonnes to 544,000 tonnes, thanks largely to the doubling of shellfish, from 113,000 tonnes to 234,000 tonnes. Total value more than doubled, from $436 million to $903 million.

In 1984, in the Maritimes and Quebec, some 14,000 boats of all sizes, mostly small, had landed an average annual catch worth $30,700. By 1999 fewer than 11,000 boats, still mostly in the small category but often bigger than before, landed an average $83,000. Even if expenses were higher and inflation had lessened the dollar's value, the fishermen were handling far more money.

As noted earlier, the Department of National Revenue compiles statistics on those people who get their single biggest share of earnings from fishing. These "main-income" fishermen tend to be full-time, professional fishermen. For the Maritimes and Quebec, the average income from all sources on such returns rose roughly from $17,700 for 16,590 fishermen (out of 31,500 registered fishermen) in 1984 to $26,700 (including $16,500 from fishing) for 14,500 fishermen (out of 29,800 registered) in 1997.

This gain only kept even with the inflation rate (goods and services worth $100 in 1984 cost $150 in 1997). Yet, odd though it may seem, many observers agreed that since the early 1980's and especially the early 1970's, a great many full-time fishermen in the Maritimes and Quebec had entered a different stage, with more money to spend, even if some of it came from Employment Insurance. One fisherman's representative noted that he started in the 1970's working for poor people; now the chief worry for many of them was taxes, including capital gains when they sold their boat and licence.[11] Others noted that for many areas, these were the first years when fishermen could live by fishing alone, without other seasonal work such as cutting pulpwood.

For some fishermen, the fishing life was still simple: go out, catch the fish, come home. For others, it became far more businesslike, with the captain worrying about fish prices, markets, financing, and where to buy or sell a quota. Particularly in Scotia–Fundy, some fishermen made close connections with larger enterprises. Processors could help not only with financing in a more expensive industry, but also with other business aspects and paper work.

Fishermen during the period became more empowered or at least entangled in management. The fisherman's licence could be a multi-page document, spelling out various rules and conditions. Integrated Fishery Management Plans could run to scores of pages.

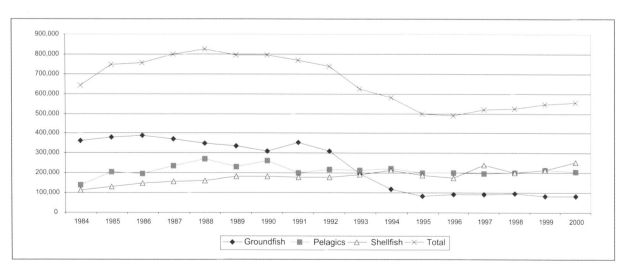

Landings (tonnes) in the Maritimes and Quebec, 1984-2000, by species group. While groundfish landings dropped, and pelagics were flat much of the time, shellfish landings grew.

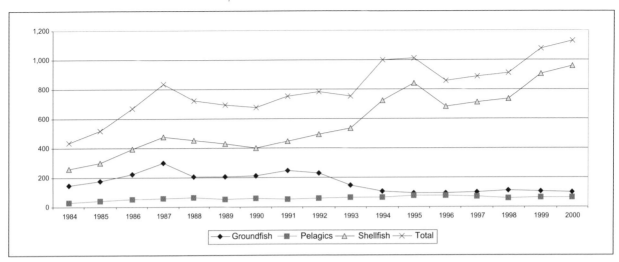

Landed value ($millions) in the Maritimes and Quebec, 1984-2000, by species group. Increasing shellfish value drove the rise in total value.

Advisory committees and organizations got involved in complicated schemes to parcel out the fish in the best way they could. Despite the nuisance factor, this new obligation represented a taking of responsibility, a step towards boat-level management of a common property.

Differences remained between areas of the Maritimes. Southwest Nova Scotia and the Bay of Fundy still manifested a more survival-of-the-fittest attitude; Cape Breton and the Gulf took a more communal approach. But in all regions, fishermen now handled more money, exercised more power, and at least in some instances had developed a higher conservation ethic. Progress was relative; many problems remained. But setting aside the great wound of the groundfish collapse and the precarious plight of a few towns like Canso, the fishery had made a good deal of headway.

That being said, the rest of the economy had expanded far more than the fishery. Other businesses, public-sector jobs, and transfer payments had in the overall picture reduced fishery dependence. As elsewhere in Canada, rural populations were losing ground to urban areas. The fishery remained important, economically, socially, and culturally. In some areas, it was still fundamental. But overall, at the turn of the millennium, the fishery in the Maritimes and Quebec looked less like the be-all and end-all, and more like a business among others.

CHAPTER 29.
Newfoundland, 1984–2000

Crab displaces cod

In Newfoundland, the most fishery-dependent province, the cod moratorium of July 2, 1992, brought changes that are still working themselves out. In a beautiful, harsh land of rock, rain, and fog, the fishing of cod had shaped the economy, society, and identity. As already seen, the groundfish collapse changed the mental landscape. The biggest groundfish company, Fishery Products International, closed most of its large plants and sold most of its trawlers. Several thousand jobs disappeared from that company alone, and many more from other groundfish enterprises.

In the early 20ᵗʰ century, D.F.O.'s Small Craft Harbours network in Newfoundland and Labrador included 378 commercial fishing harbours and 1 recreational harbour. Clockwise from top left: Jackson's Arm in White Bay; Harbour Grace, Conception Bay, boat with crab traps on dock; Port de Grave, Conception Bay; François, southwest coast. (Photos courtesy of Wayne Bungay, Bill Goulding, Bill Jenkins, and Gary Sooley, Small Craft Harbours branch, Newfoundland.)

Starting in 1997, some closed fisheries reopened with small quotas, first on the south and west coasts, then in 1999 for northern cod. On the south coast (division 3Ps), Total Allowable Catches of cod rose to 30,000 tonnes. All told, groundfish catches in the province built back to more than 60,000 tonnes. But that was only about 20 per cent of groundfish landings a decade earlier; and new cuts and reclosures would occur early in the new millennium.

In rough figures, the number of registered fishermen in Newfoundland dropped from 28,000 in 1990 to 12,900 in 1999, the number of licences from 27,000 to 21,000, and the number of boats from 17,000 to

10,000. As out-migration increased, the population dropped from 580,000 in 1993 to 534,000 in 2001. More young people stayed in school rather than taking a place in a boat or plant. Fishing was no longer such an easy way for a young fellow to pick up money (although the occupation still attracted many, with E.I. an extra inducement). Meanwhile, offshore oil and mineral development and high technology were exercising a new appeal.

Yet, despite the groundfish collapse, the waters of Newfoundland were by the mid-1990's yielding more money than ever. Landed value first dropped from $279 million in 1990 to less than $200 million in 1992, then rebounded to $346 million by 1995 and $581 million by the year 2000. This was almost a tripling of value in eight years (and far above the previous record of $293 million in 1987). As groundfish dropped to 13 per cent of total value, shellfish rose to more than 80 per cent. The main force behind the growth was snow crab.

Fishermen vie to get at crab

Even before the cod collapse, snow crab was shaping up as the great success story of the 1984–2000 period. By 1984, about 50 vessels, mostly 50–65 feet, the original "full-time" crab fleet, held licences. Landings had grown to 9,600 tonnes; value, to $6.8 million.

As inshore groundfish catches shrank in the mid-1980's, more fishermen pushed for access to crab. Minister Tom Siddon made available "supplementary" licences for vessels 35–65 feet, the idea being that crab would supplement their groundfish earnings. The total number of licences jumped to more than 600 in 1988. This meant that of Newfoundland's entire fleet between 35 and 65 feet, more than half the boats now held crab licences (although the "supplementary" licences authorized only 300 traps, compared with 800 for the "full-time" licences).

The smaller supplementary vessels generally fished close to shore, the larger ones further off with the full-time fleets. As fishermen, sometimes with provincial or federal support, explored crab in new areas around the island and in Labrador, the department granted additional licences. They totalled well over 700 by the mid-1990's.

Management measures resembled those in the Maritimes and Quebec. Limited entry had applied from 1976. D.F.O. limited the number of traps (different levels applied for different licence categories), and from 1986 applied quotas, subdividing them among different areas. To control the age at capture, the department used carapace-size limits, as well as trap-mesh sizes big enough to let females escape. Other regulations controlled seasons and limited the taking of soft-shell crab.

From 1984 to 1992, crab landings and value roughly doubled, to 16,000 tonnes and $13 million. Then the rise became even steeper. Additional effort, together with poorly understood environmental and ecological factors, increased the landings. Fishermen opened up new grounds, including some on the west coast of Newfoundland. From 1992 to 2000, landings more than tripled to 56,000 tonnes. Landed value increased an astonishing twenty-fold, to $263 million.

Meanwhile, thousands of groundfish fishermen who had lost their cod were demanding a share of the crab. This seemed to them a matter of natural justice; otherwise, a few boats would enjoy extreme earnings, while much of the newly abundant crab went uncaught. The department recognized the need for some sort of redistribution, but given the history of boom-and-bust fisheries and the unknown future of crab abundance, was wary of increasing the number of permanent licences. Instead, in 1995, D.F.O. authorized about 400 "seasonal temporary permits" for vessels less than 35 feet. Fishermen got these licences through a random draw. The different fleets subdivided quotas into individual vessel shares, usually equal.

Small-boat fishermen pressed for more from Minister Brian Tobin, and got it. From 1996 on, the department issued similar temporary seasonal permits to "core" fishermen with craft under 35 feet. This huge increase brought the total number of crab-fishing boats under 65 feet to roughly 3,300, which meant the larger part of the core fisherman fleet.

The department and advisory committees worked out arrangements for different fleets and vessel classes. "Temporary seasonal" boats less than 35 feet got 30 traps; "supplementary" boats got 150 or 300 traps, depending on their licence and size; and the original "full-time" boats got 800 traps. New zones applied for the various fleets. The full-time boats got pushed further offshore. Indeed, some of the smaller boats fished far enough off that concerns arose about safety; this brought new pressure on the department to relax the rules against increasing the size of replacement vessels.[1]

With more and more vessels getting licences, crab replaced cod as the primary source of inshore earnings. When seasonal permits began in 1995, they were supposed to apply only when quotas for the full-time and supplementary vessels exceeded the 1993 level. But at the end of the millennium, small-boat fishermen were pushing hard for permanent licences. D.F.O., its ministers, and its advisors (Max Short, a former official of the Newfoundland fishermen's union, was a ministerial advisor for much of the 1990's) faced a puzzle as in the Gulf. How many should share the wealth? What levels of revenue might be just and fair? How deeply should the department meddle in such matters?[2]

Meanwhile, industry members had invested hundreds of millions of dollars in new crab vessels and plants. Some vessels, even if 65 feet or less, now looked like big, high, squarish, seaborne armoured vehicles. Individual enterprises remained the mainstay of the fleet, some of them deeply indebted. But corporate control was creeping into the larger-vessel fleet. As in the Maritimes, it was no longer a question of old-line trawler companies, the giants of the 1970's and 1980's, taking control. Instead, it was a mixture of established companies and new investors, sometimes including fishermen themselves, wanting to extend their operations. Thirty-some crab plants were operating in Newfoundland by the year 2000. Many observers were expecting another crisis when the crab resource took a down-cycle.

Northern shrimp fishery grows

As of 1984, the primary shrimp fishery in Atlantic Canada still took place in the frigid waters of Labrador and the Davis Strait. The 12 shrimp-fishing enterprises, operating from every province except Prince Edward Island, still relied largely on chartered foreign freezer trawlers. Under departmental pressure, the companies gradually switched to Canadian boats and crews. The two Labrador co-ops kept chartering, but from other Canadian companies.

Labrador and Nunavut interests kept asking for more benefits for the people closest to the resource. In 1987, the department issued new licences to the Labrador Inuit Association and the Baffin Region Inuit Association. Two other new licences were shared between Nunavut and Quebec, and in 1991 another licence went to Newfoundland. This brought the total to 17, with Newfoundland and Labrador interests holding about half of them. Meanwhile, Enterprise Allocations began for the offshore trawlers in 1989, with each licence-holder getting an equal share in each shrimp-fishing area.

D.F.O. officials enacted typical management measures, including licence limitation, T.A.C.s, minimum trawl-mesh sizes, and Nordmore grates. Co-management became strong. The Canadian Association of Prawn Producers administered the E.A. system for the offshore trawlers, and controlled the days-on-ground in the Flemish Cap area outside the 200-mile zone. Licence-holders paid substantial fees, and also paid for observers aboard all vessels at all times.

Meanwhile, the fishery was spreading widely, in grounds ranging from eastern and western Newfoundland to Baffin Island. By the year 2000, the Total Allowable Catch of northern shrimp came to nearly 110,000 tonnes, ten times the 1984 level. As

the resource bloomed in job-hungry, cod-bereft Newfoundland, pressures arose to share the wealth.

As with crab, D.F.O. feared over-expansion, and so resorted to temporary licences. The department in 1997 gave most of the current increase in quotas to temporary new entrants, including more than 350 core fishermen in craft of less than 65 feet, and with special attention to communities such as St. Anthony on the Great Northern Peninsula. Both D.F.O. and the province sponsored experimental fishing and work on gear, to help the smaller boats get going.

Under Minister David Anderson, the department in 1997 laid out principles to govern the quota increases. Conservation would come first. A threshold level of T.A.C. (37,600 tonnes) would permit new entrants, but the additional access would be only temporary, with no permanent increase in licences. Adjacent fishermen would have priority, as would inshore fleets less than 65 feet and Aboriginal interests. In a re-echo of the Kirby report, employment would be maximized in harvesting and processing.

These guidelines were somewhat of a new departure. For years, groundfish plans had listed such considerations as adjacency and equity, but without ranking priorities. Officials and ministers generally shied away from locking themselves in. At least for shrimp, Anderson took the setting of "principles" to a new and clearer stage.

But there was still room for argument. Three years later, in 2000, Minister Herb Dhaliwal gave interests in Prince Edward Island, the only province without access to northern shrimp, a small temporary quota. Some of the proceeds were to go towards fishermen's professionalization, through an initiative led by Rory McLellan, of the P.E.I. Fishermen's Association. Newfoundland interests raised a storm, charging that the P.E.I. allocation defied the "adjacency" principle. The Newfoundland minister of fisheries, a fellow Liberal, called for Dhaliwal's resignation. Dhaliwal stuck to his guns. Meanwhile, some observers pointed out that if adjacency was the main factor, perhaps Nunavut should get more of the shrimp and turbot taken by Newfoundlanders.

The fishery kept spiralling upward. By 2000, the value of shrimp in Newfoundland reached $184 million, more than double the 1996 value and almost 50 times the 1984 value of $4 million. As of 1999, inshore vessels were taking nearly half the catch. Their share came to some 41,000 tonnes, processed on shore. Offshore vessels were taking 44,000 tonnes, frozen on board, either cooked or raw. The trawlers ranged from about 120 to well over 200 feet; some could stay at sea as long as 75 days.

Canadian producers sold most of their shrimp in Asia and Europe. Newfoundland was now taking 70 per cent of Atlantic-caught shrimp. As with crab, the thriving fishery prompted strong new investments in boats and plants, inducing fears of another race into crisis. Meanwhile, some ports saw new social divisions between better-off fishermen—holding crab or shrimp licences—and the rest.

Shrimp trawler in the ice.

Smaller fisheries plug along

The Newfoundland fishery continued to have less diversity than that of the Maritimes and Quebec. In those provinces, the three top fisheries in the year 2000—lobster, scallops, groundfish—contributed 60 per cent of landed value, with many small and intermediate fisheries making up the other 40 per cent. In Newfoundland, the three top fisheries—snow crab, shrimp, groundfish—contributed an overpowering 90 per cent of landed value. Where cod had been king, now crab contributed nearly half of landed value.

Groundfish was still important, worth nearly $80 million in the year 2000. The fishery for lumpfish roe, a new development of the 1970's and 1980's, had become notable. Lobster landings typically exceeded $20 million in the latter 1990's. Newfoundland interests also took part in the important clam fishery on the Grand Banks. Herring stocks in eastern Newfoundland had recovered some strength. But in the 1990's, the landed value of herring, capelin, swordfish, tuna, and other pelagics put together rarely equalled the value of lobster alone. And lobster was minor compared with crab and shrimp.

Meanwhile, some aquaculture development took place. Both D.F.O. and Memorial University of Newfoundland carried out research to help private efforts. After 1985, some enterprises began "growing out" cod, which were trapped and fattened up in cages for a few months until marketing.

F.F.A.W. keeps gaining

In fisheries management, the department continued doing the great bulk of the work, and the minister still had almost all the power, on paper. But in Newfoundland, the Fish, Food and Allied Workers Union (F.F.A.W.), as it renamed itself, at times appeared almost co-equal with the department. Helping to shape the Kirby restructuring and the special-aid programs, consistently backing limited entry and the "real fishermen," defending the owner–operator and separate fleet rules, negotiating prices with companies and allocations with D.F.O., pioneering the professionalization and certification of fishermen—on these and other matters, the union wielded great influence, while also carrying out extensive educational and other programs for fishermen. The union lost on some issues, for example the shrimp allocations granted by Minister Herb Dhaliwal to Prince Edward Island. But on major points it usually prevailed.

In 1987, the union had pulled out of the United Food and Commercial Workers to affiliate with the Canadian Auto Workers (C.A.W.). The C.A.W. took over representation of the Maritime plants that the Newfoundland union had organized. At the beginning of the new millennium, the Newfoundland union represented about 10,000 fishermen and another 11,000 members (some on layoff) in F.F.A.W. and C.A.W. plants.[3] The F.F.A.W. was also active nationally through the Canadian Council of Professional Fish Harvesters.

At century's end, the union remained a giant. In its three decades of life, the F.F.A.W. had equalled, if not exceeded, the achievements of Coaker in Newfoundland and of the United Fishermen and Allied Workers on the Pacific.

Fishermen, boats drop in number

In Newfoundland, the latter 1990's saw more money getting divided among fewer people. From 1984 to 2000, the fleet dropped by almost half. The number of registered fishermen fell by more than half, to 12,900 in 1999. Newfoundland now had fewer registered fishermen than Nova Scotia, and its landed value was coming close to Nova Scotia's.

Average revenue per registered fisherman before expenses increased from $6,900 for 27,600 fishermen in 1984 to roughly $45,000 for 12,900 fishermen in 1999—about the same gross revenue as in Nova Scotia, which had always led.

Average revenue per boat rose from $9,900 in 1984 to $61,000 in 1999. This still ran behind the Nova Scotia average of $127,000 per boat. (The usual caveats apply to such figures. Some boats earned a lot more money than the averages showed. Offsetting that, bigger boats might have proportionally bigger expenses.)

Incomes rose sharply, judging by National Revenue statistics for those whose single biggest source of earnings was fishing. Back in 1985, such "main-income" fishermen numbered only 9,700, out of 26,000 registered with D.F.O., and their average income was $10,500 (of which only $4,400 came from fishing). By 1997 their average income had risen to $19,000 (of which $11,600 came from fishing). The number of these "main-income" fishermen had grown only slightly, to 10,690; but they now formed a far higher proportion of registered fishermen, whose numbers had dropped to 13,300.

While trawlers declined sharply, many vessels of 65 feet and under took on a more power-packed look.

Trawlers, small-boat fleet decline

In this period, the Newfoundland trawler and small-boat fleet dropped sharply, because of the codfish collapse and government buy-back programs. Despite the shellfish build-up, the mid-size fleet also dropped.

As of 1999, compared with Nova Scotia, Newfoundland had less than half as many large vessels over 100 feet (22 vs. 54) and less than half as many mid-size vessels (1,008 vs. 2,403). In smaller boats less than 35 feet, however, Newfoundland still had more than three times the number in Nova Scotia (8,605 vs. 2,539), and nearly double the number in the Maritimes and Quebec combined.

Table 29-1. Newfoundland fleet, selected years.

	Year	Fleet
Over 100 ft.	1984	90
	1991	76
	1999	22
35–100 ft.	1984	1,438
	1991	1,342
	1999	1,008
Under 35 ft.	1984	15,020
	1991	13,678
	1999	8,605

Fewer people do better in Newfoundland

However one looks at the figures, by the turn of the millennium fewer Newfoundland fishermen were bringing in more money. It appeared that if the core fisherman policy, professionalization requirements, and the absence of groundfish kept reducing participation, the number of fishermen could drop to the 8,000–10,000 level, which had sometimes been put forward as the number of "real fishermen" in the province.

Though less marked than in the Maritimes, the trend was strong to the businessman–fisherman, spending more of his time on the phone, e-mail, and paperwork, while new technology supplied much of the fishing ability. By the end of the 1990's, more fishermen were caught up in representations for licences or quotas (although I.Q.s and I.T.Q.s were less common than in the Maritimes), complex financing arrangements, and advisory committee work.

Some 3,300 core boats with crab licences had a fair

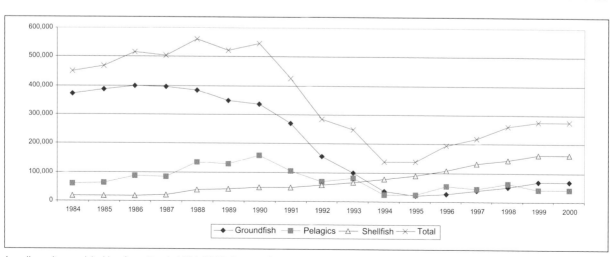

Landings (tonnes) in Newfoundland, 1984-2000, by species group.

449

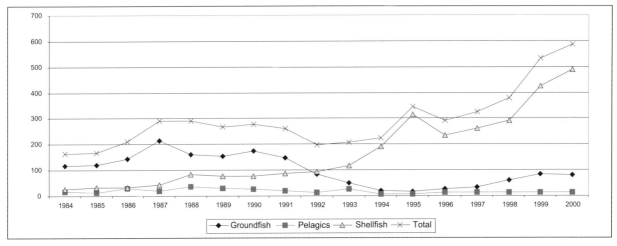

Landed values ($000's) in Newfoundland, 1984-2000, by species group.

chance at making money, as long as the stock held out. The larger of these tended to be making a lot more money than before the moratorium. The smaller ones also increased their take, although less markedly.

But many other boats lacked crab licences. In the year 2000, there were 6,300 of them, mainly non-core. Many of these enterprises were in hard shape. Fishermen lacking both crab and shrimp licences could try to survive on the remnants of groundfish, lobster in some areas, herring, scallops, lumpfish, and miscellaneous species. Or they could, as many did, tie up and find other work, while holding onto their licences and hoping for an improvement.[4]

The Atlantic industry rolls on

From the turmoil of the 1984–2000 period, the overall Atlantic industry emerged in some ways better equipped for the future. It still helped support more than a thousand communities, though for many, dependence on the sea had lessened. Although the fleet in general was still overpowered, overcrowded, and over-capitalized, it had become less numerous and more involved in management. The number of registered fishermen had dropped from 59,000 to 43,000; of boats, from 31,000 to 20,000. The future seemed to belong to the 12,000 or so officially recognized core fishermen, who headed enterprises, had an attachment to and dependence on the fishery, and held key licences for their area. Although forecasting the fishery is tempting fate, they appeared to have a good chance to survive, and to increase their incomes and voice in management.

Gains occur despite codfish collapse

How did Atlantic fishery management stack up in 1984–2000? The great cod collapse darkened the pic-

ture. Although federal fisheries had traditionally concerned itself with many matters, its first responsibility was conservation of fish. In groundfish, initial overestimates, excessive quotas compounded by politics, poor compliance, and changes in the ocean wrought catastrophe. Another historic commercial fishery, that for Atlantic salmon, came to an end, and in some areas salmon runs themselves bid fair to disappear.

That being said, Atlantic Canada had company in its problems. Various countries had fishery crises of one sort or another, including those in the New England and northeast Atlantic groundfish fisheries. In Europe at the turn of the millennium, scientists at the International Council for the Exploration of the Sea (I.C.E.S.) were calling for a cod moratorium; ministers and managers were trying to make do with lesser steps.

Canada had been a leader in limited entry, in quotas, in science, in advisory committees, and in management generally. But as fishing power increased, the game had become more dangerous than most people in industry and government thought. As the Harris report on northern cod warned, *the more you walk at the edge of a precipice, the greater the danger of falling off.* For groundfish, everybody walked at the edge too long, without knowing it.

Sectors of the industry had often complained of too-tight management and fought for relaxation of quotas. But the approach D.F.O. adopted in the 1970's was essentially right: the fishery needed comprehensive management, including co-management. The error was in the residual looseness, as fishing pressure kept creeping upward.

How well did D.F.O. and its ministers react to the great adversity of the cod crisis? In 1989, when scientists recommended cutting the northern cod catch in half, the response was far less than hindsight would deem correct. But after 1992, D.F.O. and ministers faced up to the situation, displaying compassion for the displaced industry and renewed dedication to conservation.

How well did the department do for other species, and for the Atlantic fishery in general? (Not that scientists, managers, and ministers control ocean conditions or industry attitudes, but they do influence results.) Gains in shellfish (more than tripling) and pelagics offset much of the groundfish collapse. For all Atlantic species combined, the catch decline over the period was 23 per cent in volume, bad enough but better than commonly thought. Value meanwhile roughly tripled, to $1.7 billion. Revenue per boat increased even faster, as did the value of landings per registered fisherman. The Atlantic overall was coming closer than ever to B.C. revenues. (In 1999, average revenues in British Columbia came to $90,000 per boat, compared with $72,000 on the Atlantic, and $40,000 per registered fisherman, against $37,400 on the Atlantic.)

The industry gained more of a say in controlling the fishery. Government subsidization became less of a factor, with boat-building support and the Fisheries Prices Support Board vanishing (although Employment Insurance remained). At the end of the century, the threats of over-investment and over-capacity were still present, but perhaps at a lower level.

To proclaim stability in the fishery would be foolish; environmental factors alone make it unpredictable. But at the end of the millennium, one could at least say that Atlantic fishery management had become somewhat quieter and less controversial, for several reasons. Co-management had increased. I.Q.s and I.T.Q.s had lessened overt competition: instead of trying to out-fish or out-lobby rivals, entrepreneurs could simply buy one another out. There were fewer groundfish quotas to fight about. Because shellfish fisheries were more regionalized, there were fewer clashes between local and distant fleets. There was more consciousness of the dangers of overfishing and the need for restraint.

The remaining fishery conflicts were finding less resonance in the broader public. The fishery though regionally important now made up a smaller part of the general economy. The public had to a degree lost sympathy with fishermen, whom some saw as having an excessive appetite for fish, E.I., and bailouts.

Fishery management did move forward in the period, notwithstanding the great cod debacle, and the odds improved for a decent living in the fishery.

Less industrial, more entrepreneurial

Before the Second World War, processing was relatively simple—mainly salting or canning—and marketing basic. In the post-war period, the dominant groundfish industry had turned to mass production,

Table 29-2. Earnings of "main-income" fishermen tax-filers, Atlantic and Pacific, 1984-97 ($000's).

Year	MARITIMES AND QUEBEC			NEWFOUNDLAND			BRITISH COLUMBIA		
	Total number fishermen	Average earnings fishing	Average income overall	Total number fishermen	Average earnings fishing	Average income overall	Total number fishermen	Average earnings fishing	Average income overall
1984	16,590	---	17.7	11,121	---	10.1	6,710	---	14.3
1985	17,344	13.4	19.9	9,719	4.4	10.5	7,214	14.3	21.4
1986	17,850	17.5	24.9	9,800	6.2	12.6	8,050	14.9	22.7
1987	18,590	20.2	27.9	11,810	8.9	15.8	8,490	15.4	23.3
1988	17,240	16.9	25.7	9,280	7.4	15.4	7,940	18.9	28.1
1989	16,730	14.8	23.7	9,050	6.0	14.3	7,700	15.0	25.0
1990	15,260	14.1	22.2	9,390	6.1	14.7	8,310	15.3	24.4
1991	16,150	16.4	26.2	7,620	6.0	14.9	6,620	11.4	20.9
1992	14,770	17.4	28.2	11,100	6.5	15.8	6,920	11.5	21.5
1993	14,460	16.5	27.3	11,290	12.1	17.3	8,280	14.9	25.4
1994	12,700	19.8	38.8	9,810	14.5	22.1	6,560	18.1	31.8
1995	14,680	23.3	35.8	10,890	17.4	24.1	5,780	14.5	24.2
1996	13,250	19.9	29.5	9,670	13.2	19.7	5,490	17.6	25.4
1997	14,500	16.5	26.7	10,690	11.6	19.0	10,250	8.8	15.1

often with large plants producing undifferentiated commodity products. It was often the practice for the industry to produce first, sell after.

By the mid-1980's, Clearwater and other companies were reversing the process, starting with market demand and working backwards. Giant, commodity-producing plants became less dominant. Big operators remained, but rather than running huge plants, they tended to control or at least market for a variety of medium-sized or smaller enterprises. New species or product forms such as herring and scallop roe, frozen crab sections, and sea urchins took their place in the market. The fresh-fish trade increased; boil-in-bag and other ready-made products became more common. The big restructured companies diversified into other foods as well. Free-trade arrangements with the United States gave Canadian producers more room to manoeuvre.

For small plants, marketing in the old days might have meant a trip once a year to see the Boston buyer, if that. By the end of the period it could mean product development, trade shows, and busy fax machines and computers. The trend was less marked than in Europe, where major metropolitan markets on the industry's doorstep demanded quality and encouraged product differentiation. In parts of Atlantic Canada the impulse to "catch first, sell later" was still strong. But overall, the industry had become less industrial and more entrepreneurial.

Among fishermen as well, although no one would call it a revolution, the business orientation had become stronger. A fisherman at the end of the millennium might have his steering wheel in one hand and a cell phone in the other, discussing catch composition and market conditions.

Ranking the elements of Atlantic management

Of the various elements of management, few went backwards in the years 1984–2000, and many showed improvement.

Understanding—Scientific understanding kept growing bit by bit, though major breakthroughs were scarce. As noted earlier, cutbacks from the Program Review in the mid-1990's took a toll on research. The department lost some insight into industry workings when the marketing branch shut down and the inspection corps moved out to the Canadian Food Inspection Agency. Offsetting that, the growth of advisory committees brought more fishermen and processors into direct contact with officials.

Managerial authority and capability—The Fisheries Act and other legislation continued to give the department strong powers. Although the cutbacks of the mid-1990's sliced budgets for enforcement, other changes may have helped. Fishery officers focussed more closely on police work. The number of violations detected grew. Dockside monitoring probably reduced misreporting and improved statistics and quota monitoring. Lobster fishermen in some areas took to informal self-enforcement. Still, enforcement officials and fishermen agree that a lot of poaching remains.

In Ottawa, the frequent shifting of some senior officials weakened continuity and corporate memory. But some of the new people improved dealings with the rest of government. As for fishery management itself, a strong core of dedicated fishery people remained. Co-management became stronger, with the industry taking more responsibility.

Goals—Roméo LeBlanc and the 1976 *Policy for Canada's Commercial Fisheries* had mainly wanted more fish, more money, more stability, and more power for fishermen themselves. The department in that era had made many interventions. The 1983 Kirby report had superimposed the objectives of viability, jobs, and Canadianization. But after the Conservative takeover in 1984, the government narrowed its focus, cutting back on loans, insurance, subsidies, price support, and development work. Officials talked more and more of the "core mandate" of conservation. In the latter 1990's, ministerial and departmental statements on the "fishery of the future" tended to stress economic viability, environmental sustainability, self-reliance, and resilience, together with the ever-hopeful goal of balancing the industry's capacity with the resource.

Yet, the department still got drawn into such questions as sharing the fishing among multiple users. Meanwhile, the ready availability of Employment Insurance, a factor outside the control of fishery managers, continued to draw entrants to the fishery and to subsidize their participation. The weakness of general economic development in many Atlantic areas still fostered over-dependence on the fishery.

An Atlantic Fisheries Policy Review started work in 1999. A chief objective was to define a set of guiding principles for the Atlantic fishery; this would include the clear primacy of conservation, and higher responsibilities for licence-holders in a self-reliant industry. It was also mooted that, on top of a strengthened advisory system for the fisheries, broader forums should bring in other interests: sport-fishermen, environmental groups, and so on.

Level of fishing—Since at least the early 1970's, federal fisheries had dropped the pursuit of Maximum Sustainable Yield, talking instead of "Optimum Sustainable Yield" (O.S.Y.). For finfish, officials used the $F_{0.1}$ guideline, which generally meant catching about one fish in five.

Without disavowing O.S.Y. or $F_{0.1}$, the department and the Oceans Act in the 1990's superimposed the "precautionary approach," which meant erring on the side of caution. The thinking was that when there was risk of serious or irreversible harm, lack of full scientific certainty was no reason to postpone decisions. Fisheries scientists and managers began developing "reference points" to define limits of exploitation for specific populations.

Access and allocations—In theory, access and allocation in the year 2000 remained firmly in the hands of the minister, to grant or withdraw as he pleased. In practice, licences had more and more taken on a life of their own, with transfers generally allowed, and with the government feeling obliged to pay compensation when it removed a licence. Although Individual Transferable Quotas furthered "rationalization" by reducing the number of vessels in some fisheries, they also drew charges of fostering overfishing and misreporting and of extending corporate control at the expense of the independent fisherman.

Clear progress had appeared in at least one aspect: a fisherman was far more likely in 2000 than 30 years earlier to have a share of the overall catch that he could count on. The "core fisherman" policy further buttressed the position of serious fishermen.

Fish-handling and quality—The 1970's had seen ice and containers become common. The trend continued in following years. After 1984, the government abandoned the Kirby-backed attempt to impose compulsory bleeding, gutting, and grading. Still, for much of the 1980's and 1990's, quality improved as companies became more market-oriented. Industry observers at the turn of the millennium suggested a mixed picture. Progress in some sectors seemed to have slowed or halted, especially after the Inspection service moved to the Canadian Food Inspection Agency. In other areas, gains were taking place.

Development—D.F.O. in the 1984–2000 period helped to research and develop aquaculture. But in the traditional fishery, the department now paid less attention to developing new fisheries (with some exceptions such as surf clams and silver hake), new products, and new markets. In a backhanded way, the slackening of development work showed progress. The need for D.F.O. activity was less, partly because the industry itself was more capable. Besides, governments and communities were tending to look elsewhere than the fishery for development.

The lives of fishermen—In LeBlanc's time, the department had wanted to give fishermen a better life, including, in a frequent phrase, "a voice in their destiny." As fishermen advanced, the department gave less attention to "bringing them along." No longer did officials project such ideas as catch insurance and a Fishermen's Bank. There was less talk of the fishing "way of life," more of self-reliance.

Aboriginal fishing—As the department stepped away from its godfather role with fishermen, it got further involved in the affairs of the First Peoples. In the 1970's and 1980's, the department had tried, sometimes in conjunction with Indian and Northern Affairs, to encourage Native fishermen. That work accelerated with the Sparrow decision. The department provided more food-fishing and commercial access, and worked with First Nations on training. The Marshall decision multiplied such efforts, with hundreds of millions of dollars going towards commercial licences, training, provision of wharves, and the like. In management, Native representatives began taking part in most advisory committees.

Governance—Co-management increased during the period for just about every fishery, especially for better-off ones on less-migratory stocks. That being said, such mechanisms as elections and votes are still rare, and the minister still has the ultimate say. In the industry, some people find fault with the situation; others declare the need for a strong minister.

CHAPTER 30.
On the Pacific, 1984–2000

Salmon fishery starts off strong

As with Atlantic cod in the 1984–2000 period, so with Pacific salmon: a historic fishery's initial resurgence would give way to a catastrophic decline. But at the end of the period, salmon would inspire more hope than cod.

Back at the beginning, the Pacific salmon fishery rebounded from a cyclical low in 1984 to extremely high levels in 1985. The catch value of a quarter-billion dollars was the greatest to date. The Salmonid Enhancement Program appeared to be paying off, with new hatcheries still coming on stream. Meanwhile, the habitat-protection laws, strengthened by LeBlanc's changes to the Fisheries Act, and the growing environmental consciousness offered better care for rivers and lakes. The 1985 Pacific Salmon Treaty promised better international management. In the broader economy, interest rates had dropped back, the early-1980's recession was receding, and a boom was beginning.

Two generations of Pacific coast patrol vessels: the old and new *Arrow Post*.

Living with over-capacity

Within the department at the outset of the period, some worries remained about the B.C. fleet's towering over-capacity and over-investment. But after the Pearse report's failure to reduce the fleet, no one had the appetite for major changes. Most departmental people left aside the larger picture and got on with the current job.

Minister John Fraser had criticized Pierre De Bané's fleet-reduction proposals while in opposition. As minister, he made no major initiatives for salmon. The industry still looked good under Fraser's successor, Tom Siddon. In 1988, salmon reached a landed value of more than $300 million, double the value of ten years earlier. British Columbia was back into the good times.

S.E.P. seems to help salmon runs

Always in the background during the period was the Salmonid Enhancement Program, which aimed to dou-

ble stocks to historic levels. The high mid-1980's landings seemed to reflect well on the program. By the 1990's, new building of hatcheries and spawning channels had practically ceased, but S.E.P. continued as a large operation. At its 20th anniversary in 1997, the program was operating 26 government hatcheries, 60 spawning channels, and 46 fishways. Lake enrichment was working well; staffers dumped nutrients into waters that harboured young sockeye before their move to sea. Other activities included rearing ponds; incubation boxes; modifications of water flow and temperature; and improvements to spawning gravel, streamside vegetation, and refuge areas. Officials noted that apart from lake fertilization, the enhancement techniques closest to nature, such as spawning channels and stream clearance, produced the most benefits per dollar spent.

As of 1997, S.E.P. was contracting with 21 community groups to operate smaller hatcheries. These included 13 Aboriginal communities, providing 50-odd jobs. Another 20 Native persons worked in government hatcheries. In addition, S.E.P. gave money and technical help to about 300 volunteer projects. Fifteen D.F.O. community advisors worked with some 10,000 volunteers and industry groups on such projects as building side-channels, improving water flows, stabilizing stream banks, and rebuilding estuary marshes. The educational component of the program, Salmonids in the Classroom, by 1997 had reached roughly a quarter-million schoolchildren throughout B.C. and the Yukon.

As for fish production, a department statement in 1997 said that the 600 million juveniles released each year produced four to five million fish caught, accounting for 10–20 per cent of salmon landings. Net benefits came to more than $17 million a year, and every dollar invested in the program returned about $1.74 to the economy. S.E.P. produced more than one-third of all chinook and coho in the Strait of Georgia sport-fishery. Its Babine spawning channels strongly supported commercial and Aboriginal sockeye landings on the Skeena.[1]

Salmon enhancement always had some doubters. Critics argued that heavy fishing on enhanced runs intensified the strain on natural runs in the same areas. But S.E.P., even if it had yet to match its original dreams, still seemed to most people in the mid-1990's to be a positive program. The trouble was, salmon runs were now dropping to record lows.

Salmon catches nosedive in the 1990's

Previously, in the late 1970's and early 1980's, the decline of chinook, especially in the Strait of Georgia, had caused great alarm. Closures and gear regulations had helped reverse the trend for spring-type chinook, which spawn mainly in the upper reaches of larger river systems. But in the 1990's many runs of fall-type chinook, which spawn in lower rivers and coastal streams, were in serious decline.

Coho appeared in even worse shape. Problems had started in the 1970's; the department had closed net fisheries on the Skeena and Fraser rivers. Losses had continued in the 1980's, with both fishing pressure and habitat damage drawing blame. In 1995, the department's Pacific Stock Assessment Review Committee (P.S.A.R.C.) sounded the loudest alarm yet. By 1997, the department had sharply cut commercial coho fishing in the north, and forbidden it in the south.

Chinook and coho were the foundation of the ever-growing recreational industry. As fishing lodges and charter boats multiplied, their fish were vanishing. The powerful voice of the sport-fishery advocated restraints on commercial fishing, which took well over 90 per cent of the total salmon harvest.

Sockeye and other species were also facing declines. Two incidents drew public attention. In 1992, early sockeye runs reached the upper Fraser in much smaller numbers than expected. Minister Crosbie appointed Peter Pearse, the resource economist who had written the Pearse report, and Peter Larkin, a noted biologist, to study the matter. Their report noted that near-

ly half a million sockeye "seemed to disappear" on their way upriver, "due mainly to unusually intensive fishing in the river." This in turn derived in large part from the expanded Native fisheries.

Native catches on the Fraser had already increased in the 1980's. The report noted that "it is safe to say that most of the salmon caught in the Indian fishery along the lower Fraser in recent years were sold." Then the commercial pilot projects under the 1992 Aboriginal Fisheries Strategy had increased the Native take. Since the 1992 run was large to start with, the situation was no disaster, but it was a setback that should not be repeated. Pearse's recommendations stressed such matters as commitment to conservation, working together, accountability, strict enforcement, communication, consultation, and liaison.[2]

Then came an even more notorious incident. In 1994, the year of aggressive fishing during the Canada–United States salmon dispute, the late runs of Fraser sockeye suffered from heavy fishing in Johnstone Strait, additional fishing as they approached the Fraser, and an intensive Aboriginal fishery within the Fraser, on what was by now an unexpectedly thin population. Reports emerged of 1.3 million missing sockeye salmon. Some blamed Native people; most blamed the federal fisheries department, weakened by budget cutbacks.

D.F.O. commissioned former fisheries minister John Fraser to review the situation. Fraser heated the atmosphere by reporting that "if something like the 1994 situation happens again, the door to disaster will be wide open. ... One more 12-hour opening could have virtually eliminated the late run of sockeye in the Adams River. Such an occurrence would have devastating consequences for the Pacific fishery." Fraser blamed much of the problem on a 1992–1993 Pacific Region reorganization and budget cuts, including "a sharply reduced complement of uniformed enforcement staff in 1994 (down by 47 per cent from 1989) and a reduced complement of seasonal staff."

Fraser's report spoke of "chaos," "confusion," "dysfunction," and "laxity of diligence." As well, he said, the department had over-relied on historic estimates while taking too little account of environmental changes, had shown undue optimism, and had made risky management decisions. The many recommendations included calls for better management, enforcement, communication, and co-operation; more co-management; and the establishment of a Pacific Fisheries Conservation Council.[3]

Some fishery managers considered Fraser's "12 hours from destruction" statement and the whole affair to have been an exaggeration. But it sharpened public perceptions of a conservation crisis, which was indeed approaching. For most years from 1984 on, B.C.'s salmon catches had been above 80,000 tonnes and occasionally over 100,000 tonnes. The late 1980's and early 1990's had seen the three largest Fraser runs since the Hell's Gate slide in 1913. But in 1994, B.C. salmon catches started plunging. By 1995, they were below 50,000 tonnes; by the end of the century, below 20,000 tonnes. In 1996 and 1997, the historic fisheries at Rivers Inlet and Smith Inlet, in decline since the 1970's, got closed down. Major changes took place in the processing sector, with the old, industrial canneries losing prominence.

Fishery, land, and ocean gang up on salmon

What caused the salmon decline? As on the Atlantic, scientists and managers were unable to weigh the factors precisely. Viewed through the prism of decline, many old problems that the department had fought now looked more menacing than ever. Fishing power, which Jack Davis had set out to reduce through licence limitation and buy-backs, was at its highest level ever. Despite habitat laws, policies, regulations, and a higher public consciousness, habitat degradation still seemed to be advancing as the economy grew. Salmon enhancement itself could pose dangers by supporting bigger fisheries and attracting more predators, which depleted wild stocks. Some like John Fraser said that the sharp cuts to D.F.O. budgets had weakened management and enforcement. British Columbia had more than 1,600 spawning streams; fishery officers now checked fewer of them in person.

The ocean itself had grown capricious. Fishermen have always talked of natural cycles. But for many years the struggles with more obvious threats—fishing, pollution, loss of habitat—had taken centre stage. Then, in the 1980's and 1990's, occasional El Niño events—ocean disruptions associated with warming currents off the coast of Peru—brought weather changes to large parts of the world, and warmed waters off British Columbia.

Richard Beamish, a pioneering investigator of acid rain in Canada, and other scientists at D.F.O.'s Pacific Biological Station began looking in detail at the salmon's ocean pastures. They found evidence of "regime shifts" disrupting normal patterns of feeding, predation, and survival. These were argued to have occurred in 1925, 1947, and 1977. Now British Columbia was feeling the impact of a 1989 shift that first affected more southern waters, then moved north. The "ocean regime" became a bigger consideration in management, a sobering one for those who had thought they pulled the levers of control. While on the Atlantic scientists cautiously reached a consensus that overfishing more than oceanic factors had destroyed the cod, Pacific

D.F.O.'s Pacific research trawler, the *W.E. Ricker*. Scientists laboured to sort out the impacts of fishing and other factors on the salmon decline.

scientists inched towards the opposite idea: that oceanic changes had been the primary factor for salmon, compounded by heavy fishing, which gave the stock no chance to rebound quickly.[4]

Prices plunge as aquaculture changes market

British Columbia had long enjoyed a favourable market position. About half the salmon pack went to domestic consumption, half to foreign, giving the industry a better balance than most sectors of Canada's fishery. Reliable supply and good quality had reinforced B.C.'s standing in the markets.

Now the salmon business was changing. Scarcity of wild salmon brought no increase in prices. Indeed, some species now fetched far less, as aquaculture changed world markets. Pink salmon, a less valuable species used mainly for canning, had in the 1980's fetched $1 a pound; by the end of the century the price was 15 cents.[5]

Farmed salmon from Norway, the United Kingdom, Chile (where Canadians had helped to spread aquaculture), Canada itself, and some smaller producers were overtaking wild-salmon fisheries. By 1999, world production of farmed salmon, mainly Atlantic salmon, came to nearly a million tonnes, well ahead of the wild-salmon production of less than 800,000 tonnes. British Columbia salmon farmers produced 47,000 tonnes, mostly Atlantic salmon plus chinook and a few coho; Atlantic coast farms produced another 21,000 tonnes.

That same year, British Columbia's wild-salmon fishery produced only 17,000 tonnes. This was, said a report to the fish-farming industry, "less than one per cent of total world salmon production, a significant drop from fifteen per cent of the world total in the early 1980's." The traditional B.C. wild-salmon industry had lost its clout.

Landed value of wild salmon in 1999 came to roughly $26 million. Farmed salmon provided a production value of $291 million. Industry figures said that aquaculture created 1,800 direct jobs and 1,600 indirect jobs in British Columbia. The B.C. farmed-salmon industry was a major force in U.S. markets, which took most of its exports.[6] The provincial government had applied a moratorium on new sites, but this would be lifted in the new millennium, allowing more growth.

By the early 21st century, salmon farming was drawing fierce criticism. Opponents were charging that besides being unsightly, salmon farming caused environmental harm from feed, fecal matter, sea lice, and diseases. Some critics said that farmed salmon accumulated harmful elements from their feed or from chemicals used at sites. They charged that escaped Atlantic salmon could colonize Pacific rivers, displacing the weakened native runs. Opponents also said that D.F.O. had put itself in a conflict of interest by promoting aquaculture through the Commissioner for Aquaculture Development and other efforts. Instead, it should be following the precautionary approach, to protect wild stocks from damage by aquaculture.

Round Table recommends buyback, aid

As wild stocks declined and markets plunged, the wild-salmon fishery entered another crisis. Fish were vanishing and markets plunging. The top-heavy Pacific fleet, marked by over-investment, over-building, and over-indebtedness, seemed finally about to capsize. Traditionally resilient, the B.C. fishery now faced a desperation more familiar to the Atlantic coast.

In the spring of 1995, Minister Brian Tobin launched a Pacific Round Table process. Regional director-general Louis Tousignant, an energetic administrator, pushed ahead with committees and consultations. Three panels represented the commercial gill-net, troll, and seine fleets. An overall policy panel brought together representatives of the United Fishermen and Allied Workers Union, Aboriginal groups, processors, and the minister's senior fisheries advisory group, the Pacific Regional Council. The B.C. government, under New Democratic Party premier Glen Clark, weighed in with an intensive study on the economic and community situation. Meanwhile, the recreational industry constantly pressed its views on federal fisheries and the public.

In December, the Round Table review recommended, to no one's surprise, reducing the fleet. Government should also clear up fish allocations between sectors and assist people who were displaced by fleet rationalization.

"Mifflin Plan" revises Davis Plan

In January 1996, the retired Canadian Forces admiral Fred Mifflin took over as minister and entered heavy Pacific seas. The main ideas in the departmental mind were those that had animated the Davis Plan nearly three decades earlier. The oversized fleet was threatening both resource and incomes; a buy-back and licensing changes should bring it under control.

On March 29, Mifflin announced "a comprehensive plan to revitalize the West Coast commercial salmon fishery and enhance conservation and sustainable use of the resource." On conservation, the department would continue to pursue risk-averse management; reduce the harvest of selected species; and adopt more selective, stock-specific fishing practices. The fleet needed (just as Pearse had recommended) a 50 per cent reduction in capacity, over the long term, to promote resource conservation and fishery revitalization. An $80 million "voluntary licence-retirement program" (buy-back) would take place right away. Mifflin also promised a new commercial licensing system, including higher licence fees and, in another recurring idea, charges based on landings. The latter provisions were to be delayed in light of the current poor conditions.

Area licensing returns

Area licensing had prevailed earlier in the century, only to be dismissed in the post-war period as cumber-some. Now, with a high-powered, highly mobile fleet, it again seemed useful to hinder mobility. Already the department had divided the coast by area for herring fishermen. Now the Mifflin Plan divided the coast into two areas for salmon seiners and three for gillnetters and trollers. Licence-holders would choose one area and one type of gear.

This move made plain that licences really did depend on government: those who had invested in a licence, treating it almost as private property, found its range and potentially its value reduced by a stroke of the pen. To fish another area or with different gear, a fisherman now had to acquire the licence from another licence-holder, a process known as "stacking licences." This would foster fleet reduction, by encouraging fishermen to buy one another out.

Some in the fleet supported the Mifflin Plan; after all, the fleet-reduction plan promised half as many boats, which should mean twice the fish for each. But mainly, the plan attracted criticism. Opponents feared that it would take out an unfair proportion of small boats, hand the fleet over to larger operators, and wipe out small communities, including First Nations that depended on the fishery. Department officials kept plugging along, promising that they would try to keep a balance.

At the plan's beginning in 1996, the fleet had about 4,100 salmon licences. The licence-retirement program took out nearly 800 licences, 19 per cent of the fleet, at a cost of nearly $80 million. Unlike Atlantic programs, the B.C. one concentrated on taking out actual fishing power, rather than people.[7] Meanwhile, the salmon fishery continued to worsen.

Anderson bulls it through on fleet reduction

David Anderson took over from Mifflin in June 1997. After scientists sounded another warning about coho and chinook stocks, Anderson went to cabinet for money and in June 1998 launched the Canadian Fisheries Adjustment and Restructuring program. C.F.A.R. applied to both coasts. This was the first time that the alphabet soup of 1990's programs—A.F.A.P., N.C.A.R.P., T.A.G.S., A.G.L.R.P., and so on—had reached the normally independent Pacific fishery.

C.F.A.R. devoted another $400 million to British Columbia, on top of Mifflin's $80 million. Half of it would go towards fleet reduction, the biggest such effort ever on the Pacific. By 2001, the C.F.A.R. program had removed about 1,400 salmon licences, at a cost of $192 million. More of the seine fleet than expected sold out their licences: 216 out of 500-odd, at an average cost of $436,000. From the more than 3,000 gillnet and troll licences, about 460 troll licences and 730 gillnet licences vanished. The licence retirements resulted in removing 1,007 vessels, about 30 per cent of the original fleet.[8]

In the period from 1984 to 1999, the number of boats in B.C. dropped from about 7,000 to about 3,900; the number of registered fishermen, from

Fleet shrinks in every category

Like the Atlantic coast fleet, the over-built B.C. fleet went through a resource decline and buy-back programs in the 1990's. But British Columbia, as often happened, took stronger hold of the situation. The dominant mid-size fleet dropped more than on the Atlantic. The small-boat fleet also saw a steep decline.

Table 30-1. B.C. fleet, selected years.

	Year	Fleet
Over 100 ft.	1984	38
	1991	31
	1999	19
35–100 ft.	1984	3,598
	1991	3,481
	1999	2,320
Under 35 ft.	1984	3,370
	1991	2,364
	1999	1,521

18,200 to 8,700. The vision of Pearse and others had come to pass: a fleet half its former size.

Of course, there was still enough capacity in the well-equipped fleet run by skilled fishermen to take the harvest many times over. But a smaller fleet would be easier to control for conservation purposes. And it would divide the fishery pie among fewer people, giving better prospects for decent and secure incomes, unless costs went up. (A prime expense was the buying of licences.)

"Slipper skippers" control part of fleet

Smaller-boat fishermen and some community groups had feared that the C.F.A.R. buy-back, coupled with the "stacking" of licences, would favour the big-boat fleet, specifically the seiners. In fact, the fleet-reduction program showed no big-boat bias. Even so, some still charged that the smaller vessels were suffering overmuch and losing their place. The fear was that the small operators couldn't make a living with just one area or gear type; nor could they afford to buy up licences for more. They would go broke, while those who could afford to accumulate licences and quotas would do better.

Like the Atlantic licensing system, that in British Columbia generally allowed fishermen to transfer their licences through D.F.O. to whomever they wished. But the Atlantic system had safeguards to protect the independent fleet of vessels under 65 feet in length.

British Columbia, by contrast, was wide open, particularly in the roe-herring fishery. The original limited-entry scheme had made herring licences non-transferable, though holders could lease them out. In 1989,

in the wake of a court case, the department made herring licences transferable, like those in salmon and most other fisheries. Now licences could either be sold or leased. A licence-holder had no obligation to fish the licences himself; he could lease a state privilege to others for private profit, getting a set fee or a share of the catch. Many "slipper skippers" leased out their licences, some for long periods, and at high prices. In the early 1990's, U.F.A.W.U. president John Radosevic objected, "This is a public resource and the licences are a privilege granted by the Department of Fisheries and Oceans. Then these people walk out of the door and turn a publicly-granted privilege into private gain without having to do even an hour's work."[9]

In salmon, the licence still attached to the boat; but here, too, the licence-holder, with no obligation to operate his vessel, could accumulate additional boats and licences. In those fisheries where individual transferable quotas prevailed, the fisherman could lease them out as well.

Although, back at the time of the Davis Plan, a form of gentleman's agreement had kept corporate ownership to 12 per cent or less of the salmon fleet, no regulation backed it up, and as years went by no one worried overmuch about it. Without Atlantic-style safeguards, did the feared corporate takeover occur in British Columbia? The situation at the end of the millennium was murky.

Where giant canneries used to dominate, now many types of operation shared the industry. The general opinion seemed to be that fish-buying and processing companies could usually do as well by getting fish from independents (who were either truly independent or under a financial obligation to a company). Large investors who could in theory buy up dozens or hundreds of licences (although that would probably stir up an industry and government backlash) had other uses for their money. Corporate influence over the B.C. fleet was said to be very strong among seiners, relatively light among gillnetters and trollers. In any case, the spectre of takeovers fed the fears of small operators and communities caught in the salmon decline.

B.C. industry changes shape

With catches and markets fading, the B.C. industry went through major changes. For a hundred years, large companies had dominated the scene. In industrial canneries, hundreds of workers had served on production lines. Now, with less salmon and with new labour-saving technology helping to process what remained, the processing set-up changed. Although a few major outfits still had the biggest marketing presence, the B.C. fishery at the turn of the millennium looked less industrial and more entrepreneurial.

In 1901, about 70 canneries had lined the coast; now only a handful remained. B.C. Packers, historically the largest and most famous, reduced and then, by 2000, sold off its operations, largely to Canadian Fishing Corporation. Meanwhile, new firms sprang up, selling wild salmon or other species to the food-service trade. A major factor was a shift in consumer buying habits, from canned to fresh or frozen products. A few thousand people still worked in about 190 fish processing plants, with many of the jobs now coming from farmed salmon or new fisheries, such as squid, and from new products. Seafood was British Columbia's biggest food export.[10]

Coastal, Native communities face difficulties

The B.C. coast was said to have a hundred communities with substantial fishing involvement, and some were highly dependent. Native communities in particular felt threatened by changes.

Since British Columbia joined Confederation, Indian commercial fishermen had faced many difficulties, but managed to hold a significant place in the fishery. As already noted, in the 1970's and early 1980's government had encouraged commercial fishing through initiatives including the Indian Fishermen's Assistance Program and the Northern Native Fishing Corporation. In addition, Aboriginal people paid less for salmon licences; they had been allowed free entry into the roe-herring fishery for two years after it was closed to others; they got 50 per cent of roe-on-kelp licences; and they operated many community economic development projects under the Salmonid Enhancement Program. The department funded various Native-run studies, and weighted employment opportunities towards Native people.[11] The Native Fishing Association from 1985 and the Aboriginal Fisheries Strategy from 1992 gave further aid to Native groups.

No special program for Indians came in with the Mifflin Plan. Some Native fishermen took part in the

Port Hardy, B.C., early in the 21st century. Communities along the coast felt the impact of salmon-fishery changes. (Photo courtesy of Small Craft Harbours branch, B.C.)

voluntary licence retirement (buy-back); others sold their licences to other people under the "stacking" encouraged by the plan. They still held a strong place; as of 2003, they held or exercised 27 per cent of commercial-fishing licences. Still, Native fishermen had suffered a sharp drop in the number of licences.[12] Community complaints were many.

"Adjustment and Restructuring" programs head west

As on the Atlantic, money emerged to help communities. In April 1998, Minister Anderson and cabinet colleagues announced aid programs amounting to about $14 million. Various government agencies would help fishermen and communities start new enterprises, for example in recreational fishing, alternative commercial species, tourism, or other forms of employment. Several hundred fishery-related projects were under way by mid-1999. Some funds went to help fishermen acquire additional licences through stacking.

The June 19, 1998 announcement of the Canadian Fisheries Adjustment and Restructuring (C.F.A.R.) program included another $100 million for early retirements, adjustment programs for displaced fishery workers, and community economic development. Human Resources Development Canada, Western Economic Diversification, and Indian Affairs and Northern Development were to oversee these measures. Later that year, a "vessel tie-up program" paid $6,500 to gillnetters and trollers and $10,500 to seiners who had rigged up but now couldn't make a season.

The same 1998 announcement featured a $100-million program to protect and rebuild fish habitat. Building on a previous Habitat Restoration and Enhancement Program set up under Mifflin, this new effort would spread money and jobs around to ease the crisis. It set up a permanent fund for habitat initiatives, and provided for community-based programs to protect habitat and increase public awareness. Projects went forward with many First Nations and local groups. As part of the effort, some 80 jobs came into place for community workers dealing with habitat projects around the province.[13]

Conservation closes more fisheries than ever

With evidence abounding of weak runs and ocean-regime changes, Anderson wanted more protection for fish. B.C. salmon, he said, "are our heritage, our responsibility and our legacy." Early in 1998 he set up a Coho Response Team. Severe measures followed. On most of the coast, fishing for coho salmon ceased entirely.

The June C.F.A.R. announcement gave notice of tighter restrictions and more selective fishing. Where once the department might have allowed tightly managed salmon openings, now it had less inclination to take risks. Officials under Anderson and his successor Herb Dhaliwal were moving away from "micro-management" of salmon runs, and more to a posture of "if in doubt, close it." The department above all protected escapement to the spawning grounds, keeping a foundation for recovery as ocean conditions improved.[14]

Anderson announced in 1998 the creation of the Pacific Fisheries Resource Conservation Council (P.F.R.C.C.), chaired by former minister John Fraser. (Fraser had recommended such a council in his 1994 report on Fraser River sockeye.) Like its counterpart on the Atlantic, the P.F.R.C.C. held public meetings and offered advice on research, conservation, and enhancement. Unlike the Atlantic council, the P.F.R.C.C. gave no specific annual advice on conservation measures such as Total Allowable Catches. Rather, it was to provide a long-term strategic overview.

Anderson wins P.R. battles

As Anderson pushed ahead with fleet reforms and conservation measures, time and area closures cut deeply into the commercial fishery. Cries of protest arose in the media. Fishermen complained that cutbacks were allowing salmon escapements far beyond requirements; fish were only dying uselessly in the rivers. Anderson made speeches and commissioned advertisements in the media, presenting the issues in terms of conservation. He gained strong support from sport-fishermen, who wanted restrictions on commercial fishing (and protection for their own). It became clear that the general public also sided with Anderson. The commercial fishery found itself in danger of appearing anti-conservation. Anderson emerged the winner from the media battles.

Meanwhile, the commercial salmon fishery reached new lows. Landings in 1999 were at roughly one-quarter of traditional levels. Their $25 million value was a return to 1960's dollar levels, but worth far less because of inflation.

Policy papers set new directions

Anderson was doing what De Bané had hoped to do: use a crisis, and federal injections of money, to buy change. The minister's own will was a major force. A series of policy papers starting in 1998 laid out guiding principles on major aspects of the fishery. A "New Directions" discussion document on Pacific salmon kicked off the process, followed by others on wild salmon policy, allocation, selective fishing, and improved decision-making.

The initial New Directions discussion document reaffirmed various conservation principles that had developed over time: conservation would come first, a precautionary approach would apply, the department would aim for net gain in habitat, and an ecological approach (though exact processes remained to be clarified) would prevail. As for "sustainable use" of the fish, short-term considerations would give way to long-term goals. All fishing—First Nations, recreational, and commercial—would be selective; First Nations requirements would take priority after conservation; the recreational fishery would where possible have more reliable and stable fishing opportunities; and the commercial fishery would be less dependent on salmon and more diversified.

Finally, the document said, the decision-making process would improve, with better information to the public. Government and "stakeholders" would together be responsible and accountable, with management based on "partnerships." And "enhanced community, regional and sector-wide input to decision making" would "be pursued through a structured management and advisory board system."[15]

Though much of this was old material reheated, still the New Directions document was the clearest statement on either coast for years. (A wild-salmon discussion paper supplemented it in 2000, stressing biodiversity and local populations.) Anderson's statement gave new emphasis to recreational fisheries and to wider involvement in decision-making.

A follow-up discussion paper in December 1998 laid out principles for allocation. Restating the primacy of conservation and of First Nations needs, it declared that recreational fisheries had priority for chinook and coho, and would benefit from more predictable fishing for sockeye, pink, and chum. But the commercial fishery would take at least 95 per cent of the latter three species. An extensive October 1999 policy paper reaffirmed those and related principles.[16]

"Selective fishing" brings major changes

The Pacific industry and department now undertook a major effort to modify gear and change fishing practices for selective fishing.

International estimates said that about 25 per cent of world catches—some 27 million tonnes yearly—got discarded. In Canada, various attempts at selective fishing had taken place over the years. D.F.O. and industry representatives had also worked out a Code of Conduct for Responsible Fishing. The Pacific selective-fishing program took such efforts to a new level, leaving no salmon gear untouched.

The coho crisis was a triggering factor. In a series of announcements in the spring of 1998, Anderson first tightened the coho rules, then virtually closed the fishery. There would be, almost everywhere, no retention or possession of coho. Every boat would use revival tanks, to ensure maximum survival of incidentally caught coho, which were then put back to sea. In seine fisheries, fishermen would brail salmon aboard with dipnets, rather than spilling them in over the stern. The troll and recreational fisheries would use barbless hooks. Gillnetters would fish only in daylight and make quick sets. Monitoring and logbook programs would foster compliance.

The department issued a selective fishing discussion document in May 1999, followed by consultations and a 2001 policy paper. This set out ground rules about selective fishing, whether by avoidance of the fish through timing and area restrictions (seen as the best option), avoidance by gear, release in the water, or release from the deck. More than a hundred experimental pilot projects went forward with fishermen, anglers, and First Nations. Fishermen who participated often got extra fishing allowances.

According to D.F.O. reports, it became clear that purse-seiners could cut post-release mortality of coho from the standard 25 per cent to 5 per cent by modifications in handling, such as new brailer designs. This allowed fishing in sockeye and pink fisheries that otherwise would have been closed because of coho by-catch. Salmon gillnetters in selective-fishing experiments cut coho mortality from the standard 60 per cent to as little as 5 per cent. Techniques included shorter set times and smaller mesh sizes, as well as very careful handling. Trollers also found they could improve their selectivity. The department provided gear to help First Nations catch salmon for food, social, and ceremonial purposes more selectively. Some traditional techniques, such as beach seines, fish wheels, and fish traps, returned to the scene.

The mandatory revival tanks decreed in 1998 for coho stirred high interest. Fishermen at the outset would put coho into the tanks for half an hour or an hour before releasing them, with some success. Then gillnetter Jake Fraser worked with Simon Fraser University, under department auspices, to improve the revival tanks. Survival rates increased from 60 per cent to nearly 100 per cent. The new box, seemingly able to revive the dead, got nicknamed "the Jesus Box."

Don Lawseth, who led the program for D.F.O., said the biggest effect of the program probably came through education and consciousness-raising. By the new millennium, no fishery was going ahead without careful consideration of selective fishing.

The push for conservation and selectivity caused further complaints. Trollers charged that the department was shifting too many allocations towards seiners, because of their advance in selectivity. All gear sectors complained about too many fish going uncaught.

But the Pacific efforts seemed to help. By the year 2000, certain runs were starting to resurge. The ocean was also returning to more normal conditions. Glimmers of hope were appearing in a smaller, chastened, and challenged industry. It appeared that the resilient B.C. salmon fishery might re-emerge strong.

Consultative system gropes for change

B.C. commercial fishermen had vigourous organizations, but their interests differed. As well, D.F.O. managers in British Columbia, far more than on the Atlantic, faced additional pressures from Native and environmental groups. The recreational industry, now including more than a hundred sport-fishing lodges, was demanding more attention and wielding more political influence. The interest groups were far better at articulation than compromise. The problem was to bring different sectors into some form of coherent discussion.

The earlier Minister's Advisory Council had given way in 1986 to a Pacific Regional Council. Neither of these over-arching bodies had taken firm hold. For the commercial fishery, most consultations continued to take place sector by sector, through such groups as the north and south coast salmon committees and the Herring Industry Advisory Board. Members generally came from the main organizations in the fishery, supplemented by other appointees the department deemed representative. Sport-fishing consultations generally took place separately.

Minister Fred Mifflin had spoken of a better consultation process. Then, Anderson's 1998 New Directions policy called for more public information and shared responsibility. The federal Auditor General, in a report on Pacific fisheries, offered further comments on the subject. Following all this, minister Herb Dhaliwal in 2000 commissioned a report by the University of Victoria's Institute for Dispute Resolution.

The Institute's study published in 2001 noted high levels of mistrust and made comprehensive recommendations. Besides reorganized sector committees, there should be integrated regional forums and an overriding Policy Action Committee. Public policy forums should discuss major issues—among them, the role of communities in the decision-making process. A Code of Conduct should govern consultations, and a new nomination process should choose representatives. Although there were no immediate sweeping changes, the department at the start of the new millennium appeared to be following through on some form of new system.

In the meantime, area manager Chris Dragseth and other officials worked with interest groups to develop an Aquatic Management Board for the west coast of Vancouver Island. The pilot project brought together Native, commercial, sport, community, and other interests in a 16-member board, half from government and First Nations agencies, half from the private sector.

The board was intended to operate by consensus and make recommendations to appropriate authorities.

Herring fishery seems to stabilize under tight controls

The other mainstay fisheries in the 1984–2000 period did better than salmon. Roe remained the prime product, along with a small amount of food herring. Prices bounced up and down after the early 1980's nosedive, but without causing a major crisis. At the turn of the millennium, the department still allowed no direct fishery for reduction to fish-meal, fertilizer, or aquaculture feed (although carcasses from the roe fishery went for meal). Catches were fairly stable under the 50,000-tonne level, nothing like the old post-war days when they climbed to more than 200,000 tonnes. Scientists and managers were pursuing a cautious approach, aiming to catch one fish in five. This suited the industry, which had only a limited Japanese market for roe.

The licensing system followed the trend towards individual quotas, but in a special manner. As noted earlier, back in 1981 the herring fishery had adopted a three-area system (north coast, Strait of Georgia, and west coast of Vancouver Island). Boats chose an area, but could also buy licences from other areas, or pool their catches through private arrangements. This had reduced the number of active vessels.

In 1997, a complex modification took place. Fishermen and vessel owners helped shape the plan through the Herring Industry Advisory Board. Under the new system, about 250 seine operators were assigned 55 per cent of the coast-wide catch, a long-established proportion, and some 1,260 gillnetters got 45 per cent. Meanwhile, the coast got re-divided into five herring areas.

Scientists would present their Total Allowable Catch recommendations by area, based on a 20 per cent exploitation rate. The Herring Industry Advisory Board

At the turn of the millennium, besides taking herring and reduced catches of salmon, some seiners were once again fishing pilchards. (Photo courtesy of Small Craft Harbours branch, B.C.)

would usually suggest a lower level of catch, and recommend seiner and other quotas by area.

Late in December, D.F.O. would announce catch targets by area and suggest the ideal number of vessels by area and gear type. As operators began selecting their areas, D.F.O. would make it known how many were planning to fish in specific areas. The remaining boats would adjust their plans accordingly; if too many picked one area, the average catch per boat would suffer.

Having picked an area, each operator took part in a pool, of at least four licences in the case of gillnetters; at least eight, in the case of seiners. D.F.O. allocated each pool a sub-quota proportional to the number of licences in the pool. Each pool appointed a representative to deal with D.F.O.'s on-site manager. When an area opened to fishing, D.F.O. would allow boats from each pool into the fishery, at a tightly controlled pace, until they caught up the quota for their pool. Some boats might never get to fish, yet collect their share. Other measures applied, such as dockside monitoring and industry contributions towards management costs. The wild fishery of the 1970's had become one of the most closely controlled in the country.

The fishery in the late 1990's appeared stable. Most licence-holders pursued both herring and salmon; catches were staying well within conservation bounds; the market appeared solid, though limited and subject to fluctuations; and no interest group seemed to be seeking major change, although First Nations were raising some concerns.

Spawn-on-kelp fishery stays solid

The herring spawn-on-kelp fishery continued to supply a reliable income for licence-holders. By the year 2000, more than 40 licence-holders, mostly Native, would each receive a product quota of roughly ten tons ("short tons" of 2,000 pounds) of spawn-on-kelp. Hired seiners caught the fish for impoundment. Licence-holders paid for a monitoring program, administered by the Spawn-on-Kelp Operators Association.

In 1996, however, the Heiltsuk band of Bella Bella took advantage of a Supreme Court decision, the Gladstone case, to enlarge their fishery. In their eyes, the decision gave them the right to sell unlimited quantities. As they increased their sales, other licence-holders complained that the Heiltsuk were driving down the market. D.F.O. negotiated larger quotas with the Heiltsuk. At the end of the century, the two sides appeared to be groping towards a modus vivendi.

The pilchards return

The ocean changes that decimated salmon seemed to encourage the return of pilchards (California sardines). The bonanza fishery of the 1930's and 1940's had faded in the 1950's. But by the year 2000, the biomass recovered to about a million tonnes. The great bulk of the fish remained in American waters, where the authorities were setting quotas of some 130,000 tonnes. A small number of seiners were fishing the mature pilchards that spilled over into Canadian waters. Licence-holders organized a Pacific Sardine Association under Don Pepper, which helped shape departmental policy for the emerging fishery. (Like many heads of sector associations, Pepper was ex-D.F.O.) Key features of the management plan included First Nation participation, individual vessel quotas (obtained by dividing the number of permits into the overall quota), dockside monitoring and hailing in, full observer coverage, and industry contributions to D.F.O. costs.

Halibut fishery moves to individual quotas

Of the main traditional fisheries in British Columbia, salmon plummeted in the 1984–2000 period, herring held its own, and halibut eventually began a resurgence.

For years after the 200-mile limit in 1977, the fishery had stayed in a deepening rut. As boats raced to get the biggest share, they caught up quotas faster. Conservation suffered. The fishery itself became dangerous as boats competed no matter what the weather. A fishery that once stretched over nine months could in the late 1980's last less than nine days.[17] Landings coming all at once made it harder to handle and market halibut; more fish went into the lower-value frozen market.

The 1982 Pearse report had recommended individual quotas for halibut. Lack of industry support and a change of government doomed that proposal. But by the late 1980's, industry members themselves wanted a change. In 1990–1991, the halibut fishery turned to individual transferable quotas, based mainly on catch history.

In this fishery, I.T.Q.s seem to have worked just as economists visualized them. The number of active vessels dropped from over 400 to under 300. The season lengthened from only 13 days in 1988 to 260 by 1994. As fishermen returned to fresh-fish markets, value shot up for less fish caught (7,800 tonnes worth $23 million in 1988; 6,000 tonnes worth $37 million in 1994). Halibut licence-holders elected members to a Halibut Advisory Board. Licence-holders played a stronger part in designing the management plan (which soon allowed temporary quota transfers) and paid management and enforcement costs. The fishery appeared more stable and profitable. Industry organizations took a strong role in management and covered part of the costs.

Groundfish overtake salmon

The value of groundfish landings including halibut nearly quadrupled after 1984. By 1994, it surpassed the value of Pacific salmon, a development that in earlier times would have seemed inconceivable. The trend continued; by 2000, B.C. groundfish fetched nearly triple the value of Pacific salmon, $130 million against $50 million. In some years of the 1990's, Pacific groundfish landings even surpassed those in the historic but shrunken Atlantic fishery.

The term "groundfish" in British Columbia was most commonly used for species other than halibut, such as cod, Pacific ocean perch, other rockfishes, and hake. Over-the-side sales of Pacific hake to foreign processing vessels remained a key feature of the fishery. Responding to industry pressures, the department required that at least 50 per cent of hake landings would go to Canadian processors.

In licensing, a similar progression took place as for halibut. In 1997–1998, groundfish fisheries moved to individual vessel quotas (I.V.Q.s) including transferability. The number of active vessels dropped from about 140 to less than 100. As returns improved, stories began to circulate (as for herring and salmon in some bygone years) of crew members making princely earnings. A high degree of co-management prevailed, and the industry paid a good proportion of management costs.

The inshore fishery for rockfish, however, ran into a resource crisis. The sport-fishery, with fewer salmon, added pressure to the fishery, especially in the Strait of Georgia and Johnstone Strait. Long-lived, slow-growing, and unprolific, inshore rockfish at the turn of the millennium necessitated special conservation measures, including closed areas.

Co-management rules sablefish fishery

The fishery for black cod, or sablefish, became known in the 1990's as a money-maker with a valuable product and a high degree of co-management. But it also showed that the new I.T.Q. fisheries with all their seeming strengths could have their ups and downs.

The sablefish fishery grew stronger after the 200-mile limit, as prices increased in Japan. Limited entry began in 1979 (27 trap vessels, 20 longline vessels). Total Allowable Catches governed the harvest. Vessel efficiency rose sharply. In 1981, the fleet took more than seven months to catch 2,600 tonnes; in 1989, it caught more fish (4,700 tonnes) in only 14 days.[18]

The fishermen and D.F.O. worked out an individual vessel quota (I.V.Q.) scheme beginning in 1990. Most of the 48 licence-holders joined in the Canadian Sablefish Association (C.S.A.), which entered into a contract with D.F.O. The two parties jointly developed management plans. The licence-holders paid for monitoring by observers through a contracted private firm. Although the observers lacked D.F.O.'s direct enforcement power and could only observe, record, and report, still their presence helped compliance. The C.S.A. funded research projects approved by the association and D.F.O.'s science branch. Some of these projects— biological sampling at different depths, tagging, and so on—yielded the main data for annual stock assessments. Others, such as experiments with trap-escape mechanisms for undersized fish, fostered selective fishing.

A Joint Project Agreement between the C.S.A. and the department laid out roles, responsibilities, and planned expenditures. C.S.A. expenditures paid for some or all of dockside and at-sea monitoring, biological sampling and data collection, advisory committee costs, a fishing-log program, and cost-recovery funding. Still, ocean conditions and fishing pressure combined to cause resource problems at the turn of the millennium, when the fishery faced a temporary closure, followed by an upturn.

Shellfish and the smaller fisheries

Shellfish, traditionally the smallest fishery, also pulled ahead of salmon in the 1990's, with values climbing to typically over $100 million. Shellfish became more attractive as salmon declined. About 4,000 fishermen, with 1,500 vessels, fished shellfish in the late 1990's, along with 1,900 clam diggers. The shrimp and crab fisheries used the typical management measures of quotas, licences, and a growing degree of co-management and industry funding.

A number of dive fisheries—goeduck, sea urchins, and abalone—took on new stature in the 1970's and 1980's. Like the sablefish fishery, they made use of individual quotas, high co-management, and a strong industry voice backed by industry funding.

As of the year 2000, prawns provided a landed value of $31.8 million, shrimp $6.1 million, geoduck $39.5 million, crabs $21.6 million, sea urchins $8.8 million, clams $3 million, and sea cucumber $1.7 million. Other fisheries including those for scallops and octopus brought the shellfish total to $113 million. Some of the species had been barely heard of before the Second World War. Now geoduck alone nearly equalled the value of the fabled sockeye salmon.

Changing international markets after the 200-mile limit influenced the shellfish fisheries. So did local markets in Vancouver, the third-largest city in Canada, with a multicultural population. More than on the Atlantic, a major and sophisticated market stood right alongside the fishing grounds, offering openings for many products, but also demanding exactly what it

wanted. Fewer intermediaries cluttered the road between consumer and fisherman; the market signals worked better.

As noted earlier, however, the abalone fishery ran into trouble, for several reasons. The department closed it after 1990.

B.C. fisheries turn upside down

British Columbia ended the century with its fishery world turned upside down: salmon a minor fishery for the moment, coastal communities badly damaged, Native participation in the fishery falling, and the department suffering criticism for the decline of the salmon and the cutbacks to the previous, Whitmoresque system of management. The old, industrial canneries had mostly closed. The fleet and the number of fishermen had shrunk to about half their 1984 size. Some fishermen with a good set of licences were doing well. But salmon still dominated the licensing picture with more than 2,000 licences, compared with about 1,500 for herring, and a few hundred for halibut and other groundfish, some of those now inactive. Many salmon fishermen were barely hanging on.

But salmon runs were showing signs of renewal. A new system seemed to be emerging: a smaller industry in terms of capacity; a full-scale, no-holds-barred emphasis on conservation; a clearer set of policy principles; more co-management and industry responsibility, especially in smaller fisheries; and perhaps a new consultative system that would bring more systematic representation of the commercial sector and all the interest groups.

Ranking the elements of Pacific management

In the most basic elements of management, understanding the fishery and being able to control and police it, federal fisheries in British Columbia during the 1984–2000 period struggled to hold its ground.

Scientific understanding of the nature of the resource continued to grow, although perhaps more slowly as cutbacks took their toll. D.F.O. and other researchers stimulated new thought with their work on ocean regimes. But monitoring of real-time conditions sometimes suffered from cutbacks.

The department's approach to basic regulation and enforcement went through a partial change. Under the old Motherwell–Whitmore system, fishery officers had travelled key streams every year, getting first-hand knowledge. They had dealt with the fishery in a generalist way, by everything from making school visits to assessing escapements to setting fishing boundaries and times.

With organizational reorientations in the early 1990's, fishery officers focussed more on enforcement; this, however, was offset by cutbacks in the patrol fleet and the enforcement budget in general. The department placed more responsibility on industry, through fees for dockside monitoring and growing participation

Fishery officer checking crab sizes in the early 1980's. (D.F.O. photo by Kevin McVeigh)

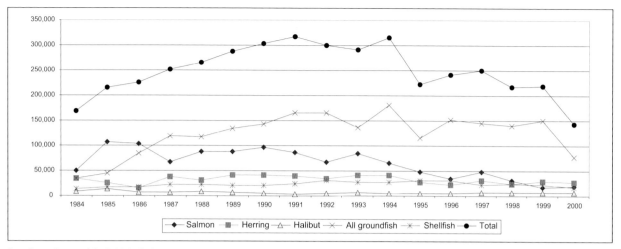

Landings (tonnes) in British Columbia, 1984-2000, total and selected species. After the mid-1980's, salmon cycles tended downwards.

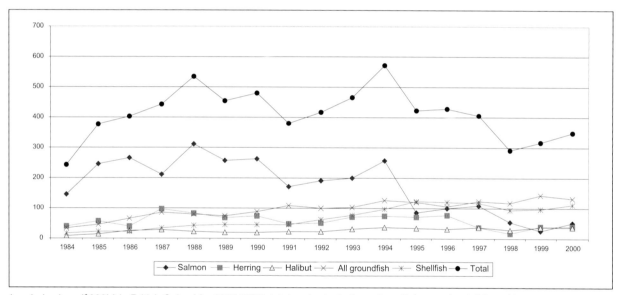

Landed values ($000's) in British Columbia, 1984-2000, total and selected species. Salmon value fell from $312 million in 1988 to less than one-sixth that level at the turn of the millennium.

in forming management plans and overseeing conservation. The system was moving by fits and starts towards a new balance between D.F.O. and industry responsibilities. Some old hands thought that the department, with its changing approaches and shifting of top managers, was losing institutional strength. Others thought the region might lead the way to something better.

After understanding and basic control, the third great and complex aspect of management is using the fish for the most benefit to the industry and society. By a rough and rocky road, British Columbia was moving towards the "fishery of the future" as enunciated by Anderson: one that was "environmentally sustainable;

economically viable; smaller than it is today; internationally competitive; and self-reliant."[19]

While restating old goals of the 1970's, these points said little about such questions as who got the fish. But the New Directions papers added clarity, for example, on the priority of Aboriginal fisheries, and the cases where recreational interests might trump commercial.

The B.C. fishery still tended to be more self-reliant than many Atlantic sectors. The rationalizer versus sympathizer conflict was less than in the East. B.C. fisheries got hit hard in the 1990's, and the fleet went through a bigger reduction than on the Atlantic. But the outcry, though strong, was less than on the

Atlantic. British Columbia's inherent goals and attitudes were more businesslike than those in the east. The higher education of the fishing workforce and the stronger surrounding economy influenced those attitudes.

Regarding the level of fishing, B.C. made a major effort to improve selectivity. For most species, as on the Atlantic, department officials tried to keep fishing at a level that would ease strain on the resource while providing enough abundance for profitable fishing. In herring, for example, both coasts aimed to take about 20 per cent of the biomass.

But for salmon, the issue posed itself differently. The department aimed mainly for proper escapements, rather than a catch percentage. The key questions were, how well could the department monitor runs, and how much safety margin should it allow?

Driven by the precautionary approach and Anderson's will, fishery openings in the late 1990's dropped sharply, producing industry complaints about fish going to waste. Seldom had the department responded so firmly to a decline. It stepped away from common practices of micromanagement and fiddling to give fishermen a break. One can argue the rights and wrongs, but the action was impressive, and paid off with healthier stocks of coho and chinook in particular.

As for access and allocations, the Pacific coast in the 1990's made far more use of I.T.Q.s. Important re-allocations took place among commercial, First Nations, and recreational fisheries. Within the commercial fleet, there were efforts to stabilize shares.

As on the Atlantic, rumours rose of wide-scale corporate control of boats and licences, but there was less of a backlash. People in British Columbia seemed more willing to take their chances and to cope in a rough-and-ready way with the existing system.

In fish-handling and quality, the department made no major direct initiatives, although I.Q.s and I.T.Q.s enabled better fishing to market. Efforts at fishery development had become, for the most part, a thing of the past. As for elevating the status of fishermen, as LeBlanc had tried to do on the Atlantic, the department and ministers felt no need for such a crusade on the Pacific; indeed, LeBlanc himself had sometimes held up the Pacific coast as a model for the East.

All told, the years 1984–2000 on the Pacific as on the Atlantic were a period not of activism, but of trying to get the basic system to work right, with more emphasis on conservation and co-management.

CHAPTER 31.
Some concluding observations

Fishery management has never been fully figured out. Otherwise, the fishery past and present would have seen more prosperity and fewer crises. About once a generation, a new trend in management seems to promise happiness, until a new set of circumstances capsizes it. But if precise understanding remains elusive, still one can offer broad observations about fisheries management in Canada.

Aircraft surveillance on the Atlantic. (Photo: Provincial Airlines Ltd.)

The fishery served Canada well

The only large industry based on hunter–gatherer methods, the fishing industry in Canada attracted the first European settlers, helped define sovereignty, supported great regions, and bred characteristics—a combination of co-operation and self-reliance—that helped form the Canadian character. If the occupation typically produced low incomes and many crises, still some people always did well, and most lived a decent life. Their chances for good incomes were best on the Pacific. Freshwater fisheries, largely outside the scope of this book, presented a mixed picture.

The biggest fishery, that on the Atlantic, was a sprawling, uncoordinated, but resilient giant. Fishermen there generally faced low earnings, instability, and in some ways, a limited horizon. Yet fishing despite its frequent desperation had a glorious, elemental appeal. Daily facing the uncertainties of fish, weather, and markets, people persevered against tough odds, creating a rich culture and close-knit coastal communities. On both coasts, the character and beauty of fishing communities were obvious to visitors; but fishing itself, the life on the water, remained largely hidden, with rewards and challenges unknown to the rest of us.

In the 1990's, the public lost some of its sympathy with fishermen. Some began to think of them as greedy fish-killers who benefitted overmuch from Employment Insurance and, on the Atlantic, from special bailouts. But the fishery always gave Canada more than it took, deserving more credit than blame. If current trends continue, the fishery could rise again in the public eye, not just as a form of local colour, but as a strong and unquestioned contributor to the economy.

The less the dependence, the better the fishery

The fishing industry tended to do best where the surrounding economy was strongest, as in British Columbia. Markets were better, over-dependence on the fishery less frequent. While old and isolated Atlantic communities sometimes got set in their ways, the later waves of settlement in British Columbia brought knowledge from modern societies elsewhere. Most Pacific fishermen lived in a smaller number of larger centres, where information circulated better than on the Atlanitc. Always allowing for exceptions, west coast fishermen tended to be less isolated and fragmented, more organized, and better educated.

Still, some Atlantic areas despite their high fishery dependence did almost as well. Southern and western Nova Scotia and the Bay of Fundy had a special set of

Surveillance in the Pacific salmon fishery.

strengths—freedom from ice, closeness to New England and West Indies markets, and a rich and varied resource. Elsewhere on the Atlantic, there were pockets of prosperity, where fish were especially plentiful or fishermen and processors particularly resourceful.

Often, however, an Atlantic fishery that started to make money would attract new entrants, who to some degree bled away profits. From the 1950's on, U.I. added to the fishery's appeal.

In recent decades, the fishing occupation has become at least somewhat less vulnerable to the pattern of over-pressure and under-profitability, because of licence limitation. As fisheries Minister Bernard Valcourt remarked in the early 1990's, "Do we want to feed two or starve five?" With controlled entry, a well-off fishery can exist amidst a poorer general economy without new entrants flooding in to sink prosperity—unless, of course, public pressure forces additional licences, as sometimes happens.

Technology shaped history

Technology drove many of the great changes in fishing history. The longline and dory intensified groundfish fishing. The gas engine was a liberator, making small-boat fishermen more independent. The purse-seine and trawler brought Canada's first great bans on modern fishing methods. Freezing technology after the Second World War brought large industrial plants to the Atlantic fishery, with Canadian factories on land and foreign factories on the water.

For all the department's limited-entry schemes and attempts to "match the fleet with the resource," fishing power grew steadily after the war. Efforts to control fishing capacity by limiting vessel size often missed the mark. Length and hold capacity could be less important than mechanization, for example with the power block. Electronics and hydraulics multiplied fishing power and helped weaken major stocks.

What about regulating technology? For a century, the department tried to foster conservation mainly by regulating gear, along with size and seasons. In the 1970's and 1980's, the regulating of technology sometimes drew scornful criticism from theorists as a primitive mechanism unsuitable for a modern industry. Yet, such regulations were often useful, though sometimes difficult to apply. When the department began supplementing "input controls" with "output controls"—quotas and the equivalent—these, too, turned out to have their drawbacks.

Do simpler systems work better?

Have simpler, passive-fishing systems worked better than complex, mobile-fishing systems? This seems to be the case at least with the collapsed groundfish and thriving lobster fisheries on the Atlantic coast.

The lobster fishery avoided the trends of mobile gear, stock assessment through virtual population dynamics, and quotas. Instead, passive-fishing traps limited effort, parcelling out fishing power in measurable units. Size limits allowed for maturation and reproduction. Meanwhile, the groundfish fishery with its more sophisticated system of high-powered fishing, complex population calculations, and subdivided quotas got partly out of control.

In the groundfish system, reliable catch statistics were supposed to support accurate stock assessments and to yield cautious quotas to be obeyed by a responsible industry. All this was to be backed up by thorough enforcement and careful monitoring of catches and related data. But in this management chain, every link had weaknesses that affected the others. Furthermore, the complexity bred a degree of isolation among different parts of the system. Fishermen, scientists, and managers were to an extent three solitudes, with too little co-operation. Thus, a system perfect on paper had weaknesses that, compounded by other factors, allowed the groundfish collapse.

What if government had managed groundfish more like lobster, applying limited entry, allowing only one restrictive type of gear, and controlling the number of units? One can argue that this style of management might have continued the good yields of the 19th and the first half of the 20th century.

In practice, though, it might have been impossible to keep the fishery in such a strait-jacket while enticing prospects of development beckoned. Allowing only a hook fishery, for example, would have left the flounder and redfish resources mainly unused, and would have held back Canada in the post-war competition against foreign fleets. Also, it would be wrong to deny the potential of population dynamics in tailoring fishing effort to the resource. As for restricting gear, even the lobster fishery with seemingly simple traps was increasing its fishing power in the late 20th century. On the Pacific coast, the ban on halibut trawling may have lessened damage but failed to guarantee abundance. And the groundfish management system could well see consistent success in future.

Still, simpler systems exert a certain appeal. The department has tried many approaches to management, all with their strengths and weaknesses; but the weaknesses were often less in the theory than in the degree of industry support and governmental follow-through. Various approaches might have worked better if applied full force. Simpler systems, if they are sensible, might have a better chance of generating support and follow-through.

Does visibility help viability?

Former fisheries minister James Sinclair said that the great advantage for fishery management in British Columbia was that "we can see the salmon." People could monitor and count them. Fishermen and the public identified with the salmon's struggle to run the gauntlet of predators, pollution, and fishermen. The Pacific salmon fishery often pioneered in management.

Lobster fishing as well takes place relatively close to shore. People can see the creatures and the traps; they consciously or unconsciously monitor the fishery. At the moment, this fishery, too, seems highly successful.

There are no rules in these matters. In Nova Scotia, the offshore scallop fishery seems a model of good management, even though it uses mobile gear, far out of sight. And Atlantic salmon, a highly visible and prized species, has declined. But, other things being equal, visibility favours sustainability.

Development work made sense, at the time

Most development work after the Second World War took place on the Atlantic. Did that work do more harm than good? It is true that development work at times pursued mirages, and at other times worked too well, fostering over-optimism, over-investment, and overfishing.

But the Atlantic fishery before the Second World War was in some ways a backward industry. To raise a voice against development in the 1940's and 1950's would have seemed foolish. From the post-war years to the early 1970's, the department and the Fisheries Research Board had notable triumphs. They spread the use of fish-finding sonar, refrigerated sea water, and everything from lobster-trap haulers to refrigerated trucks. Templeman and others took the veil off the ocean to reveal great groundfish resources. And from the 1960's to the 1980's, development work helped create the crab, shrimp, surf clam, spawn-on-kelp, and other fisheries. The promotion and marketing work that took place over many decades and the efforts to improve fish quality also had their value.

Had the department done nothing in the post-war period, would the industry have advanced just as well? It seems likely that progress would have been more fitful and foreign domination more prevalent. Federal development work in the post-war period went roughly in the right direction, even if too fast at times. It was justifiable to help put technology in Canadian hands, so the country could compete.

As doubts crept in during the later 1960's, the department cut back its fishing-development work (although boatbuilding subsidies and tax incentives lingered) and moved towards comprehensive management. The problems of northern cod, Pacific salmon, and herring on both coasts owed something to overzealous development efforts, but more to other factors.

That being said, the stronger technology which government helped introduce remained a complicating factor. Like transferable licences, new technology was easier to bring in than to curtail. And more efficient gear could temporarily mask the stock declines to which it contributed.

Regulatory approach skirts certain questions

Canada historically took a more regulatory approach to fisheries than some other nations, clearly more so than the United States. H. Scott Gordon called for a "unified directing power." Roméo LeBlanc said, "You can't manage a common-property resource with a laissez-faire attitude." Was close regulation the way to go?

Despite its failings, regulation is clearly better than no regulation. But the department never went all the way to Gordon's "unified directing power," nor did it try to shape an ideal fleet and fishery. It regulated technology less by dictating it than by hobbling it. Rarely did it explicitly address the question, what are the fisheries for? LeBlanc was a partial exception: his speeches said that the fisheries were for fishermen first, and among them for inshore fishermen first. But his 1976 policy document spoke more vaguely of "optimum benefits to society." Later efforts to define policy tended towards similar generalities.[1]

If a new, uninhabited, fish-rich Newfoundland were discovered today, how would one design its fishery for the greatest good? Open it to unrestrained competition? Hand it over to a single large corporation, crown or private, to try to create the most profits and the highest "resource rent"? Follow the Kirby principle of ensuring viability but also generating the most possible jobs? Turn the resource into a sport-fishing reserve or an ecological showplace? There is no consensus.

In theory, one could adjust the commercial fishery over time to use only the very best types of gear for conservation and cost-effectiveness, and to include the optimum number of boats and people for profitability and social strength, while leaving a safety margin to provide for flexibility and change. Among ministers, only Jack Davis ever gave much thought to the pursuit of an ideal fleet, but he abandoned the effort. LeBlanc, too, backed off from his proposal to put the large-trawler fleet in the hands of fishermen, rather than processors. The optimum fleet structure would be problematic to define and difficult to impose, and indeed would change over time.

For the most part, rather than trying to reshape the fishery in a fundamental way for a particular goal, the department operated by looking at the existing situation and making whatever adjustments seemed sensible at the time. This course of action has both advantages and drawbacks.

In licensing, for example, Canada took strong action in applying limited entry before most other countries. But it refrained from full control; instead, it let fishermen trade in licences. If, when a fisherman stopped fishing, his licence went back to the state, then the department could reduce the fleet at will, simply by issuing fewer licences. It could give fishing privileges to those applicants with the best proposals for conservation, quality, jobs, profits, or other criteria; or, it could auction them for the most money, which would return "resource rent" to the state. The industry group most responsible for triggering limited entry, B.C.'s United Fishermen and Allied Workers Union, wanted licences to go back to the state, as did the Maritime Fishermen's Union, the Levelton report, and the second Sinclair report.

Instead, Davis, LeBlanc, and their successors in effect let the fisherman control transfers, and gave licences a form of everlasting life. One can argue that this saleable asset gives the fisherman more freedom and security. One can also argue that it undercuts security, by hindering proper control of the fishery. But full licence control was a deeper intervention than ministers and officials wanted to make.

How to measure good management?

How well has the federal fisheries department managed fisheries? And how to measure success or failure? One clear indication is the protection of stocks and species against extirpation. The original Lake Ontario salmon disappeared, along with some salmon runs on each coast.[2] Among collapses short of extirpation, the commercial fishery for Atlantic salmon ended, and northern cod suffered its famous crash. But generally, Canada has preserved its resources, even if some lost strength.

It is natural to make comparisons with our giant American neighbour, and when one looks at species fished in common, Canadian management generally did better. For example, Pacific salmon in Canadian rivers survived far better than those in the northwest United States. American scallop fishermen envied the abundance of Canadian stocks just across the Georges Bank line, where different management prevailed.

What are the benchmarks apart from preservation? In the 1970's, ministers and officials often quoted the goals of volume, value, stability, and power and prosperity for fishermen. These remain a handy set of criteria. At the end of the century, the record was obviously mixed. Volume crept up to record levels in the 1980's, then fell with the Pacific salmon and Atlantic groundfish declines. In the Atlantic crisis, the department in its different aspects—scientists, managers, enforcers, ministers—played a role, and afterwards should have shone the bright light of a thorough inquiry onto the whole affair.

Harbour dredging at Port de Grave, Newfoundland. (Photo courtesy of Bill Jenkins, Small Craft Harbours branch, Newfoundland)

That being said, the federal fisheries department laboured mightily throughout its history for conservation, with a good record most of the time. It also did more than its share for value. Strenuous efforts in inspection and marketing helped haul the industry up to its present position as a good-quality, high-value producer.

What about stability? Setting aside the complex phenomena of the 1990's cod and salmon declines, the system instituted by Davis and LeBlanc—licences, quotas, zones, advisory committees—improved stability in several respects. Full-time, licensed fishermen, especially those with a portfolio of licences, in the 1990's managed to survive changes that were previously unthinkable. Paradoxically, many fishermen at the end of the century were doing better than ever. And they had more power over their fisheries.

Apart from the fishery pure and simple, fishery management gets entangled with questions of sovereignty, development, communities, and way of life. Consistently since Confederation, the federal fisheries department and ministers enhanced Canada's sovereignty. Most other meta-fishery issues, such as preserving a "way of life," escape easy analysis. The federal fisheries department waded into, or got sucked into, whatever came its way, sometimes winning praise, usually getting pilloried.

Criticism was worse because the department generally paid less attention than it should have to communications with fishermen. Dozens of studies and reports in the 20th century—for example, those of Pearse, Kirby, Haché, and Harris—called for the department to do more in the way of information, education, organization, and consultation. But such recommendations almost always got tucked into the back of a bigger report, with forceful action rare. Misunderstanding, misinformation, and mistrust caused needless fear and anger in fishing communities, and often held back co-management and progress in general.

Having said that, it may sound strange to suggest that a prime way to evaluate fishery management,

"Stream planting," as hatchery officials put salmon in a river.

apart from easily counted aspects such as volume and value, is by seeking the opinions of industry members. Even while complaining, thoughtful fishermen and processors respect good, tough, responsive, intelligent management. They factor in not only volume and money but the personal qualities of officials. They are quick to acknowledge how hard they themselves are to control. Few are satisfied during a crisis, and most are suspicious of out-of-sight administrators in Ottawa or elsewhere. But allowing for variations of time and place, they have tended to give officials, especially those with whom they can talk face to face, their solid respect. On the Atlantic, that respect continued even after the cod collapse, though tempered.

What makes a good fishery manager?

Fishery management is a game of many variables, like playing chess in the dark. Success often depends less on rules and procedures than on human qualities. In a department with many good managers, perhaps the most highly reputed was Joe Whitmore, long-time regional director on the Pacific. Whitmore exemplified three qualities—knowledge (he worked in the department more than 40 years and made it his business to listen and learn); toughness (one recalls an industry executive's comment that "when Whitmore was there, fights were fun"); and responsiveness (Whitmore faced issues on the spot, and inspired his officials to do the same).

Many top officials, and thousands more down the line, showed the same traits. The fisheries department attracts people who enjoy the tangibility of dealing with fish and fishermen, and the complexities of real life. Time and again, department officials have passed up chances to move upwards in some other agency, because they get caught up by fisheries and believe they can contribute. In that respect, the department is

almost a character test, and keeps more than its share of dedicated officials.

Do fisheries personnel, in their frequent dedication, also enjoy a feeling of power? The federal fisheries department controls the daily activities of citizens in a large industry. It says who can fish, in what kind of boat, when, where, and for what. It tells fishermen how to handle the fish and sometimes how to sell them, while making them pay for observers and dockside monitors. It affects their income. A fisherman can argue and orate at an advisory committee, but the department has the last word.

I have never met an official who acknowledged enjoying this power. Most feel chained to a desk, and more put-upon than powerful. Still, one suspects that at some level, authority forms one of the department's attractions.

Levelton's Law

How do internal bureaucratic structures and practices affect fisheries management? Especially after the 1960's, fisheries like other portfolios suffered from the governmental disease of frequent internal reorganization. As the size and subdivisions of bureaucracy increased, people could get channelled into separate units, with less intermixing than desirable.

Cliff Levelton, a senior fishery manager for many years, used to say, "Never separate the analyzers from the implementers." The fishery is never easy to grasp; it requires an overall picture, in part intuitive. Many of the best fishery managers either began as fishery officers or used some other means to gain personal experience with the fishery in all its variety, contrariness, and humanity. The more direct experience at various levels, the better.

From Governor to self-rule

The fisheries department still bears some resemblance to colonial administration. In the early 1800's, the governors of the Atlantic colonies had, at most, token elected assemblies. They ruled with appointed councils, favouring whom they pleased. By the end of that century, however, universal education, the vote, the press, and the Industrial Revolution had changed Canada into a free-market democracy.

Fishermen missed out on much of the transformation. Particularly on the Atlantic, they remained less educated than the average citizen and less organized than farmers. Information flow was fitful. There was no vote on fishery matters. For fishermen, the Industrial Revolution fully arrived only after the Second World War. Even then the minister, like the early governors, still had the last word. Although the industry or public could spark policy debate and development, the tendency was, in the end, for a small group of officials and the minister to cobble something together.

Throughout most of that long period, the department saw the fishery as a twin challenge of conserva-

Checking lobster traps on the Atlantic.

tion—mainly in inland and nearshore waters at the start—and of development. A succession of ideas seemed to offer progress: gear and season regulations for conservation; leases in eastern river fisheries and licences in the B.C. salmon fishery; hatcheries; developmental projects and subsidies; the trawler ban and fishery co-ops during the Great Depression; the development heyday after the Second World War; and, with Davis and Leblanc, comprehensive management and the 200-mile zone. The department much of the time was trying to push the industry ahead, according to its latest thinking, with sometimes unexpected results.

That was of course most true on the Atlantic. In British Columbia, industry members had more education, organization, and prosperity, and showed more leadership. Whether in licence limitation, quality improvement, international management, or salmonid enhancement, fishermen and processors were in the forefront. The B.C. industry was more assertive, and the fishery generally did better.

After the Second World War, especially under Davis and LeBlanc, Atlantic fishermen gained more organization and influence. But even today, the governor—the minister—holds unusual power.

One can argue that two main currents of the Western world shape self-control. One is freedom from government, as fostered by private enterprise and the free market. Licences and individual quotas can give enterprises a degree of freedom from arbitrary action. But it is unlikely that the common-property fishery will ever be as separate from government as most other industries.

The second approach to freedom is by control of government, through the vote, information, and all the civil engagements of society. Encouraged by LeBlanc

and to varying degrees by his successors, fishermen and other industry members, together with the department, have been moving towards co-management of one form or another. Here lies the true promise.

Calls to take all politics out of the fishery have never worked, and probably never will, as long as the resource belongs to all Canadians. What is needed is a better politics, comprising a big dose of education, democracy, and responsibility-taking, exerted through a system with the small-p political mechanisms to cope with unpredictable change. This approach appears to be developing through organizations, advisory committees, Integrated Fishery Management Plans, joint projects, and such attempts to bring the relevant interests into management.

Bringing people or their properly mandated representatives together, with shared information and responsibility, sounds banal but is basic. Co-management will still need to be coupled to authority to work; but authority without co-management will continue, too often, to fail. Fishery management is an endless struggle to square the circle of private interests and common property. The more that the people involved have a say, the better they can muddle through, as in the common property of Canada itself.

The fisheries department served Canada well

Meanwhile, the people of Canada can thank the federal fisheries department. My first term in the federal fisheries department came in the 1970's. I entered with a chip on my shoulder, objecting to what I deemed mismanagement of the Bay of Fundy herring fishery. After about six months, it dawned on me that I was among some of the best, hardest-working, most knowledgeable people I had ever met. In the Bay of Fundy, I soon saw LeBlanc, Fern Doucet, Derrick Iles, and others work with fishermen to transform the fishery for the better. I came to realize that every fishery is a separate, complex story, and the country has thousands of them, full of living, breathing people each with their own interests and problems, in an industry where many factors are beyond the managers' control. No formula applies except work, brains, and good will.

The fishery officer, the scientist, the negotiator, the habitat protector, the front-line staff in general, all belong to a special and little-understood brotherhood. Like fishermen themselves, they face a set of challenges and rewards unknown to most other people. No ministry could have had more dedicated employees, and they have done a generally commendable job for industry, communities, and country. What fish we have, we owe to them.

NOTES

Chapter 1 Notes

1. Diamond Jenness, *The Indians of Canada*, 5th ed. (Ottawa: National Museum of Canada, 1960), passim.

2. Olive Patricia Dickason, *Canada's First Nations: A History of Founding Peoples from Earliest Times* (Norman, Okla.: University of Oklahoma Press, 1992), 80. See also Hilary Stewart, *Indian Fishing: Early Methods on the Northwest Coast* (Seattle, Wash.: University of Washington Press, 1977), 161 ff.

3. See Jenness, passim; Dianne Newell, *Tangled Webs of History: Indians and the Law in Canada's Pacific Coast Fisheries* (Toronto, Ont.: University of Toronto Press, 1993), 28–45. I owe much of this chapter to Jenness, Newell, Dickason, and Stewart.

4. M.P. Shepard and A.W. Argue, *The Commercial Harvest of Salmon in British Columbia, 1820–1877* (Department of Fisheries and Oceans, Canadian Technical Report of Fisheries and Aquatic Sciences, August 1989), 2.

5. See Dickason, 63; Jenness, 59–60; Newell, 31.

6. See Jenness, 53–66; Newell 34, 35; Gordon Winant Hewest, "Aboriginal Use of Fishery Resources in Northwest North America" (diss., University of California, February 1947), 241.

7. Department of Marine and Fisheries, Annual Report for 1886, 264–5; Stewart, 46–7; Jenness, 170–3.

8. See Jenness, 59–60; Hewest, 136 ff. Dickason, 67, states that the whaling technology originated with the Aleuts and Inuit to the north. Lescarbot quotes Joseph Acosta, a Spanish Jesuit who spent years in Peru in the 16th century, about Indians leaping from their canoes to mount whales as if on horseback, using staffs to plug up the whales' blowholes, and then taking an attached line ashore to draw the whale in slowly. See Marc Lescarbot, *Nova Francia: A Description of Acadia, 1606*, trans. P. Erondelle, 1609 (London, U.K.: George Routledge and Sons, 1928), 289.

9. Cartier's account appears in J.H. Reid, Stewart Kenneth McNaught, and Harry S. Crowe, A *Source-book of Canadian History*, rev. ed. (Toronto, Ont.: Longmans Canada, Ltd., 1964), 9. See also Marcel Moussette, *Fishing Methods Used in the St. Lawrence River and Gulf* (Ottawa: National Historic Parks and Sites Branch, Parks Canada; Indian and Northern Affairs Canada, 1979), 65.

10. See Jenness, 53–66; Moussette, passim; Lescarbot, 95–6, 285; Dickason, 106–7, 167.

11. Jenness, 411.

12. Dickason, 92.

13. See Jenness 53–66, 97, 406 ff.

14. Jenness, 45, 61.

15. Newell, 42.

16. See Jenness, 53–66.

17. Roderick Haig-Brown, *The Salmon* (Ottawa: Environment Canada, Fisheries and Marine Service, 1974), 43, 48. See also Stewart, 161 ff., 174.

18. Newell, 40.

19. Jenness, 65.

20. Most of the above can be found in Newell, 40–5; Jenness, 145; Stewart, 20, 100; Marilyn G. Bennett, *Fishing and Its Cultural Importance in the Fraser River System* (Fisheries Service, Pacific Region, Environment Canada; and Union of British Columbia Indian Chiefs, April 1973), 17; Geoff Meggs, *Salmon: The Decline of the B.C. Fishery* (Vancouver/Toronto, Ont.: Douglas & McIntyre, 1991), 12.

21. Dickason, 228.

22. Jenness, 421–2, 256–7.

Chapter 2 Notes

1. Harold Innis, *The Cod Fisheries: The History of an International Economy* (New Haven, Conn.: Yale University Press, 1940), 11. Future references to "Innis" mean this work unless otherwise specified.

2. Reid et al., 8.

3. Innis, 37–8, 71.

4. Mark Kurlansky, *Cod: A Biography of the Fish that Changed the World* (Toronto, Ont.: Alfred A. Knopf Canada, 1997), 49.

5. G.B. Goode, Joseph W. Collins, R.E. Earll, and A. Howard Clark, *Materials for a History of the Mackerel Fishery* (Washington, D.C.: Government Printing Office, 1883), 217.

6. Robert W. Dunfield, *The Atlantic Salmon in the History of North America* (Ottawa: Department of Fisheries and Oceans, 1985), 12.

7. L.Z. Joncas, "The Fisheries of Canada," in *Fisheries of the World: Papers from the International Fisheries Exhibition* (London, U.K.: 1883).

8. Dickason, 95.

9. D.W. Prowse, *A History of Newfoundland from the English, Colonial and Foreign Records* (London and New York, 1895; reprint, Mika Studio, Belleville, Ont., 1972), 19.

10. See Innis, 47–9, 124–6; Shannon Ryan, *Fish out of Water: The Newfoundland Saltfish Trade, 1814–1914* (St. John's, Nfld.: Breakwater Books, 1986), xxi–xxii, 29–30, and *The Ice Hunters: A History of Newfoundland Sealing to 1914* (St. John's, Nfld.: Breakwater Books, 1994), 25–6; Jean-François Brière, "The French Fishery in North America in the 18th Century," and Roch Samson, "Good Debts and Bad Debts: Gaspé Fishers in the

19th Century," in *How Deep Is the Ocean? Historical Essays on Canada's Atlantic Fishery*, edited by James E. Candow and Carol Corbin (Sydney, N.S.: University College of Cape Breton Press, 1997).

11. Lescarbot, 286. See also Prowse, 21.

12. Innis, 23–6; Ryan, *Fish*, 31.

13. Lescarbot, 286–7.

14. Innis, 47–8.

15. Lescarbot, 287–9.

16. See Ryan, *Ice*, 65 ff.

17. Much of the above appears in Innis, 21, 47–8, 56 ff., 133; Ryan, *Fish*, 29 ff., passim.

18. Samuel Edward Dawson, *The Saint Lawrence Basin and Its Borderlands: Being the Story of Their Discovery, Exploration and Occupation* (London, U.K.: Lawrence and Bullen, 1905), 81–3. See also Innis, 14–5, 26; Prowse, 21–2.

19. Darlene Abreu-Ferreira, "Portugal's Cod Fishery in the 16th Century: Myths and Misconceptions," in *How Deep Is the Ocean? Historical Essays on Canada's Atlantic Fishery*, edited by James E. Candow and Carol Corbin (Sydney, N.S.: University College of Cape Breton Press, 1997), 35.

20. Innis, 38; Prowse, 60.

21. Innis, 92.

22. See Innis, 23; Dawson, 222–3; René Belanger, "Basques," in *The Canadian Encyclopedia*, 2ⁿᵈ ed. (Edmonton, Alta.: Hurtig Publishers, 1988).

23. Janet E.M. Pitt and Robert D. Pitt, "Red Bay," in *The Canadian Encyclopedia*, 2ⁿᵈ ed. (Edmonton, Alta.: Hurtig Publishers, 1988).

24. J.N. Tonnessen, A.O. Johnsen, and R.J. Christophersen, *The History of Modern Whaling* (Berkeley, Calif.: University of California Press, 1982), 1–2; Innis, 50; Dawson, 135.

25. Reid et al., 28.

26. Relevant information appears in Dawson, 224; Innis, 31, 39–40, 47–8.

27. I owe much of the above to Dawson, 130, 178, 224; Innis, 15–25, 31, 39–40, 44–51, 90; Tonnessen et al., 1–2; Ryan, Fish, 29 ff.

28. See Innis, 87–9, 121–7; Prowse, 187.

29. Muriel K. Roy, "Settlement and Population Growth," in *The Acadians of the Maritimes*, edited by Jean Daigle (Moncton, N.B.: Centre d'études acadiennes, 1982).

30. See Jean Daigle, ed., *The Acadians of the Maritimes* (Moncton, N.B.: Centre d'études acadiennes, 1982), 23–8; J. Bartlett Brebner, *Canada: A Modern History* (Ann Arbor, Mich.: University of Michigan Press, 1960), 66.

31. Cf. Innis, 84–6, 95, 120–9; C. Gallant and G. Arsenault, *Histoire de la pêche chez les Acadiens de l'Île-du-Prince-Édouard* (Summerside, P.E.I.: Société Saint-Thomas-d'Aquin, 1980).

32. Innis, 50, 86–91, 121–4.

33. Reid et al., 24.

34. Cf. Innis, 121–4; Moussette, 56.

35. Cf. Brebner, *Canada*, 39; Innis, 118, 120.

36. Daigle, *Acadians*, 24.

37. Innis, 183, see also 103, 131.

38. Much of the above appears in Innis, 13, 30; Dawson, 74–5, 224; Prowse, 33; Abreu-Ferreira, 34; Lunt, 361.

39. Innis, 31–8, 49–50.

40. See Ryan, *Fish*, 29, 31; Innis, 34, 50–1.

41. Dawson, 226; Tonnessen et al., 1–2.

42. See Innis, 52–74, 90, 127; Ryan, *Fish*, 32.

43. Ryan, *Ice*, 26; Innis, 95, 115–9.

44. Innis, 52, 64–5, 71, 76.

45. Most of the above can be found in W.E. Lunt, *History of England*, 4ᵗʰ ed. (New York, N.Y.: Harper and Brothers, 1957), 437–9, 492; Innis, 112–6.

46. Innis, 96, 107; Prowse, 173.

47. For the above, cf. Innis, 68–71, 75–6, 117, 134–5; Prowse, 120–1.

48. Innis, 75; Prowse, 152.

49. Dunfield, *Salmon*.

50. Cf. Goode et al., 81ff., 83, 85, 115–7, 148.

51. Innis, 75, 117.

52. Tonnessen et al., 3; Daniel Francis, *The Great Chase: A History of World Whaling* (Toronto, Ont.: Penguin Books Canada, 1990), 44–6.

53. Innis, 70–4, 116, 127.

54. Prowse, 151; Innis, 101.

55. Cf. Innis, 80–1, 101–4, 115–8, 125, 131–7.

56. Innis, 116–9, 130.

57. Prowse, 59, 95; Innis, 54.

58. For the above, cf. Innis, 54–6, 61–2; Prowse, 99 ff., 109.

59. Ryan, *Ice*, 29; cf. Innis, 57, 63–4.

60. Cf. Innis, 66–7; Prowse, 154 ff.; Ryan, *Ice*, 26.

61. James J. Talman, *Basic Documents in Canadian History* (Princeton: D. Van Nostrand, 1959), 15–9.

62. Cf. Innis, 67–70; Prowse, 110–2.

63. Ryan, *Fish*, 32; cf. Ryan, Ice, 30; Prowse, xvii, 114.

64. Most of the above comes from Ryan, *Ice*, 27; Shannon Ryan, pers. comm.; Innis, 101–2, 106.

65. Ryan, *Fish*, 32.

66. For the above, cf. Prowse, 113–4, 151–3; Innis, 98–9, 103, 151–3.

67. Prowse, 226.

68. Prowse, 154–5; Innis, 97.

69. For the above, cf. Prowse, 154–5, 167–8, 191–6; Innis, 96–9.

70. Innis, 101–2, 127.

71. For the above, cf. Prowse, 201, 274; Ryan, Ice, 31; Innis, 109, 123, 296; H.Y. Hind, *The Effect of the Fishery Clauses of the Treaty of Washington on the Fisheries and Fishermen of British North America*, part 1 (study prepared for the Fishery Commission, Halifax, N.S., 1877), 162.

72. A boat's "room" could be as much flake as would allow spreading 70 quintals of wet fish, or

40 feet of frontage, or various other definitions. Cf. Innis, 106n.

73. Cf. Innis, 102, 108–9, 317; Prowse, 253; Ryan, Ice, 32.

74. J.R. Smallwood, *The Book of Newfoundland*, vol. 3 (St. John's, Nfld.: Newfoundland Book Publishers, Ltd., 1937), 427–8.

75. Brebner, *Canada*, 70.

76. Cf. Lescarbot, 281–91; Nicolas Denys, *The Description and Natural History of the Coasts of North America (Acadia)* (New York: Greenwood Press, 1968; Champlain Society Publication, facsimile of 1908 ed.), passim.

Chapter 3 Notes

1. Innis, 149; cf. Ryan, *Fish*, 34; Prowse, 297.

2. Innis, 149–50.

3. Prowse, 273. For information on costs, earnings, and wages, see Innis, 147, 151, 181–2, 300n.

4. Innis, 151–2, 196.

5. Ryan, *Ice*, 28.

6. Innis, 147.

7. Prowse, 274; cf. Prowse, 201, 296–8, 325; Innis, 109, 147, 177, 195–6.

8. Innis, 150.

9. Great Britain, Board of Trade, *Report of the Lords Commissioners for Trade and Plantations to His Majesty, 19 December 1718*; see Innis, 157.

10. Ryan, *Ice*, 30–1.

11. Prowse, 342n; cf. Innis, 149, 198.

12. Innis, 147, 155.

13. Innis, 155.

14. Prowse, 192–200; cf. Innis, 145–7, 195–6.

15. Most of the above appears in Ryan, *Ice*, 31; Prowse, 283, 296, 324, 341; Innis, 147, 157, 195–6.

16. W.S. MacNutt, *The Atlantic Provinces: The Emergence of Colonial Society*, paperback ed. (Toronto, Ont.: McClelland & Stewart, 1968), 74.

17. Innis, 144, 153, 194, 296.

18. Cf. Andrew German, "Fishing History, 1720–1930," in *Georges Bank*, edited by Richard H. Backus and Donald W. Bourne (Cambridge, Mass.: MIT Press, 1987), 412.

19. Cf. German, 409; Innis, 160–1.

20. Innis, 200.

21. Most of the above appears in Innis, 160–6, 205; Prowse, 199; J. Bartlett Brebner, *The Neutral Yankees of Nova Scotia*, paperback ed. (Toronto, Ont.: McClelland & Stewart, Carleton Library series, 1969), 8–9; Allan Nevins and Henry Steele Commager, *A Pocket History of the United States* (New York, N.Y.: Washington Square Press, 1970), 29–30, 62.

22. For the above, cf. Francis, 55, 61–2; Tonnessen et al., 3; Prowse, 298; Innis, 191.

23. Judith Tulloch, "The New England Fishery and Trade at Canso," in *How Deep Is the Ocean?*

Historical Essays on Canada's Atlantic Fishery, edited by James E. Candow and Carol Corbin (Sydney, N.S.: University College of Cape Breton Press, 1997), 70.

24. Innis, 191, 231.

25. Hind, part 1, 166.

26. Kennedy Wells, *The Fishery of Prince Edward Island* (Charlottetown, P.E.I.: Ragweed Press, 1986), 101.

27. *Lifelines* exhibit at Canadian Museum of Civilization, Hull, Quebec, 2001. More on the walrus appears in Prowse, 336; Innis, 191, 276; Moses H. Perley, *Reports on the Sea and River Fisheries of New Brunswick* (Fredericton: Queen's Printer, 1852).

28. B.A. Balcom, *History of the Lunenburg Fishing Industry* (Lunenburg, N.S.: Lunenburg Marine Museum Society, 1977), 1–2.

29. J. Murray Beck, "Nova Scotia, 1714–1784," in *The Canadian Encyclopedia*, 2nd ed. (Edmonton, Alta.: Hurtig Publishers, 1988).

30. Frederick William Wallace, *In the Wake of the Wind Ships* (Toronto, Ont., 1927), 11–2.

31. Innis, 190.

32. Dunfield, 48.

33. Innis, 185–90.

34. For the above, cf. Prowse, 250; Innis, 138, 168–9.

35. Report quoted in Reid et al., 29. Hind, 167–8, quotes a report around 1758 of 726 decked vessels, with eight men each, and 1,555 shallops. This may have included decked vessels from France. In any case, the fishery was sizeable.

36. Craig Brown, ed., *The Illustrated History of Canada* (Toronto, Ont.: Lester & Orpen Dennys, 1987), 164.

37. Brebner, *Canada*, 66–8.

38. Roy, 134.

39. Roy, 149–53.

40. Prowse, 320.

41. G.W. Brown, *Readings in Canadian History* (Toronto, Ont., 1940), 203–4.

42. Innis, 183, 183n.

43. Hind, part 1, 166

44. Innis, 191.

45. Innis, 193–4.

46. Innis, 188, 192, 276–8.

47. Prowse, 319.

48. Cf. Innis, 193–4; Dunfield, 56; Prowse, 624.

49. Innis, 185.

50. Prowse, 321–2.

51. Cf. Brebner, *Neutral*, 8; Innis, 203.

52. Prowse, 327; cf. Innis, 193–4, 203.

53. Innis, 195 ff.

54. Innis, 151, 196, 293, 293n.

55. Hind, part 1, 117, 162–3.

56. Prowse, 344–5.

57. Innis, 209 ff., 312–6.

58. Prowse, 341 ff.; Innis, 208–9.

59. Cf. Innis, 293n.

60. Innis, 293.

61. Ryan, *Ice*, 34; Innis, 196, 293.

62. See Ryan, Ice, 49–53, 70; Prowse, 298; Sir William Mcgreggor, *Chronological History of Newfoundland Fisheries* (St. John's, Nfld., 1907; reprinted by Provincial Archives of Newfoundland).

63. Hind, part 1, 62.

64. Innis, 298, 303–5, 316 ff.

65. Innis, 204–6; United States Tariff Commission, *Treaties Affecting the Northeastern Fisheries*, Report No. 152, second series (Washington, D.C., 1994), 21.

66. Innis, 183, 185, 214–5, 293.

67. Innis, 202, 206–10; Tariff Commission, 20.

68. Innis, 220; cf. Ryan, *Fish*, 33; Tariff Commission, 20.

69. Tariff Commission, 21.

70. MacNutt, 132; cf. Innis, 232–3, 235, 247.

71. Innis, 232.

72. Innis, 230.

73. Cf. Wells, 100 ff.; Dunfield, 51.

74. Innis, 187, 230–40.

75. Innis, 188, 192, 276–8.

76. Dunfield, 39, 52.

77. Most of the above comes from Dunfield, 33, 41–3, 58; Prowse, 283, 337; Innis, 148–9, 158, 294.

78. Dunfield, 42, 48, 57.

79. See Dunfield, 53, 62, 65, 67–8.

80. Letter by Stephen J. Augustine of Fredericton to the *Telegraph-Journal*, Saint John, N.B., 3 June 1980, quoting from *Collections of the New Brunswick Historical Society* III, no. 9: 335–6.

81. Dunfield, 62, 65.

82. Prowse, 283, 337; cf. Innis, 148–9, 158, 294.

83. Perley, 88.

84. Kurlansky, 98–100.

Chapter 4 Notes

1. Brebner, *Canada*, 168–9, 182–3, 192–3, 208.

2. Innis, 307, 416.

3. Innis, 239–41.

4. Innis, 160.

5. Cf. Ryan, *Ice*, 37–8; Innis, 303, 320; MacNutt, 115–6.

6. Most of the above comes from Innis, 220–4.

7. MacNutt, 107–8.

8. Cf. Innis, 227–36, 239–40, 241–2, 416; MacNutt, 197–8, 114–5.

9. Cf. Ryan, *Ice*, 41; Innis, 223–4, 242.

10. See MacNutt, 134; Innis, 243, 245, 250, 339–41.

11. Innis, 219; Brebner, *Canada*, 196.

12. *Memorandum upon the questions at issue between Great Britain and the United States in relation to the North American fisheries, 1898.* (In the library of the Department of Fisheries and Oceans, Ottawa, Cat. No. 190902) Cf. Innis, 274 ff.

13. Cf. D.M. Johnston, *The International Law of Fisheries* (New Haven, Conn.: Yale University Press, 1965).

14. Innis, 279.

15. *Memorandum.*

16. Innis, 320.

17. Innis, 308; Perley, 51.

18. See German, passim; Innis, 225, 324–6; Hind, part 1, xiv.

19. Hind, part 1, 163.

20. Goode et al., 133.

21. Innis, 226, 348–9.

22. Innis, 222, 286, 324–5, 328, 332, 347; Hind, part 1, 140.

23. Cf. Ryan, *Fish*, 84; Innis, 253, 258; Brebner, *Canada*, 111.

24. Most of the above appears in Innis, 223, 225, 324.

25. Goode et al., 81 ff., 83, 85, 115–7. See also Innis, 225, 326; Hind, part 1, 81.

26. G.B. Goode, *The Fisheries and Fishery Industries in the United States* (Washington, D.C.: United States Commission of Fish and Fisheries, Bureau of Fisheries, 1884–1887), 428–30.

27. Innis, 225, 329.

28. Goode, 269–70; Goode et al., 77.

29. Goode et al., 81, 122–3.

30. Goode, 272. On purse-seining, see also Hind, part 1, 135, 139; German; Goode, 269–71; Hind, part 1, 135, 139.

31. Cf. Innis, 324–6, 330–1; German, passim; Hind, part 1, xiv.

32. Much of the above comes from Tonnessen et al., passim; Innis, 222; Prowse, 298.

33. Ryan, *Fish*, 91. Ryan tabulates the French bounties as of 1851.

34. Innis, 214–9; cf. also Innis, 375–83; Hind, part 1, xxi, 161–6.

35. Innis, 381.

36. Ch. de la Morandière, *Histoire de la pêche française de la morue dans l'Amérique septentrionale*, vol. III (Paris, France, 1966); Innis, 374–6.

37. De la Morandière.

38. See German; Innis, 329, 376; Balcom, *Lunenburg*, 15.

39. On Lunenburg, see Balcom, *Lunenburg*, 6–7; the rest of the above comes mainly from Innis, especially 272–6, 284–6, 347.

40. Innis, 273, 277.

41. Innis, 348.

42. Wells, 106.

43. Innis, 203, 260.

44. L.K. Ingersoll, *Report by Captain John Robb, R.N., on the state of the fisheries, the condition of the lighthouses, the contraband trade, and various other matters in the Bay of Fundy, made to His Excellency the Lieutenant-Governor, 1840* (Grand Manan Historical Society, 1965).

45. Innis, 346.

46. Innis, 331, cf. also 346.

47. Innis, 332–3.

48. Innis, 287.

49. Cf. Roch Samson, *Fishermen and Merchants in 19ᵗʰ Century Gaspé* (Ottawa: Parks Canada, 1984).

50. Innis, 278–9, 282–3.

51. Innis, 356; Perley, 39.

52. Innis, 277–8.

53. L. Bérubé, *Coup d'œil sur les pêcheries du Québec* (Sainte-Anne-de-la-Pocatière, Que.: École supérieure des pêcheries, 1941), 10–5.

54. Perley, 1, 51, 37–8, 31–2.

55. Innis, 264–6, 347.

56. Perley, 16.

57. L.W. Scattergood and S.N. Tibbo, *The Herring Fishery of the Northwest Atlantic*, Bulletin 121 (Ottawa: Fisheries Research Board of Canada, 1959).

58. James L. Warren, *Marine Sardine Industry History: An Anthology* (Maine Sardine Council, March 1986), passim.

59. Stephen A. Davis, St. Mary's University, Halifax, N.S., pers. comm.

60. Warren, passim.

61. The above comes mainly from Goode et al., 131, 239; Cicely Lyons, *Salmon, Our Heritage: The Story of a Province and an Industry* (Vancouver, B.C.: British Columbia Packers, 1969), 140; Perley, 20; Charles L. Cutting, *Fish Saving: A History of Fish Processing from Ancient to Modern Times* (London, U.K.: Leonard Hill Limited, 1955), 187–91; A. Gordon DeWolf, *The Lobster Fishery of the Maritime Provinces: Economic Effects of Regulations*, Bulletin 187 (Ottawa: Fisheries Research Board of Canada, 1974).

62. Wells, 112.

63. Cf. Cutting, 187–91.

64. Innis, 239–41, 259–65, 332–6.

65. Innis, 262–4; Balcom, *Lunenburg*, 4.

66. Wells, 109, 119.

67. Dunfield, 50.

68. Innis, 263–6.

69. Wells, 109.

70. E.E. Prince, "Fifty Years of Fishery Administration in Canada," *Transactions of the American Fisheries Society* 50 (1921): 5.

71. Goode, 441.

71. Innis, 332–4, 345–6.

72. Innis, 346.

73. Innis, 344–7; Canada, *Delimitation of the Maritime Boundary in the Gulf of Maine Area (Canada/United States of America). Annexes to the Memorial Submitted by Canada. Vol. 1. Contemporary Treaties Affecting the Northwest Atlantic Fisheries, with a Historical Introduction* (The Hague, Netherlands: International Court of Justice, September 1982), 12.

74. Tariff Commission, 54.

75. *Contemporary Treaties*, 1, 12; Tariff Commission, 42–4.

76. Innis, 348–9.

77. Tariff Commission, 46; *Memorandum,* Appendix 4.

78. Much of the above appears in Innis, 229, 249–59, 266, 337–8; J.B. Brebner, *North Atlantic Triangle: The Interplay of Canada, the United States and Great Britain* (New Haven, Conn.: Yale University Press, 1945), 111.

79. Cf. Philip A. Buckner and John G. Reid, eds., *The Atlantic Region to Confederation: A History* (Toronto, Ont.: University of Toronto Press, 1994), xv.

80. Brebner, *Canada*, 157.

81. Ryan, *Ice*, 39.

82. Ryan, *Fish*, 35, and *Ice*, 40–2.

83. Ryan, *Ice*, 93.

84. Ryan, *Ice*, 54–5; Innis, 303–5, 410–1.

85. Brebner, *Canada*, 158.

86. Prowse, 450–1, 420; Ryan, *Ice*, 52–4; Innis, 303–4, 409–11.

87. Ryan, *Ice*, 104, 445, 457.

88. Ryan, *Ice*, 100–5.

89. Innis, 264.

90. Henry D. Roberts with Michael O. Nowlan, *The Newfoundland Fish Boxes: A Chronicle of the Fishery* (Fredericton, N.B.: Brunswick Press, 1982), 28.

91. Prowse, 403–4.

92. G.M. Story, W.J. Kirwin, and J.D.A. Widdowson, eds., *Dictionary of Newfoundland English* (Toronto, Ont.: University of Toronto Press, 1982).

93. Hind, part 1, 163.

94. Ryan, Fish, 236–40; Innis, 301.

95. Innis, 307–10; Ryan, *Fish*, 240.

96. Brebner, *Canada*, 158.

97. Innis, 309.

98. Hind, part 1, 163; Balcom, *Lunenburg*, 4.

99. Innis, 273, 277, 413.

100. Innis, 308–9, 385.

101. Innis, 209, 308–9, 406–7.

102. Michael E. Condon, *The Fisheries and Resources of Newfoundland: The Mine of the Sea* (St. John's, Nfld., 1925).

103. Innis, 387.

104. Cf. D.A. Macdonald, "Social History in a Newfoundland Outport: Harbour Breton, 1850–1900," in *How Deep Is the Ocean?* Historical Essays on Canada's Atlantic Fishery, edited by James E. Candow and Carol Corbin (Sydney, N.S.: University College of Cape Breton Press, 1997), 98.

105. Innis, 306–7, 386–7; Prowse, 379.

106. Innis, 306.

107. Hind, part 1, 62.

108. Innis, 355, 391–3.

109. Hind, part 1, 81, 84.

110. W. Templeman, *The Newfoundland Lobster Fishery: An Account of Statistics, Methods, and Important Laws* (St. John's, Nfld.: Department of Natural Resources, 1941), 5.

111. Tonnessen et al., 4–5; Newfoundland Department of Marine and Fisheries, *Annual Report* for 1902, 6–7; Prowse, 298–9.

112. Commissioner of Crown Lands for the Province of Canada, *Annual Report* for 1858, 78.

113. Innis, 302, 308.

114. Innis, 386.

115. See Prowse, 431; Innis, 384–7.

116. Innis, 378, 393.

117. See Innis, 355, 391–2, 405.

118. Brebner, *Triangle*, 98.

119. Cf. Lyons, 43, 50; Shepard and Argue.

Chapter 5 Notes

1. Buckner and Reid, xii–xiii.

2. Cf. Ryan, *Fish*, 258–9.

3. The direct quote above and most other statistics come from Hind, part 1, 124, 164–6, part 2, 63–5. See also Department of Secretary of State, *Return correspondence between the Government of the Dominion and the Imperial Government on the subject of the fisheries, with other documents relating to the same, laid before the Honourable the House of Commons* (Ottawa, 1871).

4. Cf. Hind, part 1, iv–v.

5. Perley, 1.

6. D.A. Sutherland, "1810–1820: War and Peace," in *The Atlantic Region to Confederation: A History*, edited by Philip A. Buckner and John G. Reid (Toronto, Ont.: University of Toronto Press, 1994), 247.

7. Perley, 93.

8. Perley, 112.

9. Innis, 331, 350; Tariff Commission, 49 ff.; Wells, 117–9.

10. Tariff Commission, 49 ff.; W.L. Morton, *The Kingdom of Canada* (Toronto, Ont.: McClelland & Stewart, 1963), 291.

11. Innis, 350.

12. Goode et al., 251–3.

13. Ryan, *Fish*, 85.

14. Brebner, *Triangle*, 155–6.

15. Innis, 351.

16. Cf. Goode et al., 345–6.

17. MacNutt, 250. Relevant material appears in Innis, 360–1; T.F. Knight, *Report on the Fisheries of Nova Scotia* (Halifax, N.S.: A. Grant, Printer to the Queen, 1867).

18. Wells, 122; Innis, 361–2.

19. Cf. Innis, 323–4.

20. Innis, 331; German.

21. Innis, 324.

22. Perley, 15.

23. Hind, part 1, 88.

24. Goode, 440–1.

25. Hind, part 1, 121.

26. United States, Secretary of the Treasury, *Reply of the Secretary of the Treasury to the resolution of the House of Representatives of December 14, 1886* (Washington, D.C.: Government Printing Office, 1887), 54–5.

27 Cf. Hind, part 1, 59–60, 81–2; Innis, 329.

28. Innis, 329.

29. Department of Marine and Fisheries, *Annual Report* for 1867, 13.

30. Cf. Goode, 440–2; Hind, part 1, 44–5, 58–9, 119.

31. Knight, *Report on the Fisheries*, 6.

32. Department of Marine and Fisheries, *Annual Report* for 1867. See also Pierre Fortin, in *Fisheries appendices from the annual report, for 1863, of the Hon. Wm. McDougall, Commissioner of Crown Lands: statements of Fisheries Branch, Superintendents' reports, Capt. Fortin's report, synopsis of fishery overseers' reports, etc. for the year 1863* (Quebec: Hunter, Rose & Co., 1864) (in the library of the Department of Fisheries and Oceans, Ottawa, Cat. No. 223025); Hind, part 1, 163; Innis, 401 ff.

33. Perley, 11.

34. Perley, 60, 25.

35. Perley, 21.

36. Perley, 20, 97.

37. Perley, 22, 24, 5.

38. Perley, 46.

39. Perley, 98, 176.

40. Department of Marine and Fisheries, *Annual Report* for 1867, 14.

41. Perley, 16, 176.

42. Knight, *Report on the Fisheries*, part 2, 16–7.

43. Perley, 23, 22, 7.

44. Perley, 166, 176.

45. Perley, 93.

46. Perley, 104–5.

47. Perley, 39–40.

48. Perley, 18–20.

49. Pierre Fortin, 1859 report, quoted in Prince, "Fifty Years," 167. See also Fortin, in *Fisheries Appendices*; Tétu, in Department of Marine and Fisheries, *Annual Report* for 1867, 24.

50. Perley, 217.

51. I owe most of the above to Carman Bickerton, *A History of the Canadian Fisheries in the Georges Bank Area*, vol. 2 of the Canadian Counter-Memorial submitted to the International Court of Justice in the Delimitation of the Maritime Boundary in the Gulf of Maine area (Ottawa, 1983), 141–2. See also Balcom, *Lunenburg*, 14; Innis, 342, 347, 356.

52. Cf. Prince, "Fifty Years," 165.

53. Perley, 64, 89.

54. See Knight, Report on the Fisheries. Cf. also T.F. Knight, *Shore and Deep Sea Fisheries of Nova Scotia* (Halifax, N.S.: A. Grant, Printer to the Queen, 1867) and *The River Fisheries of Nova Scotia* (Halifax, N.S.: 1867).

55. Perley, 21.

56. Perley, 89, 74.

57. Perley, 86, 292; C.S. Juvet, *Licensing Policies and Procedures: Confederation–1949* (compilation in the library of the Department of Fisheries and Oceans, Ottawa, 1950, Cat. No.

208624), c. 2, 126.

58. New Brunswick Department of Natural Resources, Web-site: www.gnb.ca/0078/auction-e.asp, March 2003,

59. Perley, 128–9.

60. Department of Marine and Fisheries, *Annual Report* for 1873.

61. Perley, 177, 262, 289 ff.

62. Cf. Department of Marine and Fisheries, *Annual Report* for 1889, 101, Annual Report for 1889, 103; Prince, "Fifty Years," 165–6.

63. See Knight, *Report on the Fisheries*, 1, 23; Prince, "Fifty Years," 165; Prince calls the 1853 act "a failure."

64. Balcom, 15.

65. Innis, 375–7; Department of Marine and Fisheries, first *Annual Report*, 13; Knight, *Report on the Fisheries*, 9.

66. Innis, 355; Knight, *Report on the Fisheries*, 9.

67. Balcom, *Lunenburg*, 9; Prince, "Fifty Years," 165.

68. Goode et al., 413–5.

69. Cf. Perley, 8–9.

70. Table is from Hind, part 2, 64.

71. Hind, part 1, 169.

72. Cf. Innis, 358, 396–401; Perley, 8-9.

73. Cf. Innis, 409–11; Hind, part 2, 71.

74. Prowse, 602–3.

75. Ryan, *Ice*, 169. I owe most of this section to the same work, especially pages 112–5, 151, 156–7, 161–5, 169, 445, 457, 467. See also Hind, part 2, 60; Innis, 411.

76. Ryan, *Ice*, 329–39, 349; Smallwood, *Book of Newfoundland*, vol. 2, 1937, 100.

77. Ryan, *Ice*, 342, 350.

78. Smallwood, *Book of Newfoundland*, vol. 2, 192.

79. Innis, 398.

80. Innis, 413.

81. Ryan, *Fish*, xxiv.

82. Innis, 395, 406–7.

83. Ryan, *Ice*, 407.

84. Innis, 386–90.

85. Innis, 396, 399.

86. Hind, part 1, 107, 165.

87. Hind, part 1, 118; Macdonald, 92.

88. Hind, part 1, 125, 58.

89. Innis, 408–9; Balcom, *Lunenburg*, 11.

90. Innis, 396–7, 413.

91. Meggs, 16–7.

92. Cf. Lyons, 137–40.

93. *Report of the Select Committee on the Working of the Fishery Act* (Province of Canada, 1864).

94. Margaret Beattie Bogue, *Fishing the Great Lakes: An Environmental History, 1783–1933* (Madison, Wis.: University of Wisconsin Press, 2000), 21.

95. Cf. E.J. Chambers, *The Canadian Marine: A History of the Department of Marine and Fisheries* (Toronto, Ont., 1905), 113–4.

96. Bogue, 177–8.

97. Innis, 357–60, 407, 411–2.

98. Department of Marine and Fisheries, *Annual Report* for 1867, 25, 27.

99. Cf. A.B. McCullough, *The Commercial Fishery of the Canadian Great Lakes* (Ottawa: Environment Canada, Canadian Parks Service, 1989), 20–5; Prince, "Fifty Years"; Kenneth Johnstone, "History of Canadian Fisheries and Marine Service," historical manuscript (in the library of the Department of Fisheries and Oceans, Ottawa, 1977), 1–14; Compilation 1860, *Acts respecting fisheries, and the inspection of fish and oil: also fisheries regulations for Lower Canada* (Quebec: Stewart Derbishire & George Desbarats, 1860) (in the library of the Department of Fisheries and Oceans, Ottawa, Cat. No. 192545).

100. McCullough, 20.

101. Commissioner of Crown Lands, *Annual Report* for 1857, 58–66.

102. Commissioner of Crown Lands, *Annual Report* for 1858, 97.

103. Commissioner of Crown Lands, *Annual Report* for 1858, 11–2.

104. Commissioner of Crown Lands, *Annual Reports* for 1858 and 1859; Compilation 1860.

105. *Commissioner of Crown Lands* (1858), 70–1.

106. McCullough, 21.

107. Cf. Province of Canada, in *Fisheries appendices from the annual report, for 1863, of the Hon. Wm. McDougall, Commissioner of Crown Lands: statements of Fisheries Branch, Superintendents' reports, Capt. Fortin's report, synopsis of fishery overseer' reports, etc. for the year 1863* (Quebec: Hunter, Rose & Co., 1864) (in the library of the Department of Fisheries and Oceans, Ottawa; Cat. No. 223025); *Report of the Select Committee* (1864), d. Readers should note that this outline of the evolution of licensing power is the author's own, without benefit of legal advice.

108. N.R. MacCrimmon, J.E. Stewart, and J.R. Brett. *Aquaculture in Canada: The Practice and the Promise* (Ottawa: Environment Canada, Fisheries and Marine Service, 1974), 3.

109. Commissioner of Crown Lands, *Annual Report* for 1857.

110. Prince, "Fifty Years," 166.

111. Cf. Chalmers, 116; Johnstone, historical manuscript.

112. The above comes from Compilation 1860; *Report of the Select Committee* (1864); Compilation 1873, *The fishery acts: Dominion of Canada: consisting of: "An Act for the Regulation of Fishing and Protection of Fisheries"; "An Act respecting Fishing by Foreign Vessels," passed by the Parliament of Canada in 1868, applicable generally to the Fisheries of Canada; also certain*

Provincial Statutes continued in force in the provinces of Nova Scotia and New Brunswick (Ottawa: I.B. Taylor, 1873) (in the library of the Department of Fisheries and Oceans, Ottawa, Cat. No. 192538).

113. See Compilations 1860 and 1873 (early acts and regulations).

114. Goode, 123, 138–9.

115. Innis, 326.

116. Innis, 412–3.

117. U.S. Secretary of the Treasury, xv, 54–5; however, H.Y. Hind reported about a 50 per cent reduction over the whole period.

118. Innis, 327, 413.

119. U.S. Secretary of the Treasury, 54; Innis, 330–1.

120. Hind, part 1, v. My own great-grandfather, a Maritimer, fished at times out of Gloucester. Twice he went aboard a schooner for a voyage, didn't like the looks of the vessel, and took his seabag back ashore. Both vessels sank.

121. Hind, part 1, 169.

122. See de la Morandière.

123. Portuguese government information for Canadian Museum of Civilization, *Lifelines* exhibit (2001).

124. *Memorandum.*

125. Cf. Innis, 362; Department of Marine and Fisheries, *Annual Report* for 1867, appendix by a Mr. Johnston; Knight, *Report on the Fisheries.*

126. Cf. Brebner, *Triangle*, 182.

127. Reid et al., 205.

128. Kenneth J. Pryke, *Nova Scotia and Confederation, 1864–1874* (Toronto, Ont.: University of Toronto Press, 1979), 41–2. Standard accounts of pre-Confederation talks give little attention to fishery influence. Harold Innis, however, states that Nova Scotia's struggle against New England, France, and Newfoundland—struggles mainly connected with the fisheries—drove her to Confederation, with its greater power to make fishery regulations unimpeded by Great Britain.

129. Pryke, 24–5; Innis, 352.

130. Secretary of State, 1–3.

131. G.P. Brown, *Documents on the Confederation of British North America* (Toronto, Ont.: McClelland & Stewart, 1969), 192.

132. See Pryke, 41–2; Reid et al., 284, ff.

133. I owe the above largely to Pryke, 41–2; Innis 353–4; Goode et al., 290; Johnstone, historical manuscript; Brown, *Documents on the Confederation.*

134. Innis, 356, 364, 372–3.

135. Brebner, *Triangle*, chapter titled "Triune Understanding."

136. Innis, 385.

Chapter 6 Notes

1. Knight, *Report on the Fisheries*, 14, 33; Department of Marine and Fisheries, *Annual Report* for 1867, appendix by Johnston.

2. Department of Marine and Fisheries, Annual Report for 1868, 14–27.

3. Cf. Thomas E. Appleton, *Usque ad Mare: A History of the Canadian Coast Guard and Marine Services* (Ottawa: Department of Transport, 1968), passim. The reference to Smith's lack of popularity comes from an article by Appleton: "Our First Deputy Minister," for which I lack the full reference.

4. Prince, "Fifty Years," 167.

5. Department of Marine and Fisheries, *Annual Report* for 1868, 11.

6. Prince, "Fifty Years," 169–71.

7. See James White, "The North Atlantic Fisheries Dispute," in *Commission of Conservation Canada*, 1911 (Ottawa: The Mortimer Co. Ltd., 1911), 73; Innis, 364–5; Department of Secretary of State, 19.

8. Cf. Brebner, *Triangle*, 182–97.

9. R.S. Longley, "Peter Mitchell, Guardian of the North Atlantic Fisheries, 1867–1871," *Canadian Historical Review* 22 (1941): 390–2.

10. Department of Secretary of State, 56 ff., 64; Innis, 365.

11. Brebner, *Triangle*, 186; cf. Department of Marine and Fisheries, *Annual Report* for 1867.

12. Brebner, *Triangle*, 186.

13. Longley, 392–5; cf. Brebner, *Triangle*, chapter titled "Triune Understanding."

14. See Innis, 361–6; Brebner, *Triangle*, 186; 1898 *Memorandum.*

15. The above relies mostly on Brebner, *Triangle*, 186–96; Longley, 395–7; Innis, 366–7; Department of Secretary of State, 35.

16. Brebner, *Triangle*, 187.

17. Innis, 368.

18. Tariff Commission, 71.

19. T. Hodgins, "The Coercion of Newfoundland," in *Diplomatic North American Fisheries* (bound collection of articles, 1908–1910, in the library of the Department of Fisheries and Oceans, Ottawa).

20. Innis, 369.

21. Cf. Brebner, *Canada*, 295–7, and *Triangle*, chapter titled "Triune Understanding."

22. Cf. Innis, 370 ff.

23. See Carl M. Wallace, "Smith, Albert J.," in *The Canadian Encyclopedia*, 2nd ed. (Edmonton, Alta.: Hurtig Publishers, 1988).

24. Bogue, 210.

25. Brian Richman, Department of Fisheries and Oceans, pers. comm.

26. Most of the above comes from Department of Marine and Fisheries, *Annual Reports* for 1868, 1871, and 1880.

27. Department of Marine and Fisheries, *Annual Report* for 1868, 18, 25; cf. 34–5.

28. Compilation 1873.

29. Department of Marine and Fisheries, *Annual Report* for 1868, 19.

30. N.H. Morse and A.G. DeWolf, *Management of the Fisheries of the Saint John River System, New Brunswick, with Particular Reference to Atlantic Salmon* (Canada: Fisheries and Marine Service, 1974), 4.

31. Department of Marine and Fisheries, *Annual Report* for 1874.

32. Tariff Commission, 85.

33. Innis, 370–2. For Canadian reaction, see Department of Secretary of State, 77 ff.

34. United States, Secretary of the Treasury 55–60.

35. W.S. MacNutt, "The 1880's," in *The Canadians*, part 1, edited by J.M.S. Careless and R. Craig Brown (Toronto, Ont.: Macmillan, 1967).

36. Cf. Richard H. Gimblett, "Reassessing the Dreadnought Crisis of 1909 and the Origins of the Royal Canadian Navy," *The Northern Mariner / Le Marin du nord* 4, no. 1 (1994): 35–53.

37. 1898 *Memorandum*, 85; Department of Secretary of State 195 ff., 202 ff.

38. United States, Secretary of the Treasury, xiii.

39. Brebner, *Triangle*.

40. United States, Secretary of the Treasury, xvi.

41. Innis, 422.

42. Innis, 421; Department of Marine and Fisheries, *Annual Report* for 1888, 9 ff.

43. Innis, 423.

44. Much of the above comes from J. Tomasevich, *International Agreements on Conservation of Marine Resources* (Stanford, 1943).

45. Bogue, 217.

46. Bogue, 145, 227, 231.

47. Bogue, 309–10.

48. Bogue, 212, 240.

49. Bogue, 241–9.

50. *Dominion–British Columbia Fisheries Commission* 1905–1907, 33.

51. Bogue, 312–20.

52. Brian Richman, Department of Fisheries and Oceans, pers. comm.

53. Most of the above comes from Gimblett and from Nigel D. Brodeur, "L.P. Brodeur and the Origins of the Royal Canadian Navy," in *RCN in Retrospect*, 1910–1968, edited by James A. Boutilier (Vancouver, B.C.: University of British Columbia Press, 1982).

54. Appleton, 61–3.

55. Appleton, 84–5.

56. Department of Marine and Fisheries, *Annual Report* for 1886, 322.

57. Department of Marine and Fisheries, *Annual Reports* for 1886, 1888, and 1890. On these matters, cf. also Prince, "Fifty Years," 182; Ruth F. Grant, *The Canadian Atlantic Fishery* (Toronto, Ont.: Ryerson Press, 1934), 128.

58. Cf. Department of Marine and Fisheries, *Annual Report* for 1890, part 2, suppl. 1, 17.

59. Cf. Department of Marine and Fisheries, *Annual Report* for 1880, 17–8, 39–40.

60. Much of the above comes from Department of Marine and Fisheries, *Annual Report* for 1898; D.B. Quayle, *Pacific Oyster Culture in British Columbia*, Bulletin 169 (Ottawa: Fisheries Research Board of Canada, 1969), passim; W.B. Scott and E.A. Crossman, *Freshwater Fishes of Canada*, Bulletin 184 (Ottawa: Fisheries Research Board of Canada, 1973).

61. Cf. Kenneth Johnstone, *The Aquatic Explorers: A History of the Fisheries Research Board of Canada* (Toronto, Ont.: University of Toronto Press, 1977), 23–32.

62. Bogue, 303–4.

63. Johnstone, *Aquatic Explorers*, 17.

64. Brian Richman, Department of Fisheries and Oceans, pers. comm.

65. Department of Marine and Fisheries, *Annual Report* for 1891, lxiv ff.

66. Compilation 1894, *Consolidated Fishery Regulations: Canada 1889* (Ottawa: 1894, 1890) (in the library of the Department of Fisheries and Oceans, Ottawa, Cat. No. 192507).

67. E.E. Prince, *Fluctuations in the Abundance of Fish, in Special Reports of Professor E.E. Prince, 1894–1909* (Ottawa: Commissioner of Fisheries, 1894, 1909) (compilation in the library of the Department of Fisheries and Oceans, Ottawa, Cat. No. 192111); Department of Marine and Fisheries, *Annual Report* for 1898, lxii.

68. L.S. Parsons, *Management of Marine Fisheries in Canada* (Ottawa: National Research Council, 1993), 62–3.

69. Cf. Parsons, 58–76.

70. Cf. Department of Marine and Fisheries, *Annual Report* for 1890, Appendix 7.

Chapter 7 Notes

1. Bogue, 206–7, 223–4, 226.

2. Chambers, 115–8.

3. Bogue, 45, 54–6, 69–73, 78–9, 272.

4. McCullough, 85. McCullough's *Commercial Fishery of the Canadian Great Lakes* and Bogue's *Fishing the Great Lakes* not only illuminate the history of the Great Lakes fishery, but also provide incidents and insights relevant to Canadian commercial fisheries in general.

5. McCullough, 98.

6. Compilation 1894.

7. Prince, "Fifty Years."

8. Bogue, 235.

9. Bogue, 306–7, 319; McCullough, 96–7.

10. S.V. Ozere, "Survey of Legislation and Treaties Affecting Fisheries," in *Resources for Tomorrow: Montreal, October 22–28, 1961, vol. 2, Conference Background Papers* (Ottawa: Queen's Printer, 1961).

11. Provincial roles in more detail as of the early 21st century: In Newfoundland, the province

licenses freshwater fisheries. In the Maritimes, the provinces manage and license freshwater fisheries; they also license fishing for anadromous species in inland waters. Quebec manages and licenses freshwater, anadromous, and catadromous species. Ontario, Manitoba, and Alberta manage and license freshwater species; the federal government manages marine species in Ontario and Manitoba. Saskatchewan manages its fisheries and also deals with some legislation, except with regard to fish habitat and Aboriginal fishing. British Columbia manages and licenses freshwater species, and licenses inland salmon sport-fishing. The Yukon manages freshwater fisheries. In Nunavut and the Northwest Territories, co-management boards deal with fishery allocations and advise the Department of Fisheries and Oceans (D.F.O.) on conservation, science, and management. Nunavut and the N.W.T. also administer sport-fish licensing. (This information is from a D.F.O. brochure posted on the Internet as of 2003: "Recreational Fisheries in Canada—an Operational Policy Framework.")

12. Dave Dunn, Department of Fisheries and Oceans, pers. comm.

13. Bogue, 319–20, 324; cf. McCullough, 97–8.

14. Bogue, 366.

15. Bogue, 163–4, 330.

16. J.H. Leach and S.J. Nepszy, "The Fish Community in Lake Erie," *Journal of the Fisheries Research Board of Canada* 33 (1976).

17. Department of Marine and Fisheries, Annual Reports for 1887 and 1888.

18. Commission of Conservation Canada, 1911; Department of Marine and Fisheries, Annual Report for 1918.

19. Manitoba Fisheries Commission, interim report and final report (1910–1911) (in the library of the Department of Fisheries and Oceans, Ottawa, Cat. No. 34236), covered Saskatchewan and Alberta also.

20. *Commission of Conservation Canada, 1911.*

Chapter 8 Notes

1. A.G. Huntsman, "Edward Ernest Prince," *Canadian Field-Naturalist* 59, no. 1 (1945).

2. Hind, passim.

3. Cf. Bogue, 88–9.

4. Department of Marine and Fisheries, *Annual Reports* for 1917 and 1918.

5. H.A. Innis, "History of Commercial Fisheries," in *Encyclopedia Canadiana* (Toronto, Ont.: Grolier of Canada, 1968); see W.C. MacKenzie, "Modern Commercial Fisheries," in the same volume.

6. Innis, 428–9, 433.

7. Innis, 172.

8. The above draws heavily on Balcom, *Lunenburg.*

9. Canadian Museum of Civilization, Parks Canada, and IDON EAST, *Balancing the Scales:*

Canada's East Coast Fishery, CD-ROM (1999).

10. Department of Marine and Fisheries, *Annual Report* for 1917, 16, 28, 35–6; also *Annual Report* for 1918, 26.

11. *Commission of Conservation Canada, 1911.*

12. Hind, part 1, 199; Department of Secretary of State, 56.

13. Hind, 59–60.

14. Warren, passim.

15. The above comes largely from Warren, passim.

16. Hind, 136–9.

17. Compilation 1873.

18. Department of Marine and Fisheries, Annual Report for 1890, lxxiii; cf. *Annual Report* for 1881, xlv–xlvii.

19. E.E. Prince, *Dominion Shad Fishery Commission, 1908–1910: Report and Recommendations* (Ottawa: Government Printing Bureau, 1910), xi–xviv, xlix.

20. Department of Marine and Fisheries, *Annual Report* for 1873.

21. Joncas; Wells, 134; *Report of the Canadian Lobster Commission* (Ottawa: S.E. Dawson, 1899), 23.

22. DeWolf, 18.

23. Department of Marine and Fisheries, *Annual Report* for 1890, part 2, suppl. 1, 26–7; cf. Prince, "Fifty Years", 171.

24. DeWolf, 21.

25. The above picture emerges from DeWolf, annual reports of the period, and *Report of Commander Wm. Wakeham, Special Commissioner and Inspector of Fisheries for the Gulf of St. Lawrence, in the Lobster Industry of the Maritime Provinces and the Province of Quebec* (1909–1910).

26. N. Bourne, *Scallops and the Offshore Fisheries of the Maritimes*, Bulletin 145 (Ottawa: Fisheries Research Board of Canada, 1964), 19.

27. Most of the above comes from *Commission of Conservation Canada, 1911;* departmental annual reports; and *Dominion Shell-fish Fishery Commission: Report and Recommendations* (Ottawa, 1913).

28. E.E. Prince, *Unutilized Fishery Products in Canada* (Ottawa: Government Printing Bureau, 1907) and The *Dog-fish Plague in Canada* (Ottawa: Government Printing Bureau, 1904), both available in the library of the Department of Fisheries and Oceans, Ottawa.

29. *Balancing the Scales.*

30. *Commission of Conservation Canada, 1911,* 42.

31. Department of Marine and Fisheries, *Annual Report* for 1890, Appendix 2.

Chapter 9 Notes

1. See Innis, 413; Innis, *Commercial Fisheries.*

2. Innis, 329–30.

3. Relevant information appears in Innis,

329–30, 414; Tariff Commission, 86; C. Isham, *The Fishery Question: Its Origin, History and Present Situation: With a Map of the Anglo-American Fishing Grounds and a Short Bibliography* (New York: G.P. Putnam's Sons, 1887).

4. Cf. Innis, 354, 400, 443–6, 450; White; Newfoundland, *Report of the Commission of Enquiry Investigating the Seafisheries of Newfoundland and Labrador other than the Sealfishery* (St. John's, Nfld., 1937), 15–6.

5. Innis, 451–4.

6. Cf. Tariff Commission, xxi–xxiii; and White.

7. Innis, 399.

8. Innis, 444–6.

9. Innis, 446–9.

10. Cf. Prowse, 453; Ryan, *Fish*, 266–7; Ryan, *Ice*, 108, 318; Newfoundland, *Commission of Enquiry*; and Innis, passim.

11. Innis, 410, 446; and Albert Whiteley, *A Century on Bonne Esperance* (privately printed, no date).

12. For the above, see Innis, 374, 396–8, 456; Ian McDonald, *W.F. Coaker and the Fishermen's Protective Union in Newfoundland Politics 1908–1925* (Ph.D. diss., University of London, 1971), 16–28.

13. John McCormack, St. Joseph's, St. Mary's Bay, Nfld., pers. comm.

14. Peter Cashin, *My Life and Times, 1890-1919* (Portugal Cove, Nfld.: Breakwater Books, 1976), 47.

15. Newfoundland, *Commission of Enquiry*.

16. Innis, 396–7.

17. Newfoundland, *Annual Report* for 1902.

18. Department of Secretary of State, 56.

19. Newfoundland, *Annual Report* for 1904, 26.

20. Newfoundland, *Annual Report* for 1900, 16; Dunfield, 162.

21. Most of the above comes from Templeman, *Lobster*, especially 5–7, 28–9.

22. Most of the above comes from Shannon Ryan, *A Historical Overview of Canadian Newfoundland/World Sealing* (prepared for the Royal Commission on Seals and the Sealing Industry in Canada, 1986), passim; Ryan, *Ice* (esp. 113–7, 165–8); and *Balancing the Scales*.

23. Cf. *Balancing the Scales*.

24. Tonnessen et al., 24–5, 33–4.

25. Tonnessen et al., 5–6.

Chapter 10 Notes

1. Lyons, 137–40.

2. Meggs, 22.

3. Cf. J.L. Hart, F. Neave, and D.B. Quayle, *Brief on the Fishery Wealth of British Columbia* (Ottawa: Fisheries Research Board of Canada, 1951).

4. Meggs, 23.

5. Lyons, 171, 180, 185; Department of Marine

and Fisheries, *Annual Report* for 1887, 246.

6. Cf. Department of Marine and Fisheries, *Annual Reports* for 1917–1919.

7. Department of Marine and Fisheries, *Annual Reports* for 1826, 1887, and 1888.

8. Newell, 63.

9. Department of Marine and Fisheries, *Annual Report* for 1889, 243, 249–50.

10. Meggs, 55–6.

11. Peter H. Pearse, *Turning the Tide: A New Policy for Canada's Pacific Fisheries* (Vancouver: Commission on Pacific Fisheries Policy, 1982), 176. (This report is known as the Pearse report.)

12. Cf. Meggs, 74–80; Lyons, 255–7. Helgesen memo courtesy of federal fisheries alumnus Ron MacLeod.

13. Compilation 1873; Lyons, 149.

14. Meggs, 22; Lyons, 192.

15. Cf. Meggs, 30–4; B.A. Campbell, *An Historical Review of Developments in Salmon Licensing up until 1968*, internal paper (Ottawa: Fisheries and Marine Service, Department of the Environment, 1973), 3–4; British Columbia, *Fishery Commission Report, 1892*, passim; Department of Marine and Fisheries, Annual Reports for 1887–1890.

16. B.C. Fishery Commission (1892), 123.

17. Much of the above comes from Meggs, 34–8.

18. Lyons, 706.

19. Cf. Meggs, 38–40, 62.

20. *Dominion–British Columbia Fisheries Commission* (1905–1907), 18; Department of Marine and Fisheries, *Annual Report* for 1906, lxi.

21. Lyons, 248.

22. W.C. Duff, *British Columbia Fisheries Commission* (Ottawa: King's Printer, 1923) (the Duff report), 21–3; Meggs, 89.

23. Compilation 1906: see part 3, 149.

24. See Meggs, 89–103; Juvet, 153, 161.

25. Meggs, 72.

26. Meggs, 43.

27. Campbell, 5; Lyons, 270.

28. W.S. Evans, H.B. Thomson, and F.T. James, *Report of Special Fishery Commission, Province of British Columbia, 1917* (Ottawa, 1918) (the Evans report), 24–5; Department of Marine and Fisheries, *Annual Report* for 1917, 10; Campbell, 8.

29. Cf. Department of Marine and Fisheries, *Annual Report* for 1917, 16, 33; Campbell, 5.

30. Cf. Department of Marine and Fisheries, *Annual Report* for 1917, 10, 24–5; Meggs, 89; Lyons, 287; Campbell, passim.

31. Lyons, 175; Meggs, 86.

32. See Evans et al., 36.

33. Lyons, 577.

34. Juvet.

35. Duff, 22; cf. Meggs, 89–103; Juvet, 153, 161; Compilation 1906.

36. I owe the above mainly to Department of Marine and Fisheries, *Annual Reports* for 1888 and

1889; Lyons, 716–7; J.A. Crutchfield, *The Pacific Halibut Fishery, in The Public Regulation of Commercial Fisheries in Canada*, edited by A. Scott and P.A. Neher, Ottawa: Economic Council of Canada, 1981); Frederick William Wallace, *Roving Fisherman* (Gardenvale, Quebec: Canadian Fisherman, 1955), 256–7.

37. *Commission of Conservation Canada*, 1911,
38. Peter Rider, Canadian Museum of Civilization, provided the information on Newfoundland whalers in British Columbia.

38. Meggs, 42.

39. Meggs, 60–4.

40. I owe most of the above to A.V. Hill, *Tides of Change: A Story of Fishermen's Co-operatives in British Columbia* (Prince Rupert Fishermen's Co-operative Association, 1967); G. North and H. Griffin, *A Ripple, A Wave: The Story of Union Organization in the B.C. Fishing Industry* (Vancouver: Fisherman Publishing Society, 1974); Lyons, passim; and Meggs, passim.

Chapter 11 Notes

1. For a summary of laws and regulations in 1911, see *Commission of Conservation Canada*, 1911.

2. *Dominion–B.C. Fisheries Commission*, 46.

3. Bogue, 229.

4. Hind, part 1, 145.

Chapter 12 Notes

1. Department of Marine and Fisheries, *Annual Report* for 1917, 34–5.

2. *Commission of Conservation Canada*, 1915 (Toronto, Ont.: The Methodist Book and Publishing House, 1916).

3. Ron MacLeod, former federal fisheries official, pers. comm.

4. Ken Johnstone interview with former federal fisheries official John Lamb.

5. Department of Marine and Fisheries, *Annual Report* for 1912, 23.

6. Brian Richman, former federal fisheries official, pers. comm.

7. Eddie Moore, former fishery officer, pers. comm.

8. Department of Marine and Fisheries, *Annual Report* for 1919.

9. Evans

10. Richard W. Parisien, The Fisheries Act: *Origins of Federal Delegation of Administrative Jurisdiction to the Provinces*, report submitted to Environment Canada (Ottawa: Policy, Planning, and Research Service, 1972).

11. Department of Fisheries and Oceans, "Major federal development plan for the Quebec fisheries," news release (Ottawa, 11 July 1983).

12. Internal files provided by John Emberley, then of the Department of Fisheries and Oceans.

13. Innis, 429–30; Bérubé, 25 ff.

14. Lyons, 387–9.

15. Lyons, 389.

16. Emberley files; Ozere.

17. Cf. Lyons, 692–3.

18. Johnstone, historical manuscript, 12–8.

19. Eddie Moore, former fishery officer, pers. comm.

20. Ken Johnstone interviews with former federal officials John Lamb and Henry Mitchell.

21. Ron MacLeod, former federal fisheries official, pers. comm.

22. Most of the above comes from a Position Record Book provided me by former federal fisheries official Phil Murray.

23. I owe much of the above to R.N. Wadden, *Department of Fisheries of Canada 1867–1967* (reprinted from Fisheries Council of Canada, *Annual Review* for 1967; W.C. MacKenzie, former federal fisheries official, pers. comm.

24. Department of Marine and Fisheries, Annual Report for 1919, 22–3; Johnstone, historical manuscript, 12–8.

25. Duff, 26.

26. The above comes mostly from *The Fisheries Research Board of Canada and Its place in Canada's Scientific Development 1968–1978* (Ottawa: Fisheries Research Board of Canada, 1968); Johnstone, historical manuscript; H.B. Hachey, History of the Fisheries Research Board, Manuscript Report 843 (Ottawa: Fisheries Research Board of Canada, 1965).

27. Interview with A.G. Huntsman by J.L. Hart of the Fisheries Research Board.

28. Alfred Needler, former federal fisheries official, pers. comm.

29. Hachey (1965), manuscript, passim; Johnstone, Aquatic Explorers, passim.

30. J.L. Hart interview with A.G. Huntsman.

31. This information in part from W.E. Ricker, *The Fisheries Research Board of Canada—Seventy-five Years of Achievements* (Ottawa: Fisheries Research Board of Canada, 1975).

32. Johnstone, historical manuscript; Hachey (1965), manuscript, passim.

33. W.R. Martin, *A Summary Report of the Atlantic Groundfish Investigation* (St. Andrews, N.B.: Atlantic Biological Station, 1949), 10.

34. See Peter Jangaard article in *Fisheries of Canada* (Department of Fisheries periodical, April 1969).

35. Ricker, Fisheries Research Board.

36. Tariff Commission, 159; cf. Innis, 424.

37. Department of Marine and Fisheries, *Annual Report* for 1917, 8.

38. Tariff Commission, 150 ff; Innis, 422; MacKenzie, "Modern Commercial Fisheries."

39. Most of the above comes from Tariff Commission, 150–60; Innis, 423–43; MacKenzie,

"Modern Commercial Fisheries"; Stewart Bates, *Report on the Canadian Atlantic Sea-fishery* (Halifax, N.S.: Royal Commission on Provincial Development and Rehabilitation, 1944), 40.

40. *Report of the American–Canadian Fisheries Conference 1918* (Ottawa, 1918), 29–30.

41. Cf. Tomasevich.

42. I owe most of the above to Tomasevich, passim; Ron MacLeod, pers. comm.; International Pacific Halibut Commission Web site; Crutchfield.

43. The above comes largely from Tomasevich, 240–6, 255–6, 262; Lyons, 240–6, 255–9, 301, 677–8; MacKenzie, "Modern Commercial Fisheries"; M.P. Shepard, *History of International Competition for Pacific Salmon* Manuscript Report 921 (Ottawa: Fisheries Research Board of Canada, 1967); Duff.

44. Cf. *Commission of Conservation Canada*, 1911; American–Canadian conference.

45. Department of Marine and Fisheries, *Annual Report* for 1917, 14–5; *Annual Report* for 1918, 10–1; Evans, 47–8.

46. D. Byrne, "Practical Problems in the Fish Trade," in *Commission of Conservation Canada*, 1915 (Toronto, 1916).

47. Department of Marine and Fisheries, *Annual Report* for 1917, 173.

48. Department of Marine and Fisheries, *Annual Report* for 1918, 17.

49. J.L. Hart interview with A.G. Huntsman.

50. Wells, 161.

51. Except as noted, most of the above comes from Johnstone, historical manuscript, passim; Ken Johnstone's interview with Henry Mitchell; Nova Scotia Economic Council, *Proceedings of the Nova Scotia Fisheries Conference: Halifax, July 13–14, 1938* (Halifax, N.S.: Provincial Secretary, King's Printer, 1938), passim.

52. Cockfield, Brown & Company, *Summary of Report on the Marketing of Canadian Fish and Fish Products* (Ottawa: King's Printer, 1932), 53–4, 66. See also Innis, 442.

53. Cf. Economic Council, 78, 97; John Emberley, internal files, Department of Fisheries and Oceans; Innis, 442.

54. A.G. Huntsman, *The Processing and Handling of Frozen Fish, as exemplified by Ice Fillets*, Bulletin 20 (Biological Board of Canada, Ottawa, 1931).

55. Hart interview with Huntsman.

56. Johnstone, historical manuscript, 16-10/11, 26; John Emberley, internal files, Department of Fisheries and Oceans.

57. Jangaard.

58. Cockfield, Brown & Company.

59. Johnstone, historical manuscript, 15, 9–15; *Halifax Herald*, 22 January 1937.

60. Don Tansley, then of Department of Fisheries and Oceans, pers. comm.

61. Cf. Bérubé, 46 ff.; Bates, 33–4; C.L. Mitchell and H.C. Frick, *Government Programs of Assistance for Fishing Craft Construction in*

Canada: An Economic Appraisal, Canadian Fisheries Report No. 14 (Ottawa: Department of Fisheries and Forestry, 1970), 7.

62. Related information appears in Department of Marine and Fisheries, Annual Report for 1918, 46 ff.; *Commission of Conservation Canada, 1915; Report of Royal Commission Investigating the Fisheries of the Maritime Provinces and the Magdalen Islands* (Ottawa, 1928) (the MacLean report); Department of the Interior, Natural Resources Intelligence Service, The Utilization of Fish By-products (Ottawa, 1928).

63. *Commission of Conservation Canada*, 1911.

64. Hart interview with Huntsman.

65. MacCrimmon.

66. Cf. Ricker, *Seventy-five Years*, 18, and R.E. Foerster, The *Sockeye Salmon*, Oncorhyncus nerka, Bulletin 162 (Ottawa: Fisheries Research Board of Canada, 1968), 387.

67. Ron MacLeod, pers. comm.; cf. Foerster, 394; Lyons, 668.

68. Quayle.

69. Alfred Needler, pers. comm.

70. W.C. MacKenzie, pers. comm.

Chapter 13 Notes

1. Ryan, *Ice*, 310–6.

2. Department of Marine and Fisheries, *Annual Report* for 1919, 31–2.

3. *Commission of Conservation Canada*, 1915, 33–6.

4. *Commission of Conservation Canada*, 1915, 4.

5. Cf. Innis, 423–4; Balcom, *Lunenburg*, 45–6; Department of Marine and Fisheries, *Annual Report* for 1918, 26; Byrne, 4–6.

6. Hédard Robichaud, former fisheries official and federal Minister of Fisheries, pers. comm.

7. Cf. Innis, 423–5, 435, 468–9; Bates, passim.

8. Cf. Bates, passim; W.C. MacKenzie, "Modern Commercial Fisheries"; Innis, 426, 430–2, 442; Balcom, *Lunenburg*; Bérubé, 34 ff.

9. Halifax *Herald*, 3 April 1933.

10. B.A. Balcom, "Technology Rejected: Steam Trawlers and Nova Scotia, 1897–1933," in *How Deep Is the Ocean? Historical Eessays on Canada's Atlantic Fishery*, edited by James E. Candow and Carol Corbin (Sydney, N.S.: University College of Cape Breton Press, 1997)

11. C.R. Forrester and K.S. Ketchen, *A Review of the Strait of Georgia Trawl Fishery*, Bulletin 139 (Ottawa: Fisheries Research Board of Canada, 1963).

12. Related information appears in Innis, 433; Balcom, "Technology Rejected"; Nova Scotia Economic Council.

13. Johnstone, *Aquatic Explorers*, 157–61; W.C. MacKenzie, "Modern Commercial Fisheries"; R.A. MacKenzie, *The Canadian Atlantic Offshore Cod*

Fishery East of Halifax, Bulletin 61 (Ottawa: Fisheries Research Board of Canada, 1942); Balcom, Lunenburg, 48.

14. Alfred Needler, former official and federal deputy minister of fisheries, pers. comm.

15. Harold Thompson, *A Survey of the Fisheries of Newfoundland and Recommendations for a Scheme of Research* (St. John's, Nfld.: Newfoundland Fishery Research Commission, 1931), 50.

16. Bates, 190; Innis, 453.

17. Bates, 14, 28.

18. Halifax *Chronicle*, 30 December 1937.

19. *Canadian Fisherman*, "The National Sea Products Story," (c. 1963).

20. Halifax *Herald*, 30 December 1936; Halifax *Chronicle*, 30 December 1937.

21. Ralph Surette and Alain Meuse, Nova Scotia-based journalists, pers. comm., Nov. 2003.

22. Bates, 43; Innis, 492–3.

23. Nova Scotia Economic Council, 111.

24. Much of the above comes from Bates, 10, 167; Jean Haché, former federal fisheries official, pers. comm.; Halifax *Herald*; 7 January 1937; B.S. Keirstead and M.S. Keirstead, *The Impact of the War on the Maritime Economy* (Halifax, N.S.: Imperial Publishing Company, 1944.)

25. Johnstone, historical manuscript, 14–7.

26. *The Antigonish Movement: Yesterday and Today* (Antigonish, N.S.: St. Francis Xavier University, 1976), 4–5.

27. Innis, 420.

28. Various U.M.F. documents, 1975, 1980.

29. Halifax *Herald*, 29 November 1934.

30. Halifax *Herald*, 13 April 1935; cf. Halifax *Herald*. 12 December 1936.

31. Johnstone, historical manuscript, 15-14/16, 16-4; Innis, 441.

32. Halifax *Herald*, 7 January 1937.

33. W.J. Lever, federal fisheries, pers. comm.

34. Most of the above comes from Sue Calhoun and M. Lynk, *The Lockeport Lockout: An Untold Story in Nova Scotia's Labour History* (Halifax, N.S., 1983).

35. W.C. MacKenzie, former federal fisheries official, pers. comm.

36. For the above, cf. Bates, 14–5, 29, 34, 59; Innis, 441–2.

37. Alfred Needler, pers. comm.

38. Department of Marine and Fisheries, *Annual Report* for 1919, 33.

39. Gerald Cline, Department of Fisheries and Oceans, pers. comm.

40. Joseph Gough, "Frankenstein's Seine," in *Canadian Fisherman* magazine (October and December, 1971).

41. Gretchen Fitzgerald, *The Decline of the Cape Breton Swordfish Fishery: An Exploration of the Past and Recommendations for the Future of the Nova Scotia Fishery* (Halifax, N.S.: Ecology Action Centre, 2000); J.F. Caddy, *A Review of Some Factors Relevant to Management of Swordfish Fisheries in the Northwest Atlantic*, Technical Report 633, (Ottawa: Environment Canada, Fisheries and Marine Service, 1976).

42. Thompson, *Survey of the Fisheries*.

43. R. Reeves and E.D. Mitchell, "Beluga Whale," in *The Canadian Encyclopedia*, 2nd ed. (Edmonton, Alta.: Hurtig Publishers, 1988).

44. E.E. Prince, *The Fisheries of Canada* (Rome: Tipografia del Senato, 1913).

45. Juvet.

46. Department of Fisheries and Oceans, *The American Lobster*, leaflet in the *Underwater World series* (Ottawa, 1982).

47. Innis, 438; Newfoundland, *Royal Commission*.

48. DeWolf, 23–6.

49. H. Scott Gordon, *The Fishing Industry of Prince Edward Island* (Ottawa: Markets and Economics Service, Department of Fisheries of Canada, March 1952).

50. DeWolf, 23–6.

51. Bates, 15.

52. Cf. Department of Marine and Fisheries, *Annual Report* for 1919, 34.

53. Bourne, 19; N.S. Fisheries Conference (1938), 64.

54. I owe most of the above to Alfred Needler, former federal fisheries deputy minister, pers. comm.; Bourne; N.H. Morse, *An Economic Study of the Oyster Fishery of the Maritime Provinces*, Bulletin 175 (Ottawa: Fisheries Research Board of Canada, 1971); J.C. Medcof, *Oyster Farming in the Maritimes*, Bulletin 131 (Ottawa: Fisheries Research Board of Canada, 1961).

55. W.C. MacKenzie, former federal fisheries official, pers. comm.

56. Keirstead and Keirstead, 22–3.

Chapter 14 Notes

1. Much of the above comes from Newfoundland, *Commission of Enquiry*; Innis, 459–66; Shannon Ryan, historian, Memorial University of Newfoundland, pers. comm.

2. Ian McDonald, "Coaker and the Balance of Power Strategy: The Fishermen's Protective Union in Newfoundland," in *Newfoundland in the Nineteenth and Twentieth Centuries: Essays in Interpretation*, edited by James Hiller and Peter Neary (Toronto, Ont.: University of Toronto Press, 1980), 156.

3. W.F. Coaker, "The Bonavista Platform," in A *Coaker Anthology*, edited by Robert H. Cuff (St. John's, Nfld., 1986), 49–50; McDonald, "Balance of Power"; Innis, 463.

4. Peter Neary and Patrick O'Flaherty, eds. *By Great Waters: A Newfoundland Anthology*

(Toronto, Ont,: University of Toronto Press, 1974).

5. Cf. McDonald, diss., passim.

6. Innis, 465–6.

7. The above comes in large part from McDonald, diss., and "Balance of Power"; in lesser part, from J.R. Smallwood, *Coaker of Newfoundland: The Man Who Led the Deep-Sea Fishermen to Political Power* (London, U.K.: Labour Publishing Company, 1927); Newfoundland, Commission of Enquiry; Innis, 465–7.

8. Innis, 467–8; Barbara Jamieson and Carolyn Molson, former Newfoundland residents, pers. comm.

9. Shannon Ryan, historian, Memorial University of Newfoundland, pers. comm.

10. Newfoundland, *Commission of Enquiry*, 24.

11. Newfoundland, *Commission of Enquiry*, 19.

12. Thompson, *Survey of the Fisheries*, passim.

13. Wilfred Templeman, *Marine Resources of Newfoundland*, Bulletin 154 (Ottawa: Fisheries Research Board of Canada, 1966), 12. For the inception and the early years of the research laboratory, see Melvin Baker and Shannon Ryan, "The Newfoundland Fishery Research Commission, 1930–1934," in *How Deep Is the Ocean? Historical Essays on Canada's Atlantic Fishery*, edited by James E. Candow and Carol Corbin (Sydney, N.S.: University College of Cape Breton Press, 1997).

14. Wilfred Templeman, former federal fisheries scientist, pers. comm.

15. David Alexander, *The Decay of Trade: An Economic History of the Newfoundland Saltfish Trade, 1935–1965* (St. John's, Nfld.: Institute of Social and Economic Research, Memorial University of Newfoundland, 1977), 47–57.

16. Don Jamieson, former broadcaster and politician, pers. comm.

17. *Decks Awash* (Memorial University of Newfoundland Extension Service, February 1974).

18. Wilfred Templeman, pers. comm.

19. Sir Albert Walsh et al., *Newfoundland Fisheries Development Committee Report* (St. John's, Nfld.: Guardian, 1953) (the Walsh report), 19.

20. Wilfred Templeman, pers. comm.

21. McDonald, diss., 247.

22. Innis, 446–9.

23. Newfoundland, *Commission of Enquiry*, 23–4.

24. Melvin Baker, "Hazen Algar Russell," *Newfoundland Quarterly 89, no. 3* (Spring/Summer 1995).

25. I owe most of the above to Newfoundland, *Commission of Enquiry*; McDonald, diss.; Newfoundland Department of Natural Resources, Annual Report on the fisheries for 1942 ; former federal officials John Emberley, on inspection, and Sam Bartlett, on Newfoundland enforcement work.

26. Tariff Commission, 153–6.

27. Don Jamieson, former broadcaster and politician, pers. comm.

28. Templeman, *Lobster*, 29.

29. Templeman, *Lobster*, 19, 32.

30. Gerry Ennis, biologist, Department of Fisheries and Oceans, pers. comm.

31. Fisheries and Marine Service, *Lobster Fishery Task Force* (March 1975), 105.

32. Ryan, *Historical Overview*.

33. J.E. Candow, *Of Men and Seals: A History of the Newfoundland Seal Hunt* (Ottawa: Environment Canada, Canadian Parks Service, c. 1989), 106.

34. Candow, 106.

35. Allan P. Clarke as told to Frank E. Clarke, *My Life at the Front: The Seal Hunt in the 1940s* (http://www3.nf.sympatico.ca/feclarke/ story6.htm, June 2002).

36. Cf. Candow, 107 ff.

Chapter 15 Notes

1. Manitoba. *Report of the Commissioners appointed to investigate the fishing industry of Manitoba* (Winnipeg, Man., 1933).

2. T.A. Judson, *The Freshwater Commercial Fishing Industry of Western Canada* (Ph.D. diss., University of Toronto, Toronto, Ont., 1961), 233–41.

3. Judson, 264.

4. Judson, 260, 264–73.

Chapter 16 Notes

1. Lyons, 709.

2. Meggs, 99.

3. Shepard, *International Competition*, 5.

4. The above picture emerges from Meggs, Lyons, Campbell, and the B.C. fisheries commissions of inquiry in 1917, 1921, and 1922.

5. Cf. Department of Marine and Fisheries, *Annual Report* for 1919, 43, 49; Campbell, 13; Meggs, 103, 105; Newell, 101.

6. Campbell, 15–6.

7. Lyons, 338; Meggs, 118, 133.

8. Meggs, 118.

9. Meggs, 103–8, 115–6.

10. Ron MacLeod, former official of the Department of Fisheries and Oceans, pers. comm.; cf. Campbell, 16; Meggs, 110.

11. I owe the above to Campbell, 16–7; Duff; Hill, 5; Ron MacLeod.

12. The information on Japanese Canadian restrictions comes in part from Eddie Moore, Cliff Levelton, and Ron MacLeod, former officials of the Department of Fisheries and Oceans, pers. comm.

13. Except as noted, the above comes largely from Meggs, 120–31; Department of Marine and Fisheries, Annual Report for 1919; Duff; Campbell, 9.

14. L.A. Audette, *Claims by Pelagic Sealers Arising out of the Washington Treaty, 7ᵗʰ July 1911*, vol. 16 (Victoria, B.C., 1915).

15. Most of the organizational information comes from Lyons, 672–3, 721; she in turn drew much of it from Vic Hill.

16 Chief sources for the above are former officials Eddie Moore, Ron MacLeod, and Brian Richman, and Department of Marine and Fisheries, *Annual Report* for 1930.

17. Meggs, 116–8.

18. Information provided by Ron MacLeod.

19. Meggs, 103, 105; Department of Marine and Fisheries, *Annual Report* for 1919, 42; Newell, 101.

20. Ron MacLeod, pers. comm.

21. Ron MacLeod, pers. comm.

22. Campbell, 18–9; Meggs, 137; Lyons, 370–2.

23. Lyons, 371.

24 Ron MacLeod, pers. comm.

25. Lyons, 384.

26. Lyons, 372–4, 384; Meggs, 137.

27. Campbell, 19–20.

28. Lyons, 393.

29. Duncan Stacey, article, *The Fisherman*, periodical of the United Fishermen and Allied Workers Union (March 1981), 30.

30. Lyons, 403.

31. Eddie Moore, former fishery officer, pers. comm.

32. Jimmy Sewid, fisherman from Alert Bay area and Native leader, pers. comm.

33. Cf. Lyons, 672–3, 721; Meggs, 157–9; Hill, passim.

34. Cf. Meggs, 102–6.

35. Ron MacLeod, former federal fisheries official, pers. comm.

36. Meggs, 108–9, 122.

37. Meggs, 156; Newell, 107; Jimmy Sewid, pers. comm.

38. Meggs, 111, 154.

39. Newell, 105–6.

40. Jimmy Sewid, pers. comm.

41. Homer Stevens, fisherman and union leader, pers. comm.

42. Lyons, passim.

43. Werner Cohn, "Persecution of Japanese Canadians and the Political Left in British Columbia, December 1941–March 1942," *B.C. Studies*, No. 68 (Winter 1985–1986), 3–22.

44. Janice Palton, *The Exodus of the Japanese* (Toronto, Ont.: McClelland & Stewart, 1973), passim.

45. Ron MacLeod, former federal fisheries official, pers. comm.

46. Cf. Lyons, 716–7.

47. Crutchfield, passim.

48. Hans Underdahl, veteran halibut captain, Mrs. Underdahl, and crewman Olaf Bendikson, interview, 1979.

49. Cf. D.M. Eberts, *Report of the Hon. D.M. Eberts on Charges of Inefficiency and Irregularities by Fishery Officers in Fishery District No. 3, B.C. and into Inefficient Methods There* (Ottawa: King's Printer, 1921); Department of Marine and Fisheries, *Annual Report* for 1919, 47; A.L. Tester, *The Herring Fishery of British Columbia: Past and Present*, Bulletin 47 (Ottawa: Biological Board of Canada, 1935), 6–7.

50. Tester, 20–1.

51. J.L. Hart, *The Pilchard Fishery of British Columbia*, Bulletin 36 (Ottawa: Biological Board of Canada, 1933).

52. Gordon Gibson and Carol Remison, *Bull of the Woods* (Vancouver: Douglas & McIntyre, 1980), 53–4.

53. J.L. Hart, *Pacific Fishes of Canada*, Bulletin 180 (Ottawa: Fisheries Research Board of Canada, 1973).

54. Richie Nelson. Sr., former fish-processing company executive, pers. comm.

55. Eddie Moore, former fishery officer, pers. comm.

56. Lyons, 597.

Chapter 17 Notes

1. Cliff Levelton, former federal fisheries official, pers. comm.

2. Alfred Needler, former federal fisheries official, pers. comm.

3. Bates, 115.

4. W.C. MacKenzie and Alfred Needler, former federal fisheries officials, pers. comm.

5. Position Record Book, Nova Scotia District 1; Brian Richman, former federal fisheries official, pers. comm.

6. Miriam Wright, *A Fishery for Modern Times: The State and the Industrialization of the Newfoundland Fishery*, 1934–1968 (Don Mills, Ontario: Oxford University Press, 2000), 141.

7. Much of the above comes from Department of Fisheries, *Annual Report* for 1949–1950; W.C. MacKenzie, former federal fisheries official, pers. comm.

8. Wright, 42.

9. Statistics Canada Web-site: historical statistics of Canada http://www.statcan.ca/english/freepub/11-516-XIE/sectionn/sectionn.htm.

10. Parsons, *Management*, 537.

11. Alfred Needler, former federal fisheries official, pers. comm.

12. David Wilder, former federal fisheries official, pers. comm.

13. D.H. Cushing, "The Problems of Stock and Recruitment," in *Fish Population Dynamics*, edited by J.A. Gulland (Rochester, N.Y.: John Wiley & Sons, 1977).

14. A.G. Huntsman, "Fisheries Research in Canada," *Science* 98 (1943), 121.

15. W.E. Ricker, "Productive Capacity of Canadian Fisheries." Background paper for

Resources for Tomorrow conference, Montreal, 22–28 October 1961; published as Circular No. 64 (Nanaimo, B.C.: Fisheries Research Board of Canada, June 1961).

16. Cf. Parsons, *Management*, 536–9; Ralph Halliday, federal fisheries official, pers. comm.

17. W.C. MacKenzie, former federal fisheries official, pers. comm.

18. Frank McCracken, former federal fisheries official, pers. comm.

19. Ron MacLeod, former federal fisheries official, pers. comm.

20. Ralph Halliday, former federal fisheries official, pers. comm.

21. Department of Fisheries and Oceans, Scotia–Fundy Region, *Fish Counting: The Report of the Working Committee on the Catch and Effort Statistics Module in Scotia–Fundy Region* (12 November 1982).

22. W.C. MacKenzie, former federal fisheries official, pers. comm.

23. H.S. Gordon, "The Economic Theory of a Common-Property Resource: The Fishery," *Journal of Political Economy* 62 (1954).

24. W.C. MacKenzie, former federal fisheries official, pers. comm.

25. H.S. Gordon, former federal fisheries official, professor emeritus, Queen's University and University of Indiana, pers. comm.

26. Department of Fisheries, *Annual Report* for 1949–1950.

27. William W. Warner, *Distant Water* (Penguin Books, 1983), 32–44.

28. Cliff Levelton, former federal fisheries official, pers. comm.

29. Ozere.

30. Cliff Levelton, former federal fisheries official, pers. comm.; Department of Fisheries, news release (12 April 1957).

31. W.E. Ricker and Ron MacLeod, former federal fisheries officials, pers. comm.

32. James Sinclair, former Minister of Fisheries, pers. comm.

33. Shepard.

34. Alfred Needler, former federal fisheries official, pers. comm.

35. B.A. Applebaum, *Law of the Sea—The Exclusive Economic Zone* (Canadian Museum of Civilization Web-site: www.civilization,ca/hist/lifelines/apple1e.html, 20 September 2002, 6.

36. Cf. Wright, 133–8.

37. W.C. MacKenzie, former federal fisheries official, pers. comm.

38. Walsh et al.

39. Much of the above comes from Mitchell and Frick, passim; R.W. Crowley, B. MacEachern, and R. Jasperse, "A Review of Federal Assistance to the Canadian Fishing Industry, 1945–1990," in *Perspectives on Canadian Marine Fisheries Management*, edited by L.S. Parsons and W.H. Lear, Canadian Bulletin of Fisheries and Aquatic Sciences, No. 226 (Ottawa: National Research Council and Department of Fisheries and Oceans, 1993).

40. Cf. Wright, 140; Department of Fisheries, *Trends in the Development of the Canadian Fisheries*, background document for fisheries development planning (Department of Fisheries, Economic Services, April 1967).

41. Department of Fisheries, news release (6 January 1959).

42. John Emberley and Ronald Bond, former federal fisheries officials, pers. comm.; Johnstone, historical manuscript, 17–5.

43. Crowley, 345–6.

44. Fisheries Council of Canada, *Centennial Annual Review* for 1967.

45. Department of Fisheries, *Federal–Provincial Conference on Fisheries Development*, background papers (Ottawa, 20–24 January 1964), 77–9; John Emberley, former federal fisheries official, pers. comm.

46. W.C. MacKenzie, former federal fisheries official, pers. comm.

47. W.C. MacKenzie, pers. comm.

48. Fisheries Council of Canada. *Centennial*; W.C. MacKenzie, pers. comm.

49. W.C. MacKenzie, former federal fisheries official, pers. comm.

50. Cf. J.W. Pickersgill, *Seeing Canada Whole: A Memoir* (Markham, Ont.: Fitzhenry and Whiteside, 1994), 418–20.

51. Wright, 141

52. Fisheries Council of Canada, *Selected Programs of Federal Financial Assistance* (Ottawa, December 1969).

53. *Canadian Fishing Report* (June 1982).

54. Carl Sollows, former federal fisheries official, pers. comm.

55. Department of Fisheries, news release (April 1960).

56. Wright, 140.

57. Atlantic Development Board, *Fisheries in the Atlantic Provinces* (Ottawa, 1969), 33–4.

58. Environment Canada, Fisheries and Marine Service, *Policy for Canada's Commercial Fisheries* (Ottawa, May 1976), 26.

59. Environment Canada, Fisheries and Marine Service, 12.

60. C.L. Mitchell, Canada's *Fishing Industry: A Sectoral Analysis* (Ottawa: Department of Fisheries and Oceans, 1980).

Chapter 18 Notes

1. Cliff Levelton, former federal fisheries official, James Sinclair, former Minister of Fisheries, pers. comm.

2. W.C. MacKenzie, former federal fisheries official, pers. comm.

3. Parsons, *Management*, 175.

4. J. Proskie and J.C. Adams, *Survey of the Labour Force in the Offshore Fishing Fleet, Atlantic Coast* (Fisheries Service, Economic Branch, 1971), 4.

5. Most of the above comes from Mitchell and Frick, passim; Crowley et al., passim; W.C. MacKenzie, former federal fisheries official, pers. comm.

6. Mitchell and Frick, 26–7.

7. Atlantic Development Board.

8. W.H. Lear and L.S. Parsons, "History and Management of the Fishery for Northern Cod in NAFO Divisions 2J, 3K and 3L," in *Perspectives on Canadian Marine Fisheries Management*, edited by L.S. Parsons and W.H. Lear, Canadian Bulletin of Fisheries and Aquatic Sciences, No. 226 (Ottawa: National Research Council and Department of Fisheries and Oceans, 1993). 64.

9. Parsons, *Management*, 111.

10. R.R. Logie, "Trends in the Atlantic Coast Fisheries of Canada," *Transactions of the American Fisheries Society* 97 (January 1968).

11. Alfred Needler, former federal fisheries official, pers. comm.

12. W. Templeman, *Redfish Distribution in the North Atlantic*, Bulletin 130 (Ottawa: Fisheries Research Board of Canada, 1959), 31.

13. Frank McCracken, former federal fisheries official, pers. comm. Cf. also Department of Fisheries, *Proceedings: Canadian Atlantic Offshore Fishing Vessel Conference, Montreal, February 7–9, 1966* (Ottawa: Department of Fisheries, 1966), 13 ff.

14. Hédard Robichaud, former fisheries official and Minister, pers. comm.

15. Medford Matthews, fisherman, pers. comm.

16. Cliff Levelton, former federal fisheries official, pers. comm.

17. Audber Matthews, fisherman, pers. comm.

18. Scott Parsons, former federal fisheries official, pers. comm.

19. Cf. Fitzgerald; Caddy.

20a. Cf. DeWolf, 23–6, 33.

20b. J.D. Pringle and D.L. Burke, "The Canadian Lobster Fishery and Its Management, with Emphasis on the Scotian Shelf and the Gulf of Maine," in *Perspectives on Canadian Marine Fisheries Management*, edited by L.S. Parsons and W.H. Lear, Canadian Bulletin of Fisheries and Aquatic Sciences, No. 226 (Ottawa: National Research Council and Department of Fisheries and Oceans, 1993).

21. Cf. Parsons, *Management*, 106.

22. David Wilder, former federal fisheries official, pers. comm.

23. Pringle and Burke, 97, 113.

24. Cf. Parsons, *Management*, 170.

25. Department of Fisheries and Forestry, news release (20 January 1969).

26. J.B. Rutherford, D.G. Wilder, and H.C. Frick, *An Economic Appraisal of the Canadian Lobster Fishery*, Bulletin 157 (Ottawa: Fisheries Research Board of Canada, 1967).

27. Parsons, *Management*, 171.

28. Department of Fisheries and Forestry, news releases (20 January and 27 February 1969).

29. Cliff Levelton, former federal fisheries official, pers. comm.

30. Frank McCracken, former federal fisheries official, pers. comm.

31. Cf. G.M. Hare and D.L. Dunn, "A Retrospective Analysis of the Gulf of St. Lawrence Snow Crab (*Chionoecetes opilio*) Fishery, 1965–1990, in *Perspectives on Canadian Marine Fisheries Management*, edited by L.S. Parsons and W.H. Lear, Canadian Bulletin of Fisheries and Aquatic Sciences, No. 226 (Ottawa: National Research Council and Department of Fisheries and Oceans, 1993), 178–80; Department of Fisheries and Forestry, *Proceedings: Meeting on Atlantic Crab Fishery Development, Fredericton, N.B., March 4–5, 1969* (Ottawa: Department of Fisheries and Forestry, 1969).

32. Much of the above comes from H. Lear, "The Management of Canadian Atlantic Salmon Fisheries," in *Perspectives on Canadian Marine Fisheries Management*, edited by L.S. Parsons and W.H. Lear, Canadian Bulletin of Fisheries and Aquatic Sciences, No. 226 (Ottawa: National Research Council and Department of Fisheries and Oceans, 1993).

Chapter 19 Notes

1. Wright, 23–33.

2. P.D.H. Dunn, *Fisheries Re-organization in Newfoundland*, radio address, 21 January 1944 (in the library of Memorial University of Newfoundland, St. John's, Nfld.: Cat. No. SH 229 D86).

3. H.V. Dempsey, "Fisheries," in *Canada One Hundred 1867–1967* (Ottawa: Dominion Bureau of Statistics, 1967).

4. Don Jamieson, former broadcaster and politician, pers. comm.

5. The background information on Newfoundland comes in part from Ken Johnstone interview with John Lamb, former federal fisheries official; Newfoundland Fisheries Board, *Annual Report* for 1942; John Emberley and Sam Bartlett, former federal fisheries officials, pers. comm.

6. Wright, 34–5.

7. Aidan Maloney, former Minister of Fisheries for Newfoundland and federal fisheries official, pers. comm.

8. Department of Fisheries, news release (7 February 1966); Walsh et al., 30, 51 ff.

9. Cf. Parzival Copes, *The Development of the Newfoundland Fishing Economy* (Burnaby, B.C.: Simon Fraser University, Dept. of Economics and Commerce, 1971).

10. Alexander, *Decay*, 141.

11. Wright, 59–66, 71, 80.

12. Wright, 66–86.

13. Wright, 85.

14. Government of Newfoundland and Labrador, *Resettlement Statistics*, Industry Canada Web-site: http://collections.ic.gc.ca/placentia/newpag.htm (10 December 2003).

15. Alexander, *Decay*, passim; W.C. MacKenzie, former federal fisheries official, pers. comm.

16. *Canadian Fishing Report* (October 1983).

17. Alexander, *Decay*, 137; Wright, 71.

18. Wright, 118.

19. Wilfred Templeman, former federal fisheries official, pers. comm.

20. Templeman, *Redfish*.

21. Templeman, *Marine Resources*; Ricker, Seventy-five Years of Achievements., passim.

22. Wright, 108–9.

23. Templeman, *Marine Resources*, 2, 23; Wright, 142–3.

24. Wright, 114–5.

25. Scott Parsons, former federal fisheries official, pers. comm.

26. Wright, 99–100.

27. Wright, 103.

28. Wright, 94–109.

29. Wright, 114.

30. Wright, 117–22.

31. J.R. Smallwood, former premier of Newfoundland, pers. comm.

32. Wright, 145–6.

33. Scott Parsons, pers. comm.

34. Jerry Ennis, Department of Fisheries and Oceans, pers. comm; Templeman, *Lobster* (1941).

35. Much of the above comes from Candow; A.H. Malouf, *Seals and Sealing in Canada: Report of the Royal Commission*, Vol. 1 (Ottawa: Minister of Supply and Services, 1986); Department of Fisheries and Forestry, news releases (19 June 1968 and 15 October 1969).

36. W.C. MacKenzie, former federal fisheries official, pers. comm.

Chapter 20 Notes

1. Hart, *Pacific Fishes*, 122.

2. W.E. Ricker, former federal fisheries official, pers. comm.; Lyons, passim.

3. Much of the above comes from Lyons, passim.

4. James Sinclair, former Minister of Fisheries, pers. comm.

5. Cliff Levelton, former federal fisheries official, pers. comm.

6. Fleet information largely from Cliff Levelton and Ron MacLeod, former federal fisheries officials, pers. comm.

7. Homer Stevens, former U.F.A.W.U. official, pers. comm.

8. Campbell, 21.

9. Allen Wood, former federal fisheries official, pers. comm.

10. James Sinclair, pers. comm.

11. Richie Nelson, B.C. Packers, pers. comm.

12. Cliff Levelton, former federal fisheries official, pers. comm.

13. Rod Hourston, former federal fisheries official, pers. comm.

14. Most of the above comes from Ron MacLeod, former federal fisheries official, pers. comm.

15. Cliff Levelton, former federal fisheries official, pers. comm.

16. M.C. Healey, "The Management of Pacific Salmon Fisheries in British Columbia," in *Perspectives on Canadian Marine Fisheries Management*, edited by L.S. Parsons and W.H. Lear, Canadian Bulletin of Fisheries and Aquatic Sciences, No. 226 (Ottawa: National Research Council and Department of Fisheries and Oceans, 1993), 251; Ron MacLeod, former federal fisheries official, pers. comm.

17. Dave Denbigh, former federal fisheries official, pers. comm.

18. James Sinclair, former Minister of Fisheries, pers. comm.

19. Ricker, "Productive Capacity."

20. Meggs, 164–6.

21. S. Sinclair, *Licence Limitation: British Columbia: A Method of Economic Fisheries Management* (Department of Fisheries, 1960), 140–5; Parsons, Management, 160.

22. Brian Richman and Cliff Levelton, former federal fisheries officials, pers. comm.

23. D.N. Outram, *Canada's Pacific Herring* (Ottawa: Department of Fisheries, 1965).

24. Much of the above comes from Rod Hourston, Cliff Levelton, and Ron MacLeod, former former federal fisheries officials, pers. comm.; Parsons, *Management*, 99; M. Stocker, "Recent Management of the British Columbia Herring Fishery," in *Perspectives on Canadian Marine Fisheries Management*, edited by L.S. Parsons and W.H. Lear, Canadian Bulletin of Fisheries and Aquatic Sciences, No. 226 (Ottawa: National Research Council and Department of Fisheries and Oceans, 1993).

25. I owe the above largely to various papers by K.S. Ketchen, of the Pacific Biological Station; Lyons, passim; and Homer Stevens, U.F.A.W.U., pers. comm.

26. Cf. Parsons, *Management*, 94–5.

Chapter 21 Notes

1. Fisheries and Marine Service, speaking notes on main estimates for Standing Committee on Fisheries and Forestry, 1974.

2. Cf. Parsons, *Management*, 538.

3. Roméo LeBlanc, former federal minister, pers. comm.

4. Ken Lucas, former federal fisheries official, pers. comm.

5. Cf. Parsons, *Management*, 26 ff.

6. Cf. Parsons, *Management*, 310.

7. Cf. Parsons, *Management*, 65.

8. Derrick Iles, former federal fisheries official, pers. comm.

9. In this and preceding sections on international matters, I owe much to Bob Applebaum, who served many years in Department of Fisheries and Oceans's international branch.

10. Parsons, *Management*, 66.

11. Warner, 74.

12. Roméo LeBlanc, speech to Halifax Board of Trade, 29 January 1979.

13. Department of Fisheries and Oceans, news release, 22 February 1973.

14. Roméo LeBlanc, speech to Halifax Board of Trade, 16 January 1979.

15. Roméo LeBlanc, speech to Halifax Board of Trade, 16 January 1979.

16. Department of Fisheries and Oceans, news release, 14 November 1979.

17. Brian Richman and Steve Tilley, Department of Fisheries and Oceans, provided much of the above information.

18. Department of Fisheries and Oceans, news release, 11 April 2002.

19. *Canadian Fishing Report*, December 1983.

20. Ron MacLeod, former federal fisheries official, pers. comm.

21. Roméo LeBlanc, speech to the Atlantic Provinces Economic Council, Halifax, 22 October 1974.

22. Roméo LeBlanc, speech to the Fisheries Council of Canada, Halifax, 5 May 1975.

23. Roméo LeBlanc, speech to Halifax Board of Trade, 26 January 1979.

24. Environment Canada, Fisheries and Marine Service, news release, 13 August 1976; Richard Cashin, *Charting a New Course: Towards the Fishery of the Future*, Report of the Task Force on Incomes and Adjustment in the Atlantic Fishery (Ottawa: Department of Fisheries and Oceans, 1993), 162. (the Cashin report).

25. Much of the above comes from Crowley et al.

26. Roméo LeBlanc, speech to Halifax Board of Trade (26 January 1979).

27. Cf. Parsons, *Management*, 26 ff.

Chapter 22 Notes

1. W.C. MacKenzie, former federal fisheries official, pers. comm.

2. Department of Fisheries, news release, 17 July 1968.

3. Cf. Alexander, *Decay*, 137 ff.

4. Wright, 123.

5. Department of Fisheries and Forestry, news releases, 28 November and 19 December 1969.

6. Robin Molson, former federal fisheries official, pers. comm.

7. *Canadian Fishing Report*, October 1983.

8. William E. Schrank, "Extended Fisheries Jurisdiction: Origins of the Current Crisis in Atlantic Canada's Fisheries," *Marine Policy* 19, no. 4 (1995).

9. Cf. Parsons, *Management*, 176–8.

10. Environment Canada, *Policy*, 48–9.

11. Roméo LeBlanc, speech to Halifax Board of Trade, Halifax, N.S., 26 January 1979.

12. Roméo LeBlanc, speech to Halifax Board of Trade, Halifax, N.S., 26 January 1979.

13. Department of Fisheries and Oceans, *Quality Excellence in the 1980's*, booklet (Ottawa, 1980); Greg Peacock, Department of Fisheries and Oceans, pers. comm.

14. Quotes from LeBlanc are from the same Halifax Board of Trade speech cited earlier, which happened to summarize many of LeBlanc's views.

15. Roméo LeBlanc, speech to Halifax Board of Trade, Halifax, N.S., 26 January 1979.

16. Information on F0.1 comes in part from Parsons, *Management*, 55–6, 66; in part from Ralph Halliday, federal fisheries scientist, pers. comm.

17. A.C. Finlayson, *Fishing for Truth* (St. John's, Nfld.: Memorial University of Newfoundland, 1994), 135–42; Steve MacPhee, former regional director of science with Department of Fisheries and Oceans, pers.comm.

18. H.B. Nickerson and Sons Ltd. and National Sea Products Ltd., *Where Now?*, booklet (c.1978).

19. *Canadian Fishing Report*, July 1979.

20. Cf. Halifax *Chronicle Herald*, 12 July, 5 and 6 September 1977; Toronto *Globe and Mail*, 1 September 1977.

21. Roméo LeBlanc, speech to Fisheries Council of Canada, Québec, Que., 3 May 1978.

22. Roméo LeBlanc, speech to Halifax Board of Trade, Halifax, N.S., 26 January 1979.

23. Scott Parsons, former federal fisheries official, pers. comm.

24. Cf. Parsons, *Management*, 284–5.

25. Roméo LeBlanc, speech to St. John's Rotary Club, St. John's, Nfld., May 1977.

26. Art May, speech to Fisheries Council of Canada, St. John's, Nfld., 1983.

27. Ralph Halliday, federal fisheries scientist, pers. comm.

28. Parsons, *Management*, 124–5.

29. *Canadian Fishing Report*, July 1979.

30. Lennox O. Hinds and James Trimm, "Utilization of Catch Now Discarded at Sea," in *Proceedings: Government–Industry Meeting on the Utilization of Atlantic Marine Resources, Queen Elizabeth Hotel, Montreal, P.Q., February 5–7, 1974* (Environment Canada, Fisheries and Marine Service, 1974).

31. Parsons, *Management*, 176.

32. Roméo LeBlanc, speech to Rotary Club,

Yarmouth, N.S., 28 November 1977.

33. *Canadian Fishing Report*, February 1981.

34. Charles Gaudet, Department of Fisheries and Oceans, pers. comm.

35. Cf. M.J.L. Kirby, *Navigating Troubled Waters*, Report of the Task Force on Atlantic Fisheries (Ottawa, 1982) (the "Kirby report"), 107; Environment Canada, *Policy*; Mitchell (1980).

36. Part of this information appears in Kirby, 31.

37. Louise Hebert, *Great Expectations: The Atlantic Fish Processing Sector* (Moncton, N.B.: Department of Fisheries and Oceans, Economic Research and Analysis Division, Gulf Region, 1989).

38. Cf. Parsons, Management, 180–4; Stephen Kimber, *Net Profits: The Story of National Sea* (Halifax, N.S.: Nimbus, c. 1989), passim.

39. C.R. Levelton, *Towards an Atlantic Commercial Fisheries Licensing System* (Ottawa: Department of Fisheries and Oceans, 1979).

40. *Canadian Fishing Report*, March 1981.

41. Parsons, *Management*, 131–2.

42. Parsons, *Management*, 201–3.

43. Parsons, *Management*, 206–7; Brian Lester, Department of Fisheries and Oceans, pers. comm.

44. Jim Jones, Department of Fisheries and Oceans, pers. comm.

45. *Canadian Fishing Report*, Feb. 1982; Ralph Halliday, federal fisheries scientist, pers. comm.

46. Miscellaneous 1981 documents provided by Don Tansley, former deputy minister of Fisheries and Oceans.

47. *Canadian Fishing Report*, June 1982.

48. Parsons, *Management*, 70–2.

49. Relevant information on the above appears in *Canadian Fishing Report*, February, April, June 1983.

50. *Canadian Fishing Report*, June 1983; L.S. Parsons, "Shaping Fisheries Policy: the Kirby and Pearse Inquiries—Process, Prescription and Impact," in Parsons and Lear, *Perspectives*.

51. Peter John Nicholson, statement at Fisheries Council of Canada annual meeting, 1993.

52. *Canadian Fishing Report*, August 1981.

53. L.S. Parsons and W.H. Lear, eds., *Perspectives on Canadian Marine Fisheries Management*, Canadian Bulletin of Fisheries and Aquatic Sciences, No. 226 (Ottawa: National Research Council and Department of Fisheries and Oceans, 1993), 77; Ralph Halliday, federal fisheries scientist, pers. comm.

54. Parsons, "Shaping Fisheries Policies," 408.

55. Ralph Halliday, federal fisheries scientist, pers. comm.

56. Kirby, 123–5, 242.

57. Scott Parsons, former federal fisheries official, pers. comm.

58. *Canadian Fishing Report*, July 1984.

59. Cf. Parsons, *Management*, 127–8.

60. *Canadian Fishing Report*, July 1984.

61. *Canadian Fishing Report*, June 1982.

62. Parsons, *Management*, 364–75.

63. *Canadian Fishing Report*, December 1983.

64. Parsons, *Management*, 370.

65. *Canadian Fishing Report*, February, August 1984; Kimber, 253–6.

66. *Canadian Fishing Report*, July 1984.

67. Parsons, *Management*, 375.

68. *Canadian Fishing Report*, April 1984.

69. *Canadian Fishing Report*, April 1984. Much of the above comes from material provided by Ken Jones, Department of Fisheries and Oceans, and from Lear.

70. D.L. Dunn, *Canada's East Coast Sealing Industry 1976*, Fisheries and Marine Service, 1977).

71. Department of Fisheries and Forestry, news release, 15 October 1969.

Chapter 23 Notes

1. Cf. Silver Donald Cameron, *The Education of Everett Richardson: The Nova Scotia Fishermen's Strike*, 1970–71 (Toronto: McClellan and Stewart, 1977).

2. Hare and Dunn, 180.

3. *Canadian Fishing Report*, February 1981.

4. *Canadian Fishing Report*, November 1982, January 1983.

5. Carl Myers, Department of Fisheries and Oceans, pers. comm.; Rory McLellan, P.E.I. Fishermen's Association, pers. comm.; *Canadian Fishing Report*, July, October, November 1979.

6. Parsons, *Management*, 196.

7. *Canadian Fishing Report*, August, September 1983.

8. Iles, T.D., "The Management of the Canadian Atlantic Herring Fisheries," in *Perspectives on Canadian Marine Fisheries Management*, edited by L.S. Parsons and W.H. Lear, Canadian Bulletin of Fisheries and Aquactic Sciences, No. 226 (Ottawa: National Research Council and Department of Fisheries and Oceans), 145; Kirby, 331–7.

9. *Canadian Fishing Report*, September 1983.

10. Fisheries and Marine Service, *Lobster*, 21–2.

11. Parsons, *Management*, 170–3.

12. Much of the above appears in Parsons, *Management*, 174–5.

13. Much of the above comes from Douglas Pezzack, D.F.O., pers. comm.

14. Al Youngren, American fisherman, pers. comm.

15. P.R. Toews, *The Canadian Atlantic Shrimp Fishery: Prospects for Development* (Ottawa: Department of Fisheries and Oceans, 1980).

16. I owe most of the above to Hare and Dunn, 177–92.

17. D.B. Lister, *A Brief Review of Salmon Enhancement Programs in Pacific and Atlantic Canada* (Vancouver: Department of the Environment, Fisheries and Marine Service, 1974).

18. Richard Saunders, former federal fisheries scientist, pers. comm.

Chapter 24 Notes

1. Parsons, *Management*, 26–34.
2. Gordon Inglis, *More Than Just a Union: The Story of the NFFAWU* (St. John's: Jesperson Press, 1985), as quoted on Fish Food and Allied Workers Union Web site, 2003.
3. *Canadian Fishing Report*, February 1983.
4. The above comes largely from Jerry Brothers, Department of Fisheries and Oceans, pers. comm.
5. Parsons, *Management*, 303–8.
6. Cf. Finlayson, 134–43; Parsons, *Management*, 584.
7. Most of the above comes from National Revenue income statistics; and from Cashin, 162.

Chapter 25 Notes

1. Campbell, 111.
2. Department of Fisheries, news release, 6 September 1968.
3. Department of Fisheries, news release, 21 November 1968; Ron MacLeod, former federal fisheries official, pers. comm.
4. Department of Fisheries, news release, 23 October 1968.
5. Sinclair, 140–5.
6. Campbell, 111–2.
7. Meggs, 181.
8. W.C. MacKenzie, former federal fisheries official, comments that the argument over licensing the vessel or the man was somewhat misplaced; the ultimate objective was to control fishing power, which implied licensing or otherwise controlling the fishing enterprise, that is, the decision-making entity.
9. Campbell, 243–51.
10. Campbell, 257–69.
11. Cf. Campbell, 343–4.
12. Campbell, 434–5.
13. Department of Fisheries, news release, 21 November 1968.
14. Department of Fisheries, news release, 6 January 1970.
15. Campbell, 406–7.
16. Pearse, 99–100.
17. Cf. Parsons, 160–3.
18. Campbell, 409–29.
19. Cf. Parsons, 163; Campbell, 430–66.
20. Jack Davis, speech to United Fishermen and Allied Workers Union, 28 January 1974.
21. Dick Crouter and Ron MacLeod, former federal fisheries officials, pers. comm.
22. Max Stocker, "Recent Management of the British Columbia Herring Fishery," in *Perspectives on Canadian Marine Fisheries Management*, edited

by L.S. Parsons and W.H. Lear, Canadian Bulletin of Fisheries and Aquatic Sciences, No. 226 (Ottawa: National Research Council and Department of Fisheries and Oceans, 1993), 274.
23. *Canadian Fishing Report*, November 1983.
24. Chris Newton, former federal fisheries official, pers. comm.
25. Dennis Brock, former federal fisheries official, pers. comm.
26. Brian Richman, former federal fisheries official, pers. comm.
27. The above relies heavily on Stocker, "Recent Management"; *Canadian Fishing Report*, November 1983; Don Pepper, former federal fisheries official, pers. comm.
28. Ron MacLeod, former federal fisheries official, pers. comm.
29. Ricker, *Productive Capacity.*
30. Lister.
31. Ron MacLeod, former federal fisheries official, pers. comm.
32. Jack Davis, speech to United Fishermen and Allied Workers Union, 28 January 1974.
33. Pearse, 48.
34. Pearse, 49.
35. Ron MacLeod, former federal fisheries official, pers. comm.
36. Cf. Pearse, 49–57.
37. Kate Glover, former federal fisheries official, pers. comm.
38. S. Sinclair, *A Licensing Fee System for the Coastal Fisheries of British Columbia*, unpublished report for Department of Fisheries and Oceans (1978) (available in the library of the Department of Fisheries and Oceans, Ottawa).
39. Garrett Hardin, "Essays on Science and Society, Extensions of The Tragedy of the Commons," in *Science*, Vol. 280, Number 5364 (1 May 1998), 682–3.
40. Charles Friend, former federal fisheries official, pers. comm.
41. Parsons, *Management*, 165.
42. Pearse, 94.
43. *Canadian Fishing Report*, February 1984.
44. Cf. Parsons, *Management*, 404–7.
45. Ron MacLeod, former federal fisheries official, pers. comm.
46. Cf. *Canadian Fishing Report*, October 1982.
47. Parsons, *Management*, 95.
48 Parsons, *Management*, 95, 122–3.
49. Ben Muse, *Management of the British Columbia Abalone Fishery* (Alaska Commercial Fisheries Entry Commission, May 1998).
50. Ron MacLeod, former federal fisheries official, pers. comm.
51. Ron MacLeod, pers. comm.
52. Ron MacLeod, pers. comm.; Pearse, 102.
53. This information comes from background material provided by the Department of Fisheries and Oceans for the Pearse commission.
54. Information on the Northern Native Fishing

Corporation largely from Parsons, *Management*, 422–3; Ron MacLeod, former federal fisheries official, pers. comm.

55. Ron MacLeod, pers. comm.

56. *Canadian Fishing Report*, February 1983, May 1984.

57. Kirby, 25.

Chapter 26 Notes

1. Parsons, *Management*, 72–3.

2. Parsons, *Management*, 474.

3. Office of the Auditor General of Canada, *1988 Report of the Auditor General of Canada* (Ottawa, 1988).

4. Office of the Auditor General of Canada, 1997 *Report of the Auditor General of Canada* (Ottawa 1997), ch. 14.

5. Department of Fisheries and Oceans, *Challenges Facing the Fishery Sector: Report to First Ministers* (Ottawa, 1986).

6. Parsons, "Shaping Fisheries Policy," 393.

7. Department of Fisheries and Oceans, Atlantic *Fisheries Development: 1983/1984 Annual Project Review* (Ottawa, 1984).

8. Jerry Conway, Department of Fisheries and Oceans, pers. comm.

9. Cf. Parsons, *Management*, 426.

10. Department of Fisheries and Oceans, *Challenges*.

11. Department of Fisheries and Oceans, "Canada's Foreign Fisheries Relations Policy," Internet: http://www.dfo.mpo.gc.ca/communic/fish_man/forfish/b64/sld001/htm (accessed 9 May 2002). The main items were small quotas for France and for a few joint-venture arrangements.

12. Hon. R. Thibault, Minister of Fisheries and Oceans, statement, "Canada speaks with united voice at NAFO meeting," 17 September 2002.

13. Much of the above comes from Applebaum.

14. Bob Applebaum, former federal fisheries official, pers. comm.

15. Cf. Parsons, *Management*, 308–31; Department of Fisheries and Oceans, *Challenges*, 1986.

16. Brian Richman and Ron MacLeod, former federal fisheries officials, pers. comm.

17. Steve Tilley, Department of Fisheries and Oceans, pers. comm.

18. R. Halliday and A. Pinhorn, "North Atlantic Fishery Management Systems: A Comparison of Management Methods and Resource Trends," *Journal of Northwest Atlantic Fishery Science* 20 (1996): 29–30.

19. Canadian Council of Professional Fish Harvesters, About the Council, Internet: http://www.ccpfh-ccpp.org/eng/faccveil-html (accessed 2002).

20. Bob Applebaum, former federal fisheries official, pers. comm.

21. Department of Fisheries and Oceans, news release, 4 July 1995.

22. Bob Applebaum, former federal fisheries official, pers. comm.

23. Department of Fisheries and Oceans, news release, 3 March 1995; David Bevan, Department of Fisheries and Oceans, pers. comm.

24. Scott Parsons, former federal fisheries official, pers. comm. Another source relates that during a tense meeting with Foreign Affairs officials, a Department of Fisheries and Oceans enforcement official blocked the door so that no one would leave until everyone, including cautious diplomatic types, had signed an agreement for action.

25. Robert Thibault, Minister of Fisheries and Oceans, statement to the Northwest Atlantic Fisheries Organization, 17 September 2002.

26. The European Union signed the United Nations Fish Agreement (U.N.F.A.) in December 2003. Meanwhile, Canada in November 2003 ratified the United Nations Convention on the Law of the Sea (U.N.C.L.O.S.). The existence of U.N.F.A. in effect plugged a loophole and encouraged Canada's ratification of U.N.C.L.O.S.

27. Department of Fisheries and Oceans, news release, 28 Feb. 1995.

28. Scott Parsons, former federal fisheries official, pers. comm.

29. Department of Fisheries and Oceans, news release and backgrounder, 28 February 1995.

30. These figures are from the government financial estimates: *Fisheries and Oceans Canada: Performance Report for the Period Ending March 31, 2001*, Internet: http://www.tbs-sct.gc.ca/rma/dpr/00-01/F&Ocoodpre.pdf.

31. Scott Parsons, former federal fisheries official, pers. comm.

32. Department of Fisheries and Oceans, news release, 28 February 1995.

33. Scott Parsons, pers. comm.

34. Department of Fisheries and Oceans, PowerPoint presentation: "Action on the Water," 2000.

35. C. Prud'homme, Department of Fisheries and Oceans, pers. comm.

36. Roméo LeBlanc, speech to the 1975 Fisheries Council of Canada meeting. On LeBlanc and co-management, cf. John Kearney, "The Transformation of the Bay of Fundy Herring Fisheries, 1976–1978: An Experiment in Fishermen–Government Co-Management.," in *Atlantic Fisheries and Coastal Communities: Fisheries Decision-Making Case Studies*, Edited by Cynthia Lawson and Arthur J. Hanson (Halifax, N.S.: Dalhousie University Ocean Studies Programme, 1984), 175–6.

37. Department of Fisheries and Oceans, Canada's Oceans Act, Internet: http://www.pac.dfo-mpo.gc.ca/oceans/Oceans%20Act/oceanact_e.html (accessed 6 June 2002).

38. Department of Fisheries and Oceans, Gulf

Region, *First Indian Food Fisheries Conference, Gulf Region* (Memramcook, N.B.: 1982).

39. Jack Stagg, formerly of Department of Indian and Northern Affairs and Department of Fisheries and Oceans, pers. comm.

40. Department of Fisheries and Oceans, *Challenges*, 128.

41. Department of Fisheries and Oceans, *Challenges*, 128.

42. Scott Parsons, former federal fisheries official, pers. comm.

43. Cf. Office of the Commissioner for Aquaculture Development, *Canadian Aquaculture Industry Profile* (14 December 2001), Internet: http://ocad-bcda.gc.ca/eaquaculture.html (accessed 7 February 2002).

44. Ken Lucas, former federal fisheries official, pers. comm.

Chapter 27 Notes

1. *Canadian Fishing Report*, February 1983, February 1984.

2. L. Burke, *Experience with Individual Quota and Enterprise Allocation (IQ/EA Management in Canadian Fisheries, 1972–1994* (Halifax, N.S.: Department of Fisheries and Oceans, Policy and Economic Branch, 1994).

3. Fisheries Council of Canada, *Building a Fishery That Works: A Vision for the Atlantic Fisheries* (Ottawa: Fisheries Council of Canada, 1994).

4. Cf. Canadian Council of Professional Fish Harvesters, *Preliminary Response of the Canadian Council of Professional Fish Harvesters to the Atlantic Fisheries Policy Review* (12 March 2001), Internet: http://www.ccpfh-ccpp.org/eng/polc-cpp.html (accessed 13 June 2002).

5. Late in 2003, after lobbying by fishermen's organizations, Fisheries and Oceans Minister Robert Thibault announced a series of consultations on "preserving the independence of the inshore fleet in Canada's Atlantic fisheries."

6. Standing Senate Committee on Fisheries and Oceans, *Straddling Fish Stocks in the Northwest Atlantic* (Ottawa: Senate, June 2003), 19–22; Parsons, *Management*, 275.

7. Information on the federal stake in restructured plants appears in Parsons, *Management*, 375. Quote is from Senate Standing Committee on Fisheries, *The Atlantic Groundfish Fishery: Its future* (Ottawa: Senate, December 1995). Instructive tables on plant numbers and capacity in 1981 and 1991 appear in the Cashin report, 146-7.

8. Cf. Parsons, *Management*, 285–7.

9. Kirby, 14–5, 208–11.

10. Greg Peacock, Department of Fisheries and Oceans, pers. comm.

11. Part of the above information relating to draggers comes from GTA Fisheries Consultants,

The Mobile Gear Groundfish Fleet under 65 ft. in Atlantic Canada (report prepared for Department of Fisheries and Oceans, 1988).

12. Richard Cashin, former president of Newfoundland fishermen's union, pers. comm.

13. J.R. Angel, D.L. Burke, R.N. O'Boyle, F.G. Peacock, M. Sinclair, and K.C.T. Zwanenburg, *Report of the Workshop on Scotia–Fundy Groundfish Management from 1977 to 1993*, Canadian Technical Report of Fisheries and Atlantic Sciences No. 1979 (1994), 107–8.

14. Scott Parsons, former federal fisheries official, pers. comm.

15. Scott Parsons, pers. comm.

16. Scott Parsons, pers. comm.

17. Scott Parsons, pers. comm.

18. Cf. Lear and Parsons, "Northern Cod," 79–80.

19. Ralph Halliday, former federal fisheries official, pers. comm.

20. Jake Rice, Department of Fisheries and Oceans, pers. comm.

21. Lear and Parsons, 80.

22. Cf. Parsons, *Management*, 379.

23. Scott Parsons, former federal fisheries official, pers. comm.

24. Scott Parsons, pers. comm.

25. Leslie Harris, *Independent Review of the State of the Northern Cod Stock*, report to the Hon. Tom Siddon (February 1990), 2–3.

26. Harris, 23, 7.

27. Cf. Lear and Parsons, 82–3.

28. Cf. Angel et al., 43–9; J.-E. Haché, former federal fisheries official, pers. comm.

29. Parsons, *Management*, 378.

30. Angel et al., 75.

31. Jorgen Hansen, Department of Fisheries and Oceans, pers. comm.

32. Cf. John C. Crosbie, *No Holds Barred: My Life in Politics* (Toronto: McClelland & Stewart, 1977), 380–5; Parsons, *Management*, 383.

33. Department of Fisheries and Oceans, *Government Assistance to the Fishing Industry: Canada* (paper presented in 1998 to the Organization for Economic Co-operation and Development, AGR/FI/RD(98)39; hereafter referred to as OECD, 1998).

34. Testimony before Standing Committee on Fisheries and Oceans (15 March 2002).

35. The above draws heavily on contemporary documents of the Canadian Atlantic Fisheries Scientific Advisory Committee and Atlantic Groundfish Management Plans published by the Department of Fisheries and Oceans; Lear and Parsons, 80–4; Scott Parsons, former federal fisheries official, pers. comm.

36. Crosbie, 377–81.

37. Angel et al., 58–60; D.L. Burke, R.N. Boyle, P. Partington, and M. Sinclair, eds., *Report of the Second Workshop on Scotia–Fundy Groundfish Management*, Canadian Technical Report of

Fisheries and Aquatic Sciences No. 2001 (1996), 45.

38. Cf. K.F. Drinkwater, "A Review of the Role of Climate Variability in the Decline of Northern Cod," *American Fisheries Society Symposium* 32 (2002).

39. Jake Rice, Department of Fisheries and Oceans, pers. comm.

40. Cf. Harris, 2 ff.

41. Parsons, *Management*, 579.

42. This information is from Sherwood–Meder Marketing Consultants, *Survey of South-Western Nova Scotia Fishing Vessel Owners*, report commissioned by Carl Myers (Halifax, N.S.: Department of Fisheries and Oceans Communications, 31 March 1989). 43. Cf. Department of Fisheries and Oceans, *Resource Prospects for Canada's Atlantic Fisheries, 1989–1993* (1988), 1.

44. Cf. Parsons, *Management*, 577, 582–4.

45. Parsons, *Management*, 575.

46. Cf. Finlayson, 96–8, 112–4.

47. Gus Etchegary, formerly of Fishery Products International, as quoted by reporter Jim Wellman.

48. Henry Lear, Department of Fisheries and Oceans, pers. comm.

49. Cf. Angel et al.; Burke et al.

50. Herb Dhaliwal, speech to the Atlantic Provinces Economic Council, 24 March 2000.

51. Scott Parsons, former federal fisheries official, pers. comm.

52. Office of the Auditor General of Canada, *1993 Report of the Auditor General of Canada* (Ottawa, 1993), ch. 15, 396.

53. Department of Fisheries and Oceans, *Government Assistance*

54. Department of Fisheries and Oceans, *Government Assistance*; Tim Hsu, Department of Fisheries and Oceans, pers. comm. The department's Government Assistance document gave the picture up to 1996; some A.F.A.P. expenditure continued after.

55. Office of the Auditor General, *1993 Report*, 409.

56. Office of the Auditor General, *1993 Report*.

57. Human Resources Development Canada, Evaluation and Data Development, Evaluation of the Atlantic Groundfish Strategy (TAGS) (March 1998).

58. Much of the above comes from an evaluation of the Northern Cod Adjustment and Recovery Plan carried out for Department of Fisheries and Oceans by Gardner, Pinfold Consulting Economists, Halifax, N.S.

59. Task Force on Incomes and Adjustment in the Atlantic Fishery, *Incomes and Adjustment in the Atlantic Fishery* (Ottawa, December 1992).

60. Cashin report, 74, 194–5.

61. Cashin report, 9.

62. Pat Chamut, Department of Fisheries and Oceans, pers. comm.

63. Cashin report, 56.

64. Office of the Auditor General, *1997 Report*.

65. Pat Chamut, Department of Fisheries and Oceans, pers. comm.

66. Cashin report, 146–7.

67. I owe part of the above to Greg Peacock and Pat Chamut, Department of Fisheries and Oceans, pers. comm.

68. Much of the above comes from GTA Consultants, *Evaluation of the Atlantic Groundfish Strategy (DFO components)*, report prepared for Department of Fisheries and Oceans, Review Directorate (December 1998), Internet: http://www.dfo-mpo.gc.ca/communic/cread/reports/98-99/tags98/index_e.html.

69. Department of Fisheries and Oceans, Review Directorate, *Evaluation of the Canadian Fisheries Adjustment and Restructuring Program Licence Retirement Programs* (May 2002), Internet: http://www.dfo-mpo.gc.ca/communic/cread/reports/02-03/cfar/idex_e.html (accessed January 2003).

70. Task Force on Incomes and Adjustment in the Atlantic Fishery.

71. Cf. Department of Fisheries and Oceans, *Government Assistance*.

72. Ken Jones and Julia Barrow, Department of Fisheries and Oceans, pers. comm.

73. Standing Committee on Fisheries and Oceans, *The Seal Report* (Ottawa, 1990).

74. A.H. Malouf, *Seals and Sealing in Canada: Report of the Royal Commission*, Vol. 1 (Minister of Supply and Services, 1986).

75. Standing Committee on Fisheries and Oceans.

76. Standing Committee on Fisheries and Oceans.

77. The above information comes largely from Standing Committee on Fisheries and Oceans. On the interaction of seals and fish, cf. various Fisheries Resource Conservation Council (F.R.C.C.) reports available on the council's Web-site (e.g., *Uncharted Waters*, the F.R.C.C.'s annual report on groundfish conservation for the year 2000).

Chapter 28 Notes

1. Maurice Beaudin, *Towards Greater Value: Enhancing Eastern Canada's Seafood Industry* (Moncton, N.B.: Canadian Institute for Research on Regional Development, 2001).

2. Information from Clearwater's Internet Web-site (2002).

3. Much of the above comes from Doreen Liew and Maurice Bourque, *Trends in Landed Value and Participation in the Maritime Region's Fisheries* (Dartmouth, N.S.: Department of Fisheries and Oceans, 1997).

4. The above comes in part from Liew and Bourque.

5. In the case of Gulf crab, a decision in 2003 by Minister Robert Thibault, a Nova Scotian who suc-

ceeded Herb Dhaliwal in 2002, allowed permanent small-boat licences, causing violent protests from established crabbers. Boats got burnt on the Caraquet waterfront.

6. Liew and Bourque.

7. Department of Fisheries and Oceans, Gulf Region, *General Overview of the Gulf Fisheries 1987–1999* (Moncton, N.B., 2001).

8. The latter statistics are from the New Brunswick government Web-site in 2002 <http://www.gnb.ca/AFA-APA>.

9. Nova Scotia government Web-site in 2002 <http://gov.ns.ca/nsaf/aquaculture/stats>.

10. Statistics from Nova Scotia and P.E.I. (<http://www.gov.pe.ca>) government Web-sites in 2002.

11. Rory McLellan, P.E.I. Fishermen's Association, pers. comm.

Chapter 29 Notes

1. In May 2003, Minister Robert Thibault announced a more flexible approach, allowing fleets to propose changes to replacement rules, provided they respected conservation, brought no increase in the fleet's overall capacity, and followed other guiding principles.

2. As in the Gulf, Minister Robert Thibault in 2003 converted temporary seasonal permits to regular licences.

3. This information is from the Canadian Auto Workers Web-site (2002).

4. In 2003, those fishermen in Newfoundland and the Quebec North Shore who still depended to some degree on groundfish took another blow. The department, after cautiously reopening small cod quotas on the northeast and northwest coasts and in the southern Gulf, reclosed them.

Chapter 30 Notes

1. Department of Fisheries and Oceans, Pacific Region, news release, 30 September 1997.

2. P.H. Pearse and P.A. Larkin, *Managing Salmon in the Fraser* (Vancouver, B.C.: Department of Fisheries and Oceans, 1992).

3. J.A. *Fraser, Fraser River Sockeye 1994 : Problems and Discrepancies* (Ottawa: Fraser River Sockeye Public Review Board, 1995), xi–xv, 58.

4. Cf. Pacific Fisheries Resource Conservation Council, *Ocean Habitat*, Internet: http:///www.fish.bc.ca/html/fish2115.htm (accessed 2003).

5. John Sutcliffe, Canadian Council of Professional Fish Harvesters, pers. comm.

6. David Egan, Pricewaterhouse Cooper, Presentation to the B.C. Salmon Farmers Association Annual General Meeting, Vancouver,

B.C., 15 June 2000, Internet: http://www.salmon-farmers.org (accessed 10 April 2002).

7. Department of Fisheries and Oceans, Review Directorate, A*udit of Pacific Commercial Salmon Licence Retirement Program—November 1997*, Internet: http://www.dfo-mpo.gc.ca/communic/cread/reports/97-98/retirement/index_e.html.

8. Department of Fisheries and Oceans, Review Directorate, *Evaluation.*

9. This information comes from a 1994 departmental review of licensing issues.

10. B.C. Ministry of Agriculture, Food and Fisheries Web-site, 2002.

11. Pierre De Bané, speech to the Native Brotherhood of British Columbia, 25 November 1983.

12. Donald M. McRae and Peter H. Pearse, *Treaties and Transition: Towards a Sustainable Fishery on Canada's Pacific Coast* (Vancouver: Fisheries and Oceans Canada and the B.C. Minister Responsible for Treaty Negotiations and Minister of Agriculture, Food and Fisheries, April 2004), 11; Allen Wood, former federal fisheries official, pers. comm., from a study for the Native Brotherhood.

13. Part of the above comes from Department of Fisheries and Oceans news releases of 1 and 15 September 1999.

14. Pat Chamut, Department of Fisheries and Oceans, pers. comm.

15. Department of Fisheries and Oceans, *A New Direction for Canada's Pacific Salmon Fisheries*, discussion paper and associated news release and backgrounders, 14 October 1998.

16. Department of Fisheries and Oceans, *Wild Salmon Policy: Discussion Paper* (March 2000); *Allocation Framework for Pacific Salmon 1999-2005* (discussion paper, December 1998); *An Allocation Policy for Pacific Salmon* (policy paper, 22 October 1999). A revised wild salmon policy document appeared in 2004.

17. Parsons, *Management*, 216–7.

18. Pearse, 125–6; Parsons, *Management*, 217–8.

19.. David Anderson, speech to the Standing Committee on Fisheries and Oceans, 28 January 1998.

Chapter 31 Notes

1. Parsons, *Management*, 58-74.

2. Jake Rice, Department of Fisheries and Oceans, pers. comm.

GLOSSARY

Anadromous species
Fishes such as salmon that spawn in fresh water but spend much of their adult lives in salt water.

Boat
See vessel.

Bultow
Old name for longline.

Bye-boat
A boat kept in Newfoundland by early British fishermen who fished at Newfoundland in season.

Cuddy
A small vessel or boat's forward, underdeck accommodation, and storage space, often with bunks and a stove.

Drag, dragger
A large, bag-shaped net (or "trawl") pulled along the ocean bottom, with its mouth held open, by a groundfish dragger. Fishermen also use drags of various types for scallop, shrimp, and other species. In federal fisheries management, "dragger" usually applies to vessels less than 100 feet long, and "trawler" to longer vessels (although scallop vessels, even over 100 feet, generally get called "draggers.") See also trawl.

$F_{0.1}$
A technical term used in stock assessments. $F_{0.1}$ corresponds to a level of fishing that produces somewhat less than maximum sustainable yield from a stock, but gives more fish per unit of effort and reduces the danger of overfishing. For many groundfish stocks, $F_{0.1}$ means catching about two fish of every ten, each year.

Fathom
Six feet.

Flake
A device (e.g., stage, platform, or slotted metal tray) on which fish are laid to dry.

Floater
Traditionally, a cod-fishing vessel that moved from place to place along the northeast Newfoundland or Labrador coast during the season.

Green fishery
Traditionally, a fishery for cod sold wet-salted rather than dried.

Groundfish
Generally, whiter fleshed species, such as cod and haddock, that live near the ocean bottom.

Handline
To fish using hook and line.

High-liner
A top fisherman (said to come from the higher mark left by the water on the side of his laden boat).

Inshore
A term originally associated with small, often open boats that fished close to shore and tied up every night. In federal fisheries terminology, however, the term has sometimes meant vessels as large as 65 feet long. See also midshore, offshore.

Jig
An unbaited hooking device that can have several barbs; the fisherman jigs it up and down in the water.

Livier
One who lives on the Labrador coast.

Longline
A long line with many attached shorter lines (often called snoods or gangions) carrying hooks.

Midshore
A loose term that came into currency on the Atlantic n the 1970's and 1980's for vessels between about 45–50 and 100 feet. Most belonged to independent operators; some belonged to corporate fleets. See also inshore, offshore.

Mile
In this book, generally a nautical mile, which is 6,080 feet or 1,853.2 metres. Canada's 200-mile zone equals about 371 kilometres.

Offshore
A loose term usually associated with the fishery by larger vessels, most often owned by companies rather than individuals, able to stay at sea for days or weeks at a time, and often fishing 12 or more miles offshore. In federal fisheries statistics, the term "offshore" has meant different things at different times: vessels over 25.5 gross tons, vessels over 65 feet, or vessels over 100 feet. See also inshore, midshore, inshore.

Overfishing
Can mean yield or growth overfishing (fishing hard enough to reduce yield; fishing less would let the fishermen catch more), recruitment overfishing (fishing so hard that it threatens the very reproduction of the stock), economic overfishing (high fishing effort cutting profits below what they could be), or overrunning quotas or other conservation regulations.

Pelagic species
Darker-fleshed species, such as herring, swordfish, and tuna, that live near the ocean's surface.

Pickled fish
Fish cured in a salty brine inside a container.

Planter
A settler in the New World.

Purse-seine
See seine.

Quintal
A hundredweight (112 pounds), a measure commonly used for dried salted cod. A quintal is one-twentieth of a long ton (2,240 pounds).

Recruitment
The joining of young fish to the parent stock of those big enough to catch ("fishable stock").

Room
An old term in Newfoundland for working space along the shore, for drying fish and keeping equipment.

Salt bulk
Piles of salt cod before they are dried.

Salt fish
Any of various forms of salted fish, whether dry (e.g., saltfish) or wet (e.g., pickled fish, brine-cured fish).

Saltfish
Any of various cures of dried salted cod.

Scalefish
Usually means Atlantic groundfish other than cod, but can also take in cod, herring, etc.

Seine
A net with floats on the top and sinkers on the bottom, pulled to encircle fish. In beach-seining, common up into the 20th century, fishermen pulled the net to shore, using the beach and the bottom to prevent the fish from escaping. In purse-seining in open water, fishermen bring the ends of the net together and use a "purse-line" to close up its bottom.

Stage
In Newfoundland, a wharf or platform used for landing and handling fish.

Stationer
Traditionally, a fisherman who located in season at a particular spot on the northern Newfoundland or Labrador coast.

Stock
A rather loose term for a group of fish defined by fishery scientists on the basis of shared areas, behaviours, and biological characteristics.

Wet-salted
Fish salted but not dried.

Tierce
An old term of volume, between a hogshead and a barrel.

Ton, tonnage (for ships and boats)
Said to have originally been related to the tun, a large cask; tonnage indicated how many tuns a ship could carry. The tun eventually became fixed at 252 gallons, which weighed about 2,240 pounds: a "long ton." Meanwhile, various and sometimes confusing ways evolved to measure vessel tonnage. For more than a century, however, "gross tonnage" has meant the volume of major, specified enclosed spaces, as measured in units of 100 cubic feet (e.g., a fish hold that happened to be 10 feet _ 10 feet _ 10 feet would equal 1,000 cubic feet, or 10 tons). There are other ways to measure vessel tonnage; for example, displacement and deadweight tonnage relate to weight. In this book, "tonnage" normally means gross tonnage, as measured at the time referred to. There is no hard and fast relation between tonnage and length. In Atlantic Canada in the 1960's, craft of ten or more tons were likely to be 40 or more feet long, and those of less tonnage likely to be shorter. A vessel of 25 gross tons was likely, and one of 50 tons almost certain, to be more than 50 feet. Since then, vessels have tended to get bulkier, with more tonnage for a given length. A 60-footer built in the 1990's might have twice the tonnage of one built in the 1960's.

Ton (weight)
Can mean 2,000 pounds (short ton) or 2,240 pounds (long ton).

Tonne
A metric tonne, 1,000 kilograms, 2.204.6 pounds.

Trawl
A longline, a string of traps, or an otter trawl: that is, a large bag-shaped net towed to capture fish. "Trawler" means a larger vessel using an otter trawl. Most such trawls get pulled along the bottom; midwater trawls fish higher in the water column. See also drag.

Troll
To pull a hook or hooks through the water from a moving boat.

Vessel

Until after the First World War, commonly meant a larger, decked craft with enclosed spaces. A "boat" meant a small, open craft. In following years, the term "boat" often included somewhat larger craft, which might have at least a partial deck. After the Second World War, the federal fisheries department for a time classified craft over ten tons (which were generally at least 40 feet long) as vessels, and smaller craft as boats. But by then, the term "vessel" was seeing less use; the trend was to call fishing craft of whatever size "boats." In this book, before the Second World War, the terms "vessel" and "boat" generally carry their original meaning; after, they are often used indiscriminately.

BIBLIOGRAPHY
Government publications and documents

Canada and the Province of Canada

Angel, J.R., and D.L. Burke, R.N. O'Boyle, F.G. Peacock, M. Sinclair, and K.C.T. Zwanenburg. *Report of the Workshop on Scotia–Fundy Groundfish Management from 1977 to 1993.* Canadian Technical Report of Fisheries and Atlantic Sciences No. 1979 (1994).

Applebaum, B.A. *Law of the Sea—The Exclusive Economic Zone.* Canadian Museum of Civilization Web-site: *www.civilization.ca/hist/lifelines/apple1e.html* (accessed 20 September 2002).

Appleton, T.E. *Usque ad Mare: A History of the Canadian Coast Guard and Marine Services.* Ottawa: Department of Transport, 1968.

Atlantic Development Board. *Fisheries in the Atlantic Provinces.* Ottawa, 1969.

Audette, L.A. Claims by Pelagic Sealers Arising out of the Washington Treaty, 7th July 1911, vol. 16. Victoria, B.C., 1915.

Auditor General of Canada. *Reports.*

Balcom, B.A. *History of the Lunenburg Fishing Industry.* Lunenburg, N.S.: Lunenburg Marine Museum Society, 1977.

Balcom, B.A. *The Cod Fishery of Isle Royale, 1713–58.* Ottawa: Environment Canada, Parks Canada, 1984.

Bennett, Marilyn G. *Indian Fishing and Its Cultural Importance in the Fraser River System.* Fisheries Service, Pacific Region. Environment Canada and Union of British Columbia Indian Chiefs. April 1973.

Bickerton, Carman. *A History of the Canadian Fisheries in the Georges Bank Area.* Vol. 2 of the Canadian Counter-Memorial submitted to the International Court of Justice in the Delimitation of the Maritime Boundary in the Gulf of Maine area. Ottawa, 1983.

Bourne, N. *Scallops and the Offshore Fisheries of the Maritimes.* Bulletin 145. Ottawa: Fisheries Research Board of Canada, 1964.

British Columbia. *Fishery Commission Report,* 1892.

Burke, L. *Experience with Individual Quota and Enterprise Allocation (IQ/EA Management in Canadian Fisheries, 1972–1994.* Halifax, N.S.: Department of Fisheries and Oceans, Policy and Economic Branch, 1994.

Burke, D.L., R.N. Boyle, P. Partington, and M. Sinclair, eds. *Report of the Second Workshop on Scotia–Fundy Groundfish Management.* Canadian Technical Report of Fisheries and Aquatic Sciences No. 2100 (1996).

Caddy, J.F. *A Review of Some Factors Relevant to Management of Swordfish Fisheries in the Northwest Atlantic.* Technical Report 633. Ottawa: Environment Canada, Fisheries and Marine Service, 1976.

Campbell, B.A. *An Historical Review of Developments in Salmon Licensing up until 1968.* Internal paper, Fisheries and Marine Service, Department of the Environment, 1973.

Canada. *Delimitation of the Maritime Boundary in the Gulf of Maine Area (Canada/United States of America).* Annexes to the Memorial Submitted by Canada. Vol. 1. *Contemporary Treaties Affecting the Northwest Atlantic Fisheries, with a Historical Introduction.* The Hague, Netherlands: International Court of Justice, September 1982.

Canadian Fishing Report. Various issues. July 1979–August 1984.

Canadian Museum of Civilization, Parks Canada, and IDON EAST. *Balancing the Scales: Canada's East Coast Fishery.* CD-ROM. 1999.

Candow, J.E. *Of Men and Seals: A History of the Newfoundland Seal Hunt.* Ottawa: Environment Canada, Canadian Parks Service, c. 1989.

Cashin, Richard. *Charting a New Course: Towards the Fishery of the Future.* Report of the Task Force on Incomes and Adjustment in the Atlantic Fishery. Ottawa: Department of Fisheries and Oceans, 1993. (The Cashin report.)

Chambers, E.J. *The Canadian Marine: A History of the Department of Marine and Fisheries.* Toronto, Ont., 1905.

Cockfield, Brown & Company, Ltd. *Summary of Report on the Marketing of Canadian Fish and Fish Products.* Ottawa: King's Printer, 1932.

Commission of Conservation Canada, 1911. Ottawa: The Mortimer Co., Ltd., 1911.

Commission of Conservation, 1915. Toronto, Ont.: The Methodist Book and Publishing House, 1916.

Commissioner of Crown Lands for the Province of Canada. *Annual Reports.*

Commission to Inquire into the Herring and Sardine Industry of the Bay of Fundy, as well as into the Ravages of the Dog-fish and the General Condition of the Lobster Fishery at the Magdalen Islands, St. Mary's Bay and the Bay of Fundy. 1903–1905.

Commission to Investigate a Scheme of Scientific Fishing for Sea Herring in the Atlantic Waters of the Dominion, 1915. (The Hjort expedition.)

Commission to Investigate Grievances and Complaints existing in regard to Salmon and Lobster Fisheries in Gloucester County, New Brunswick, 1904.

Compilation 1860. Acts respecting fisheries, and the inspection of fish and oil: also fisheries regulations for Lower Canada. Quebec: Stewart Derbishire & George Desbarats, 1860. In the library of the Department of Fisheries and Oceans, Ottawa, Cat. No. 192545.

Compilation 1873. The fishery acts: Dominion of Canada: consisting of: "An Act for the Regulation of Fishing and Protection of Fisheries"; "An Act respecting Fishing by Foreign Vessels," passed by the Parliament of Canada in 1868, applicable generally to the Fisheries of Canada; also certain Provincial Statutes continued in force in the provinces of Nova Scotia and New Brunswick. Ottawa: I.B. Taylor, 1873. In the library of the Department of Fisheries and Oceans, Ottawa, Cat. No. 192538.

Compilation 1894. *Consolidated Fishery Regulations: Canada 1889.* Ottawa: 1894, 1890. In the library of the Department of Fisheries and Oceans, Ottawa, Cat. No. 192507.

Compilation 1906. Acts and Regulations re Fisheries. In the library of the Department of Fisheries and Oceans, Ottawa, Cat. No. 190391.

Crowley, R.W., B. McEachern, and R. Jasperse. "A Review of Federal Assistance to the Canadian Fishing Industry, 1945–1990." In *Perspectives on Canadian Marine Fisheries Management.* Edited by L.S. Parsons and W.H. Lear. Canadian Bulletin of Fisheries and Aquatic Sciences, No. 226. Ottawa: National Research Council and Department of Fisheries and Oceans, 1993.

Debate on the fisheries bill, of the Hon. Alex Campbell, Commissioner of Crown Lands, on the 9t[h] and 10th of March, 1865. Quebec: Daily News Office, 1865.

Dempsey, H.V. "Fisheries." *In Canada One Hundred 1867–1967.* Ottawa: Dominion Bureau of Statistics, 1967.

Department of Fisheries. *Annual Reports.* 1930–1968.

Department of Fisheries. "Federal–Provincial Conference on Fisheries Development." Background papers. Ottawa, January 20–24, 1964.

Department of Fisheries. *Proceedings: Canadian Atlantic Offshore Fishing Vessel Conference, Montreal, February 7–9, 1966.* Ottawa: Department of Fisheries, 1966.

Department of Fisheries. *Trade News.* Periodical published by Department of Fisheries. Ottawa, 1948–66.

Department of Fisheries. *Trends in the Development of the Canadian Fisheries.* Background document for fisheries development planning. Department of Fisheries, Economic Services, April 1967.

Department of Fisheries and Forestry. *Proceedings: Meeting on Atlantic Crab Fishery Development.* Fredericton, N.B.: March 4–5, 1969.

Department of Fisheries and Oceans. *A Station in History: Commemorating the 75ᵗʰ Anniversary of the Pacific Biological Station.* Nanaimo, B.C.: Pacific Biological Station, Department of Fisheries and Oceans, 1983.

Department of Fisheries and Oceans. *Quality Excellence in the 1980's.* Booklet. Ottawa, 1980.

Department of Fisheries and Oceans. *The American Lobster.* Leaflet in the Underwater World series. Ottawa, 1982.

Department of Fisheries and Oceans. "Major federal development plan for the Quebec fisheries." News release, Ottawa, 11 July 1983.

Department of Fisheries and Oceans. *Atlantic Fisheries Development: 1983/1984 Annual Project Review.* Ottawa, 1984.

Department of Fisheries and Oceans. *Challenges Facing the Fishery Sector: Report to First Ministers.* Ottawa, 1986.

Department of Fisheries and Oceans. *Resource Prospects for Canada's Atlantic Fisheries, 1989–1993.* 1988.

Department of Fisheries and Oceans. *Government Assistance to the Fishing Industry:* Canada. Paper presented at a Workshop on Seafood Inspection, 21–23 January 1998, Paris. Document AGR/FI/RD(98)39. Paris: Organisation for Economic Co-operation and Development, 1998.

Department of Fisheries and Oceans. *Canada's Foreign Fisheries Relations Policy.* Internet: *http://www.dfo.mpo.gc.ca/communic/fish_man/forfish/b64/sld001.html* (accessed 9 May 2002).

Department of Fisheries and Oceans. *Canada's Oceans Act.* Internet: *http://www.pac.dfo-mpo.gc.ca/oceans/Oceans%20Act/oceanact_e.html* (accessed 6 June 2002).

Department of Fisheries and Oceans, Gulf Region. *First Indian Food Fisheries Conference,* Gulf Region. Memramcook, N.B., 1982.

Department of Fisheries and Oceans, Gulf Region. *General Overview of the Gulf Fisheries 1987–1999.* Moncton, N.B., 2001.

Department of Fisheries and Oceans, Review Directorate. Audit of Pacific Commercial Salmon Licence Retirement Program—November 1997. Internet: *http://www.dfo-mpo.gc.ca/communic/cread /reports/97-98/retirement/index_e.html.*

Department of Fisheries and Oceans, Review Directorate. *Evaluation of the Canadian Fisheries Adjustment and Restructuring Program Licence Retirement Programs.* May 2002. Internet: *http://www.dfo-mpo.gc.ca/communic /cread/reports/02-03/cfar/idex_e.html* (accessed January 2003).

Department of Fisheries and Oceans, Scotia–Fundy Region. *Fish Counting: The Report of the Working Committee on the Catch and Effort Statistics Module in Scotia–Fundy Region.* 12 November 1982.

Department of Marine and Fisheries. *Annual Reports.* Ottawa, 1868–1914, 1921–1930.

Department of Naval Service. *Annual Reports.* 1914–1920.

Department of the Interior, Natural Resources Intelligence Service. *The Utilization of Fish By-products.* Ottawa, 1928.

Department of the Secretary of State. *Return correspondence between the Government of the Dominion and the Imperial Government on the subject of the fisheries, with other documents relating to the same, laid before the Honourable the House of Commons.* Ottawa, 1871.

DeWolf, A. Gordon. *The Lobster Fishery of the Maritime Provinces: Economic Effects of Regulations.* Bulletin 187. Ottawa: Fisheries Research Board of Canada, 1974.

Dominion–British Columbia Boat-Rating Commission, 1910.

Dominion–British Columbia Fisheries Commission, 1905–1907.

Dominion Shell-fish Fishery Commission: Report and Recommendations. Ottawa, 1913.

Dorion-Robitaille, Y. *Captain J.E. Bernier's Contribution to Canadian Sovereignty in the Arctic.* Ottawa: Department of Indian and Northern Affairs, 1978.

Duff, W.C. *British Columbia Fisheries Commission.* 1922. (The Duff report.)

Dunfield, Robert W. *The Atlantic Salmon in the History of North America.* Ottawa: Department of Fisheries and Oceans, 1985.

Dunn, D.L. *Canada's East Coast Sealing Industry 1976.* Fisheries and Marine Service, 1977.

Eberts, D.M. *Report of the Hon. D.M. Eberts on Charges of Inefficiency and Irregularities by Fishery Officers in Fishery District No. 3, B.C. and into Inefficient Methods There.* Ottawa: King's Printer. 1921.

Environment Canada, Fisheries and Marine Service. *Policy for Canada's Commercial Fisheries.* Ottawa, May 1976.

Evans, W.S., H.B. Thomson, and F.T. James. *Report of Special Fishery Commission, Province of British Columbia, 1917.* Ottawa, 1918. (The Evans report.)

Finn, D.B., C.R. Molson, R.W. Bedard, and W.S. Posthumus. *Commission of Enquiry into the Atlantic Salt Fish Industry, 1965.* Canada: Nethercut and Young, 1965.

Fisheries and Marine Service. *Lobster Fishery Task Force.* First report. March 1975.

Fisheries of Canada (1966–71). Periodical published by Department of Fisheries and Department of Fisheries and Forestry.

Fisheries Research Board of Canada. *The Fisheries Research Board of Canada and Its place in Canada's Scientific Development 1968–1978.* Ottawa: Fisheries Research Board of Canada, 1968.

Fisheries Resource Conservation Council. *A Conservation Framework for Atlantic Lobster: Report to the Minister of Fisheries and Oceans.* Report No. FRCC95.R.1, November 1995

Foerster, R.E. *The Sockeye Salmon,* Oncorhyncus nerka. Bulletin 162. Ottawa: Fisheries Research Board of Canada, 1968.

Forrester, C.R., and K.S. Ketchen. *A Review of the Strait of Georgia Trawl Fishery.* Bulletin 139. Ottawa: Fisheries Research Board of Canada, 1963.

Fortin, Pierre. *Annual Reports, in the Journal of the Legislative Assembly of Canada,* 1852–1857.

Fortin, Pierre. In *Fisheries appendices from the annual report, for 1863, of the Hon. Wm. McDougall, Commissioner of Crown Land: statements of Fisheries Branch, Superintendents' reports, Capt. Fortin's report, synopsis of fishery overseers' reports, etc. for the year 1863.* Quebec:

Hunter, Rose & Co., 1864. In the library of the Department of Fisheries and Oceans, Ottawa, Cat. No. 223025.

Fortin, Pierre. *Reports to the Legislative Assembly of Canada.*

Fraser, J.A. *Fraser River Sockeye 1994: Problems and Discrepancies.* Ottawa: Fraser River Sockeye Public Review Board, 1995.

Frick, H.C. *Economic Aspects of the Great Lakes Fisheries of Ontario.* Bulletin 149. Fisheries Research Board of Canada, 1965.

Georgian Bay Fisheries Commission, 1905–1908.

Gordon, H. Scott. *The Fishing Industry of Prince Edward Island.* Markets and Economics Service, Department of Fisheries of Canada, March 1952.

Gough, Joseph. "A Historical Sketch of Fisheries Management in Canada." In *Perspectives on Canadian Marine Fisheries Management.* Edited by L.S. Parsons and W.H. Lear. Canadian Bulletin of Fisheries and Aquatic Sciences, No. 226. Ottawa: National Research Council and Department of Fisheries and Oceans, 1993.

Gough, Joseph. *Fisheries Management in Canada, 1880-1910.* Canadian Manuscript Report of Fisheries and Aquatic Sciences; 2105. Halifax: Department of Fisheries and Oceans, 1991.

GTA Consultants. *Evaluation of the Atlantic Groundfish Strategy (DFO components).* Report prepared for the Department of Fisheries and Oceans, Review Directorate (December 1998). Internet: *http://www.dfo-mpo.gc.ca/communic/read/ reports/98-99/tags98/index_e.html.*

GTA Fisheries Consultants. T*he Mobile Gear Groundfish Fleet under 65_ in Atlantic Canada.* Report prepared for Department of Fisheries and Oceans, 1988.

Haché, J.-E. *Report of the Scotia-Fundy Groundfish Task Force.* Ottawa: Department of Fisheries and Oceans, 1989.

Hachey, H.B. *Oceanography and Canadian Atlantic waters.* Bulletin 134. Fisheries Research Board of Canada, 1961.

Hachey, H.B. *History of the Fisheries Research Board.* Manuscript Report 843, Ottawa: Fisheries Research Board of Canada, 1965.

Haig-Brown, R.L. *Canada's Pacific salmon.* Department of Fisheries. Reprinted from *Canadian Geographical Journal,* 1952.

Haig-Brown, Roderick. *The Salmon.* Ottawa: Environment Canada, Fisheries and Marine Service, 1974.

Hare, G.M., and D.L. Dunn. "A Retrospective Analysis of the Gulf of St. Lawrence Snow Crab (*Chionoecetes opilio*) Fishery, 1965–1990." In *Perspectives on Canadian Marine Fisheries Management.* Edited by L.S. Parsons and W.H. Lear. Canadian Bulletin of Fisheries and Aquatic Sciences, No. 226. Ottawa: National Research Council and Department of Fisheries and Oceans, 1993.

Harris, Leslie. *Independent Review of the State of the Northern Cod Stock.* Report to the Hon. Tom Siddon, February 1990.

Hart, J.L. *The Pilchard Fishery of British Columbia.* Ottawa: Biological Board of Canada, 1933.

Hart, J.L., F. Neave, and D.B. Quayle. *Brief on the Fishery Wealth of British Columbia.* Fisheries Research Board of Canada, 1951.

Hart, J.L. *Pacific Fishes of Canada.* Bulletin 180. Ottawa: Fisheries Research Board of Canada, 1973.

Healey, M.C. "The Management of Pacific Salmon Fisheries in British Columbia." In *Perspectives on Canadian Marine Fisheries Management.* Edited by L.S. Parsons and W.H. Lear. Canadian Bulletin of Fisheries and Aquatic

Sciences, No. 226. Ottawa: National Research Council and Department of Fisheries and Oceans, 1993.

Hebert, Louise. *Great Expectations: The Atlantic Fish Processing Sector.* Moncton, N.B.: Department of Fisheries and Oceans, Economic Research and Analysis Division, Gulf Region, 1989.

Hind, H.Y. *The Effect of the Fishery Clauses of the Treaty of Washington on the Fisheries and Fishermen of British North America.* Parts I and II. Study prepared for the Fishery Commission. Halifax, N.S., 1877.

Hinds, Lennox O., and James Trimm. "Utilization of Catch Now Discarded at Sea." In *Proceedings: Government–Industry Meeting on the Utilization of Atlantic Marine Resources, Queen Elizabeth Hotel, Montreal, P.Q., February 5–7, 1974.* Environment Canada, Fisheries and Marine Service, 1974.

Hourston, A.S., and C.W. Haegele. *Herring on Canada's Pacific Coast.* Canadian Special Publication of Fisheries and Aquatic Sciences 48. Ottawa: Department of Fisheries and Oceans, 1980.

Human Resources Development Canada. *Evaluation of the Atlantic Groundfish Strategy (TAGS).* Human Resources Development Canada, Evaluation and Data Development, March 1998.

Iles, T.D., "The Management of the Canadian Atlantic Herring Fisheries," in *Perspectives on Canadian Marine Fisheries Management,* edited by L.S. Parsons and W.H. Lear, Canadian Bulletin of Fisheries and Aquactic Sciences, No. 226. Ottawa: National Research Council and Department of Fisheries and Oceans, 1993.

Jenness, Diamond. *The Indians of Canada,* 5th ed. Ottawa: National Museum of Canada, 1960.

Johnson, W.W. *Development of stern drum seining in Canada.* Fisheries and Marine Service, 1975.

Johnstone, Kenneth. *The Aquatic Explorers: A History of the Fisheries Research Board of Canada.* Toronto, Ont.: University of Toronto Press, 1977.

Johnstone, Kenneth. "History of Canadian Fisheries and Marine Service." Historical manuscript in the library of the Department of Fisheries and Oceans. Ottawa, 1977.

Juvet, C.S. "Licensing Policies and Procedures: Confederation–1949." Compilation in the library of the Department of Fisheries and Oceans, Ottawa, 1950, Cat. No. 208624.

Kennedy, W.A. *A History of Commercial Fishing in Inland Canada.* Manuscript Report Series. London, Ont.: Fisheries Research Board of Canada, 1966.

Kennedy, W.A. *The first Ten Years of Commercial Fishing on Great Slave Lake.* Bulletin 107. Fisheries Research Board of Canada, 1956.

Kirby, M.J.L. *Navigating Troubled Waters.* Report of the Task Force on Atlantic Fisheries. Ottawa, 1982. (The Kirby report.)

Knight, T.F. *Report on the Fisheries of Nova Scotia.* Halifax, N.S.: A. Grant, Printer to the Queen, 1867.

Knight, T.F. *Shore and Deep Sea Fisheries of Nova Scotia.* Halifax, N.S.: A. Grant, Printer to the Queen, 1867.

Knight, T.F. *The River Fisheries of Nova Scotia.* Nova Scotia, 1867. Halifax, N.S.: A. Grant, Printer to the Queen, 1867.

Lear, W.H. "The Management of Canadian Atlantic Salmon Fisheries." In *Perspectives on Canadian Marine Fisheries Management.* Edited by L.S. Parsons and W.H. Lear. Canadian Bulletin of Fisheries and Aquatic Sciences, No. 226. Ottawa: National Research Council and Department of Fisheries and Oceans, 1993.

Lear, W.H., and L.S. Parsons. "History and Management of the Fishery for Northern Cod in NAFO Divisions 2J, 3K and 3L." In *Perspectives on Canadian Marine Fisheries Management.* Edited by L.S. Parsons and W.H. Lear. Canadian Bulletin of Fisheries and Aquatic Sciences, No. 226. Ottawa: National Research Council and Department of Fisheries and Oceans, 1993.

Leim, A.H., and W.B. Scott. *Fishes of the Atlantic Coast of Canada.* Bulletin 155. Fisheries Research Board of Canada, 1966.

Levelton, C.R. *Towards an Atlantic Commercial Fisheries Licensing System.* Ottawa: Department of Fisheries and Oceans, 1979.

Liew, Doreen, and Maurice Bourque. *Trends in Landed Value and Participation in the Maritime Region's Fisheries.* Dartmouth, N.S.: Department of Fisheries and Oceans, 1997.

Lister, D.B. *A Brief Review of Salmon Enhancement Programs in Pacific and Atlantic Canada.* Vancouver, B.C.: Department of Fisheries and Oceans, 1974.

Low, A.P. *Report on the Dominion Government Expedition to Hudson Bay and the Arctic Islands on Board the D.G.S. Neptune, 1903–1904.* Ottawa, 1906.

MacCrimmon, H.R., J.E. Stewart, and J.R. Brett. *Aquaculture in Canada: The Practice and the Promise.* Ottawa: Environment Canada, Fisheries and Marine Service, 1974.

MacKenzie, R.A. *The Canadian Atlantic Offshore Cod Fishery East of Halifax.* Bulletin 61. Ottawa: Fisheries Research Board of Canada, 1942.

Malouf, A.H. *Seals and Sealing in Canada: Report of the Royal Commission,* Vol. 1. Minister of Supply and Services, 1986.

Martin, W.R. *A Summary Report of the Atlantic Groundfish Investigation.* St. Andrews, N.B.: Atlantic Biological Station, 1949.

Materials on 1865 fisheries bill: the debate and the act. In the library of the Department of Fisheries and Oceans, Ottawa. Cat. Nos. J102 A2 1865 and SH 3L7. C21 A121 1865.

McCullough, A.B. *A Select, Annotated Bibliography on the History of the Commercial Fisheries of the Canadian Great Lakes.* Ottawa: Parks Canada, 1986.

McCullough, A.B. *The Commercial Fishery of the Canadian Great Lakes.* Ottawa: Environment Canada, Canadian Parks Service, 1989.

McRae, Donald M., and Peter H. Pearse. *Treaties and Transition: Towards a Sustainable Fishery on Canada's Pacific Coast.* Vancouver: Fisheries and Oceans Canada and the B.C. Minister Responsible for Treaty Negotiations and Minister of Agriculture, Food and Fisheries, April, 2004.

McKay, W., and Ouellette, K. *A review of the British Columbia Indian Fishermen's Assistance Program 1968/69 – 1977/78.* Indian and Northern Affairs Canada, Policy Research, and Evaluation Group, 1978.

Medcof, J.C. *Oyster Farming in the Maritimes.* Bulletin 131. Ottawa: Fisheries Research Board of Canada, 1961.

Memorandum upon the questions at issue between Great Britain and the United States in relation to the North American fisheries, 1898. In the library of the Department of Fisheries and Oceans, Ottawa, Cat. No. 190902.

Mensikai, S.S. *Plant Location and Plant Size in the Fish Processing Industry of Newfoundland.* Canadian Fisheries Reports No. 11. Ottawa: Department of Fisheries and Forestry, 1969.

Mitchell, C.L. *Canada's Fishing Industry: A Sectoral Analysis.* Ottawa: Department of Fisheries and Oceans, 1980.

Mitchell, C.L., and D.B. McEachern. *Developments in the Atlantic Coast Herring Fishery and Fish Meal Industry, 1964–1968.* Canadian Fisheries Reports No. 16. Ottawa: Department of Fisheries and Forestry, August 1970.

Mitchell, C.L., and H.C. Frick. *Government Programs of Assistance for Fishing Craft Construction in Canada: An Economic Appraisal.* Canadian Fisheries Report No. 14. Ottawa: Department of Fisheries and Forestry, 1970.

Morse, N.H. *An Economic Study of the Oyster Fishery of the Maritime Provinces.* Bulletin 175. Ottawa: Fisheries Research Board of Canada, 1971.

Morse, N.H., and A.G. DeWolf. *Management of the Fisheries of the Saint John River System, New Brunswick, with Particular Reference to Atlantic Salmon.* Halifax, N.S.: Department of Economics, Dalhousie University. Canada: Fisheries and Marine Service, 1974.

Moussette, Marcel. *Fishing Methods Used in the St. Lawrence River and Gulf.* Ottawa: National Historic Parks and Sites Branch, Parks Canada; Indian and Northern Affairs Canada, 1979.

Office of the Auditor General of Canada. *1988 Report of the Auditor General of Canada.* Ottawa, 1988.

Office of the Auditor General of Canada. *1993 Report of the Auditor General of Canada.* Ottawa, 1993.

Office of the Auditor General of Canada. *1997 Report of the Auditor General of Canada.* Ottawa, 1997.

Office of the Commissioner for Aquaculture Development. *Canadian Aquaculture Industry* Profile, 14 December 2001. Internet: *http://ocad-bcda.gc.ca /eaquaculture.html* (accessed 7 February 2002).

Outram, D.N. *Canada's Pacific Herring.* Ottawa: Department of Fisheries, 1965.

Outram, D.N., and R.D. Humphreys. *The Pacific Herring in British Columbia Waters.* Pacific Biological Station, 1974.

Ozere, S.V. "Survey of Legislation and Treaties Affecting Fisheries." In *Resources for Tomorrow: Montreal, October 22–28, 1961.* Vol. 2. Conference Background Papers. Ottawa: Queen's Printer, 1961. pp. 797–805.

Pacific Fisheries Resource Conservation Council. *Ocean Habitat.* Internet: *htpp://www.fish.bc.ca/ html/fish2115.htm* (accessed 2003).

Parisien, Richard W. *The Fisheries Act: origins of federal delegation of administrative jurisdiction to the provinces.* Report submitted to Environment Canada: Policy, Planning and Research Service, Ottawa, 1972.

Parsons, L.S. Management of Marine Fisheries in Canada. Ottawa: National Research Council, 1993.

Parsons, L.S. "Shaping Fisheries Policy: The Kirby and Pearse Inquiries—Process, Prescription and Impact." In *Perspectives on Canadian Marine Fisheries Management.* Edited by L.S. Parsons and W.H. Lear. Canadian Bulletin of Fisheries and Aquatic Sciences, No. 226. Ottawa: National Research Council and Department of Fisheries and Oceans, 1993.

Parsons, L.S., and W.H. Lear, eds. *Perspectives on Canadian Marine Fisheries Management.* Canadian Bulletin of Fisheries and Aquatic Sciences, No. 226. Ottawa: National Research Council and Department of Fisheries and Oceans, 1993.

Pearse, Peter H. *Turning the Tide: A New Policy for Canada's Pacific Fisheries.* Vancouver, B.C.: Commission on Pacific Fisheries Policy. 1982. (The Pearse report.)

Pearse, P.H., and P.A. Larkin. *Managing Salmon in the Fraser.* Vancouver, B.C.: Department of Fisheries and Oceans, 1992.

Pelagic Sealing Commission, 1913–1916.

Prince, E.E. "The Biological Board of Canada." In *4th Annual Report of the Commission of Conservation*. Ottawa: Commission of Conservation, 1913.

Prince, E.E. *The Dog-fish Plague in Canada*. Ottawa: Government Printing Bureau, 1904. In the library of the Department of Fisheries and Oceans, Ottawa.

Prince, E.E. *Dominion Shad Fishery Commission, 1908–1910: Report and Recommendations*. Ottawa: Government Printing Bureau, 1910.

Prince, E.E. *Fluctuations in the Abundance of Fish*. In *Special Reports of Professor E.E. Prince, 1894–1909*. Ottawa: Commissioner of Fisheries, 1894, 1909. Compilation in the library of the Department of Fisheries and Oceans, Ottawa, Cat. No. 192111.

Prince, E.E. "The Fisheries of Canada." In *30th Annual Report*, Department of Marine and Fisheries. Ottawa, 1898.

Prince, E.E. *Unutilized Fishery Products in Canada*. Ottawa: Government Printing Bureau, 1907. In the library of the Department of Fisheries and Oceans, Ottawa.

Pringle, J.D., and D.L. Burke. "The Canadian Lobster Fishery and Its Management, with Emphasis on the Scotian Shelf and the Gulf of Maine." In *Perspectives on Canadian Marine Fisheries Management*. Edited by L.S. Parsons and W.H. Lear. Canadian Bulletin of Fisheries and Aquatic Sciences, No. 226. Ottawa: National Research Council and Department of Fisheries and Oceans, 1993.

Proceedings: Canadian Atlantic Herring Fishery Conference, Fredericton, N.B., May 5–7, 1966. Sponsored by Federal–Provincial Atlantic Fisheries Committee. Canadian Fisheries Reports No. 8. Ottawa: Department of Fisheries, 1967.

Proskie, J., and J.C. Adams. *Survey of the Labour Force in the Offshore Fishing Fleet, Atlantic Coast*. Fisheries Service, Economic Branch, 1971.

Proulx, Jean-Pierre. *Whaling in the North Atlantic: From Earliest Times to the Mid-19th Century*. Ottawa: Parks Canada, 1986.

Province of Canada. In *Fisheries Appendices from the Annual Report, for 1863, of the Hon. Wm. McDougall, Commissioner of Crown Lands: Statements of Fisheries Branch, Superintendents' reports, Capt. Fortin's Report, Synopsis of Fishery Overseer's Reports, etc. for the Year 1863*. Quebec: Hunter, Rose & Co., 1864. In the library of the Department of Fisheries and Oceans, Ottawa, Cat. No. 223025.

Quayle, D.B. *Pacific Oyster Culture in British Columbia*. Bulletin 169. Ottawa: Fisheries Research Board of Canada, 1969.

Report of Commander Wm. Wakeham, Special Commissioner and Inspector of Fisheries for the Gulf of St. Lawrence, in the Lobster Industry of the Maritime Provinces and the Province of Quebec, 1909–1910.

Report of the American–Canadian Fisheries Conference 1918. Ottawa, 1918.

Report of the Canadian Lobster Commission. Ottawa: S.E. Dawson, 1899.

Report of the Commissioner of Crown Lands of Canada. Toronto, Ont.: John Lovell, 1856–58.

Report of the Commission of Inquiry into Freshwater Fish Marketing, 1965–66. (The McIvor report.)

Report of the Dominion Fishery Commission on the Fisheries of the Province of Ontario, 1892–1893.

Report of the Royal Commission Investigating the Fisheries of the Maritime Provinces and the Magdalen Islands. Ottawa: 1928. (The Maclean report.)

Report of the Royal Commission on Price Spreads, 1934–1935.

Report of the Select Committee on the Working of the Fishery Act. Province of Canada, 1864.

Ricker, W.E. "Productive Capacity of Canadian Fisheries." Background paper for Resources for Tomorrow conference, Montreal, 22–28 October 1961. Circular No. 64. Nanaimo, B.C.: Fisheries Research Board of Canada, June 1961.

Ricker, W.E. *Synopsis of Achievements of the Fisheries Research Board of Canada and its Predecessors, 1898–1972*. Fisheries Research Board of Canada, 1970.

Ricker, W.E. *The Fisheries Research Board of Canada — Seventy-five Years of Achievements*. Ottawa: Fisheries Research Board of Canada, 1975.

Rigby, M.S., and A.G. Huntsman. *Materials Relating to the History of the Fisheries Research Board of Canada (Formerly the Biological Board of Canada) for the Period 1898–1924*. Manuscript Report 660. Fisheries Research Board of Canada, 1958.

Royal Commission on Canada's Economic Prospects: The Commercial Fisheries of Canada, 1955–1957.

Rutherford, J.B., D.G. Wilder, and H.C. Frick. *An Economic Appraisal of the Canadian Lobster Fishery*. Bulletin 157. Fisheries Research Board of Canada, 1967.

Ryan, Shannon. *A Historical Overview of Canadian Newfoundland/World Sealing*. Prepared for the Royal Commission on Seals and the Sealing Industry in Canada, 1986.

Samson, Roch. *Fishermen and Merchants in 19th Century Gaspé*. Ottawa: Parks Canada, 1984.

Scattergood, L.W., and S.N. Tibbo. *The Herring Fishery of the Northwest Atlantic*. Bulletin 121. Ottawa: Fisheries Research Board of Canada, 1959.

Scott, A., and P.A. Neher, eds. *The Public Regulation of Commercial Fisheries in Canada*. A study prepared for the Economic Council of Canada. Ottawa, 1981.

Scott, W.B., and E.J. Crossman. *Freshwater Fishes of Canada*. Bulletin 184. Ottawa: Fisheries Research Board of Canada, 1973.

Senate Standing Committee on Fisheries. *The Atlantic Groundfish Fishery: Its future*. Ottawa: Senate, December 1995.

Sergeant, D.E. *Harp Seals, Man and Ice*. Ottawa: Department of Fisheries and Oceans, 1991.

Shepard, M.P. *History of International Competition for Pacific Salmon*. Manuscript Report 921. Fisheries Research Board of Canada, 1967.

Shepard, M.P., and A.W. Argue. *The Commercial Harvest of Salmon in British Columbia, 1820–1877*. Department of Fisheries and Oceans, Canadian Technical Report of Fisheries and Aquatic Sciences, August 1989.

Sherwood–Meder Marketing Consultants. *Survey of South-Western Nova Scotia Fishing Vessel Owners*. Report commissioned by Carl Myers. Halifax, N.S.: Department of Fisheries and Oceans Communications, 31 March 1989.

Sinclair, S. *Licence Limitation: British Columbia: A Method of Economic Fisheries Management*. Ottawa: Department of Fisheries, 1960.

Standing Committee on Fisheries and Oceans. *The Seal Report*. Ottawa, 1990.

Standing Senate Committee on Fisheries and Oceans. *Straddling Fish Stocks in the Northwest Atlantic*. Ottawa: Senate, June 2003.

Stocker, M. "Recent Management of the British Columbia Herring Fishery." In *Perspectives on Canadian Marine Fisheries Management*. Edited by L.S. Parsons and W.H. Lear. Canadian Bulletin of Fisheries and Aquatic Sciences, No. 226. Ottawa: National Research Council and Department of Fisheries and Oceans, 1993.

Sutterlin, A.M., E.B. Henderson, S.P. Merrill, R.L. Saunders, and A.A. MacKay. *Salmonid Rearing Trials at Deer Island, New Brunswick, with Some Projections on Economic Viability.* Canadian Technical Report of Fisheries and Aquatic Sciences 1011. Ottawa: Department of Fisheries and Oceans, 1981.

Swann, L.G. *A Century of B.C. Fishing.* Ottawa: Department of Fisheries, 1958. Reprint from *Trade News.*

Symposium on Policies for Economic Rationalization of Commercial Fisheries. Special edition of the *Journal of the Fisheries Research Board of Canada* 36, no. 7 (July 1979). Notable articles by A.W.H. Needler, A. Scott, G.A. Fraser, and P.H. Pearse.

Task Force on Incomes and Adjustment in the Atlantic Fishery. *Incomes and Adjustment in the Atlantic Fishery.* Ottawa, December 1992. (The Cashin discussion paper.)

Taylor, F.H. *Life History and Present Status of British Columbia Herring Stocks.* Bulletin 143. Fisheries Research Board of Canada, 1964.

Taylor, J. Garth. " Inuit Whaling Technology in Eastern Canada and Greenland." In *Thule Eskimo Culture: An Anthropological Perspective.* Edited by Allan P. McCartney. National Museum of Man Mercury Series, Archaeological Survey of Canada Paper No. 88, 1979.

Templeman, W. *Redfish Distribution in the North Atlantic.* Bulletin 130. Ottawa: Fisheries Research Board of Canada, 1959.

Templeman, Wilfred. *Marine Resources of Newfoundland.* Bulletin 154. Ottawa: Fisheries Research Board of Canada.

Tester, A.L. *The Herring Fishery of British Columbia: Past and Present.* Bulletin 47. Ottawa: Biological Board of Canada, 1935.

Thompson, P. "Institutional Constraints in Fisheries Management." *Journal of the Fisheries Research Board of Canada* 31, no. 12 (1974).

Tibbo, S.N., L.R. Day, and W.F. Doucet. *The Swordfish (Xiphias gladius L.), Its Life-History and Economic Importance in the Northwest Atlantic.* Ottawa: Department of Fisheries, 1961.

Toews, P.R. *The Canadian Atlantic Shrimp Fishery: Prospects for Development.* Ottawa: Department of Fisheries and Oceans, 1980.

Wadden, R.N. *Department of Fisheries of Canada 1867–1967.* Reprinted from the 1967 Annual Review of the Fisheries Council of Canada.

Walsh, Sir Albert, et al., *Newfoundland Fisheries Development Committee Report.* St. John's, Nfld.: Guardian, 1953. (The Walsh report.)

White, James. "North Atlantic Fisheries Dispute." In *Commission of Conservation, 1911.* Ottawa: Commission of Conservation, 1911.

Other government publications and documents

Bates, Stewart. *Report on the Canadian Atlantic Sea-fishery.* Halifax, N.S.: Royal Commission on Provincial Development and Rehabilitation, 1944.

Dunn, P.D.H. *Fisheries Re-organization in Newfoundland.* Radio address, 21 January 1944. In the library of Memorial University of Newfoundland, St. John's, Nfld., Cat. No. SH 229 D86.

Earll, R.E. "The Sardine Industry." In *The Fisheries and Fishery Industries of the United States.* U.S. Fish Commission, 1887.

Goode, G.B., Joseph W. Collins, R.E. Earll, and A. Howard Clark. *Materials for a History of the Mackerel Fishery.* Washington, D.C.: Government Printing Office, 1883.

Goode, G.B. *The Fisheries and Fishery Industries of the United States.* Washington, D.C.: United States Commission of Fish and Fisheries. Bureau of Fisheries, 1884–1887.

Government of Newfoundland and Labrador. *Resettlement Statistics.* Industry Canada Web-site: *http://collections.ic.gc.ca/placentia/newpag.htm* (accessed 10 December 2003).

Great Britain. Board of Trade. *Report of the Lords Commissions for Trade and Plantations to His Majesty,* 19 December 1718.

Great Lakes Fishery Commission. *Lake Huron: The Ecology of the Fish Community and Man's Effects on It.* Ann Arbor, Mich., 1973.

Great Lakes Fishery Commission. *Lake Michigan: Man's Effects on Native Fish Stocks and Other Biota.* Ann Arbor, Mich., 1973.

Great Lakes Fishery Commission. *Commercial Fish Production in the Great Lakes 1867–1977.* Ann Arbor, Mich., 1979.

Great Lakes Fishery Commission. *Review of Fish Species Introduced into the Great Lakes, 1819–1974.* Ann Arbor, Mich., 1985.

Ingersoll, L.K. *Report by Captain John Robb, R.N., on the state of the fisheries, the condition of the lighthouses, the contraband trade, and various other matters in the Bay of Fundy, made to His Excellency the Lieutenant-Governor, 1840.* Grand Manan Historical Society, 1965.

Jensen, Albert C. *A Brief History of the New England Offshore Fisheries.* U.S. Bureau of Commercial Fisheries, 1967.

Manitoba. *Report on the Commercial Fishing Industry of Manitoba.* Winnipeg, Man.: H.C. Grant, 1938.

Manitoba. *Report of the Commissioners Appointed to Investigate the Fishing Industry of Manitoba.* Winnipeg, Man., 1933.

Manitoba Fisheries Commission. 1910–1911. Interim report and final report. In the library of the Department of Fisheries and Oceans, Ottawa, Cat. No. 34236.

Mcgreggor, Sir William. *Chronological History of Newfoundland Fisheries.* St. John's, Nfld., 1907. Reprinted by Provincial Archives of Newfoundland.

Muse, Ben. *Management of the British Columbia Abalone Fishery.* Alaska Commercial Fisheries Entry Commission, May 1998.

New Brunswick Department of Natural Resources Web-site: *www.gnb.ca/0078/auction-e.asp* (accessed March 2003).

Newfoundland. *National Fisheries Development: A Presentation to the Government of Canada by the Government of Newfoundland,* February 1963.

Newfoundland. *Report of the Commission of Enquiry Investigating the Seafisheries of Newfoundland and Labrador Other than the Sealfishery.* St. John's, Nfld., 1937.

Newfoundland Department of Marine and Fisheries. *Annual Reports.*

Newfoundland Fisheries Board. *Annual Report* for 1942.

Newfoundland Fisheries Commission. *Report and Recommendations of the Newfoundland Fisheries Commission to the Government of Newfoundland, April 1963.* St John's, Nfld., 1963.

Nova Scotia Department of Fisheries. *Sea, Salt and Sweat.* Halifax, N.S., 1977.

Nova Scotia Economic Council. *Proceedings of the Nova Scotia Fisheries Conference: Halifax, July 13–14, 1938.* Halifax, N.S.: Provincial Secretary, King's Printer, 1938.

Pearson, John C. *The Fish and Fisheries of Colonial North America: A Documentary History of Fishery Resources of the United States and Canada.* United States National Marine Fisheries Service, 1972.

Perley, Moses H. *Reports on the Sea and River Fisheries of New Brunswick.* Fredericton, N.B.: Queen's Printer, 1852.

Quebec. *The Lower North Shore,* 1984.

Sabine, Lorenzo. *Report on the Principal Fisheries of the American Seas.* Washington, 1853.

Saskatchewan. *Report of the Royal Commission on the Fisheries of Saskatchewan.* Regina, Sask., 1947.

Stacey, D.A. *Sockeye and Tinplate: Technological Change in the Fraser River Canning Industry, 1871–1912.* British Columbia Provincial Museum, 1982.

Templeman, Wilfred. *The Newfoundland Lobster Fishery:*

An Account of Statistics, Methods, and Important Laws. St. John's, Nfld.: Department of Natural Resources, 1941.

Thompson, Harold. *A Survey of the Fisheries of Newfoundland and Recommendations for a Scheme of Research.* St. John's, Nfld.: Newfoundland Fishery Research Commission, 1931.

United States. Secretary of the Treasury. *Reply of the Secretary of the Treasury to the resolution of the House of Representatives of December 14, 1886.* Washington, D.C.: Government Printing Office, 1887.

United States Tariff Commission. *Treaties Affecting the Northeastern Fisheries.* Report No. 152, second series. Washington, D.C., 1944.

Watt, John W. *A Brief Review of the Fisheries of Nova Scotia.* Nova Scotia Department of Trade and Industry, 1963.

Non-government publications

Abreu-Ferreira, Darlene. "Portugal's Cod Fishery in the 16th Century: Myths and Misconceptions." In *How Deep Is the Ocean? Historical Essays on Canada's Atlantic Fishery.* Edited by James E. Candow and Carol Corbin. Sydney, N.S.: University College of Cape Breton Press, 1997. pp: 31–44.

Alexander, David. "Development and Dependence in Newfoundland, 1880–1970." *Acadiensis* 4 (1974): 3–31.

Alexander, David. "Newfoundland's Traditional Economy and Development to 1934." *Acadiensis* 5 (1976): 56–78.

Alexander, David. "The Political Economy of Fishing in Newfoundland." *Journal of Canadian Studies* 11 (Feb. 1976).

Alexander, David. *The Decay of Trade: An Economic History of the Newfoundland Saltfish Trade, 1935–1965.* St. John's, Nfld.: Institute of Social and Economic Research, Memorial University of Newfoundland, 1977.

Andersen, R. *Social Organization of the Newfoundland Banking Schooner Cod Fishery, circa. 1900–1948.* St. John's, Nfld.: Department of Anthropology, Memorial University of Newfoundland, 1980.

Anspach, Rev. Lewis Amadeux. *A History of the Island of Newfoundland.* London, U.K.: privately printed, 1819.

Backus, Richard H., and Donald W. Bourne, eds. *Georges Bank.* Cambridge, Mass.: MIT Press, 1987.

Baker, Melvin, "Hazen Algar Russell." *Newfoundland Quarterly* 89, no. 3 (Spring/Summer 1995): 35–7.

Baker, Melvin and Shannon Ryan. "The Newfoundland Fishery Research Commission, 1930–1934." In *How Deep Is the Ocean? Historical Essays on Canada's Atlantic Fishery.* Edited by James E. Candow and Carol Corbin. Sydney, N.S.: University College of Cape Breton Press, 1997.

Balcom, B.A. "Technology Rejected: Steam Trawlers and Nova Scotia, 1897–1933." In *How Deep Is the Ocean? Historical Eessays on Canada's Atlantic Fishery.* Edited by James E. Candow and Carol Corbin. Sydney, N.S.: University College of Cape Breton Press, 1997.

Bartlett, Capt. Robert A. *The Log of Bob Bartlett.* New York, 1928.

Bartlett, Capt. Robert A. "The Sealing Saga of Newfoundland." *National Geographic,* July 1929.

Beaudin, Maurice. *Towards Greater Value: Enhancing Eastern Canada's Seafood Industry.* Moncton, N.B.: Canadian Institute for Research on Regional Development, 2001.

Beck, J. Murray. "Nova Scotia, 1714–1784." In *The Canadian Encyclopedia.* 2nd ed. Edmonton, Alta.: Hurtig

Publishers, 1988.

Belanger, René. "Basques." In *The Canadian Encyclopedia.* 2nd ed. Edmonton, Alta.: Hurtig Publishers, 1988.

Bérubé, L. *Coup d'œil sur les pêcheries du Québec.* Saiute-Anne-de-la-Pocatière, Que.: École supérieure des pêcheries, 1941.

Bishop, Morris. *Champlain, the Life of Fortitude.* New York: A.A. Knopf, 1948.

Blake, Raymond B. *From Fishermen to Fish: the Evolution of Canadian Fishery Policy.* Contemporary Affairs Series, Number 7. Toronto: Irwin Publishing Company, co-published by Canadian Institute of International Affairs, 2000.

Bogue, Margaret Beattie. *Fishing the Great Lakes: An Environmental History, 1783–1933.* Madison, Wis.: University of Wisconsin Press, 2000.

Boreman, John, Brian S. Nakashima, James A. Wilson, and Robert L. Kendall. *Northwest Atlantic Groundfish: Perspectives on a Fishery Collapse.* Bethesda, Md.: American Fisheries Society, 1997.

Brebner, J. Bartlett. *North Atlantic Triangle: The Interplay of Canada, the United States and Great Britain.* New Haven, Conn.: Yale University Press, 1945.

Brebner, J. Bartlett. *Canada: A Modern History.* Ann Arbor, Mich.: University of Michigan Press, 1960.

Brebner, J. Bartlett. *The Neutral Yankees of Nova Scotia.* Paperback ed. Toronto, Ont.: McClelland & Stewart, Carleton Library series, 1969.

Brière, Jean-François. *The French Fishery in North America in the 18th Century.* In *How Deep Is the Ocean? Historical Essays on Canada's Atlantic Fishery.* Edited by James E. Candow and Carol Corbin. Sydney, N.S.: University College of Cape Breton Press, 1997. pp. 47–64.

Brodeur, Nigel D. "L.P. Brodeur and the Origins of the Royal Canadian Navy." In *RCN in Retrospect, 1910–1968.* Edited by James A. Boutilier. Vancouver, B.C.: University of British Columbia Press, 1982.

Brown, Cassie. *Death on the Ice.* Toronto, Ont., 1972.

Brown, Craig, ed. *The Illustrated History of Canada.* Toronto, Ont.: Lester & Orpen Dennys, 1987.

Brown, G.P. *Documents on the Confederation of British North America.* Toronto, Ont.: McClelland & Stewart, 1969.

Brown, G.W. *Readings in Canadian History.* Toronto, Ont., 1940.

Brown, J.J. *Ideas in Exile: A History of Canadian Invention.* Toronto, Ont., 1967.

Buckner, Philip A., and John G. Reid, eds. *The Atlantic Region to Confederation: A History.* Toronto, Ont.: University of Toronto Press, 1994.

Busch, Briton Cooper. *The War Against the Seals.* McGill–Queen's University Press, 1985.

Byrne, D. "Practical Problems in the Fish Trade." In *Commission of Conservation, 1915.* Toronto, Ont., 1916.

Calhoun, Sue, and M. Lynk. *The Lockeport Lockout: An Untold Story in Nova Scotia's Labour History.* Halifax, N.S., 1983.

Cameron, Silver Donald. *The Education of Everett Richardson: The Nova Scotia Fishermen's Strike, 1970–71.* Toronto, Ont.: McClelland and Stewart, 1977.

Canadian Council of Professional Fish Harvesters. *Preliminary Response of the Canadian Council of Professional Fish Harvesters to the Atlantic Fisheries Policy Review.* March 12, 2001. Internet: *http://www.ccpfh-ccpp.org/eng/polccpp.html* (accessed 13 June 2002).

Canadian Council of Professional Fish Harvesters. "About the Council." Internet: *http://www.ccpfh-ccpp.org/eng/fac-cveil-html* (accessed 2002).

Canadian Fisherman. "The National Sea Products Story." *Canadian Fisherman* (Gardenvale, Que.), c. 1963.

Candow, James E., and Carol Corbin, eds. *How Deep Is the Ocean? Historical Essays on Canada's Atlantic Fishery.* Sydney, N.S.: University College of Cape Breton Press, 1997.

Careless, J.M.S., and R. Craig Brown, eds. *The Canadians*, part 1. Toronto, Ont.: MacMillan, 1967.

Carrothers, W.A. (with foreword by H.A. Innis). *The British Columbia Fisheries.* Toronto, Ont.: University of Toronto Press, 1941.

Cashin, Peter. *My Life and Times, 1890–1919.* Portugal Cove, Nfld.: Breakwater Books, 1976.

Chantraine, P. *The Living Ice: The Story of the Seals and the Men Who Hunt Them in the Gulf of St. Lawrence.* Translated by David Londell. Toronto, Ont.: McClelland & Stewart, 1983.

Clarke, Allan P., as told to Frank E. Clarke. *My Life at the Front: The Seal Hunt in the 1940s.* Internet: *http://www3.nf.sympatico.ca/feclarke/story6.htm* (accessed June, 2002).

Coaker, W.F. *The History of the Fishermen's Protective Union of Newfoundland.* St. John's, Nfld.: Union Publishing Co., 1920.

Coaker, W.F. "The Bonavista Platform." In *A Coaker Anthology.* Edited by Robert H. Cuff. St. John's, Nfld., 1986. pp. 49–51.

Cohn, Werner. "Persecution of Japanese Canadians and the Political Left in British Columbia, December 1941–March 1942." *B.C. Studies*, No. 68 (Winter 1985–1986): 3–22.

Condon, Michael E. *The Fisheries and Resources of Newfoundland: The Mine of the Sea.* St. John's, Nfld., 1925.

Copes, Parzival. *The Development of the Newfoundland Fishing Economy.* Burnaby, B.C.: Department of Economics and Commerce, Simon Fraser University, 1971.

Crosbie, John C. *No Holds Barred: My Life in Politics.* McClelland & Stewart, 1977.

Crutchfield, J.A. "The Pacific Halibut Fishery." In *The Public Regulation of Commercial Fisheries in Canada.* Edited by A. Scott and P.A. Neher. Ottawa: Economic Council of Canada, 1981. Study No. 2.

Cushing, D.H. "The Problems of Stock and Recruitment." In *Fish Population Dynamics.* Edited by J.A. Gulland. Rochester, N.Y.: John Wiley & Sons, 1977.

Cutting, Charles L. *Fish Saving: A History of Fish Processing from Ancient to Modern Times.* London, U.K.: Leonard Hill Limited, 1955.

Daigle, Jean, ed. *The Acadians of the Maritimes.* Moncton, N.B.: Centre d'études acadiennes, 1982.

Dawson, Samuel Edward. *The Saint Lawrence Basin and Its Borderlands: Being the Story of Their Discovery, Exploration and Occupation.* London: Lawrence and Bullen, 1905.

de la Morandière, Ch., *Histoire de la pêche française de la morue dans l'Amérique septentrionale.* Vol. III. Paris, 1966.

Denys, Nicolas. *The Description and Natural History of the Coasts of North America (Acadia).* New York: Greenwood Press, 1968; Champlain Society Publication, facsimile of 1908 ed.

Dickason, Olive Patricia. *Canada's First Nations: A History of Founding Peoples from Earliest Times.* Norman, Okla.: University of Oklahoma Press, 1992.

Drinkwater, K.F. "A Review of the Role of Climate Variability in the Decline of Northern Cod." *American Fisheries Society Symposium* 32, 2002.

England, George Allan. (Reprint of *Vikings of the Ice*, New York, 1924). *The Greatest Hunt in the World.* Montreal, Que., 1969.

Fairley, H.S. *Canadian Federalism, Fisheries and the Constitution: External Constraints on Internal Ordering.* LL.M diss. Harvard University, Cambridge, Mass., 1980.

Finlayson, A.C. *Fishing for Truth.* St. John's, Nfld.: Memorial University of Newfoundland, 1994.

Fisheries Council of Canada. *Building a Fishery That Works: A Vision for the Atlantic Fisheries.* Ottawa: The Council, 1994.

Fisheries Council of Canada. *Centennial Annual Review*, 1967.

Fisheries Council of Canada. *Selected Programs of Federal Financial Assistance.* Ottawa, December 1969.

Fitzgerald, Gretchen. *The Decline of the Cape Breton Swordfish Fishery: An Exploration of the Past and Recommendations for the Future of the Nova Scotia Fishery.* Halifax, N.S.: Ecology Action Centre, January 2000.

Francis, Daniel. *The Great Chase: A History of World Whaling.* Toronto, Ont.: Penguin Books Canada, 1990.

Gallant, C., and G. Arsenault. *Histoire de la pêche chez les Acadiens de l'Île-du-Prince-Édouard.* Summerside, P.E.I.: Société Saint-Thomas-d'Aquin, 1980.

German, Andrew W. "History of the Early Fisheries: 1720–1930." In *Georges Bank.* Edited by Richard H. Backus and Donald W. Bourne. Cambridge, Mass.: MIT Press, 1987.

Gibson, Gordon, and Carol Remison. *Bull of the Woods.* Vancouver, B.C.: Douglas & McIntyre, 1980.

Gimblett, Richard H. "Reassessing the Dreadnought Crisis of 1909 and the Origins of the Royal Canadian Navy." *The Northern Mariner / Le Marin du nord* 4, no. 1 (1994): 35–53.

Gordon, H.S. "The Economic Theory of a Common-Property Resource: The Fishery." *Journal of Political Economy* 62 (1954): 124–42.

Gough, Joseph. "Frankenstein's Seine." *Canadian Fisherman.* October and December 1971.

Gough, Joseph. "The Canadian Fishing Industry and Georges Bank." In *Georges Bank.* Edited by Richard H. Backus and Donald W. Bourne. Cambridge, Mass.: MIT Press, 1987.

Gough, Joseph. "Fisheries History." In *The Canadian Encyclopedia.* 2nd ed. Edmonton, Alta.: Hurtig Publishers, 1988.

Grant, Ruth F. *The Canadian Atlantic Fishery.* Toronto, Ont.: Ryerson Press, 1934.

Gushue, Raymond. "The Territorial Waters of Newfoundland." *Canadian Journal of Economics and Political Science* 10, no. 3 (August 1942).

Halliday, R., and A. Pinhorn. "North Atlantic Fishery Management Systems: A Comparison of Management Methods and Resource Trends." *Journal of Northwest Atlantic Fishery Science* 20 (1996).

Hardin, Garrett. "Essays on Science and Society, Extensions of The Tragedy of the Commons." *Science 280*, no. 5364 (1 May 1998).

H.B. Nickerson and Sons Ltd., and National Sea Products Ltd. *Where Now?* Booklet. c. 1978.

Hewest, Gordon Winant. *Aboriginal Use of Fishery Resources in Northwest North America.* Ph.D. diss., University of California, February 1947.

Hill, A.V. *Tides of Change: A Story of Fishermen's Co-operatives in British Columbia.* Prince Rupert Fishermen's Co-operative Association, 1967.

Hiller, James and Peter Neary. *Newfoundland in the Nineteenth and Twentieth Centuries: Essays in Interpretation.* Toronto, Ont., 1980.

Hodgins, T. "The Coercion of Newfoundland." In *Diplomatic North American Fisheries.* Bound collection of articles, 1908–1910. In the library of the Department of Fisheries and Oceans, Ottawa.

Huntsman, A.G. "Edward Ernest Prince." *Canadian Field-Naturalist* 59, no. 1 (1945).

Huntsman, A.G. "Fisheries Research in Canada." *Science* 98 (1943): 17–122.

Inglis, G. *More Than Just a Union: The Story of the NFFAWU.* St. John's, Nfld.: Jesperson Press, 1985.

Innis, H.A. "History of Commercial Fisheries." In *Encyclopedia Canadiana.* Toronto, Ont.: Grolier of Canada, 1968.

Innis, Harold. *Select Documents in Canadian Economics History, 1497–1783.* Toronto, Ont., 1929.

Innis, Harold. *The Cod Fisheries: The History of an International Economy.* New Haven, Conn.: Yale University Press, 1940.

Isabella, Jude. "A Turbulent Industry." *Journal of the Maritime Museum of British Columbia* 45 (Spring 1999).

Isham, C. *The Fishery Question: Its Origin, History and Present Situation: With a Map of the Anglo-American Fishing Grounds and a Short Bibliography.* New York: G.P. Putnam's Sons, 1887.

Jensen, Albert C. *The Cod.* New York, 1972.

Joncas, L.Z. "The Fisheries of Canada." In *Fisheries of the World: Papers from the International Fisheries Exhibition.* London, 1883.

Johnston, D.M. *The International Law of Fisheries.* New Haven, Conn.: Yale University Press, 1965.

Judson, T.A. The Freshwater Commercial Fishing Industry of Western Canada. Ph.D. diss. University of Toronto, Toronto, Ont., 1961.

Kearney, John. "The Transformation of the Bay of Fundy Herring Fisheries, 1976–1978: An Experiment in Fishermen–Government Co-Management." In *Atlantic Fisheries and Coastal Communities: Fisheries Decision-Making Case Studies.* Edited by Cynthia Lawson and Arthur J. Hanson. Halifax, N.S.: Dalhousie University Ocean Studies Programme, 1984. pp. 165–203.

Keirstead, B.S., and M.S. Keirstead. *The Impact of the War on the Maritime Economy.* Halifax, N.S.: Imperial Publishing Company, 1944.

Kimber, Stephen. *Net Profits: The Story of National Sea.* Halifax, N.S.: Nimbus, c. 1989.

Kurlansky, Mark. *Cod: A Biography of the Fish That Changed the World.* Toronto, Ont.: Alfred A. Knopf Canada, 1997.

Leach, J.H., and S.J. Nepszy. "The Fish Community in Lake Erie." *Journal of the Fisheries Research Board of Canada* 33 (1976): 622–38.

Le Moyne d'Iberville, Pierre. "Memoir concerning the coas of Florida and a part of Mexico" (n.d.). In MacKirdy, K.A., J.S. Moir, and Y.F Zoltvany, *Changing Perspectives in Canadian History: Selected Problems.* Don Mills, Ont., 1971.

Lescarbot, Marc. *Nova Francia: A Description of Acadia, 1606.* Translated by P. Erondelle, 1609. London, U.K.: George Routledge and Sons, 1928.

Logie, R.R. "Trends in the Atlantic Coast Fisheries of Canada." *Transactions of the American Fisheries Society* 97 (January 1968).

Longley, R.S. "Peter Mitchell, Guardian of the North Atlantic Fisheries, 1867–1871." *Canadian Historical Review* 22 (1941): 389–402.

Lounsbury, R.G. *The British Fishery at Newfoundland, 1634–1763.* New Haven, Conn. and London, 1934.

Lunt, W. E. *History of England.* 4th ed. New York: Harper and Brothers, 1957.

Lyons, Cicely. *Salmon, Our Heritage: The Story of a Province and an Industry.* Vancouver, B.C.: British Columbia Packers, 1969.

Macdonald, D.A. "Social History in a Newfoundland Outport: Harbour Breton, 1850–1900." In *How Deep Is the Ocean? Historical Essays on Canada's Atlantic Fishery.* Edited by James E. Candow and Carol Corbin. Sydney, N.S.: University College of Cape Breton Press, 1997.

MacKean, R., and R. Percival. *The Little Boats: Inshore Fishing Craft of Atlantic Canada.* Fredericton, N.B., 1979.

MacKenzie, W.C. "Modern Commercial Fisheries." In Encyclopedia Canadiana. Toronto, Ont.: Grolier of Canada, 1968.

MacKenzie, W.C., and J. Gough. "Fisheries." In *The Canadian Encyclopedia.* 2nd ed. Edmonton, Alta.: Hurtig Publishers, 1988.

MacKirdy, K.A., J.S. Moir, and Y.F. Zoltvany. *Changing Perspectives in Canadian History: Selected Problems.* Don Mills, Ont., 1971.

MacNutt, W.S. "The 1880's." In *The Canadians.* Part 1. Edited by J.M.S. Careless and R. Craig Brown. Toronto, Ont.: Macmillan, 1967.

MacNutt, W.S. *The Atlantic Provinces: The Emergence of Colonial Society.* Paperback ed. Toronto, Ont.: McClelland & Stewart, 1968.

McDonald, Ian. *W.F. Coaker and the Fishermen's Protective Union in Newfoundland Politics 1908–1925.* Ph.D. diss., University of London, London, U.K., 1971.

McDonald, Ian. "Coaker and the Balance of Power Strategy: The Fishermen's Protective Union in Newfoundland." In *Newfoundland in the Nineteenth and Twentieth Centuries: Essays in Interpretation.* Edited by James Hiller and Peter Neary. Toronto, Ont.: University of Toronto Press, 1980. pp. 148–80.

Meggs, Geoff. *Salmon: The Decline of the B.C. Fishery.* Vancouver, B.C., and Toronto, Ont.: Douglas & McIntyre, 1991.

Mercer, M.J. "Relations between Nova Scotia and New England 1815–1867: With Special Reference to Trade and the Fisheries." M.A. diss., Dalhousie University, Halifax, N.S., 1938.

Morton, W.L. *The Kingdom of Canada.* Toronto, Ont.: McClelland & Stewart, 1963.

Neary, Peter and Patrick O'Flaherty, eds. *By Great Waters:*

Newfoundland Anthology. Toronto, Ont.: University of Toronto Press, 1974.

Nevins, Allan, and Henry Steele Commager. *A Pocket History of the United States.* New York: Washington Square Press, 1970.

Newell, Dianne. *Tangled Webs of History: Indians and the Law in Canada's Pacific Coast Fisheries.* Toronto, Ont.: University of Toronto Press, 1993.

North, G., and H. Griffin. *A Ripple, A Wave: The Story of Union Organization in the B.C. Fishing Industry.* Vancouver, B.C.: Fisherman Publishing Society, 1974.

Palton, Janice. *The Exodus of the Japanese.* Toronto, Ont.: McClelland & Stewart, 1973.

Pepper, D.A. Men, Boats and Fish in the Northwest Atlantic: An Economic Evaluation. Ph.D. diss., University of Wales, 1978.

Pickersgill, J.W. *Seeing Canada Whole: A Memoir.* Markham, Ont.: Fitzhenry and Whiteside, 1994.

Pitt, Janet E.M., and Robert D. Pitt. "Red Bay." In *The Canadian Encyclopedia.* 2nd ed. Edmonton, Alta.: Hurtig Publishers, 1988.

Prince, E.E. *The Fisheries of Canada.* Rome: Tipografia del Senato, 1913.

Prince, E.E. "Fifty Years of Fishery Administration in Canada." *Transactions of the American Fisheries Society* 50 (1921): 163–86.

Prowse, D.W. *A History of Newfoundland from the English, Colonial and Foreign Records.* London and New York, 1895. Reprint, Mika Studio, Belleville, Ont., 1972.

Pryke, Kenneth J. *Nova Scotia and Confederation, 1864–1874.* Toronto, Ont.: University of Toronto Press, 1979.

Reeves, R., and E.D. Mitchell. "Beluga Whale." In *The Canadian Encyclopedia.* 2nd ed. Edmonton, Alta.: Hurtig Publishers, 1988.

Reid, J.H.; Stewart Kenneth McNaught, and Harry S. Crowe. *A Source-book of Canadian History.* Rev. ed. Toronto, Ont.: Longmans Canada, Ltd., 1964.

Roberts, Henry D., with Michael O. Nowlan. *The Newfoundland Fish Boxes: A Chronicle of the Fishery.* Fredericton, N.B.: Brunswick Press, 1982.

Roy, Muriel K. "Settlement and Population Growth." In *The Acadians of the Maritimes.* Edited by Jean Daigle. Moncton, N.B.: Centre d'études acadiennes, 1982.

Ryan, Shannon. *Fish out of Water: The Newfoundland Saltfish Trade,* 1814–1914. St. John's, Nfld.: Breakwater Books, 1986.

Ryan, Shannon. *The Ice Hunters: A History of Newfoundland Sealing to 1914.* St. John's, Nfld.: Breakwater Books, 1994.

Ryan, S., M. Drake, and C. Andrews. *Seals and Sealers: A Pictorial History of the Newfoundland Seal Fishery: Based on the Cater Andrews Collection.* St. John's, Nfld.: Breakwater Books, 1987.

Ryan S., and L. Small. *Haulin' Rope and Gaff: Songs and Poetry in the History of the Newfoundland Seal Fishery.* St. John's, Nfld.: Breakwater Books, 1978.

Samson, Roch. "Good Debts and Bad Debts: Gaspé Fishers in the 19th Century." In *How Deep Is the Ocean? Historical Essays on Canada's Atlantic Fishery.* Edited by James E. Candow and Carol Corbin. Sydney, N.S.: University College of Cape Breton Press, 1997.

Schrank, William E. "Extended Fisheries Jurisdiction: Origins of the Current Crisis in Atlantic Canada's Fisheries." *Marine Policy* 19, no. 4 (1995): 285–99.

Skogan, J. Skeena. *A River Remembered.* Vancouver, B.C.: B.C. Packers, 1983.

Smallwood, J.R. *Coaker of Newfoundland: The Man Who Led the Deep-Sea Fishermen to Political Power.* London: Labour Publishing Company, 1927.

Smallwood, J.R. *The Book of Newfoundland.* Vols. 2 and 3. St. John's, Nfld., 1937.

Smith, Nicholas. *Fifty-two Years at the Labrador Fishery.* London, 1936.

Stansby, M.E. *Industrial Fishery Technology. A Survey of Methods for Domestic Harvesting, Preservation, and Processing of Fish Used for Food and for Industrial Products.* Huntington, N.Y.: Robert E. Krieger Publishing Company, 1976.

Stewart, Hilary. *Indian Fishing: Early Methods on the Northwest Coast.* Seattle, Wash.: University of Washington Press, 1977.

Story, G.M., W.J. Kirwin, and J.D.A. Widdowson, eds. *Dictionary of Newfoundland English.* Toronto, Ont.: University of Toronto Press, 1982.

Sutherland, D.A. "1810–1820: War and Peace." In *The Atlantic Region to Confederation: A History.* Edited by Philip A. Buckner and John G. Reid. Toronto, Ont.: University of Toronto Press, 1994.

Talman, James J. *Basic Documents in Canadian History.* Princeton: D. Van Nostrand, 1959.

Tansill, C.C. *Canadian–American Relations, 1875–1911.* New Haven, Conn.: 1943.

The Antigonish Movement: Yesterday and Today. Antigonish, N.S.: St. Francis Xavier University, 1976.

Tomasevich, W.E. *International Agreements on Conservation of Marine Resources.* Palo Alto, Calif.: Stanford University Press, 1943.

Tonnessen, J.N., A.O. Johnsen, and R.J. Christophersen. *The History of Modern Whaling.* Berkeley, Calif.: University of California Press, 1982.

Tulloch, Judith. "The New England Fishery and Trade at Canso." In *How Deep Is the Ocean? Historical Essays on Canada's Atlantic Fishery.* Edited by James E. Candow and Carol Corbin. Sydney, N.S.: University College of Cape Breton Press, 1997.

Wallace, Carl M. "Smith, Albert J." In *The Canadian Encyclopedia.* 2nd ed. Edmonton, Alta.: Hurtig Publishers, 1988.

Wallace, Frederick William. *In the Wake of the Wind Ships.* Toronto, Ont., 1927.

Wallace, Frederick William. *Wooden Ships and Iron Men.* Boston, Mass., 1931.

Wallace, Frederick William. *"Roving Fisherman." Canadian Fisherman,* 1955.

Warner, William W. *Distant Water.* Penguin Books, 1983.

Warren, James L. *Marine Sardine Industry History: An Anthology.* Maine Sardine Council, March 1986.

Wells, Kennedy. *The Fishery of Prince Edward Island.* Charlottetown, P.E.I.: Ragweed Press, 1986.

Western Fisheries. 15th anniversary issue. Vancouver, B.C., September 1979.

Whitelaw, W.M. *The Maritimes and Canada Before Confederation.* Oxford, U.K.: Oxford University Press, 1934.

Whiteley, Albert S. *A Century on Bonne Esperance: The Saga of the Whiteley Family.* Ottawa: privately printed, 1977.

Wilbur, J.R., and E. Wentworth. *Silver Harvest: The Fundy Weirmen's Story.* Fredericton, N.B., 1986.

Wright, Miriam. *A Fishery for Modern Times: The State and the Industrialization of the Newfoundland Fishery, 1934–1968.* Oxford University Press

INDEX